AF327924

Introduction to
High-Temperature Superconductivity

SELECTED TOPICS IN SUPERCONDUCTIVITY

Series Editor: Stuart Wolf
Naval Research Laboratory
Washington, D.C.

CASE STUDIES IN SUPERCONDUCTING MAGNETS
Yukikazu Iwasa

INTRODUCTION TO HIGH-TEMPERATURE SUPERCONDUCTIVITY
Thomas P. Sheahen

A Continuation Order Plan is available for this series. A continuation order will bring delivery of each new volume immediately upon publication. Volumes are billed only upon actual shipment. For further information please contact the publisher.

Introduction to High-Temperature Superconductivity

Thomas P. Sheahen
Western Technology Incorporated
Derwood, Maryland

Plenum Press • New York and London

Library of Congress Cataloging-in-Publication Data

Sheahen, Thomas P.
    Introduction to high-temperature superconductivity / Thomas P.
  Sheahen.
        p.   cm. -- (Selected topics in superconductivity)
    Includes bibliographical references and index.
    ISBN 0-306-44793-2
    1. High temperature superconductivity.   I. Title.   II. Series.
  QC611.98.H54S48   1994
  537.6'23--dc20   94-34415
                                                                      CIP

ISBN 0-306-44793-2

©1994 Plenum Press, New York
A Division of Plenum Publishing Corporation
233 Spring Street, New York, N.Y. 10013

Printed in the United States of America

Foreword

High-temperature superconductivity (HTSC) has the potential to dramatically impact many commercial markets, including the electric power industry. Since 1987, the Electric Power Research Institute (EPRI) has supported a program to develop HTSC applications for the power industry. The purpose of EPRI is to manage technical research and development programs to improve power production, distribution, and use. The institute is supported by the voluntary contributions of some 700 electric utilities and has over 600 utility technical experts as advisors.

One objective of EPRI's HTSC program is to educate utility engineers and executives on the technical issues related to HTSC materials and the supporting technologies needed for their application. To accomplish this, Argonne National Laboratory was commissioned to prepare a series of monthly reports that would explain the significance of recent advances in HTSC. A component of each report was a tutorial on some aspect of the HTSC field. Topics ranged from the various ways that thin films are deposited to the mechanisms used to operate major cryogenic systems. The tutorials became very popular within the utility industry. Surprisingly, the reports also became popular with scientists at universities, corporate laboratories, and the national laboratories. Although these researchers are quite experienced in one aspect of the technology, they are not so strong in others. It was the diversity and thoroughness of the tutorials that made them so valuable. The authors spent many hours with leading experts in each topic area and went through a painstaking review process to ensure that the information in the tutorials was complete, concise, and correct.

The tutorials that were originally published by EPRI in a newsletter format have evolved into many of the chapters of this book. Hopefully the value that we tried to provide for our member utilities with these tutorials will also benefit the entire industry through the publication of this book. Utility engineers and electric equipment manufacturers will benefit from the chapters describing the theory and characteristics of the HTSC materials. Scientists working with the materials will appreciate the chapters that discuss the engineering of the various applications that will make use of the HTSC materials.

Because of the HTSC's potential for a strong impact on business and society, it is important that new and working engineers become knowledgeable in the technology. This book will become an invaluable resource for understanding the fundamental characteristics of the materials and how they can be used.

Donald W. Von Dollen

Electric Power Research Institute

v

Preface

High-Temperature Superconductivity (HTSC) is most certainly a multidisciplinary field. Drawing from physics, mechanical engineering, electrical engineering, ceramics, and metallurgy, HTSC spans nearly the entire realm of materials science. No one is expert in all these disciplines; rather, each researcher brings a special expertise that is complemented by the skills of colleagues. Therefore, it is necessary for each to obtain a modest understanding of these allied specialities.

This book tries to present each of those disciplines at an introductory level, with the goal that the reader will ultimately be able to read the literature in the field. Recognizing that there is no need to read introductory material in your own specialty, the chapters were organized with the expectation that each reader would skip part of the book. As a consequence, some repetition occurs in places; for example, Josephson junctions are introduced in both Chapter 5 and Chapter 13. On the expectation that most engineers will be interested in only a few of the applications, the later chapters are designed to stand alone.

In various places, numerical values are given for certain quantities of interest. In a fast-moving field like HTSC, it is impossible to be absolutely up-to-date with the latest numbers. It would be missing the point to dwell on numerical values. Rather, the intent of the book is to convey a general understanding of the accomplishments, problems, and motivations that lead researchers to try various ways of improving the HTSC materials.

OUTLINE

The HTSC field is also quite large, and conceptually splits nicely into applications directed toward carrying electrical power and applications directed toward electronic circuits. This book deals primarily with the former. Electronic applications, including the very broad field of thin-film superconductors, are given very little attention. This is because the book grew out of a series of reports prepared at Argonne National Laboratory for the Electric Power Research Institute (EPRI), during the period of rapid development in HTSC from 1988 to 1992. EPRI's interest in power applications drove the choices of reporting topics, and consequently determined the scope of this book.

There are five major divisions of the text:

1. Conventional Superconductivity—This part describes the present-day playing field on which HTSC is striving to compete.

2. Properties of the HTSCs—This series of chapters describes what we know about the basic physics, chemistry, and materials science of these compounds. Because

of the complexity and interrelatedness of several different fields here, this was the most difficult portion of the book to unify into a coherent presentation.

3. Carrying Electricity—These chapters deal specifically with those aspects of HTSC that relate to making wire and conducting electricity. Because of the very rapid pace of research and development in the HTSC field, and the likely success of some of the government–industry partnerships carrying it out, this is the portion most likely to be in need of revision soon.

4. Near-Term Applications—The known needs of the electric power industry are featured here, in a series of chapters that each focus on one specific application of HTSC. These could plausibly be termed the practical applications.

5. Futuristic Applications—The HTSC field has a lot of room to grow, and in these chapters we peer over the horizon for potential future uses of HTSC. A modest amount of speculation is in order here, and if some exceptional breakthrough occurs tomorrow, some of these applications may move into the practical category.

Of course, for a full understanding it is best to read all five parts. However, Parts 4 and 5 can be read without having a detailed knowledge of all that went before. In general, *no* single chapter in the book is so pivotal that it absolutely *must* be read. From the outset, I aimed for a reader whose other demands preclude reading everything.

Thomas P. Sheahen

Acknowledgments

Every author is always indebted to his colleagues, and so it is a standard custom in the scientific literature to say thanks for many helpful discussions. That is not enough here. The long hours put in by many friends and professional colleagues (heavily, but certainly not exclusively, at Argonne National Laboratory) are deserving of much greater recognition.

First of all, several chapters are co-authored with researchers who are more skilled than I in the pertinent subject matter. My role here was often to integrate their work into the overall presentation of the book.

Second, at the outset I certainly did not know all the various required disciplines. I had to be tutored in the subject matter of each report to EPRI. After that, my written drafts had to be reviewed, corrected, and critiqued both for factual accuracy and for clarity of presentation. In assembling and updating the tutorials to make chapters for the book, I continued to rely very heavily on the patience and generosity of many colleagues. A lot of very fine people took time away from their own pursuits in order to help me succeed.

Foremost among my collaborators at Argonne National Laboratory was Dr. Robert F. Giese; we worked together in preparing the series of EPRI reports for more than 4 years. Those reports were each roughly equivalent to the size of one chapter here. Bob's contributions have been very great indeed.

From the beginning, the primary source of up-to-date information about what was taking place in the HTSC field was *High T_c Update*, featuring the "Note Bene" section written by John Clem of Iowa State University. The guidance through the very extensive literature provided in this way was indispensable to the completion of our reports.

Alan Wolsky supervised the EPRI project, and Bobby Dunlap and Roger Poeppel read and critiqued each of the EPRI reports. Much of the clarity of presentation of various topics originated in the reviews and discussions that were held with them.

Many other Argonne scientists contributed to my education in the HTSC field, and several reviewed individual chapters, which resulted in the elimination of a number of errors and mistaken concepts. In this regard I am particularly grateful to Howard Coffey, Steve Dorris, George Crabtree, John Hull, Jim Jorgensen, Dick Klemm, Hagai Shaked, J.P. Singh, and Jack Williams.

Colleagues at the National Institute of Standards and Technology deserve recognition, both for educating me on various subjects and for critiquing portions of the manuscript. Chapter 3 on refrigeration follows very closely the work of Ray Radebaugh; he could easily be called a co-author. Others who provided in-depth consultation include Frank Biancaniello, John Blendell, Steve Frieman, George Mattingly, Steve Ridder, and Bob Roth.

Stuart Wolf of the Naval Research Laboratory worked very hard to raise my level of knowledge of the theoretical aspects of HTSC. Two British scientists (whom I have never met) have taught me a lot: J. E. Gordon and Martin N. Wilson have written books of such clarity that I can only cite the old slogan "Imitation is the sincerest form of flattery" to acknowledge my debt to them. I would have fallen far behind in my knowledge of wire development were it not for the continuous help of Alex Malozemoff and Bart Riley of American Superconductor Corp., and of Pradeep Haldar and Lech Motowidlo of Intermagnetics General. Roger Koch of IBM straightened out my understanding of flux pinning considerably. Xingwu Wang of Alfred University clarified conventional SMES and its applications to the electric utility sector. Mas Suenaga of Brookhaven explained ac losses, and Yuki Iwasa of M.I.T. helped me to understand stability in the HTSCs. Jerry Selvaggi of Eriez Magnetics and Gene Hirschkoff of Biomagnetics Technologies each patiently explained their devices to me. Eddie Leung of Martin Marietta corrected several lapses in my grasp of fault current limiters.

These are but a few examples of the countless sources of help—*interdisciplinary* help—from which I have benefitted en route to writing this book.

Another 20 or more researchers from national laboratories, universities, and corporations have reviewed individual chapters, and have explained and clarified one point or another. In short, this effort has received a lot of support from friends who saw the value in it. I am very grateful to all my colleagues who have helped me to get it right. To the extent that errors remain in the text, I personally have to take the responsibility for them.

This book would not have been completed without the strong and direct encouragement and support of Jim Daley of the U.S. Department of Energy and Don Von Dollen of EPRI. Their unfailing confidence made it possible to get through some very difficult aspects of the work.

I also wish to thank all those researchers who generously gave permission for me to reproduce their original figures, and frequently took the trouble to provide me with pristine copies. On the subject of actually preparing the manuscript, special thanks go to Erika Shoemaker of Argonne for guiding me through a series of word-processing hurdles, and to Laurie Culbert for turning many sketches into excellent figures. Finally, I greatly appreciate the generosity of Charlie Klotz of Argonne in providing me with support services during the later stages of writing the book.

Thomas P. Sheahen

Contents

Part I. Superconductivity

Part II. High-Temperature Superconductivity (HTSC) Basic Properties

Part III. Carrying Electricity

Chapter 16. Wire

Thomas P. Sheahen and Alan M. Wolsky

Chapter 17. Protecting Against Damage

Thomas P. Sheahen and Robert F. Giese

Chapter 18. AC Losses

Part IV. Electric Power Applications of HTSC

Chapter 19. Transmission Lines

John S. Engelhardt, Donald Von Dollen,
Ralph Samm, and Thomas P. Sheahen

Chapter 20. Levitation

John R. Hull and Thomas P. Sheahen

Chapter 21. Superconducting Magnetic Energy Storage

Susan M. Schoenung and Thomas P. Sheahen

Chapter 22. Electric Motors

Howard E. Jordan, Rich F. Schiferl, and Thomas P. Sheahen

Chapter 23. Fault Current Limiters

Robert F. Giese, Magne Runde, and Thomas P. Sheahen

V. Future Possibilities

Chapter 24. New Refrigerators

Chapter 25. Applications to Measurement and Process Control

Chapter 26. High Magnetic Fields

Chapter 27. Organic Superconductors

Chapter 28. Aerospace Applications

Appendix A. Measurement of Critical Current 555

Appendix B. Magnetic Measurements Upon Warming or Cooling 563
Donn Forbes and John R. Clem

1

SUPERCONDUCTIVITY

1

Introduction and Overview

The field of superconductivity, once a mere laboratory curiosity, has moved into the realm of applied science in recent years. Even more applications may become possible because of the discovery of ceramic superconductors, which operate at comparatively "high" temperatures.

1.1. SUPERCONDUCTORS

What is a *superconductor*? For most materials, which are *normal* conductors, whenever electrical current flows, there is some resistance to the motion of electrons through the material. It is necessary to apply a voltage to keep the current going, to replace the energy dissipated by the resistance. Ordinary copper wire in a house is a good conductor, with only a little resistance; the filament in a light bulb has a high resistance, and generates so much heat that light is given off. Electronics is based on components in which the resistance changes under control of an input voltage; these components are made of semiconductors. A superconductor, in contrast, is a material with no resistance at all.

A lot of metals, but not all, show modest electrical resistance at ordinary room temperatures, but turn into superconductors when refrigerated very near to absolute zero. The first metal discovered to be a superconductor was mercury,[1] soon after the invention (in 1908) of a cryogenic refrigerator that could attain the temperature at which helium becomes a liquid: $4.2 \text{ K} = -452°\text{F}$. In the subsequent 60 years, many more superconductors were found at these very low temperatures. By the 1960s, certain alloys of niobium were made that became superconductors at 10–23 K. It was generally believed on theoretical grounds that there would be no superconductors above 30 K.

Since a superconductor has no resistance, it carries current indefinitely without requiring voltage or an expenditure for electricity. Once the current is started, it continues for "geological" time durations, provided that the superconductor is kept cold. For many years, the requirement of refrigeration to extremely low temperatures had the effect of confining superconductivity to the realm of research laboratories. The cost of running a superconducting persistent current loop is simply the cost of refrigeration, which in most cases means the cost of purchasing liquid helium—about $7 per liter.

Electromagnets are the most important application of current loops, but it is expensive to run a large electromagnet built out of ordinary wire like copper. By the 1970s, it became cost effective (in some cases) to pay the price for refrigerating a superconductor instead of paying the utility for electricity lost through resistance. In this way an industry evolved, in

3

which large superconducting magnets were used in certain applications. One familiar use was in hospitals, where Magnetic Resonance Imaging (MRI) has become a standard diagnostic tool for scanning the body to see what is wrong inside. The cost of running such a device is far less than "exploratory surgery."

1.2. HIGH-TEMPERATURE SUPERCONDUCTORS

There would be a lot more practical uses for superconductivity if it weren't for the very high cost of liquid helium coolant. Any gas will liquefy at sufficiently low temperatures; for example, oxygen becomes liquid at 90 K and nitrogen at 77 K. It is far less costly to liquefy these gases than to liquefy helium. Liquid nitrogen sells for about six cents per liter (in truckload quantities); moreover, it has a much greater cooling capacity than liquid helium. For any application in which liquid nitrogen can replace liquid helium, the refrigeration cost will be about 1000 times less.

There are several ceramics, based on copper oxide, which remain superconducting near 100 K. For example, the compound yttrium barium copper oxide (YBCO) has been found to be superconducting up to 92 K. This may not seem like a "high" temperature to most people, but to the engineers figuring the cost of refrigerants, it is high enough: liquid nitrogen is sufficient to cool YBCO into its superconducting range. Additional important ceramic superconductors include BSCCO (bismuth strontium calcium copper oxide) and TBCCO (thallium barium calcium copper oxide); and HBCCO (mercury barium calcium copper oxide). The latter has the highest critical temperature of superconductivity, $T_c = 133$ K = $-220°$F. Table 1.1 presents the chemical formulas and T_c values for each of these compounds.

The ceramic superconductors of greatest interest are very anisotropic compounds; that is, their properties are quite different in different crystalline directions. For that reason, researchers take considerable pains to obtain good grain alignment within any finite-sized sample. Figure 1.1 is a drawing of the molecular structure of YBCO. The structure is essentially that of a sandwich, with planes of copper oxide in the center, and that is where the superconducting current flows. The compounds BSCCO and TBCCO are even more pronounced in their anisotropy; indeed, very little current can flow perpendicular to the copper oxide planes in those lattices.

The role of the elements other than copper and oxygen is secondary. In YBCO, yttrium is only a spacer and a contributor of charge carriers; indeed, nearly any of the rare earth

Table 1.1. HTSC Materials

Name	Formula[a]	Transition Temperature
Yttrium barium copper oxide	$Y_1Ba_2Cu_3O_7$	92 K
Bismuth strontium calcium copper oxide (2223 phase)	$(Bi,Pb)_2Sr_2Ca_2Cu_3O_x$	105
Thallium barium calcium copper oxide (1223 phase)	$Tl_1Ba_2Ca_2Cu_3O_y$	115
Mercury barium calcium copper oxide (1223 phase)	$Hg_1Ba_2Ca_2Cu_3O_y$	135[b]

[a]The subscripts (2223, etc.) provide a shorthand code for naming each generic compound.
[b]Under extremely high pressure, $T_c > 150$ K has been observed.

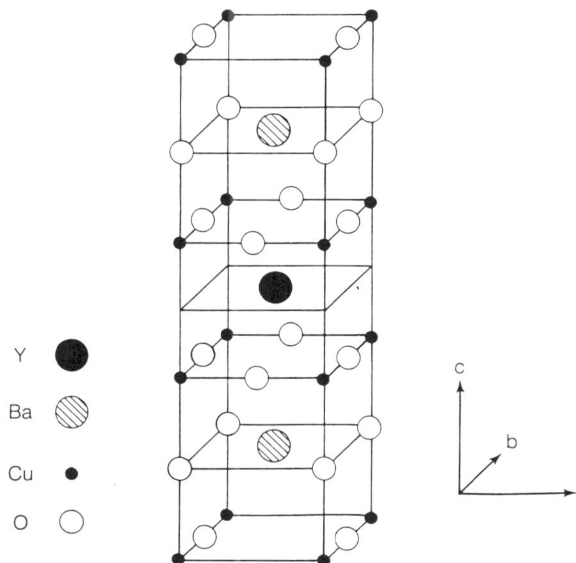

Figure 1.1. Unit-cell structure of YBaCuO superconductor.

elements (holmium, erbium, dysprosium, etc.) can be substituted for yttrium without changing the transition temperature T_c significantly. Often the formula is written as $(RE)_1Ba_2Cu_3O_7$, to emphasize the interchangability of other rare earths (RE) with yttrium.

The bismuth compounds exhibit the interesting property of being *micaceous*; that is, they are like mica. The crystal lattice shears easily along the bismuth oxide planes, and this allows BSCCO to be deformed and shaped with less difficulty than the other ceramic superconductors. This advantage has led researchers to invest more effort in making wire out of BSCCO: lengths of over one kilometer have been made so far.

Unfortunately, the new high-temperature superconductors have two major drawbacks: they are very brittle (like most ceramics), and they do not carry enough current to be very useful. One problem is that of brittleness. Ceramics are by nature brittle, and so is copper oxide. The idea of making wire out of ceramics would be a subject of derision, were it not for the example set by fiber optics. It is true that if one makes a strand of sufficiently tiny diameter, then a cable made from such strands can have a bending radius of a few centimeters without over-straining the individual strands. For the high temperature superconducting materials, the engineering task of overcoming brittleness is proving more difficult than it was for fiber optics.

A more important drawback is that the magnetic properties of these materials are substantially different from conventional metallic superconductors. The workhorse material of low temperature superconducting magnets, niobium-titanium (NbTi), allows lines of magnetic flux to penetrate in such a way that these lines tend to stay put: the phenomenon is known as *flux pinning*. By contrast, the exceptional crystalline structure of the copper oxide superconductors causes the magnetic flux lines to fragment (they become shaped like sausages), and hence they move around readily, thus dissipating energy and defeating the advantage of superconductivity. In one of those perverse conspiracies of nature, the crystalline properties that offer the best chance to circumvent the brittleness problem are the very same properties that tend to degrade flux pinning.

1.3. HISTORY

Before continuing with what HTSCs may lead to, it is appropriate to look back and see what they have come from. The history of high-temperature superconductivity as a field distinct from ordinary superconductivity is very brief. It began in late 1986 when news spread that J. George Bednorz and Karl Müller of the IBM research laboratory in Zurich, Switzerland, had reported[2] the observation of superconductivity in lanthanum copper oxides doped with barium or strontium at temperatures up to 38 K. This caused tremendous excitement because 38 K was above the ceiling of 30 K for superconductivity that had been theoretically predicted almost 20 years earlier (and which had become an unquestioned belief among scientists and engineers interested in superconductivity).

Once the barrier was broken, hundreds of scientists rushed to try various chemical compounds to see which one would give the highest T_c. In March 1987, the American Physical Society meeting included a session dealing with new discoveries in superconductivity. That session, which lasted all night, had over 1000 people trying to squeeze in the doors of the meeting room, and would later be remembered as "the Woodstock of physics." At that point, the compound yttrium barium copper oxide ($Y_1Ba_2Cu_3O_7$, or just YBCO for short) took center stage,[3] because of it's high value of $T_c = 92$ K.

Subsequently, attention was focused on copper oxides, and before long the compound bismuth lead strontium calcium copper oxide was found[4] with $T_c = 105$ K. That was followed by the discovery[5] in 1988 of thallium barium calcium copper oxide, with $T_c = 125$ K. Almost five years elapsed before the mercury compounds[6] boosted the T_c record to 133 K. Under extremely high pressure,[7] T_c can be pushed over 150 K.

As soon as *one* superconductor had reached a temperature above 77 K, the era of high-temperature superconductivity had arrived. Some observers believed that room-temperature superconductors were just around the corner waiting to be discovered. A number of exuberant articles appeared in the popular press extolling the many ways our lives would change. Others realized the stunning advantage associated with having superconductors near 100 K and turned their attention to studying and improving the properties of the compounds already discovered.

Moreover, all previous (low-temperature) superconductors require expensive ($7 per liter) liquid helium to cool them to around 4 K. Also, substantial skill and training is required to transfer liquid helium from one container to another without freezing the apparatus. Consequently, only rarely has conventional superconductivity emerged from the physics lab. In the meantime, anyone can pour liquid nitrogen, so a major obstacle to using superconductors in practical applications vanishes if they can operate above 77 K.

These features of superconductivity, well known in 1987, have provided the driving force to sustain superconductivity research ever since. The payoff has been so great that many researchers have devoted major resources to pursuing practical applications. Of course, the path toward high-temperature superconductivity has never been all roses, and the research community has had to sustain itself through several early disappointments. The bubble generated by the popular press didn't exactly burst, but deflated around 1990. The early exuberance was replaced by the sober realization that there are many serious obstacles to overcome in physics, materials science, and mechanical and electrical engineering before these new superconductors find widespread practical application. Serious research managers do not expect to see any large-scale applications until the twenty-first century. Some early applications to delicate sensors and electronic devices are beginning to appear in the mid-1990s.

1.4. SUPERCONDUCTING MAGNETS

A leading use of superconductors is to produce high magnetic fields. Magnetic fields exceeding 10 T have been produced in a handful of laboratories, but have never been employed either in health care (MRI scans, for example) or in industry. The potential applications for higher magnetic fields are just beyond the horizon, and therefore subject to speculation. The idea of using very high magnetic fields (> 30 T) to separate industrial chemicals, thus retrieving value from a waste stream and reducing pollution, is a very attractive concept. However, such mundane considerations as the structural integrity of the supporting framework must be brought into the engineering design, because high magnetic fields exert very great forces, and no one has yet built a large-scale magnet of such magnitude. Optimistic recognition of possibilities needs to be tempered with cautious engineering pragmatism about what can actually be accomplished at a low cost. If the price of an entire magnetic system is too high, no one will buy the device and the application will not come into widespread use.

Meanwhile, interest has increased in applications of low-temperature superconductors; and the possibility of using the ceramic copper oxide superconductors at low (4 K) or intermediate (20–30 K) temperatures is worth considering. Conventional low-temperature superconductors are often used in magnets running at 4 K, but they lose their superconductivity in high magnetic fields, typically above 6 T (= 60,000 gauss); although niobium tin (Nb_3Sn) will remain superconducting even out to 10 or 15 T. The ceramic superconductors do much better. Bismuth strontium calcium copper oxide (BSCCO) carries adequate current and remains superconducting well above 20 T, at 20 K. Therefore, the best way to obtain very high magnetic fields is to use the ceramic superconductors at low temperatures.

Of course, in order to wind a coil to produce a magnetic field, the first prerequisite is to make long lengths of wire from the copper oxide superconductors; thus, the application to high magnetic fields awaits the development of a reliable wire-manufacturing technique. There is no guarantee of ultimate success here, which is why ceramic superconductivity remains a research field.

1.5. WIRE MAKING

The *critical current density* J_c (current per cross-sectional area, A/cm^2) is the major electrical parameter of a superconductor's performance. Therefore, the main focus in HTSC research today is on trying to make wire with high J_c. There are four distinct categories of obstacles to be overcome:

- Large currents in magnetic fields
- Fabricating uniform long lengths of wire
- Mechanical properties
- Joining and contact techniques

Each of these obstacles contains subcategories by which R&D activities can be classified. Here we touch on only the first two.

1.5.1. Large Currents in Magnetic Fields

Ceramic oxide wires present two problems that were not encountered in the earlier development of low-temperature, intermetallic wires. The first is due to the granular nature of these materials. Very large currents can flow within grains, but grain boundaries impede the current flow between grains. It is necessary to achieve very good alignment between adjacent grains in order to circumvent this problem. Methods have been developed both to align grains and to provide "clean" grain boundaries, but these processing methods still need improvement.

The second problem occurs when current is passed through HTSC wires (even when grains are aligned and grain boundaries are clear) but the operating temperature exceeds a certain value. This temperature may be as low as 30 K for some materials and as high as 90 K for others.[8] It is known from development of LTSC wires that high transport current required pinning of the magnetic flux lines that penetrate the material.[9] *Lorentz forces*, proportional to both current and magnetic field strength, will move the flux lines unless the flux lines are sufficiently pinned. Flux line movement causes losses (which may exceed that of copper resistance) even in the presence of superconductivity.

1.5.2. Fabricating Uniform Long Lengths

For the HTSCs, none of the ordinary standard methods of making wire have proved successful. It is not easy to make wire from the ceramic superconductors. To circumvent the problem of brittleness, it is customary to sheath the ceramic material with some ductile metal (usually silver) that is readily handled in wire fabrication equipment. Figure 1.2 is an illustration of one typical process. The raw ingredients are oxide powders of the key elements (in this illustration, BSCCO is being made). These are treated at high temperatures to make a powder of the superconducting compound. It is often helpful to substitute different elements. $(Bi_xPb_{1-x})_2Sr_2Ca_2Cu_3O_y$ is known (by its subscripts) as 2223, where partial substitution of lead for bismuth is understood.

The powder is next packed into a tube of typically a half-meter length and an 8-mm diameter (see Figure 1.2). The wire-making process of drawing, rolling, or swaging follows, leaving a final shape well below 1 mm in diameter but very long. To restore the ceramic core to the superconducting state, it is necessary to heat treat it further, at perhaps 800–900°C. Finally, the wire must be annealed in oxygen very slowly (typically 100 hours) in order to allow oxygen atoms to slowly recover their proper positions in the crystal lattice. Without this step, only a small percentage of the material would be superconducting, and the wire would not be useful for carrying current.

Sumitomo Electric Corp. in Japan was the first company to make over 100 m lengths of wire. Subsequently, companies in Europe and the United States also made lengths over 100 m, and now the competition is intense. Questions of manufacturability, bending radius, and insulation are being explored, demonstrating that companies consider wire-making to be more than just a research venture.

There are many different varieties of processing techniques, the details of which are proprietary within each organization. Sumitomo also made the first multifilament strands by a repeated rolling and annealing process, packing as many as 1,296 fibers into a wire.[10] By 1993, American Superconductor Corp. used a metal precursor process[11] to surpass that

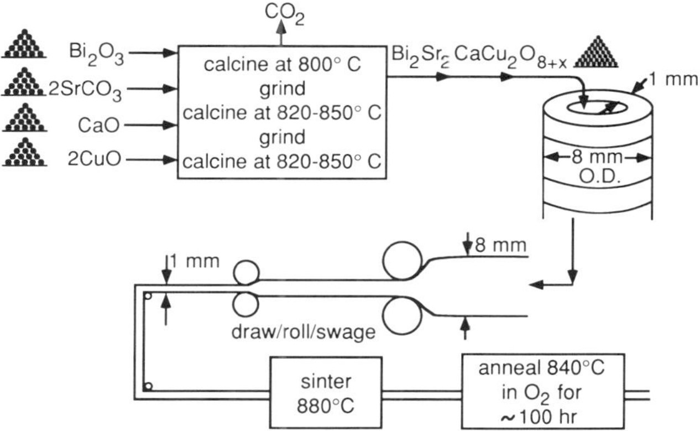

Figure 1.2. Schematic of BSCCO Powder-In-Tube wire-making process.

record. Making a long wire that contains thousands of ultrafine filaments is everyone's goal, because this will allow greater flexibility of the overall wire without cracking the internal filaments.

Certainly BSCCO wire processing is the most advanced, but BSCCO suffers from *flux lattice melting* at modest temperatures (≈ 30 K) when a magnetic field is present. Therefore, BSCCO is *not* going to be the 77 K wire that will revolutionize the industry. There is similar effort to make wire out of the thallium compounds, which perform better in magnetic fields. Once copious quantities of the mercury compounds are available, wire-making efforts will presumably begin there as well.

1.6. ELECTRIC POWER APPLICATIONS

Most electrical applications depend on high values of J_c. With one exception, HTSC electric power applications require coils (or magnets) able to provide strong magnetic fields (2 to 10 T). These coils—in large motors, generators or magnetic storage systems—will require several kilometers of high-performance wire. (The one exception is power transmission and distribution cables, where the magnetic field is low.)

Engineers will always make trade-offs that make technical and economic sense; one of these, historically true, is that higher current density or higher magnetic field is more valuable than higher operating temperature. Operating temperatures of 77 K or higher are preferable for all applications and essential for some, but most large applications are expected to be economically feasible at intermediate temperatures (20–40 K) using new types of cryogenic refrigerators. In fact, current, temperature and magnetic field trade off among one another. The goal of research is to raise the operating envelope of these three parameters, so that trade-offs can occur over a wider range.

The total current I flowing in the wire must be large for power applications. Also, the overall length L of the wire usually must be long for practical applications. In fact, the product of current times length, $I*L$, is a very useful way to capture the "size" concept for a particular application. There are other parameters, notably temperature and magnetic field, but for our immediate purpose, we set all these aside, focusing on the relation between J_c and $I*L$, which is important for applications.

Figure 1.3 displays a collection of applications on a graph where the vertical axis is J_c and the horizontal axis is $I*L$. This log–log plot enables us to place widely disparate applications on the same graph.[12] The more difficult ones appear at the upper right and the simpler applications toward the lower left. The figure shows that a current density of 10,000 A/cm^2 is required in nearly all cases of importance to the electric power industry.

The location of the oval associated with each application is determined by its typical design criteria. As a simple example, for a short transmission line 50 m long and carrying 10 kA, $I*L = 500,000$, obviously; and if one assumes a cross section of 1 cm^2, then the critical current density must be about 10,000 A/cm^2. The other applications are more complicated, but the same idea is used in placing them on the plot.

Progress is shown growing from the lower left corner. The earliest results were on very small samples (small L) with poor current characteristics, but there has been substantial progress since 1988. The first application reached was that of current leads for LTSC magnets, meaning short wires that carry current to a superconducting magnet cooled by liquid helium at 4.2 K, such as used in magnetic resonance imaging (MRI). In fact, a team of Westinghouse and Argonne National Laboratory[13] produced leads that carry 2000 A. That was a good start, but the next nearest application demanded that $I*L$ improve by a factor of 100.

1.7. OTHER DEVICES

Fortunately, not all potential applications of high-temperature superconductivity are associated with high currents and high magnetic fields. Other unique properties, generally valid at 77 K, promise some entirely different applications. First, these materials can switch from the superconducting state to the nonsuperconducting state in 10^{-12} sec, about 1000 times faster than silicon. On the face of it, this suggests that computers made from superconductors might be 1000 times faster than computers based on silicon chip technology. No one expects to gain that entire advantage, but substantial improvements in speed seem assured. Research strives to make hybrid circuits, combining the best features of silicon technology and superconducting technology, on a single chip.

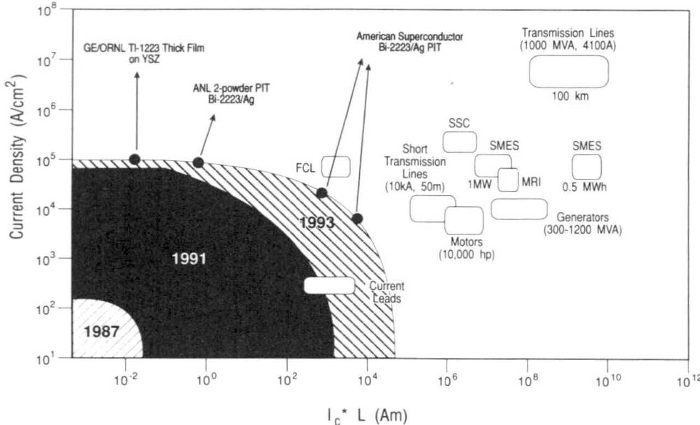

Figure 1.3. Log–log plot relating current density J_c to product of current I times wire length L, to show level of anticipated difficulty for various applications.

Second, the property of magnetic field repulsion by superconductors (known as the *Meissner effect*) opens the door to using high-temperature superconductors as a bearing material. The familiar photo of a small magnet floating in air above a disk of YBCO immersed in liquid nitrogen demonstrates the concept. A magnetic material will stand away from a superconductor. Therefore, it is possible to build a bearing surface with absolutely no contact between pieces. In test rigs, rotational speeds of 240,000 rpm are achievable, because of the negligible friction. Space applications come to mind for such bearings because in the weightlessness of space they do not need to carry heavy loads. Industrial (heavy-loaded) applications will be slower to appear, because hybrid magnet-and-superconductor combinations will be needed to carry the weight.

This same principle is the basis of an energy storage device. It is well known that electric power plants face their peak demand from customers in the late afternoon, but have excess generating capacity in the hours between midnight and dawn. If electricity could be generated at night and stored for half a day, the power plant would be much more efficient. One way to store energy is to make a flywheel spin rapidly; but energy is gradually lost to friction in the flywheel's bearings. With high-temperature superconductors employed as bearings, the efficiency of flywheel energy storage can improve dramatically. Figure 1.4 shows one typical configuration.

Yet another useful characteristic of these materials is that a superconductor reflects electromagnetic waves perfectly. When the interior walls of a closed chamber are coated with superconducting material, the resonance properties of the box improve tremendously. Microwave resonators, which already have a number of room-temperature applications, perform much better when coated with films of ceramic superconductors. The economic trade-off, comparing the value of sharper "Q" of the resonant cavity versus the cost of refrigeration, will determine how widespread this application will ultimately be. In the past, the cost of cooling to 4 K was prohibitive, but having to cool only to 77 K is a much smaller cost penalty.

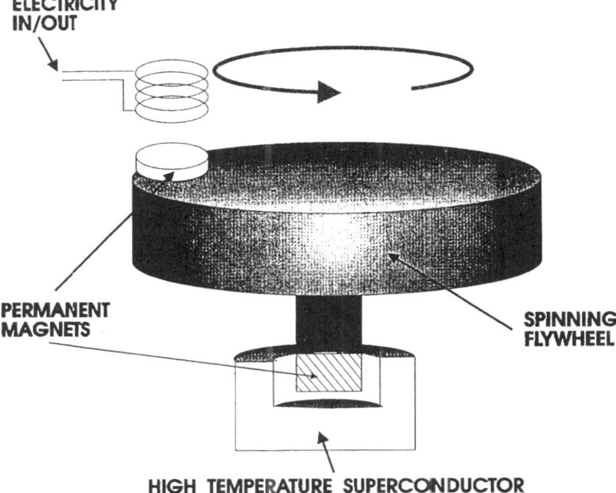

Figure 1.4. Conceptual drawing of a flywheel supported by a superconducting levitation bearing, for storage of power.

1.8. FUTURE OPPORTUNITIES AND CHALLENGES

The U.S. Department of Energy sponsors research at the National Laboratories, including some on applications of high-temperature superconductors. Their object is to develop the technical capability for industry to produce a wide range of advanced energy-efficient products: transmission and distribution cables, SMES (superconducting magnetic energy storage), motors, and generators. This is the major federal effort on energy applications of HTSC.

The research program definitely is an evolving one.[14] The focus today is on making HTSC wire, which is essential to everything downstream. Indeed, without uniform long lengths of high performance HTSC wire, there can be no HTSC electric power devices.

Likewise, the particular compounds of greatest interest have evolved over time, too. In 1987–1989, tremendous attention was given to YBCO, and thus its properties were measured in greater detail than the other compounds. In 1989–1991, led by wire-making accomplishments reported from Japan,[15] BSCCO was the subject of greatest interest. The thallium compounds, TBCCO, were given very little attention prior to 1991, because of fears that thallium (a relatively volatile heavy metal) was extremely toxic, and therefore was dangerous to have in the laboratory. By 1992, the rather limited progress with YBCO and BSCCO encouraged more researchers to take a fresh look at the thallium compounds. Then in 1993 HgBaCaCuO came along. No one can say whether the thallium or mercury compounds will eventually be more suitable for wire than BSCCO or YBCO.

Simultaneously, interest has grown in new refrigeration methods to produce temperatures intermediate between that of liquid helium (4 K) and liquid nitrogen (77 K). Liquid neon (28 K) is an unlikely candidate, because it is so expensive and scarce that it would have to be contained in a closed-cycle system, not allowed to boil off. Liquid hydrogen (20 K) has already been put to use in bubble chambers for physics research, but it can explode if ignited, and hence may be too dangerous for widespread applications. Engineers are hopeful of finding new types of refrigerators that will reach intermediate temperatures without paying the penalty (in thermodynamic efficiency) associated with cooling all the way down to 4 K.

Meanwhile, the aura of attention has given a boost to low temperature superconductivity. Storage of electricity via superconducting magnets was demonstrated years ago on a small scale; that is now being scaled up. Magnetically levitated trains, already demonstrated in Germany and Japan, may be built in America using liquid helium refrigeration (an Orlando to Disney World line is proposed). Major accelerators for physics research are under construction around the world. These are all projects in the several billion dollar range. Without the excitement of the new discoveries of ceramic superconductors, they might still be on the drawing boards.

The entire applications program is motivated by the realization that electric energy savings could be realized throughout all sectors of the economy if HTSCs were to "come true." The rosy predictions are by no means false; rather, it is a very difficult and challenging task to work through (or around) all the obstacles to implementing HTSCs in "the real world."

BIBLIOGRAPHY

The majority of this chapter first appeared as the article *Ceramic Superconductors* by T. P. Sheahen in *Magill's Survey of Science: Applied Science Series*, copyright © 1993, and is reprinted by permission of the publisher and copyright holder, Salem Press, Inc.

John Bardeen, "Historical Introduction," in *Theories of High-Temperature Superconductivity* (Addison-Wesley, Reading, MA: 1988). This chapter is much more readable than a standard textbook on theory of superconductivity; it provides a number of important and interesting details about the period 1986–1987, when the ceramic superconductors were first discovered.

S. J. Dale, S. M. Wolf, and T. R. Schneider, *Energy Applications of High-Temperature Superconductivity, Volume I: Extended Summary Report*, Report ER-6682, February 1990. (Request copy from Electric Power Research Institute, Research Reports Center, P.O. Box 50490, Palo Alto, CA 94303). This report goes into more technical detail on several of the specific devices that were briefly described above, and contains a number of explanatory drawings.

Robert M. Hazen, *The Breakthrough: The Race for the Superconductor* (Summit: 1988). This book, written by an active participant in the early research pertaining to high-temperature superconductors, conveys the excitement of the rush to understand these new materials and helps the reader understand why scientists were so surprised by these materials.

U.S. Congress, Office of Technology Assessment, *Commercializing High-Temperature Superconductivity*, OTA-ITE-388 (U.S. Government Printing Office, Washington, D.C.: June 1988). This report, directed toward the nonspecialist, provides an overview of the most likely applications as perceived in 1987 at the outset of the research activity.

REFERENCES

1. H. K. Onnes, *Leiden Comm.* **120b, 122b, 124c** (1911).
2. J. G. Bednorz and K. Mueller, *Z. Phyzik* **B64,** 189 (1986).
3. M. K. Wu *et al., Phys. Rev. Lett.* **58,** 908 (1987).
4. H. Maeda *et al., Japanese J. Appl. Phys.* **27,** L209 (1988).
5. Z. Z. Sheng and A. M. Hermann, *Nature* **332,** 55 (1989).
6. A. Schilling *et al., Nature* **363,** 56 (1993).
7. M. Nunez-Regueiro *et al., Science* **262,** 97 (1993); and C. W. Chu *et al., Nature* **365,** 323 (1993).
8. D. H. Freedman, *Science* **255,** 158 (1992).
9. M. Tinkham, *Introduction to Superconductivity* (Krieger Publishing Co.. Malabar, FL: 1980).
10. K. Sato *et al., IEEE Trans. Magn.* **MAG-27,** 1231 (1991).
11. A. Otto *et al., IEEE Trans. Appl. Superconductivity* **3,** 919 (1993).
12. Y. S. Cha and J. R. Hull, private communication.
13. J. L. Wu *et al., IEEE Trans. Magn.* **MAG-27,** 1861 (1991).
14. J. G. Daley and T. P. Sheahen, *Proc. Amer. Power Conf.*, Chicago, 1992
15. H. Mukai, *Proc. Third Int'l. Symp. Supercond.* (Sendai, Japan, November 6–9, 1990).

2

Magnetism and Currents in Superconductors

The first concept that comes to mind upon hearing the word superconductor is zero resistance. However, there is another equally significant aspect of superconductors that is less widely recognized—the unique magnetic properties of superconductors. This chapter introduces the reader to the magnetic side of superconductivity. The presentation here emphasizes the macroscopic aspects of the subject; it deals with magnetic fields and currents in the familiar (older) superconductors, deferring any mention of the new high-temperature superconductors (HTSCs).

Inevitably, it is necessary to decide what material to include at what point in a presentation, and what to leave out. In this explanation of "old superconductivity," the classical-physics tools of thermodynamics and Maxwell's equations are used. Not only is this chapter limited to the low-temperature superconductors, it does not use the quantum-mechanical explanation for superconductivity, known as the BCS theory. Here we only mention the BCS theory, waiting until Chapter 6 to present more detail. (The BCS theory is not mandatory for describing the observed behavior of superconductors of practical interest to engineers.)

Even so, certain supplementary concepts are indispensable. Thus, in this chapter the concepts of coherence length and penetration depth are introduced, but not really justified on any theoretical basis. That decision represents a trade-off between being thorough on the one hand and being clear about a limited amount of content on the other. Flux quantization and the notion of vortex lines are essential to understanding magnetism within superconductors. The prevention of vortex-line motion, known as flux pinning, is a key design goal in making practical superconducting wire.

One concept having practical consequences for current flow in superconductors is the Bean Critical State Model, which is described here. We also distinguish between Type I and Type II superconductors; only the latter carry high currents, and hence all practical wire is made from Type II materials.

2.1. ORIGINS OF SUPERCONDUCTIVITY

Perhaps the least celebrated similarity between the new and old superconductivities is that both were discovered empirically at a time when theory predicted no such phenomenon. The discovery in 1911 of superconductivity[1] is by now a familiar story; however, it is not

15

widely remembered that H. K. Onnes's experiments were directed toward finding a steady rise in electrical resistivity with decreasing temperature. (Prevailing theory at the time held that the free electrons in the metal would eventually freeze out at sufficiently low temperatures.) Most semiconductors show rising resistivity as the temperature falls; indeed, germanium is commonly used as a low-temperature thermometer because of its steeply rising resistivity. Metals, on the other hand, level off to a low value of resistivity near absolute zero, mainly due to impurities.[2] In some metals (i.e., ones that are poor conductors at room temperature), the resistivity suddenly vanishes at very low temperatures, and the material becomes superconducting.[3]

A second important discovery about superconductors is the *Meissner effect*,[4] which was found experimentally in 1933 without any theoretical basis. A metal expels any magnetic

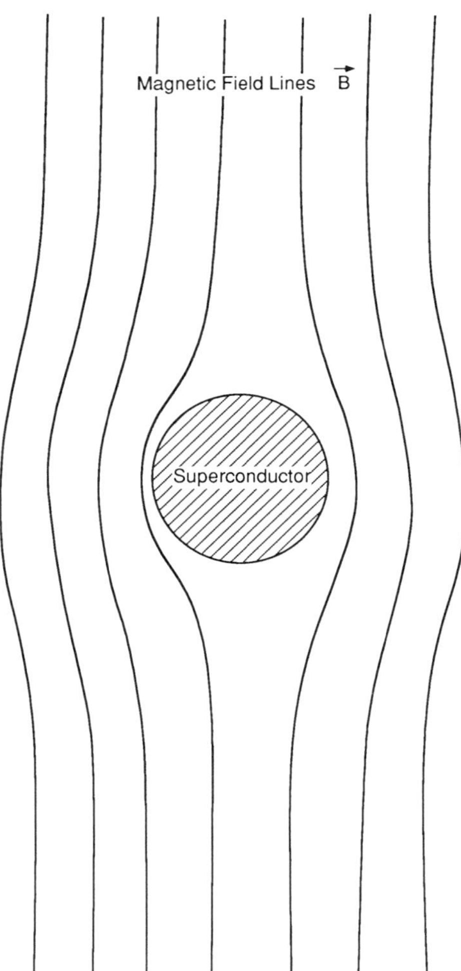

Figure 2.1.   The Meissner effect in a superconductor, showing the distorted magnetic field lines.

field inside it when it cools through T_c and becomes superconducting. By expelling the field and thus distorting nearby magnetic field lines, as shown in Figure 2.1, a superconductor will create a strong enough force field to overcome gravity. This gives rise to the memorable photos of a small magnet floating freely above a cooled block of superconductor.

Superconductivity remained an empirical science for several decades. After quantum mechanics was introduced in the late 1920s, theorists gradually began to suspect that superconductivity and superfluidity were quantum phenomena, and semi-empirical theoretical rules for superconductors were developed in the 1940s. Shortly after World War II, isotopes of various elements became available, and soon the isotope effect[5] was discovered. Here the transition temperature T_c varies as $1/M^{1/2}$, where M is the mass of an isotope of a particular element. This pointed to the importance of lattice vibrations (whose frequency would be proportional to $M^{1/2}$) in mediating superconductivity.

In the 1950s, it was gradually understood that the principal mechanism was a coupling between the electrons and lattice vibrations. This culminated in the Bardeen–Cooper–Schrieffer (BCS) theory of 1957.[6] It took 45 years to develop this theory, but it proved to be a very good theory indeed. By the early 1960s, superconductivity was considered to be a "mature" science and attention shifted to engineering applications.

2.2. THE MEISSNER EFFECT

Because the zero-resistance feature of superconductors was discovered first, it is widely believed that this is the most fundamental property of superconductors. Actually, the Meissner effect is of equal or greater significance, and plays a central role in the magnetic phenomena associated with superconductivity.

As stated above, the Meissner effect is the expulsion of a magnetic field from within a superconductor. It is important to be precise here. This expulsion is different from merely not letting in an external field; any metal with infinite conductivity would do the latter. If a magnetic field is already present, and a substance is cooled through T_c to become a superconductor, the magnetic field is expelled. The significance of the difference is that the Meissner effect cannot be explained merely by infinite conductivity.[7] Rather, it is necessary to develop a totally different picture of what is going on inside the superconductor.

No superconductor can keep out very strong magnetic fields. In fact, at any temperature (below the transition temperature T_c, of course), there is some magnetic field of sufficient strength such that the Meissner effect can be overcome and superconductivity vanishes. This is known as the *critical magnetic field* and is denoted by $H_c(T)$. At zero temperature, the upper limit of critical magnetic field is $H_c(0) = H_0$. At T_c, the critical magnetic field goes to zero: $H_c(T_c) = 0$. It is desirable to find superconductors with high critical field values, and these are generally associated with materials having a high T_c value.

A typical *type I superconductor* excludes all magnetic fields below H_c and admits magnetic fields without hindrance when H exceeds H_c. This behavior is termed *perfect diamagnetism*. In any material, the applied magnetic field $\mathbf{H}$ is related to the magnetization $\mathbf{M}$ and the magnetic induction $\mathbf{B}$ by the simple relation[8]

$$\mathbf{B} = \mu_0(\mathbf{H} + \mathbf{M}) \tag{2.1}$$

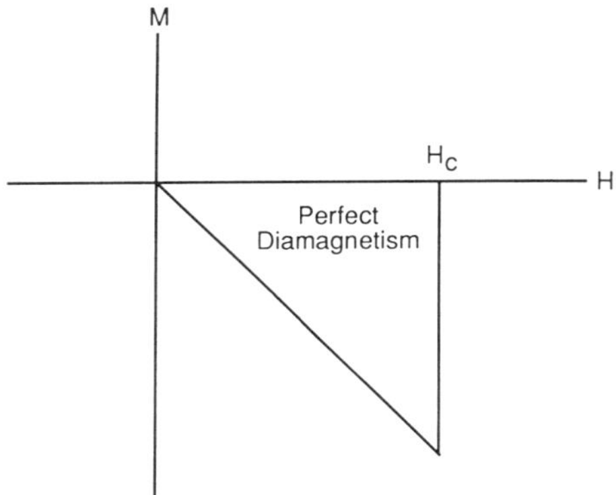

Figure 2.2. The variation of magnetization **M** with applied magnetic field **H** for a simple (pure element) superconductor. The material shows perfect diamagnetism below H_c. Some authors plot $-$**M** in such drawings, thus shifting the picture to the first quadrant.

In a perfect diamagnet, **M** = $-$**H**, so that $B = 0$. This exact cancellation is shown in Figure 2.2. For any value of H, there is exactly one corresponding value of **M**, and **B** is either zero or μ_0**H**. This holds true regardless of the path by which the magnetic field was imposed.

From the time of Onnes's original discovery until the Messner effect was reported in 1933, no one thought that the superconducting state was a thermodynamic equilibrium state. The single-valuedness of the *M–H* curve came as a shock, and demonstrated that the transition from the normal to the superconducting state represents a phase transition.

The route to understanding this phenomenon relies upon remembering the thermodynamic principle that in nature, the free energy is always minimized at equilibrium.[9] With this in mind, the superconducting state must have the lowest free energy in its temperature range, and the normal state must have the lowest free energy at higher temperatures. The pathway by which magnetic fields intervene must have to do with free energy minimization.

The free energy is equal to the work done to achieve a particular thermodynamic state. In the case of a superconductor, the condition that **B** = 0 requires it to have a magnetization **M** = $-$**H**. The work done on a superconductor moved from infinity to a position **r** near a permanent magnet[7] is the integral over $-$**M** $\cdot$ *d***H**, and this is also the increment in free energy *dF*. To calculate the difference in free energy between a superconductor in a zero applied field and in any other applied field, one can easily carry out the integral to find

$$F_s(H) - F_s(0) = \mu_0 H^2/2 \qquad (2.2)$$

The free energy curve for the superconducting state is a simple parabola in H.

Meanwhile, the normal state has no special magnetic properties; ignoring minuscule susceptibility, we set **M** = 0 throughout. Then, whatever the free energy of the normal state might be, it is the same in any applied magnetic field:

$$F_N(H) - F_N(0) = 0 \qquad (2.3)$$

The free energy of the normal state is a flat line, which will be crossed by the parabola of the superconducting state at some value of magnetic field H, as shown in Figure 2.3.

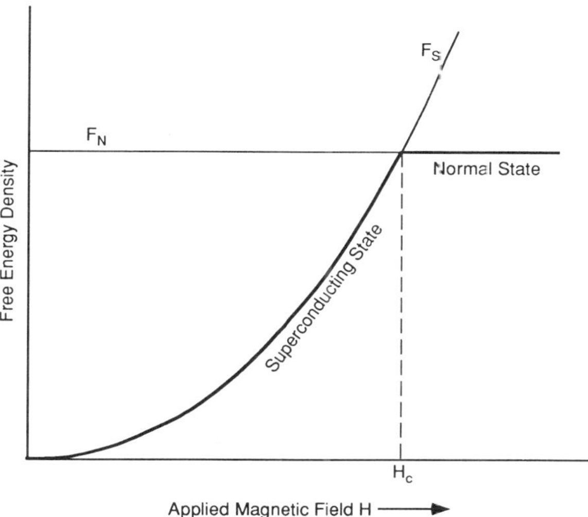

Figure 2.3.   Free energy curves for both normal and superconducting states; the point of crossover defines the critical magnetic field H_c. This schematic relationship holds for any temperature below T_c, although the numerical values of F and H vary with temperature. (From Ref. 7; reprinted by permission of John Wiley & Sons, Inc.)

The normal and the superconducting states will be in equilibrium when their free energies are equal; that is, at the *critical magnetic field* H_c, $F_N(H_c) = F_s(H_c)$. At that point, utilizing equations (2.2) and (2.3) quickly produces an expression for the difference in free energy ΔF between the normal and superconducting states:

$$\Delta F = F_N(H = 0) - F_s(H = 0) = \mu_0 H_c^2/2 \tag{2.4}$$

The value of H_c determined in this way is called the *thermodynamic critical field*.

This can all be experimentally verified via an entirely different pathway.[10] The entropy of any system is the derivative of the free energy $S = -(\partial F/\partial T)_V$, and the specific heat is $C_V = T(\partial S/\partial T)_V$. But the specific heat is a readily measured quantity (which will be discussed more fully in Chapter 17). For low-temperature superconductors, the normal state can be produced by applying a strong magnetic field, so specific heat data can be obtained in both states over the entire temperature range. Therefore, by starting with experimental specific heat data and integrating twice, the free energy can be recovered for each state, and the difference ΔF can be calculated at any temperature. When compared with independent measurements of $H_c(T)$, the agreement is excellent.

The Meissner effect is a very important characteristic of superconductors. Among the consequences of its linkage to the free energy of the superconductor are the following facts: (a) the superconducting state is more ordered than the normal state; (b) only a small fraction of the electrons in a solid need participate in superconductivity; (c) the phase transition must be of second order; that is, there is no latent heat of transition in the absence of any applied magnetic field; and (d) superconductivity involves excitations across an energy gap. These all proved to be important clues for understanding the fundamental nature of superconductivity.[11]

2.3. THE LONDON EQUATION

The Meissner effect could not be explained by any conventional model of electricity in solids, but a bold hypothesis was put forth by F. and H. London[12]: Since current flows unimpeded within a superconductor, let there be circulating currents inside the superconductor which set up a magnetic field that exactly cancels the magnetic field being applied externally. The form required for such a circulating current turns out to be surprisingly simple; we follow Kittel's presentation[7] here.

Recalling that magnetic field $\mathbf{B}$ is related to the vector potential $\mathbf{A}$ by $\mathbf{B} = \nabla \times \mathbf{A}$, the London hypothesis makes the current density $\mathbf{j}$ linearly proportional to $\mathbf{A}$:

$$\mathbf{j} = \frac{-1}{\mu_0 \lambda_L^2} \mathbf{A} \tag{2.5}$$

This so-called London equation is dramatically different from the normal Ohm's law, $\mathbf{j} = \sigma \mathbf{E}$. (The proportionality constant seems a bit contrived; it will become apparent soon.) From here on, Maxwell's equations do the rest.[8] The vector potential can be exchanged for the magnetic field by taking the curl of both sides and obtaining

$$\nabla \times \mathbf{j} = \frac{-1}{\mu_0 \lambda_L^2} \mathbf{B} \tag{2.6}$$

But we know from Maxwell's equations that, in the absence of a time-varying electric field,

$$\nabla \times \mathbf{B} = \mu_0 \, \mathbf{j} \tag{2.7}$$

taking the curl of this equation, we have

$$\nabla \times \nabla \times \mathbf{B} = \mu_0 \nabla \times \mathbf{j} \tag{2.8}$$

Now, another Maxwell equation says $\nabla \cdot \mathbf{B} = 0$, so this reduces to

$$-\nabla^2 \mathbf{B} = \mu_0 \nabla \times \mathbf{j} \tag{2.9}$$

and invoking equation (2.6) above, this yields

$$\nabla^2 \mathbf{B} = \frac{1}{\lambda_L^2} \mathbf{B}. \tag{2.10}$$

The only *constant* solution inside the superconductor must be $\mathbf{B} = 0$, which is another way of saying that magnetic fields are excluded. The variable solution has the general form

$$B(x) = B_0 \exp\left(-x/\lambda_L\right) \tag{2.11}$$

This explains the contrivance of the proportionality constant relating j to A. The value λ_L is called the *London penetration depth*,[12] and will be discussed more fully in Section 2.5 below.

The hypothesis of circulating shielding currents thus give a concise account of the Meissner effect; $\mathbf{j} \propto \mathbf{A}$ is all that is needed. Years later, when the BCS theory came along and justified the London equation, the issue was settled satisfactorily.

2.4. TYPE I AND TYPE II SUPERCONDUCTORS

So far, we have discussed the temperature and magnetic field properties of superconductors, but have not touched upon the current flowing in them. That mirrors actual history.
Prior to about 1960, superconductors were interesting from the point of view of physics, but
had no practical applications because they couldn't carry any significant amount of current.
Only when a new class of superconductors was discovered did practical applications become
possible. The two classes are distinguished as type I and type II superconductors, also known
as *soft* and *hard* superconductors, because of the dramatic difference in their magnetic and
current-carrying properties. There are such enormous differences between J_c in the two types
that an entire industry[13] is based on type II superconductors, while type I superconductors
have only very limited applications.

The current density (j or J) is current divided by the cross-sectional area through which
it flows; it is usually given in amps per centimeter squared. Just as superconductors have a
critical temperature T_c and a critical magnetic field H_c, so too do they have a critical current
density J_c as well. That there must be some upper limit to the current density in a
superconductor is required[3] by the relationship between current and magnetic field; for a
wire of radius a carrying current I, the magnetic field at the surface is $I/2\pi a$. The current
cannot exceed the amount that produces a critical magnetic field H_c at the superconductor,
which implies a critical current $I_c = 2\pi a H_c$, and $J_c = 2H_c/a$. For real superconductors, the
actual critical current density is less than this upper limit and the actual current is limited by
other physical mechanisms.[14]

For a type I superconductor, critical current is simply a consequence of the critical
magnetic field H_c. Since H_c is low in type I superconductors, their critical current densities
J_c are likewise low. This is why type I superconductors have not been of interest to the electric
utilities or magnet builders.

In a type II superconductor, the relationship is much more complicated; indeed, Figure
2.4 shows the critical surface in the 3-dimensional space of temperature, magnetic field, and
current.[15] This is known as a THJ plot, after the three axes. The critical current is no longer
related in a trivial way to the magnetic field.

The response to an applied magnetic field is quite different in the two cases. In Section
2.2 above, the behavior described is that of a type I superconductor: there is exact cancellation
of an applied magnetic field **H** by an equal and opposite magnetization **M**, resulting in **B** =
0 inside the superconductor. Above the critical field, superconductivity vanishes. It is all very
simple.

In type II superconductors, the Meissner effect is partially circumvented. The magnetic
field starts penetrating into the material at a lower critical field H_{c1}. Penetration increases
until at the upper critical field H_{c2} the material is fully penetrated and the normal state is
restored. Figure 2.5 shows this behavior, in which **M** rises to a negative maximum at H_{c1},
but then **M** retreats as flux lines begin to penetrate. The cancellation of **H** by **M** is no longer
perfect, and **B** is finite within the superconducting material.

Thus we seem to have a major violation of the principle that superconductors exclude
magnetic fields, for obviously magnetism and superconductivity co-exist in a type II
material. To understand this we must introduce the concept of a *coherence length*, within
which superconductivity takes place.

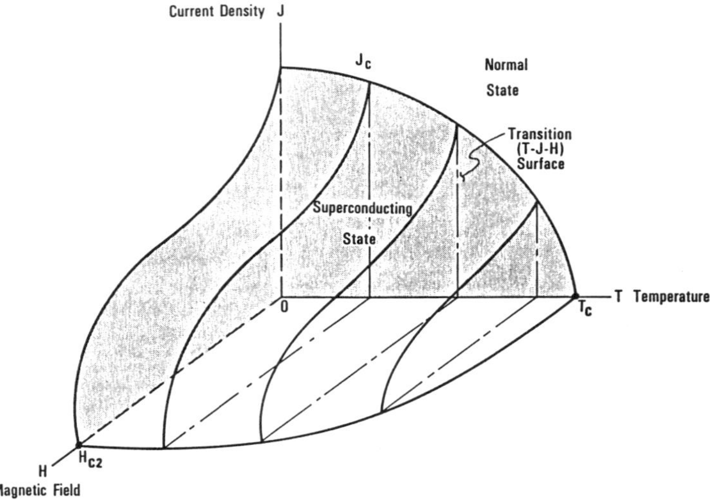

Figure 2.4. The relationship between temperature, magnetic field, and critical current in a superconductor. Anywhere inside the surface, the material is superconducting. (From Ref. 15.)

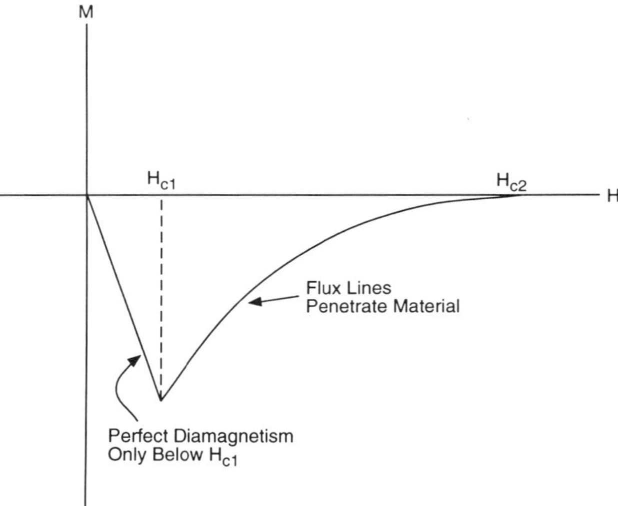

Figure 2.5. Magnetization as a function of applied magnetic field for an ideal type II superconductor. Above H_{c1} magnetic flux begins to penetrate the material until at H_{c2} there is no magnetization left at all, and the material returns to the normal state.

2.5. PENETRATION DEPTH AND COHERENCE LENGTH

In Section 2.3 above, the solution to the London equation showed that a magnetic field could penetrate a little way into a superconductor; equation (2.11) says that the field falls off exponentially over a mean distance λ_L, known as the *penetration depth*. Typically this is less than 0.05 μm, and so a macroscopic sample of a superconductor could safely be said to entirely exclude magnetic fields. But what happens on a scale smaller than λ?

A central contribution to the theory of superconductivity was made by Ginsburg and Landau,[16] who introduced the notion of a *coherence length*, generally denoted by ξ. By the time of their hypothesis, superconductivity was agreed to be an interaction among electrons, and so it was natural to imagine this interaction occurring within some limited distance. Basically, ξ is a measure of how likely it is that a pair of electrons will interact with each other.

The Ginsburg–Landau equations were the first use of quantum-mechanical wave functions to describe superconductivity, and were clearly a major theoretical step forward. Several years later, Abrikosov[17] showed how type II superconductivity arises from the Ginsburg–Landau model. Soon after the BCS theory[6] appeared, Gorkov[18] derived the Ginsburg–Landau equations from BCS. In this way, the governing theory of type II superconductivity became the GLAG theory, for Ginsburg–Landau–Abrikosov–Gorkov. We will return to the concept of a coherence length again in Chapter 6. For now, however, we need only note that the key to understanding how superconductivity and magnetism can co-exist lies in the relationship between penetration depth λ and coherence length ξ.

The *intrinsic coherence length* ξ_0 can be calculated[19] from the Ginsburg–Landau model, and often exceeds 0.3 μm in type I superconductors.[7] For a pure metal, the actual coherence length is about the same, but in alloys or impure compounds it is much smaller, because the mean free path for electrons is smaller. Type II metals fall into this category. Therefore, the superconducting properties of a material can be changed by altering the electron mean free path, such as by introducing lattice defects. The exploitation of this principle has led to a number of advances in engineering the best materials for practical superconductors.

For convenience of categorizing superconductors, the Ginsburg–Landau ratio is defined as $\kappa = \lambda/\xi$. In type I superconductors $\kappa < 1$; that is, the coherence length is larger than the penetration depth. The fundamental difference in type II superconductors is that this relationship is reversed, i.e., $\kappa > 1$. (Actually, the breakpoint comes at $\kappa = 0.707$, a minor distinction.)

A comparison of λ and ξ in type I and type II superconductors is shown in Figure 2.6. For type I ($\kappa < 1$), any magnetic field will not penetrate far enough to affect the electrons within a coherence length. On the other hand, for type II ($\kappa > 1$), superconductivity is confined to within such a short coherence length that it can still live with a nearby magnetic field that has penetrated the material.

2.6. FLUX QUANTIZATION

Looking at Figure 2.5, it is evident that above H_{c1} the magnetic field penetrates the superconductor. There must be some reason why it is energetically favorable to have this. Before searching for that reason, it is first necessary to introduce one more fact—the quantization of magnetic flux.

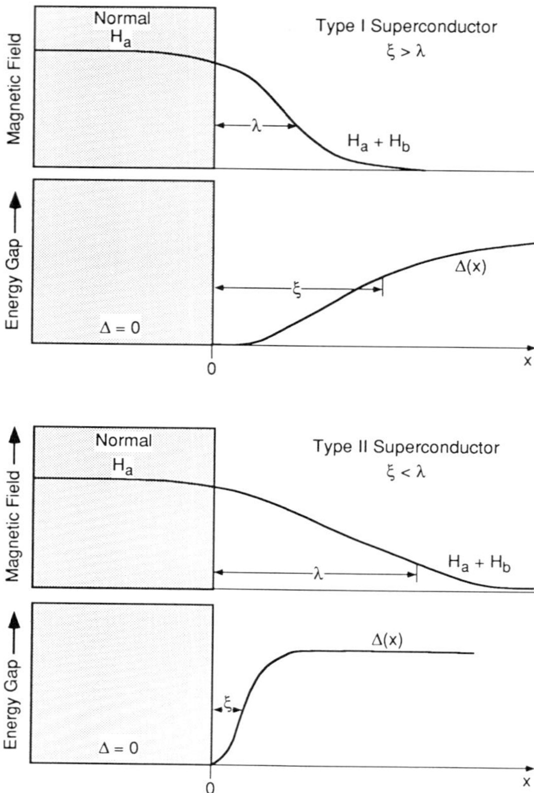

Figure 2.6. The magnetic field and the energy-gap parameter $\Delta(x)$ at the boundary between superconducting and normal regions, for both type I and type II superconductors. The differences are due to differences in penetration depth λ and coherence length ξ. (After Ref. 7.)

Just as it took some centuries for scientists to realize that electric charge was quantized (the electron), so it came as a surprise to find that magnetic flux lines also have discrete values. In the mid 1960s,[20] individual magnetic flux lines were identified by allowing very fine particles of iron to settle on the surface of a superconductor in a magnetic field. This is called a "decoration" experiment. Heavy black dots collected wherever a flux line emerged from the surface of the material. Very recently, quantized flux lines were studied by passing a beam of neutrons through a niobium crystal and detecting their scattering angles.[21] Figure 2.7 shows the pattern in which individual magnetic flux lines penetrate a type II superconductor.

Flux quantization has been observed for many type II superconductors, and for the HTSCs as well, but for type I superconductors, there is no such pattern. This is a very clear difference between type I and type II. The explanation is quantum mechanical, involving the phase of a wave function around a loop. The derivation is presented in a variety of books (see, for example, Ref. 7), and yields the *flux quantum*, or *fluxoid*,

$$\Phi_0 = h/2e = 2.0678 \times 10^{-15} \text{ Webers} \tag{2.12}$$

Figure 2.7. Neutron scattering pattern of a niobium crystal, showing that magnetic vortex lines are quantized. In this computer-drawn plot, the intensity of scattered neutrons has been mapped into a gray-scale. The six-fold pattern of first-order peaks has the strongest intensity, with higher order peaks having weaker intensity. (From Ref. 21.)

where h is Planck's constant and e is the charge on the electron. The important consequence of this quantization of flux is that any magnetic field penetrating a superconductor will be quantized.

In fact, the magnetic field penetrates one flux-quantum at a time, each one being a single normal vortex that stands surrounded by superconducting material. That is exactly what Figure 2.7 shows. Moreover, the flux lines array themselves into a regular pattern, known as the *Abrikosov lattice*.[22] This configuration is very important for *flux pinning*, which determines how much current a superconductor can ultimately carry without dissipating energy. In Section 2.8 we discuss this in more detail; in Chapter 14 we consider the role of flux pinning for the HTSCs. For now, it suffices to think about only one single flux line.

2.7. THE VORTEX STATE

To understand the reason why a state of mixed normal and superconducting regions can co-exist, we turn once again to thermodynamics and consider the free energy. Whenever there is a surface between two phases of a material, there will be some associated surface energy. If the surface energy is positive, the material will minimize the surface area by having only one borderline. If the surface energy is negative, it will be energetically advantageous to have many borders between regions, and the equilibrium configuration will be one of mixed normal and superconducting zones.

For any superconductor, when a magnetic field is first applied to it, the Meissner effect will try to expel that field. As the field is increased, it costs more energy to expel it, and the free energy difference between normal and superconducting states diminishes. A type I superconductor always has a positive surface energy, so it expels a field as much as it can, until the normal and superconducting free energies are equal, beyond which point superconductivity vanishes.

By contrast, a type II superconductor expels a field for a while (up to H_{c1}), but as the applied magnetic field increases further the surface energy associated with the border[3] becomes negative, so it is energetically favorable to let some of the material switch to the normal state in order to accommodate the flux lines of the field. Small circular regions form, which are normal inside but have superconducting currents circulating outside to shield the magnetic field—hence the name *vortex state*.

The first penetration of flux lines occurs at H_{c1}, where a single flux quantum Φ_0 occupies a core of approximate radius equal to the coherence length ξ. However, the spacing between them approximately equals the penetration depth λ. Thus the measured value of H_{c1} offers a way to estimate λ:

$$\Phi_0 = H_{c1}\,\pi\lambda^2 \tag{2.13}$$

In the same way, we can estimate the coherence length ξ by examining the final state of the material just before it goes normal. At H_{c2}, the maximum number of flux lines have penetrated; there is no room for any more. The cores are packed tightly, only ξ apart. This condition corresponds to:

$$\Phi_0 = H_{c2}\,\pi\xi^2 \tag{2.14}$$

(When this treatment is done rigorously, the only change is a minor numerical factor.)

Note from equations (2.13) and (2.14) that the ratio of H_{c2} to H_{c1} is $\lambda^2/\xi^2 = \kappa^2$. In extreme type II superconductors (of which the HTSCs are examples), κ is very great and H_{c2} can be many Tesla, even though field penetration begins at H_{c1} of a small fraction of a Tesla.

It is helpful to relate all this to some real type II materials, such as NbTi and Nb_3Sn. There, lines of magnetic flux penetrate the superconductor anytime the magnetic field exceeds H_{c1}, usually around 0.01 T, but superconductivity is not destroyed until the magnetic field exceeds H_{c2}, of the order of 10 T or more. In this mixed state, large currents (perhaps 10^6 A/cm^2) flow in the presence of large magnetic fields (perhaps 6T) at 4.2 K. As Figure 2.4 displays, both J and H must diminish to zero if the superconductor is operated very near its critical temperature.

One final point is that the *thermodynamic critical field* H_c is the geometric mean of the two critical fields:

$$H_c^2 = H_{c1}H_{c2} \qquad\qquad (2.15)$$

Furthermore, the Ginsburg–Landau ratio enters in the form $H_{c1} = H_c/\kappa 2^{1/2}$ and $H_{c2} = 2^{1/2}\kappa H_c$. For a type II superconductor, the only way to determine H_c independently is via calorimetry,[7] integrating specific heat data to obtain the free energy.

2.8. CURRENT FLOW IN SUPERCONDUCTORS

The individual lines of magnetic flux penetrating the superconductor each act as a small independent vortex, forming a cylindrical region of normal material. It is fair to ask whether these normal cylinders get in the way of the superconducting current. Fortunately, the cylinder radius is only of the order of the coherence length. In a type II superconductor, the coherence length is smaller than the London penetration depth, and vortices are repelled from one another. Consequently, the normal cylinders are confined to small local areas, and there is still a superconducting path through the material. The current will pass unimpeded as long as that current does not disrupt the flux vortices.

There is a natural tendency for exactly such a disruption to occur. A magnetic flux line with a current passing nearby feels a force (the Lorentz force) that tries to move it sideways ($\mathbf{F} = \mathbf{J} \times \mathbf{B}$). The sideways motion of a flux line through the lattice dissipates energy and is therefore equivalent to resistance. If flux lines always moved, superconductors would not carry current without resistance. Fortunately, something else happens. A flux line can be stopped in its sideways motion by grain boundaries, impurities, or many other kinds of impediments. The flux line is said to be pinned. A nearby flux line, about one penetration length away, feels the influence of the pinned line. Thus, when another moving flux line bumps into that one, it too stops moving, and soon the traffic jam of flux lines form a lattice, none of which can move unless they all move at once. The word frozen is used to denote that the lattice of flux lines behaves very much like a solid, with mutual repulsion holding the lines a fixed distance apart.

In type II superconductors the critical current J_c depends on the detailed history of preparation. For example, it has become good engineering practice to introduce many defects into the metal in order to offer plentiful pinning sites. By enhancing the pinning strength, the allowable current is increased. For the workhorse of the superconductivity industry, NbTi and its composites with copper, the metallurgy of optimizing pinning to improve J_c has been studied extensively,[13] and is still being improved.

There is always some small resistance, caused by incomplete pinning, in a type II superconductor at nonzero temperatures. Historically, the value of this resistance has been negligibly small at 4 K, so the distinction between true zero and finite resistance has not usually been important. Only in applications such as MRI (magnetic resonance imaging) has this *flux creep* become important. In Chapter 14, we shall see that in HTSCs the finite energy dissipation due to flux motion is significant at practical temperatures.

2.9. THE BEAN CRITICAL STATE MODEL

When current flows inside a superconductor, how is its current density distributed within the material? In a type II superconductor, in its vortex state, there is some penetration of the magnetic field. Although it is quantized into flux lines, we can treat it classically for purposes of examining the macroscopic current. The current and magnetic field are, of course, intimately related.

Consider a wire in which the current flows in the z-direction: $\mathbf{J} = (J_r, J_\theta, J_z) = (0, 0, J_z)$. What does its magnetic field B look like under these circumstances? From Maxwell's equations we have $\nabla \times \mathbf{B} = \mu_0 \mathbf{J}$, so $\nabla \times \mathbf{B}$ has only a z-component. The radial component of $\mathbf{B}$ must be zero, or else magnetic fields would be diverging from within the wire, i.e., $\nabla \cdot \mathbf{B} \neq 0$, violating Maxwell's equations. The θ-component of $\mathbf{B}$ is all that is left. It cannot vary with θ, because of the circular symmetry of the wire. In fact, it can only vary with r (not z), since the current is constant down the wire and there is no difference from one point along z to the next. We thus conclude

$$\mathbf{B} = (0, B_\theta(r), 0) \tag{2.16}$$

and it immediately follows that $\partial B_\theta(r)/dr = J_z$. The same symmetry conditions on θ and z ensure that the current density varies only with radius: $J_z = J_z(r)$. B will have its maximum value at the surface, with lesser values interior to the wire.

Despite the presence of a magnetic field, the Meissner effect has not simply gone away. In fact, the attempt by the superconductor to enforce the Meissner effect ultimately determines the distribution of current within the wire.

The inner axis of the wire is the most protected location, from the point of view of the Meissner effect; near the surface is the easiest-penetrated region. To enlarge that protected region of zero field as much as possible, it makes sense for the current to flow preferentially near the outer surface of the wire. In response to even the tiniest applied potential, a current will start to flow on the surface. There being no resistance in the superconductor, the current will start to become infinitely large, in an extremely shallow surface layer. However, as J exceeds J_c in that shallow layer, it becomes normal, J decays to J_c, and superconductivity is restored.

The *Bean critical state model*[23] introduces the hypothesis that whenever a current I_0 flows in a superconducting wire, the current density at the outside will immediately jump up to the critical current value J_c. To transport more total amps, the shallow layer thickens, again carrying J_c, but over a larger cross-section. In this way, the current is always flowing with the critical current density J_c, but only the outer portion of the superconductor carries a current other than zero. The total current I_0 flows in an annulus near the outside. $J = J_c$ at the outer edge $r = a$, and also inward to another radius $r = b$ such that

$$I_0 = J_c \, \pi(a^2 - b^2) \tag{2.17}$$

The inner edge of the annulus is at

$$b = a\left(1 - \frac{I_0}{J_c \pi a^2}\right)^{1/2} \tag{2.18}$$

The Bean model has been verified by experimentally measuring currents at various depths within a superconductor.[24]

The Bean critical state model has several interesting consequences. The foremost of these is that, although any part of the wire is *able* to carry current density J_c, only that fraction of the wire that is *needed* to get the job done is actually put to work. Note that when $I_0 = j_c \pi a^2$, the *entire* wire is carrying the critical current J_c. Any further increase in I causes resistance to appear, and current no longer flows without requiring energy.

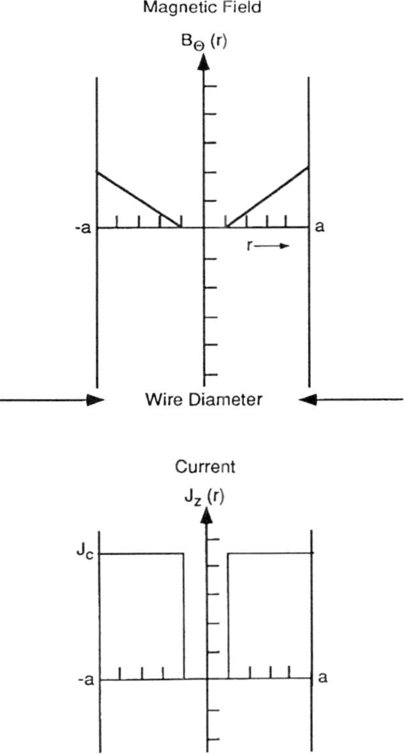

Figure 2.8. Magnetic field and current distribution according to the Bean critical state model.

In the annular ring where current J_c is flowing, $\partial B_\theta(r)/\partial r = J_c$, a constant, and is zero elsewhere. Thus, B varies linearly with r. The shapes of E and J are sketched in Figure 2.8.

It bears mentioning that whenever a magnetic field is present in a superconductor, the critical current density J_c is less than its value in zero field. This means that J_c itself varies through the wire, being higher in the interior. The effect of this dependence on applied field is to cause $J_z(r)$ to decrease gently with radius r, because the magnetic field is greater near the outer surface. This in turn changes the expected shape of the magnetization curve M versus H. This is inconsequential to the niobium superconductors, but may make a difference[25] in the HTSCs, for which the critical current falls off sharply even in small magnetic fields.

In later chapters, we will employ the Bean model to describe current flow in a variety of situations, including flow of magnetic flux lines, AC losses, etc.

2.10. HYSTERESIS IN SUPERCONDUCTORS

The quasi-static (including 60 Hz) magnetic behavior of materials is represented by plotting the magnetization **M** versus the applied magnetic field **H**. When **M** depends only upon the present value of **H** and on nothing else, then the material returns to its initial magnetic state when the applied field returns to its initial state and no energy is dissipated. In practice, copper behaves this way. In particular, the magnetization of copper is essentially zero for all applied magnetic fields. Iron, cobalt and nickel behave differently. Their magnetization depends on their histories as well as on the present value of the magnetic field.

After the applied magnetic field is cycled many times, the magnetization of iron responds by following the familiar hysteresis curve, with two possible values of $\mathbf{M}$ for each value of $\mathbf{H}$. After $\mathbf{H}$ is cycled once, $\mathbf{M}$ does not return to its initial value. Instead, magnetic energy has been converted to heat. This conversion is irreversible or dissipative. It accounts for the need to cool transformers. Whenever $\mathbf{M}$ is a single-valued function of $\mathbf{H}$, the material will conserve, not dissipate, magnetic energy when $\mathbf{H}$ is cycled. On the other hand, when $\mathbf{M}$ is not a single-valued function of $\mathbf{H}$, the material (iron or superconductor) will heat up when $\mathbf{H}$ is cycled. Just as magnet iron is called hard when it has a significant hysteresis loop, so a superconductor is called hard when it has a significant hysteresis loop.

In superconductors, hysteresis (with concomitant irreversibility) arises from flux pinning. When there is no flux pinning, there is likewise no irreversibility; that is, the magnetic behavior of the superconductor is perfectly reversible. As depicted in Figure 2.2, this is the case for type I superconductors. In an idealized type II superconductor, without any flux pinning, the behavior is also reversible, with $\mathbf{B}$, $\mathbf{H}$ and $\mathbf{M}$ uniquely related to one another. Referring back to Figure 2.5, if the applied magnetic field $\mathbf{H}$ is increased from zero to H_{c2}, and then decreased, the line will be exactly retraced as $\mathbf{H}$ decreases.

However, no real material gives the exact retracing indicated by the idealized curve. Structural imperfections or chemical impurities prevent flux lines from moving freely through the crystal; this is called pinning. By the 1960s, the niobium compounds were well known to pin flux lines at sites of crystal imperfections, and the practical goal of material engineers has been to introduce as many pinning sites as possible in order to permit high currents to flow under high magnetic fields.

In an attempt to understand the phenomenon of pinning better, Farrell *et al.*[26] went the other way, studying special lead–indium alloys deliberately contrived to minimize pinning (an ultrasoft superconductor). The best they could do looked like Figure 2.9: having increased H to H_{c2}, the decreasing-H path of $\mathbf{M}$ versus H deviated gradually from the upward path; the discrepancy was greatest near H_{c1}. When the applied field $\mathbf{H}$ was reduced entirely to zero, there was a small amount of flux trapped in the material, causing a small residual

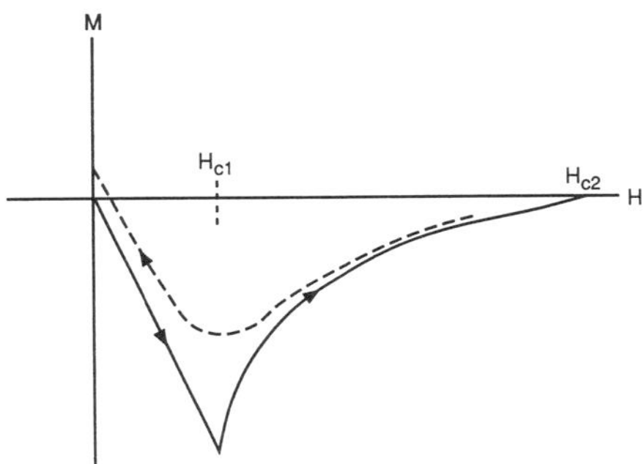

Figure 2.9. Near-perfect canonical type II magnetization behavior with almost no pinning. Because of flux pinning, the return path of magnetization (when field H is first raised to H_{c2} and then decreased) differs from the forward path. A small amount of magnetization remains even at the zero applied field.

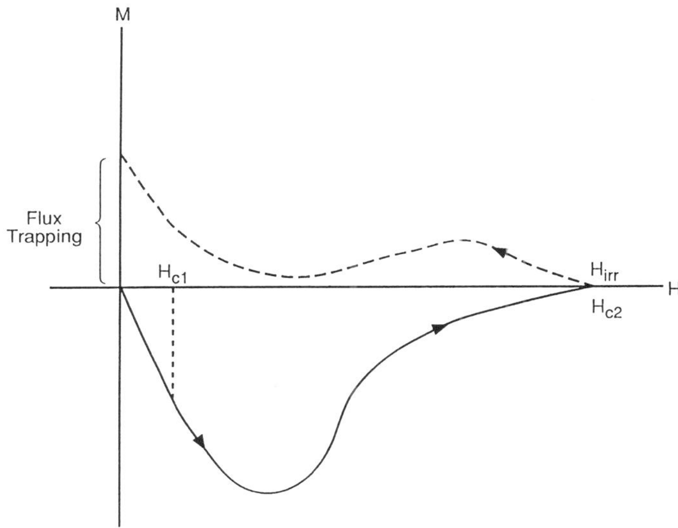

Figure 2.10. Realistic type II hysteresis with pinning. A well-engineered type II superconductor has very great flux pinning, allowing J_c to be large and causing a considerable excursion of the return path away from the forward path in the magnetization diagram. The field at which irreversibility sets in, H_{irr}, is virtually the same as H_{c2}.

magnetization **M**, much like a permanent magnet. By 1970, J. E. Evetts[27] had shown that this condition was attributable to pinning at phase boundaries, especially the surface. Thus, some flux pinning is practically unavoidable in type II superconductors, and a great deal of pinning is desirable for most practical applications.

A more realistic type II superconductor has a more complicated magnetic history, which is sketched in Figure 2.10. First, there is no sudden change in magnetization **M** as **H** passes through H_{c1}—only the deviation from linearity of the **M** versus **H** curve tells that the diamagnetism is no longer perfect and that some amount of flux is penetrating the material. Eventually, of course, increasing H leads to extensive flux penetration, **M** decreases, and at H_{c2} penetration is complete, $\mathbf{B} = \mu_0\mathbf{H}$ and the material goes normal. In the superconducting state, much of the flux is pinned inside the superconductor. As **H** is decreased, **B** remains elevated, and since $\mathbf{B} = \mu_0(\mathbf{H} + \mathbf{M})$, this means that **M** rises to a positive value. Further decline in **H** is accompanied by changes in **M** that leave **B** finite even when $\mathbf{H} = 0$, which is due to flux trapped in the material. This permanent magnetism will remain until the temperature is changed.

We shall take up this topic again in Chapter 14, which is concerned with flux motion in the HTSCs.

2.11. PRACTICAL SUPERCONDUCTING WIRE

The parameters of temperature, magnetic field, and current trade off against one another in any superconductor. We would like to have a single material that maximizes all three, but nature is never that cooperative. Furthermore, there are additional trade-offs to be made: in manufacturing wire with good stability and AC loss properties, it is necessary to construct a composite of many fine filaments of superconductor embedded in copper. For this, we prefer a ductile metal over a brittle crystalline material.

The clearest example of this competition comes in comparing NbTi with Nb_3Sn. Niobium-tin has better thermal, electric and magnetic properties (higher T_c, H_c, J_c) than niobium titanium,[13] but Nb_3Sn is an *A-15 compound*, which is brittle. Therefore, it is difficult and expensive to form into wire. NbTi, on the other hand, is a metal alloy having typical ductility. Actually, NbZr was preferred at first, but it did not extrude well without breaking. Consequently, the material of choice for most superconducting wire applications continues to be the nonbrittle NbTi (with T_c only 10 K). One leading maker of superconducting wire is Intermagnetics General Corporation. The multifilamentary wire used in MRI applications, in which NbTi strands are embedded in a copper matrix and then extruded into wire, is the major product of Intermagnetics General. NbTi constitutes about 95% of their business, while Nb_3Sn is still in the pilot-line stage 30 years after its discovery.

The difficulty of manufacturing Nb_3Sn explains the difference.[28] Intermagnetics General makes Nb_3Sn wire via their internal tin process: They begin with strands of niobium and dip them in tin. A collection of these strands is then embedded in bronze (bronze is a copper-tin alloy). This wire is extruded and wrapped into a coil or whatever shape is required of the final configuration. At this point it is still not superconducting, but it is no longer going to be bent, twisted, extruded or otherwise strained. The final stage is to heat treat the entire assembly at 700°C, and thereby cause some of the niobium and tin to diffuse together and form Nb_3Sn. The brittle superconducting substance is formed in its final geometrical shape. When the coil (or whatever shape) is immersed in liquid helium, the superconducting current flows in the many thin-walled cylinders of Nb_3Sn surrounding each niobium core. Figure 2.11 is a photograph of such wire.

Intermagnetics General also makes superconducting tape by a similar process: a strip of niobium is dipped in tin, thus coating both sides. A layer of copper (for conductivity in the normal state, so-called *quench protection*) is laminated to one side, and a layer of steel (for structural strength) is laminated to the other. Then the whole sandwich is heat treated under pressure; once again, diffusion leads to the formation of two thin regions of Nb_3Sn. After cooling, the tape is dipped in solder for ease of wiring connections.

Clearly, difficulties of this sort motivate the search for alternative materials. Several other A-15 compounds of niobium are also superconducting. V_3Ga has slightly higher H_{c2} and J_c than Nb_3Sn; V_3Ga was used routinely in the inner core of composite magnets to achieve higher fields. Nb_3Ge has $T_c = 23$ K, which is appreciably higher than the 18 K of Nb_3Sn. V_3Ga has been used routinely in composite magnets, but otherwise attempts to make practical wire from these compounds have been disappointing. Still, research continues. Here is one typical example:

The applied superconductivity group at Frascati in Italy strives to make wire from Nb_3Al. They hope that Nb_3Al will prove superior to Nb_3Sn and enable manufacture of modest-size magnets whose fields are greater than 10 T. The manufacturing process steps are as follows: (a) wrap a thin cylinder of copper with a "jelly roll" of interleaved niobium and aluminum foils; (b) drill 19 holes in a copper billet—each parallel to the long axis of the billet—then insert one jelly roll cylinder into each hole; (c) carry out all extrusion and related metal work so that each jelly roll becomes a hollow prefilament; (d) coat with a glass or ceramic insulation; (e) heat treat, at ~800°C, to promote the chemical reaction forming Nb_3Al; (f) add a resin to fill up space left vacant by the insulation; and (g) heat treat at 125°C to cure the resin. So far, Frascati has been unable to produce Nb_3Al in commercial quantities.

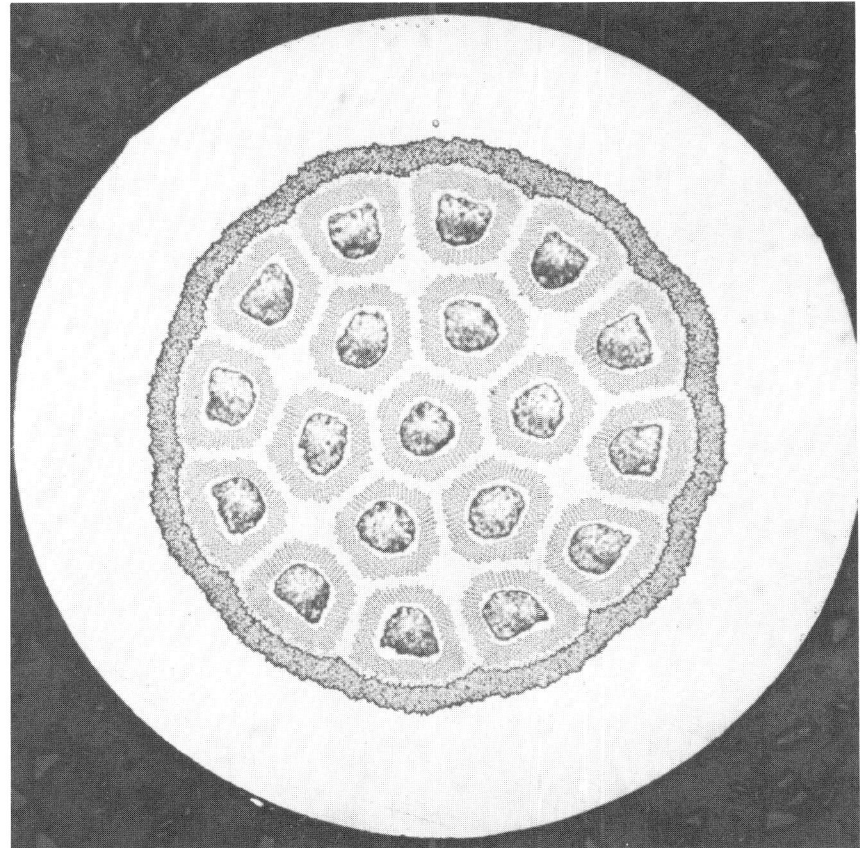

Figure 2.11.   Photograph of Nb_3Sn wire made by Intermagnetics General Corporation.

Their continuing effort is motivated by the higher strain limit (0.6% compared to 0.2% for Nb_3Sn) and the fact that short samples of Nb_3Al have twice the J_c of Nb_3Sn in fields of 10 T at 4.2 K.

This example is neither the best nor the worst case, but is illustrative of the struggle to make real wire from compounds that are basically brittle materials. The point is that some applications need high magnetic fields and high currents so badly that enormously expensive wire can be justified.

The value of $J_c = 150,000$ A/cm^2 in 6 T magnetic field became the benchmark for NbTi wire. However, the present state of the art exceeds 300,000 A/cm^2. For the Superconducting Super Collider (SSC), the designers pushed to this limit: the magnet carried a specification of 274,000 A/cm^2. Typical magnets, once constructed, suffer about 6% defective filaments, so 300,000 A/cm^2 delivers an *apparent* J_c of about 280,000 A/cm^2.

The record continues to advance gradually. In 1983, Furukawa Corporation in Japan obtained 383,000 A/cm^2 in NbTi wire, but in only a 5 T field. This gain was primarily due to an improved manufacturing process that allowed precise control of the NbTi filament diameter: the filaments, embedded in a copper matrix, were less than 1 μm in diameter, which contrasted with the previous best diameter of 5 μm. The most important aspect of maximizing J_c is to manufacture filaments of small (micron-sized) diameter. Furukawa's ability to extrude a copper/NbTi composite wire to such dimensions is certainly both an advance and a challenge; within a few years (1992), several manufacturers worldwide announced submicron filamentary NbTi wire.

Continuing improvements in related manufacturing technologies, such as photolithography and precise etching, are making it easier to make submicron-width conducting paths on circuit boards. There are theoretical reasons to believe that if paths are narrowed to less than 0.1 μm (1000 Å), still further improvements in J_c can be expected.

2.12. SUMMARY

This chapter has covered the magnetic and current-carrying properties of conventional (low-temperature) superconductors. (Many of these features are retained in the HTSCs as well.) The penetration depth is a unique property of superconductors that follows from the Meissner effect.

The difference between type I and type II superconductors can be understood by introducing the coherence length ξ and comparing it to the penetration depth λ. The Ginsburg–Landau ratio is $\kappa = \lambda/\xi$, and in simple (type I) superconductors $\kappa < 1$. However, when the coherence length is short compared to the penetration depth, $\kappa > 1$ and it becomes possible for lines of magnetic flux to penetrate into the material without destroying superconductivity. This is called the vortex state. Such type II superconductors are able to carry high currents in strong magnetic fields and consequently have converted a laboratory curiosity (superconductivity) into a practical engineering tool. The Bean critical state model is used to relate magnetism to current flow in the vortex state.

The interaction between flowing current and lines of magnetic flux will dissipate energy unless the flux lines are prevented from moving (flux pinning), and there is an art to achieving this goal, which involves the deliberate introduction of defects into the metal. Manufacturers of superconducting wire for practical applications—mostly in magnets—utilize a variety of processing techniques to optimize the current and magnetic properties of these materials.

Jumping ahead many chapters, it may be observed that all the troubles that arise in making LTSC wire are even worse with the HTSCs. The basic principles that govern type II superconductivity continue to hold in this new field.

REFERENCES

1. H. K. Onnes, *Comm. Phys. Lab. Univ. Leiden* **120B**, 3 (1911).
2. N. F. Mott and H. Jones, *Theory of the Properties of Metals and Alloys* (Clarendon Press, Oxford: 1936).
3. M. Tinkham, *Introduction to Superconductivity* (McGraw-Hill, New York: 1975).
4. W. Meissner and R. Ochsenfeld, *Naturwiss.* **21**, 787 (1933).
5. E. Maxwell, *Phys. Rev.* **78**, 477 (1950).
6. J. Bardeen, L. N. Cooper, and J. R. Schrieffer, *Phys. Rev.* **108**, 1175 (1957).
7. C. Kittel, *Introduction to Solid-State Physics*, 6th ed. (Wiley, New York: 1986).
8. J. D. Jackson, *Classical Electrodynamics* (Wiley, New York: 1962).
9. C. Kittel, *Elementary Statistical Physics* (Wiley, New York: 1961).
10. C. Kittel and H. Kroemer, *Thermal Physics*, 2nd ed. (W.H. Freeman, San Francisco: 1980).
11. C. Kittel, *Quantum Theory of Solids*, Ch. 8 (Wiley, New York: 1963).

12. F. and H. London, *Proc. Royal Soc. A* **149**, 71 (1935); and F. London, *Superfluids*, Vol. I (Wiley, New York: 1950).
13. E. W. Collings, *Applied Superconductivity, Metallurgy, and Physics of Titanium Alloys* (Plenum, New York: 1986).
14. M. N. Wilson, *Superconducting Magnets* (Oxford Univ. Press, New York: 1983).
15. L. Goodrich and S. L. Bray, *Cryogenics* **30**, 667 (1990).
16. V. L. Ginsburg and L. D. Landau, *Zh. Eksp. Teor. Fiz.* **20**, 1064 (1950).
17. A. A. Abrikosov, *Zh. Eksp. Teor. Fiz.* **32**, 1442 (1957).
18. L. P. Gorkov, *Zh. Eksp. Teor. Fiz.* **36**, 1918 (1959).
19. P. G. deGennes, *Superconductivity of Metals and Alloys* (Benjamin, New York: 1966).
20. U. Essmann and H. Trauble, *Phys. Lett.* **24A**, 526 (1967).
21. J. W. Lynn *et al.*, *Phys. Rev. Lett.* **72**, 3413 (1994).
22. A. A. Abrikosov, *Fundamentals of the Theory of Metals* (North-Holland, Amsterdam: 1988).
23. C. P. Bean, *Rev. Mod. Phys.* **36**, 31 (1964).
24. H. T. Coffey, *Cryogenics* **7**, 73 (1967).
25. G. R. Kumar and P. Chaddah, *Phys. Rev. B* **39**, 4704 (1989).
26. D. E. Farrell, S. Chandrasekhar, and H. V. Culbert, *Phys. Rev.* **177**, 694 (1969).
27. J. E. Evetts, *Phys. Rev. B* **2**, 95 (1970).
28. E. Gregory, in *Encyclopedia of Materials Science and Engineering* (edited by R. W. Cahn), Vol. 2, pp. 1080–1086 (Pergamon, Elmsford, New York: 1990).

3

Refrigeration

Cooling something to 77 K is conceptually simple: immerse the object in liquid nitrogen. No one worries about where liquid nitrogen comes from; it is a by-product of oxygen production, and costs 6¢/liter in truckload quantities. However, the field of superconductivity has long been relegated to the research laboratory explicitly because it is so expensive to reach 4 K, the temperature of liquid helium. The purpose of this chapter is to explain how very low temperatures are achieved.

In this chapter, we first cover the basic principles governing refrigeration systems. This presentation closely follows the work of Ray Radebaugh of NIST.[1] From elementary thermodynamic principles, we build up to describe the way helium is liquified and introduce specific types of refrigerators. After that, we enumerate the economic factors relating to superconductivity. Finally, we describe typical large-scale cooling systems, with examples drawn from industrial and research facilities.

The possibility of using the HTSCs at intermediate temperatures in between 4 K and 77 K has received increasing attention because of the properties of the high-temperature superconductors. To carry this out, special refrigeration systems are required to establish an operating environment at some temperature such as 35 K. In Chapter 24 we will return to that subject and consider some novel heat-removal schemes that are being developed into practical refrigerators.

3.1. THERMODYNAMIC PRINCIPLES

All heat transfer systems are governed by the laws of thermodynamics: the first law states the conservation of energy; the second law specifies that entropy always increases; the third law says that the heat capacity of any material goes to zero as the temperature nears absolute zero; and the fourth law says that absolute zero is unattainable. A popular mnemonic device for students is: "You can't win; you can't even break even; things will get worse before they get better; and who says things will get better."

Here we are concerned primarily with the second and third laws. The object of a good refrigeration system is to minimize the increase in entropy, that is, to do the least work required to remove a certain amount of heat,[2] and to optimize the overall system efficiency even though the cold end of the system suffers from lower efficiency.

Figure 3.1 introduces the basic concepts of a refrigerator. Some amount of work W is input at the warm end in order to remove an amount of heat Q from the cold end. The overall process leads to a net increase in total heat, but all of it is rejected into the surroundings at

37

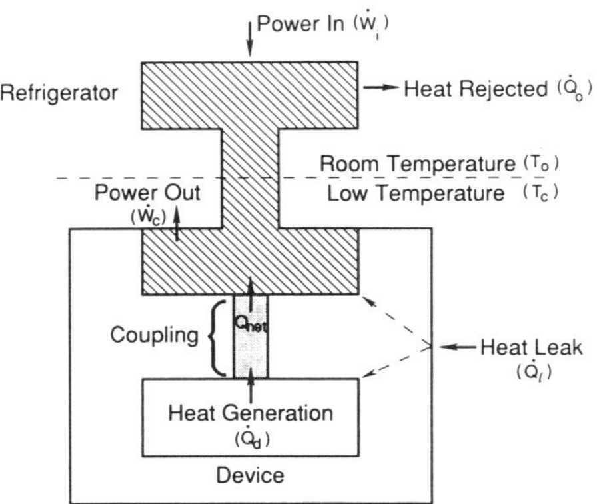

Figure 3.1. Schematic of a general refrigeration system and the source of heat and power. Heat leaks should be thermally anchored to the refrigerator to prevent a heat flow (and resulting ΔT) across the coupling. (Courtesy R. Radebaugh, NIST.)

the warm end. The important achievement is that the cold end gets colder, and for this we are willing to pay some price.

The essential elements of any refrigeration system are shown in the schematic of Figure 3.1. For our purposes, the *heat generation device* is typically a superconductor in which AC losses, excursions, and transients (such as that due to the sudden motion of a flux-line bundle) are the only sources of heat. The heat leak comprises radiation input from warmer surrounding walls, conduction of heat along electrical leads that service the apparatus, and heat conduction through supporting structures attached to the external surroundings. The refrigerator itself (shown shaded) operates between the ambient and the temperature of the superconductor. The coupling can be anything from a copper bar (known as a *cold finger*) to flowing helium gas. For most realistic applications of superconductors (e.g., power lines, energy storage, etc.), the superconductor is kept in good thermal contact with the cold end of the refrigerator, and all considerations of heat transfer are between the two ends of the refrigerator. This situation is our topic here.

The first important distinction between types of refrigerators is that of open versus closed systems. Denoting heat, internal energy, and work, respectively, by Q, U, and W, entropy by S, and temperature by T, Figure 3.2 defines the relationships: in a closed system, only heat and work cross the system boundary, and $dQ = dU + dW$. In an open system, mass flow dm crosses the boundary, bringing in enthalpy (h) and taking it away. In that case, for a steady-state system ($dU = 0$) when the mass flow is constant, we have $dQ = dW + (h_o - h_i)dm$. It is noteworthy that in a *reversible* process, $dQ = T\,dS$. This is relevant in some magnetic refrigeration systems. However, for gas refrigerators utilizing expansion valves, the process is *irreversible*.

The *efficiency* η of a refrigerator is defined with respect to an ideal *Carnot cycle*. It is a measure of the irreversibility of the process. The ratio of heat removed (Q_{net}) to work input (W) can be written as

$$\frac{Q_{net}}{W} = \eta \frac{T_{cold}}{T_{hot} - T_{cold}} \qquad (3.1)$$

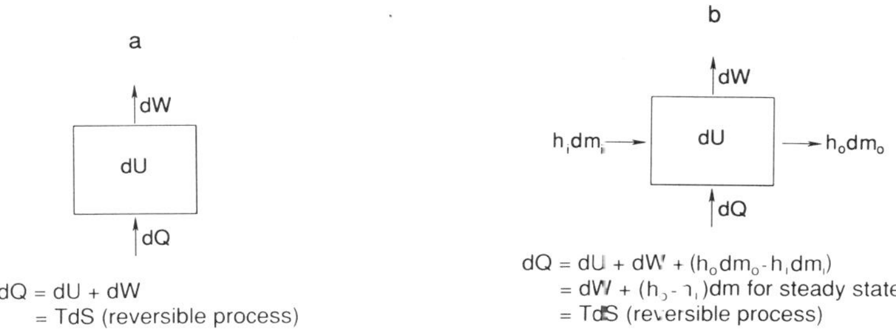

Figure 3.2. Defining relationships for (a) closed and (b) open thermodynamic systems.

where T_{hot} and T_{cold} are the upper and lower temperatures of the cooling system. Incidentally, *specific power* is defined as the inverse of this ratio, W/Q_{net}. In an ideal refrigerator, $\eta = 1$ and we have a Carnot cycle, but all real refrigerators are less efficient. Figure 3.3 shows the scatter in efficiency of typical real systems, and tells what can be expected in various cases. Low-wattage refrigerators (such as in spacecraft) are often only about 1% efficient, whereas major facilities (accelerator-sized) may approach 30%.

Of course, these figures are still relative to the Carnot efficiency, which itself decreases with decreasing temperature. It is useful to plot the *specific power* required to reach a required low temperature; this appears in Figure 3.4. (Every real refrigerator lies to the upper right of the line for $\eta = 100\%$ Carnot.) The advantage of liquid nitrogen temperatures compared to helium temperature operation is immediately obvious from Figure 3.4. It is a factor of 24 easier to get to 77 K than to 4.2 K, even in an ideal system. Some large superconducting systems run at 1.8 K, where the Carnot efficiency is worse yet. It becomes even more

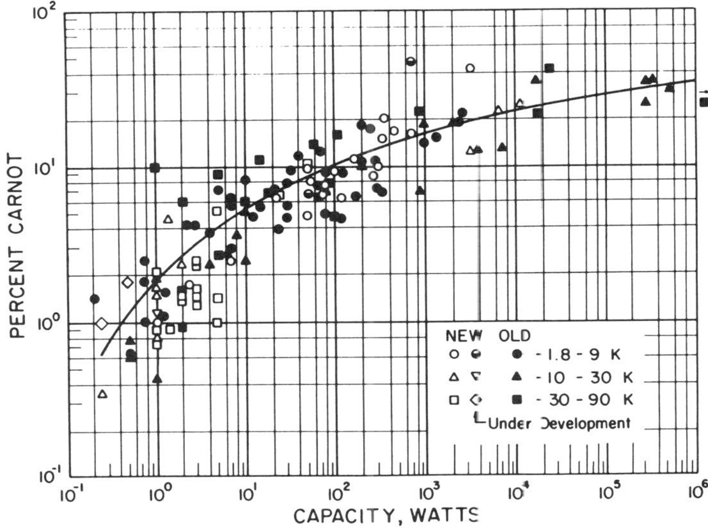

Figure 3.3. Coefficient of performance relative to Carnot for many refrigerators. (Courtesy R. Radebaugh, NIST.)

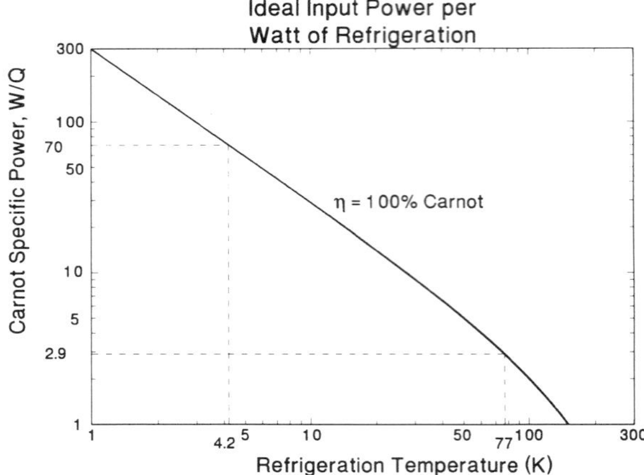

Figure 3.4.   The minimum power input required to produce 1 W of refrigeration at various temperatures. (Courtesy R. Radebaugh, NIST.)

important to strive for high efficiency in such systems. To gain better conductor performance, designers sometimes are forced to accept the lower efficiency that comes with lower operating temperatures.

3.2.   GAS REFRIGERATORS

Just as there is no realizable ideal Carnot cycle, so, too, there is no truly ideal gas. It is the deviations from ideal-gas behavior that makes it possible to build refrigerators that liquefy gases. The cornerstone of gas refrigerators is the *Joule–Thompson effect*, and this plays an important role in the following discussion.

A standard refrigeration cycle is shown[3] in Figure 3.5. An automobile air conditioner is a good practical example. Starting with a gas at room temperature, work is input at the

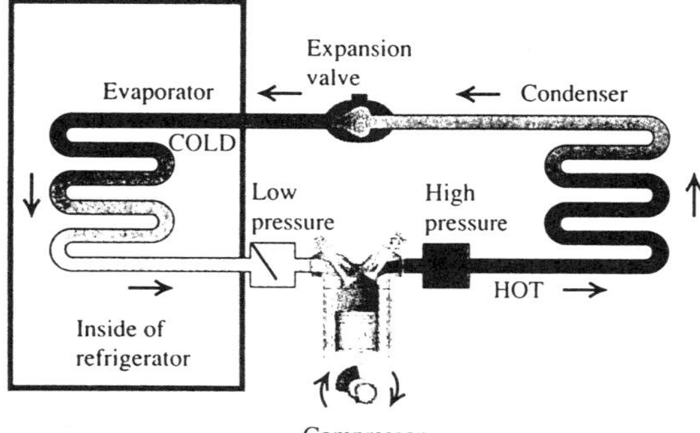

Figure 3.5.   Major components of a refrigeration system. A high-pressure fluid (usually a liquid) is expanded through a throttling valve, becoming a much cooler gas. Heat is removed from the lower reservoir by this gas. The gas thus warmed flows into the compressor, where work is added; a hot, high-pressure gas exits. In the condensor, the gas cools and a greater amount of heat is rejected into the high-temperature reservoir. That "reservoir" is often the ambient surroundings. (From Ref. 3; ©1992 by Addison-Wesley Publishing Co., Inc. Reprinted by permission of the publisher.)

compressor, leaving the gas both hot and pressurized. A condenser (e.g., the automobile's radiator) reduces the temperature to near ambient. Next, the high-pressure gas enters a throttling valve, and as it expands both pressure and temperature drop. Now this cool fluid (whether gas or liquid) is able to remove heat from the surroundings (e.g., the passenger compartment of the car). In doing so, it rises in temperature on its way back to re-enter the compressor.

A gas liquefier is highly reminiscent of this. Figure 3.6(a) shows the temperature–entropy diagram corresponding to the compression–expansion cycle of Figure 3.6(b). This is called the *Linde–Hampson* liquefaction system, and is a simplified representation of real liquefiers.[4] By following Figures 3.6(a) and (b) together, we can relate the changes in the

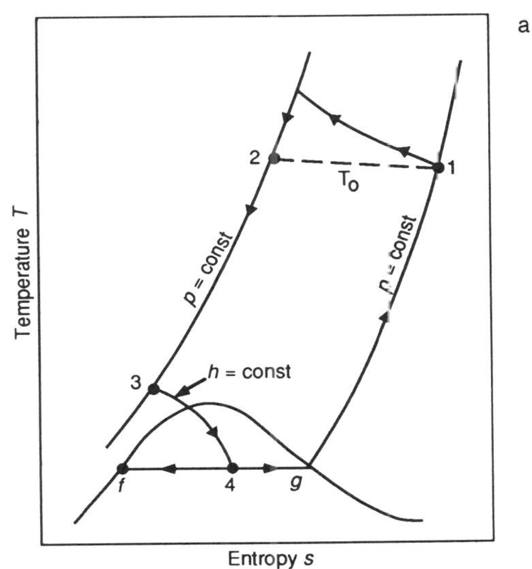

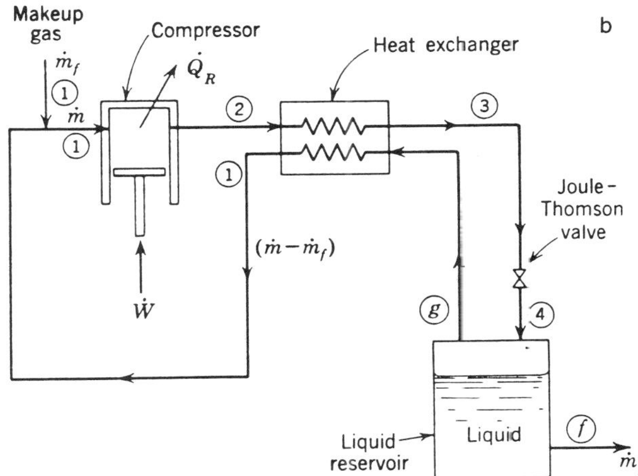

Figure 3.6. (a) The Linde–Hampson cycle. The state-points refer to the numbered points in part (b), which is the Linde–Hampson liquefaction system. (From Ref. 4.)

thermodynamic properties to the movement through the components of the refrigeration system. Starting at point 1, the gas (e.g., ambient air) is compressed and run through a radiator to reject heat Q_r. At point 2 the high pressure gas has returned to ambient temperature. Next, it enters a heat exchanger which reduces its temperature to point 3. The heat exchanger is a counterflow device that allows some of the gas above the liquid pool to return to ambient temperature, taking away the heat from the incoming high-pressure gas in the exchanger. The step from point 3 to point 4 is taken at the Joule–Thomson valve, a throttling valve which reduces both temperature and pressure, expanding the gas into a thermodynamic region [Figure 3.6(a)] where both liquid and gaseous phases co-exist. The liquid phase (f) is collected in a pool; the gas (g) returns to ambient via the heat exchanger.

The key step is in the throttling valve. The valve is simply an obstruction, like a partially open faucet, and the gas velocity changes very little across the valve. There is dissipation going through the valve, so entropy goes up while enthalpy stays the same: $dh = 0$. However, from thermodynamics we know that

$$dh = T\,dS + v\,dP \qquad (3.2)$$

where v is the molar volume and P is the absolute pressure. Therefore having $dh = 0$ requires that

$$dS = -v\,dP/T$$

Since pressure drops, dP is negative and dS is positive; disorder increases while energy is merely dissipated. So far we have not said what the temperature does at this expansion valve.

To determine the temperature change, we must inquire about the thermodynamic quantity $(\partial T/\partial P)_h$, i.e., the rate of change of temperature with pressure *when enthalpy is held constant*. (Constant enthalpy is different from constant entropy processes.) To calculate this quantity (which is also known as the Joule–Thomson coefficient), we convert it to familiar quantities:

$$\mu_{JT} = \left(\frac{\partial T}{\partial P}\right)_h = -\left(\frac{\partial T}{\partial h}\right)_P\left(\frac{\partial h}{\partial P}\right)_T$$

We then note that $(\partial h/\partial T)_P = C_P$, and $(\partial h/\partial P)_T = v - T(\partial v/\partial T)_P$. Combining these, we find

$$\mu_{JT} = \frac{1}{C_P}\left[T\left(\frac{\partial v}{\partial T}\right)_P - v\right] \qquad (3.3)$$

If this quantity is positive, the gas will cool when the pressure is reduced without changing enthalpy h. If μ_{JT} is negative, the gas will heat up upon *isenthalpic* expansion. For refrigeration, it is necessary to work only with gases that cool on expansion.

The exact numerical value of the Joule–Thomson coefficient depends upon the equation of state of the gas. For a perfect gas, $Pv = RT$; in that case, $(\partial v/\partial T)_P = v/T$, and inserting this into the expression for μ_{JT} results in $\mu_{JT} = 0$. Thus, a perfect gas does not change temperature upon isenthalpic expansion. This is consistent with the notion that the internal energy of a perfect gas is a function of temperature alone, so if $dh = 0$, $dT = 0$ as well. By contrast, a gas obeying the more realistic van der Waals equation of state gives a finite Joule–Thomson coefficient. This is due to the influence of intermolecular forces in real gases.

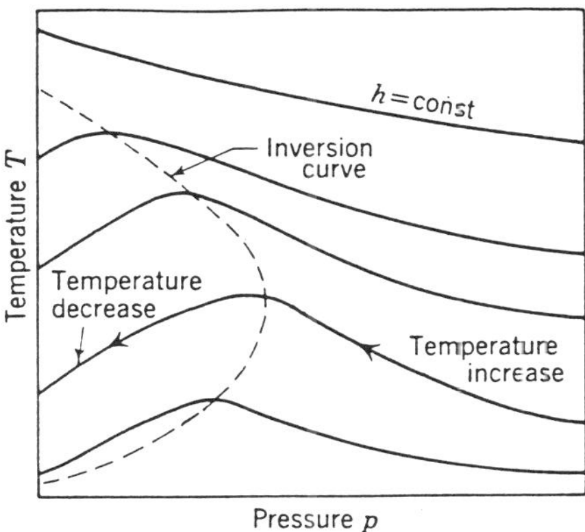

Figure 3.7. Pressure–temperature plot of a real gas. Solid lines are lines of constant enthalpy. Dashed line is the inversion curve. (From Ref. 4.)

For any individual gas, it is possible to draw lines of constant enthalpy on a pressure–temperature diagram; Figure 3.7 is typical. Any isenthalpic expansion will move the system along one of these solid lines. (Incidentally, for an ideal gas, the graph would be boring: straight horizontal lines, giving the same enthalpy no matter what the pressure.) One can plot the locus of points for which $(\partial T/\partial P)_h = 0$, and this (dashed line in Figure 3.7) is called the *inversion curve*. In the temperature–pressure regime to the left of the inversion curve, refrigeration is possible. A gas that starts off extremely hot must first be reduced in temperature to lie within its inversion curve before it can be liquefied.

Here is where the most important difference appears in cryogenic refrigeration relevant to superconductors: nitrogen has a maximum inversion temperature of 621 K, whereas helium's maximum is below 50 K. Thus, helium qualifies as extremely hot at room temperatures and even at 77 K. Neon has a maximum inversion temperature of 250 K (and hydrogen 205 K), so it is necessary to construct a multistage refrigerator in order to reach 4 K. First, nitrogen is cooled from room temperature to 77 K. Then neon or hydrogen is placed in thermal contact with the liquid nitrogen. The cool hydrogen can then be liquefied, reaching about 20 K. Helium gas exchanges heat with that second liquid bath to get below its inversion temperature of 45 K. At last, the helium can be liquefied using the Joule–Thomson principle. This is illustrated in Figure 3.8.

It would be desirable to span the intermediate temperature regime with mixtures of nitrogen–neon or similar combinations. However, these liquids are immiscible, and nitrogen solidifies well above the temperature where neon or hydrogen become liquid. There is no "antifreeze" with which to establish a liquid bath at some intermediate temperature.

3.3. CRYOGENIC REFRIGERATORS

Clearly, if gases are to exchange heat in steps en route to cryogenic temperatures, we should reject heat to surroundings at the highest allowable temperatures (typically, room temperature) because thermodynamic efficiency continues to deteriorate as we go to lower and lower temperatures. Thus, in a helium liquefier, we first bring the helium gas down to 77 K before cooling it to hydrogen temperature. Similarly, additional small steps are

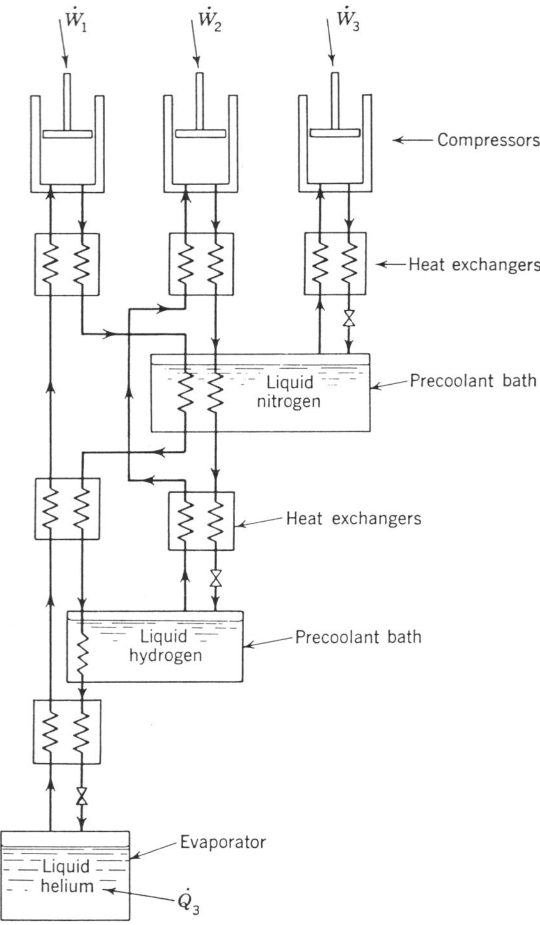

Figure 3.8. Three-stage Joule–Thomson liquid helium refrigerator. (From Ref. 4.)

introduced throughout the cooling process, in order to construct a path to the lowest temperature that stays as close as possible to the ideal Carnot cycle behavior all along the way.

There are two basic types of heat exchangers: *recuperative* and *regenerative*. In a *recuperative* system, the working fluid flows continuously in one direction, completing the thermodynamic cycle as it does so; the gas refrigerators discussed above are examples of recuperative systems. Heat is rejected to the outside world, and the gas stream recuperates to its original condition. In a *regenerative* heat exchanger, the flow of heat reverses periodically; heat is stored during the first half-cycle, and rejected during the second half-cycle. The *heat wheel*, widely used in industrial metal-treating processes, is an example of this. One way to run a cryogenic regenerative system is to collect heat in a working fluid at low temperature (e.g., helium gas) and then use a *displacer* to transport that fluid to a place where it can reject its heat to the surroundings. Both kinds of heat exchangers have their advantages and disadvantages. Table 3.1 displays these in a simple plus-or-minus format.

Table 3.1. Advantages and Disadvantages of
Cryocooler Type

	Recuperative	Regenerative
Reliability	0–	0+
Efficiency	–	+
Vibration	+	–
Transport refrigerant	+	–
Simple heat exchanger	–	+

It is instructive to consider how the simple Joule–Thomson configuration is modified in real refrigerators. The diagram of Figure 3.6(b) is redrawn in Figure 3.9(a); instead of a pool of collected liquid, the low end is depicted as simply the input of heat Q at temperature T; at the high end, heat Q_o is rejected to the surroundings at temperature T_o. This is a recuperative system. Figure 3.9(b) is a schematic of the Brayton cycle, a modification which includes an *expansion engine* operating at an intermediate temperature. Still another modification is the Claude cycle, which adds another heat exchanger at intermediate temperature, diverts part of the flow through an expansion engine, and expands only enough gas through the Joule–Thomson valve to remove the heat Q arising from the low-temperature end of the cycle.

The *Collins liquefier*, invented in the 1940s, is a variant of the Claude cycle: at the outlet of the Joule–Thomson valve, the temperature and pressure conditions are sufficiently reduced that liquid helium collects in a pool. Later on, it is drawn off through a transfer tube into a dewar, whence it is taken to the application venue. This Collins process is the workhorse of the liquid helium industry. In Section 3.6 below we shall present a detailed description of an actual large-scale helium liquefier and cooling system.

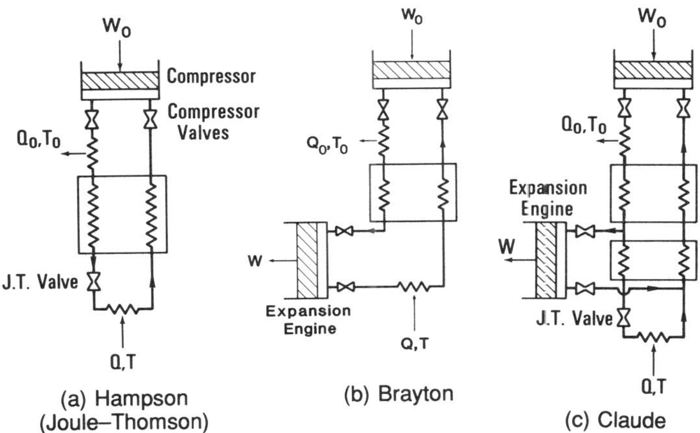

Figure 3.9. Three types of recuperative refrigerators. (a) In the Joule–Thomson cycle, a recuperative heat exchanger spans the temperature range from room temperature to some low temperature. Parts (b) and (c) are two additional recuperative refrigerators which rely on an expansion engine for cooling. The expansion engine may be either a reciprocating type with inlet and outlet valves as shown here or it may be an expansion turbine which does not require valves.

Regenerative refrigerators contain a *displacer* which is linked to a *regenerator*. The purpose of the displacer is to cycle the gas between the hot and cold ends of the refrigerator and to maintain the proper phase relationship with the oscillating pressure generated by the oscillating piston. Figure 3.10(a) is a schematic of the *Stirling cycle*, and Figure 3.10(b) displays a variant known as the *pulse tube refrigerator*.

The Stirling cycle itself has different configurations, such as the two shown in Figure 3.11. In the simpler configuration, there are moving parts at low temperatures, which means a maintenance headache. The mechanical couplings present in the integral Stirling cycle cause vibration and wear. To circumvent this, it is only necessary to produce a cyclical variation in the pressure of the working fluid by some other means. The split-Stirling cycle uses the piston motion to vary the pressure in a tube of the gas, eliminating the mechanical coupling and concomitant vibration.

In the pulse-tube refrigerator [Figure 3.10(b)], the solid displacer of the Stirling cycle is completely eliminated, replaced with a segment of gas that separates the hot end from the cold end of the pulse tube. As the gas in the pulse tube is compressed and heated, a portion of this gas flows through the orifice, rejecting heat on the way as it passes through the hot

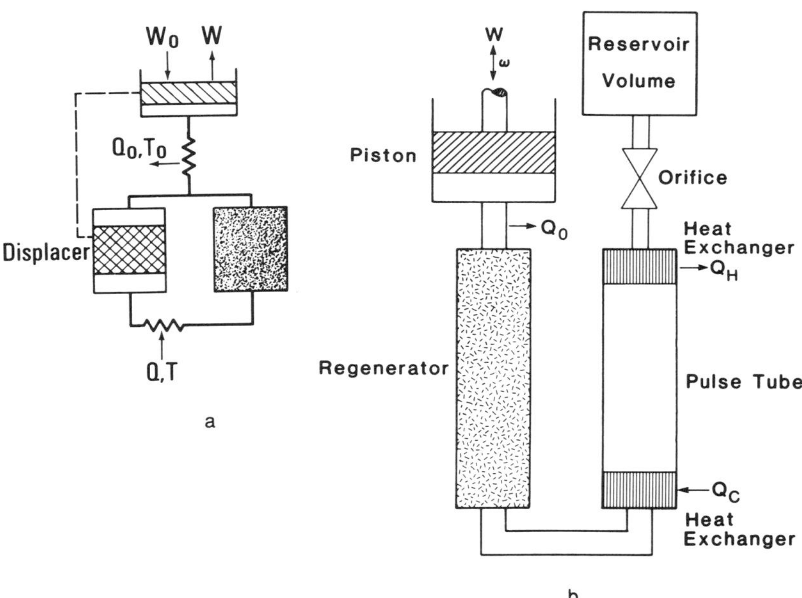

Figure 3.10. Two radically different types of regenerative refrigerators. (a) The displacer in the Stirling refrigerator moves synchronously with the piston but leads it by about 90° in phase. The pressure difference between the two ends of the displacer is nominally zero, but in practice a small difference occurs to overcome pressure drop in the regenerator. The work input W_0 to the piston occurs during the compression part of the cycle and the work W is recovered during the expansion part of the cycle. The working fluid undergoes a pressure oscillation with a typical pressure ratio of 2. Helium gas is nearly always used as the working fluid in regenerative refrigerators. The heat of compression Q_0 is dissipated at room temperature T_0, and the heat Q is absorbed at the low temperature T. (b) In the pulse-tube refrigerator, the orifice provides the proper phase shift in the mass flow rate to simulate the Stirling cryocooler. The solid displacer of the Stirling cryocooler is replaced with a segment of gas which separates the hot and cold ends of the pulse tube. (Courtesy R. Radebaugh, NIST.)

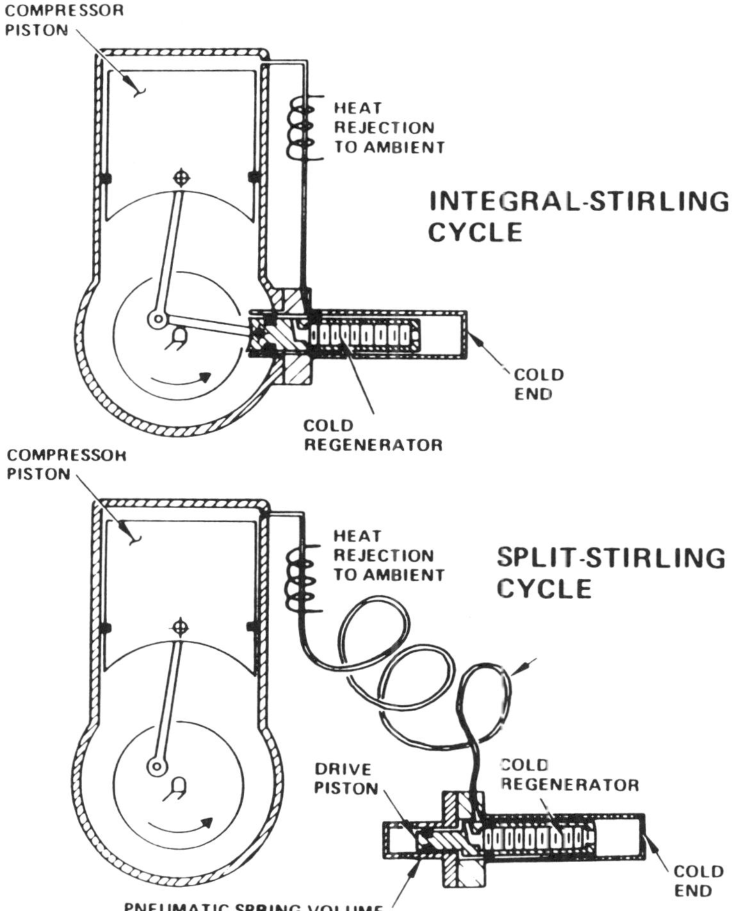

Figure 3.11. A comparison of the integral Stirling cycle with the split Stirling cycle. In the integral Stirling refrigerator the displacer motion is brought about by a mechanical linkage to the crankshaft, whereas in the split Stirling refrigerator the displacer motion is caused by the oscillating pressure in the working fluid. The advantage to the split version is that the vibration of the compressor is not transmitted to the cold end.

heat exchanger. When the gas in the pulse tube is expanded and cooled, gas flows from the reservoir through the orifice and forces the cold gas in the pulse tube to pass through the cold heat exchanger, where it absorbs heat.

There are several other types of refrigerators used in both large and small systems. Examples include the Solvay and the Vuilleumier systems, and the Gifford–McMahon system, which is described in detail in Chapter 24. For now, we only note that there have been dramatic improvements in recent years.[5] Pulse-tube refrigerators, for example, have seen their lowest attainable temperatures (with one stage) drop from 124 K to 23 K since 1984. Table 3.2 is a chart showing this improvement. If the copper oxide superconductors are to operate near 35 K, pulse-tube refrigerators should be considered to get there. At the present time, these are one-of-a-kind systems, still much closer to a research project than to a commercial product.

Table 3.2. Pulse Tube Temperatures

Temperature	Type	Investigator	Year
195 K	Resonant	Wheatley	1985
165 K	Basic (no valve)	Gifford	1967
124 K	Basic (valve)	Longsworth	1967
105 K	Orifice (valves and air)	Mikulin *et al.*	1984
60 K	Orifice (no valves and He)	Radebaugh	1985
39 K	Orifice (split)	Radebaugh	1990
33 K	Orifice (integral)	Tward *etal.*	1990

The merit of each of the choices of refrigerators is subject to constraints of reliability and original cost, but all other things being equal, the efficiency of operation is usually the dominant factor in the long run. A good way to compare the choices is by comparing their *specific powers*. This is defined as the ratio of work in over work out. Figure 3.12 compares various systems designed to reach 80 K. Specific power ratios of about 100 to 1 are typical. If 80 K were the final target temperature, Stirling cycle refrigerators would win out; but reliability and other criteria must be weighed more heavily for lower temperatures. Pulse-

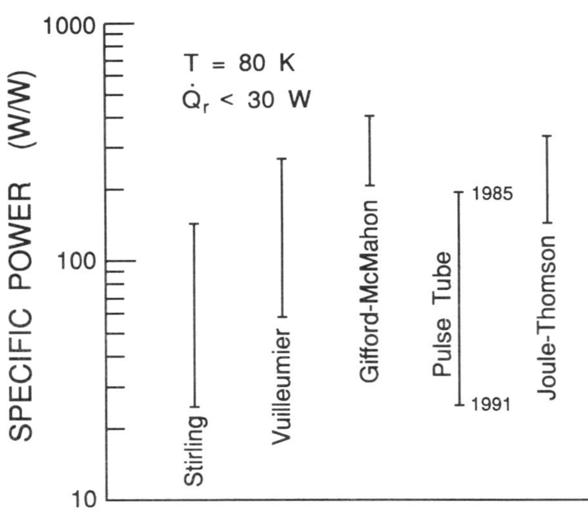

Figure 3.12. Specific Power: The ratio of input power to the refrigeration power at 80 K for various types of cryocoolers. Manufacturers are steadily improving performance: e.g., for pulse tubes, values of W_{input}/W_{output} down to about 25 have been obtained only recently. (Courtesy R. Radebaugh, NIST.)

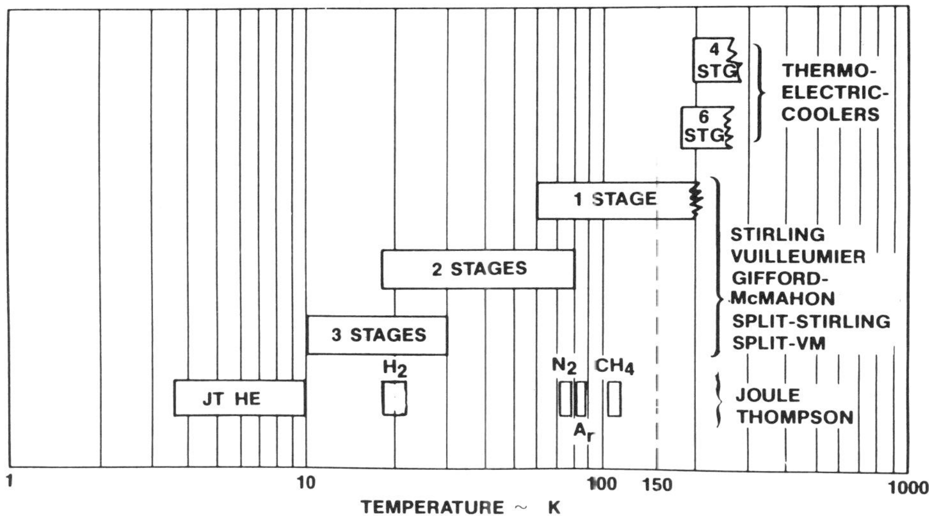

Figure 3.13. The number of stages required to reach various temperatures.

tube devices, which had specific powers of about 200 early in their development, have been advanced rapidly and have recently achieved specific powers as low as 25, which is comparable to the best Stirling refrigerators.

To go lower in temperature requires more stages, as discussed above at the end of Section 3.1. Figure 3.13 shows how many stages are needed to reach various temperatures. In a recuperative system, the final step to 4.2 K is usually taken by a Joule–Thomson expansion of helium. Regenerative systems have generally been used only at higher temperatures. The most common configuration for liquefying helium today is a combination of a Gifford–McMahon upper end and a Claude–Collins lower end, culminating in a Joule–Thomson final stage. However, improvements in regenerative systems are steadily being made. Recently, some Gifford–McMahon refrigerators have reached 4 K without the use of a Joule–Thomson stage. Moreover, by employing three consecutive stages of pulse-tube refrigerators, a temperature of 3.6 K has been reached.[6]

3.4. EXTREME LOW TEMPERATURE REFRIGERATION

How do you get below 4.2 K? First, remember that 4.2 K is the equilibrium temperature of liquid helium at one atmosphere pressure. At lower pressure, the equilibrium temperature is lower. Therefore, the simplest thing to do is start with a bath of 4.2 liquid helium and pump a partial vacuum on it. With a large diffusion pump, it is possible to get to 0.8 K. The popular intermediate temperature of 1.8 K is commonly reached by vacuum pumping. It is worth noting that at 2.17 K, liquid helium becomes a superfluid; this imposes requirements of extra care to avoid vacuum leaks in a system, because the superfluid helium can penetrate where no ordinary gas can go. Nevertheless, using proper care, large-scale cryogenic systems have been operated satisfactorily below 2 K.

There are alternate ways to reach very low temperatures.[7] The isotope ^{3}He boils at a lower temperature than conventional ^{4}He, and so it can reach 0.3 K by vacuum pumping. However, ^{3}He is terribly expensive, so it is never vented to the atmosphere. Mixtures of ^{3}He and ^{4}He are used in a *dilution refrigerator* to achieve still lower temperatures.[8] Finally, to get down very close to absolute zero, the technique of *adiabatic demagnetization* is employed.

3.4.1. Adiabatic Demagnetization

In adiabatic demagnetization we make use of the thermodynamic fact that as disorder increases, entropy rises. If we prevent the system from exchanging heat with its surroundings, then $dQ = 0$. Therefore, if the entropy is to go up within the system, heat must be removed elsewhere within the closed system. The *magnetocaloric effect* is the operating principle of any magnetic refrigerator. When a magnetized material is removed from a magnetic field, if it is kept isolated from its surroundings so that no heat transfer takes place, it will cool as the lattice of aligned spins relax to a disordered state. Thermodynamically, the substitution of magnetic energy for the term $-v\,dP$ is a very successful analogy: the specific volume v is replaced by the magnetic moment per unit mass, and pressure P is replaced by $-\mu_0 H$, where H is the applied magnetic field and $\mu_0 = 4\pi \times 10^{-7}$. The basic thermodynamic equation is

$$dU = T\,dS + \mu_0 v H \cdot dM \tag{3.4}$$

and the change in entropy is given by

$$dS = (\partial S/\partial T)_H dT + (\partial S/\partial H)_T \partial H \tag{3.5}$$

Now $(\partial S/\partial T)_H = C_H/T$, where C_H is the specific heat at constant magnetic field. There is a useful *Maxwell equation* for the second term:

$$(\partial S/\partial H)_T = \mu_0 v (\partial M/\partial T)_H$$

In the adiabatic case, $dS = 0$, which immediately requires that

$$dT = -(T/C_H)\,\mu_0 v\,(\partial M/\partial T)_H\,dH \tag{3.6}$$

The change in temperature ΔT is simply the integral of this dT. Since the term $(\partial M/\partial T)_H$ is generally a function of both H and T, this integral isn't always easy to carry out. Fortunately, for several important cases the dependence is simple enough to allow us to see the outcome.

To build a refrigerator, ΔT should be as large as possible. It helps if C_H is very small. Here is how a magnetic refrigerator typically works in practice: The apparatus to be cooled is sealed inside a vacuum can, which is placed inside another chamber of liquid helium, as shown in Figure 3.14. Equilibrium is established at the lowest temperature conveniently attainable by the helium, typically below 1 K. The apparatus contains a fairly large crystal of a paramagnetic salt, such as cerium magnesium nitrate. A magnetic field is applied, thus lining up the spins in the paramagnetic salt. This tends to heat up the salt, but that heat is carried away by an exchange gas (helium) circulating around it. Just before the cooling is about to begin, that exchange gas is pumped away, so that the only thermal contact with the surroundings is through fine silk threads holding the apparatus in place. This establishes the adiabatic condition $dQ = 0$.

The magnetic field is then shut off. With no reason to remain aligned, the spins gradually become randomized, thus increasing their entropy; but to supply the heat needed for that

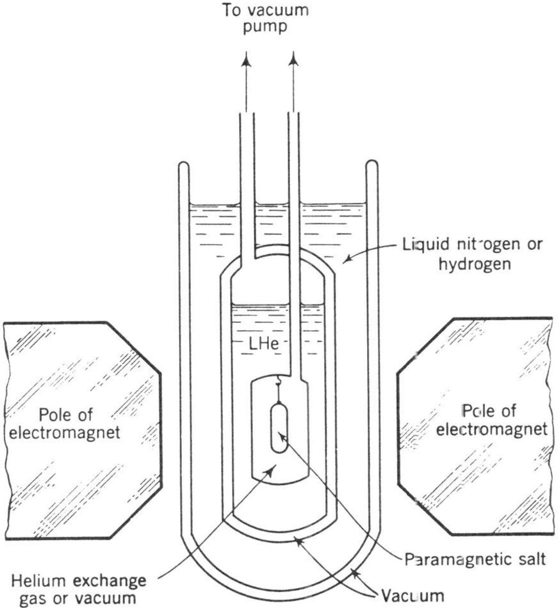

Figure 3.14. Apparatus for carrying out an adiabatic demagnetization process. (From Ref. 4.)

randomization, the entire isolated apparatus must cool down. We find that the relevant parameter is the *magnetocaloric coefficient* $\mu_M = (\partial T/\partial H)_s$. This coefficient is always positive, which means that decreasing the magnetic field always leads to a decrease in temperature. Temperatures below 0.001 K have been reached by exploiting this technique.

Paramagnetic materials are used in magnetic refrigerators to achieve extremely low temperatures. Above 0.05 K, the magnetization $M(T)$ shows a simple $1/T$ dependence, known as the *Curie law*:

$$M(T) = N\mu^2 H/3k_B T \tag{3.7}$$

Defining the parameter $a = \mu H/k_B T$, we can rewrite this

$$M(T) = \mu Na/3$$

This is actually an approximation to the *Langevin function*

$$M(T)/\mu N = L(a) = \coth(a) - 1/a \tag{3.8}$$

and at extremely low temperatures this function does *not* go to infinity but saturates. This is because μN is the maximum possible magnetic moment, when all spins are aligned.

When this $M(T)$ function is inserted into the expression for dT, the result is

$$dT = \frac{1}{C_H} \mu_0 v \, \frac{N\mu^2}{3k_B T} H \, dH \tag{3.9}$$

and for paramagnetic materials $\mu^2 = g^2 J(J + 1)\,\mu_B^2$, where μ_B is the *Bohr magnetron*, and g is the *gyromagnetic ratio* $\cong 2$. Whenever the field is low or the temperature is high, the Langevin function simplifies to give the $1/T$ behavior. Anytime a is < 2 this is valid, and typical values of a are about 0.3.

It would be nice if adiabatic demagnetization were useful at temperatures well above absolute zero. Next we describe one possible path to making this come true.

3.4.2. Nano-Composite Refrigeration

This section describes an advance in the field of adiabatic demagnetization that may extend the range of usefulness of this refrigeration technique. We have discussed paramagnetic materials, where the Curie law $[M(T) \sim H/T]$ governs the rate of cooling during adiabatic demagnetization. Similarly, a ferromagnet behaves according to the *Curie–Weiss law*:

$$M(T) \sim H/(T - T_{\mathrm{cu}}) \tag{3.10}$$

where T_{cu} is the *Curie temperature* at which the material becomes a ferromagnet. This means that instead of $dM(T)/dT$ peaking at $T = 0$ as in a paramagnet, its maximum is at $T = T_{\mathrm{cu}}$. Furthermore, this yields for the cooling upon demagnetization

$$dt_{\mathrm{ferro}} = \left(\frac{\mu_0 v}{C_H}\right)\frac{N g^2\, J(J + 1)\mu_B^2 T}{3 k_B (T - T_{\mathrm{cu}})^2}\, H\, dH \tag{3.11}$$

in complete analogy with the expression for cooling of a paramagnet. Evidently the cooling effect will be greatest near the Curie point T_c when a ferromagnetic material is substituted for the paramagnet. To exploit the magnetocaloric effect in some specific temperature range, it is only necessary to pick a ferromagnet with its transition in that range.

This much is background, and is generally known, but is not particularly useful. However, what is new is the development of materials ("magnetic nanocomposites") that combine the best features of both paramagnetism and ferromagnetism. These materials, developed and tested at the National Institute of Standards and Technology (NIST), offer the promise of greatly enhanced refrigeration effects.

A nanocomposite is a composite of immiscible metals and nonmetals. The size scale of the constituents is 1 to 20 nm. (Co-depositing from vapor, or sol-gel methods are typical ways to make them.) A typical phase diagram is given in Figure 3.15. It shows that as the volume-fraction of silica increases, there arises a region of *superparamagnetism*, or clusters of paramagnetism. The N spins comprising the lattice associate into n clusters, each having a regional magnetic moment

$$\mu_{\mathrm{region}} = (N/n)\,\mu$$

This causes the net magnetic moment to become

$$M(T) = n\,\mu_{\mathrm{region}}^2 H/3 k_B T \tag{3.12}$$

What is remarkable about this is that (N/n) now appears as a squared quantity! What had been simply N for a simple paramagnet becomes $n\,(N/n)^2$ for a superparamagnet, and this can be huge if $N \gg n$. The magnetic moment goes up by orders of magnitude, and the implication for refrigeration is that the factor dT does the same. The size of the clusters matters: if they get so big as to form *domains*, the advantage is lost.

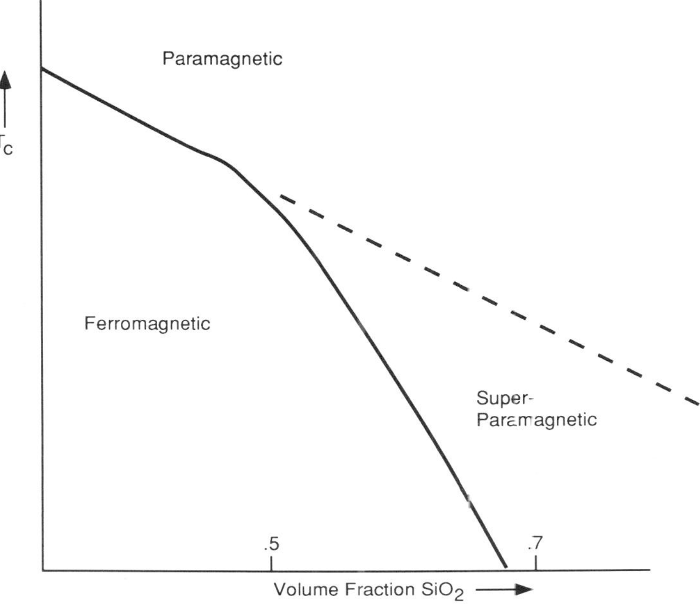

Figure 3.15. Phase diagram for nanocomposite of nickel-silica showing not only a paramagnetic-to-ferromagnetic phase transition but a new region in which the composite exhibits superparamagnetism.

When these superparamagnets interact with one another, it is similar to ferromagnetism, but with the interaction peaking at some temperature T_I. We then have

$$dT_{\text{nano}} = \left[\frac{\mu_0 v}{C_H}\right]\left[n\left(\frac{N}{n}\right)^2\right]\frac{\mu^2 T}{3k_B(T - T_I)^2} H\, dH \qquad (3.13)$$

which again features the factor $n\,(N/n)^2$. For typical values of N and n, dT is enhanced by a factor of 100 or more.

At the NIST, Robert Shull and co-workers[9] built a calorimeter to test this idea. First, a paramagnet of gadolinium gallium garnet (GGG) was operated in a field of 5 T, to verify that the apparatus worked correctly (a standard precaution in low-temperature heat-transfer measurements); their data agreed with others', thus enhancing confidence. Next a composite made of 11% iron and silica gel was treated with H_2 gas to make a superparamagnetic material. At 70 K in a 5 T field, the ΔT is 3 K. For this material, the *maximum* magnetocaloric effect is expected to occur around 30 K—right in the most interesting region for HTSCs being operated at intermediate temperatures.

Recalling that the efficiency of most gas refrigerators is far below the Carnot efficiency (see Figure 3.3), there is an urgency associated with finding new means of cooling. Adiabatic demagnetization has been used at extremely low temperatures simply because nothing else works in that range. It is currently being pursued for a number of space applications at temperatures around 20 K using gadolinium compounds and superconducting magnets. This new technique offers the potential for lower cost and higher performance, possibly without the need for superconducting magnets. In the near future, NIST and others will fine-tune their composite materials, attempting to maximize the magnetocaloric effect. If they are successful, we can look forward to more efficient and more reliable refrigeration systems operating in an important temperature regime.

3.5. ECONOMIES OF SCALE

So far, we have only described how refrigerators work, and not how one chooses one type over another. The evolving design of major liquefaction machines has been a series of trade-offs between throughput, thermodynamic efficiency, maintenance and reliability, and cost of materials of component assemblies. For spacecraft, the severe cost penalty associated with weight is an additional important tradeoff.

An even more fundamental question is whether to utilize superconductors in accomplishing some goal. It sometimes happens that a particular goal is so highly valued that almost any refrigeration price is acceptable. At other times, the prohibitive cost of the cryogenics causes designers to set superconductivity aside. For spacecraft, the former is usually the case. For the electric power industry, those who advocate superconductivity must remain cognizant of the need to keep refrigeration costs down. The imperative of system reliability shifts the balance among the trade-offs between temperature, capital cost, maintenance, and operating cost.

The economy of scale associated with large refrigerators is very great, and plays an important role in determining the size of major industrial facilities. Returning once again to Figure 3.3, we can see that to achieve even 10% of Carnot efficiency, the cooling capacity generally needs to exceed 100 W; that in turn implies about 100 kW input power. Roughly speaking, an increase of one order of magnitude in efficiency comes only after three orders of magnitude increase in cooling capacity. Careful scrutiny of Figure 3.3 reveals a cluster of 30%-efficient refrigerators near the 20 K range having over 100,000 W cooling power: these are NASA's hydrogen liquefiers.

Another point that emerges (by comparing square blocks and round dots in Figure 3.3) is that liquid nitrogen systems consistently out-perform liquid helium systems of comparable cooling capacity. Recalling that all the points on this chart are relative to Carnot efficiency, and observing the dependence of Carnot efficiency upon temperature displayed in Figure 3.4, it becomes clear that there is a tremendous opportunity for improvement in absolute efficiency whenever liquid nitrogen cooling is sufficient.

It is exactly this point that motivates the interest in HTSC today: there are a few things that can be done economically using liquid helium; there are lots of things that could be done economically with liquid nitrogen. With nitrogen, you can downsize the economic break-even point on the wattage scale. "Superconductivity in the home" was a popular buzzword because of this downsizing.

What developments in cooling can we expect due to HTSC? In the past, refrigeration efficiency has never been an important criterion where liquid nitrogen is concerned. The process for making liquid oxygen encourages development of efficient refrigerators, but liquid nitrogen is just a by-product. Sold for 6 cents per liter in truckload quantities, liquid nitrogen is commonly vented to the atmosphere. For many large-scale utility applications, especially where reliable delivery by truck is an issue, it may be economical to recycle liquid nitrogen.

As soon as intermediate temperatures (e.g., 30 K) are taken seriously, liquid neon comes to mind, but it is far too expensive to throw away. Hence, closed-loop systems that recycle the neon are of interest. Better yet, it is possible to reach 20 K with certain two-stage helium gas systems, as Figure 3.13 indicates. The increased reliability associated with fewer stages is an advantage. In the past, refrigerators were generally designed to reach much lower temperatures, and so comparatively little effort went into optimizing a system for an end point near 20 K. What has changed now is that the collection of trade-offs associated with HTSC includes the possibility of having high critical current at intermediate temperatures. This creates an incentive to develop efficient refrigerators for that regime.

3.6. OPERATING PRACTICAL REFRIGERATORS

To give a connection with the real world of practical refrigeration, we present three full-scale examples, which have had several years of operating experience. The first accompanies a magnetic separation unit used in a remote factory location, and therefore must exhibit excellent reliability. The second is the refrigerator for the "Atlas" facility at Argonne National Laboratory. The third is the major cryogenic system serving the Fermilab synchrotron. These examples bring out the principles discussed in the previous sections. However, one very important point for practical refrigerators, which is certainly not apparent from theory, is the absolute requirement for fail-safe performance. Neither a major research laboratory nor an industrial facility can afford to be hit with unscheduled down-time due to a malfunctioning refrigerator. Consequently, these real cryogenic systems include a number of exceptional features in their design.

To place these systems in perspective, their cooling capacities are compared with several large utility applications. The magnetic separator is served by a refrigerator that liquefies 25 liters per hour (l/hr) of helium, although the heat load is normally far less. The system is overdesigned, but that's the best way to serve industrial needs when engineers are far away. Utilities are usually nervous in the absence of such overdesign, so this system is a good paradigm. The Atlas system provides 700 W of refrigeration at 4 K. Imagining a future utility power plant, a 300-MW LTSC generator would require approximately 1 kW of cooling at 4 K (due primarily to heat leaks). This is roughly the size of the Atlas refrigeration system. The large Fermilab system is capable of providing 24 kW of refrigeration at 4.5 K, plus approximately 700 l/hr of liquid helium to cool the magnet current leads. By contrast, a large-scale (10,000 MVA) superconducting AC transmission line over 100 km long would require approximately 300 kW of cooling capacity near 5 to 7 K. A cryogenic system that large has not yet been built. Unlike the other applications, approximately 75% of the total heat load would be due to internally generated voltage- and current-related losses.

3.6.1. Magnetic Separator Cooling System

Our first example is conceptually the simplest and closest to the helium liquefiers described in Section 3.3 above. The Eriez Magnetics Corporation sells complete magnetic-separation systems to industry, and their magnets are made of superconducting wire. In Chapter 4 we discuss the process in more detail; here it suffices to observe that their typical cryogenic system has to liquefy 25 liters per hour (l/hr) of helium, which is plenty to service an equivalent refrigeration heat load of 86 W. This in itself does not impose severe design requirements, but the need for reliability in a system running at a remote factory site dramatically affects the engineering choices.

Table 3.3. Cryogenic System Characteristics

Coolant	Liquid helium bath
Temperature	4.5 K
Pressure	130 kPa (18.85 psia)
Precoolant	Liquid nitrogen
Temperature	80 K (144°R)
Pressure	137 kPa (20 psia)
Heat Load: Steady State for Vapor-Cooled Loads	4.0 l LHe/h
Heat Load (1 min Ramp)	
Coil	900 J
Cryostat	4200 J
Liquid and Helium Capacity	
Magnet	200 l
Standpipe	32 l (normal level)
Dewar	4000 l

The operating configuration of the entire magnetic separator imposes[10] four basic requirements:

- Ability to cool the system down from ambient to operating temperature;
- Steady-state operation, including periodic ramping;
- Capability to warm the system in a controlled manner;
- Capability for continuing magnet operation during refrigerator maintenance.

The characteristics of the system are shown in Table 3.3. A block diagram of the cryogenic system appears in Figure 3.16. It is worth walking through this diagram, in order to compare this real system with the idealized versions of Section 3.3.

A compressor (A) begins the process by pumping room-temperature helium gas from 1 atm up to 15 atm. The first striking feature is the presence of a duplicate compressor (B), which allows the refrigeration system to continue running during maintenance to compressor A. This kind of redundancy is a hallmark of industrial equipment; its cost is fully justified by the very great premium placed on avoiding down-time in a factory. The two examples that follow will display even more sophisticated ways of preventing down-time.

The compressed helium gas has oil (from the compressor) removed at C, and then enters the *cold box D*. A supply of liquid nitrogen from storage dewar I cools the compressed gas to 80 K, after which a Claude-cycle engine cools it further. Next, liquid helium falls into a dewar E, while rather cold helium gas exits upward to return to the cold box. From the dewar E, helium flows through a standpipe F to cool the magnet G. Heat losses cause helium to evaporate and return to the cold box and compressor; however, there is a large buffer tank H (a standard 15,000-gallon propane tank) to reduce pressure fluctuations in the gas stream, particularly at the input to the compressor. The suction pressure is maintained at a preset pressure by the gas management system J.

The cold box is the heart of the system and deserves closer scrutiny. Figure 3.17 is a more detailed schematic of the lower part of the system; obviously, it is more complex than the idealized Claude cycle of Figure 3.9(c). The designers' own description is abbreviated

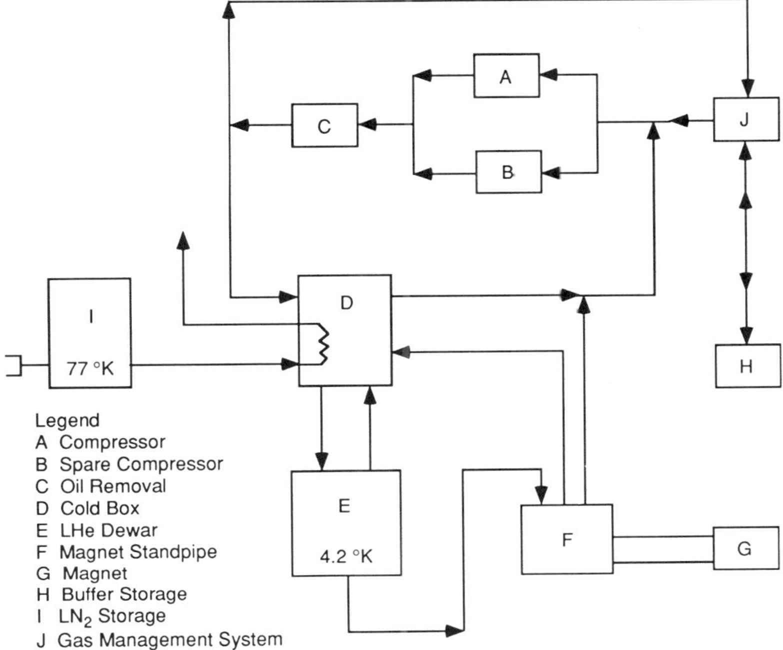

Figure 3.16. Block diagram of the cryogenic system for the Eriez magnetic separator.

here[10]: There are six heat exchangers in three bundles, two adsorbers at 80 K, one reciprocating expansion engine, and one Joule–Thomson expansion valve. At each exchanger, both the temperature and the pressure of the helium drop. The ideal design gives a large ΔT and a small ΔP, so that relatively cold helium gas arrives at the Joule–Thomson valve still at high pressure. A good expansion engine optimizes that trade-off. In this engine, the observed efficiency is approximately 74%. (Either overexpansion or underexpansion would tend to increase the irreversibility of the cycle, and that would show up as less efficiency.) Beyond the J-T valve, sudden expansion produces liquid helium at 4.5 K, corresponding to a pressure of about 1.2 atm. Needless to say, all gas-phase helium leaving the J-T valve returns to the compressor via heat exchangers, where it helps to cool incoming compressed gas as it gradually warms up.

Again, reliability was a prominent design factor. This Claude cycle, with a single expansion engine, was chosen to keep the system and controls as simple as possible. After all, the setting is an industrial plant, not a research laboratory. A two-engine cycle, such as a Collins liquefier uses, would have higher efficiency (i.e., lower operating cost), but reliability and simplicity outweigh that criterion. As experience builds, future refrigerators might take this route. However, when one is selling the first of a kind, building credibility with the customer is everything, and consequently complexity must be held to a minimum.

The system designers report very favorable experience[10]:

The system has been designed to maximize on-stream time, with components backing up other components. . . . The backup is accomplished as follows: (1) Loss of the compressor can be offset by the spare compressor; (2) Loss of the expander can be offset by consumption of some liquid

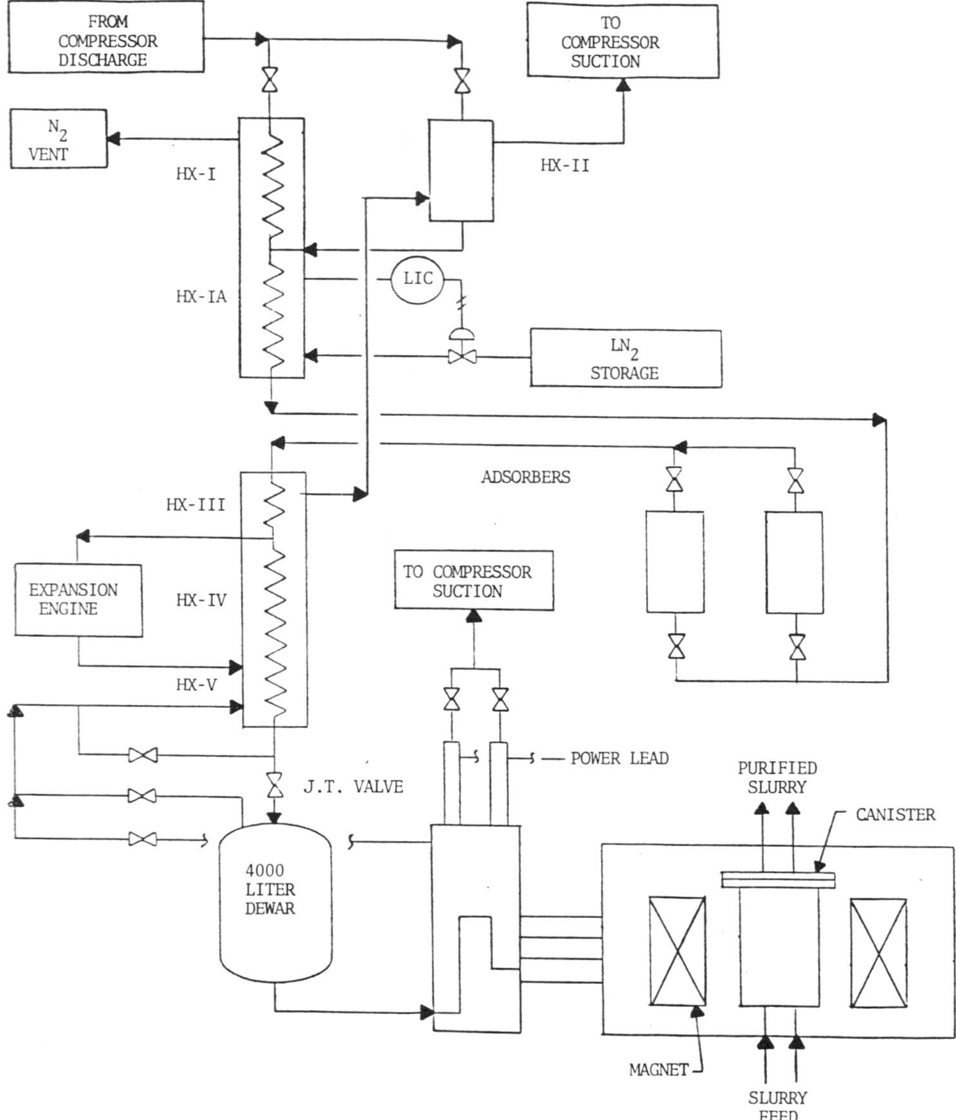

Figure 3.17. Schematic of the cold box, dewar, standpipe, and magnet of the Eriez separator.

helium . . .; (3) Loss of the refrigerator cold box is offset by the vaporization of liquid helium without recuperation of the sensible refrigeration.

The first superconducting magnetic separator has been in operation since May 1986 with an on-stream factor of 99% (8,500 hours in the first year). There have been frequent shutdowns due to thunderstorm power outages. Each one requires some time for recovery. . . .

These authors go on to report on the economics of their system. Suffice it to say that the same customer subsequently bought more such units, a clear criterion of success. Annual savings of about \$130,000 on the factory's electric bill easily justified the higher initial capital cost for superconducting magnets. Reliable refrigeration to 4 K is the cornerstone of that success.

3.6.2. Argonne "Atlas" System

"Atlas" is a superconducting, heavy-ion, linear accelerator at Argonne National Laboratory. Atlas was the first accelerator to use superconducting RF cavities, began operation in 1978, and is still operating, producing heavy-ion projectiles for impact on fixed targets.

As with any major cryogenic system, reliability is the principal design constraint. In the Atlas cooling system, reliability is enhanced by flexibility, redundancy, and excess capacity. There are 3 refrigerators, 11 helium compressors, approximately 100 m of coaxial liquid helium transfer lines, 3 thousand-liter dewars, and 76 liquid helium valves that deliver steady-state flowing liquid helium to 16 beam-line cryostats. The cryogenic system has a cooling capacity of 700 W of liquid helium at 4.6 K, approximately 50% to 100% more than required.

Figure 3.18 is a schematic of the basic interconnection and control features of the Atlas cryogenic system.[11] The cryogenic system serves three separate elements: (1) the injector, (2) the booster, and (3) the Atlas linac cryostats. A complicated distribution system is required to provide independent service to each of these elements. On the main outgoing line there is a pair of tees and valves with which to direct flow to and from each cryostat. The bypass valves can be fully closed to force the entire flow through the cryostats or may be throttled to provide only the required flow, the rest being bypassed to downstream components. With this arrangement, any cryostat can be removed from the system without affecting operation of the others. The three dewars serve several functions: (1) interconnection between the three refrigerators, (2) pool-boiling heat exchangers, (3) surge tanks, (4) phase separators for the gas return lines, and (5) control elements for maintaining levels and pressures.

The control system manages several functions. First, the controls of the liquid helium system must adapt easily to large and sudden changes in heat load; these occur frequently due to changes in operating conditions of the linac. Second, the controls must be able to provide cooling to parts of the linac while other parts are out of service. Finally, the linac must be kept cold continuously for long periods of time, preferably years.

Operating experience with the Atlas cryogenic system has been quite good. Approximately 15% of beam down time is due to cryogenic failures. Most of these are due to power glitches which trip the interlocks on the refrigeration systems. Failure of the screw compressors is the next biggest source of down time. However, because of redundancy (11 compressors) and approximately 50% excess capacity, compressors are repaired at the time of failure; preventive maintenance is not required. The 3 refrigerator systems shown in Figure 3.18 are all commercial units; "2800" and "1630" are model numbers. One is a reciprocating unit, and the other two are low-maintenance, turbo-expander units. For repair, these units go back to their manufacturer.

The system maintains an inventory of 1,500–2,000 liters of liquid helium. Still, there are losses: helium losses occur on the warm side at a rate of approximately 1.25 l/hr. The system is topped off with liquid helium supplied by a vendor every few weeks. Delivery of liquid helium is more expensive than using gaseous helium, but use of liquid helium eliminates many impurities (especially neon), which more than justifies the additional cost. Meanwhile, the liquid nitrogen supply is replenished three times per week; each truck delivers 2,000–2,500 gallons (approximately 7,600–9,500 liters).

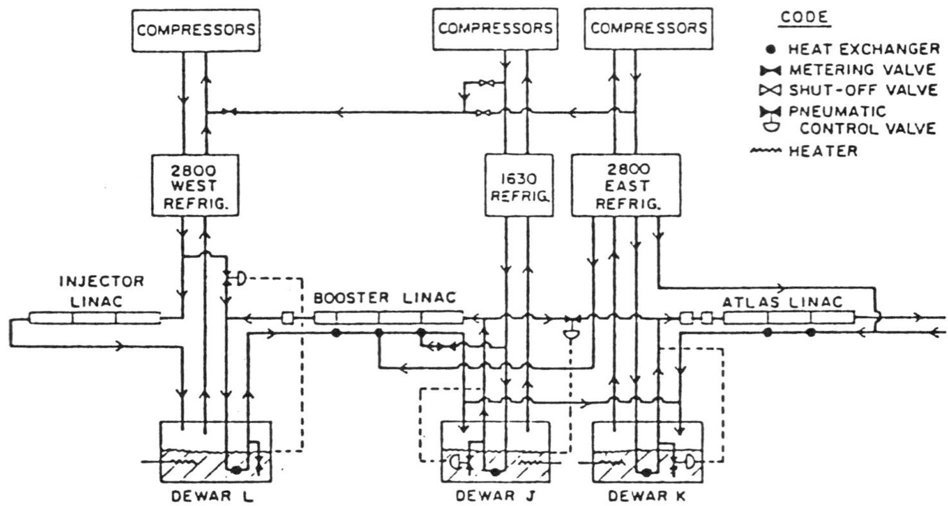

Figure 3.18.   Schematic of basic interconnections and control features of the Atlas cryogenic system under normal operating conditions.

Redundancy is a key characteristic of the Atlas system. By starting with a large number of elements (three refrigerators, seven 100-kW screw compressors, and four reciprocating 25-kW reciprocating compressors), and then adding an interconnected system of valves and controls, Atlas achieves flexibility in matching refrigeration capacity to varying heat loads. One large compressor would degrade reliability. This arrangement also allows for continued operation of the linac at reduced energy levels in the case of refrigerator or compressor maintenance or failures. Most important, it permits keeping the linac cold for long periods of time in spite of such failures, thus minimizing the loss of accelerator running time.

3.6.3. *Fermilab Tevatron Cryogenic System*

The size and complexity of cryogenic systems for major research facilities has reached well beyond the Atlas system. (Atlas is big enough to illustrate complexity, but is still small enough to describe here.) For example, Fermilab is a national laboratory for high-energy physics experiments using a proton accelerator called the Tevatron. Figure 3.19 is a familiar photo of the site. The Tevatron is a 2-km-diam synchrotron containing nearly 1,300 cryogenic components, primarily superconducting magnets. The cryogenic system at Fermilab provides 24 kW of refrigeration, making it about 30 times bigger (in capacity) than the Atlas system. It cools about 1,000 magnets and 300 other devices.

Cooling for the Tevatron is supplied by a hybrid system consisting of a *central helium liquefier* (CHL) connected to 24 satellite refrigerators by a 7-km liquid helium and liquid nitrogen transfer line.[12] The CHL occupies one building; it contains three 2,000-HP helium compressors; a helium cold box; a 90,000 kg/day nitrogen reliquefier; and various storage tanks. Not surprisingly, the system is much more complicated than the illustrations of Figures 3.8, 3.9, and 3.10. It can deliver up to 5,000 liters per hour of liquid helium at approximately 300 kPa absolute (100 kPa = 1 atm).

Figure 3.19. Aerial photograph of the Fermi National Accelerator Laboratory.

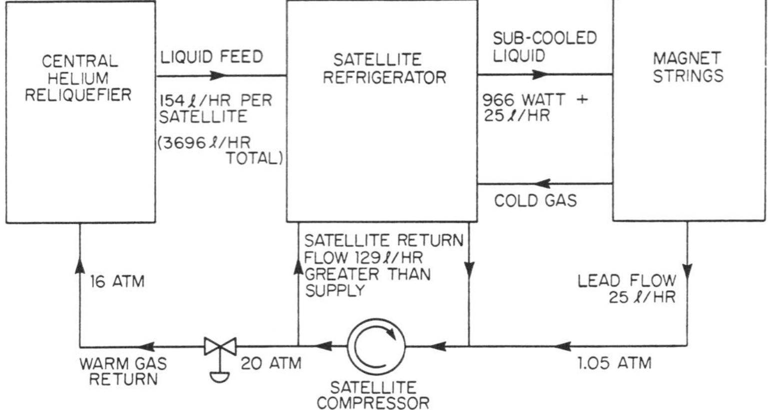

Figure 3.20. Helium flows for a single Fermilab satellite refrigerator system. (From Ref. 13.)

This system combines the advantages of a single central facility with those of individual, stand-alone units distributed around the Tevatron accelerator ring. Figure 3.20 shows the flows for a single satellite. The entire flow system is closed. Liquid helium from the CHL supplies 24 similar refrigerators, and each in turn provides cooling for magnets, current leads, and various experimental devices. Each satellite was designed to produce 1 kW of refrigeration plus 25 l/hr of liquid helium to cool the magnet current leads.

Why was the system built this way? The CHL has the high efficiency associated with large components, but the requirements for distribution of cryogenic liquids and electric power to the service buildings located around the ring are reduced.[13] This system provides reliability through a combination of redundancy and storage of large inventories of liquid helium (63,000 l), gaseous helium (5100 kg at 1700 kPa) and liquid nitrogen (254,000 l). Moreover, the system is designed to permit flexible operations, having an extensive set of control valves, regulators, and sensors; plus, it has a control system which can be operated under computer control, or under a computer/manual combination, or completely manually.

Planned maintenance is an adjunct of reliability. Only two compressors are required to satisfy the liquid helium requirements of the Tevatron. Because the third compressor can be held in reserve, maintenance requirements are severely reduced. The major sources of failure are bearings and valves which can be replaced in a relatively short period of time. This eliminates the need for most preventive maintenance. Wear on the piston rings requires major maintenance every three to four years. Since this wear causes a gradual decrease in compressor efficiency and not catastrophic failure, this maintenance can be scheduled for normal Tevatron shut-down periods.

This system has been in operation since 1983, and valuable operating experience has led to improvements. Even in the first two years, the CHL was available for 97% of the time the Tevatron was scheduled to run. Initially, contamination due to dust, water, and nitrogen was the biggest source of down-time.[14] Over time, modifications have overcome such problems; for example, in the oil removal system, the seal between the oil and water regions has been improved. Fortunately, liquid helium itself is not subject to contamination, because all other materials are solids by temperatures low enough for helium to be a liquid.

Reliability considerations at Fermilab have led even to nitrogen self-sufficiency. In routine industrial use, liquid nitrogen is delivered in big trucks periodically. At Fermilab, the liquid nitrogen dewars require topping off every four hours. During periods of bad weather, particularly in winter, failure to receive delivery has caused some shut-downs. This has prompted construction of a nitrogen reliquefier, simply for the sake of reliability. Fermilab estimates that at this scale of production, the cost of reliquefying nitrogen is a break-even venture. Projecting to other industries, it is unlikely than anything but a utility would operated on a large enough scale to warrant this.

Because of its excellent performance record, the cryogenic system at Fermilab has been designated[15] an "International Historic Mechanical Engineering Landmark" by the American Society of Mechanical Engineers (ASME). Still, it would be misleading to assert that all is perfect in this or any other refrigeration system. Here, stuck relief valves continue to be a nuisance. Nevertheless, based upon a number of years of successful operation, Fermilab concludes that careful attention to design of the cryogenic system (including redundancy, large inventories of cryogens, excess cooling capacity, and a flexible, well-monitored system) can lead to high levels of reliable operation. This provides a basis for optimism about future large industrial refrigeration systems.

3.7. SUMMARY AND CONCLUSIONS

The ceramic oxide superconductors may eventually prove to be quite useful in the intermediate temperature range around 30 K. For this to happen, it will be necessary for engineers to move beyond the elementary level of cooling via a liquid nitrogen bath, and instead design systems that include more complex refrigeration systems.

This chapter covers the principles of operation, as well as the chief practical features, of cryogenic refrigerators in common use today. The most striking characteristic of any practical system is that it functions at an efficiency far below the ideal Carnot efficiency, as shown in Figure 3.3. It is fair to say, therefore, that there is a long way to go in improving refrigerators. Here is clearly a business opportunity for an innovative refrigerator manufacturer.

Standard refrigerators for very low temperature applications involve expansion–cooling of gases, utilizing the Joule–Thomson effect. The physics and thermodynamics of typical refrigeration cycles are well established, but the practical implementation of real hardware into a cooling system is still an engineering art. The performance is fundamentally limited to no better efficiency than that of a Carnot Cycle, and this in itself carries a serious cost penalty when very low temperatures (4.2 K) must be achieved. Moreover, the actual efficiency of real refrigerators is often only 10% of the Carnot efficiency. Thus, there is plenty of room for improvement as the state of the art advances.

Cooling to 4.2 K used to be a matter of filling a laboratory dewar with liquid helium, and then minimizing the heat leak associated with small-scale scientific apparatus. However, as superconductivity becomes more commonplace in commercial and industrial applications, the need for intermediate-sized refrigerators (that do something other than liquefy great quantities of helium) is increasing. Because of the scarcity of liquid helium, many applications (such as magnetic resonance imaging units in Third World hospitals) require closed-loop refrigeration. Scheduled maintenance is certainly possible, but there is a premium on reliable operation in the field. Industrial separators that utilize superconducting magnets likewise cannot afford to fail unexpectedly. Consequently, makers of refrigerators are considering new materials and new ways to improve efficiency that would have escaped attention only ten years ago.

Spacecraft requirements are also helping to advance the state of the art in refrigerators. With no scheduled maintenance permitted, reliability carries such a high premium that cooling systems are usually redundant, in spite of the cost of the weight for the duplicate system. It is possible to pursue radical new principles of heat exchange and refrigeration, even sacrificing efficiency, if reliability can be improved. The interest in pulse-tube systems, devoid of moving parts at low temperatures, is explained by the quest for reliability.

Large refrigerators are more efficient than small ones; for example, a liquid helium machine might have 20% of Carnot efficiency while a spacecraft cooler typically has 2%. This has important consequences for the decision of whether to utilize HTSCs or LTSCs (or neither!) in electrical power applications such as SMES. In very large systems, the fractional cost associated with cooling decreases, but in smaller units the cooling cost may destroy the economic gain sought in the first place.

There is a strong market pull toward better cryogenic systems, and the next decade should see considerable advances in refrigeration technology. Continued expansion of superconducting magnets into medical applications, underground transmission lines, SMES and other large-scale applications, all call for better refrigerators. With so much gain in efficiency still waiting to be captured, it is plausible to anticipate a very competitive market ahead.

REFERENCES

1. R. Radebaugh, NIST-Boulder, "Refrigeration Systems," Lecture at NATO Advanced Study Institute, 15 Sept. 1990.
2. M. Tribus, *Thermostatics and Thermodynamics*, pp. 342–347 (Van Nostrand, New York: 1961).
3. H. D. Young, *University Physics*, 8th Edition (Addison-Wesley, Reading, Mass.: 1992).
4. R. F. Barron, *Cryogenic Systems*, pp. 60–82 (Oxford Univ. Press, London: 1985).
5. R. Radebaugh, "Progress in Cryocoolers," in: *Applications of Cryogenic Technology* 10, ed. by J. P. Kelley, pp. 1–14 (Plenum Press, New York: 1991).
6. Y. Matsubara and J. L. Gao, Cryogenics **34**, 259 (1994).
7. O. V. Lounasmaa, *Experimental Principles and Methods Below 1 K*, (Academic Press: 1974).
8. C. Kittel and H. Kroemer, *Thermal Physics*, 2nd ed., Chapter 12 (W. H. Freeman: 1980).
9. R. D. Shull *et al.*, Materials Research Soc. Symposium Proc. **286**, 449 (1993).
10. A. J. Winters, Jr. and J. A. Selvaggi, Chemical Engineering Progress, January 1990, pp. 36–40.
11. J. M. Nixon and L. M. Bollinger, "The Liquid Helium System of Atlas," (report by Atlas Group at Argonne).
12. G. A. Hodge *et al.*, "Fermilab Central Helium Liquifier Operations," in *Advances in Cryogenic Engineering*, R. W. Fast, Ed., **29**, 461 (Plenum Press, New York: 1984).
13. C. H. Rode, *et al.*, "Operation of Large Cryogenic Systems," *IEEE Trans. Nucl. Science* **NS-32**, 3557 (1985).
14. R. J. Walker, *et al.*, "Recent Operating Experience with the Fermilab Central Helium Liquefier," in *Advances in Cryogenic Engineering*, R. W. Fast, Ed., **31**, 647 (Plenum Press, New York: 1986).
15. See *Cold Facts*, newsletter of the Cryogenic Society of America, Fall 1993 issue.

4

Industrial Applications

Conventional superconductors are already being used to a limited extent by industry. The inconvenience of supplying liquid helium has been the major obstacle so far, but as closed-cycle refrigeration systems become more reliable, the advantages associated with superconductors will lead to greater use.

The role of economics must not be underestimated when thinking about industrial applications. It is easy enough to do a cost/benefit analysis on a proposed new piece of equipment, presuming that the new equipment works correctly. That is the normal basis on which vendors of machinery sell their wares to factories and mills. However, the buyer of such machinery, the factory, often looks suspiciously at such cost/benefit analyses, because of fears that the new equipment may not work as advertised. Therein lies the conflict between buyer and seller. The factory manager knows that "down time" is extremely expensive, and usually writes a contract with the vendor that contains severe financial penalties if the new equipment should force the production lines to go down.

Every college student who works in a laboratory with liquid helium quickly learns how temperamental equipment becomes at very low temperatures: Tiny stress cracks lead to vacuum-seal leaks, etc., and consequently down time is substantial. Anyone can pour liquid nitrogen, but liquid helium requires care and training to handle. Generally, over the years, the risk has been too high to capture the imagination of practical-minded factory managers. Therefore, superconductivity has never really had an opportunity to show what it can do for industry.

Proponents of superconductivity will argue their case by asking their audiences "When's the last time you bought freon for your refrigerator?" This viewpoint presupposes the existence of a reliable closed-cycle refrigerator, albeit for liquid helium. As we saw in Chapter 3, today such systems exist at prices that can be factored into the overall cost projections for an industrial application. This fact in itself is a major step toward enabling superconductivity to reach industry.

Clearly, a second major step will be to have devices commercially available that operate at 77 K instead of 4 K, thus relaxing the refrigeration requirements; that is the whole point of the high-temperature superconductors. Nonetheless, some low-temperature superconducting devices already have entered the industrial arena, and are performing well. As the record of reliability and cost-effectiveness grows, we expect more interest to follow, and subsequent innovators will propose still more industrial uses for superconductivity. This chapter describes existing devices that point toward the potential that lies ahead.

4.1. POWER QUALITY CONDITIONING IN FACTORIES

The ordinary electricity that comes out of the wall is one of the most reliable things in the world, but it is not perfectly reliable: there are momentary power outages, voltage sags, and assorted glitches on the line all the time. Consumers of electricity rarely notice these, unless their digital clocks are suddenly blinking "12:00" for no apparent reason. Authors of books are equally oblivious to power disturbances, right up until their word processors suddenly go off when they haven't done a "save" recently.

4.1.1. Critical Conditions

However, certain factories suffer serious losses in production due to very minor imperfections in the power supply, and so they look for a way to improve the quality of power reaching their machinery. Each industry has its own level of tolerance pertaining to power quality. For instance, "heavy" industry (steel, oil refining, etc.) was not bothered by minor glitches in the past, but with modern factories running under elaborate electronic control

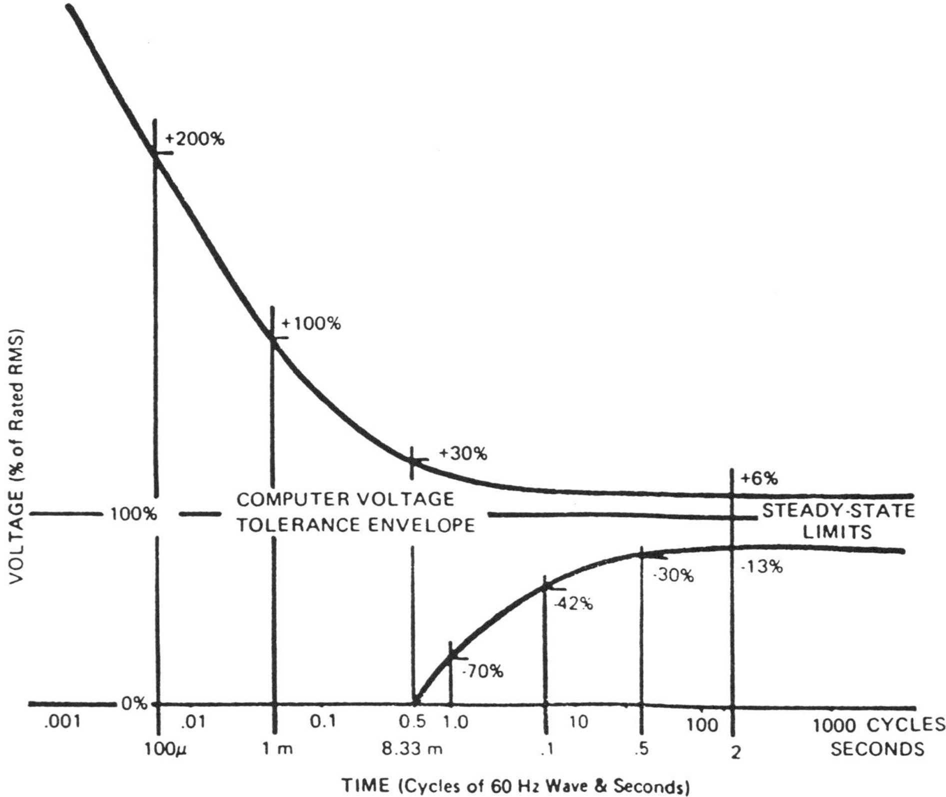

Figure 4.1. Typical computer system voltage tolerance envelope (adapted from IEEE standard 446). Two logarithmic horizontal scales are shown: cycles or seconds. Thus, for 1 millisecond, the system must tolerate a voltage 100% above the nominal voltage; for 1 cycle, it must tolerate a deviation of –70%; and for 1/10 second, a deviation of –42%. For any time duration, however long, the system must withstand swings of +6% or –13%.

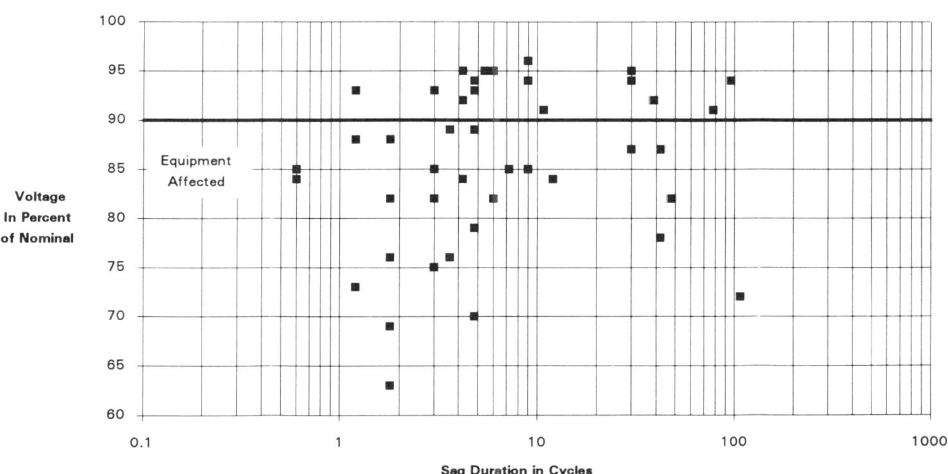

Figure 4.2. Occurrences of voltage sags in an industrial plant. (From Ref. 1.)

systems, even very large industrial plants are vulnerable to interruptions, sometimes with very expensive consequences. For example, Intel in Albuquerque cites losses of $100,000 annually due to infrequent power problems that affect integrated-chip manufacturing lines.

One typical graph is the voltage tolerance envelope used by the computer industry (Figure 4.1). Any computer is expected to withstand excursions within that envelope. However, it is not difficult to imagine a voltage sag lasting only 1/4 of a (60 Hz) cycle that would exceed the envelope, possibly either causing a trip-out or even damaging equipment. Voltage sags of 15% or more are reasonably commonplace (and routinely described as due to "squirrels and birds"). One study[1] compiled the data portrayed in Figure 4.2, showing the depth of sags and their duration at a typical factory.

The "obvious" solution is to use uninterruptable power supplies, which generally means batteries, to power critical portions of the factory. Unfortunately, batteries drain too quickly to be satisfactory and are seldom properly maintained in factories—hence the search for alternative ways to guarantee high power quality. There are also "active" power-quality conditioning devices, but they have not yet been widely evaluated.

4.1.2. Small SMES

The concept of superconducting magnetic energy storage (SMES) is one possible answer to the power-quality problem. Figure 4.3 sketches the basic concept involved: source and load interface with the SMES through an AC/DC power-conditioning system, and the energy storage itself consists of a DC persistent-current flowing through a superconducting coil in a liquid helium bath. Basically, all a SMES unit does is store energy in the magnetic field of a coil; when superconducting wire is used for the coil, there are no resistive losses in it, and a DC current will persist for as long as the coil is kept cold. (AC current in a SMES would be lossy and inefficient.)

Often SMES is imagined as a very large energy storage system that takes advantage of an economy of scale; but SMES certainly can be used on a small scale as well. One company,

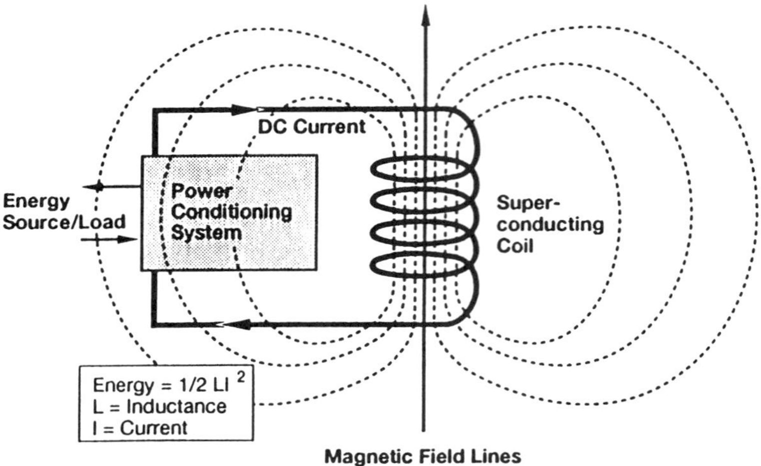

Figure 4.3. Basic concept of superconducting magnetic energy storage (SMES).

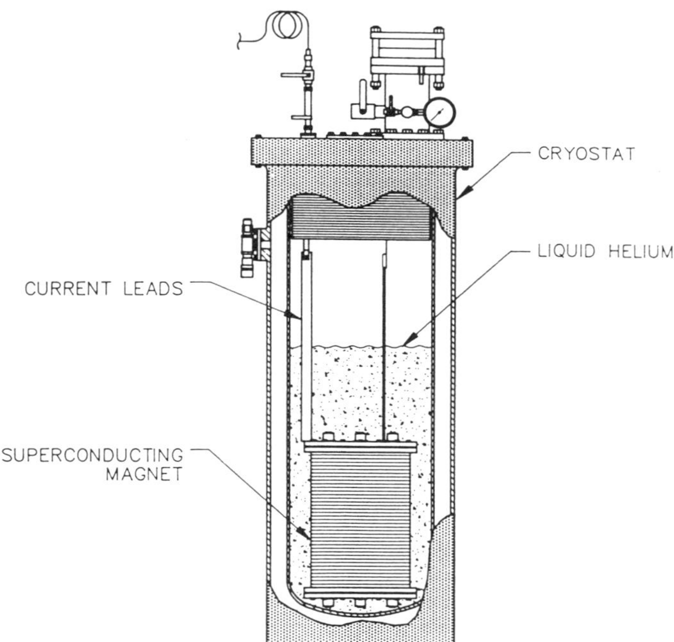

Figure 4.4. Magnet and cryostat assembly in Superconductivity, Inc.'s superconducting storage device (SSD).

Superconductivity, Inc., of Madison, Wisconsin, offers such a device for sale. The evaluative question asked by industrial customers is, "Does the value of better power quality in my application justify the expenditure for a small SMES unit?"

The Superconductivity, Inc., power-conditioning unit (called an SSD) comes in a semi-trailer, and can be treated as a "black box" by the customer.[2] A typical unit capable of delivering 750 kW for 2 sec costs almost $1 million for a complete package. In addition to running the load in the first place, power from the utility also drives the SMES power supply and the refrigerator. During operation, the SSD uses about 50 kW of power continuously. (At $0.05/kWh, that runs to over $20,000 annually.) Clearly, the SSD is for applications where power quality really counts. The example of Figure 4.2 is one such case.

Figure 4.4 is a diagram of the SSD apparatus, a magnet immersed in a helium dewar inside a vacuum jacket. Note that current leads leave the cryogenic bath; these wires will dissipate energy resistively, and hence the stored energy will eventually be lost unless it is renewed by an external power supply. Of the 50 kW drawn by the unit, some power goes to this purpose while most power runs the liquid helium refrigerator. Figure 4.5 shows the actual apparatus.

Figure 4.6 indicates one way the SSD is connected in a typical industrial setting. Power to the load is AC, but the SMES coil carries DC current, so there is a mismatch. Under normal conditions, the switch inside the voltage regulator is closed and the DC current circles idly through the coil in the cryostat; this is the "standby" mode. However, whenever there is a momentary failure in the utility power source, the solid-state isolation switch opens, and the voltage across the capacitor bank starts to fall, as energy stored in the capacitors is used to supply the load. The GTO ("gate turn-off") thyristor switch in the voltage regulator opens and closes at up to 400 Hz, to send pulses of current from the magnet to the capacitor bank.

The result is that voltage is maintained across the capacitors within very tight tolerances, allowing the inverter to supply constant AC power to the load. The output waveform is by no means perfect, because of serious harmonic distortion originating in the inverter; but that's better than suffering a complete power outage. Actual data from a test appears in Figure 4.7, and shows how a series of pulses leave the superconducting coil during a half-second outage. The power to the load remains very constant, which, of course, is what the customer is paying for.

The current in the superconducting coil has been depleted (by about 35% in Figure 4.6), but that can be restored by the power supply as soon as the utility outage is over. If the outage lasts longer than the storage capacity of the coil (typically about 2 sec), at least there is enough time for other protective equipment to trip and shed the load without damage. There are factories containing certain critical equipment where just a single protective event in 5 years justifies the entire cost of the SSD system.

One point worth noting is that the superconducting coil will not pour current back into a fault condition at the utility. An isolation switch (see Figure 4.6) ensures that stored power will only go to the load. There are other possible installation configurations, but they are beside the point.

The essential message here is that SMES already exists and serves the needs of industry. It is more expensive than battery-based alternatives; but it is subject to rational cost/benefit analysis, and may be justified in selected cases. The obvious advantage of changing to liquid nitrogen conditions would be to lower the cost, and thus shift the cost/benefit equilibrium point, leading to greater use by industry.

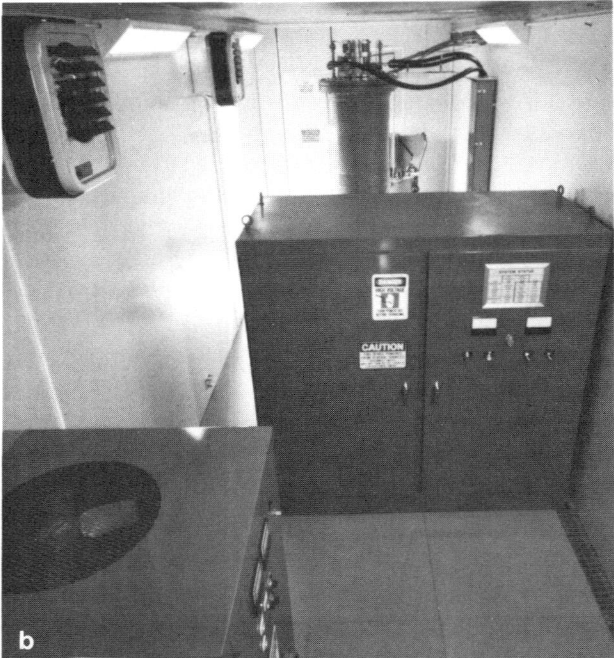

Figure 4.5. Photographs of Superconductivity Inc.'s SSD: (a) Trailer parked outside a factory; cryostat is visible through the doorway. (b) Interior of trailer, showing cryostat in background, voltage regulator in middle, and magnet charger in foreground.

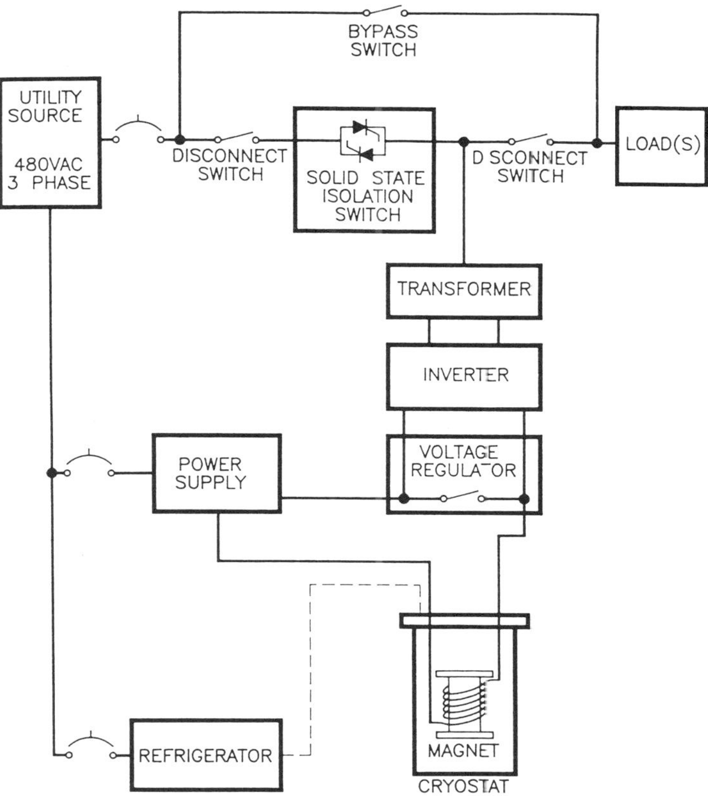

Figure 4.6. Shunt–connected SSD, typical of installed configuration.

4.2. MAGNETIC SEPARATION

SMES is not the only present-day use of superconductivity by industry. Because superconductors enable high magnetic fields to be sustained, almost any application that utilizes a high magnetic field is eligible for superconductivity. Here we describe one example where superconductivity appears in an unusual industrial setting.

4.2.1. Separation Principles

In many high-magnetic field applications, e.g., measurement systems such as MRI, some of the available field strength must be sacrificed in the interest of achieving uniformity and stability. However, there are also "brute force" applications of magnetic fields that can have great commercial value. For industry, the case of magnetic separation is a good example.

The goal of industrial separation processes is to make use of strong fields in the mainstream of production. Conventional electromagnets separate iron from aluminum or copper scrap; magnetic fields are not generally strong enough to divert impurities from water

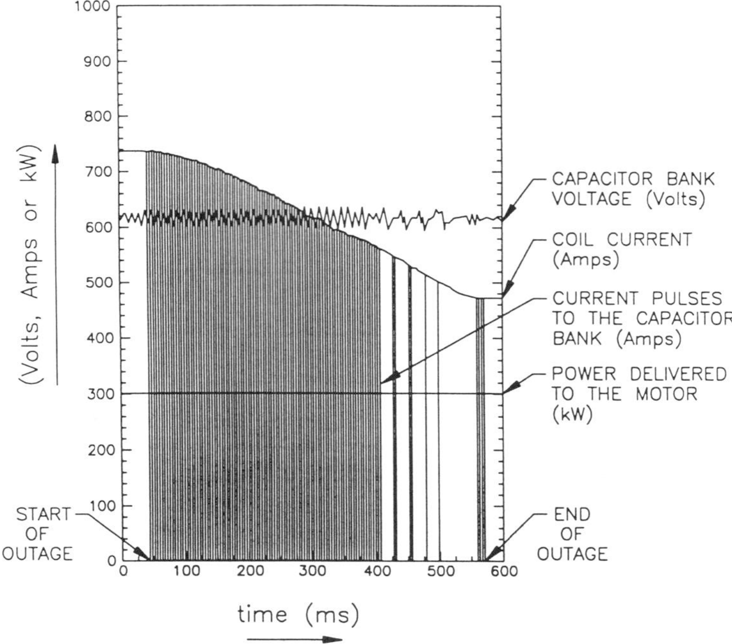

Figure 4.7. Test results of October 1990 for SSD operation. Note especially the very flat line for power delivered to the motor.

or flue gas streams. However, a superconducting magnet of 20 T would be a very effective separation device. Even if it were nonuniform, variable and suffered brief intermittent outages, it would still be useful.

It is easy to construct an illustration of the principle: In a chemical plant, it is frequently required to separate a stream containing two or more useful products. Referring to Figure 4.8, a step could be incorporated where one of the components was "tagged" with a magnetic element through a chemical reaction. (Chelating comes to mind as a mechanism for this step.) Then the mixed stream would enter the magnet, where the magnetic tag would be diverted to one side, taking one of the chemicals with it. A subsequent step would remove the tag and restore the now-purified stream to its former state. The tagging compound would be recycled to rejoin the original mixed stream ahead of the magnet.

The whole process is reminiscent of the way lime (CaO) in a paper pulp mill plays a crucial role in a peripheral loop that recovers NaOH from Na_2CO_3. Whether such a method is implemented by a factory depends on the economics of the several steps involved. Superconducting magnets must compete with conventional electromagnets. The point is that high magnetic fields make it possible to consider the option.

Another industrial application of magnets is the use of magnetic field *gradients* for separation purposes. (A classic physics experiment of the 1920s was the Stern–Gerlach experiment, in which atoms with electron spins up were separated from spin-down atoms by passing them between specially shaped magnet poles that gave a uniform gradient to their

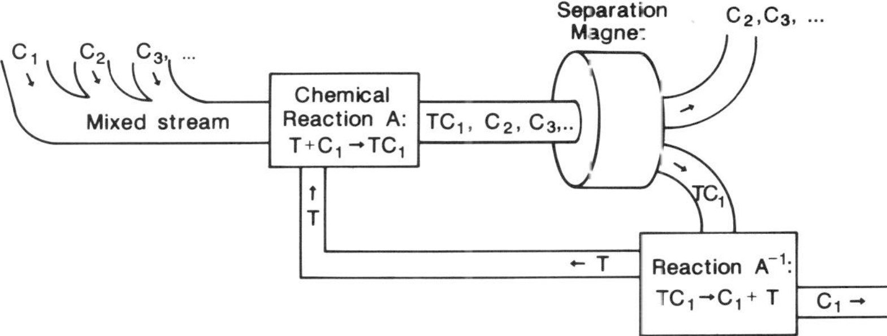

Figure 4.8. General schematic of magnetic separation process. T denotes a magnetic tag that is attached only to chemical C_1 in reaction A; reaction A^{-1} removes that tag. C_1 is separated out while T is recycled repeatedly.

magnetic field. The experiment became a precursor of quantum mechanics by showing that electron spin had only two possible values.) In industry today, there are plenty of chemicals that are distinguishable by their magnetic moments—ortho and para hydrogen immediately come to mind—and strong magnetic gradients offer a way to sort them out.

In liquifying air, oxygen will collect on the poles of a nearby magnet, while nitrogen will not. Although this has been observed for nearly a century, the magnetic attraction is so weak that other means of separating oxygen from nitrogen have so far won out economically. Powerful high-field magnets may change that economics. Since the steel industry is the biggest user of oxygen,[3] it would benefit by a cheaper oxygen separation technique.

4.2.2. Eriez Magnetic Separator

The task of purifying ores is eligible for magnetic gradient separation, and there is an application of superconducting magnets operating today that does this. Eriez Magnetics Corporation of Erie, Pennsylvania, makes a magnetic-gradient separator[4] for the mining industry which incorporates a superconducting magnet. Figure 4.9 is a photograph of one of their units. Figure 4.10 shows the general layout of the device.

These are *very* big magnets—definitely not a research facility! Eriez had made conventional electromagnets of this size for some time. It was not uncommon to have a 600 kW power level for them. Around 1980, electricity costs began to rise sharply; for example, the demand charge in some areas rose from $2.50 per peak kW to $12.00. When attention fell on the high cost of running a conventional electromagnet, Eriez began designing a superconducting magnet. After a test period, the first superconducting unit became operational in 1986. In this unit, the entire steel chamber is 7 ft high, of which the iron-mesh zone is 20 in. The energy storage is 3.5 MJ and the inductance is 10 Henrys.

Eriez's initial customer was in the kaolin industry, which needs to remove iron oxide from titanium dioxide ore in order to produce whitener for paper. The process is as follows: a slurry of kaolin ore flows into the 7-ft bore of the magnet, which is blockaded by a huge "Brillo pad" of magnetic stainless-steel mesh. The steel wires create steep magnetic field gradients in their vicinity, thus attracting and holding iron oxide particles. The exiting flow is a milky color, instead of orange-brown from iron oxide. Every half-hour or so, depending on the slurry concentration, the magnet is shut down and the steel mesh is back-flushed.

Figure 4.9.   Eriez Magnetics's superconducting magnetic separator in a kaolin mill.

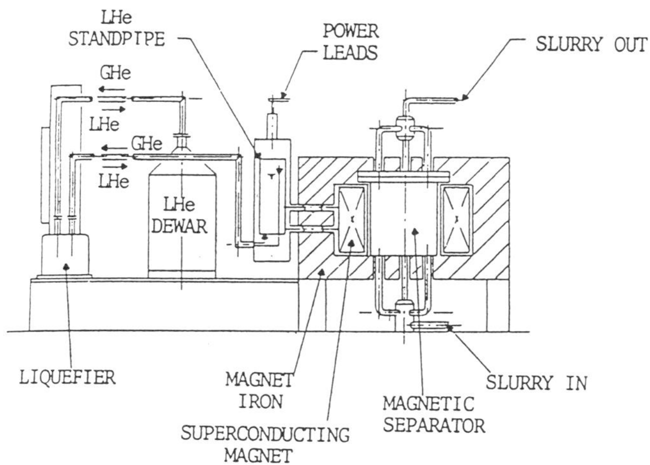

Figure 4.10.   Conceptual layout of an Eriez separator.

Table 4.1. Magnetic Separator

Stored energy	7.6 MJ
Current	860 A
Inductance	21 Henry
Ramp rate	1 min up, 1 min down; 3 cycles per hour
Charging voltage	300 V
Coil type	Superconducting solenoid
Orientation	Vertical axis
Cooling	Liquid helium bath, 4.5 K (nominal)
Heat load (calculated)	
Steady-state operation Vapor-cooled leads	4.0 l/hr
For a 1-min ramp	
Coil	1.6 KJ
Cryostat	15 KJ
Liquid helium capacity	430 l
Conductor	
Material	NbTi/Cu
Configurations	Flattened one-level cable
Coil configuration	Double-pancake wound
Number of layers	28
Number of turns per layer	36
Winding I.D.	$130\frac{3}{8}$-inch
Winding O.D.	$140\frac{3}{8}$-inch
Coil height	20 in.
Coil weight	9300 lb
Axial force constant	144,380 lb/in. (calculated)
Radial force constant	14,000 lb/in. (calculated)
Helium vessel inner tube O.D.	130 in.
Helium vessel outer tube I.D.	143.75 in.

The economics driving this technology[5] are easy to understand: in a conventional water-cooled copper magnet, typically 300 kW are dissipated by I^2R losses, and all this must be carried away by water. A 30-hp pump is required just to circulate the water through a heat exchanger. In a superconducting magnet, the only power required is for the helium liquefier, which is 60 kW. Thus, the net savings is 240 kW—worth over \$100,000 per year under typical factory parameters.

The parameters of a factory also drive the technical design of the magnet.[6] Down-time is so expensive that the Eriez superconducting magnet was colossally overdesigned. As described in Chapter 3, there are two helium compressors, for redundancy; even those are oversized. The full magnetic field is only 2 T, but this suffices to saturate the iron wires that create the magnetic field gradients needed for separation. Because there is no need for high Fermilab-type fields, parsimony of the conductor area is not an issue. Therefore, for total safety and stability, the ratio of copper to NbTi in the wire is 39 to 1. The magnet never quenches and is designed not to "train" either.[4] Testing is done at 400 A, but only 20 A is needed to achieve 2 T. These design parameters have a "business" explanation: Eriez's entire credibility and hopes for future growth are tied to the fail-safe image of this magnet, so they are taking no chances.

The fact that this device works at all is a major engineering achievement; that it works in a remote field setting, far from any science laboratory, is a stunning accomplishment.

The satisfied customer soon had Eriez build a second one for another plant, this time with a 10-ft bore to accommodate greater throughput. The second unit stores 7 MJ and weighs 1 million pounds.[7] The parameters of this magnet are listed in Table 4.1. As one might expect, the liquid helium consumption is substantially higher, although not quite double that of the first unit. However, the liquid helium consumption is actually lower, owing to design improvements learned en route to the first unit. (The magnet-support structure was refined.) Here, as in many other cryogenic applications, the cooling load depends mostly on the support structure, the current lead-ins, and peripheral activities; it depends very little upon the size or current in the cooled magnet itself.

There is more to the story of this unusual magnet. For the 7-ft-diam. bore magnet, the actual separation process requires back-flushing three times per hour, so the magnet must be shut off after 15 min or so, and then recharged 5 min later. This could waste a lot of electricity if the entire magnet power were merely dissipated. Eriez found that the most cost-effective strategy was to send electricity back to the utility, so they invented a clever circuit that does so while discharging the magnet. Called the *bipolar power supply*, it ramps the magnetic field down and sends the power back to the electric utility; upon restart of the separation process, the magnetic field ramps back up. They cycled the magnetic field 25,000 times in the first two years. Unfortunately, eddy-current losses in the metal dewar and current leads dissipate 40% of the energy; so their system has plenty of room for improvement.

Still, the decision to return power to the utility rather than store it is important: the two-quadrant power supply circuit is an innovation that gives Eriez an advantage over potential competitors in this market. In a future scenario in which many industrial customers do the same thing, the utility will not be hurt as long as the customers don't all send power back at the same time. The need for storage (on either side of the meter) depends upon multicustomer changes in demand. The leading example of this, which motivates much of the research on utility-side SMES, is the diurnal variation in electricity demand.

There is another economic benefit that accrues to the customer by using a superconducting magnet. In a conventional magnet, the I^2R losses trade off against the ability to separate materials, which increases linearly with B or I. Beyond a certain point it doesn't pay to separate further. By contrast, a superconducting magnet can run "flat out" for the same refrigeration cost; therefore, lower-grade ores can be treated economically, which in turn extends the raw materials supply at no extra cost. This will be discussed further in Chapter 26.

4.3. UTILITY-BASED SMES

Our third major example of actual superconducting devices is drawn from the experience of the Bonneville Power Administration during the early 1980s. Although this system is not running today, its story is an excellent illustration of the way concepts are tested under real-world conditions.

In contrast to a critical-unit protection device for industry, a utility-based superconducting magnetic energy storage (SMES) system strives to assure energy supply over sustained periods of time. Twelve hours of energy storage is a design goal of contemporary SMES designers. We defer discussion of such storage devices until Chapter 21, because no such equipment has yet been built. Here we describe a small SMES unit that was used to improve the stability of Bonneville's electrical power grid.

Figure 4.11. Small SMES for utility line stabilization at Tacoma, Washington. (Bonneville Power Administration photo.)

The basic concept of a stabilizing SMES is about the same as any other SMES. However, the storage capability need not be great; the unit must be able to deliver several megawatts for a few seconds; so its total storage is in the kilowatt-hour range—about a factor of 10 more than the device of Section 4.1.

This unit (Figure 4.11 is a photograph) was built at the Tacoma, Washington, substation of Bonneville Power Administration (BPA) during 1982. It took 3 months to cool down to 4.5 K and first was operated in February 1983. Testing proceeded until November 1983, after which it operated well for 3 months. During that time it ran over 1000 hours and exchanged energy with the system over 1 million times. Many details are provided in papers by the project team.[8,9] Here it suffices to state that the SMES components worked very well, meeting all design criteria, but the refrigerator was the weak link in the chain. An inordinate amount of manpower went into coddling the refrigerator, to which utility engineers were not accustomed. Here is one description of the circumstances[10]:

"Our nemesis was the refrigerator system," said [BPA manager Barry L.] Miller. "There were two reasons why. First, the refrigerator was a very sophisticated piece of equipment. Second, as with any R&D project where you are stretching beyond your grasp, we were way down on the learning curve. By the time we figured out how to make the refrigerator work, the project was almost ready to shut down. Most of the damage to the refrigerator, however, came through misuse. That particular refrigerator was simply never intended for a place like

Tacoma. It was a sealed unit, designed to be started up once and left to run. Its gas-bearing turbines, for example, were the size of hen's eggs and spun at about 250,000 rpm. It didn't take much to unbalance them. A speck of dirt would do it. It also proved to be hypersensitive to line voltage fluctuations, which would shut it down in a calamitous way. . . .''

Practical engineers with field experience have heard similar stories throughout their careers. The simple reality is that when technology is transferred from the laboratory to the real world, things like this happen. It is essential to remember that what is "proven technology" in the laboratory is only at the "beginning experiment" stage for a utility system. Consequently, a retrospective view of the installation and performance of this SMES regards the outcome as a successful experiment because it didn't do any harm and useful information was obtained to guide future designs.

In the intervening years, substantial improvements have been made in refrigerators, and so if this experiment were repeated today it would suffer far fewer problems. Because *one* utility showed that superconductivity is *not* incompatible with the electric power industry, it is now possible to consider other, more advanced, forms of SMES for various utility purposes. Chapter 21 covers some of those designs for the future.

4.4. OTHER APPLICATIONS

Two other areas of possible applications are worth discussing, even though neither is a plausible user of HTSCs in the near future.

4.4.1. Accelerator Magnets

Superconducting magnets have only a few markets to date, such as magnetic resonance imaging in hospitals and high-energy physics accelerators. The Superconducting Super Collider (SSC) was to use conventional LTSC magnets rather than wait for the development of suitable HTSC magnets.

Progress in LTSC magnets for accelerators has been steady and impressive. As recently as 1975, the CERN magnets were simply copper, and were limited to 2 T. The Tevatron at FermiLab subsequently transformed the idea of superconducting magnets into a practical reality. Problems of operational stability, cryogenic environment and mechanical structure were overcome en route to building that facility. Today, all accelerators routinely use superconducting magnets. (If copper magnets were selected for the SSC, it would have taken 14 GW to run it!)

An important factor limiting the magnetic field of an accelerator is the difficulty of making wire. The HTSCs are already notoriously difficult to form into wire, but it is less well known that niobium-tin (Nb_3Sn) also poses serious manufacturing problems. As discussed in section 2.11, Nb_3Sn is brittle, and despite its relatively high T_c of 18 K, it has found few applications. The non-brittle niobium-titanium alloy (NbTi, with T_c only 10 K) is preferred for all applications except where very high fields are mandatory.[11] The leading American maker of either kind of superconducting wire is Intermagnetics General Corp., and Figure 2.11 shows a cross section of their Nb_3Sn wire. Imagine the difficulty of maintaining a uniform cross section of that complexity for the many miles of wire needed to wind a magnet. From a manufacturing point of view, NbTi is far easier, and hence less expensive.

At CERN in Switzerland, the Large Electron-Positron (LEP) collider is running with 2000 superconducting magnets in a tunnel 27 km long, located 150 meters underground. The LHC (Large Hadron Collider) is the next generation accelerator, and will occupy the same tunnel. It will have a magnetic field of 10 tesla, and will operate at 1.8 K, to enable the beam to reach 8 TeV energy. Alsthom Corporation in France developed LTSC wire with filament diameters below 1 μm, and Alsthom-Intermagnetics S.A. (a joint venture) is the likely supplier of the magnets for the LHC.

Today's planned timetable for building accelerators, plus the stringent requirements for reliability, combine to eliminate HTSC magnets from contention for the LHC. The additional cost of helium refrigeration (compared to nitrogen) is smaller than either the cost of waiting a few years or the cost of the down-time associated with a system that is inoperative 10% of the time. Furthermore, the extreme cold of the helium jacket has become an essential component of the vacuum-pumping required in the tube that carries the particle beam. Any liquid nitrogen system would have to spend additional money on active vacuum pumps to evacuate the tube.

4.4.2. Power Electronics

The purpose of power electronics is usually to switch large currents without having any moving mechanical parts; a transistor that changes states from "on" to "off" is the heart of such devices. The three figures of merit are (a) speed of opening and closing; (b) energy dissipation in either the "on" or "off" state; and (c) losses during "standby."

There is still plenty of room for improvement in efficiency of power electronics, so, naturally, speculation arises as to the potential gain from using superconductors. But the "silicon" industry offers very formidable competition.

Power electronics is often cited as a potential application of HTSCs, because the HTSCs can be switched from superconducting state to normal state cleanly and in very short times: a magnetic pulse can convert the material from the superconducting state to the normal state in less than 10^{-9} seconds. The HTSCs are also expected to have very low "forward" voltages, a perennial problem for conventional power electronics. The "on" state dissipates no energy as a superconductor, while the "off" state dissipation is related to the normal-state resistivity. Therefore, very high resistivity in the ceramic oxides assures relatively low off-state dissipation.

However, think of the amount of current that must be handled in a typical power electronics application, such as an adjustable speed drive. Recall that the energy cost of refrigeration is a "standby" cost—it's the same no matter which position the switch is in. The cable that brings electricity into the cryostat is typically copper, and in power equipment that copper will have a comparatively large diameter. Accompanying that will be a large heat leak, and this refrigeration cost must be charged against the efficiency of any superconducting power electronics device.

What is the competition for HTSCs in this field? Commercial products (nonrefrigerated) available today include Toshiba's 5000 V, 5000 A thyristor, with a switching time of 10 msec. Its disadvantage is that it has several volts forward drop in the "off" state. (Here is where HTSCs might be useful.) On the other hand, the Power Electronics Applications Center in Oak Ridge, Tennessee, reports that devices now emerging from the laboratory switch 100 A in 1 msec, with a forward drop of only 1 to 1.5 V, while withstanding 500 to 1000 V. Therefore, the HTSCs will only be an improvement if they can get below 0.2 V forward drop or below 0.1 msec switching time. In other words, HTSC power electronics will face serious competition for market share from silicon devices.

4.5. SUMMARY

The foremost obstacle to using superconductivity in industry is the difficulty associated with liquid helium. Until very recently, refrigeration systems have not been robust enough to persuade industrial managers to give superconductivity a try. However, now that equipment vendors can reliably predict performance, it is possible to believe cost/benefit calculations without assigning enormous uncertainty factors. That in turn means that for selected process steps, which justify very expensive technology, low-temperature superconductivity can be competitive.

This chapter reviews several examples where the necessary conditions hold to make superconductivity a cost-effective option. In every case, the decision to supplant conventional technology with superconductivity came only after someone looked with an open mind at the overall task to be done, so as to permit consideration of a nonstandard approach to the problem.

Looking downstream, the continuation of this kind of thinking will make superconductivity an industrial reality in the future. It is possible to enumerate a catalog of creative ideas for industry, but the point is well made by the few cited here. Engineers need to ask creative questions about their processes in order to discover how to best use new technologies.

Notice that a creative new idea can be tried out in a factory even today using low-temperature superconductivity. If it proves feasible, it will add to the market pull driving the development of high-temperature devices. From the point of view of an R&D agenda, it is entirely reasonable to test new ways to use magnetism, in anticipation of high-field, high-temperature equipment someday in the future.

REFERENCES

1. J. D. Lamoree, *Analysis of Power Quality Concerns (at Industrial Sites)*, Report to Central Hudson Gas & Electric, by Electrotek Concepts, Inc. (21 August 1991).
2. C. C. DeWinkel and P. F. Koeppe, "Superconducting Technology Offers Ride-Through Capability for Large Industrial Critical Process Loads," in *Proc. American Power Conference*, p. 1252 in Vol. 2, published by Illinois Institute of Technology (1992).
3. *Steel at the Crossroads*, American Iron & Steel Institute (1981).
4. J. A. Selvaggi, P. C. Vander Arend, and J. Colwell, in *Advances in Cryogenic Engineering* **33**, 53–60 (Plenum Press, New York: 1988).
5. A. M. Wolsky, R. F. Giese, and E. J. Daniels, *Sci. Am.* **260**, 61 (1989).
6. A. J. Winters, Jr., and J. A. Selvaggi, *Chem. Eng. Prog.*, 36–40 (January 1990).
7. J. A. Selvaggi, T. Kranyecz, and S. Bell, "A 124 warm bore superconducting ironclad high-gradient magnetic separator," Eriez internal report (unpublished).
8. J. D. Rodgers, R. I. Schermer, B. M. Miller, and J. F. Hauer, *Proc. IEEE* **71**, 1099 (1983).
9. J. F. Hauer and H. J. Boenig, *IEEE Trans. Power Systems* **PWRS-2**, (2), 443 (1987).
10. Bonneville Power Administration, *Engineering Review* pp. 22–35 (Spring 1986).
11. E. Gregory, "Multifilamentary Composite Superconductor Design and Fabrication" pp. 1080–1086 in *Encyclopedia of Materials Science and Engineering*, R. W. Cahn, ed., (Pergamon Press, New York: 1990).

5

Sensitive Applications

The use of superconductors in very large magnets is impressive, both for the size of the devices and for the money saved in electricity. However, there are additional applications where superconductors are not mere substitutes for conventional magnets but where the unique properties of superconductors permit their use in very delicate applications. Several of these are presently in operation, using low-temperature superconductors. The most familiar use of superconductivity is in magnetic resonance imaging (MRI), which has become (in only a relatively few years) a commonplace hospital diagnostic.

This chapter covers those applications of superconductivity that both provide either the *production* of very precise magnetic fields or the *measurement* of very small magnetic fields, and which are in use today. The future possible applications of high-temperature superconductors to these technologies is a separate question; such substitution may save money eventually, but will not revise the way superconductors are used. For now, attention is focused on the uniqueness of superconductors and their use in practical applications.

5.1. NUCLEAR MAGNETIC RESONANCE IMAGING (MRI)

The most familiar application of superconducting technology is nuclear magnetic resonance imaging (MRI). MRI is a noninvasive technique for seeing inside the body which uses no ionizing radiation. Almost unheard of in 1980, MRI has now found widespread use in diagnosing injuries to bones and joints and detecting tumors. Today, doctors regard MRI as a primary diagnostic instrument for many diseases that change the anatomy. MRI magnets are now running all day long, 7 days per week, in hospitals throughout the world. A typical MRI exam costs nearly $1,000. However, most neurological disorders show "normal" MRIs, so there is a need for additional diagnostics.

MRI is the one place where the public enjoys the benefits of superconductivity, although most people are oblivious to that aspect of it. Incidentally, the word "nuclear" was dropped from the name to minimize apprehension among the public. Once inside the machine, the patient's greatest apprehension is claustrophobia. This is mitigated by cleverly placing a small prism-mirror above the patient's eyes, allowing a view outside into the room. (It's a boring view: the patient's knees and feet, and the wall across the room; but at least it shows that there *is* an outside, thus reducing claustrophobia.) A typical hospital MRI test lasts about 1 hour.

Nuclear magnetic resonance detects the position of hydrogen nuclei (protons), primarily in water. The hydrogen nucleus has a magnetic moment. When subjected to a DC magnetic

81

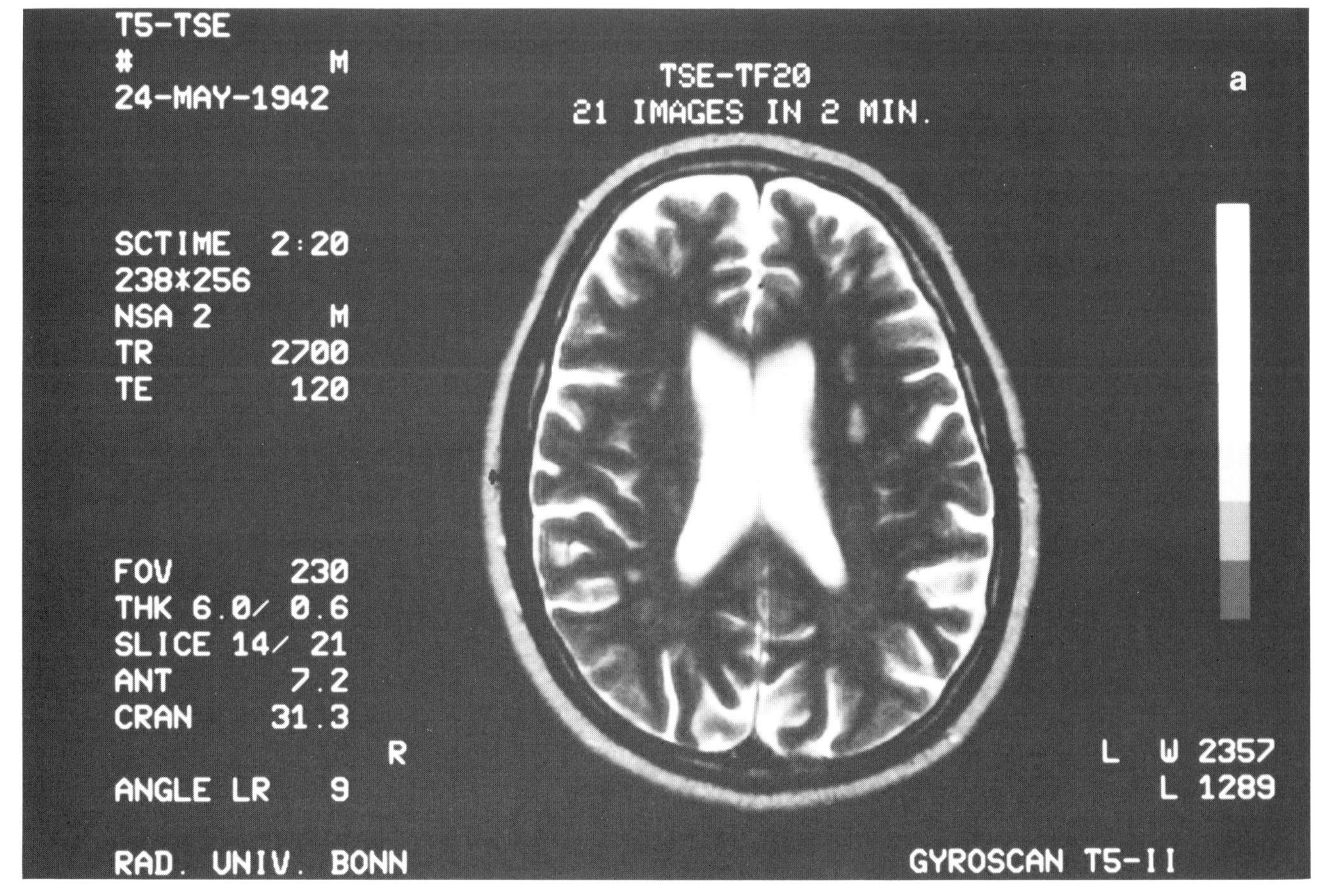
T5-TSE
M
24-MAY-1942
TSE-TF20
21 IMAGES IN 2 MIN.
a
SCTIME 2:20
238*256
NSA 2   M
TR   2700
TE   120
FOV   230
THK 6.0/ 0.6
SLICE 14/ 21
ANT   7.2
CRAN   31.3
R
ANGLE LR 9
RAD. UNIV. BONN
GYROSCAN T5-II
L W 2357
L 1289

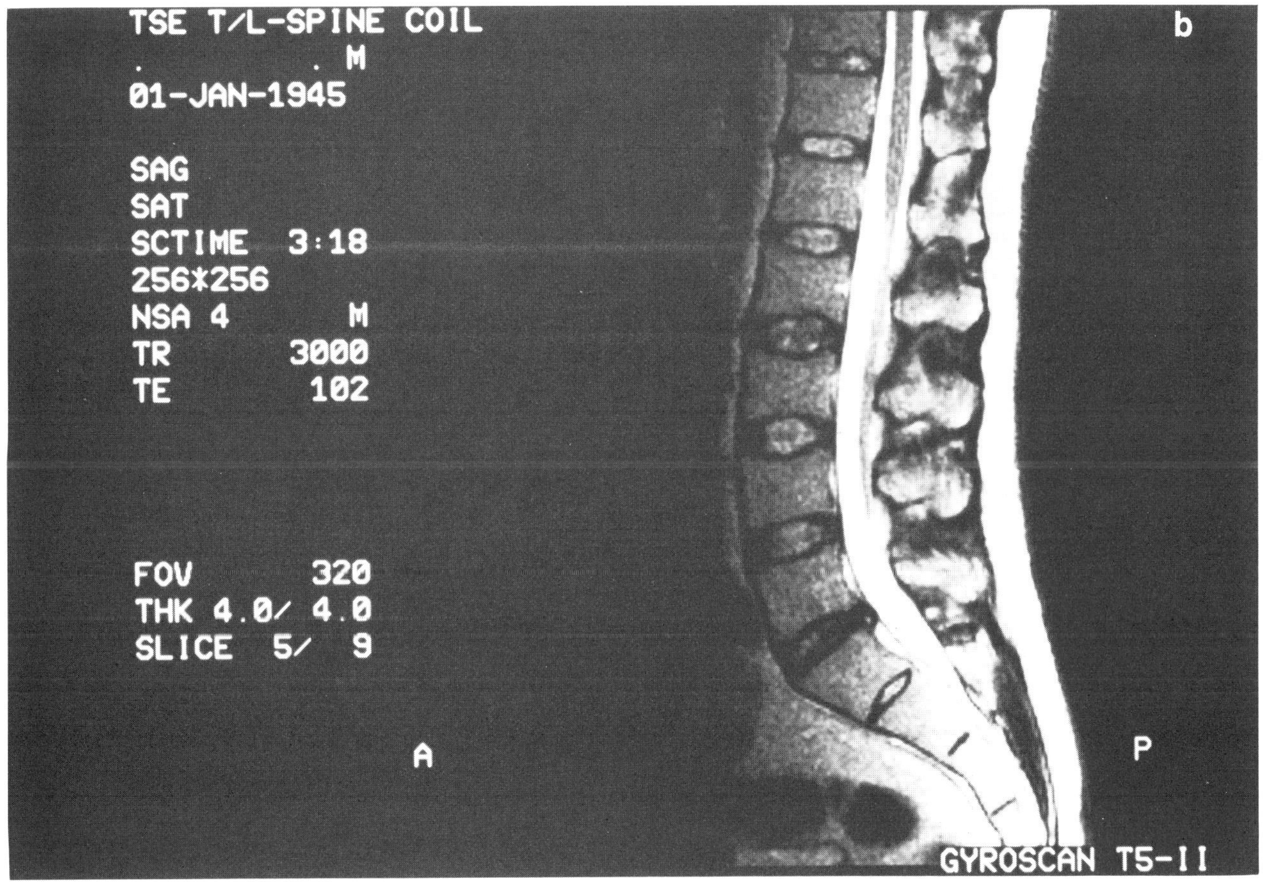

Figure 5.1. Nuclear magnetic resonance images (MRI) of (a) a brain and (b) a spine. (Photos courtesy of Phillips Medical Systems, Inc.)

field, the magnetic moments almost align, and *precess* about the direction of the magnetic field at a characteristic frequency called the *Larmor frequency*, which is proportional to the applied magnetic field. A second applied magnetic field, this one small and at high (rf) frequency, causes the hydrogen nuclei to flip from a parallel to antiparallel orientation (the higher-energy state). Removal of the rf field causes the nuclei to relax to their ground state (the lower-energy, parallel orientation), emitting an rf signal that is detected. Mathematical processing of that signal, especially Fourier transforms, yields information not only about the distribution of protons but also about the chemical environment surrounding the protons.[1] A variety of clever modifications[2,3] have advanced this type of analysis to a very sophisticated art.

If instead of using a constant DC magnetic field, a small gradient is superimposed on top of the constant field, the Larmor frequency will vary with position. By analyzing the emitted rf signal with respect to frequency, good spatial resolution can be obtained. On the other hand, if the gradient is irregular or the field is unsteady, the resolution is degraded. The entire procedure requires a substantial amount of computing power, but the result is a map of hydrogen in the body. If done correctly, the map is very detailed and provides a picture so accurate that in many cases exploratory surgery may be unnecessary. Figure 5.1 shows images of a brain and a spine produced by MRI.

Field strengths of 0.5 to 2.0 T are common in MRI machines presently in service. If this were the only requirement, then MRI would have been put into practice long ago. (There are permanent-magnet MRIs as well as electromagnetic MRIs.) However, there is a corollary need for extremely stable magnetic fields, both in time and in space. The *persistent currents* present in superconducting magnets provide this stability, where no conventional electromagnet could even come close. The time variation of a modern MRI magnet is about 1 part in 10^9, and its spatial variation is about 1 part in 10^5. Without this stability, MRI pictures would be diffuse, unfocused, and of limited diagnostic use.

The difficulty of controlling spatial variation should not be underestimated. An ideal solenoid has a uniform magnetic field inside it, but all real solenoids have fringing effects that must be taken into account. Manufacturers of MRI magnets add extra windings or small pieces of steel at specific points along the magnet to "shim" the field for maximum uniformity. These shims are used to account for the effects of manufacturing tolerances and of nearby magnetic objects on the uniformity of the central field. The placement and number of turns in these windings is a closely guarded trade secret. Furthermore, the ambient background field varies enormously near structural steel, so a magnet must be shimmed to correct for that if it is to be uniform to 10^{-5}.

One good example of a commercial success is the T-5 magnet, shown in Figure 5.2. It is made by Intermagnetics General Corporation for inclusion in the Phillips MRI system. The magnetic field (0.5 T) is homogeneous to within a few parts per million in a spherical volume of 45 cm diameter, which is large enough for head or body imaging. The key design factors for this magnet are homogeneity, persistence, weight and size, and decreased fringe fields. Low operating cost is also important. The unit weighs only 6000 lbs, so no special floor reinforcement is required at the hospital site, and the field outside the magnet is less than 5 gauss.

Some research MRI units are operating at field strengths up to 4 T. The bore size is being increased to accommodate large patients including professional athletes, who often suffer damage to muscle tissue, ligaments, and joints. Previously, arthroscopic surgery was required just to look for such damage. Now MRIs perform this function noninvasively. With greater

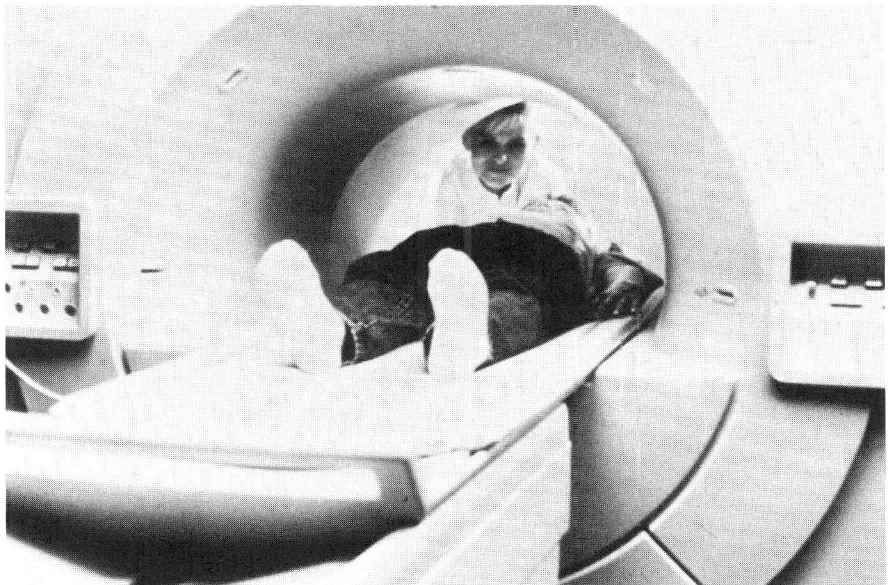

Figure 5.2. An MRI magnet in a typical hospital configuration. With a large magnet bore, the patient feels less claustrophobic. (Photo courtesy of Phillips Medical Systems, Inc.)

sensitivity, MRIs are also used for angiography (determining the flow of blood). Although the resolution is not as good as conventional, invasive angiography, MRIs are able to produce a three-dimensional image which can be displayed and rotated by a computer.

After a decade of hospital experience, MRI has developed into an intricate and sophisticated science. One exceptional feature is that the rate of relaxation of the nuclear spins can change with certain conditions, notably the chemistry of the surrounding body tissue. This can be an important clue to a medical condition; therefore, measuring different relaxation rates allows one to see different conditions within the body. As doctors gain a better understanding of these variations, the diagnostic value of MRI increases.

However, price is a limiting factor. Most modern MRI systems use two-stage cryocoolers to reduce helium losses below 0.1 l/hr, so the refrigeration costs are small relative to other MRI-associated costs such as the diagnostic room, rf power supply and detection system, computer analysis system, and medical staff. An MRI unit typically costs $1.2 to $2.2 million, so it is a major capital investment for any hospital. The most expensive components are the superconducting magnets, costing $200,000–$500,000. Moreover, a maintenance contract for the system can exceed $100,000/year; and the liquid helium supply costs perhaps $10,000 annually. Add in doctors' fees and staff costs, and these numbers result in the patient being billed over $1,000—provided that the number of patients is large enough to keep the MRI unit busy full time. In rural areas, a MRI might be housed in a semi-trailer, which travels cyclically among several hospitals in the region over a two-week period.

MRI as a medical diagnostic has value all over the world, but high costs retard its use in the Third World. (In places where liquid helium is not available, a more expensive closed-cycle refrigerator must be included with a superconducting unit.) The search for a cheaper form of MRI has led toward lower magnetic fields, even though it then becomes necessary to sacrifice some signal-to-noise ratio in the images. Often, it suffices to view a region of the body with less than state-of-the-art precision, and compared to the alternatives—exploratory surgery, or no action at all—even this much is a great improvement.

Permanent magnets generally are able to produce a field of 0.15 to 0.2 T over a suitable volume, and they have no current flow and hence no refrigeration or power consumption. Using good computer algorithms, a decent picture is obtainable. Lower magnetic fields have the disadvantage that the signal-to-noise ratio is lower. Still, permanent-magnet MRI is the fastest-growing segment of the industry.

With more sophisticated computer software, magnets as low as 0.02 T can present a useful MRI image. At that low field, copper-coil magnets are back in the picture,[4] and the cost of refrigeration is averted, but power consumption increases. Reducing the field from 0.5 T to 0.02 T increases the noise by a factor of 70. However, by using a fast pulse repetition rate and carefully controlling field inhomogeneities, the noise can be reduced to a factor of 8, which is good enough for some diagnostics. Moreover, low-field MRI is sometimes more sensitive to characteristics of certain body tissues than is standard MRI.[4] Researchers are finding that mixing various combinations of magnetic field and pulse rate gives additional useful information about the patient. Low-field techniques definitely have a place in the future of MRI.

As concern grows over the possible effects of magnetic fields, magnetic shielding of MRI units becomes more important. Their is still some debate as to whether active or passive magnetic shields are best, but the designs for both are quite advanced, as in IGC's T5. Passive shielding is simple: it utilizes iron to return the flux lines outside the main coils, but iron contributes significant weight to the system. An active shield involves placing superconducting coils outside the main coils, with currents flowing in the opposite direction from the main coils, so as to cancel the magnetic field outside the magnet. This requires more superconductor for the main coils. Both methods give approximately the same magnetic field uniformity, and therefore the same spatial resolution for the MRI.

5.2. SUPERCONDUCTING QUANTUM INTERFERENCE DEVICES

The other major practical use of superconductors is in detecting very small magnetic fields. Superconductors open the door to many applications of magnetism, spanning over 12 orders of magnitude. Not only can superconductors be used to generate magnetic fields greater than $10\,T\,(10^{+5}$ gauss), they can detect magnetic fields below $10^{-14}\,T$. This remarkable sensitivity is achieved by *s*uperconducting *qu*antum *i*nterference *d*evices (SQUIDs).

The underlying principle of a SQUID is tunnelling, a quantum-mechanical effect which (in superconductors) produces the Josephson effect.[5] Since we have not yet discussed the theory of superconductors, it is impossible in this chapter to explain how tunnelling takes place, and hence how a SQUID works. In Chapter 13, the behavior of Josephson junctions is presented in the light of contemporary theory of the high-temperature superconductors, and the very important concept of *weak link behavior* is discussed there. For now, suffice it to say that a SQUID is an extremely sensitive means of detecting a magnetic field. This fact can be exploited to collect data of a type not available by any other means.

Just as electric charge is quantized in multiples of e, so also magnetic flux is quantized in units of $hc/2e$ (h = Planck's constant; c = speed of light). This unit is customarily denoted by Φ_0, the *flux quantum*. The quantization of individual lines of flux will prove to be

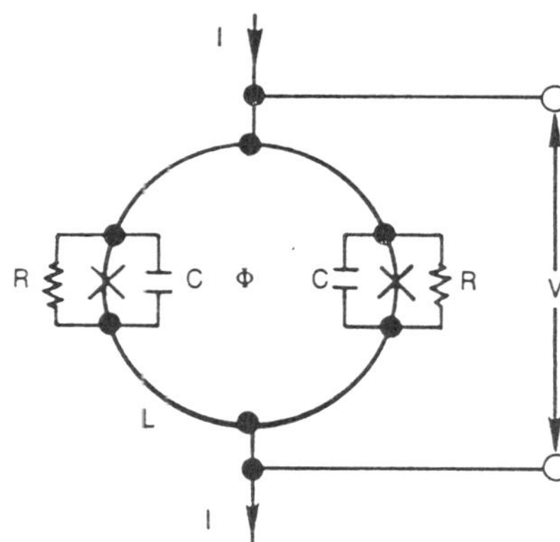

Figure 5.3. Circuit diagram representing a SQUID. The two Josephson junctions are identical; the self-capacitance is denoted by C and a shunt resistor is denoted by R. A DC bias current I flows through the device, which outputs a voltage V due to detection of the magnetic flux Φ.

important later on; in the meantime, any magnetic field can be expressed as a multiple of Φ_0. When a flux line penetrates a conducting loop, it will induce a current in that loop.[6] A SQUID is a loop formed of two Josephson junctions, and Figure 5.3 presents the standard circuit diagram that represents it. (A Josephson junction is denoted by an x, with resistance and capacitance shunting the junction.) Flowing through the circuit is a DC bias current such that the current density J is just below J_c. Then any additional induced current generates a voltage across the SQUID. This makes it possible to detect even a single flux quantum under laboratory conditions. Commercially available SQUIDs are much less sensitive, but still they can detect energy levels of about 1000 h.

In conventional SQUIDs, thermal fluctuations determine the noise level,[7] and therefore SQUIDs normally are operated near absolute zero, to achieve maximum sensitivity. Where there is plenty of room, the SQUID is simply kept in a liquid helium bath; where space is tight, very cold helium gas is piped to the vicinity of the SQUID, maintaining it at perhaps 10 K. Clearly, if the material warms enough to go normal, it ceases to be a SQUID. For some applications, this requirement of extreme cold is prohibitive; in this case one must use a *flux-gate magnetometer*, or *Hall effect magnetometer*, with orders of magnitude less sensitivity. In other cases, a slight sacrifice in sensitivity is acceptable, and the SQUID can be run at temperatures moderately above 4 K. Looking ahead to HTSCs, a SQUID operating at 77 K could be expected to have thermal noise some 20 times greater.[8]

A typical operating configuration of a commercial SQUID is shown in Figure 5.4. The signal under study is detected by a pickup coil and is coupled inductively to the SQUID itself via an input coil.[9] The accompanying electronics detects the change in the SQUID current and feeds back a current to offset it. That current passes through a large resistor, and the voltage thus produced is big enough to give an appreciable signal. The gain of this amplifier can easily be 10^6.

Usually, the surroundings contain nuisance magnetic fields, which must be screened out. To this end, the pickup coil is not just a wire loop, but instead is a *gradiometer*, as shown in Figure 5.5. Here the same wire forms two loops, one running in the opposite direction to the other. The separation between loops is called the *baseline*. A magnetic field produced by a source far away will cause equal and opposite currents, and no net signal will reach the SQUID. The earth's field, for example, is screened out this way. On the other hand, a magnetic field from a nearby source (within a few baseline lengths) will cause unequal

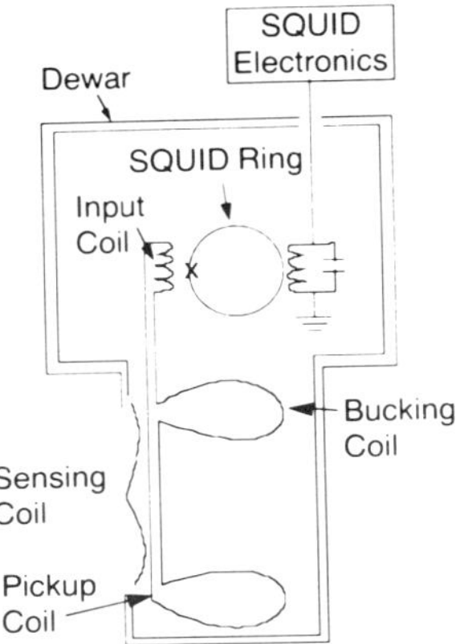

Figure 5.4. Typical configuration of a commercial SQUID detector. The sensing coil is a combination of pickup coil and bucking coil, which couples to the SQUID itself through an input coil. Electronics at room temperature feeds back a balancing current, and produces the output voltage from the device. (From Ref. 9.)

currents in the loops, and thus generate an output signal. Depending on the source being studied, gradiometers can be first-order or second-order, to enhance the degree of selectivity, as needed. The single loop picks up the tiniest signals, but it can easily be fooled by extraneous fields such as a spectator wearing a wristwatch. The first-order gradiometer is the commonest type of input to most SQUID measurements.

5.2.1. Corrosion Detection by SQUIDs

As an example of a process important to industry that may be influenced by SQUID technology, consider the case of corrosion detection. Economic costs associated with corrosion in the United States were estimated during the 1970s at \$70 billion.[10] Clearly, corrosion detection and mitigation is an important industrial goal.

In an electrochemical cell (such as an automobile battery), a current flows from anode to cathode, through an ionic liquid (an acid). Ions leave the anode and go into solution, thus diminishing the mass of the plates. When a metal corrodes, the same process takes place, but

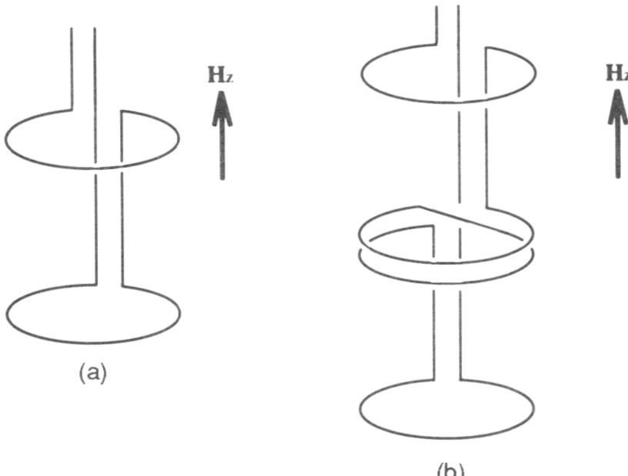

Figure 5.5. Loop configurations for the sensing coils associated with a SQUID. (a) A first-order gradiometer in which two oppositely wound coils discriminate against constant magnetic fields. (b) A second-order gradiometer in which three coils combine to screen out both constant fields and their first derivative.

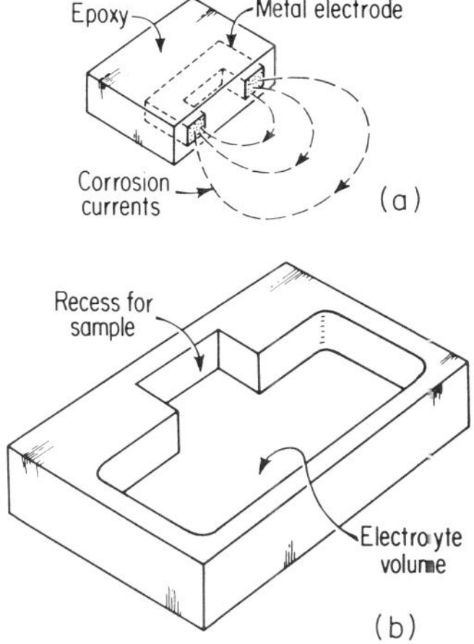

Figure 5.6. Corrosion detection apparatus employing SQUIDs, used by Bellingham *et al.*[12] (a) sample; (b) enclosure.

without yielding the useful power derived from a battery. Generally with the passage of time two different places on the same metal will alternate in the roles of anode and cathode, and the current between them will reverse, taking metal away from each.

To minimize the destructive effects of these corrosion currents, one tries to have the currents flow only one way, by offering a "sacrificial" metal anode that can be eaten away without harm. The most familiar example is the practice of putting plates of zinc on the hulls of steel ships: the expendable zinc preferentially corrodes away as current flows through sea water, but the steel does not corrode.

The extremely tiny currents of corrosion generate tiny magnetic fields, which until now have escaped detection. However, with SQUIDs, it is now possible to measure those fields. Under laboratory conditions, MacVicar and colleagues[11] used the configuration shown in Figure 5.6 to detect fields in the range of 10^{-11} T. (Their detector noise level was below 10^{-13} T, but background fields due to the metal itself were in the 10^{-12} T range.) In a typical test cell, fields up to 10^{-9} T were observed 3 cm from the corroding surface. Their data are also clear enough to see the reversal of polarity that takes place over several minutes. In fact, their studies revealed a pattern of chaos in the reversal of currents with time.[12]

The key point is that SQUIDs offer a noncontacting means of observing corrosion long before visible deterioration takes place. It is fair to assert that there are numerous other industrial processes similarly open to investigation with SQUIDs, but the need for liquid helium cooling has kept them from being economically justified thus far.

5.3. BIOMAGNETISM

For laboratory purposes, SQUIDs are certainly useful, but operating them is so costly that they are little-used in everyday commercial science. However, the one major exception is where human health is concerned, because the cost is small compared to the value of the information obtained. MRI has not only fully justified its cost, it has revolutionized medical diagnostic techniques; it is now standard procedure. The area of biomagnetism promises a similar evolution for the future diagnosis of brain and neurological afflictions.

5.3.1. Magnetic Properties of Neurons

Any place a current flows, there will be a magnetic field. In the body, neurons and muscle fibers both generate currents when they are activated. The term *neuron firing* refers to the sudden passage of a pulse of current along a neuron. That sets up a magnetic field because a finite cluster of charge has moved a finite distance. A neuron is typically 2 mm long and 2 μm in diameter. The current flowing when a single neuron fires cannot be detected; the field generated is too small even for a SQUID.

However, in the brain neurons are clustered together, somewhat aligned in certain patterns, and act synchronously in groups. When a group of 10,000 parallel-aligned neurons all fire simultaneously, the net current is sufficiently intense to generate a magnetic field detectable outside the skull,[9] using SQUIDs. The neurons parallel to the surface of the skull generate the detectable field; neurons perpendicular to that surface have their magnetic fields concealed within the cranium. Fortunately, the majority of the brain's neurons are of the former variety and give a signal. The important factor is that many of the neurons in a particular region are strongly interconnected, and fire simultaneously. Typically, a volume of 0.1 mm^3 can generate a distinguishable magnetic field.

Incidentally, the return current from neurons, which is more diffuse and travels outside the neuron cells, does not create a large enough field to detect magnetically. Nevertheless, because there are billions of neurons in the brain, *electroencephalography* (EEG) can detect the average current through conventional electrodes. Unfortunately, the widely varying electrical resistivity of various parts of the brain and skull obscure the meaning of EEG readings[9]; by contrast, magnetic data is comparatively unaffected by the surroundings, which have uniform magnetic permeability.

A neuron cluster does not have to be right near the surface to be detectable. Neuron clusters deep within the brain that act simultaneously behave like a large dipole, and hence give a detectable signal. By assuming that a dipole produced the observed field, it is possible to work backward from the observations to reconstruct the position of the original dipole. Therefore, systematic scanning combined with a model of brain functioning and some serious number crunching can determine the origin of magnetic fields almost anywhere within the brain.

5.3.2. Magnetoencephalography (MEG)

The combination of scanning, modeling and software is known as *magnetoencephalography* (MEG). This medical diagnostic uses SQUIDs to detect the magnetic field arising from currents within the brain, and produces a map of the brain's magnetic activity. Like MRI, this is a noninvasive technology which relies on the extreme sensitivity of a SQUID to detect very faint magnetism, arising from either regular or abnormal conditions within the brain.

Figure 5.7 shows a scale comparing various biological magnetic fields.[9] The biological field strengths range from ten thousand to a billion times smaller that the earth's magnetic field. This application is extremely difficult; typically, the magnetic fields range from 50 femtotesla (5×10^{-14} T) up, detectable above a noise level of about 5 femtotesla. Deliberately evoked sensory responses can be repeated and averaged, so signal-to-noise is less of a

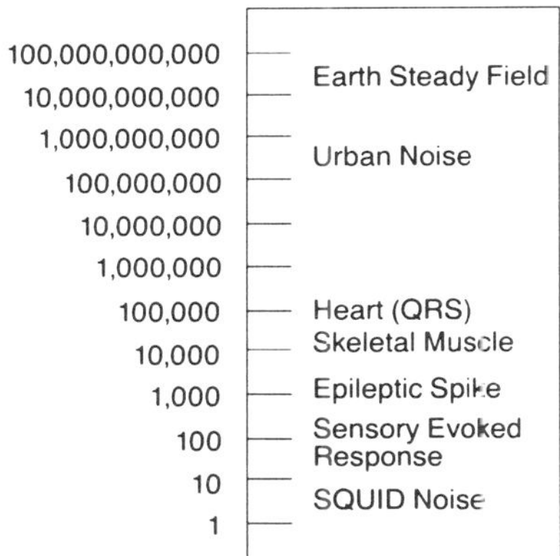

Figure 5.7. Comparative strengths of biological and environmental magnetic fields. (Courtesy of Biomagnetic Technologies, Inc.)

problem for that subset of potential observations. Still, for most of the brain, not only are the fields extremely tiny but the value of the data is limited, because a model of the underlying neural activity is required if one is to explain the observed magnetic fields.

MEG requires extensive precautions to eliminate the effects of external magnetic fields. As explained above in Section 5.2, gradiometers (which are sensitive to gradients in magnetic fields but not constant magnetic fields) are employed to reduce the effects of the earth's magnetic field and urban magnetic noise.

Also, since the field strength falls off rapidly with distance from the neuronal source (typically as the inverse square of the distance), it is important to get the pickup coils as close as possible to the magnetically active regions. Considerable engineering goes into designing the geometrical arrangement of sensors. Figure 5.8 illustrates the problem: each coil must be cooled near absolute zero while being positioned within centimeters of the brain cells. It would be very nice to have a flexible shape for the array of coils, in order to accommodate different-shaped heads. In any real configuration, such as that shown in Figure 5.9, design trade-offs create some limitations. Biomagnetic Technologies, Inc., names their MEG system MAGNES®. They use a curved array which closely follows the contour of the head.

In clinical practice the refrigeration is no special problem. Twice a week someone must top up the dewar with liquid helium, but the technicians who operate the system treat it as just another medical diagnostic device. As with an MRI unit, the hospital personnel are almost oblivious to the presence of liquid helium. (Also like an MRI, an MEG unit costs about $2.5 million.)

So far MEG has been used primarily as a research tool.[13,14] For example, Siemens has made clinical trials using a planar array of SQUIDs to diagnose heart disease. (The magnetic signals from the heart are comparatively strong.) In some applications, the clinical results of the noninvasive MEG technique give results identical to conventional invasive techniques. Because of reassuring experiences like that, MEG is rapidly emerging into accepted medical practice.[15]

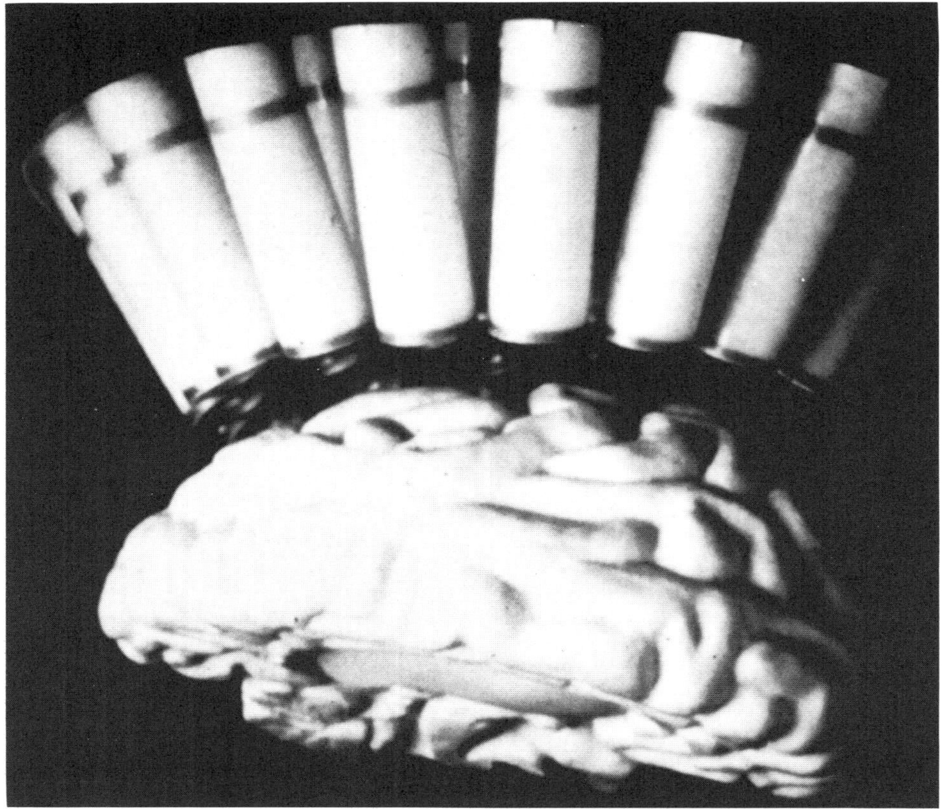

Figure 5.8. By fitting the concave surface of the coil array to the convex curvature of the brain, each input coil is positioned close to a magnetic source. (Photo courtesy of Biomagnetic Technologies, Inc.)

A great deal of benefit can be obtained without knowing the details of exactly what is going on in the brain. The value of good MEG measurements is truly stunning: Figure 5.10 is an MEG image of a patient's brain in which a tumor lies close to important blood vessels and close to the sensory cortical region that processes sensations of the hand and face. When someone is about to have a brain tumor removed, MEG scanning beforehand can show the surgeon where the motor nerves are. The surgeon can then plan the surgical procedure to minimize the risk of inadvertent injury. (Cutting motor nerves can leave the patient paralyzed.) Once the operation is underway and the craniotomy has been done, the surgeon can compare the visible blood vessels to the MEG image prepared earlier to further optimize the safety and success of the surgery.

The computer graphic of Figure 5.10 is actually MEG data superimposed on a substrate provided by MRI. Here, the MRI and the MEG are complementary to each other, and together provide more detailed information about the brain than either could alone. Biomagnetic Technologies, Inc., has coined the phrase magnetic source imaging[®] (MSI) to refer to the combination of MEG with MRI as a diagnostic tool.

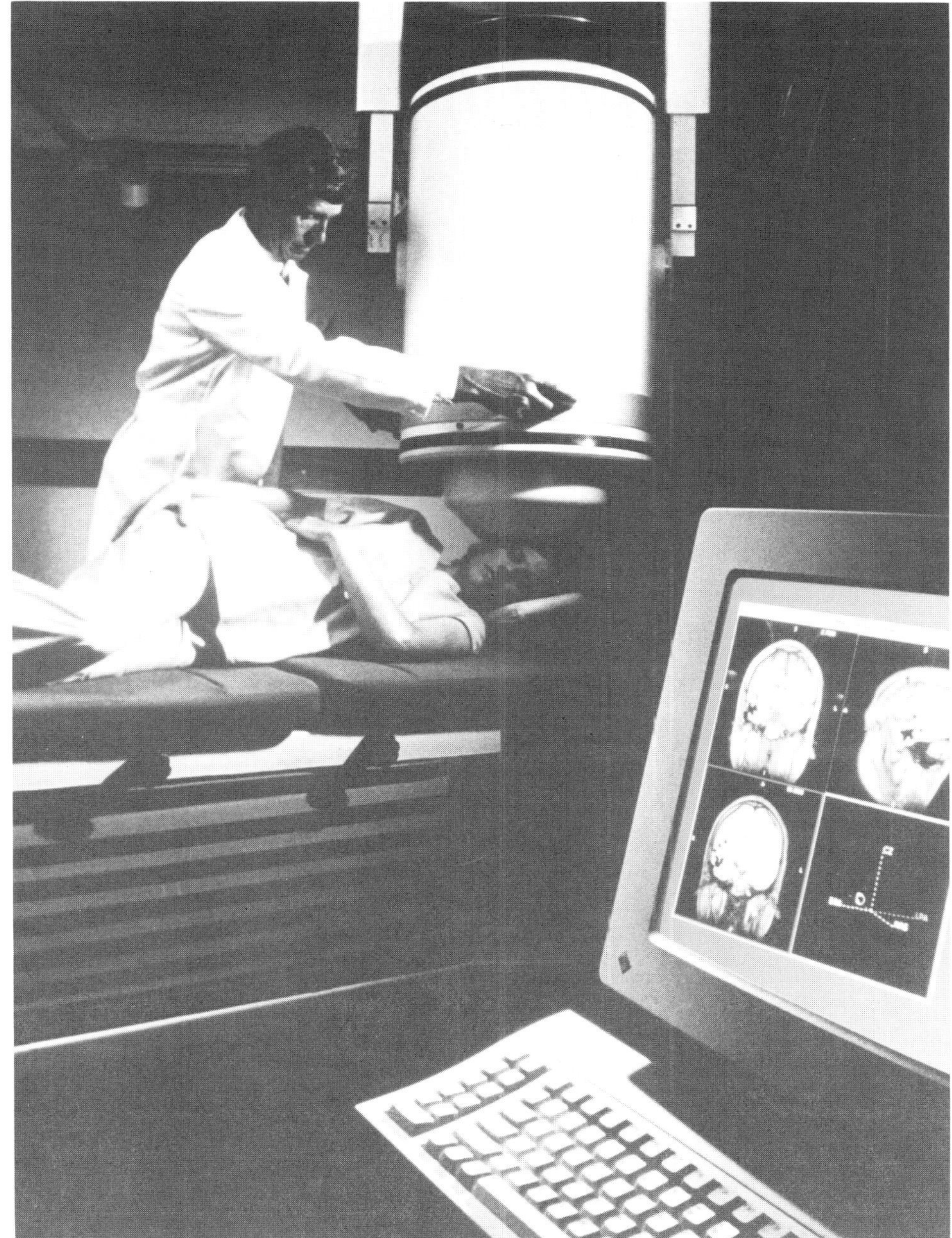

Figure 5.9. MEG system in clinical use. (Photo courtesy of Biomagnetic Technologies, Inc.)

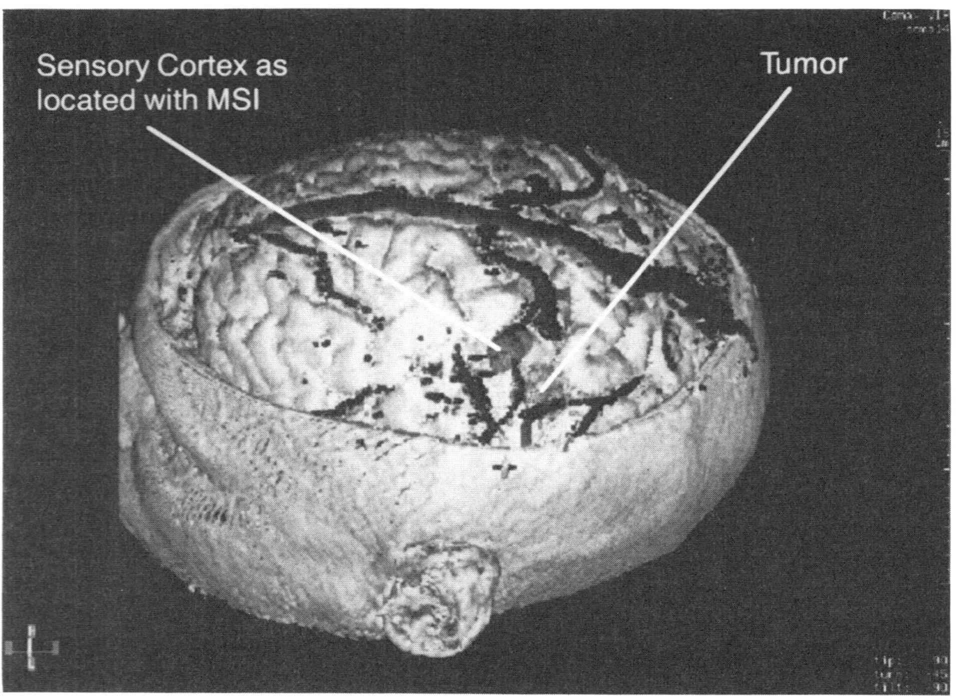

Figure 5.10. MEG map of magnetic activity from the brain. The substrate, or background, is derived from MRI data. The MEG clearly shows blood vessels, a tumor, and a key sensory cortex area. The designation MSI stands for Magnetic Source Imaging®. (Computer graphic courtesy of Biomagnetic Technologies, Inc.)

There are many other conditions where MEG is helpful. For example, in the case of epileptic patients, the magnetic signals are relatively strong (a factor of 10 higher than normal), and hence MEG provides a map that can guide the surgeon to the affected area. The Scripps Research Institute in San Diego has pioneered this kind of surgery.[16]

Among other things, MSI has shown that the brain can compensate for damage by shifting the locale where certain functions are performed. Recovering stroke victims regain their motor ability slowly as these functions move to new positions; MEG can observe over time as the position changes. MSI has also been used to study the phenomenon of phantom limbs, where people with amputations feel some sensation that seems to come from the missing limb when someone touches their face. As surprising as this may be, it shows that brain activity is very complex. In the absence of MSI, it is unlikely that these observations would have been made by any other means.

It is noteworthy that in addition to mapping brain activity, this technology can detect magnetized contaminants (e.g., metals) in the lungs and other organs, the position of metallic tracers introduced into the body, cardiograms, fetal heart activity, His-Purkinje activity (cardiac electrical system), and indeed any phenomenon where many neurons fire simultaneously. The limits of usefulness of biomagnetic measurements are set not by the technology of the apparatus, but by the pace at which doctors understand the significance of these new diagnostic measurements, and adapt their medical practice to take advantage of them.

5.4. FUTURE OUTLOOK

It is easy to jump to the conclusion that all these applications are wonderful and that they will all get much easier once HTSC arrives. The reality is appreciably different: LTSC has a very solid future all by itself, and where refrigeration is not a large component of the overall cost there is less incentive to introduce HTSCs.

5.4.1. Medical Diagnostics

Magnetic resonance imaging is now a standard diagnostic tool, and its commercial success is not in doubt. In a typical year, 1000 units are sold at a rough cost of $1–$2 million each. Will the same thing happen in magnetoencephalography? The commercial market is there: with total health care costs of $500 billion annually, neurological problems account for about 25% of those costs.

The experience of Biomagnetic Technologies is well worth noting: this company is the successor to SHE, Inc., a manufacturer of low-temperature physics apparatus. They make not only the SQUID array but the entire medical system and packaging as well. The only component they buy is the computer with which to analyze the data. It is more accurate to say Biomagnetic Technologies is in the medical instrument business now, rather than in the low-temperature physics business: they have phased out all their other products in order to concentrate on this medical technology.

However, MEG is not yet a commercial success. Since this is an experimental technique, insurance companies are hesitant to cover the cost of MEG, so reimbursement comes only slowly, justified on a case-by-case basis. Very few hospitals can afford an expensive diagnostic tool when the patients can't pay for its use. Until the medical literature contains enough examples of the value of MEG, the economics will remain unfavorable.

It is doubtful that HTSCs would impact the future of MEG very much. Recalling that SQUID noise depends on temperature, the higher noise level in a SQUID operating at 77 K (rather than 4.2 K) may be so detrimental as to offset the savings in refrigeration costs. If a HTSC SQUID had to be run at 4.2 K anyway, there would be little motivation to use HTSC materials.

On the other hand, HTSCs could serve as pickup coils for either the rf signal of an MRI or for MEG. The pickup coils for an MRI are usually copper. An HTSC pickup coil would have a better signal-to-noise ratio than copper coils at 77 K. In fact, exactly that application is to be served by the *Super*Sensor™, made by Superconductor Technologies, Inc. This HTSC coil is a simple plug-in replacement for a conventional copper coil, and by April 1994 the *Super*Sensor™ was close enough to market to merit being "filed" with the Food and Drug Administration.[17]

The pickup coil in an MEG system is usually a LTSC operating at 4 K. An HTSC coil operating at 77 K could be located closer to the head. The decrease of about 5 mm in cryostat wall thickness due to operating at 77 K instead of 4.2 K helps to improve spatial resolution and sensitivity. However, the real advantage may be in the ability to make a flexible cryostat wall (liquid helium cryostats require rigid vacuum chambers), permitting the MEG pickup coils to more closely follow the contours of different-sized heads.

5.5. SUMMARY

Superconductivity is currently in use in a variety of ways where very stable magnetic fields are required, or where very tiny magnetic fields are to be detected. The exceptional stability that comes from *persistent currents*, which are unique to superconductors, has made possible the medical diagnostic known as magnetic resonance imaging (MRI). No conventional electromagnet could have delivered the kind of sharp images necessary for the medical profession to place confidence in MRI as a means of visualizing the human body.

Another unique property of superconductors is the Josephson effect, which is the principle underlying superconducting quantum interference devices (SQUIDs). SQUIDs can detect magnetic fields of a few femtoTesla (10^{-15} T), which makes them the most sensitive instruments known. There are countless applications to measurements having economic importance to be explored with SQUIDs. In this chapter, we describe one of these: the detection of corrosion currents. The point is that there are many other potential measurements of the same sort which have not yet been studied because the requirement for liquid helium cooling precludes the use of SQUIDs in all but very high-value measurements.

Perhaps the highest-value measurements known are associated with the health of human beings, and SQUIDs find an important application in the diagnostic known as magneto-encephalography (MEG). The name is reminiscent of electroencephalography (EEG), but the data comes from magnetic fields generated within the human brain, through the action of neurons firing synchronously. Diverse conditions such as epilepsy, stroke, and brain tumors all present magnetic signals which can be detected via MEG; doctors are still learning how to interpret the measurements.

In the future, we can look forward to wider use of superconductivity in applications where sensitive control or measurement of magnetic fields is required. The value added in many cases is sufficiently great that low-temperature superconductivity is cost effective, and there is no need to wait for HTSCs to reach the marketplace. Still, if and when HTSCs become available, they will further extend the range of applicability of superconductors.

REFERENCES

1. R. R. Ernst and W. A. Anderson, *Rev. Sci. Instruments* **37**, 93 (1966).
2. R. Kaiser, *J. Magn. Res.* **3**, 28 (1970).
3. J. Dadok and R. F. Sprecher, *J. Magn. Res.* **13**, 243 (1974).
4. D. Kleppner, "MRI for the Third World," *Phys. Today*, 9–10 (March 1992).
5. C. Kittel, *Introduction to Solid-State Physics*, 6th Ed. (Wiley, New York: 1986).
6. M. Tinkham, *Introduction to Superconductivity* (Krieger Publ. Co., Malabar, FL: 1980).
7. S. Foner and B. Schwartz, eds., *Superconductor Materials Science* (Plenum Press, New York: 1981).
8. J. E. Zimmerman *et al.*, *App. Phys. Lett.* **51** (8), 617 (1987).
9. *Biomagnetism: A Primer*, Biomagnetic Technologies, Inc. (San Diego, CA: 1991).
10. L. H. Bennett *et al.*, "Economic Effects of Metallic Corrosion in the United States," National Bureau of Standards Special Publication 511-1 (1978).
11. J. G. Bellingham, M. L. A. MacVicar, and M. Nisenoff, *IEEE Trans. Magn.*, **MAG-23** (2), 477 (1987).
12. J. G. Bellingham *et al.*, *J. Electrochem. Soc.* **133**, 1753 (1986).
13. C. Pantev *et al.*, *Neurobiology* **88**, 8996 (1991).
14. C. Gallen *et al.*, "Detection and Localization of Delta Frequency Activity in Human Strokes," in *Biomagnetism: Clinical Aspects*, M. Hoke *et al.*, eds. (Elsevier, New York: 1992).
15. "Medical News and Perspectives," *J. Amer. Medical Assn.* **263**, 623 (1990).
16. C. Gallen *et al.*, "Functional Imaging Guided Neurosurgery," Preprint.
17. F. Reynolds, Superconductor Technologies, Inc., Private communication.

6

Basic Concepts of Theory of Superconductivity

This chapter deals with the BCS theory of superconductivity. The emphasis here is on explaining the concepts, perhaps at the expense of the formalities. In the interest of brevity, no derivations are presented; many excellent books[1] offer that level of treatment. Moreover, although this book focuses on high-temperature superconductors (HTSCs), this chapter is limited to the aspects of theory that were known (and trusted) before 1986. In Chapter 12, the BCS theory is extended into the realm of HTSCs. (It is fair to state here that the BCS theory is still applicable to the HTSCs, but with modifications.)

This chapter provides the background that nonspecialists need to deal with certain aspects of superconductivity. The first aim is to rise above the level of buzzwords. Phrases like *intrinsic vs. extrinsic*, the presence of "magic numbers" (e.g., the *BCS ratio*), and unfamiliar concepts (such as *anisotropy of the energy gap*) really do matter at a fundamental level and have great downstream impact on the possibility of eventually obtaining useful products from these materials.

When the BCS theory first appeared[2] in 1957, an *energy gap* Δ was an essential component of the theory. Additionally, electrons were said to travel in *Cooper pairs*, staying within a *coherence length* ξ of each other. Magnetic fields enter a superconducting material only to a *penetration depth* λ, generally smaller than ξ. With suitable approximations, some formidable mathematics reduced to a tractable model that explained a lot of data. Here, we will present the major elements of this theory.

No theory is better than the data upon which it is based, and tunneling experiments supply the most crucial data. In general, whenever quantum mechanics is involved, tunneling can serve as a very powerful probe of physical phenomena. (Tunneling is the basic phenomenon on which devices such as scanning tunneling microscopes depend.) Tunneling is important for studying superconductors because it is a quantum-mechanical condition which presents a measurable output, and superconductivity is a quantum-mechanical phenomenon in the first place. The point of tunneling experiments is twofold: to test the validity of the BCS theory, and to determine the most important property of a superconductor.

In the last 36 years, considerable advances in theory have been made,[3] and new compounds have been found as well, most notably the ceramic HTSCs. Type II superconductors, in which the coherence length ξ is shorter than the penetration depth λ, were unfamiliar in 1957. Today, all the interesting superconductors, including the HTSCs, are type II. The Eliashberg[4] modification of BCS is now the standard theory, wherein the energy gap is determined by an integral equation that can only be calculated numerically. Exposition at

97

that level is beyond our scope and is not required to appreciate the general behavior of superconductors.

How well do theory and experiment agree? Surprisingly, the observed behavior is not terribly different from the predictions of the ideal BCS model. Fortunately, the original BCS model, with its relatively simple energy gap, yields a sufficiently accurate picture for general understanding. With respect to the HTSCs, it turns out that the BCS theory is still good and no new mechanism for superconductivity is warranted, although some modifications are required to explain what we know today. Although a number of exotic theories have been introduced to cope with unexpected observations about the HTSCs, none of them do as well as BCS, properly understood. Based on tunneling and similar experimental inputs, our state of knowledge today is compatible with BCS.

There are several essential building blocks to the BCS theory of superconductivity which will be assembled in the following sections. We begin with a discussion of how lattice vibrations in solids determine the phonon spectrum. This is followed by the concept of the Fermi level. Modest alterations in the number of electrons near the Fermi level can greatly affect the properties of a material. This leads to the notion of *Cooper pairing* and the *energy gap*, which are the crucial elements of the BCS theory of superconductivity. Following that, tunneling experiments are introduced, because they measure the properties of the energy gap. Finally, we set the stage for returning in Chapter 12, where the BCS theory will be applied to the HTSCs.

6.1. LATTICE VIBRATIONS

Metals and alloys have been studied for centuries using phenomenological theories, and a great deal of competence in solid-state engineering is attainable without explicit knowledge of quantum mechanics. However, in order to understand superconductivity it is necessary to grasp certain quantum aspects of solid-state physics.[5]

Solids are made up of regular, periodically repeating patterns of atoms formed into a *lattice*. A vibration of the lattice is called a *phonon*. Because the lattice is regular and periodic, one *unit cell* is interchangeable with another, and the lattice vibrations can propagate from one cell to the next without change. Indeed, there is little point in describing such motion in terms of the physical space that holds the individual atoms. Rather, the description of phonons is made in terms of *momentum space*, also known as *k-space*, because the letter k is used to denote momentum throughout solid-state physics: $\mathbf{k} = m\mathbf{V}$, and the energy is

$$E = \frac{1}{2}mV^2 = \frac{k^2}{2m}$$

As always in quantum mechanics, $E = \hbar\omega$, and ω is the frequency of any wave. One hardly needs a *k-space* in which to describe so simple a concept. However, in a solid, the energy-momentum relation is not so simple, and hence there is merit in a description that allows a more general functional relationship $E = f(k)$, which is generally known as a *dispersion relation*.

In a solid, the phonons *could* follow the simple form $E = k^2/2m$, except for several intervening factors. Figure 6.1(a) shows that simple parabolic form. Real solids, however, are made up of a finite number of atoms spaced a finite distance apart, and this introduces modifications into the dispersion relation. First and foremost, there is no way the wavelength

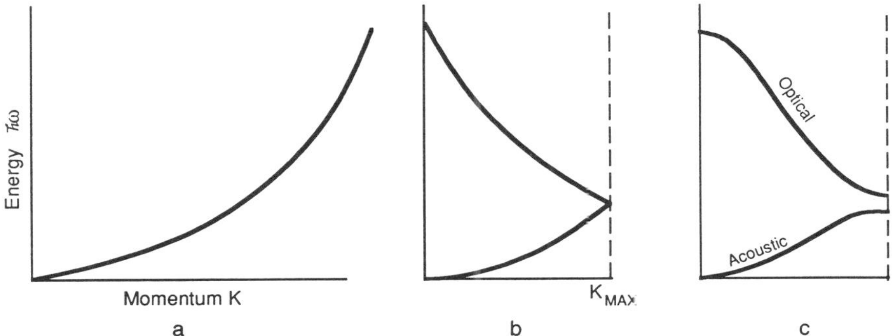

Figure 6.1. Energy-momentum relations for lattice vibrations in a solid. (a) The free-particle formula $E = k^2/2m$. (b) At a maximum allowable value of k (minimum wavelength corresponding to a half-wave per lattice constant), the curve folds back upon itself; the dashed line defines the boundary of the Brillouin zone. (c) The high-frequency vibrations are called the optical branch of the phonon spectrum; low-frequency vibrations comprise the acoustic branch.

of any vibration can be less than twice the interatomic spacing. When half the atoms are moving up while half are moving down, the shortest wavelength has been reached. In fact, if *all* the atoms were at the positive peak of their wave motion all at once, the situation would be indistinguishable from no motion at all, or infinite wavelength; such a wave motion would carry zero momentum.

Thus, the dispersion relation has to reflect back toward zero momentum for phonons with energies sufficient to drive the atomic motions faster than some cutoff wavelength λ_{min} corresponding to some momentum k_{max}. This is depicted in Figure 6.1(b). It may seem bizarre to have waves of higher energy (ω) even as the momentum decreases. To grasp this, imagine two atoms in the lattice moving directly toward each other, oscillating out of phase: their net momentum is zero despite their considerable energy.

The momentum space in the range $\pm k_{max}$ is known as the *Brillouin zone* after Leon Brillouin, a pioneering solid-state theoretician.[6] Furthermore, symmetry conditions force the phonon spectra to intersect the values of $k = 0$ and $k = k_{max}$ with a zero derivative ($d\omega/dk = 0$), and hence a realizable phonon spectrum must take on the shape of Figure 6.1(c). With the lattice vibrations thus divided along two allowable energy-momentum curves, we conventionally term the higher-frequency oscillations *optical phonons* and the lower-frequency oscillations *acoustic phonons*.

As solids deviate from simple cubic structure, the smooth curves of Figure 6.1(c) do not necessarily occur, and for extremely complex solids (the HTSCs are such examples) the optical and acoustic branches might have considerable structure as they traverse the Brillouin zone. This structure greatly affects those properties of the solid that depend heavily upon the lattice, such as the specific heat, the thermal conductivity, and so forth. Superconductivity in a solid also depends intimately on the details of the lattice phonons.

6.2. THE FERMI LEVEL

So far, we have not discussed the role of electrons at all, and yet they are extremely important for superconductivity. In an atom, each electron occupies a particular state which has a corresponding energy level. If two or more states have exactly the same energy level, this condition is termed *degeneracy*. We do not customarily employ the word degeneracy to describe the situation where a roomful of gas molecules all have identical energy levels among the myriad atoms; each molecule is a separate *system*, because they are so far apart. Figure 6.2(a) indicates the energy levels of a typical atom.

As atoms come closer together, they start to interact, and the electrons of one perturb their neighbors. This leads to a splitting of energy levels, and when several levels are closely spaced to begin with, the perturbed levels overlap and smear out. This condition is depicted in Figure 6.2(b).

Next, when all the atoms are assembled into very close proximity in a solid, then the interactions between them become dominant. In something like copper, for example, the tightly bound 1*s* electrons close to the nucleus retain their specific energy levels because those electrons are oblivious to other nearby atoms. They do not participate in most of what

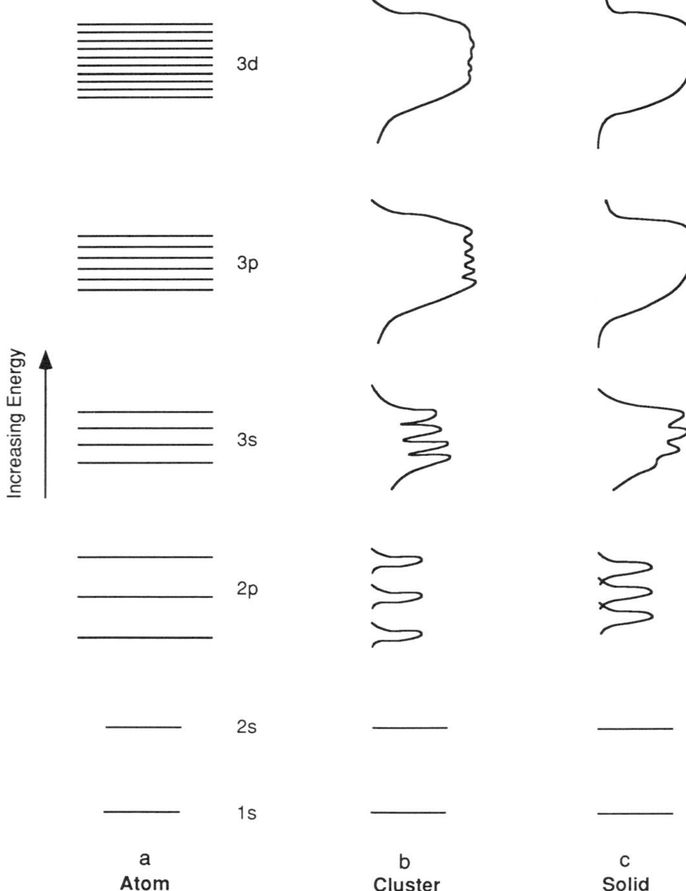

Figure 6.2. Evolution from energy *levels* in a single atom to energy *bands* in a solid. (a) A single atom has well-defined energy levels for each shell. (b) As atoms come together, interactions among the outer electrons cause the higher levels to broaden but leave the innermost electronic shells unaffected. (c) In a solid, the interactions are so comprehensive that the individual levels are no longer discernable, and yet the innermost shells are still unaffected.

is called solid-state physics and for our purposes may be ignored. Meanwhile, the outer electrons interact with neighboring atoms, and give each solid its unique characteristics. These electrons are able to move about throughout the entire solid and are no longer confined to one atom. Under such circumstances, the particular atomic states occupied by all the outer electrons in the solid get mixed together, as do their corresponding energy levels. The outcome is an enormous number of incredibly closely spaced energy levels—a continuum, effectively—that forms a band. The collection of all such levels is called the *Fermi sea*. Just as interacting atoms split up degenerate energy levels into multiple new levels, Figure 6.2(c) shows how energy bands eventually result for a solid.

It is a general principle of physics that objects seek their lowest allowable energy level, and so do electrons. Electrons are fermions and obey Fermi–Dirac statistics,[7] which means no two can occupy exactly the same state. They fill the available states from the bottom up, and the highest filled state is called the Fermi level E_f. The probability of finding an electron in energy state E is given by

$$\frac{1}{\exp[(E - E_f)/k_B T] + 1} \qquad (6.1)$$

Whenever $k_B T$ is small compared to E_f (and it always is, in a solid), this probability function drops precipitously from nearly 1 to nearly 0 as E passes through E_f. We use k_B to denote Boltzmann's constant.

Whether that Fermi level falls in between energy bands or in the middle of a band profoundly affects the properties of the solid. Again using copper as an example, where the $3d$ shell for an atom is not completely filled, it is no surprise that the $3d$ band for the solid is likewise unfilled. Indeed, for any solid to be a metal, the number of electrons has to be enough to leave the outermost band partially unfilled. Otherwise, when a band is entirely filled and the Fermi level falls in between bands, the solid is an insulator. This is so because it would require a great deal of energy to boost an electron all the way from the highest filled state (in one band) to the lowest available state (in the next higher band).

6.3. THE DENSITY OF STATES

If a solid block contains approximately 10^{22} atoms, it will have about 10^{22} $1s$ electronic energy states, 10^{22} $2p$ states, 10^{22} $3d$ states, and so forth. The inner-shell electrons are oblivious to their neighboring atoms, and so all 10^{22} $1s$ electron states have virtually the exact same energy. However, it is the interactions among atoms that make a solid different from a gas; and the interacting outer electrons (such as those in the $3d$ state in copper) lose the uniqueness of any affiliation with a "home" atom. Concomitantly, their energy levels are not all identical but are smeared out across a band. The band may contain 10^{22} total states, but that fact is less interesting than the density of the available states across the entire band. Do the states accumulate preferentially at the top of the band? At the bottom? Are they uniformly distributed across the band? Knowing the density of states in a solid is the key to understanding several of its properties, including those associated with superconductivity.

In the simplest model of a normal metal with free electrons in an outer band, the density of states are distributed as

$$N(e) \propto e^{1/2}$$

We thus have a continuum of available states that rises gently with increasing energy. The Fermi level E_f lies somewhere inside this band, and the states are filled up to E_f and empty above E_f—except in the very narrow range $\pm k_B T$ away from E_f. In that range the electrons are free to hop among partially filled energy levels. If anything happens to change the number of available levels (the density of states) quite near the Fermi level, the properties of the metal may change dramatically.

The deliberate substitution of other elements (known as *doping* in small quantities but *alloying* in large quantities[8]) is done with the express intention of altering some important property. The entire semiconductor industry is built on this principle.[9] Solid pure silicon is

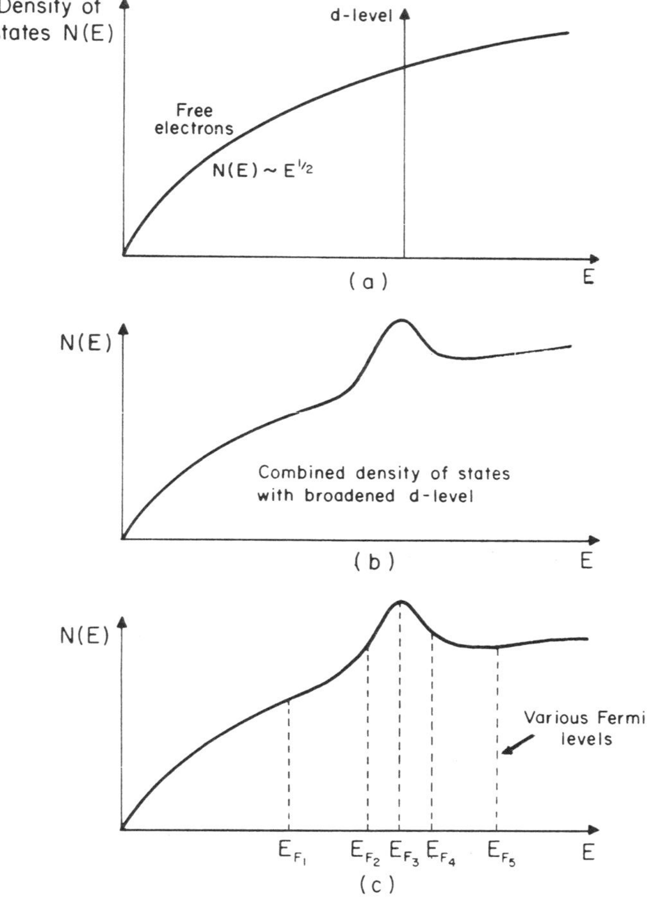

Figure 6.3. The effect of alloying on the density of states of a metal: (a) A free-electron gas (solvent metal) and a transition-metal impurity; the transition metal has a very sharp d-level. (b) When the transition metal is dissolved in the solvent, lifetime effects broaden the d-level, leaving a net density of states as shown. (c) Depending on the valence and concentration of the impurity, the Fermi level may or may not fall within the d-band region of the density of states. If it does (E_{F_2}, E_{F_3}, E_{F_4}), a net magnetic moment will form.

an insulator; its Fermi level lies in between bands, and it is a very poor electrical conductor. However, by deliberately adding a small percentage of impurity atoms that have extra electrons (e.g., phosphorus), the Fermi level can be pushed upward until it lies just a bit below an empty band. It is at this point that the factor $k_B T$ comes into play: The Fermi probability function, equation (6.1), allows a very small number of electrons to hop upward into the hitherto empty band, where they can then act as *conduction electrons*, just like those in a metal. Applying a voltage greatly changes the number of such conduction electrons, allowing a large current to flow. The silicon so treated is called a semiconductor because of this.

The formation of a magnetic moment within a conventional nonmagnetic metal is a second example of manipulating the density of states near the Fermi level. For example, only ½% manganese in copper increases the electronic specific heat by 600%; this is entirely due to a sharp increase in the density of states. The overriding factor is the relative position of the impurity's energy level in the d shell, the "d-level," in relation to the Fermi level of the host metal.

Observe Figure 6.3(a): an atom with a sharp d-level is placed in a host composed only of free electrons. The d-level will be broadened [Figure 6.3(b)] by "lifetime" effects. The Fermi level of the host material then is taken to lie at one of the five levels indicated in Figure 6.3(c). If the Fermi level falls at E_{f1}, the d-shell is completely empty and no magnetic moment forms; if it falls at E_{f2}, a few of the d-levels are occupied and a weak moment forms; at E_{f3} the d-level is half-filled and the magnetic moment is strongest; at E_{f4} the d-shell is almost filled and the moment weakens; at E_{f5} the shell is completely filled and no net magnetic moment forms. The density-of-states $N(E)$ is far bigger at the center of the d-level than it would be in the absence of the magnetic impurities. Accordingly, those properties which depend upon the density-of-states at the Fermi level should be severely changed, even for a small concentration of impurities, if the impurities have their energy level near the host's Fermi level.

The preceding paragraphs may at first appear to be a digression, but they are relevant to superconductivity because they point out how very modest alterations near the Fermi level can greatly affect the properties of a material. In a superconductor, the density of states right at the Fermi level is modified is a most unusual way.

6.4. PAIRING IN SUPERCONDUCTORS

Superconductivity occurs due to an interaction between electrons, which is transmitted by phonons. BCS theory[2] says that electrons travel in pairs, interacting with each other through lattice vibrations, or phonons: an electron having spin up and forward momentum **k** pairs with a spin-down electron traveling in the opposite direction with momentum −**k**. (Boldface type denotes vectors.) This is called Cooper pairing.

Either electron left to itself would collide with lattice atoms frequently and bounce around erratically; the dissipated energy constitutes electrical resistance. In a normal metal, exactly that happens. However, when paired, it is necessary for both electrons to undergo exactly the same scattering events, or else the pair will break. The probability of simultaneous identical scattering events is negligible, and there is an energy cost associated with breaking a pair. Therefore, if the pairing energy is sufficiently strong, the electrons retain their paired motion, and upon encountering a lattice atom do not scatter. Under these circumstances, energy is not dissipated, and there is no electrical resistance. When conditions change such that electrons can obtain the required pair-breaking energy, the material returns to its normal state and electrical resistance also returns.

The real mystery is why this paired state is energetically favorable. This leads into concepts unique to quantum mechanics, which are nonexistent in classical physics. When the theory of solids is approached using the tools of field theory,[1] transport phenomena (especially resistivity) depend on electron–phonon scattering; this scattering appears as a matrix element for processes in which electrons either absorb or emit phonons, and the resulting change in momentum of the electrons constitutes the scattering event. It is simpler to characterize this by a mean free path l, which expresses the average distance an electron travels between collisions. The quantity $1/l$ is proportional to the scattering probability.

The *coherence length* ξ is a similar concept. The paired electrons interact with each other over a finite distance, which may be different for every pair. Still, the average spatial extent of a pair must be at least as big as the uncertainty principle limit ($\Delta x \Delta p > h/2\pi$), and since the superconducting electrons have energies within $k_B T_c$ of the Fermi level, the size of the pair must be about $\Delta x > h v_F / 2\pi k_B T_c$, where v_F is the Fermi velocity. This is called the *intrinsic coherence length* ξ_0. It is analogous to the mean free path for normal electrons. In fact, these two add like resistors in a parallel circuit, to yield the actual coherence length

$$1/\xi = 1/\xi_0 + 1/l$$

In a very pure superconductor like gallium, the mean free path l can be centimeters long, so the coherence length reaches its intrinsic value, perhaps $1/3$ μm. In alloys like NbTi, l is so short that ξ is in the 100-Å range. For now, the central point is that the paired electrons interact with each other over a finite range within the crystal.

In 1950, the first inkling of the modern theory of superconductivity was presented by Frohlich,[10] who observed that there is another effect of the electron–phonon interaction: Suppose one electron emits a phonon and is scattered, and then this phonon is absorbed by a second electron; the net result is that the two electrons have interacted, with the phonon transmitting the interaction. This Frohlich interaction is akin to the sonic boom of classic acoustics: the velocity of a typical electron is the Fermi velocity, which is about 100 times the speed of sound in the metal. Therefore, as an electron moves through the metal, a narrow cone of phonons trails it. It is important to realize that these phonons are *virtual* phonons rather than *real* phonons; the distinction can be seen by thinking of a real phonon as one that has already been given off, with a certain change in the momentum and energy of the electron, while a virtual phonon is one that is available for only those interactions that occur off the energy shell, i.e., that leave the total momentum and energy of the two interacting electrons unchanged.

One loose end that needs to be cleared up pertains to the matter of paired electrons carrying current. It is a fair question to ask, How can any current flow if the pairs have equal and opposite momentum? To overcome this paradox, it is necessary to look closely at the difference between *having a certain momentum* and *moving in a particular direction*. Classically, there is no distinction. But in a solid where waves propagate through consecutive unit cells, something different can happen. When a particle of momentum **k** crosses a unit cell, its wave function is multiplied by an amount $\exp(i2\pi)$, which equals one, thus producing no net change in the wave function. A particle crossing in the opposite direction changes only by $\exp(-i2\pi) = 1$. The paired momenta are defined with respect to a Brillouin zone, and can jump from one zone to another without loss of pairing. Similarly, the positions are defined with respect to a unit cell, and motion from one cell to another leaves the wave function unchanged. Electrons are paired in k-space and may be quite far apart in position-space. To introduce an electron from one side of a solid or remove it from the other does not upset the pairing. Therefore, superconductivity persists and real current flows.

6.5. THE SUPERCONDUCTING ENERGY GAP

Thermodynamically, superconductivity occurs because the normal Fermi sea distribution is unstable in an electron gas interacting via this pairing mechanism. For a strong enough interaction (a large coupling constant in the matrix element representing the phonon-induced interaction) a different distribution of electrons gives a lower energy. Denoting by subscript k the factors associated with the pair having momentum k, the superconducting distribution function is closely akin to the conventional Fermi distribution:

$$f_k = \frac{1}{\exp\,(E_k/k_BT) + 1} \tag{6.2}$$

where the "zero" of energy has been shifted to coincide with the Fermi level, and

$$E_k^2 = e_k^2 + \Delta(T)^2 \tag{6.3}$$

That is, the energy spectrum contains a gap, the width of which varies with temperature. The value e_k is the single-electron energy with respect to the normal-state Fermi surface, and $\Delta_k(T)$ is the energy gap.

Mediated by a phonon, the interacting electrons overcome their customary Coulomb repulsion and experience a net attraction to one another; hence, they exist in a lower-energy state when paired. Once in that lower-energy state they tend to remain that way. No electron can leave its pair within the condensed state until it has first acquired sufficient energy to cross the gap. At modest temperatures, the available thermal energy makes this a negligible restriction; but at low temperatures, the energy gap is equivalent to a major barrier. Because of this, pairing (and hence superconductivity) is preserved only at low temperatures.

The way the gap changes with temperature is very important for superconductivity, as shown in Figure 6.4. The solid line is the curve

$$\delta^2 = \cos\,(\pi t^2/2) \tag{6.4}$$

where we have introduced the commonplace shorthand $t = T/T_c$ and $\delta = \Delta(T)/\Delta(0)$. This function remains quite close to $\delta = 1$ for $t < 0.5$, and falls off at higher temperatures. Experimental data for the energy gaps of elementary superconductors all lie near it.[11] This similarity of behavior from one superconductor to another is an example of the law of *corresponding states*.[12] It is also found for many of the thermodynamic parameters of superconductors and is one of the most convincing aspects of the BCS theory: it makes sense out of a lot of data.

As the temperature rises, thermal energy eventually overcomes the pairing interaction, and so superconductivity vanishes as the gap goes to zero. The energy gap also falls off with increasing magnetic fields. Figure 6.5 depicts the joint behavior of Δ with H and T. (Figure 6.4 is essentially the mirror image of the right-hand face of Figure 6.5.)

It is not too surprising that the key to understanding superconducting behavior in a material is to understand its energy gap. To go further, it is necessary to explain how tunneling occurs.

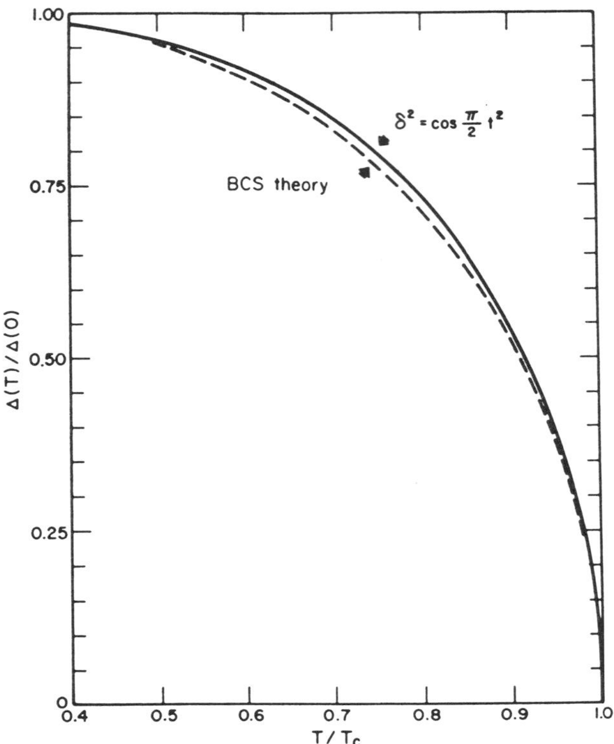

Figure 6.4. Energy gap as a function of temperature in the BCS theory of superconductivity. T/T_c is defined as t, and $\Delta(T)/\Delta(0)$ is defined as δ.

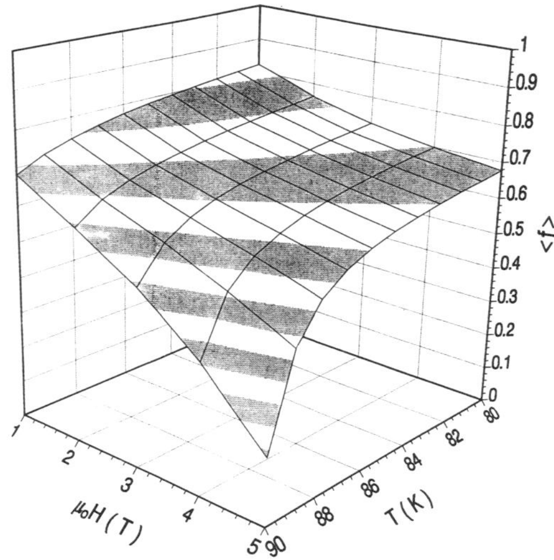

Figure 6.5. Variation of energy gap with both temperature and magnetic field. (Courtesy D. Finnemore, Ames Laboratory.)

6.6. THE GAP AND TUNNELING

Tunneling experiments are of great interest explicitly because they measure the density of states adjacent to the gap. Tunneling is strictly a quantum-mechanical phenomenon, because in classical physics a particle cannot penetrate a barrier. Yet there is a finite probability of finding a quantum-mechanical particle beyond a barrier.

A moving particle is described by its wave function ψ,

$$\Psi(\mathbf{x},t) = A(\mathbf{x},t) \exp\left[i(\mathbf{k}\cdot\mathbf{x})\right] \exp\left(i\,\omega t\right)$$

where $\mathbf{x}$ is the position vector and $\mathbf{k}$ is momentum. When the particle encounters a barrier as shown in Figure 6.6, the potential energy inside the barrier exceeds the total system energy outside the barrier, in which case the particle cannot enter the barrier, classically speaking. In quantum mechanics, however, there is a finite probability of the particle penetrating. Consequently, the amplitude A is attenuated by the barrier, but if the barrier is reasonably thin, a finite amplitude exists on the far side of it. The probability of finding a particle at any point is simply $|A|^2$, and thus the particle may sometimes pass through a barrier that would totally stop it, classically. In this way, current can flow from one metal to another through an insulating layer. This is the principle that underlies the scanning tunneling microscope and other instruments as well.

In a normal metal, the Fermi energy lies somewhere out in the middle of a partially filled energy band, and thus the density of states in the vicinity of the Fermi level has no unusual properties. The current simply increases linearly with voltage, according to Ohm's law, and so tunneling measurements detect a nearly flat density of states. In a superconductor, however, the existence of an energy gap centered on the Fermi level forces profound changes.

The energy relationship of a superconducting pair, equation (6.3), forces a gap to be present in the density of states, right at the Fermi level. We denote the normal density of states right at the Fermi level by $N(0)$ or N_0. The density of states in the superconducting state N_s is zero for energies closer to the Fermi level than Δ, and all the states that would have normally fallen into that energy district have been displaced to lie slightly above or below the gap. Comparing the two states, we have

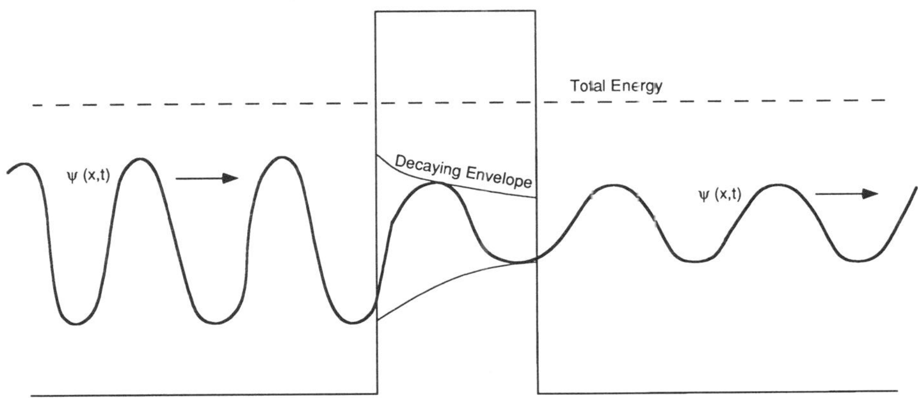

Figure 6.6. Quantum-mechanical penetration of a barrier by a moving particle.

$$\frac{N_s(e)}{N_s(0)} = \frac{e}{(e^2 - \Delta^2)^{1/2}} \quad \text{for } e > \Delta \tag{6.5}$$

where the Fermi energy is defined as zero. This has the unusual consequence of creating a singularity in $N_s(e)$ right at the gap edge, $e = \Delta$. Figure 6.7 shows this density of states[3]: Part (a) is the large-scale picture, reminiscent of Figure 6.3; part (b) is a close-up of the region

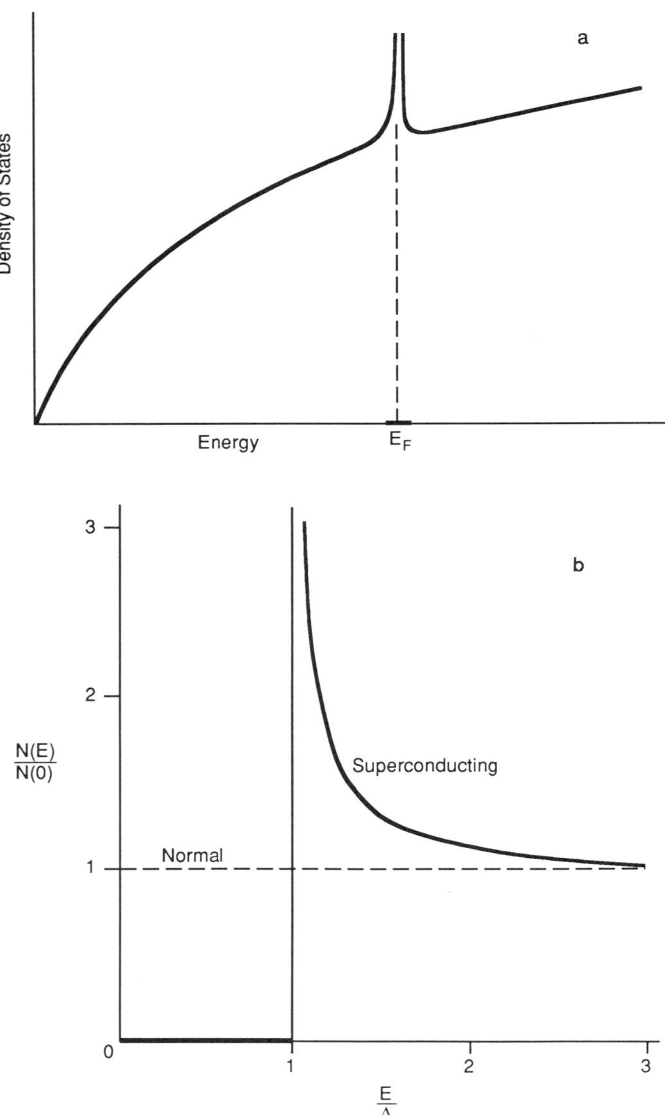

Figure 6.7. (a) Density of states in a superconductor, showing a very narrow gap located right at the Fermi level E_F. (b) Close-up of narrow region just above the Fermi level (taken as $E = 0$ here), comparing the superconductor's density of states (solid) with that of a normal metal (dashed line).

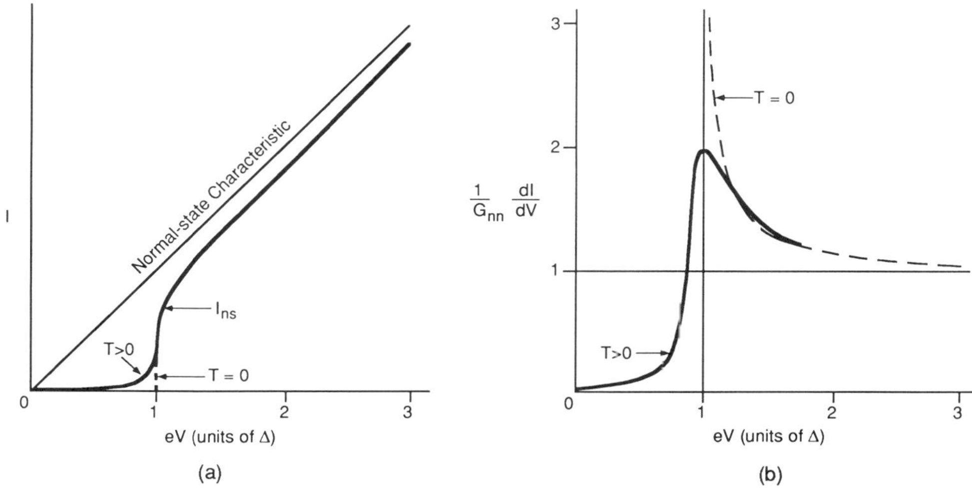

Figure 6.8. (a) Current–voltage characteristic for superconductor showing the sudden change when applied voltage reaches the value of the energy gap. (b) First derivative of the current–voltage characteristic, both at zero temperature (dashed line) and at higher temperature (heavy solid line). (From Ref. 3.)

just above the Fermi level. Now when tunneling takes place, the current–voltage characteristic is quite different. At any voltage less than Δ, no current flows because there are no occupiable states for electrons of that energy. As soon as V exceeds Δ, the current rises steeply and approaches the normal-metal I-V characteristic asymptotically. This is portrayed in Figure 6.8(a). The sharp rise in current is attributable to the great number of available states for electrons of that energy.

As we have seen, Δ varies with temperature. More important, our extremely simple model changes at finite temperatures. The distribution of actual energies occasioned by thermal motion leads to a smearing out of any sharp corners in the measured I-V curves. This stands out particularly clearly in Figure 6.8(b), which is just the derivative of Figure 6.8(a). The normal state has a perfectly flat constant derivative, while the superconducting state has a narrow peak at $eV = \Delta$.

Experimentally, the location of that narrow peak (as the voltage is increased continuously) gives the measured value of Δ. As the temperature increases above zero, the value of Δ declines (as predicted by Figure 6.4), but thermal smearing also makes the exact location of the peak less distinct, so the error brackets on Δ increase as well. This is of minor importance near 4 K, but (for eventual applications to the HTSCs) it is much more troublesome near 77 K.

Of still greater interest is the data beyond the peak corresponding to Δ. Comparing Figures 6.7(b) and 6.8(b), we see that the derivative of the I-V curve is the same as the density of states. This is only rigorously true at zero temperature, and at finite temperatures thermal smearing degrades the precise relationship slightly. (Again, at 4 K, the distinction is minor, but at 77 K it must be taken into account when analyzing data.) Nevertheless, tunneling experiments are the best way to determine the density of states in a superconductor. The discovery and exploitation of this technique led to the Nobel Prize in Physics for Ivar Giaever[13] of General Electric in the 1960s.

6.7. CONSEQUENCES OF THE BCS EQUATIONS

We have not yet explained why anyone should care about tunneling results. The reason lies in the very close relationship between the basic mechanism of superconductivity and these experiments. For LTSCs, it is tunneling that certifies the validity of BCS. For the HTSCs, the validity of the BCS mechanism has been questioned by some researchers, and it is hoped that tunneling experiments can clear up the controversy.

The BCS theory starts with the hypothesis of Cooper pairs of electrons interacting via the Frohlich mechanism[10] of electron–phonon coupling, and goes on to ask about the energy of such a system. The details are beyond our scope here, but minimization of the free energy leads to a nonlinear integral equation for the gap $\Delta_k(T)$:

$$\Delta_k(T) = -\frac{1}{2}\sum_{k'} V_{kk'} \frac{\Delta k'(T)}{E_{k'}} \tanh \frac{E_{k'}}{2k_BT} \tag{6.6}$$

The interaction $V_{kk'}$ actually includes the screened Coulomb repulsion between electrons as well as the electron–phonon interaction. If any realistic form for the interaction, such as that given by Eliashberg,[14] were put into equation (6.6), machine computation would be necessary to find $\Delta_k(e_k, T)$. In order to obtain analytic solutions of this integral equation, BCS completely ignored the Coulomb interaction and approximated the electron–phonon interaction by the extremely simple form:

$$V_{kk'} = -V \quad \text{for } | e_k | \text{ and } | e_{k'} |, \text{ both } < \hbar\omega_c$$

$$= 0, \text{ otherwise}$$

where V is a constant greater than 0. Furthermore, in the BCS model, Δ_k is a constant at each temperature up to a distance $\hbar\omega_c = k_B\theta_D$ from the Fermi surface and is zero at a greater distance. (θ_D is the Debye temperature.) Under this restricted model, the summation of equation (6.6) is confined to a narrow band of states within $\pm \hbar\omega_c$ of the Fermi energy. Then, recalling that for a free-electron gas the density of states $N(e') \sim (e')^{1/2}$, we see that it will change very little over the narrow range of integration. Accordingly, we approximate the density of states by a constant, $N_0 \equiv N(E_f)$. Converting the sum to an integral by the usual substitution, this BCS integral equation then appears in the form

$$\Delta(T) = \frac{N_0V}{2} \int_{-\hbar\omega_c}^{+\hbar\omega_c} de \frac{\tanh(E/2k_BT)}{[e^2 + \Delta^2(T)]^{1/2}} \Delta(T) \tag{6.7}$$

We have pulled one other trick in writing equation (6.7): the gap Δ is shown as only a function of temperature, not of energy as it was a moment earlier in equation (6.6). This, too, is a simplification used by BCS; it was necessary to keep the theory tractable and exempt from machine computation. When Δ can be taken out of the integral, equation (6.7) further simplifies (writing hyperbolic tangent in terms of exponentials) to

$$\frac{1}{N_0V} = \int_0^{\hbar\omega_c} \frac{\exp(E/k_BT) - 1}{\exp(E/k_BT) + 1} \frac{de}{E} \tag{6.8}$$

where $E^2 = e^2 + \Delta(T)^2$, as before.

At last we reach a point of genuine simplification: At 0 K, this integral is easy to carry out and yields for the zero-point energy gap:

$$\frac{\Delta_0}{\hbar\omega_c} = \frac{1}{\sinh(1/N_0V)} \approx 2\exp(-1/N_0V) \qquad (6.9)$$

as long as the product N_0V is small. At the other temperature extreme (T_c) the energy gap Δ vanishes, in which case equation (6.8) simplifies again, and there emerges the relationship

$$k_BT_c/\hbar\omega_c = 1.14\exp(-1/N_0V) \qquad (6.10)$$

Equations (6.9) and (6.10) combine to give the BCS ratio $2\Delta_0/k_BT_c = 3.53$. This "magic number" has received tremendous emphasis as the benchmark of an ideal BCS superconductor; it is a case of overemphasis. When N_0V is no longer small, the 3.53 number drifts upward and correction terms are required,[15] but these need not concern us here.

By the mid-1960s, it became customary to regard the BCS theory as applying only in the limit of very weak coupling. For strong-coupling cases, extensive calculations based on a Green's function approach[16] to the interaction had been performed, and this explained some very intricate tunneling characteristics.[17] More-or-less correct results for thermal conductivity,[18] ultrasonic attenuation,[19] and other difficult problems were obtained by such calculations. There is little doubt of the fundamental correctness of this approach; however, it has three disadvantages: (1) an involved calculation is needed for every property of every superconductor, because no single calculation predicts results for several superconductors; (2) the error brackets are rather large[20] on these theoretical calculations, since several unknowns enter each calculation in a crucial way; and (3) little physical insight into the interaction mechanism is gained by a procedure[21] that numerically generates a complex and intricate energy gap function.

Machine calculations of great complexity may give a correct description of a physical phenomenon and still leave only a vague notion of the true physics underlying the problem. In superconductivity it is desirable to get as much as possible out of the BCS theory before seeking numerical solutions for individual cases. It makes sense to try the same thing for the HTSCs that worked so well for the LTSCs. The fascination with exotic theories of HTSC peaked in 1987 and has since given way to an effort to retain the basic BCS mechanism while adjusting the model to account for effects of anistropy and more than one energy gap. For the HTSCs, the agreement between tunneling data and the BCS model is now very encouraging.

6.8. EXPERIMENTAL CONSIDERATIONS

The experimental confirmation of the BCS theory occurs first and foremost through tunneling data. The density of states is related to the derivative of the I vs. V curve, so the measurement to be made is the conductance, $G = dI/dV$. If experimental data were perfect, the density of states would drop right out, and it could be compared with equation (6.5) to see how well it fits a simple theory.

When things work well, tunneling is a powerful probe of superconductors. However, it is a surface probe, and if the surface is dirty or irregular, the results are very poor. Consequently, tunneling measurements require great care if they are to be valid. (This is a particular problem with HTSCs, where the surfaces are known to be poor.) Three types of

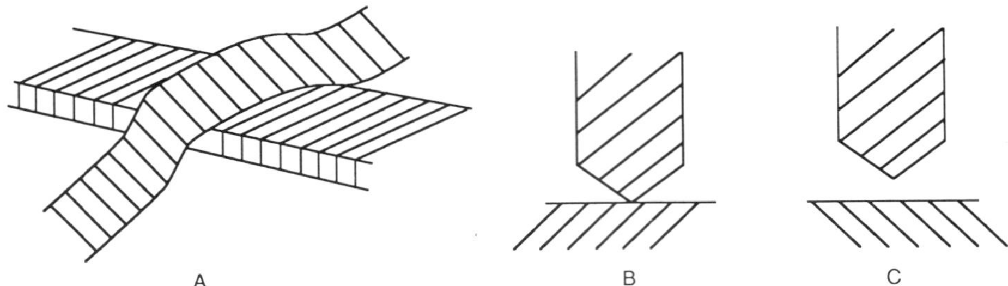

Figure 6.9. Three different geometries used in tunneling experiments: (a) Overlapping thin films; (b) surface-contacting; (c) noncontacting.

tunnel geometries appear in Figure 6.9. Because of surface dirt, oxidation, and so on, no one has yet been able to make a good tunnel junction of type A above 25 K. In type B, the probe is used to penetrate below the surface layer. Type C is very hard to control experimentally. Therefore, most tunneling data has been obtained using type B.

The derivative of I vs. V is determined as follows: a small AC current is superimposed on the DC current, and a phase-locked detector is used to isolate the AC signal produced as V is increased. The AC component might be about 1 mV peak-to-peak, while V is scanned from 0 to 50 mV. The signal detected at twice the operating frequency is d^2V/dI^2, and this is very clean. Integrating the data once produces G, which is proportional to $N(e)$.

It is important to remember that one must measure the conductance in both the superconducting state and the normal state in order to form the ratio expressed by equation (6.5). Fortunately, the normal state produces pretty boring data in the range of interest. Generally, noise is not too much of a problem out to about 50 mV, and hence the data is acceptably good over most of the range. However, if there is structure present in the density of states beyond 50 mV (typically at the 1% level), it will likely be swamped by the noise.

Simple superconductors sometimes give nearly textbook perfect[22] density of states from tunneling data, as shown in Figure 6.10. However, in other cases there is normal-state

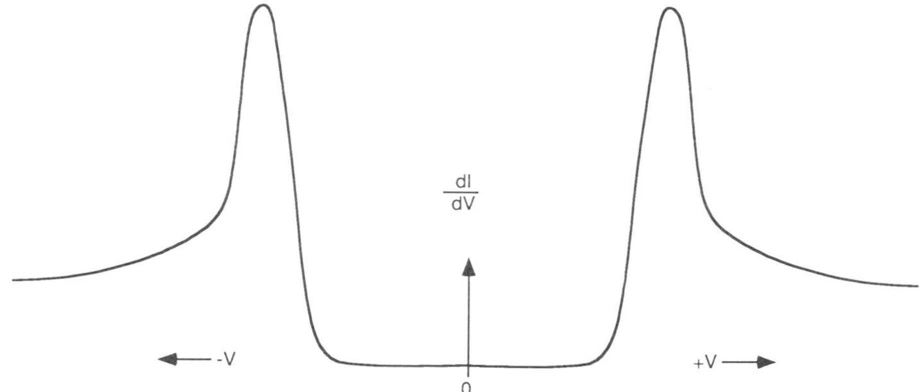

Figure 6.10. Tunneling data for a typical ideal BCS superconductor.

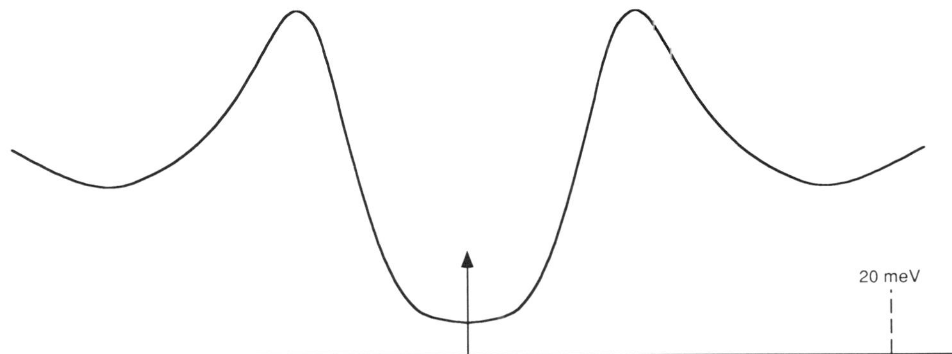

Figure 6.11. Tunneling data showing superposition of normal-state conductance with superconducting tunnelling.

conductance along with superconducting tunneling, in which case the data looks like Figure 6.11. Then it is necessary to separate those two contributions in order to recover the density of states. Furthermore, when the interaction strength N_0V is relatively strong, even in type I superconductors like mercury and lead, it is no longer easy to convert raw data into the density of states.

6.9. ANALYSIS OF DATA

Because real superconductors deviate substantially from the simple BCS model, it is usually necessary to use machine computations to recover density-of-states information from experimental data. If we retreat to equation (6.6), with experimental data in hand, the first restriction that falls is the notion of holding Δ constant for all choices of $\mathbf{k}'$. In fact, even for the same numerical magnitude of k' values, different directions give different values of Δ. This condition is called *anisotropy of the energy gap*. Moreover, the interaction strength $V_{kk'}$ is not constant either, and doesn't suddenly cut off at $\hbar\omega_c$ as in the original BCS theory. In fact, it really doesn't make sense to separate the interaction strength from the density of states, so the notation N_0V is not used any more, having been supplanted by the *interaction parameter* denoted by λ.

Furthermore, what tunneling determines is actually the product of the electron density of states and the phonon density of states as a function of energy. In the simple BCS model, that distinction can be ignored, because the phonon spectrum has no peaks or other important variation with energy. That applies to most weak-coupling type I superconductors, but as soon as strong-coupling occurs between electrons and phonons, a correction is required. The expression for the electron density of states, equation (6.5), must be revised to allow the energy gap Δ to be a complex function of energy,[23] and the observed density of states is only the real part of the modified equation (6.5). Specifically, we have for the ratio of conductances in the superconducting and normal states

$$\frac{\sigma_s}{\sigma_n} = \mathrm{Re}\left\{\frac{e}{[e^2 - \Delta(e)^2]^{1/2}}\right\}$$

The integral that evolves from equation (6.6) is much more cumbersome than equation (6.7) for the BCS case. Among other things, the effect of Coulomb repulsion is included in the

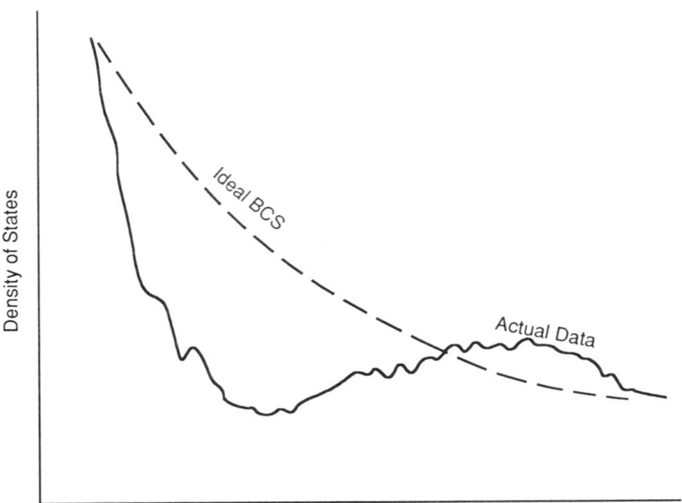

Figure 6.12. Density of states for niobium, calculated from tunneling data, and showing how a strong-coupling superconductor deviates from the simple BCS model.

modern treatment. At the end of the pipeline, a density-of-states factor (commonly labeled $\alpha^2 F[E]$) is found by inverting the governing integral equation, starting from measured I-V curves.[24]

Other measurements can help to refine this information. For example, neutrons do not interact with electrons, but scatter off the lattice ions, and thus can be used to independently determine the phonon density of states. Armed with such data, it is possible to decouple the phonons from the electron density of states. This is particularly helpful when the phonon spectrum contains sharp peaks, as in strong-coupling superconductors.

If the data is good, the outcome might look like Figure 6.12, which depicts the density of states in niobium, a strong-coupling superconductor. Once the density of states has been retrieved from the tunneling data, it is a straightforward computational integration to obtain the coupling-strength parameter λ, which in the simple BCS case is identical with $N_0 V$. Niobium, for example, shows $\lambda = 1.1$ and $2\Delta_0/k_B T_c = 3.9$—results known from the early 1960s.[25,26] In fact, numerical integration of the BCS equation (6.8) yields[15] $2\Delta_0/k_B T_c = 3.85$ for the choice $N_0 V = 1.1$. This demonstrates that despite its intricate phonon structure and density of states, niobium is still well-described within the BCS theory.

Early thin-film data on niobium showed that there is anisotropy in $\alpha^2 F$, even in a cubic compound like niobium: the [100] and [110] crystalline directions showed slightly different phonon spectra. Needless to say, if niobium is that complex, the highly anisotropic copper oxide superconductors are much worse.

6.10. SUMMARY

This chapter introduced a number of basic concepts that are used in the highly successful BCS theory of superconductivity. The Fermi level, the density of states, Cooper pairing, and so on, all are important components of our modern understanding of superconductivity. The density of states is a pivotal concept here because there is a gap in the density of states located right at the Fermi level; the gap is an essential characteristic of a superconductor. Tunneling experiments investigate the energy gap and the surrounding density of states. Using the BCS theory, tunneling data is readily converted into the density of states.

Throughout the 1960s and 1970s, a tremendous amount of data could be explained by minor modifications and generalizations of the BCS theory, and that collective experience made it seem quite incontrovertible. The BCS theory was so successful that predictions based on it were readily believed. For example, one prediction about phonon-coupled superconductivity was that T_c could not exceed 30 K. Consequently, the discovery of high-temperature superconductivity in late 1986 came as an immense surprise to the community of scientists and engineers conversant with this field.

It is very much to the credit of experimental researchers that even very good theories are challenged. However, as we shall see in Chapter 12, it is also to the credit of the BCS theory that it can still account for the main features of the HTSCs.

REFERENCES

1. C. Kittel, *Quantum Theory of Solids* (Wiley, New York: 1963).
2. J. Bardeen, L. N. Cooper, and J. R. Schrieffer, *Phys. Rev.* **108**, 1175 (1957).
3. M. Tinkham, *Introduction to Superconductivity* (Krieger Publ. Co., Malabar, FL: 1980).
4. G. M. Eliashberg, *Soviet Phys.* JETP **29**, 1298 (1969).
5. C. Kittel, *Introduction to Solid-State Physics*, 6th Ed. (Wiley, New York: 1986).
6. L. Brillouin, *Wave Propagation in Periodic Media* (Dover, New York: 1960).
7. L. D. Landau and L. M. Lifshitz, *Statistical Physics* (Pergamon Press, London: 1958).
8. E. W. Collings, *Applied Superconductivity, Metallurgy and Physics of Titanium Alloys* (Plenum Press, New York: 1986).
9. G. Carter and W. A. Grant, *Ion Implantation in Semiconductors* (Wiley, New York: 1976).
10. H. Frohlich, *Phys. Rev.* **79**, 845 (1950).
11. T. P. Sheahen, *Phys. Rev.* **149**, 368 (1966).
12. P. G. deGennes, *Superconductivity of Metals and Alloys* (Benjamin, New York: 1966).
13. I. Giaever, *Phys. Rev. Lett.* **5**, 464 (1960).
14. G. M. Eliashberg, *Soviet Phys.* JETP **2**, 696 (1960).
15. T. P. Sheahen, *Phys. Rev.* **149**, 370 (1966).
16. Y. Wada, *Phys. Rev. A* **135**, 1481 (1964).
17. G. J. Culler *et al., Phys. Rev. Lett.* **8**, 399 (1962).
18. L. Tewordt, *Phys. Rev.* **129**, 657 (1963).
19. L. P. Kadanov and I. I. Falko, *Phys. Rev. A* **136**, 1170 (1964).
20. J. W. Garland, Jr., *Phys. Rev. Lett.* **11**, 1111, 1114 (1963).
21. J. C. Swihart *et al., Phys. Rev. Lett.* **14**, 106 (1965).
22. D. R. Tilley and J. Tilley, *Superfluidity and Superconductivity*, 2nd. ed. (Adam Hilger, Bristol: 1986).
23. J. R. Schrieffer *et al., Phys. Rev. Lett.* **10**, 336 (1963); *Phys. Rev.* **148**, 263 (1966).
24. W. L. McMillan and J. M. Rowell, *Phys. Rev. Lett.* **14**, 108 (1965).
25. H. A. Leupold and H. D. Boorse, *Phys. Rev. A* **134**, 1322 (1964).
26. P. Townsend and J. Sutton, *Phys. Rev.* **128**, 591 (1962).

7

The New Superconductors

This chapter is about the history of high-temperature superconductivity (HTSC). It covers a fairly short period of time, because the first HTSC was discovered only in 1986 by Bednorz and Muller[1] of IBM's Zurich Research Lab. Nevertheless, since then the way in which successive events have unfolded presents a most interesting picture of contemporary scientific research in the dynamic world of government sponsorship and industrial competition. Here, we try to synopsize the major events and weave a thread of continuity among them to explain how the HTSC industry has come to where it is today.

High-temperature superconductivity arrived on the scene as a total surprise. It was widely believed to be impossible. A brief account of the major early discoveries leads to a discussion of the extensive hype that then infected the field. More sobering is the genuine progress made through careful research worldwide; however, only a few of the highlights are mentioned here. Participation by government, in America as well as in Japan, has given a considerable boost to this field; our perspective on this is presented. As a result, a viable industry based on HTSC wire, devices, and applications is emerging, and so we offer a view of that development.

The reader must appreciate that it is impossible for any participant to be totally objective and unbiased in recounting contemporary history. Accordingly, this chapter should be viewed as *one* view of what has taken place and not necessarily the eventual final view of historians of science.

7.1. WHY IT WAS "IMPOSSIBLE"

One memorable characteristic of HTSCs is that their discovery was rewarded with the Nobel Prize in physics[2] for 1987. Even more remarkable is that the elapsed time between discovery and prize was shorter for HTSC than for any other Nobel Prize ever given. That happened because the experimental discovery of HTSCs completely overthrew a firmly established body of theory, one that had served so well that many thought of superconductivity as a closed (or "mature") field of science. The excitement generated by HTSCs was certainly due to their potential applications, but it was equally due to the way in which scientists were reminded that experimental surprises can occur at any time.

To understand why this discovery was so revolutionary, it is necessary to understand why it was such a "sure thing" that the highest transition temperature T_c could not go above 30 K. This was one of the major secondary results following upon the BCS[3] theory; it was such a clear and easy-to-grasp conclusion that no reasonable person (skilled in superconductivity) could possibly doubt it. The major paper on the subject was that of W. L. McMillan in 1968.[4] With the advantage of hindsight from a 1990s perspective, one can see where certain assumptions (eminently plausible and universally applicable in the 1960s) sent the theory down a path that would later prove to be a limited perspective, from which layered superconductors, including the copper oxides, are exempt. In fairness, it should be noted that McMillan was killed in a traffic accident in the 1970s; had he lived, he would most probably have been among the first to see the implications of the Bednorz–Muller discovery, and would have removed the limitations from his own theory to accommodate the higher T_c values.

As we saw in Chapter 6, a key result of BCS is the formula for the transition temperature

$$T_c = 1.14 <\omega> \exp(-1/N_0V) \tag{7.1}$$

where $<\omega>$ is the average phonon energy,[†] N_0 is the density of states at the Fermi level, and V is the pairing potential arising from the electron–phonon interaction. More sophisticated treatments[5] that followed BCS converted N_0V to the more generalized λ, which accounted for a number of effects (Coulomb screening, etc.) that had been skipped in the original BCS theory. Equation (7.1) suggests that no matter how large $N_0V = \lambda$ becomes (i.e., no matter how strong the coupling), there will be an upper limit on T_c set by the average phonon energy $<\omega>$. On that basis, the pathway to high T_c is via higher phonon energies.

It is not that simple, however; the coupling constant λ is itself depends on the phonon energy. McMillan started from the refined BCS theory and derived a formula [his equation (18)] that gave T_c as a function of λ. Based on data available at that time, and limiting his treatment to metals and alloys, McMillan selected the relation

$$\lambda \approx C/M<\omega^2> \tag{7.2}$$

where $<\omega^2>$ is the mean-square average phonon energy, M is the mass, and C is a catch-all constant, fixed within a given class of materials. This was a critical step, and McMillan wrote, "The most important [refinement needed] is to test the relationship between the coupling constant and the phonon frequencies for a wider range of materials. . . ."[4] Nonetheless, this particular choice for λ as a function of ω, upon insertion into a more precise version of equation (7.1), led to the upper limit

$$T_c^{max} = (C/2M)^{1/2} \exp(-3/2) \tag{7.3}$$

Within a class of materials (C fixed), how close one could get to T_c^{max} depended on λ:

$$T_c/T_c^{max} = (2/\lambda)^{1/2} \exp(1/2 - 1/\lambda) \tag{7.4}$$

This has a broad maximum in T_c for $\lambda \approx 2$, and to reach that it is desirable to *decrease* the average phonon energy, contrary to the superficial appearance of equation (7.1).

McMillan went on to compute the maximum T_c for each class of materials. For the class akin to V_3Si, he obtained $T_c^{max} = 40$ K, but warned that a lattice instability prevents formation of any compound with so high a value. Scaling by the square root of the mass ratio, he found $T_c^{max} = 28$ K for the Nb_3Sn class of materials. Rounding that off, 30 K became the "magic number" that stuck in nearly everyone's mind for years to come.

[†]Comparing with Eq. (6.10), the factors of $h/2\pi$ and k_B have been suppressed here, being incorporated into the units of phonon energy (temperature).

Several years later, in 1975, Allen and Dynes[6] re-examined McMillan's treatment and showed that the broad maximum in T_c near $\lambda \approx 2$ was spurious. They presented a new T_c formula valid for asymtotically large $\lambda > 10$:

$$T_c \approx 0.15(\lambda <\omega^2>)^{1/2} \tag{7.5}$$

However, equation (7.2) shows that $\lambda<\omega^2> = C/M$. Allen and Dynes replaced McMillan's materials-dependent constant C with the parameter η, a purely electronic property independent of lattice dynamics. It was believed that η dropped off for reasons of *covalent instability*. Allen and Dynes went on to speculate about the behavior of η, including the suggestion of $\eta_{max} \approx 2.4$ in a Tl-Pb-Bi system. Recognizing their own many approximations, Allen and Dynes did not try to calculate any specific values for T_c^{max} , but used empirical T_c data to aid their estimation of λ and η.

Within McMillan's paper,[4] certain cautions were expressed: "We have neglected the anisotropy of the energy gap . . ."; "We have extrapolated the theoretical formula [equation (18)] for T_c vs. λ, which was derived for $\lambda \leq 1$, to larger values of λ. The errors are probably not serious . . ."; "We have made one special assumption by using the phonon density of states for niobium. This introduces important errors only for the strong-coupled ($\lambda > 1$) superconductor with a wildly different phonon spectrum."

What we have, in fact, in the HTSCs is a "wildly different phonon spectrum," as will be explained in Chapter 12. For now, it must suffice to note that McMillan obviously understood just how limited his theory was. Unfortunately, not enough others did. Allen and Dynes[6] wrote, "McMillan made suitable cautionary and qualifying remarks which have been largely forgotten with time." People gave up hope that T_c would ever exceed 30 K.

By their discovery, Bednorz and Muller proved again that experiment always prevails over theory. In early 1988, during a coffee break in a conference, I posed the question to Muller: "Since you knew of the theoretical upper limit of 30 K, why did you keep on looking?" Muller replied "I asked the theorists to explain to me how they got that limit, and when they were all done, I didn't understand it, so I went back to work in the lab." That answer may be the perfect prescription for how to win a Nobel Prize.

7.2   THE DISCOVERIES OF 1986–1987

In 1986, the first indication of superconductivity above 30 K was found in barium-doped La_2CuO_4,[1] as shown in Figure 7.1. The first reaction of most practitioners of superconductivity was to think, "There must be a new mechanism, because phonon-driven superconductivity is impossible at so high a temperature." With the 30 K barrier broken, the race was on to find still higher T_cs. The first step upward[7] was via strontium substitution: $La_{2-x}Sr_xCuO_4$ gave $T_c = 38$ K.

Meanwhile, for over two decades investigations had continued on the electronic properties of materials under high pressure. Paul Chu and colleagues at the University of Houston soon found[8] that high pressure could increase T_c to 50 K. Then they started trying other combinations of atoms of different sizes and valences, in order to simulate pressure via chemical substitution. One of the variations was to substitute yttrium into the perovskite structure of $BaCuO_3$. To everyone's surprise, this compound ($YBa_2Cu_3O_7$) went superconducting[9,10] at 92 K. Immediately thereafter, hundreds of research laboratories joined the search for other high-temperature superconductors. Fairly early in 1987, it became clear that nearly any rare earth element had the same effect as yttrium, yielding a T_c around 90 K.

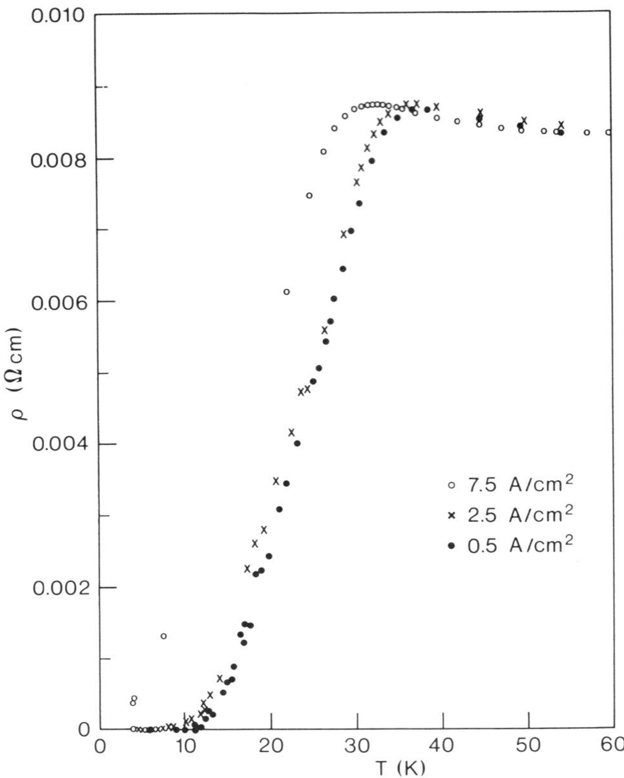

Figure 7.1.   Data by Bednorz and Muller[1,2] from 1986 for LaBaCuO, showing decline in resistivity below 30 K.

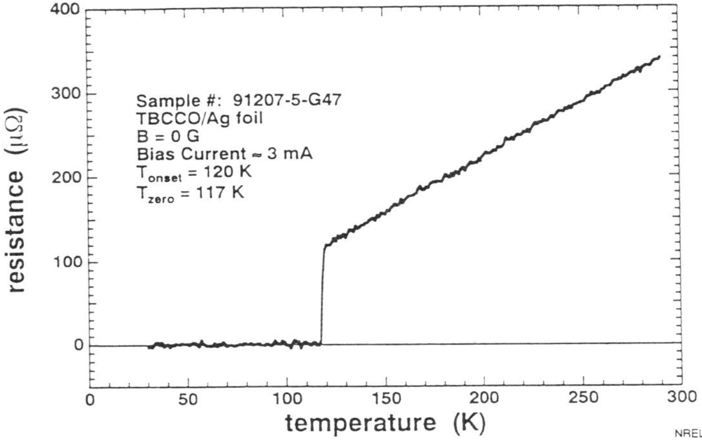

Figure 7.2.   Plot of resistivity vs. temperature for TBCCO made in 1992. (From Ref. 11.)

The copper oxides proved to be more complicated than previously known superconductors. Ideally, the resistivity of a superconductor drops from a finite value to zero at one particular temperature. Such a graph is shown[11] in Figure 7.2 for TBCCO; however, that data was taken in 1992. The more typical result of 1987 had data all over the map, with T_c varying greatly depending upon how the sample was annealed after sintering. The role of oxygen was not well understood in 1987, and so many combinations of sintering temperature and annealing temperature were tried. Where partial superconductivity was present it was attributed to forming mixtures of the proper phase ($Y_1Ba_2Cu_3O_{7-\delta}$) with assorted other phases.

With all this variability, it is little wonder that there were many conflicting reports of much higher T_c measurements, some nearing room temperature. The poor reproducibility of many of these results was eventually attributed to the changing phases of the compounds, associated with a loss of oxygen atoms.

Another major surprise was the anomalous behavior observed when the HTSCs were placed in a magnetic field. Instead of T_c merely shifting downward as H increased, as per the familiar case of LTSCs, the transition broadened, as shown[12] in Figure 7.3. The transition width ΔT_c was about 3 K in a 1 T field, and nearly 15 K in 9 T. Eventually, this condition was attributed to motion of magnetic flux lines within YBCO, known as *giant flux creep*.[13] This was a little-known phenomenon from LTSC that was greatly magnified in the HTSC case. However, in early 1987, the difference between these observations and what was expected fueled additional speculation that there was an entirely new mechanism of superconductivity at work.

The tremendous advances of 1987 were almost totally empirical, with theorists struggling to sort out the better speculations. A number of exotic theories were proposed to account for the data; but in many cases the data were inferior because of poor sample quality. Again, this problem was corrected eventually, but not soon enough to avert a wholesale trashing of the BCS theory. Remember, the BCS theory had been "proved wrong"—T_c was above 30 K—and theorists, sharing the enthusiasm of experimenters, wanted to be the first to come up with the correct new theory.

As mentioned in Chapter 2, the phase space for superconductivity can be represented by a surface in three dimensions: the THJ surface, for temperature, magnetic field, and critical

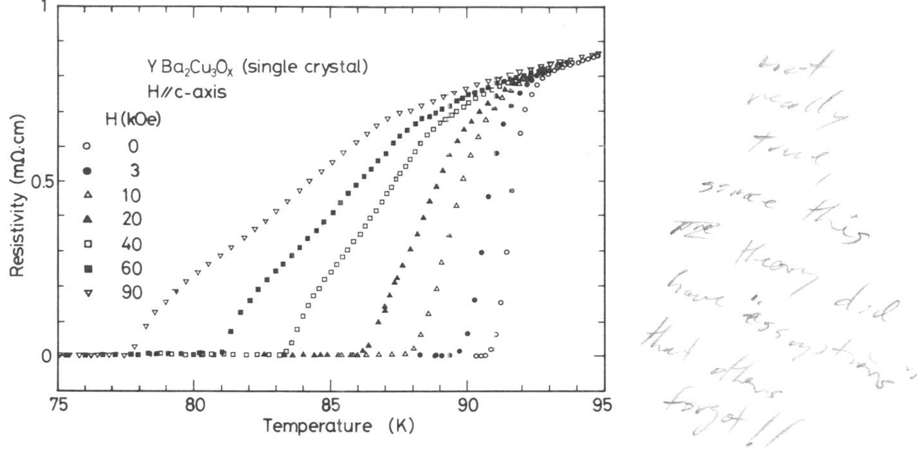

Figure 7.3. Variation of superconducting transition width in a magnetic field. (From Ref. 12.)

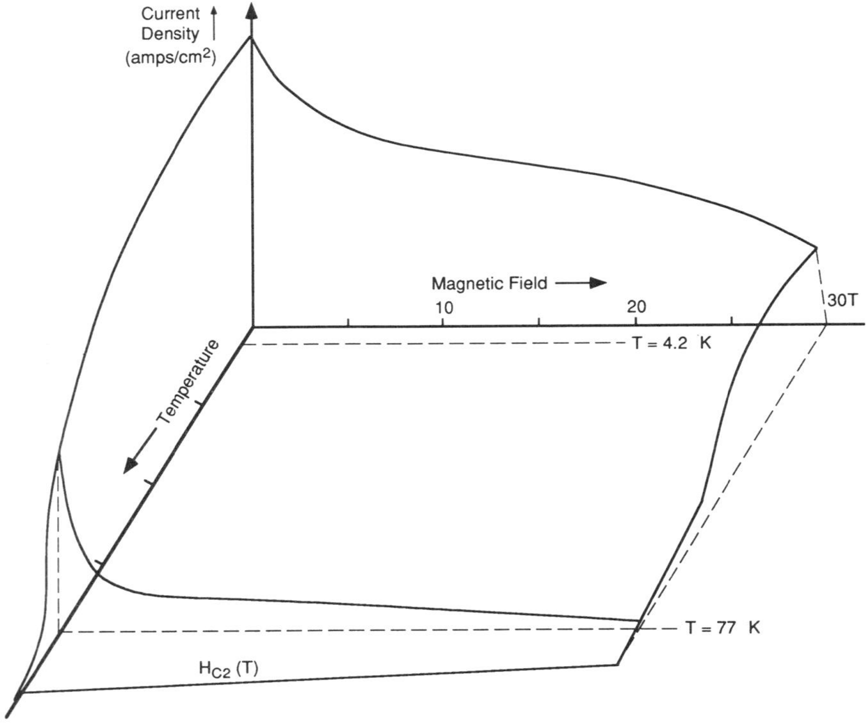

Figure 7.4. Temperature–field–current plot for a typical HTSC.

current density. This has a convex shape for ordinary LTSCs, allowing all three to be fairly large at the same time, as shown in Figure 2.4. In HTSCs, however, the THJ surface was shaped more like Figure 7.4. There was a very steep decline in J_c for even a small applied magnetic field, a condition that was soon attributed to contamination at the grain boundaries in the sintered samples of YBCO. Because high currents and high magnetic fields generally go together, this was evidently a severe weakness of the new superconductors. Therefore, very early on, it became an engineering goal to improve the grain structure of HTSC samples in order to maintain high J_c even in high fields. This goal would prove to be far more elusive than anyone imagined in 1987.

Meanwhile, regardless of the mechanism or details of HTSC, the fact that it existed at all beckoned scientists to consider the possible applications of this amazing new technology. The value was obvious at once: 77 K is much more accessible than 4.2 K. Liquid helium is hard to handle, and requires careful training of a physics graduate student to avoid freezing the pipes shut during transfer from one container to another. On the other hand, anybody can pour liquid nitrogen.

The U.S. Department of Energy (DoE) and the Electric Power Research Institute (EPRI) jointly sponsored an effort by researchers from eight National Laboratories to identify the best applications of HTSC and calculate their probable energy savings. The first report of that group,[14] printed in 1987, included both technical and economic analyses, and greatly increased the level of interest by industry, which in turn kept the level of public and congressional[15] interest high. Subsequently, more detailed technical analyses for certain specific applications were carried out. That report[16] explained the operation and energy-saving potential of a number of devices, and is still a useful reference today.

However, even in the early days there was a sober recognition that there were major obstacles, and the research would be long term.[17] Others[18] cautioned that *low*-temperature superconductivity (LTSC) would continue to dominate the applications for many years to come.

7.3. HYPE

Because a room-temperature superconductor was considered one of the holy grails of physics research, it is little wonder that the level of action in the early days of HTSC resembled an Indiana Jones movie. The excitement among scientists naturally spilled over into the public forum.

7.3.1. Scientific Speculation

To the experienced researcher, the jumble of incoherent results about HTSCs was a normal step in the initial phases of any new branch of materials science. However, an unusually large amount of media coverage brought into the public eye the kind of stumbling that is normally buried in conferences of technical specialists.[19] Because some preliminary (nonrepeatable) indications of superconductivity were seen at 240 K, there was optimism that a room-temperature superconductor was just over the horizon, and the furor intensified.

The *New York Times, Wall Street Journal*, and many others got on the bandwagon, as did countless magazines (*Time* and *Newsweek* come to mind). "Our lives have changed!" screamed the headlines. The journalists were seldom able to do more than quote the scientists, and the quotes frequently appeared garbled, because of a confusion between optimistic *wishes* and actual *accomplishments*. Today, many spokesmen for HTSC don't like to be reminded of their early quotations. All of us *wanted* it to be true so much that we often responded to questions in an ambiguous way. Journalists, trying to sell a few more copies, naturally chose to print the more flamboyant interpretation.

Also at play was scientific ignorance on the part of the scientists, many of whom were operating with one foot on firm ground and the other in unfamiliar territory. The HTSC field is truly interdisciplinary, and experts in one field fell victim to their own misunderstanding of other fields. Here is one typical example[20] from a 1987 report:

> What is important in all this theory is that there definitely is some other mechanism for superconductivity besides phonon-coupled electron pairs. At the moment, this new mechanism appears only in certain ceramic copper oxide compounds. It is entirely plausible that when our understanding of the phenomenon improves, it will be possible to make other superconducting compounds that are free of the difficulties associated with ceramics. Because of this, we can optimistically say that the age of "new" superconductors has arrived.

There are no lies in this; just enthusiasm coupled with a shortage of knowledge. This same condition prevailed for several years in both the public forum and in scientific circles.

7.3.2. Press Exuberance

Businessmen who are trying to raise venture capital have a very difficult line to walk, because investors will hold back if they learn of every consecutive obstacle along the way.

Still, it is tempting to let media errors go uncorrected when it results in an optimistic forecast. In fairness to the press, some of the confusion probably arose because technical entrepreneurs benefited by allowing ambiguity to grow.

It is instructive to examine the pathway by which *wishes* about products were converted into implied *promises*; two examples are enough to typify the pattern.

CPS Superconductor Corporation, a subsidiary of Ceramics Process System Corporation, was headed by Peter Loconto. A July 1988 report[21] said:

> CPS Superconductor has succeeded in creating HTSC wire of unlimited lengths, produced at the extremely high speed of 10 feet per minute.
>
> [They] will soon be able to ship HTSC wire "in high volume, produced at rapid rates." That would make it the first American company to ship such wire in quantity.
>
> Loconto says that current density is substantially higher than has been previously reported by other companies.... [They] have been designated to receive a DARPA award ... to construct a 75 KW electric motor.
>
> Loconto says that he is constrained by SEC regulation from releasing more detailed information about the company's wire advances until a venture capital financing, now under way, is completed.

In this 400-word article, it was never mentioned that the "wire" was the green phase of YBCO, which is not superconducting. It must be formed into its final shape and then sintered in order to convert it to the superconducting state, after which it is extremely brittle. In January 1989, Loconto told me that the reporters changed his "YBCO" to "HTSC," and that his company never claimed to have wire in production quantities. At that time, short lengths of CPS's sintered superconducting YBCO-123 wire carried only 900 A/cm^2. CPS had a big inventory of green-state YBCO "thread."

Perhaps this was all explained after the venture capital financing was completed.

By the end of 1990, Sumitomo Electric Industries had built a small coil of BSCCO, which operated at 4.2 K. This was a finite step along the path of research. Their biggest real accomplishment was to make multifilament wire that could be wrapped into a coil shape without breaking. However, the coverage in *Nikkei Superconductors* stressed that it was a "one tesla" coil. But scrutiny of Sumitomo's figures revealed that the magnetic field produced at the center axis of their "0.9 T coil" was actually 0.5120 T, and that the maximum field observed anywhere was 0.8950 T, just inside the inner radius.

The *Nikkei Superconductors* article went on to quote a "magnet researcher" who compared this performance with NbTi magnets:

> Although the claimed values of I_c and J_c of the Bi system wire are about 1/5 to 1/2 those of commercially available NbTi wires, values 4 to 5 times as high as the announced values have been reported for shorter wires. Improvements in long wire production technology will soon increase I_c and J_c values.

Certain phrases here demand attention: "for shorter wires" and "will soon increase." Experience suggests that such accomplishments do not come forth trivially, but only after much hard work. The actual achievement by Sumitomo was finite; but it certainly didn't merit the next assertion: "This clearly calls for immediate development of appropriate wire winding technologies."

This sort of ebullience kept public (and government) enthusiasm high from 1987 to 1990, but in the long run serious scientists and engineers recognized the very difficult obstacles that needed to be overcome. The hypemeisters found themselves backpedaling with, "Well, we didn't *exactly* say that . . .," and soon they lost credibility.

7.3.3. Entertainment

Sometimes the hype was all in fun. At the American Physical Society meeting in both March 1988 and March 1989, some people published (anonymously) a takeoff on the *National Enquirer* called the *SuperCONducting Enquirer*, with the emphasis on the "CON." Because the typeface was incredibly tiny, it carried the logo "Sponsored in part by the American Association of Optometrists." Advertisements included a scientific-paper-writing service, which charged $2000 to write a *Physical Review Letter* article based on a HTSC experiment, $4000 to write up a *correct* experiment, and $35 to write a theoretical paper.

Headlines read, "Elvis discovered 1-2-3 in 1963, was sworn to secrecy by space aliens." It contained a collection of brief press-release-sounding announcements, such as that Michael Jackson had bought the rights to commercialize Y3CO from IBM. My own favorite was a parody on the way less brittle materials were being developed, albeit at the cost of a lower transition temperature. It seems that by substituting niobium into YBCO, they had been able to make a far more flexible wire. The new compound had the formula $Nb_1Nb_2Nb_3Nb_7$, and it's only deficiency was that $T_c \approx 9$ K, "but we're working on that."

To my knowledge, no one ever took credit for this excellent spoof. It was very helpful in enabling researchers to relax and remember that careful research eventually wins out over press-release science.

The hype must never be allowed to detract from the fact that in HTSC, we are definitely looking at a major breakthrough. The clarity with which some applications of HTSC can be recognized is a clear light at the end of a very long tunnel that sustains HTSC research even to the present time.

7.4. REAL PROGRESS

By 1989, the HTSC industry had settled down into the hard work phase. Samples of good quality were universally available, so the papers appearing in scientific journals became more reliable. The exotic theories declined in popularity and the BCS theory regained prominence, with modifications to accommodate the unique character of the copper oxide layered superconductors. Such concepts as *flux lattice melting*[22] and *weak links* were introduced and shown to be relevant. In the chapters ahead, we will examine those various steps forward in a systematic way.

What everybody wanted, of course, was wire so they could build devices to implement the promise of superconductivity. By far the biggest disappointment associated with HTSC has been the difficulty of making wire. Certainly there has been real progress over the years. In 1990, it was possible to write,[23] "There is no *wire* available today made of HTSC. By *wire* we mean something flexible that can carry at least 10,000 A/cm^2." Progress on YBCO came to a standstill, but BSCCO did better. Figure 7.5 is a chart[24] prepared by Sumitomo Electric Industries showing the improvement in J_c over time of their BSCCO wire. In more recent years, several different manufacturers have made BSCCO wire in lengths > 100 m.

The most encouraging characteristic of charts such as Sumitomo's is that BSCCO appears likely to find a niche at temperatures below 30 K. The upper limit of NbTi coils is usually around 7 T. Every indication so far is that the bismuth compounds will retain superconductivity out beyond 20 or 30 T when cooled to 4.2 K. Thus, for high-field applications, it seems probable that BSCCO will find a home.

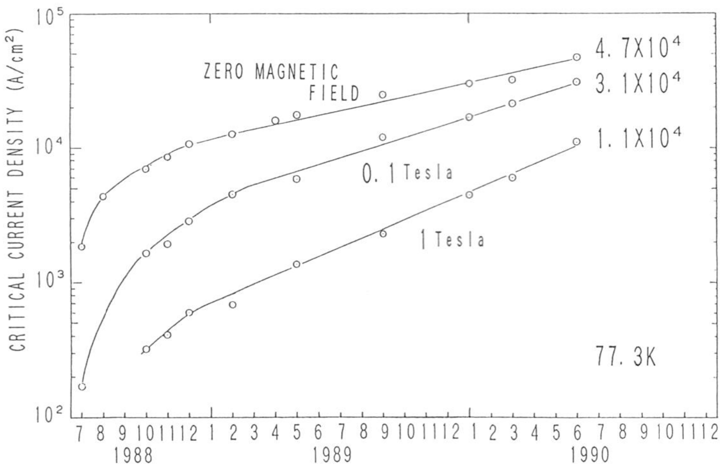

Figure 7.5. Improvement in J_c over time. (Courtesy Sumitomo Electric Industries, Ltd.)

On the other hand, BSCCO appears unsuitable for 77 K applications requiring large magnetic fields (e.g., generators). It is necessary to achieve strong pinning at 77 K, and YBCO or TBCCO appear more likely to achieve that goal. Unless a means can be found to increase J_c at 77 K, high field applications (such as motors) will not be feasible using BSCCO. However, BSSCO might be suitable for 77 K applications that only produce small magnetic fields, such as electric power transmission.

Thallium compounds (especially TBCCO-1223) are the subject of intense study now that TBCCO starting powder is more readily available compared to 1990. Still, there is no guarantee that acceptable wire can ever be made; the mechanical properties of TBCCO are much poorer than BSCCO.

The mercury compounds are still in their infancy because of the great difficult of making the material in the first place. The susceptibility data shown[25] in Figure 7.6 is reminiscent of the early data for previous copper oxide superconductors. Compounds with *two* copper oxide layers (Hg–Ba–Ca–Cu–O) have T_c up to 135 K.[26] Furthermore, it has been shown that under very high pressure,[27,28] the mercury-based compounds with three copper oxide planes have $T_c \approx 155$ K. The next question is whether substituting another element such as strontium for barium can result in an equally high T_c value at normal pressure. This is an exact analogy of the situation during 1987.

Questions about the practical applications of the HTSCs have to do with the temperature range near 20–30 K. It is thermodynamically much easier to reach such temperatures than to reach 4.2 K; the cost of refrigeration is considerably smaller, and therefore the economic advantage is worth pursuing. But what about the economic penalties of working with these materials? Can inexpensive wire ever be made? We don't know yet.

One "sure thing" application of HTSCs is as a transition wire for conventional LTSC magnets. In their present configuration, LTSC magnets are fed by copper wires carrying perhaps 2000 A. This results in a typical heat load into the helium bath of about 4 W, which may cause half the total helium loss in a well-designed system.

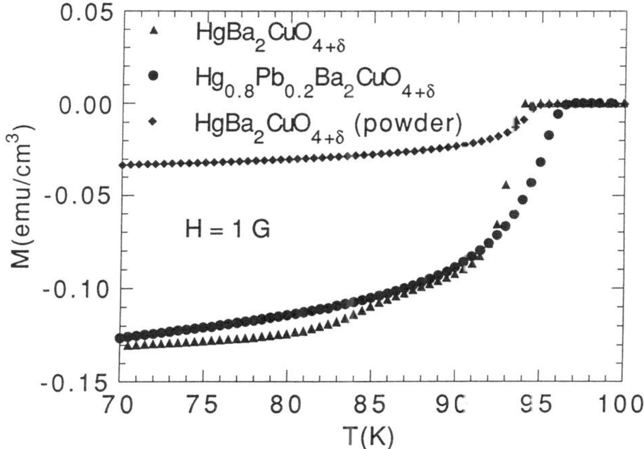

Figure 7.6. Susceptibility of HgBaCuO-related compounds. (From Ref. 25.)

It is possible to replace part of the copper leads (the portion that runs from the helium bath out to the collar on the dewar where the temperature is 77 K) with leads made of YBCO or other HTSC materials. For very modest parameters (2000 A implies J_c only 350 A/cm^2), the much lower thermal conductivity of the HTSCs compared to copper results in a calculated heat load of only 0.7 W to the helium bath. This load includes the ohmic contacts at the junction to the NbTi wire.[29] Argonne engineers first demonstrated[30] the performance of such current leads in 1989.

7.5. GOVERNMENT'S ROLE

One calm voice in all the excitement of 1987 was that of the Electric Power Research Institute, which took a long-range view[31]:

News of the discovery of HTSCs has been greeted with a mixture of caution and excitement: caution born of long experience with complex technologies, including superconductivity; excitement stemming from the potential that better superconducting materials might have for fundamentally changing electric power systems. Although the technical challenge is an engineer's delight, the succession of unsubstantiated or difficult-to-reproduce claims is a source of concerned skepticism for planners.

. . . "We're prepared to roll the dice in this area," says Richard Balzhiser, EPRI's executive vice-president for R&D. "All research involves risk, but I believe the potential importance of HTSCs to the electric power industry certainly justifies committing significant resources, even at this early stage in the game. We have to proceed cautiously, however, beginning with a realistic evaluation of the potentials and seeking partners among other funding agencies. The payoff will not come quickly, but it may be quite substantial."

The key phrase here is "seeking partners among other funding agencies." Because EPRI believed in the promise of HTSC, so did a presidential commission, the Congress, Defense Advanced Research Projects Agency (DARPA), and the Department of Energy (DoE). From the outset, HTSC research contained a high degree of private-sector participation.

Two special circumstances boosted progress in HTSC research. First, during the 1980s, Congress had passed a series of laws trying to enhance the transfer of technology outward from the many federal research laboratories. The *Omnibus Trade and Competitiveness Act* of 1988 officially established the government as a major player in HTSC research. Joint public–private cooperative agreements soon blossomed, and applied research in HTSC got underway in the very first example of a new philosophy of partnership in government-sponsored research. Superconductivity Pilot Centers were established at three National Laboratories, with the explicit charge of forming agreements with industry to develop commercial applications of superconductivity.

The second favorable occurrence was that during the 1980s the federal government took several major oil companies to court for overcharging people for gasoline during the 1970s. The settlement of that issue led to sizable fines (revenues) being forwarded to each state to be used for energy conservation purposes. HTSC burst onto the scene at exactly the right moment, and in several states a major share of that windfall was spent on superconductivity research. New York, Texas and Illinois—three of the most populous states—all established major HTSC research centers.

However, in order to team with a national lab on superconductivity research, a partner was required to put up an equal share of the cost. Normally, such money would be hard to find. Fortunately, monetary support by EPRI, as well as the supporting state revenues from the oil overcharges, were allowed to count toward those matching funds. The effect of this was to create a viable community of researchers in this new HTSC field, spanning universities, federal research labs and industry. The Superconductivity Pilot Centers took the initiative to establish working relationships with over 60 different companies in the following four years.

From the outset, government sponsorship of HTSC research was co-ordinated, but there was some inevitable overlap; alongside DoE and NSF (National Science Foundation), DARPA sponsored a broad program in YBCO technology. By 1990, The National Commission on Superconductivity decided[32] that DoE should take the lead on electric power applications, and DARPA should take the lead on electronic applications. This gave DoE the "bulk" franchise, and DARPA the "thin film" territory. Thereafter, the contracts originating from each agency reflected that specialization of interests. This book is an outcome of the electric-power side of that split.

The compendium *Federal Research Programs in Superconductivity* periodically summarizes the levels of spending by the several agencies of the U.S. government. Their December 1992 report[33] reveals that the Departments of Energy and Defense are the leading funders of both low- and high-temperature superconductivity research, with NSF in third place. The total federal commitment for FY 1992 accumulated to $\approx$ 362 million dollars ($M362); but that includes $M116 for procurement of magnets for the Superconducting Super Collider (SSC). Therefore, the R&D total is about $M246. Figure 7.7 shows the spending trend over recent years. The gentle peak in the LTSC budget in 1991 was because SSC-related research at DoE peaked and then declined as the SSC went into procurement of its magnets. Meanwhile, total HTSC funding has grown steadily[34] since its inception in 1987; for FY 1993 it exceeded $M150. What is not shown in this cumulative graph is the continually evolving breakdown between categories of basic research, technology development, and prototype demonstration. Also, the level of funds-matching by industry does not appear here.

The Superconductivity Technology Program within DoE is aimed explicitly at applications for utilities and stresses the formation of partnerships between industry and National Labs to bring research results to commercial fruition.[35] Their view of the progression of a

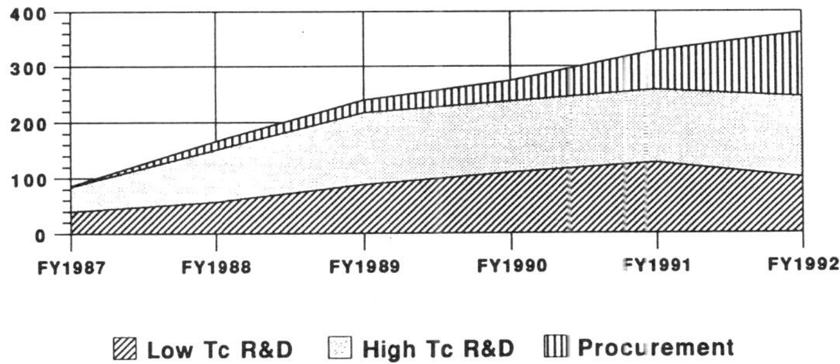

Figure 7.7. Total federal spending on superconductivity ($ million).

technology is captured in Figure 7.8, which shows how spending proceeds as an innovation moves from the laboratory bench to the marketplace. With DoE sponsorship in the range of $M22 annually, and industry matching part of that, this program is America's largest comprehensive effort to develop new HTSC applications.

It would be terribly myopic to suggest that American research stands by itself in the HTSC field, but space does not allow a discussion of international programs. As one example, it is noteworthy that in Japan, sponsorship by MITI more than doubled since 1988, and since 1991 the support increased about 20% (to $M144), although nearly all of that increase came in 1992. Because of different accounting methods (Japanese figures are exclusive of research salaries), the two countries' programs are actually about equal.[36] Moreover, in America government provides about 2/3 of the total support of superconductivity, whereas in Japan the industrial contribution is larger and government provides only 1/3.

A more significant comparison is to ask how the money is spent. In Japan, the bulk of the funds go toward precommercialization research and product development efforts, whereas in America the emphasis is on generic research.[36] Moreover, American R&D had been heavily defense oriented in the past, but is now shifting toward commercial applications. It will be interesting to watch how this change affects HTSC support. Clearly,

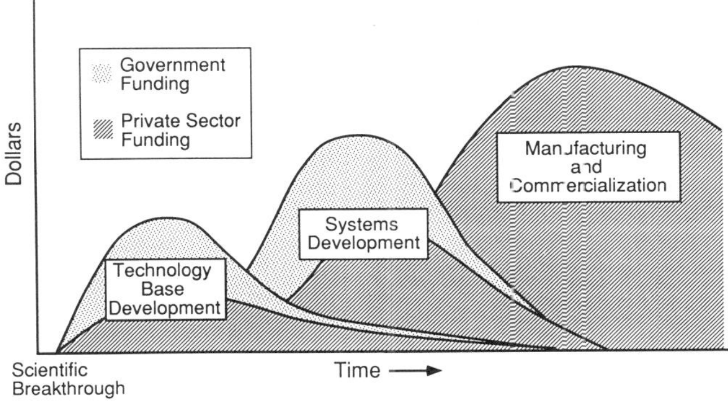

Figure 7.8. Conceptual drawing of how development proceeds, with funding shifting from government to private sources and emphasis changing in the direction of manufacturing and commercialization.

international competition and cooperation will combine to keep the research effort strong throughout the world.

7.6. DEVELOPMENT OF AN INDUSTRY

In a relatively short time, a viable HTSC industry has come into existence. Plenty of companies are actively involved in making thin-film devices, but here we deal only with the "wire" side of the industry. Company size ranges from major well-established firms (e.g., Siemens, Sumitomo Electric Industries) through existing LTSC wire-makers who wish to participate in the HTSC market (e.g., Intermagnetics General) to small start-up firms dedicated exclusively to making HTSC wire (e.g., American Superconductor). Competitors of all sizes see enormous potential profits if they can make practical wire in quantities and at a manageable cost. There are many more companies waiting for someone to sell them wire, who will then build electric motors, energy storage units, and other applications of HTSC technology.

It is instructive to look at one start-up firm as a paradigm for how this entire industry is growing. American Superconductor Corporation (ASC) came into existence in April 1987 with the intent of commercializing the brand new opportunities in HTSC. MIT Professor Greg Yurek and his associates began with their only real asset being some promising patents (licensed from MIT) on processes to make wire. They were determined to make "technology transfer" come true. ASC proceeded to raise venture capital, first from Pirelli Cable Co. of Italy, and later by publicly selling stock (NASDAQ: AMSC). Arrangements with Inco Alloys International and Hoescht AG were subsequently added. On the technical side, ASC entered into cooperative research efforts with all three of the Superconductivity Pilot Centers, and thus has leveraged a considerable amount of National Lab effort into their wire-making endeavors. They have chosen to specialize in BSCCO wire, which will likely find application at temperatures < 20 K, for reasons which will become clear in Chapter 16. Another infusion of venture capital ($\approx$ \$28 million) was obtained in early 1994, in order to scale-up their process for efficient production. There is no guarantee that this start-up will ultimately be a success story.

Whether large companies or small, American, European or Japanese, a lot of obstacles and disappointments have been encountered since the optimistic days of 1987. Cleverness, persistence and hard work are the essential qualities of those still active in the wire business.

Certain elements of good business sense have a timeless quality to them. For over 100 years, E. I. duPont de Nemours & Cie. have commercialized technology with a corporate philosophy that may be stated as follows: "You have got to get the *materials* right in the first place, or else the product will never be right downstream." This is a very powerful principle, one that provides excellent guidance to commercial-minded researchers in the HTSC field. Alan Lauder of duPont has pointed out that the technical details really constitute the "easy" challenges, whereas the "hard" challenges are those of integration. He asks, How do we get the necessary close cooperation between materials engineers and the system designers? There are lots of different possibilities, and no single approach will be universally applicable. It will be necessary to follow many different avenues. No company can do it all, and *no nation* can do it all either. With limited resources, we have to leverage them by forming partnerships and collaborations in a large cooperative program.

In Japan, the legendary MITI helped several companies form ISTEC, a cooperative R&D consortium focused on HTSC commercialization. The European Community formed a looser team of corporations interested in HTSC. In America, a trade association was formed to coordinate and promote HTSC. Named the Council on Superconductivity and American Competitiveness (CSAC), it helped to pass legislation enabling greater cooperation between National Labs and American industry. After several years without commercial products appearing, these groups decided it was in their mutual best interests to enhance communications with each other.

In May 1992, CSAC hosted the first annual International Superconductivity Industrial Summit (ISIS) in Washington, D.C. The following year they met in Japan. The ISIS meetings offer a somewhat unique perspective on where industrial leaders from diverse countries see superconductivity research going. In 1992, the central point that stood out clearly was this: industry is attentive to *low*-temperature superconductivity (LTSC), and is making plans based on that. HTSC is strictly *downstream*, perhaps *futuristic*. Nearly every speaker described plans for specific applications (using niobium-based superconductors) and then added at the close that "high-temperature superconductors might someday be used." Thus, industrial participants in the superconductivity industry are moving ahead without depending on the eventual success of the copper oxide superconductors.

The bar chart of Figure 7.9 was prepared[37] by Sumitomo Electric Industries and uses a creative means of displaying the state of progress in superconductivity applications. They evaluate the degree of success on a scale of 1 to 10, and then draw a bar to denote what's been accomplished as a fraction of what's needed. With adjacent bars for the United States and Japan pertaining to 13 different categories, one can see at a glance the enormous gulf between LTSC and HTSC progress in the several categories. Evidently, HTSC is most promising in the electronics area. (It is noteworthy that Sumitomo considers the Japanese Maglev as only through the "research" stage.) This chart is certainly conservative; in America, we would typically claim that several HTSC areas are well beyond the beginning of the study phase, and we regard the NbTi magnets made for industry by Eriez Magnetics (see Chapter 4) as beyond the "pilot plant" stage. Nevertheless, this kind of chart is an extremely useful way to compare the two country's efforts.

How big a market will superconductivity become? Again, estimates by Sumitomo in Japan are fully plausible: in 1990, the market depended on liquid helium at 4.2 K and was worth $0.4 billion/year; by 2000, it should be liquid hydrogen at 20 K, worth $3.9 billion/year; and by 2010, liquid nitrogen at 77 K, with a market value of $7.8 billion/year. If conditions of no refrigeration at all ever prevail, the market would be worth $78 billion.

The market size for superconductivity is a matter of considerable debate. Actually, LTSC has made relatively little penetration into copper systems, and it is fair to wonder how HTSCs can expect to do better. New technology is seldom cheap, and it must pay for itself promptly or it will not be used. For major investments, capital leveraging plays a very central role today. On one hand, utilities are normally willing to front the money necessary for a company to buy and install a piece of energy-saving equipment, and this has changed the economic picture dramatically. On the other hand, when one tries to sell to a utility, that very conservative industry will resist change until the device is installed at *no risk* and *no cost*. Thus, the innovator takes all the risk and must spread the cost over a longer time frame than normal. No single company is willing to assume that risk until the technology is clearly viable. The result is that corporations, in any country, look to government for support throughout the early commercialization phases of product development.

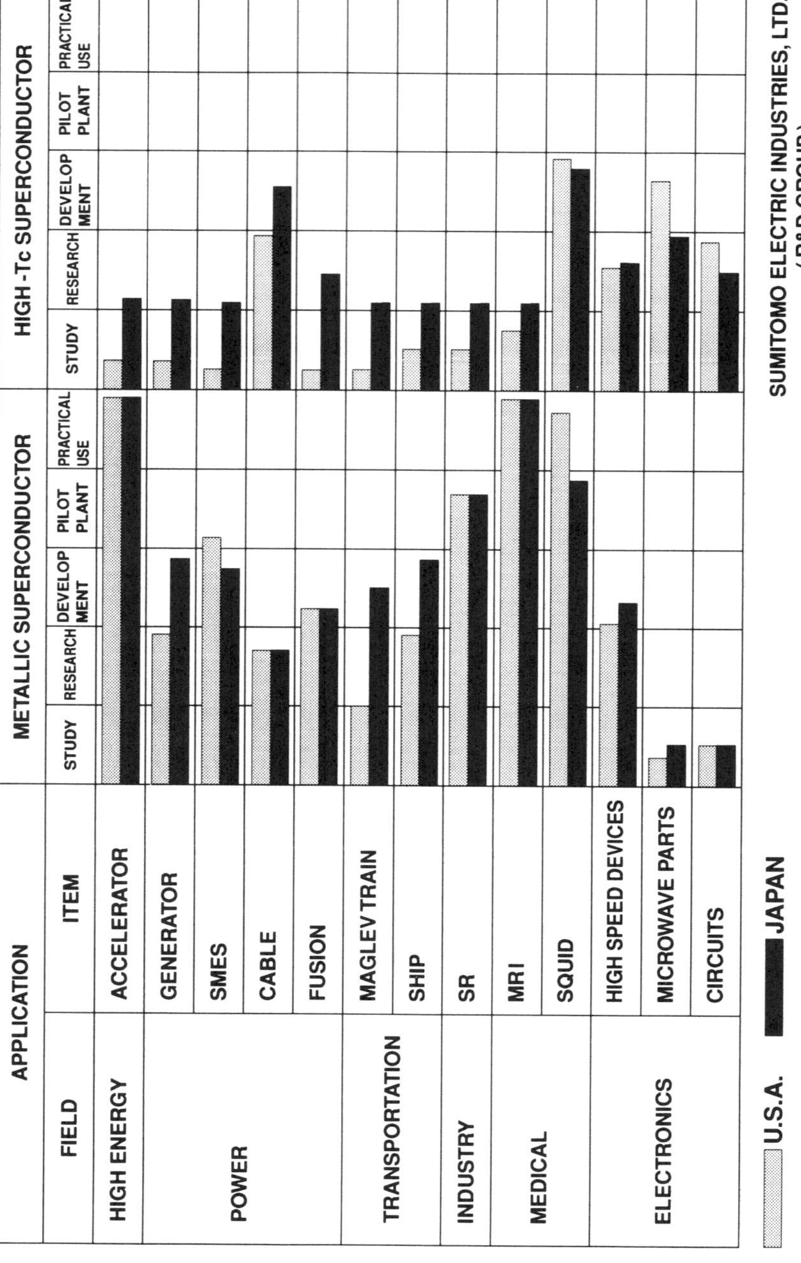

Figure 7.9. Bar chart depicting progress in superconducting applications, as of late 1993. (Courtesy of Sumitomo Electric Industries, Ltd.)

7.7. SUMMARY

This chapter has presented a view of the evolution of HTSC from a research idea into a commercial industry. The transition from the early days of furious try-anything research to the steady growth of a new industry has taken only a few years. The speed of modern communications has certainly helped to accelerate the pace of development, but so has the new climate of government–industry cooperation. It is hoped that the HTSC research effort will someday be remembered as an outstanding example of the right way to do technology transfer.

The research path for HTSC has proven more difficult than the public expected in 1987, and the early optimism has been replaced with a sober commitment to continued R&D, usually in a partnership mode. With the spotlight off, scientists and engineers are working together to identify and solve critical problems. It is clear that progress has been made over the first six years—substantially more than realistic researchers had expected.

In July 1987, President Reagan gave a speech[38] to the assembled community of HTSC workers, in which he characterized the three stages of development as: "(1) It won't work. (2) You'll never make a nickel on it. (3) I told you it was a great idea all along." The HTSC field is presently somewhere between (1) and (2). There is absolutely no guarantee that (3) will ever happen. But people are not giving up on HTSC, despite the setbacks. The potential applications are too valuable a prize to abandon in mid-stream.

REFERENCES

1.   J. G. Bednorz and K. A. Muller, *Z. Physik B* **64**, 189 (1986).
2.   J. G. Bednorz and K. A. Muller, *Rev. Mod. Phys.* **60**, 585 (1988).
3.   J. Bardeen, L. N. Cooper, and J. R. Schreiffer, *Phys. Rev.* **108**, 1175 (1957).
4.   W. L. McMillan, *Phys. Rev.* **167**, 331 (1968).
5.   G. M. Eliashberg, *Soviet Phys. JETP* **11**, 696 (1960); **12**, 1000 (1961).
6.   P. B. Allen and R. C. Dynes, *Phys. Rev. B* **12**, 905 (1975).
7.   J. G. Bednorz, K. A. Muller, and M. Takashige, *Science* **236**, 73 (1987).
8.   C. W. Chu *et al., Science* **235**, 567 (1987); *Phys. Rev. Lett.* **58**, 405 (1987).
9.   M. K. Wu *et al., Phys. Rev. Lett.* **58**, 908 (1987).
10.   P. H. Hor *et al., Phys. Rev. Lett.* **58**, 911 (1987).
11.   R. N. Bhattacharya, P. A. Parilla, and R. D. Blaugher, *Physica C* **211**, 475 (1993).
12.   Y. Iye *et al., Japanese J. Appl. Phys.* **26**, part 2, L1057 (1987).
13.   Y. Yeshuran and A. P. Malozemoff, *Phys. Rev. Lett.* **60**, 2202 (1983).
14.   A. M. Wolsky *et al.,* "Advances in Applied Superconductivity: Goals and Impacts: A Preliminary Evaluation," Argonne report to DoE (Sept. 25, 1987).
15.   U.S. Congress, Office of Technology Assessment, *High-Temperature Superconductivity in Perspective,* OTA-E-440 (Washington D.C.: U.S. Government Printing Office, April 1990).
16.   S. J. Dale *et al.,* "Energy Applications of High-Temperature Superconductors, Vol I: Extended Summary Report," EPRI report #ER-6682 (February 1990).
17.   A. M. Wolsky, R. F. Giese, and E. J. Daniels, "The New Superconductors: Prospects for Applications," *Sci. Am.* **260** (2), 60–69 (February 1989).
18.   S. Foner and T. P. Orlando, *Tech. Rev.* **91** (2), 36 (1988).
19.   A. L. Robinson, "A Superconductivity Happening," reported in *Science* **235**, 1571 (1987).
20.   T. P. Sheahen, "Industrial Superconductivity," Report to Dept. of Energy Office of Industrial Programs (October 1987).
21.   *Cambridge Report on Superconductivity* (July 1988).
22.   D. Nelson and S. Seung, *Phys. Rev. B* **39**, 9153 (1989).
23.   Argonne monthly report to EPRI on Superconductivity Progress (February 1990).
24.   K. Sato *et al., IEEE Trans. Magn.* **MAG-27**, 1231 (1991).

25. U. Welp *et al., App. Phys. Lett.* **63**, 693 (1993).
26. A. Schilling *et al., Nature* **363**, 56 (1993); L. Gao *et al., Physica C* **213**, 261 (1993).
27. C. W. Chu *et al., Nature* **365**, 323 (1993).
28. M. Nunez-Regueiro *et al., Science* **262**, 97 (1993).
29. F. J. Mumford, *Cryogenics* **29**, 206 (1989).
30. J. L. Wu *et al., IEEE Trans. Magn.* **MAG-27**, 1861 (1991).
31. "Pursuing the Promise of Superconductivity," *EPRI J.* **12** (6), 4–15 (1987).
32. "Report of the National Commission on Superconductivity," issued August 7, 1990.
33. "Federal Research Programs in Superconductivity," compiled by COMAT Working Group on Superconductivity, FCCSET Report (December 1992).
34. "Federal Research Programs in Superconductivity," compiled by COMAT Subcommittee on Superconductivity, FCCSET Report (March 1989).
35. J. G. Daley and T. P. Sheahen, *Proc. Amer. Power Conf.*, Chicago (1992).
36. K. Ott, "Surveying the Budget Landscape," in *Superconductor Industry* (Washington, D.C.: CSAC, Spring 1993), pp. 7–9.
37. T. Nakahara, Sumitomo Electric Industries, presentation at First International Superconductivity Industry Summit (ISIS) conference (May 1992).
38. R. W. Reagan, presentation at Federal Conference on Commercial Applications of Superconductivity, Washington D.C., July 28–29, 1987.

II

HIGH-TEMPERATURE SUPERCONDUCTIVITY (HTSC) BASIC PROPERTIES

8

Structure

An important fraction of the early research on HTSCs was devoted to elucidating their structure. Knowledge of structure (the location of atoms) is a prerequisite for understanding the properties of the new superconductors. In this chapter, we explain the terms most often used when describing the new superconductors and then go on to discuss their structure. We also describe the relation between structure and properties.

As a preliminary, note that there are essentially *no* important structural effects in LTSC type I superconductors. There the coherence length ξ is much longer[1] than the penetration depth λ, and the range of paired electrons may span many grains within a bulk sample. When a huge single crystal (typical dimensions of centimeters) is grown, the superconducting properties are indistinguishable from polycrystalline material of the same element.[2,3] Moreover, LTSC type II superconductors are only weakly influenced by structure. Here, the coherence length is short compared to the penetration depth, and the material is a mixture of regions of normal and superconducting phases. The critical current J_c of certain superconducting alloys[4] can be enhanced by cold-working the material, thus introducing defects into the lattice, which in turn serve as flux-pinning sites and allow more current to pass, all without making any significant change in T_c.

Macroscopic samples of materials with cubic unit cells are expected to have isotropic properties—both normal properties (e.g., dielectric strength and electrical resistance) and superconducting properties (e.g., λ and ξ) are expected to be the same in every direction. By contrast, in HTSCs the structure is a major determinant of both the mechanical and electrical properties, especially superconductivity. The observed anisotropy in the superconducting properties of $Y_1Ba_2Cu_3O_{7-x}$ the bismuth, mercury, and the thallium compounds reflect the structure of their crystals.

The entire field of layered[5] superconductors is a new discipline in which superconductivity is nearly a 2-dimensional phenomenon, taking place in layers within a 3-dimensional crystal. Our purpose in this chapter is to introduce the reader to the essential concepts of structure that make the HTSCs so different and so interesting.

8.1. TERMINOLOGY

A number of new terms are used to describe the structure of crystals.[6,7] The smallest distance, describing the structure, is the *interatomic spacing*. This is the distance from the center (nucleus) of one atom to the center (nucleus) of its nearest neighbor. For most elemental solids, this distance lies between 1 and 2 Å ($1–2 \times 10^{-10}$ m).

137

A *unit cell* is the smallest polyhedron containing the pattern of atoms that (when repeated billions of times) constitutes a perfect single crystal. When a crystal is composed of only one element (e.g., diamond is composed of carbon), the dimensions of the unit cell are on the order of the distances between atoms and their nearest neighbors, the interatomic spacing. However, when the crystal is composed of several elements (e.g., NaCl or $YBa_2Cu_3O_{7-x}$) its unit cell contains at least one of each kind of atom. Thus, the unit cell is bigger than in the mono-atomic case. Trillions of these unit cells repeat to form a very small crystal, known as a *grain*. Millions of grains are usually required to make a macroscopic sample.

When processing methods are well developed, it is possible to make a single very large grain, known as a *single crystal*. The place where two grains come together is naturally called the *grain boundary*. If there is extraneous material that does not belong in the unit cell then, as the grain forms, this extraneous material will tend to collect at the grain boundaries. In superconductors, this debris often forms a very thin insulating barrier between adjacent grains. Some extraneous atoms do get caught within the grain; these are called *interstitials* (see below).

Many familiar materials have a cubic unit cell. For example, in cesium chloride, a single atom of cesium lies at the center of a cube and 8 chlorine atoms are at the corners. This configuration is called *body-centered cubic* and is shown in Figure 8.1(a). The chemical formula is *not* $CsCl_8$, because each chlorine atom participates in 8 other unit cells, so only 1/8 of each chlorine atom counts in forming one unit cell. Another very similar unit cell is the *face-centered cubic* cell, in which atoms lie at the *center* of each face on a cube, as illustrated in Figure 8.1(b). Ordinary table salt, NaCl, has this structure.

Unit cells can form in many shapes other than cubes. For example, the extremely pure element gallium forms a hexagonal unit cell, even though all atoms are chemically identical. Many minerals form unit cells that combine one element with a common substructure of others. For example, potassium silicate (widely used in making glass) finds the silicon enclosed in a tetrahedron of four oxygen atoms, and that tetrahedron bonds to the potassium as a unit. Throughout nature, small polyhedrons are readily identifiable within unit cells. When the unit cell has dimensions of different lengths, it is said to be anisotropic, and the axes of the cell are labeled *a, b,* and *c*. By convention, *c* is the longest axis; *a* is the shortest. In many crystals, *a* and *b* are the same length; in some crystals *b* differs only very slightly in length from *a*. In the latter case, as the unit cell is repeated again and again during crystal growth, it is quite easy for the *a* and *b* axes to become reversed. There is then a slight change in the direction of the orientation. Not enough to be considered an entirely new grain, this phenomenon is called *twinning*.

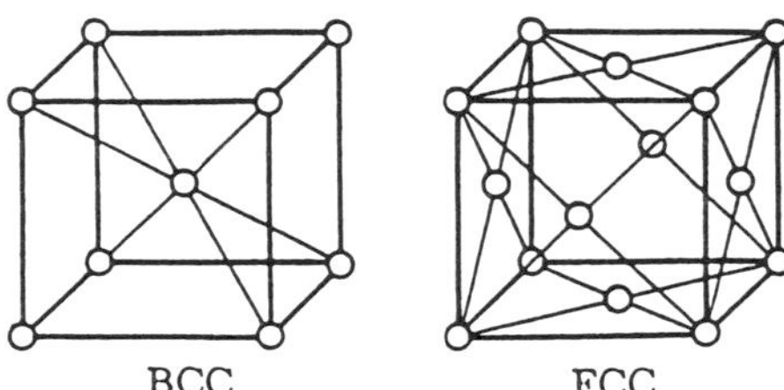

Figure 8.1. Cubic crystal structures: (a) body-centered cubic; (b) face-centered cubic.

Every distinct type of unit cell has its own name. If all the lattice dimensions are equal (i.e., $a = b = c$) and all the corner angles are 90°, the crystal is *cubic*. If one lattice dimension is longer than the others ($a = b$ but $a \neq c$) while the angles are still 90°, the crystal is called *tetragonal*. When all three of a, b, and c are different but the angles are still 90°, the crystal is *orthorhombic*. Although there may be only a small geometric difference between a tetragonal and an orthorhombic crystal of the same material, the crystals may have strikingly different properties. The *tetragonal* crystals of $YBa_2Cu_3O_{7-x}$ are not superconductors, whereas *orthorhombic* crystals of the same material are superconductors.

As noted above, the structure of an ideal or perfect crystal results from the flawless repetition of the crystal's unit cell. In reality, the repetition is not flawless; instead, there are defects. They influence the properties of real crystals. Defects are of two kinds: Chemical defects, in which a lattice site is occupied by the wrong kind of atom, and structural defects, in which atoms are located in unusual places (such atoms are called interstitials) or where a location is not occupied by any atom (such locations are called *vacancies*). Combinations of these types of defect occur; for example, an alien element or impurity can be in an unusual place. Sometimes these impurities are introduced deliberately, in which case their introduction is called *doping*. Often, the purpose of doping is to increase or decrease the number of conduction electrons in the material; this is a particular goal of doping when manufacturing semiconductors.

Another common type of defect is a *dislocation*, in which the repetition of the unit cell is interrupted in a regular way, not by an isolated impurity, interstitial, or vacancy. A dislocation may be a whole line or plane of atoms that is out of joint. Crystals containing dislocations usually have much less strength than crystals without dislocations.

8.2. HTSC CRYSTAL STRUCTURES

The HTSCs are variations of the crystal type known as perovskites.[8] Perovskites are minerals whose chemical formula is ABO_3. They can have several crystal structures, including a simple cubic structure. However, the new superconductors do not have the simple formula ABO_3. Instead they usually have more than one kind of A-atom. It is conducive to understanding these structures to examine models (or at least pictures) of the various crystals.[9]

8.2.1. The Perovskite Cubic Superconductor $Ba_{1-x}K_xBiO_3$

The simplest HTSC is $Ba_xK_{1-x}BiO_3$ ($T_c = 30$ K) in which barium (Ba) and potassium (K) are both the "A-atoms." It is *not* a copper oxide. This superconductor has a cubic crystal structure; Figure 8.2 shows that structure.[9] The bismuth atom is at the center of a cube, while either barium or potassium atoms occupy the corners; on each face is an oxygen atom. Each of the six oxygens are shared by two unit cells; each of the eight bariums (or its substitute potassium) are shared by eight unit cells.

As in most of the other HTSCs, the value of the subscript x is crucial. A condition for superconductivity to occur here is that $x > 0.25$. Chapter 10 is devoted to explaining the ramifications of various amounts of doping, sometimes with several different elements. It should be noted that which site the doping takes place on also matters. In Ba-K-Bi-O, the substitution takes place on the A site of the perovskite unit cell; the same holds true for several other HTSC compounds.

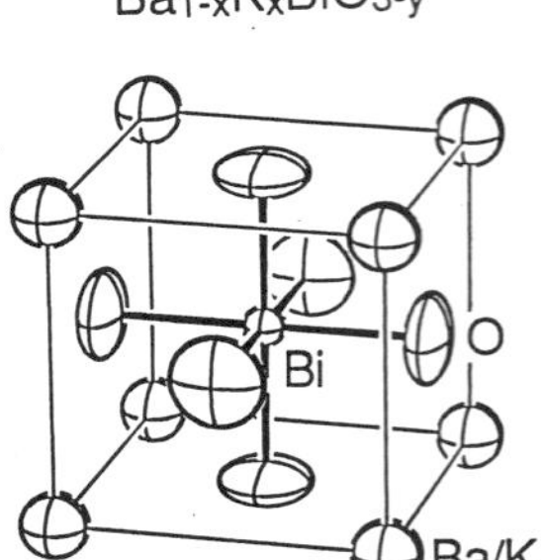

Figure 8.2. Structure of the cubic $(Ba_{1-x}K_x)BiO_{3-y}$.

Barium–potassium–bismuth–oxide is not anisotropic, and therefore it is simpler to study than most HTSCs. In particular, it cannot be a *layered* superconductor. By measuring a number of the superconducting properties of this compound, investigators have found that the BCS theory is still valid for this oxide superconductor. This lends credibility to the assertion that the BCS theory, suitably modified for anisotropy, will also explain the other HTSCs. For exactly this point, Ba-K-Bi-O has played an important role in the relatively brief but fast-moving history of these materials.

8.2.2. The Perovskite "Layered" Compound $La_{2-y}Sr_yCuO_4$

The original HTSC discovered by Bednorz & Muller[10] was lanthanum barium copper oxide, but because of higher T_c values there is greater interest in its close relative, lanthanum strontium copper oxide. Figure 8.3 shows $La_{2-y}Sr_yCuO_4$ and permits the introduction of a few more important terms. This compound is often termed the 2-1-4 structure, because it has 2 lanthanums, 1 copper, and 4 oxygens. (Some of the lanthanum is replaced by strontium.) Upon examining the unit cell, several features can be detected. First, Figure 8.3 reveals that the basic 2-1-4 structure is doubled to form a unit cell. Therefore a more proper label might be 4-2-8. The reason for this doubling is that every other CuO_2 plane is offset by one-half a

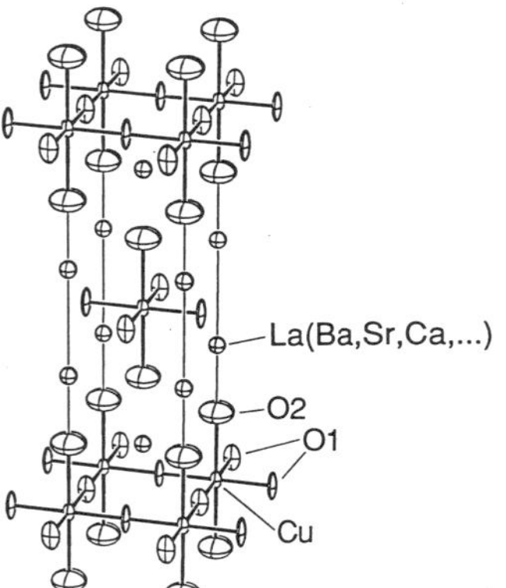

Figure 8.3. Structure of the single layered $La_{2-x}Sr_xCuO_4$ compound. The unit cell here is doubled, because every other CuO_2 plane is offset by half a lattice constant. The designation O1 and O2 denote oxygens associated with the copper and lanthanum atoms, respectively.

lattice constant, so that the unit cell would not be truly repetitive if we stopped counting after one cycle of the atoms.

Close scrutiny of Figure 8.3 reveals a total of nine copper atoms: four at the top, each shared by eight adjoining unit cells; four similar ones at the bottom; and one exactly in the middle of the drawing, not shared by any other unit cell. Summing the fragments, 4/8 + 4/8 + 1 = 2, giving two coppers in one unit cell. Similarly, there are eight atoms of lanthanum (or its substitute strontium) along the edges of the unit cell, each shared by four unit cells; plus two others (one near the top and one near the bottom) contained entirely within this unit cell. The net lanthanum content is four in each unit cell. The many oxygen atoms can be summed in the same way; Figure 8.3 contains 46 oxygen atoms, 24 of which lie entirely outside the unit cell; summing the fragments gives a total of eight oxygens participating in this unit cell.

This may seem tedious, but is indispensable for understanding the more complicated thallium and bismuth compounds.

The 2-1-4 compounds have only one CuO_2 plane. Looking at the exact center of Figure 8.3, the CuO_2 plane appears as one copper atom surrounded by four oxygen atoms, with one LaO plane above the CuO_2 plane and one below it. These LaO planes are said to be *intercalated*. Obviously, the entire structure is layered.

The CuO_2 plane is termed the *conduction plane*, and superconductivity takes place within it. The intercalated planes are called *charge-reservoir layers*. When the intercalated plane contains mixed-valence atoms, electrons are drawn away from the copper oxide planes, leaving holes to form the pairs needed for superconductivity. This mechanism is known as a *charge-transfer model*. The possible choices for placing specific atoms at the various sites opens the door into the topic of doping, i.e., changing the electron concentration. This is a very broad subject, which greatly affects superconductivity in these materials. We defer further discussion of this topic until Chapter 10 so as to concentrate in this chapter on structure alone.

8.2.3. $YBa_2Cu_3O_{7-x}$

The first superconductor found[11] with $T_c > 77$ K, and subsequently the most widely studied HTSC, is yttrium barium copper oxide ($Y_1Ba_2Cu_3O_{7-x}$), commonly termed "1-2-3." Its structure appears in Figure 8.4. It is related to the perovskite structure as follows: by tripling the perovskite (ABO_3) unit cell and substituting one yttrium atom for every third barium atom, the formula $Y_1Ba_2Cu_3O_9$ results. However, a little more than two oxygen vacancies are required for superconductivity. The formula can be thought of as $Y_1Ba_{3-1}Cu_3O_{9-2-x}$. The unit cell is orthorhombic—almost but not quite three cubes stacked upon one another.

A key feature of this unit cell is the presence of two layers or planes of CuO_2. Other HTSCs (the bismuth and thallium compounds) also form crystals which are relatives of the basic perovskite structure, again featuring layers of CuO_2. There is widespread agreement that the superconductivity takes place in the CuO_2 planes. One particularly noticeable feature in Figure 8.4 is that the two copper oxide planes are separated by a single yttrium atom; the yttrium plane contains no oxygens. Basically, this is because Y has a valence of +3, as contrasted with the +2 of Ba. Because of the *bond sum rule* for charge balancing, each Y coordinates with eight oxygens (valence $= -2$), located in the layers above and below the Y atom. With no oxygens in the yttrium layer, the formula is $Y_1Ba_2Cu_3O_7$ instead of $Y_1Ba_2Cu_3O_9$.

The role of yttrium is very minor: it just pries the two CuO_2 layers apart. When yttrium is replaced by many of the lanthanide series of rare-earth elements, there is no appreciable

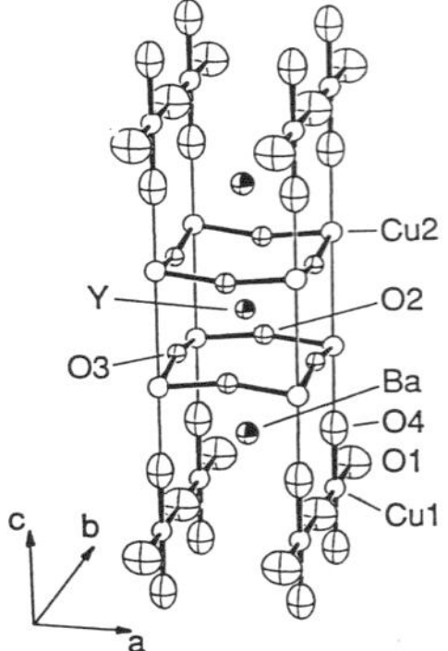

Figure 8.4. Structure of the double layered $YBa_2Cu_3O_7$ compound. This drawing has Y at the exact center and Cu atoms on the four edges. Oxygen atoms play four different roles in this unit cell, depending on position; also, there are two different roles for copper atoms. The designations O1, . . ., O4 and Cu1, Cu2 identify each type of atom.

change in superconducting properties. Thus, the yttrium (or other choice) serves only as a spacer—a "shim" between CuO_2 layers.

Outside this sandwich (but still within the unit cell) is a BaO plane. Referring to Figure 8.4, this means one barium atom surrounded by four oxygens along the edges of the unit cell. Finally, at the top (or bottom) of each unit cell is a copper oxide region that has certain oxygens missing. Since this does not qualify any longer as a CuO_2 plane, it is known as a copper oxide *chain*. Figure 8.4 shows the single Y atom and the two Ba atoms; the copper oxide planes near the middle each contribute one net copper atom (their four coppers are shared with four unit cells), and the top and bottom chains each contribute 1/2 coppers (four copper atoms shared with eight unit cells). Hence the name "1-2-3."

The additional nomenclature O1, O2, O3, O4 has been introduced in Figure 8.4 to help distinguish the four different roles played by oxygen atoms. O1's occur in the copper oxide chains; O2's lie in the CuO_2 planes; O3's likewise occur in the CuO_2 planes, but they are in line with the O1's in the chains above and below; O4's are associated with barium atoms. Likewise, we distinguish between Cu1's in the chains and Cu2's in the planes.

The missing oxygens are very important in $Y_1Ba_2Cu_3O_{7-x}$. The subscript x in the formula indicates that a fraction of the conventionally expected oxygens are missing. T_c maximizes near 92 K when $x = 0.15$; should $x = 0.50$, superconductivity goes away. Figure 8.4 depicts $Y_1Ba_2Cu_3O_7$ ($x = 0$). Note the missing oxygens along the a direction at the top and bottom of the unit cell. (If these oxygens were not missing, the formula would be $Y_1Ba_2Cu_3O_8$.) The missing oxygens result in the lattice parameters $a \neq b$, and the unit cell is orthorhombic. As x increases from zero, oxygen vacancies appear in the chains. At $x = 0.50$, there is equal probability for vacancies to occur along the a and b directions, causing the unit cell to have square symmetry. When that happens, the lattice parameter $a = b$, and the crystal is tetragonal. In 1987, when it was learned that the tetragonal phase is *not* superconducting, theorists came up with a "chain" theory of HTSC, which later was abandoned.

Typical $Y_1Ba_2Cu_3O_{7-x}$ dimensions are compared with those[4] of NbTi in Table 8.1.

8.2.4. Alternative Drawings

There is another, entirely different way to draw these crystals, using polyhedrons of copper oxide. Six oxygens surrounding a copper atom form an 8-sided octagon, and these attach to various barium and yttrium atoms. Figure 8.5 presents the YBCO unit cell in that form. Historically, the different types of drawings have been a matter of choice, with most

Table 8.1 Comparison of Characteristic Dimensions for Low-Temperature and High-Temperature Superconductors (meters)

Characteristic	NbTi Alloys[a]	$YBa_2Cu_3O_{7-x}$
Interatomic spacing	$\sim 1.5 \times 10^{-10}$	$\sim 2 \times 10^{-10}$
Unit cell dimensions[b]	$\sim 3 \times 10^{-10}$	$a = 3.8 \times 10^{-10}$
		$b = 3.9 \times 10^{-10}$
		$c = 12 \times 10^{-10}$
Maximum grain size	1×10^{-5}	1×10^{-4}
Penetration depth λ^c	16×10^{-8}	9×10^{-8}
		80×10^{-8}
Coherence length[d,e,f]	34×10^{-10}	$4\text{-}8 \times 10^{-10}$
		$15\text{-}50 \times 10^{-10}$
Mean free path of conduction Electrons in normal state	7×10^{-10}	8×10^{-9}

[a]55% Nb by weight, 39% Nb by mole fraction
[b]NbTi is not a crystal. Typical dimensions of volume containing minimum stoichiometric ratio of atoms. The density of this NbTi alloy is 7.26 gm/cm^3 (7.26 $\times 10^3$ kg/m^3).
[c]For applied field in the c direction.
[d]For applied field in the ab plane.
[e]Measured in c direction.

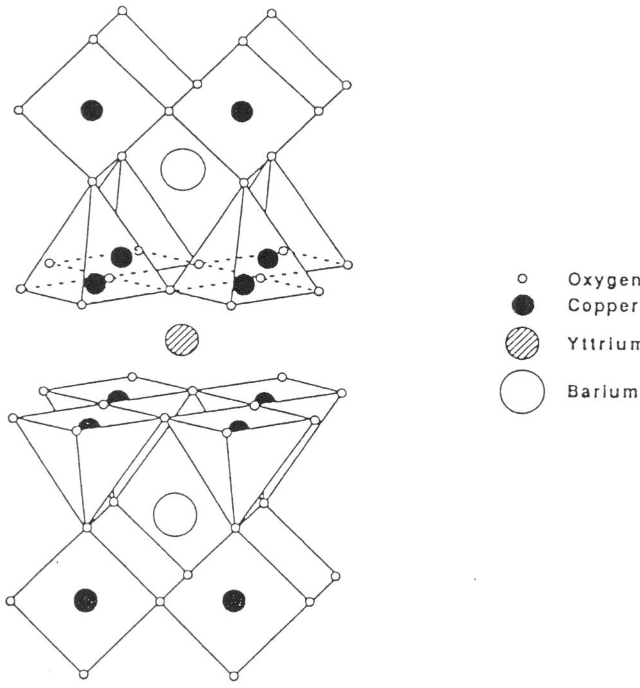

Figure 8.5. Alternative type of diagram of $YBa_2Cu_3O_7$, using polyhedrons to denote relationships of oxygen atoms to the metal atoms. In this form, the absence of oxygens on the yttrium plane stands out. (Some of the copper atoms are hidden in this type of drawing.)

geologists and chemists preferring the polyhedral method (Figure 8.5), whereas physicists prefer Figure 8.4. This reflects the interests of each group: chemists care about thermal and chemical features of the compounds, and the polyhedrons help to understand these characteristics; physicists, who measure X-ray diffraction or neutron diffraction patterns, are mainly interested in knowing where the atoms are located. Both kinds of drawings are useful because they emphasize different features of the same materials.

8.3. TWINNING

In YBCO, the slight difference in length between the *a*-axis and the *b*-axis of the unit cell breaks the symmetry of the crystal lattice. During crystal growth, it is energetically quite easy for propagation to switch from *a* to *b*. This is called *twinning*, and results in a minor irregularity in an otherwise uniform single crystal. Twinning occurs haphazardly as a single crystal grows. In polycrystal YBCO, each individual grain may contain a number of places where the change in propagation occurred, known as *twin boundaries*.

There is reason to believe that twin boundaries affect flux pinning. Figure 8.6 displays a vortex line going through YBCO crystals with twin boundaries in different orientations. It may be speculated than when twin boundaries are perpendicular to the direction of vortex motion, they might inhibit vortex motion; whereas with twin boundaries parallel to the vortices, the vortices might move easily *along* the boundaries, but not *across* them.

In order to explore this idea experimentally, it is necessary to compare data on both twinned and untwinned crystals. The quest for twin-free crystals led researchers from Argonne National Laboratory to seek a means to remove twin boundaries within a single crystal. Ulrich Welp et al.[12] devised a way to literally squeeze the twin boundaries from the lattice at elevated (400 °C) temperature without losing oxygen from the crystal. Independently, Deborah Kaiser and colleagues at the National Institute of Standards and Technology (NIST) also de-twinned crystals of YBCO.[13,14]

Anticipating a subject covered in Chapter 11, note that the flexural strength of YBCO is about 10^8 N/m^2; the grains shear apart at higher stresses. This applies to polycrystal material at room temperature. When a single crystal is heated to about 400 °C and squeezed with a pressure of 10^7 or 10^8 N/m^2, slight rearrangements of the atoms take place. In particular, the only difference between the *a*- and *b*-directions in the unit cell is the placement of one oxygen atom along the cell edge in the *b*-direction and the absence of that oxygen in the *a*-direction. Under pressure, that atom will jump diagonally across the unit cell, relocating

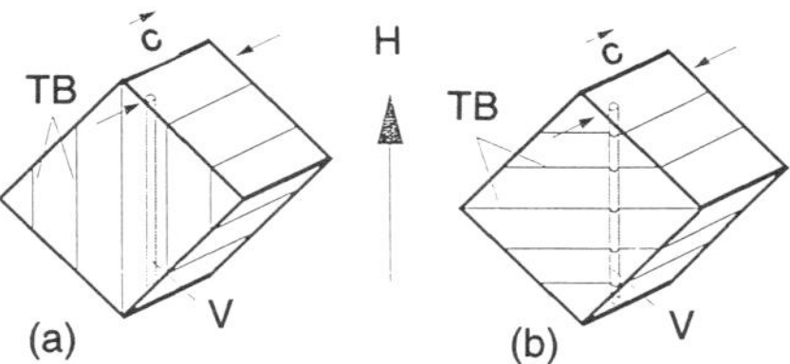

Figure 8.6. A vortex line (*V*) penetrating YBCO crystals with different twin-boundary (*TB*) orientations. (a) Twin boundaries parallel to the vortex; (b) twin boundaries perpendicular to the vortex. Motion of vortex lines will presumably differ in these different geometries.

in the other position. The result is to change the *b*- to the *a*-direction. A twin boundary goes away by this change.

Figure 8.7 shows before-and-after pictures of a YBCO crystal subjected to this squeezing. The change in twin boundaries is easy to see. Although this process cannot convert polycrystals to single crystals, it is clear that control of twin boundaries is possible. The

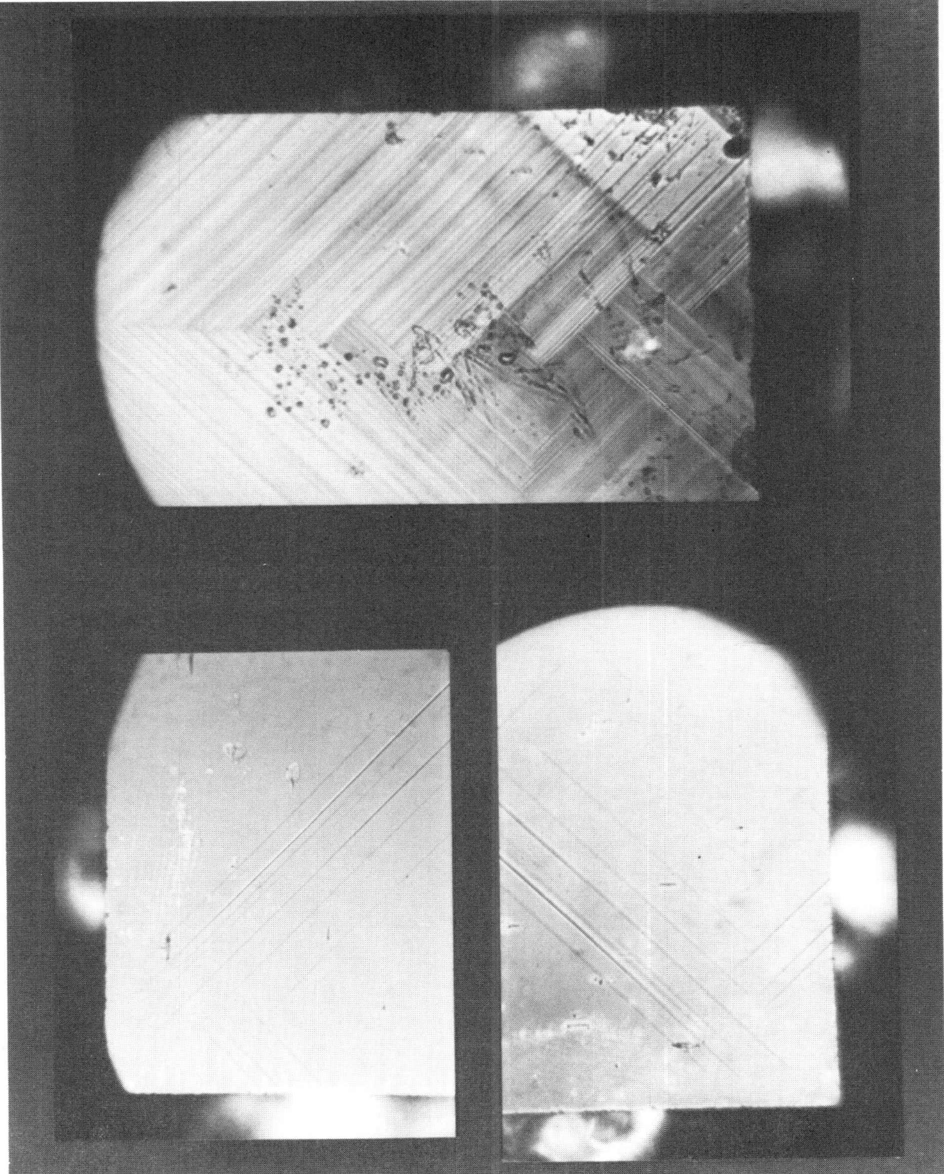

Figure 8.7.   Before-and-after polarized-light photos of YBCO crystal subjected to high pressure in order to remove twin boundaries. (From Ref. 12.)

relevance of twin boundaries to flux pinning, which has an important influence on critical current density J_c, will be addressed in Chapter 14.

8.4. THALLIUM, MERCURY, AND BISMUTH COMPOUNDS

The copper oxide superconductors containing layers of thallium oxide, mercury oxide, or bismuth oxide offer generally higher transition temperatures than YBCO, substantially exceeding 100 K in some configurations. Because all three types of compounds have very similar structures, it is convenient to focus the discussion on just one type, the thallium-based series Tl-Ba-Ca-Cu-O, or in shorthand, TBCCO. The mercury- and bismuth-based HTSCs are nearly identical.

8.4.1. Diverse Unit Cells

The structure of the thallium compounds is reminiscent of simpler HTSCs, but the very large number of atoms involved obscures that fact. Among other things, confusion is introduced because TBCCO may contain many intercalated layers, the exact role of which are not understood. Figure 8.8 presents three different examples of a thallium compound, with one, two, or three layers of CuO_2. Each contains one net layer of thallium oxide (TlO), together with two barium oxide (BaO) layers; calcium spacers lie between each CuO_2 plane.

Figure 8.9 is even worse-looking, because it contains *two* layers of TlO. Still, in all cases the fundamentals are the same. The entire issue is one familiar to any teenager working for McDonalds who must assemble either a Big Mac® or a double cheeseburger or a regular hamburger on a moment's notice. It is simply a matter of stacking consecutive intercalated

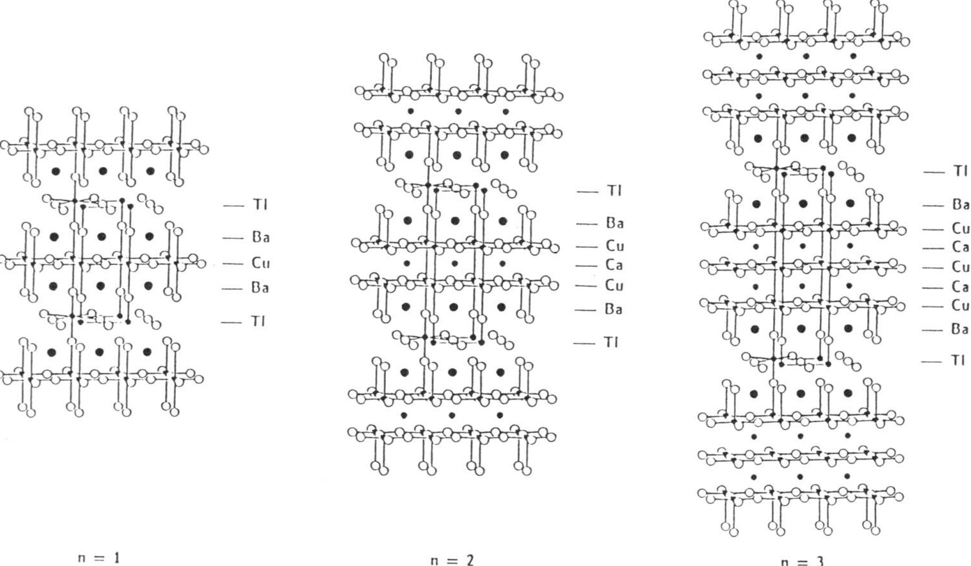

Figure 8.8.   Structure of the multilayered $Tl_mBa_2Ca_{n-1}Cu_nO_{2n+m+2}$ or $Bi_mSr_2Ca_{n-1}Cu_nO_{2n+m+2}$ compounds for $m = 1$.

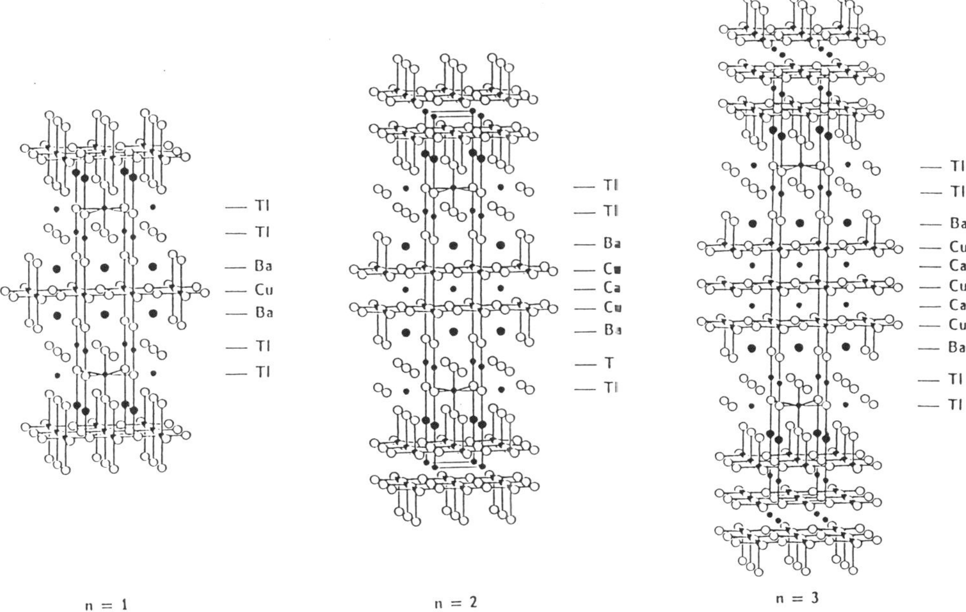

Figure 8.9. Structure of the multilayered $Tl_mBa_2Ca_{n-1}Cu_nO_{2n+m+2}$ or $Bi_mSr_2Ca_{n-1}Cu_nO_{2n+m+2}$ compounds for $m = 2$.

planes according to some pattern. You still have a piece of ground meat between slices of bread. In exactly the same way, each of these configurations has planes of CuO_2 pried apart by spacers (made of calcium here, not yttrium).

The intercalated planes of BaO and TlO take electrons out of the central sandwich, leaving holes behind to form superconducting pairs. The exact mechanism of this is not clear, but one significant observation is that T_c increases with increasing number of CuO_2 planes. Even outside the thallium family, this seems to be a general rule, since the single-layer 2-1-4 compounds have $T_c < 40$ K, the double-layer YBCO has $T_c = 92$ K, and triple-layer compounds have $T_c > 100$ K.

The general formula here is $Tl_mBa_2Ca_{n-1}Cu_nO_{2n+m+2}$, where n denotes the number of Cu atoms and m is the number of Tl atoms. The subscripts of the metals are used to create the nomenclature: The three structures shown in Figure 8.8 are known as the 1201, 1212, and 1223 compounds; Figure 8.9 shows the 2201, 2212, and 2223 compounds. Note that when $n = 1$, there is *no* calcium present. Usually the T_c of such compounds is too low to draw interest, but in the mercury-containing compounds (to which we return in Section 8.6.2), the 1201 phase has the surprisingly high value of $T_c = 95$ K.

The "bismuth compounds" refer to very similar structures, with thallium replaced by bismuth, and barium replaced by strontium: Bi-Sr-Ca-Cu-O, or BSCCO. Of course, there are differences in T_c, H_c, and J_c, as well as in normal-state properties, but the structural features are essentially the same. Meanwhile, in bismuth, mercury, and thallium compounds, the lattice constants $a = b$ and so there is no twinning within a crystal.

8.4.2. Syntactic Crystals

As any one of the thallium compounds grows by replicating a unit cell (say 2212, for example), the growth in the a- and b-directions can be expected to match layer-for-layer as the crystal expands. However, the growth in the c-direction (called *stacking*) is not especially constrained, and it is easy for an alternate structure (say 1223) to begin growing. After a while, the pattern might shift to yet another choice, including perhaps 2212 again. This phenomenon is termed *syntactic* growth. It creates all kinds of difficulties in trying to read x-ray diffraction patterns and reconstructing the crystal dimensions. The nuisance effect that this has on the thallium compounds is mitigated by the close proximity of all the T_c's to one another, but still it does make it difficult to compare the results of different research laboratories.

8.4.3. Effects of Layer Spacing

Improved understanding of the role of interlayer coupling suggests that thallium compounds hold a distinct advantage over bismuth compounds where flux motion is concerned. This, in turn, has important consequences for current-carrying capacity.

In the various compounds known as 2212, 1223, 2223, and so on, when identical unit cells are stacked one upon another, the repeat distance (denoted by s) is usually around 15 Å. For two (or three) CuO_2 layers that are close together, the thickness of that group is usually 3–6 Å, and is denoted by d_s. Then the distance between adjacent CuO_2 multilayers is the remaining amount, typically 9–12 Å, and is denoted by d_i. With the superconductivity carried in the CuO_2 planes, the remaining layers within the span d_i are insulating, and hence communication between CuO_2 multilayer groups is minimal.

Consequently, one may expect a very steep increase in the resistivity for current flowing in the c-direction of the different phases 1212, 2212, and so on. Table 8.2 displays the parameters s, d_s, and d_i for selected thallium and bismuth compounds. Comparing thallium 1212 with 1223 shows that d_i is the same in both cases, because no extra insulating layers have been introduced, and so both s and d_s increase together. Consequently, the c-direction resistivity should be the same for either 1212 or 1223, but for 2212 it should be higher. Exactly that has been found by Kim *et al.*[15] Figure 8.10 verifies that prediction. Meanwhile, BSCCO compounds have larger values of d_i, and going from 2212 to 2223 does nothing to decrease d_i.

More important than the c-axis resistivity, the distance (d_i) between adjacent CuO_2 multilayers affects the way that magnetic flux vortices are pinned in these compounds. It is necessary to delay explaining this point until Chapter 14, where flux line motion is treated in detail. For the present, we simply note that TBCCO will postpone *flux lattice melting* better than BSCCO in a magnetic field at 77 K, because of the relative vertical spacings between the superconducting layers. Moreover, since this is an intrinsic property of the unit

Table 8.2 Structural Data for the Highly Anisotropic HTSCs

Sample	S(Å)	d_s	d_i
Tl-1212	12.75	3.18	9.57
Tl-1223	15.87	6.28	9.59
Tl-2212	14.7	3.15	11.59
Bi-2212	15.45	3.16	12.29

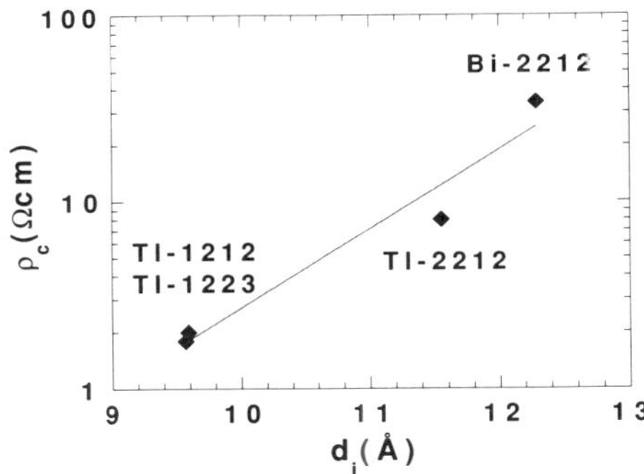

Figure 8.10. Logarithm of the c-axis resistivity ρ_c plotted against d_i. This approximately exponential dependence of ρ is expected for tunneling between CuO_2 layers (from Ref. 15).

cell of each material, it cannot be circumvented by creating additional pinning sites. (That is commonly done by introducing defects, to obstruct motion of flux lines.) Therefore, the thallium compounds appear to be a much better bet for achieving acceptable performance at 77 K.

On the other hand, BSCCO is much more micaceous than TBCCO. As a result, researchers attribute their success in making wire—long lengths of BSCCO with good grain alignment—to the easy-slip properties of the BSCCO lattice. This property is directly related to the large spacing between the CuO_2 multilayers (d_i), and in TBCCO that spacing is not large. Thus, the very property that makes TBCCO encouraging from the flux-pinning point of view simultaneously is discouraging from the mechanical-deformability point of view.

8.5. LAYERED STRUCTURES AND ANISOTROPY

Because of the layered structure of the HTSC compounds, electrons move easily in some directions, and with difficulty in others. This gives rise to the property known as *anisotropy*.

8.5.1. Effective Mass

To begin, recall Newton's second law: $\mathbf{F} = m\mathbf{a}$. Within a crystal, sometimes the very same magnitude of force F applied along different axes produces different amounts of electron motion (different $\mathbf{a}$). To account for this anomaly, we say that the *effective mass* changes with crystal direction. Instead of being a single-valued scalar m, the effective electron mass becomes a *tensor*. In the most general case, the vector $\mathbf{a}$ no longer needs to be aligned with the vector $\mathbf{F}$. For our purposes with HTSCs, $\mathbf{a}$ and $\mathbf{F}$ both have the same direction, and so the effective mass is taken to be a diagonal tensor. Furthermore, it is highly accurate to set $m_a = m_b$ so that the only difference is between m_a and m_c.

Anisotropy is defined by the ratio of the effective mass of the electron in the various directions: $(m_c/m_a) = \gamma^2$. In YBCO, the *effective mass ratio* $\gamma^2 \approx 30$, but in bismuth compounds it is greater, and in thallium compounds greater still—perhaps 10^5. Such a large anisotropy means that electrons can barely move in the c-direction, and the superconductor is effectively two-dimensional.

This condition of anisotropy was totally foreign to previously known LTSCs. (In fact, even though the crystalline axes were of different lengths, the momentum space of the superconductor could be isotropic; the element gallium is an example of this. Where superconducting electrons are concerned, the shape of the momentum space is what matters.) With HTSCs, in which the planes of copper oxide carry the supercurrent, anisotropy must be taken into account.

A widespread failure to grasp the significance of anisotropy caused a lot of early HTSC experiments to be incorrectly interpreted. Samples of sintered materials, with randomly oriented grains, were thought to be truly representative materials, and when unexpected measurements resulted, elaborate theories were constructed to explain the data. After a few years, the role of anisotropy became clear and most of the convoluted explanations vanished.

8.5.2. Penetration Depth and Coherence Length

Above we described the size and shape of the atomic arrangement in HTSCs. These sizes should be borne in mind when considering the lengths over which the superconducting properties can change by a factor of ~3. In any superconductor, there are two important characteristic lengths, which were introduced in Chapter 2. One length, the *Ginzburg–Landau penetration depth* λ, sets the scale for the exponential decay of the magnetic induction as one moves into the superconductor away from a normal boundary. The other length, the *Ginzburg–Landau coherence length* ξ, sets the scale for spatial variations in the density of superconducting charge carriers (e.g., electron-pairs or hole-pairs). Unlike the NbTi alloys, these lengths depend on direction in the HTSCs.

For the HTSCs in general, the coherence length is very small (roughly a few interatomic spacings) and the penetration depth is very large (several thousand interatomic spacings). Per the definitions in Chapter 6, this makes the HTSCs extreme type II superconductors. But there is a further anomaly associated with ξ and λ. Table 8.1 gives some typical numbers for HTSCs. The important point to notice is that ξ and λ are different in the a and c crystallographic directions. Since these lengths appear in the form $\exp[-x/\xi]$ or $\exp[-x/\lambda]$ in the formulas affecting superconducting properties, it is no surprise that their modest degree of anisotropy (a factor of 5 or so) is magnified into major changes in the superconducting properties of the materials in different directions.

8.5.3. Grain Alignment

It is a good approximation to imagine the separated CuO_2 planes as skating rinks for superconducting pairs. Cooper pairs of charge carriers move relatively freely within these planes, but only with difficulty perpendicular to them. With this image, we can see that when mismatched crystal grains come in contact (the definition of a grain boundary), there will be an interruption in the easy flow of current.

All bulk samples of HTSCs suffer from a barrier effect at the grain boundaries; the superconductivity is carried through an interwoven network of Josephson junctions, in which each grain boundary functions as the insulating layer. This is termed *weak-link* behavior, and Chapter 13 is devoted to that topic. Here, we merely note that if impurities are eliminated from the boundaries, the insulators might be minimized, but the change in orientation between adjoining CuO_2 layers or planes will *not* be eliminated. Clearly, this creates an incentive to grow large crystals with perfect alignment.

One place that such growth is possible is in thin films. (However, almost all HTSC thin films are mosaics, not a single thin crystal). Here, the underlying substrate creates a preferred direction of growth because the lattice constants of the substrate are fixed; as the superconductor is deposited, it grows in a direction such that its dimensions along the substrate surface match up. For example, a substrate of MgO will cause a different growth pattern from that of a SrTiO$_3$ substrate. For most thin-film experiments, it is desirable to grow the film such that the c-axis is perpendicular to the substrate. In that case, the a-b plane is in the plane of the film, and higher current flows in such a sample. The directional mismatch between grains is likewise in the a-b plane, whereas the c-axis remains well-aligned over many adjacent grains.

The strong interest in thin films in part reflects the belief that what we learn from films is applicable to bulk samples. It is generally conceded that alignment of crystal directions is indispensable to carrying high current in HTSCs. It was in thin films that the highest J_c values were obtained, establishing a benchmark for what might be attainable in larger single crystals.

8.5.4. Measurement of Extreme Anisotropy

The measurement of anisotropy is a matter of some delicacy. It requires a sample whose crystalline alignment is known perfectly, and most early measurements on bulk samples were compromised by mismatched grain alignment. Eventually a value for YBCO was determined: $m_c/m_a = \gamma^2 = 30$. For the multilayer HTSCs, the anisotropy was so great that very sophisticated experimental techniques were required. It is helpful to examine one such experiment, in order to appreciate just how critical grain alignment can be.

The thallium compounds are the most anisotropic HTSCs. This was proven experimentally by Farrell *et al.*[16] using a torque-measuring apparatus that can determine the anisotropy of superconducting ceramic crystals. The technique of torque magnetometry measures the twisting of a quartz rod attached to a crystal, which is induced when an applied magnetic field interacts with circulating currents in the sample. Farrell used a feedback/control method to prevent the crystal from actually rotating, and measured the torque required to maintain the null position. This apparatus was refined to the point of measuring angular differences as small as 0.001°.

The angular dependence of the torque of a small single crystal of a 2212 compound is shown in Figure 8.11. The angle θ is the angle between the c-axis and the applied magnetic field **H**. The most interesting behavior is that very near to 90°, which is shown close-up in Figure 8.11(b). The torque remains quite large until about 89.8°, and then falls precipitously to zero at exactly 90°. Using data like this, values of $\gamma = 210$ or $\gamma = 350$ emerge, which means the effective mass ratio ($\gamma^2 = m_c/m_a$) is about 10^5. This is the largest anisotropy yet measured by this technique, and shows the extreme two-dimensionality of these superconductors.

Earlier measurements of anisotropy gave much smaller values. This is due[16] to tiny misalignment of grains in previous samples. From Figure 8.11, it is evident that even 1° misalignment of *some* grains in a macroscopic sample would throw off the anisotropy measurement. By previous standards, samples aligned to within 1° were considered well-aligned, but that is not good enough for compounds as anisotropic as the thallium HTSCs.

Thin-film experiments are helpful in understanding this condition. Using a thallium thin film prepared by Kampwirth and Gray at Argonne, Farrell obtained[17] a value of $\gamma^2 \sim 10^4$ by this torque-measurement technique; this confirmed what the Argonne researchers had measured for γ independently.[18]

The message of all this is that measurements on highly anisotropic HTSCs depend critically upon alignment. This contrasts with the modest stringency-of-alignment require-

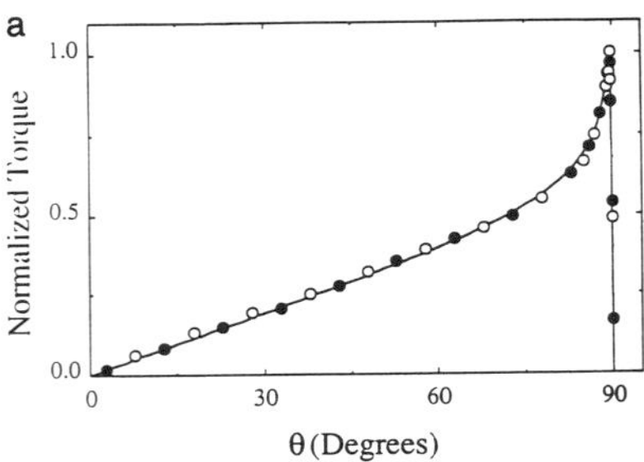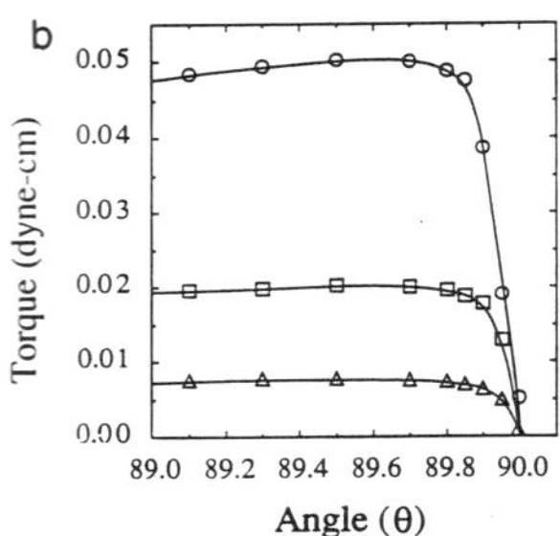

Figure 8.11.   Angular dependence of the normalized torque observed at $T = 85$ K and in a magnetic field of 1 T. (a) Full range of θ; (b) the high-angle region. The magnetic field was held constant at $H = 1$ T. Circles, squares, and triangles represent data taken at $T = 95$, 100, and 105 K, respectively. The curves through the data are a guide to the eye (from Ref. 16).

ment for the less anisotropic YBCO ($\gamma^2 \sim 30$) and with the LTSCs, which are commonly isotropic. The confinement of superconductivity within the CuO_2 planes is very pronounced in thallium compounds, giving a behavior that is almost entirely two-dimensional. Since the BCS theory was devised for standard three-dimensional materials, some surprises may well be expected in such radically different superconductors.[5]

8.6.   OTHER OXIDE SUPERCONDUCTORS

There are many other types of ceramic oxide superconductors, each having a distinct structure. Unfortunately, owing to their relatively low transition temperatures, most of them have received little attention. This may not be entirely wise. With the passage of several years since the original discoveries, it is turning out that the superconductor that is easiest to make into wire (BSCCO) carries very little current at 77 K, and many engineers are contemplating operating their devices at 20 K, in order to meet BSCCO halfway. If, in that milieu, another superconductor with $T_c \approx 50$ or 60 K were found to carry adequate current, it might become the material of choice.

It is impossible to tell whether anyone is pursuing such an approach, because any smart wire manufacturer would keep secret any progress being made on such a material. Anyone who finds a way around the tremendous difficulties associated with turning a brittle ceramic into wire has a gold mine in hand. Consequently, there is relatively little to be said about other ceramic oxide compounds. To illustrate how diverse these materials can be, we cite just a few examples which have sufficiently unusual structures to be interesting.

8.6.1.   Lead-Substituted Compounds

The most widely known variations on HTSCs involve the partial substitution of lead on the A-site in either TBCCO or BSCCO, such that the formula begins ($Tl_xPb_{1-x}\cdots$). For reasons that will become clearer after Chapters 9 and 10, it is easier to make these compounds than the pure thallium or bismuth compounds. However, there is no structural change associated with these compounds.

On the other hand, the construction of a double-sandwich of lead oxide planes with copper oxide planes is a comparatively unique structure. In 1988, researchers at AT&T Bell

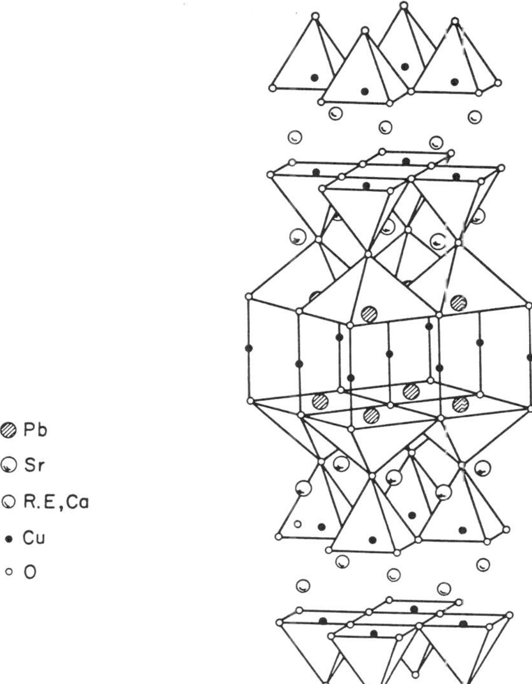

Figure 8.12.   Structure of PbO/CuO/PbO sandwich compound, with emphasis given to the polyhedrons related to each metal atom in this configuration.

Labs[19] discovered a family of HTSCs of the form $Pb_2Sr_2ACu_3O_{8+y}$, where "*A*" can be any of several rare-earth elements: Y, La, Eu, Pr, etc., perhaps mixed with Ca or Sr. The structure of such compounds is shown in Figure 8.12; the unique feature is in the central layers where superconductivity is carried: two lead oxide layers enclose a copper oxide plane. This sandwich of PbO/CuO/PbO plays a role similar to that of the TlO layers in Figure 8.9. The highest T_c obtained in this series of compounds was 68 K (in $Pb_2Sr_2Y_{0.5}Ca_{0.5}Cu_3O_8$), which is sufficiently high to confirm the layered picture of how supercurrents flow, but still too low to be of commercial interest.

8.6.2. Mercury Compounds

In Section 8.4.1 we remarked that compounds such as TBCCO-1201 (i.e., no calcium and only one CuO_2 plane) were of little interest because T_c was generally too low. The analog of this compound using mercury in place of thallium is $Hg_1Ba_2Ca_0Cu_1O_{4+\delta}$ or simply $HgBa_2CuO_{4+\delta}$. It has T_c = 94 K, which is exceptionally high for a single-CuO_2 layer compound.[20] Figure 8.13 displays its unit cell. The mercury oxide plane is in the middle (four Hg's are shared between four cells), and the copper oxide plane is at the top (or bottom) of the unit cell. The BaO plane is far from flat, being bowed toward the CuO_2 planes.

Continuing the analogy with TBCCO, a combination of the structures Hg-1212 with Hg-1223 became superconducting at 133 K.[21] That reinforced the idea that the more CuO_2

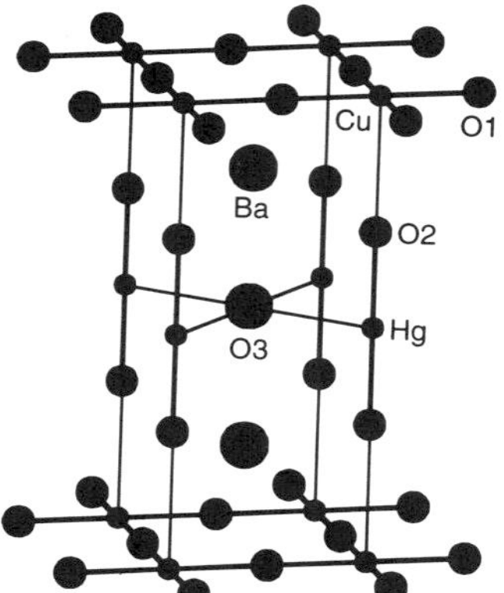

Figure 8.13. Structure of $HgBa_2CuO_4$, with Hg placed at the exact center, O on the edges, and the CuO_2 planes at top or bottom of the unit cell. O_1 denotes oxygens in the CuO_2 planes; O_2 is associated with the barium plane; O_3 is in the mercury plane. (From Ref. 24.)

layers, the higher is T_c; indeed, the different phases show $T_c = 94$ K, 128 K, and 134 K for Hg-1201, Hg-1212, and Hg-1223, respectively. Furthermore, under extremely high pressure, T_c in the Hg-1223 phase rises[22,23] to over 150 K, which tells us that T_c is very sensitive to small changes in the structure.

However, a small discrepancy in oxygen content (δ) makes all the difference. By changing the annealing cycle so as to draw excess oxygen away,[24] T_c drops to 59 K in $HgBa_2CuO_{4+\delta}$ (Hg-1201). This occurs because the oxygen–mercury bond is quite different from the oxygen–thallium bond, and the presence of additional oxygen changes the charge-reservoir layers dramatically.[25] This condition has drawn researchers' attention to the importance of oxygen defects in sustaining superconductivity.[26] Chapter 10 contains a discussion of the ways in which oxygen vacancies affect superconductivity.

8.6.3. Oxycarbonates

An entirely different form of substitution is to introduce a carbonate group (CO_3) into the conventional copper oxide configuration. The first of these[27] had CO_3 substituted for copper, in the form $YBa_4Cu_{2.27}(CO_3)_{0.53}O_{7.08}$. A variant of the 1-2-3 configuration was constructed by Miyazaki *et al.*,[28] based on $YSr_2Cu_3O_{7-\delta}$ with CO_3 substituted for one of the CuO_2 planes. A French group[29] directed by Bernard Raveau determined the formula $Y_4Sr_8Cu_{11}CO_3O_{25}$ for this compound. They also found[30] the quite similar compound $Y_{1.6}Ca_{0.4}Ba_4Cu_5CO_3O_{11}$, and the pair of these compounds led them to propose the general formula

$$(Y,Ca)_n(Ba,Sr)_{2n}Cu_{3n-1}(CO_3)O_{7n-3}$$

thus implying a whole family of compounds with diverse values of n. Unfortunately, not many of these ceramic oxides are superconductors, so attention turned to substituting carbonates into the thallium compounds.

Raveau's group synthesized[31] $Tl_{0.5}Pb_{0.5}Sr_4Cu_2(CO_3)O_{7-\delta}$, and obtained $T_c = 70$ K, which gives immediate encouragement to look further. This formula looks reminiscent of the 1-2-2-3 TBCCO formula, $(Tl,Pb)_1Ba_2Ca_2Cu_3O_9$, with strontium substituted on both the barium and the calcium sites, and one of the coppers replaced by CO_3. However, by using high-resolution electron microscopy, they concluded that the more likely structure is an intergrowth of two crystalline structures: double rock salt layers [(TlO)(SrO)] and single

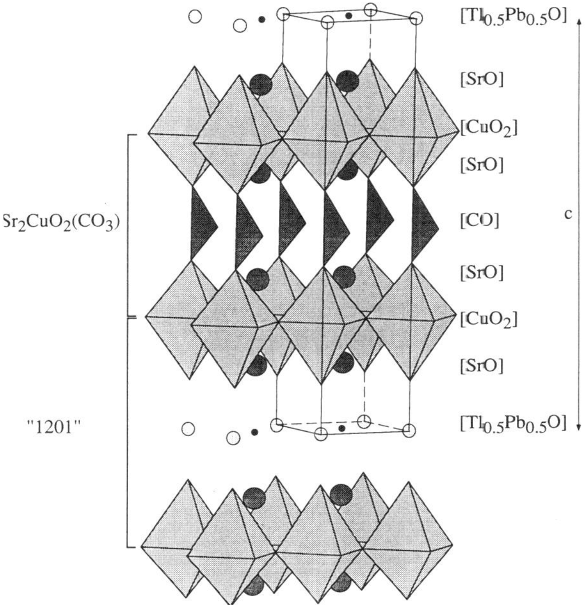

Figure 8.14. Representation of the structure of the tentative proposed model for Tl$_{0.5}$Pb$_{0.5}$Sr$_4$Cu$_2$(CO$_3$)O$_7$. Open circles = oxygen, solid circles = metals. The position of the oxygen atoms at the level of carbon is only a guide.

perovskite layers [SrCuO$_3$], linked through single carbonate layers. The composite structure is shown in Figure 8.14, where the upper portion of the unit cell is essentially Sr$_2$CuO$_2$(CO$_3$) and the lower portion is Tl$_{0.5}$Pb$_{0.5}$Sr$_2$CuO$_5$. Several other variants on this general scheme have been made,[32] and typically the T_c values fall below 77 K.

There is no fundamental reason to restrict substitutions only to carbonates, and the same research team has experimented with substituting nitrates[33] into the lattice. The compound YCaBa$_4$Cu$_5$(NO$_3$)$_{0.3}$(CO$_3$)$_{0.7}$O$_{11}$ yielded indications of superconductivity up to 82 K. This is very encouraging, because the equivalent oxycarbonate is not a superconductor at all. In this compound, the structure is the same as in the oxycarbonate, because the nitrate group is so similar to the carbonate group. However, the density of charge carriers is different, and this can affect superconductivity profoundly. Since Chapter 10 is devoted almost entirely to examining the many ways in which substitution of elements changes the electronic properties of the HTSCs, we leave this subject at this point.

The overall message to be derived from the existence of these HTSCs with alternate structures is that the common feature of all these superconductors is their copper oxide layers. Despite many different configurations of the surrounding atoms, this feature remains consistent from one compound to another.

8.7. SUMMARY AND FORECAST

This chapter dealt with the structure of the HTSCs. After introducing certain requisite terminology, we presented drawings of the unit cells of several different ceramic oxides. The unit cells are not perfectly symmetric, which has important consequences for superconductivity. In particular, the modest anisotropy of the crystal lattice is magnified many times into a severe anisotropy of the electronic properties; superconductivity is virtually two-dimensional in the HTSCs. This is quite different from conventional superconductors, for which crystal orientation matters very little.

The HTSCs have in common the presence of copper oxide layers, with superconductivity taking place between these layers. The more layers of CuO_2, the higher is T_c. However, there are trade-offs: for example, in comparing BSCCO and TBCCO, the very same structural property that makes BSCCO easy to deform and shape into wire also gives it weak flux pinning; TBCCO is the opposite, with good flux pinning but poor deformability. Choices like this seem to abound in HTSC.

No comprehensive theory predicts the properties of the HTSCs from knowledge of their structure, but the BCS theory still accounts for superconductivity in these materials. Later chapters will examine in more detail certain other aspects of superconductivity which are related to structure, and which have been briefly touched upon here:

- The type and location of particular atoms affects the electron density in the copper oxide planes, and doping of the crystal lattice is used to enhance T_c and other properties of the HTSCs. Chapter 10 covers this topic.
- Defects, particularly dislocations, affect the mechanical properties of the superconductors as described in Chapter 11.
- The presence of CuO_2 layers in the crystal structure is necessary for superconductivity above 77 K. Chapter 12 brings out some of the theoretical problems peculiar to layered superconductors.
- Both the volume between grains and the debris filling it substantially reduce the transport current density (intergrain J_c) to levels well below the magnetization current density (intragrain J_c), due to weak-link behavior. This is explained in Chapter 13.
- Structural anomalies, such as twinning, layer spacing, and defects play an important role in flux pinning, which is the subject of Chapter 14.

Before getting into any of these matters, however, it is first necessary to recognize that the HTSCs are very complex chemicals, not readily found in nature. The principles by which these compounds are formed is discussed in the following chapter on phase equilibrium. Following Chapter 14, we return to the very important matter of how to *make* HTSC materials in Chapter 15, and then how to turn it into wire in Chapter 16.

REFERENCES

1. C. Kittel, *Introduction to Solid-State Physics*, 6th edition (Wiley, New York: 1986).
2. For example, compare N. E. Phillips, *Phys. Rev. A* **134**, A385 (1964) with W. D. Gregory *et al., Phys. Rev.* **150**, 3770 (1966).
3. T. P. Sheahen, *Phys. Rev.* **149**, 370 (1966).
4. E. W. Collings, *Applied Superconductivity, Metallurgy, and Physics of Titanium Alloys* (Plenum Press, New York: 1986).
5. R. A. Klemm, *Layered Superconductors* (Oxford University Press, New York: in press).
6. L. H. Van Vlack, *Elements of Materials Science and Engineering*, 4th ed. (Addison-Wesley, Reading MA: 1980).
7. B. D. Cullity, *Elements of X-Ray Diffraction* (Addison-Wesley, Reading, MA: 1974).
8. R. M. Hazen, *Sci. Am.*, pp. 74–81 (June 1988).
9. I. K. Schuller and J. D. Jorgensen, *MRS Bulletin*, pp. 27–30 (January 1989).

10. J. G. Bednorz and K. Muller, *Z. Physik B* **64**, 189 (1986).
11. M. K. Wu *et al., Phys. Rev. Lett.* **58**, 908 (1987).
12. U. Welp *et al., Physica C* **161**, 1 (1989).
13. D. Kaiser *et al., J. Materials Research* **4**, 745 (1989).
14. W. Wong-Ng *et al., Phys. Rev. B* **41**, 4220 (1990).
15. D. H. Kim *et al., Physica C* **177**, 431 (1991).
16. D. E. Farrell *et al., Phys. Rev. B* **42**, 6758 (1990).
17. K. E. Gray, R. T. Kampwirth, and D. E. Farrell, *Phys. Rev. B* **41**, 819 (1990).
18. J. H. Kang *et al., Appl. Phys. Lett.* **52**, 2080 (1988).
19. R. J. Cava *et al., Nature* **336**, 211 (1988).
20. S. N. Putalin *et al., Nature* **362**, 226 (1993).
21. A. Schilling *et al., Nature* **363**, 56 (1993).
22. C. W. Chu *et al., Nature* **365**, 323 (1993).
23. M. Nunez-Regueiro *et al., Science* **262**, 97 (1993).
24. J. L. Wagner *et al., Physica C* **210**, 447 (1993).
25. P. G. Radaelli *et al., Physica C* **216**, 29 (1993).
26. J. D. Jorgensen *et al.*, "Roles of oxygen defects in copper oxide superconductors," presented at conference
     on *Physics and Chemistry of Molecular and Oxide Superconductors*, Eugene, OR (July 27–31, 1993).
27. C. Greaves and R. P. Slater, *Physica C* **175**, 172 (1991).
28. Y. Miyazaki *et al., Physica C* **198**, 53 (1992).
29. B. Domenges *et al., Physica C* **207**, 65 (1993).
30. B. Raveau *et al., Physica C* **209**, 153 (1993).
31. M. Huve *et al., Physica C* **205**, 219 (1993).
32. A. Maignan *et al., Physica C* **208**, 149 (1993).
33. A. Maignan *et al., Physica C* **208**, 116 (1993).

9

Phase Equilibrium

Phase equilibrium is a very important aspect of the study of HTSCs because it lies at the core of preparing the basic materials. Understanding phase diagrams helps understand the melting reaction for solidification, which in turn helps grow single crystals. Also, it helps discern the consequences of mixing powders inhomogeneously. Early samples of YBCO (and other HTSCs) contained a number of poorly understood phases, and consequently the experimental data obtained was not representative of the true compound and only impaired progress toward understanding superconductivity in these materials.

In terms of practical applications, such as making wire for devices, phase diagrams are entirely relevant. The slow progress toward wire that can carry high current may be traced directly back to reactions and compatibility relationships that stand out on the phase diagram. "You can't get there from here" is not just a quip, but a statement with a clear explanation, to be found within the phase diagram.

Phase equilibrium is a difficult subject. Consequently, we put considerable effort into establishing basic concepts and terminology. We begin with elementary material about phase diagrams in general, using the one-component example of water as an illustration. Advancing from one-component to two- and three-component systems, we introduce a series of concepts that appear prominently in the phase diagrams pertaining to YBCO and BSCCO.

As the number of elements in these compounds grow, so do the complexity of their phase diagrams. It takes considerable preparation to reach a position from which we can cogently discuss the very complex phase diagram of YBCO. Indeed, these materials are so new and so complex that good phase-equilibrium information is only now beginning to be published; it will be a while before anyone can seriously claim to understand the phase equilibrium of the HTSCs.

9.1. INTRODUCTION TO PHASE DIAGRAMS

This first cornerstone principle to be recognized is that phase diagrams only deal with equilibrium conditions. There are a great many compounds in everyday use that are not in equilibrium, e.g., glasses, in which the rate of formation is paramount. That brings in the entirely separate subject of kinetics, with its many ramifications. Here we only state a caution: it is important to remember the limited nature and the usefulness of phase diagrams. Among chemists and ceramists, whenever a nonequilibrium process is used to make a particular material, the use of a phase diagram to describe it must be carefully qualified.

159

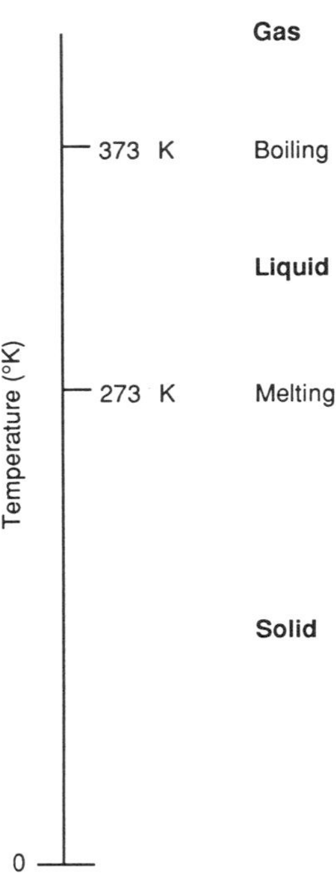

Figure 9.1. The single-component system, water, showing its melting and boiling temperatures at sea level.

To start, we consider the example of pure water. It is a familiar substance, and so its phase diagram is an ideal vehicle for introducing terminology and concepts pertinent to any phase diagram. The point here is not to learn about water but to apply these concepts to a study of the phases of HTSCs.

Figure 9.1 shows how water behaves at a pressure of one atmosphere. As temperature (on the vertical axis) rises, water first melts at $T = 273$ K (0°C) and subsequently boils at $T = 373$ K (100°C). These transitions are determined by the *free energy*, which is always a minimum when a system is in equilibrium. Above 100°C, H_2O has a lower free energy in the gas phase than in the liquid phase. All phase transitions occur because of this principle.

Actually, Figure 9.1 is really a slice through the *phase space* for a one-component system: pressure is held constant ($P = 1$ atm). We know that at high altitudes, water boils at lower temperatures, so if we had taken a different slice we would have shown a different temperature as the boiling point. Thus, we infer that even a one-component fluid system has a phase diagram involving two state variables, commonly T and P.

The first step upward is to a two-dimensional plot where T and P both vary for this one-component system; Figure 9.2 is an example for the case of water. Figure 9.2(a) is drawn to scale and identifies the separate regions in which water is a solid, liquid, or gas. The substance always minimizes its free energy and seeks its lowest possible chemical potential (μ_ϕ). The transition from one phase to another occurs when two phases have the *same* chemical potential. Figure 9.2(b) is not to scale[1]; it shows both the *triple point*, at which all three phases are in equilibrium, and the *critical point*, above which it doesn't matter whether it's called a gas or a liquid—the phase is both disordered and dense.

Continuing further, this phase diagram can be expanded to three dimensions by showing volume explicitly. In Figure 9.3 we still have a one-component system (water), but with three variables depicted: pressure, temperature, and volume. For any choice of P, T, and V the point of lowest chemical potential lies somewhere on the surface in Figure 9.3. The behavior of the substance under small changes of P, T, or V is determined by the slope of the surface nearby.

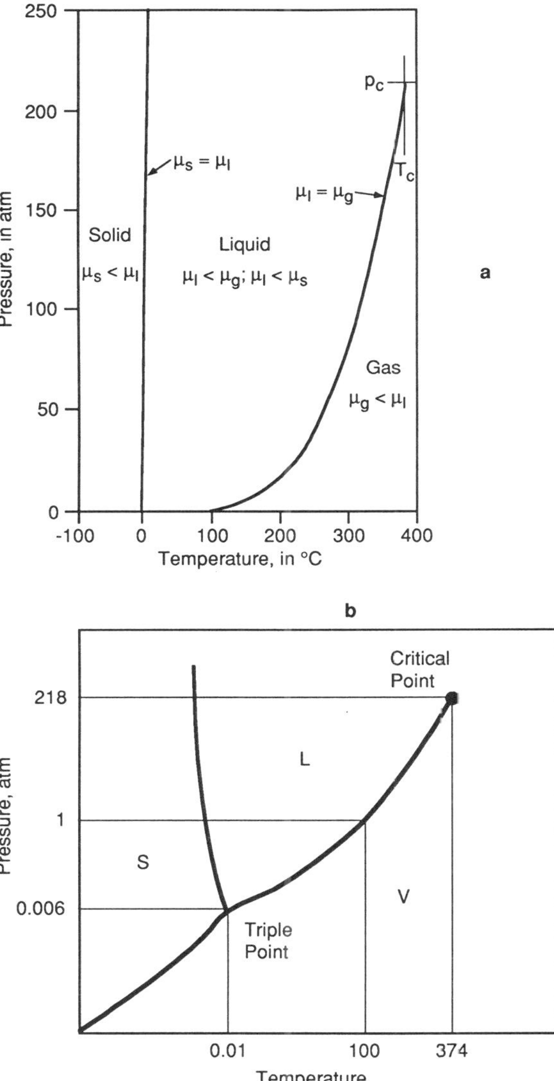

Figure 9.2. The pressure–temperature phase diagram for water. (a) Drawn to scale, this shows that phase boundaries occur at points of equal chemical potential. Between liquid and solid, the boundary is not perfectly vertical but slopes slightly to the left. (From *International Critical Tables* 3, and P. W. Bridgman, *Proc. Am. Acad. Sci.* **47**, 441 (1912)). (b) Not-to-scale drawing to emphasize the triple point and critical point. [From F. W. Sears *et al., University Physics*, 7th ed., p. 398 (Addison-Wesley, Reading, MA: 1987). Reprinted by permission of the publisher.]

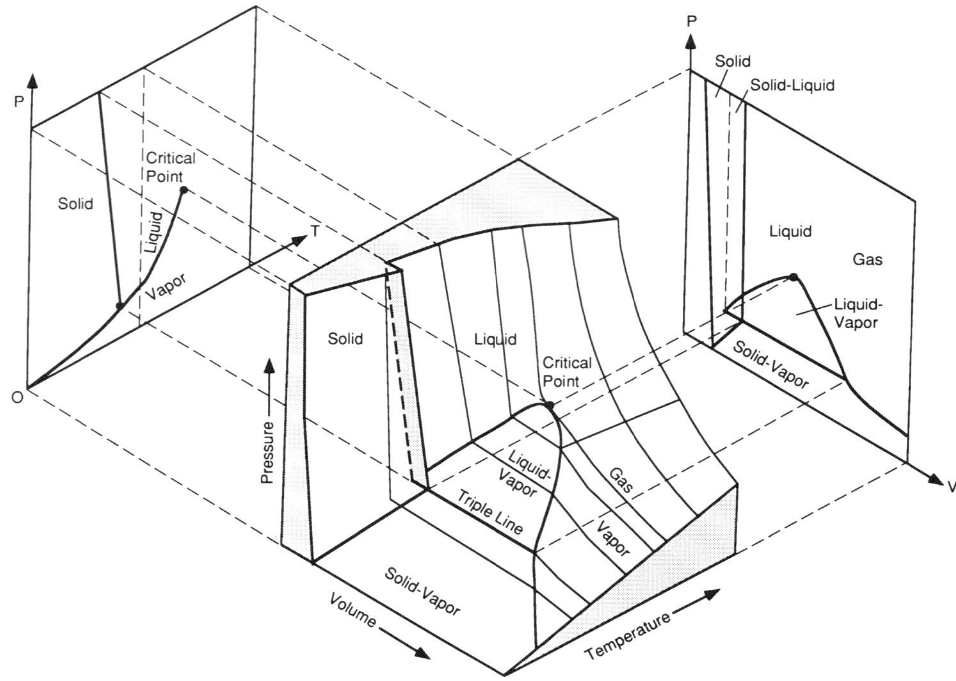

Figure 9.3. Projection of phase space for water, with variables *P*, *T*, and *V*. Again, the boundaries between zones on the surface are lines of equal chemical potential for the adjoining phases. (After Ref. 1.)

Imagine a horizontal plane slicing through Figure 9.3. With *P* held constant at 1 atm, as *T* increases from a very cold origin, ice expands a little as it warms. When it comes to the melting temperature of 0°C, its volume shrinks a little; at still higher temperatures the volume of the liquid water increases until it reaches 100°C. At this point there is a great increase in volume as it boils into the gas phase. Above that temperature, the volume expands more, eventually following the perfect gas law, $PV = NRT$. A second example is that of a constant-volume slice: if ice warms up without being allowed to expand, the pressure goes way up, usually until the container gives way. This is known as a *frost heave* and is known to break highway pavement.

The vertical "walls" on the periphery of Figure 9.3 are projections onto two dimensions. The left wall, a *P-T* plane, is just like Figure 9.2(b). Volume is not constant in this drawing, so it is not a constant-volume slice. Similarly, the back wall is a *P-V* plane, a projection of the various features that occur at assorted temperatures. A slice at constant temperature,[2] an *isotherm*, appears in Figure 9.4. It is a far simpler drawing showing how at constant temperature, reducing the pressure eventually causes the liquid to boil.

As we go on to systems with two, three, or four components, it is well to bear in mind that these other state variables remain a hidden part of the story, not displayed in the two-dimensional slices that appear on paper. Generally speaking, since we are dealing with solids and liquids from here on, the volume changes so little that it can be dropped as a

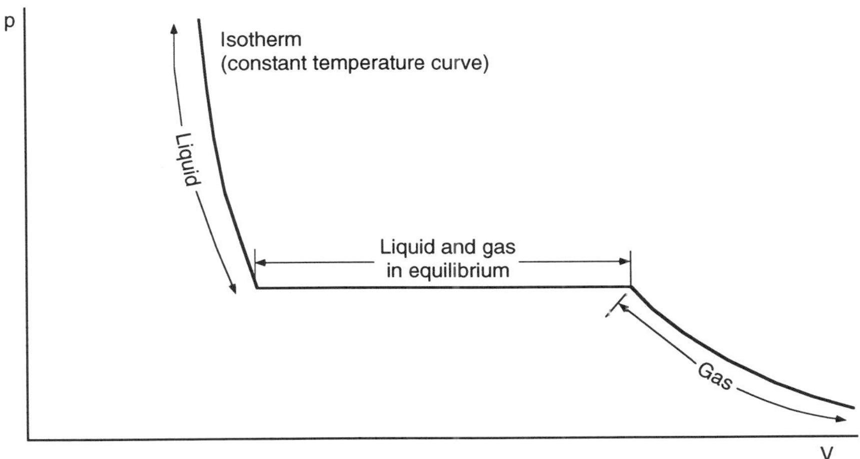

Figure 9.4. An *isotherm* for water, showing that volume expands with declining pressure, including a major expansion (boiling) at a suitably low pressure. (From *Thermal Physics* by Kittel and Kroemer. Copyright © 1980 by W.H. Freeman & Co. Reprinted with permission.)

variable. Moreover, pressure is far less important than temperature, until we get into very advanced considerations of the multicomponent HTSCs.

9.2. TWO-COMPONENT PHASE DIAGRAMS

9.2.1. Solid Dissolved in Water

Two-component systems raise the level of complexity. Returning to Figure 9.1, consider mixing salt (NaCl) with water. From high school chemistry, recall that this raises the boiling point and lowers the freezing point. Figure 9.5 expands Figure 9.1 in the direction of this second component. At the other extreme of pure salt and no water, the freezing point is far higher, and the boiling point of molten NaCl is out of sight. Thus, there must be some transition that occurs at some mixture of H_2O and NaCl.

There is. Up to a point, NaCl dissolves in water; the free energy of the solution gets lower; and the freezing point declines as well. However, there comes a point of saturation, at 8.17% (by mole) NaCl. What happens at salt concentrations higher than that? Figure 9.6 displays the sequence of events for salt and water of one specific proportion. The downward arrows indicate the progressions as a hot liquid mixture is cooled. At some temperature T_1, salt crystals start to form, and these precipitate out of solution, that is, the system breaks into two phases in equilibrium with each other. The phases continue to segregate as the temperature is lowered. As the remaining liquid becomes depleted of salt, its composition follows the downward-sloping line to the left, as indicated by the arrows.

Meanwhile, up at higher temperatures, the combination of NaCl and H_2O behaves in a simple way. When you heat salt water to the boiling point, what leaves is pure water, and the solution that stays behind is more concentrated salt water. This *phase separation* is exactly the process of distillation by which undesirable salts and other contaminants are left behind

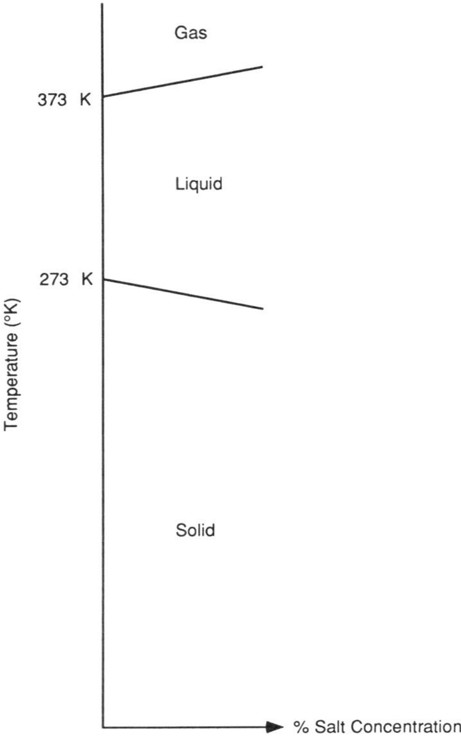

Figure 9.5. Adding salt to water makes it a two-component system. The boiling point of the mixture rises, and the melting point falls, when NaCl is first added to water.

as pure water is drawn off. Ultimately, all that is left is a dry bed of salt. This can be very large, as in western Utah.

This is a good place to bring up the existence of nonequilibrium phase mixtures, which become quite important in certain HTSC processes. It is possible to create a *supersaturated solution* as follows: Referring again to Figure 9.6, consider the salt water solution at some temperature above T_1. If the temperature is lowered below T_1 fairly quickly (without disturbing the liquid solution), it will remain entirely liquid, in a metastable state (a supersaturated solution), until *nucleation* begins, usually at an impurity site. Only then will NaCl crystals start to form. Actually, it is a lot of fun for children to do this with sugar in water, and the large sugar crystals so formed make a delightful candy, especially if colored.

Narrowly speaking, this supersaturated state shouldn't exist because it is not an equilibrium state. Clearly, however, nature tolerates nonequilibrium conditions, and many useful products depend on things remaining out of equilibrium for extremely long times. Makers of steel, glass, and other products all have exploited the kinetics of phase formation to achieve such states. It is an engineering art to see how far you can push a multicomponent system to obtain desired characteristics in a final compound that will remain in a nonequilibrium state. Such strategies are an active part of HTSC research.

9.2.2. Binary Materials

Turning our attention now to ceramics and metals, temperature is the most important state variable in a phase diagram. Imagine starting at a high temperature, in which the system

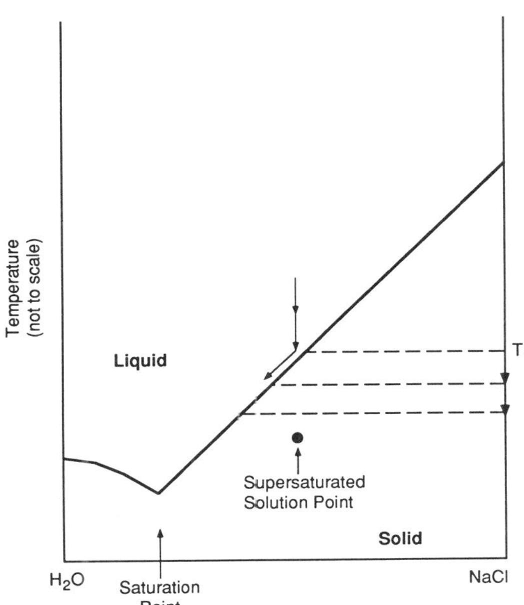

Figure 9.6. Two-component phase diagram to illustrate saturation of water with salt. When a rich solution is cooled (dashed vertical arrow), eventually solid NaCl precipitates out, while the remaining liquid retains less salt as it continues to cool. The supersaturated condition can be reached by cooling the solution quickly, before nucleation of crystals begins.

is entirely in a liquid state. Normally, when a hot molten metal cools it reaches the freezing temperature and then solidifies. Several significant events occur at the melting temperature:

First of all, the liquid metal contains a *heat of fusion* which must go into the surroundings before the temperature can drop any further. (When going up in temperature, it is this same heat of fusion that keeps the temperature from rising until the solid is completely melted— which is why ice cubes are put into drinks.) This heat of fusion is a quantity of energy associated with the disorder present in a liquid; a noncrystalline substance is less orderly than its crystalline form and requires additional energy. Thus, the removal of this energy is a major step in solidification.

Second, solidification begins at nucleation sites (generally impurities or container surfaces, but not always), and lots of small crystals grow at once. Solidification is complete when they meet and fill all space. These individual crystals are called grains. In conventional solidification, things that are foreign to the crystals being formed (impurities and precipitates) are preferentially driven out of the crystals. They accumulate at the boundaries between grains because there is nowhere else to go.

Under very specialized conditions, a solid can grow from only a single nucleation site, and thus only a single crystal will result. The value of some gems depends on the perfection of the crystalline structure. For metals, single crystals are used mainly as research specimens. Typically, as metals solidify, the grain size is around a millimeter.[3] There are reasons why either larger or smaller grains are desirable. For practical alloys, it is generally more desirable to minimize the grain size. For example, in superconducting wire, NbTi carries much higher current if it is made up of very small grains.

Third, in a mixture of two or more elements, the equilibrium percentage of mixing differs between the liquid and solid state. In many mixtures, the process of solidification is accompanied by a separation of the components into two phases, one phase rich in the first

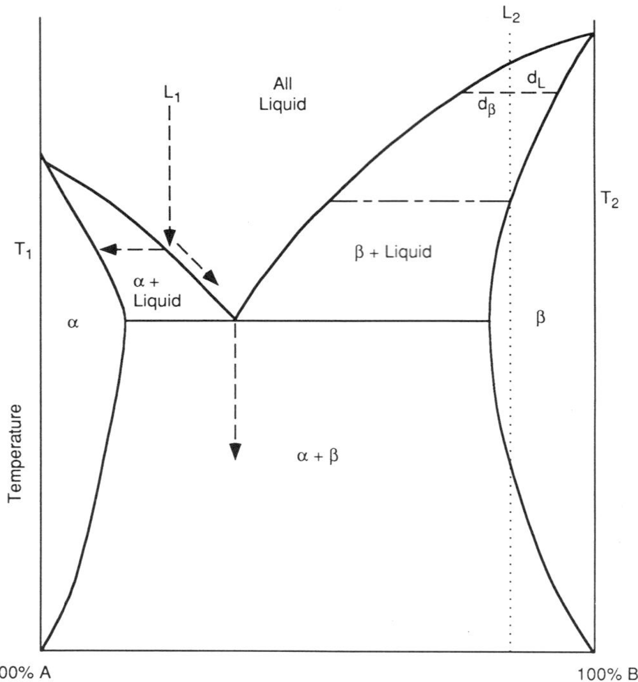

Figure 9.7. Phase diagram for a typical binary mixture. At the left is 100% component A, at the right is 100% component B. Phase α is rich in component A; β is rich in B. The path of the dashed arrows and other marked features are explained in the text.

constituent and the other rich in the second. A typical phase diagram for a binary mixture is shown in Figure 9.7. The solid phases are commonly designated by Greek letters. Cooling from a hot liquid of initial composition L_1, nothing happens until temperature T_1 is reached.

The dividing line is known as the *liquidus*. After that the composition begins to change: as one solid phase (α, predominantly A) segregates out, the remaining liquid becomes slightly richer in component B. So it continues as the temperature drops, with more solid (α, rich in A) forming at the same time as the remaining liquid phase gets richer in component B. Eventually a composition and temperature are reached at which the B-rich liquid also solidifies. That is termed the *eutectic point*.

The final solid is a conglomerate of two solid phases, α and β. The β phase is all of one composition (that of the eutectic temperature, on the horizontal line in Figure 9.7). The α phase includes a range of compositions formed at various temperatures during cooling. While this may be typical of real metals, it is not actually equilibrium. To reach true equilibrium, the solid α phase would have to evolve (through solid-state diffusion) to the α composition of the eutectic temperature. Very rarely is that degree of equilibrium achieved.

9.2.3. The Phase Rule

All this is made easier to grasp by using the *phase rule*, which relates the number of coexisting phases (P) and the number of components (C) to the degrees of freedom (F):

$$F = C - P + 2$$

However, one of those degrees of freedom is usually taken up by fixing the pressure, so sometimes ceramists and metallurgists change the 2 to a 1 and set aside pressure. Referring to Figure 9.7, when in the all-liquid state (only one phase) of this two-component system, both the temperature and the composition are free to change (as well as the pressure): $P = 1$, $C = 2$, $F = 3$. When two phases coexist, such as in the $\alpha + L$ region, F drops to 2—pressure, and either temperature or composition of one phase.[*] Selecting the temperature in a two-phase region fixes the composition of both phases; selecting the composition of either phase (here, α or liquid) fixes the temperature. At the eutectic point, where α, β, and the liquid are all in equilibrium, $C = 2$, $P = 3$ and $F = 1$; but pressure takes up that degree of freedom, so neither the composition nor the temperature is free to vary.

Going back to Figure 9.2, the same rule applies to a one-component system. The triple point ($C = 1$, $P = 3$) is an *invariant* point because $F = 0$: neither pressure nor temperature is free to change. Along the liquid–vapor equilibrium line, specifying either temperature or pressure fixes the other one. Above the critical point, there is only one phase, so both T and P are free to vary. The phase rule becomes more valuable in multicomponent systems, where it helps to keep track of the relations in mixtures containing various liquid and solid phases.

Certain cherished notions from elementary chemistry vanish in a two-component system. There is no longer *one* melting temperature; solidification occurs over a temperature range. Also, the A-rich phase is not of one exact chemical constituency; it can have a range of percentages of A and B in it. As the temperature decreases, the precipitating solid changes chemical composition continuously as it moves along the solidus line. Moreover, the final product depends intimately on the starting composition: as any particular starting mixture cools to form a solid, there are places on the phase diagram that cannot be reached.

So far, the role of kinetics has not been mentioned because phase diagrams deal with equilibrium configurations. Kinetics refers to the rate at which changes take place, and in the real world kinetics can be very important. For example, using a high-speed quench from the liquid to trap certain components in the proper ratio is common in steelmaking. You wind up with a metastable mixture, and subsequent mild heating (annealing) produces the desired properties. The phase rule is evaded by departing from equilibrium. In fact, the art of kinetics is so extensively developed and practiced that here we can only acknowledge its existence.

9.2.4. Eutectics and Peritectics

Referring back to Figure 9.6 and the water–salt system, the melting temperature reaches a minimum at –21.2°C, which is its eutectic point. Figure 9.6 could easily be converted into a true phase diagram by adding the lines that separate regions such as "ice + brine," "brine + NaCl," and so forth.

The tin–lead system, shown in Figure 9.8, is a good practical example of a real two-component system and is reminiscent of Figure 9.7. The eutectic composition is 26.1 atomic % lead, or 38% by weight. Radio solder, usually marked 60/40 in stores, is quite close to that composition, which melts/solidifies at 183°C and thus is preferred to avoid damaging electrical parts when soldering. For plumbing connections, 50/50 tin–lead solder (actually 36.4% lead by weight) is preferred, because the lead-rich phase (around 75% lead, per Figure

---

[*]If the composition of *one* coexisting phase is fixed, the other one will automatically be fixed by conservation principles.

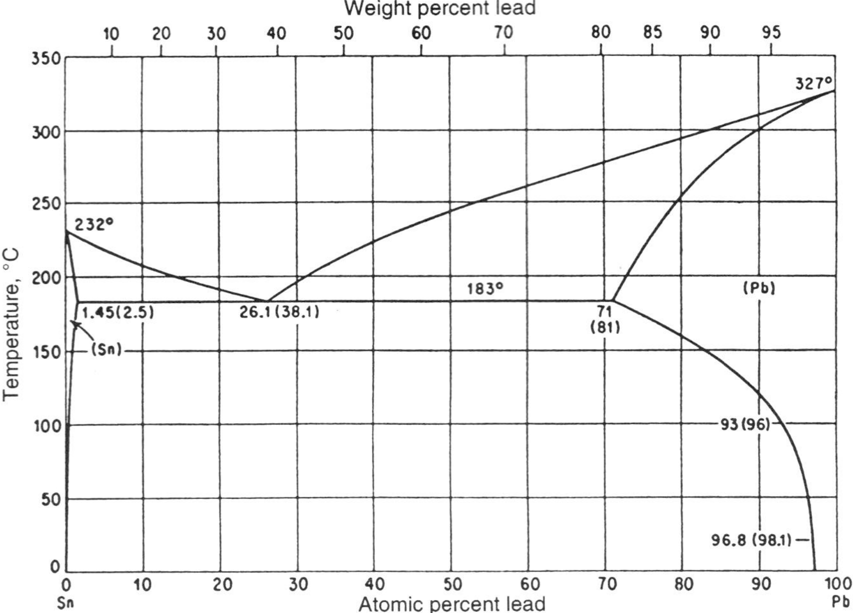

Figure 9.8. Actual phase diagram for tin–lead system. Either atomic percent or weight percent can be used for the horizontal axis. The eutectic point is denoted by the vertical arrow, at 26.1 atomic % lead = 38 weight % lead. (From *Thermal Physics* by Kittel and Kroemer. Copyright © 1980 by W.H. Freeman & Co. Reprinted with permission.)

9.8) adds strength to the joint, and copper pipe can easily tolerate the temperature. For repairing automobile body damage, a 30/70 tin/lead weight ratio is used, because it forms a slurry that can be worked into position over time using a spatula. In each case the composition is selected in order to gain some advantage.

This general kind of eutectic behavior is fairly common in binary alloys, although there may be different phases than illustrated in Figures 9.7 and 9.8. For example, the solder used on gold jewelry is a eutectic made of 69% gold and 31% silicon, which melts at 370°C. No mixed-composition phases form; hence, upon freezing, it separates into phases of silicon alone and gold alone, which occupy the same space. Other, more complex binary systems have many phases. In every case, the determining factor is the minimization of the free energy, and the phase rule is obeyed.

A very different kind of phase equilibrium occurs for a *peritectic* composition. Figure 9.9 shows such a phase diagram. This is easier to grasp by considering a rising temperature situation. The vertical line at the lower left shows a compound of composition A$_2$B. Upon heating, when it reaches its peritectic point it does not simply turn into a liquid; rather, it separates into a mixture of a liquid and pure solid A. That is, the free energy balance shifts so that the original solid A$_2$B vanishes all at one temperature, and above that temperature there is equilibrium between a liquid and a different solid phase, neither of which have the original composition. This is called *incongruent* melting.

Running the process in the opposite direction, if a hot liquid starting with two parts A to one part B is cooled, the first thing that happens is that pure A solidifies, and the liquid

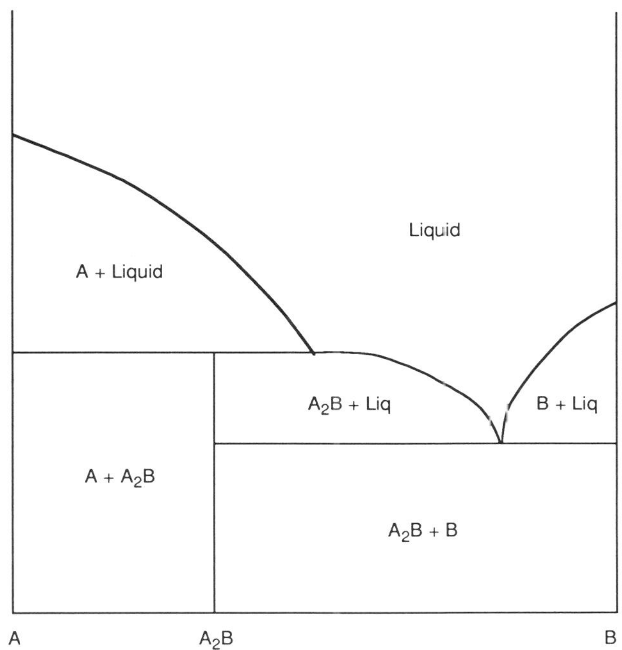

Figure 9.9. Phase diagram for a two-component system having a *Peritectic* transition. The vertical line 1/3 across from the left denotes the compound A₂B. At a certain temperature, this compound is in equilibrium with a liquid and the solid phase of pure A.

becomes richer in B. Upon reaching the peritectic temperature, the liquid by now is 50/50 A/B, and it then reacts with solid A to form the new solid A₂B.

9.2.5. The Lever Rule

So far, our discussion has focused on the left side of Figure 9.7. Consider now the composition {A = 15%, B = 85%}, indicated by the vertical dotted line L_2 on the right. Upon cooling to the liquidus on the right-hand side, phase β begins precipitating and the remaining liquid will enrich in A. However, this liquid will never reach the eutectic point, because it will be all gone—all turned into β—by the time the temperature reaches T_2. This outcome is determined by how much of A was there in the first place (only 15%), and its quantitative description is given by the *lever rule*.

At any temperature above T_2 in the "β + liquid" zone, imagine a lever balancing on a fulcrum that coincides with the dotted line of L_2. In order to balance it, use the same principle as on a playground seesaw: The amount of liquid times its lever arm length must equal the amount of β times its arm length. In terms of the lever arms labeled on Figure 9.7, we have

$$\%L = d_L/(d_L + d_\beta) \quad \text{and} \quad \%\beta = d_\beta/(d_L + d_\beta)$$

Near the top of the dotted line for L_2, at temperatures just below the liquidus, only the first small traces of β form, and nearly all the mixture remains in the liquid state because d_β (the arm length to the liquid state) is tiny compared to d_L (the arm length over to β). At a slightly lower temperature, the percentage going into β increases. As drawn on Figure 9.7, the split is about 50/50. Eventually, the mixture cools to T_2, where the length of lever arm d_L shrinks to zero. At that point the lever rule says that the amount remaining as liquid goes to zero as well.

The lever rule can be applied in any two-phase region. Later on, the lever rule will be very helpful in understanding why some methods of producing YBCO yield so little material of the useful composition (YBCO-123).

9.3. THREE-COMPONENT (TERNARY) PHASE DIAGRAMS

The principles described above continue to govern the behavior of more complex systems. Free-energy minimization determines the equilibrium state, and equality of chemical potentials sets the point of phase transitions. The phase rule still holds, but now $C = 3$. Temperature is still the most important state variable. Solidification still involves separation into different phases, eutectics, heat of fusion, and so on. The big change is in the variety of phase transitions available to the system. The dimensionality of phase space increases with each new chemical constituent, and the increasing complexity becomes harder to represent on a sheet of paper.

When yet another component is added to the phase diagram, even with pressure and volume suppressed, the first question is "Where do we put it?" In order to represent any composition (i.e., any percentages of components A, B, and C), the phase diagram has a triangular-shaped base[4] as depicted in Figure 9.10. With the three corners labeled for components A, B, and C, the grid lines are labeled from 0% to 100% of each. At vertex A the mixture is 100% component A, and at the farthest side from A the compound is 0% A; and so forth. In Figure 9.10, the point I is 40% A, 20% B, and 40% C. If, for example, the three components are copper, zinc, and tin, we have the phase diagram of brass. If they are iron, chrome, and nickel, the phase diagram pertains to stainless steel.

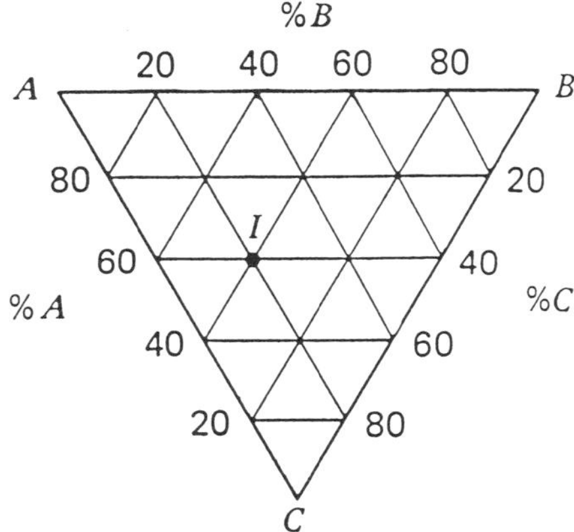

Figure 9.10. Triangular grid for a general three-component system A, B, and C. The point identified as I has the composition $A = 40\%$, $B = 20\%$, and $C = 40\%$. (From Ref. 4. Reprinted by permission of John Wiley & Sons, Inc.)

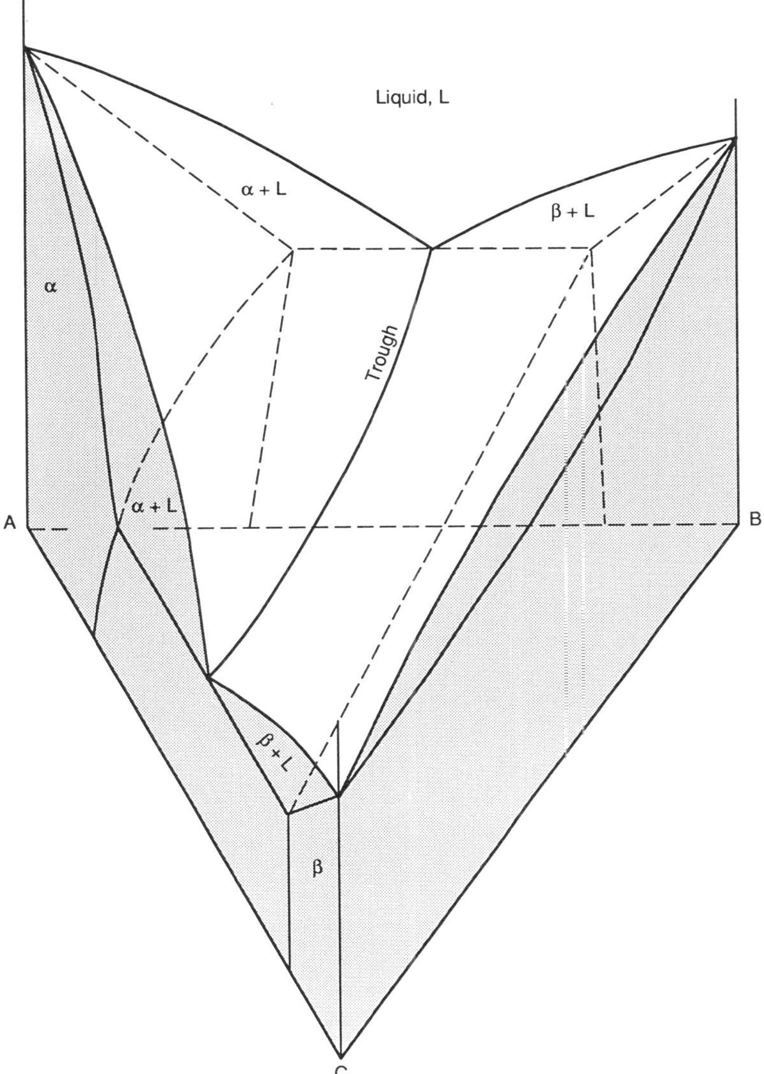

Figure 9.11. Three-dimensional phase diagram for a certain three-component system, where the z-axis is temperature and the triangular grid forms the base in the x-y plane. (After Ref. 4.)

Beginning with a triangular base, a three-dimensional phase diagram can be constructed for a typical ternary mixture by showing temperature as the z-axis (Figure 9.11). This is a fictitious metal, intended only to convey certain concepts visually. It will be immediately apparent that there are several different solid phases, and these can be in equilibrium with the liquid in selected temperature and composition ranges.

The increasing generalization from lower dimensionality has important consequences, which follow directly from the phase rule. In a one-component system, there is one temperature of melting. In a two-component system, there is a liquidus *line*, and both the temperature and the concentration ratio of the system vary together. In a three-component system there is a liquidus *surface* in phase space, so that the melting temperature depends on *both* concentration ratios.

The most important consequence of increasing the number of components is to make the definition of the cooling path require more compositional variables.[5] The end result depends on both the starting ratio of elements and the pathway of cooling from the liquid mixture. We see this again and again in high-temperature superconductivity. When experimenters describe their observations about some superconducting property of YBCO, it is of utmost importance for them first to describe exactly how they made their samples.

For the system of Figure 9.11, this point is illustrated in Figure 9.12. Here a liquid metal alloy starts cooling from a high temperature with a composition L_1. Eventually, it comes to the liquidus line for phase α at temperature T_1. A slice through Figure 9.12 (an isothermal plane) is shown in shading. At this temperature, the solid phase α starts to precipitate out with a composition α_1, which is rich in component A. As cooling continues, the remaining liquid becomes richer in components B and C, moving downward along the liquidus surface toward the *cotectic* trough.

A top-down view of the isothermal slice is shown in Figure 9.13. Here the liquidus and solidus lines for both phases are shown in order to bring out another point (which gets to be particularly significant in treating HTSCs): There are *tie lines* connecting the various compositions of liquid and solid phases that are in equilibrium at this temperature. For example, L_1 connects to α_1. These tie lines always lie in the isothermal planes and determine the composition of the phase that solidifies from the liquid. It works in the opposite direction, too. If a solid of composition α_1 is heated until melting begins, the first liquid to appear will have composition L_1. This is one reason why the properties of HTSCs depend on their processing path.

It is worth noting that very often ternary phase diagrams are presented as triangles akin to Figure 9.13, but with many temperatures superimposed together. In a manner analogous to the projections on the "walls" of Figure 9.3, such triangular drawings are not restricted to being isothermals. Features whose equilibrium occurs at various temperatures are commonly shown on the same drawing. Therefore, they conceal the pathways of varying composition below the liquidus that a system follows to its equilibrium.

Just as a slice can be made at any temperature, so the phase diagram can be sliced along any composition line. In fact, it need not be along a line that keeps one phase constant. (The two outer walls visible in Figure 9.12 are slices with $B = 0$ and $A = 0$, which are pure binary phase diagrams.) Figure 9.14 shows a slice that runs from the composition $\{A = 80\%, B = 20\%\}$, to the composition $\{C = 60\%, B = 40\%\}$. The shading deliberately omits the liquid region, for easy visualization, but, of course, the slicing plane continues upward in temperature. This kind of a cut is called a *pseudo-binary* phase diagram. It is often convenient to utilize these in order to focus on one particular aspect of the phase space.

Moreover, in analogy with the way many temperatures may be superimposed on a single triangular drawing, it is common practice to superimpose various constituencies upon a single plane such as in Figure 9.14. However, great care is required in doing so, because some complex chemistry may be misrepresented that way. Nonetheless, *pseudo-binary* diagrams are widely used and are indispensable for understanding complex systems.

Returning to the pathway of Figure 9.12, what happens as the temperature continues to drop? Figure 9.15 continues the story. On the one hand, the liquid gradually changes composition from L_1 to L_2 as it moves down the liquidus surface for phase α; at the same

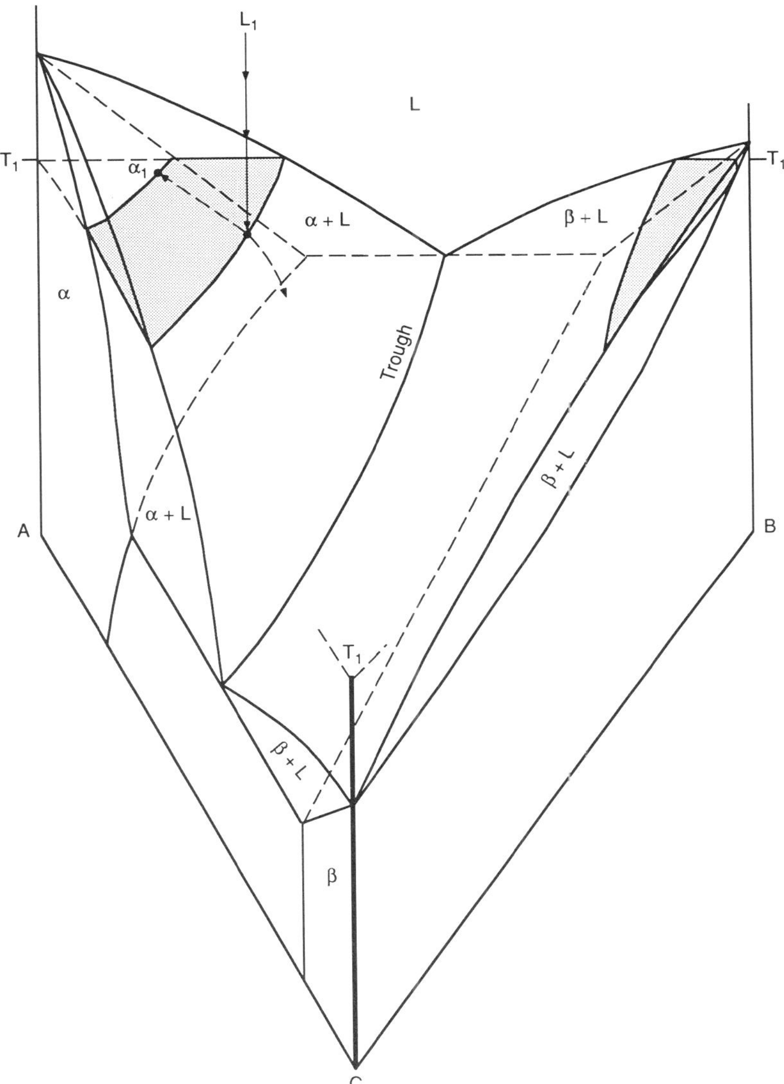

Figure 9.12. Figure 9.11 redrawn with background construction lines suppressed, and a slice through the drawing taken at temperature T_1. When the liquid mixture (with starting composition L_1) is cooled, it reaches a point on the liquidus line from which a phase rich in component A (phase α) solidifies. With still further cooling the composition of the remaining liquid changes, becoming richer in *B* and *C*. (After Ref. 4.)

time, the composition of the solid phase is changing from α_1 to α_2. Upon reaching the trough, an isothermal slice (shaded) shows that the liquid is in equilibrium with two different solid phases (α and β) having compositions α_2 and β_2, respectively. Notice what this means: this is a cotectic point, at which two different solids are forming. The only constraint upon composition is that the average of the solid compositions that precipitate lie along the line

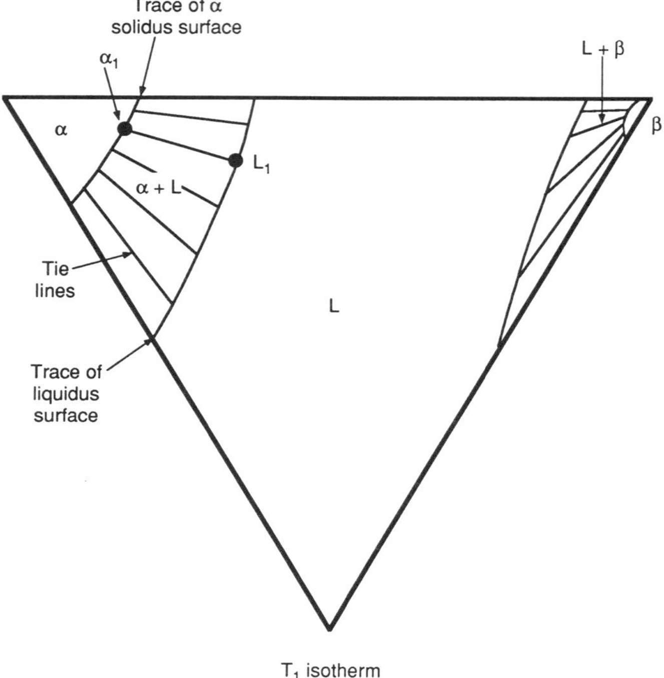

Figure 9.13. The isotherm for temperature T_1 corresponding to Figure 9.12. This is *not* a projection onto the base plane, but a single-temperature phase diagram. The *tie lines* from any particular initial composition determine the composition of the solid phase precipitated. In this illustration, solid α_1 is formed from liquid L_1. (From Ref. 4. Reprinted by permission of John Wiley & Sons, Inc.)

connecting α_2 and β_2. But either of these are light in component C compared to the liquid from which they are freezing.

Thereafter, further cooling will require the remaining liquid to become richer in C, moving down the trough toward composition L_3. Even there, the compositions of the solid precipitates are richer in A and B than the liquid composition L_3, as shown by the lower shaded plane in Figure 9.15. *The liquid has not frozen at one temperature, but over a range of temperatures, as well as a range of compositions!* Even if one started with a liquid of composition L_2 instead of L_1, the final liquid does not have that composition. When everything is frozen, the average composition equals the starting composition, but the compositions of the individual phases can be widely different. They are, however, fixed at equilibrium by the tie lines (not shown), which describe the compositions of α and β that coexist with the liquid at various temperatures.

Figure 9.15 is contrived to have L_2 lie directly above the line connecting α_3 and β_3 (at point X). Therefore, the equilibrium final state (starting from L_2) would be solid phases α_3 and β_3. But that is not necessarily what forms in real processes, because ternary alloys do not always freeze in an equilibrium manner. Kinetics must also be considered. The old slogan "You can't get there from here" is often very relevant in studying multicomponent phase diagrams.

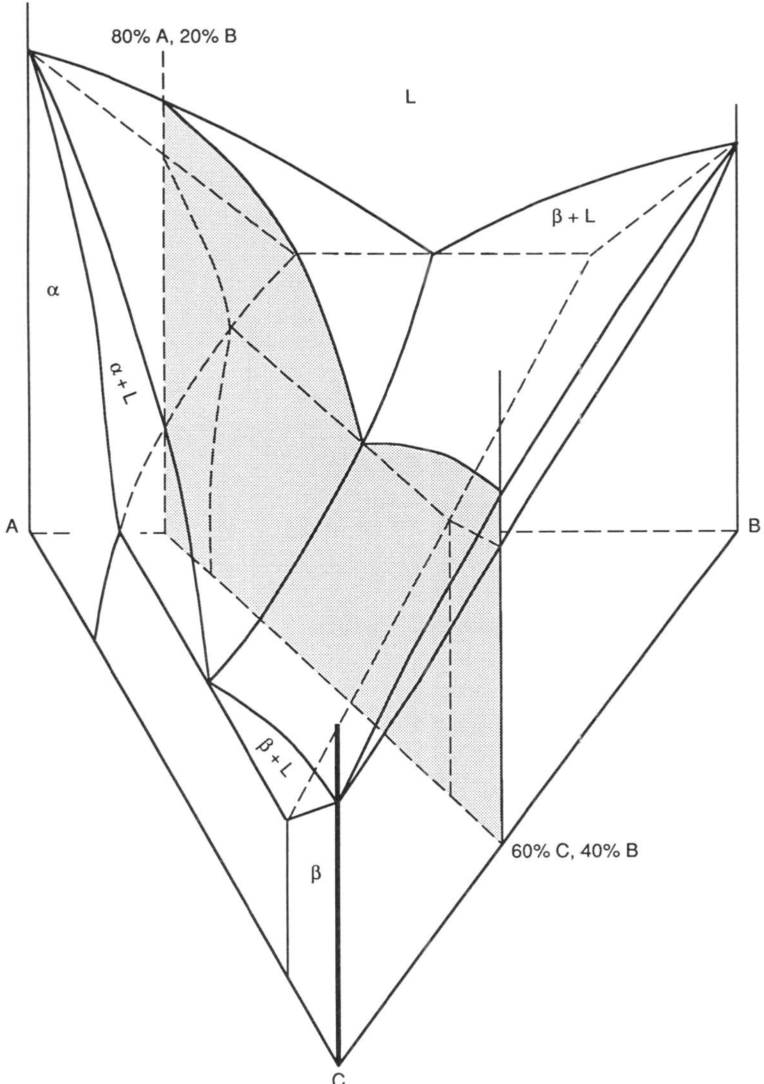

Figure 9.14. A *pseudo-binary* phase diagram is a plane sliced through a phase plot, in order to emphasize some aspect of the behavior. The shaded plane has composition varying from {80% *A*, 20% *B*} to {60% *C*, 40% *B*}.

9.4. PHASE DIAGRAM FOR YBCO

After all those preliminaries, we are at last ready to discuss the phase diagram of YBCO. The objective in what follows is to convey an appreciation of why it is so difficult to make HTSCs and to indicate how ceramists are utilizing phase diagrams to choose plausible paths toward that goal. Even the various methods of making wire (powder-in-tube, etc.) from good starting materials involve excursions to high temperatures, so phase diagrams have a most practical application.

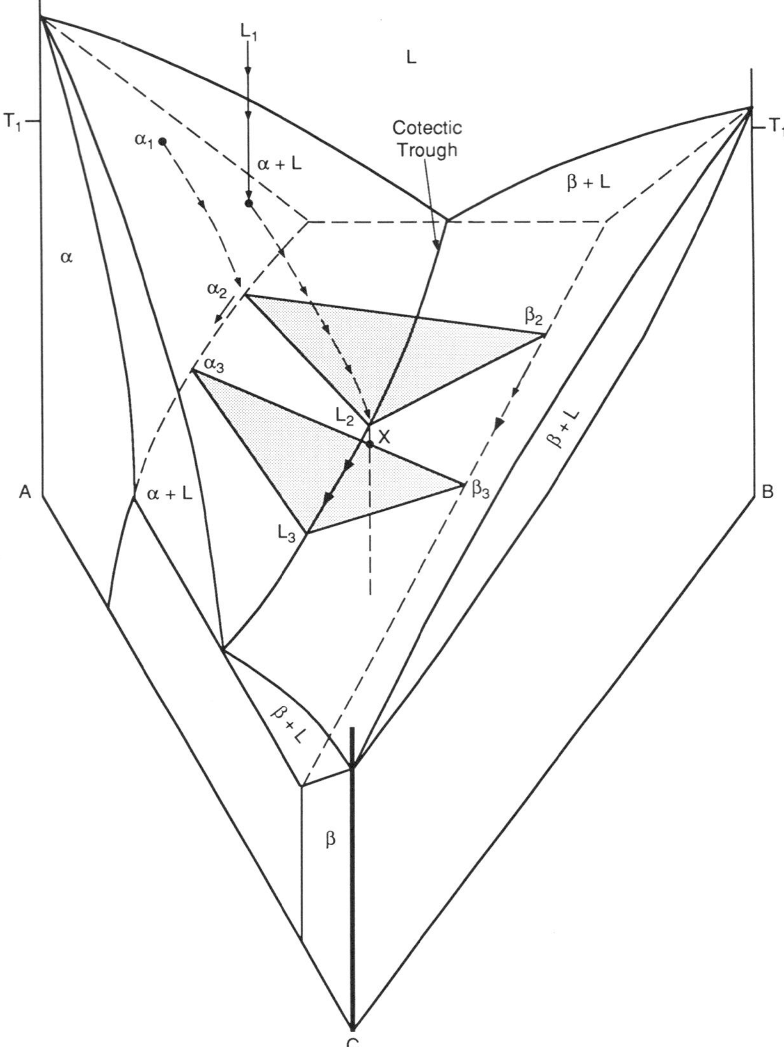

Figure 9.15. After further cooling and precipitation of more of phase α, the liquid composition follows the arrows and reaches the *cotectic trough* at a temperature T_2 with a composition L_2. The upper shaded triangle is the isotherm for T_2; it shows that both phases α and β form. Still further cooling leads to composition L_3, and the lower triangle designates the new compositions of the solid phases formed. The point X is discussed in the text. (After Ref. 4.)

First of all, although YBCO contains four elements, it is regarded as a three-component system by using the oxide compounds as the pure components. Thus, pure component A is Y_2O_3, component B is BaO (or sometimes $BaCO_3$), and component C is CuO. (Actually, the effective valence of the copper ion varies from one to three during formation of the superconductor, so it should be CuO_x.) The superconducting compound is $Y_1Ba_2Cu_3O_7$. If life were simple, we could mix the right proportions of each, heat it, and obtain the

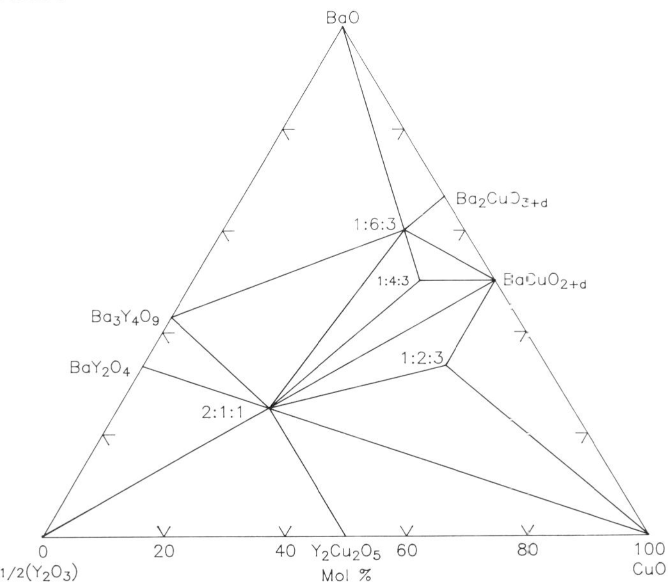

Figure 9.16. Triangular phase diagram for YBCO, with important phases at 900°C labeled. Each tie line connects two (or more) phases that exist in equilibrium at this temperature. [From F. Abbattista *et al., Materials Chem. Phys.* **21** (5), 521 (1989).]

superconductor. Regrettably, all the complexities described above come into play en route to forming YBCO-1-2-3.

The most familiar diagrams published are top-down projections of the three-component plane. Figure 9.16 is typical.[6] It is an isotherm at 900°C (in air) and it includes the tie lines connecting the most important phases. (The composition of the phases labeled 1:4:3 and 1:6:3 are still controversial, and may contain CO_2 as a component.) Mixtures such as liquid plus CuO, which form at much higher temperatures, do not appear here. The ternary eutectic may actually occur slightly below 900°C in air, giving rise to a transient liquid that results in liquid phase sintering, often found in YBCO ceramics. The 1-2-3 phase ($Y_1Ba_2Cu_3O_7$) lies well over to the right. It is composed of three parts CuO, two parts BaO, and one-half part Y_2O_3. The oxygen present in such a mixture adds up to only 6.5 parts oxygen, and so there is normally a final anneal in air or oxygen in order to bring the oxygen content up to 7.

Many other equilibrium mixtures and compounds are locatable on the triangle, such as $BaCuO_2$, halfway down the line connecting BaO with CuO, and directly opposite the Y_2O_3 vertex. The 2-1-1 phase of YBCO, also known as the green phase, which is not superconducting, is lower left of center. Had the corners been labeled in some other order, this diagram would rotate or reflect about its center; that is not significant.

However, the trick of reflection is useful for visualizing special characteristics of the phase diagram. Consider the rhombus on Figure 9.16 that connects four points in the lower right region: the 2-1-1 phase (Y_2BaCuO_5), the 0-1-1 phase ($BaCuO_2$), the 0-0-1 phase (CuO), and the 2-0-2 phase ($Y_2Cu_2O_5$). The 1-2-3 phase lies out in the middle of that district. It is helpful to construct a three-dimensional representation to portray temperature variations in

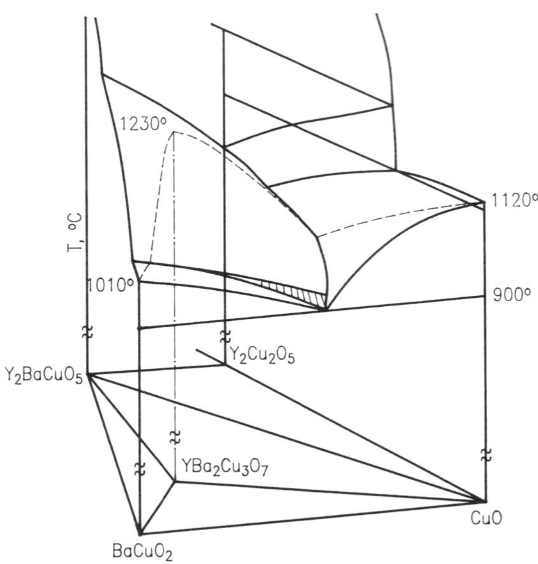

Figure 9.17. Phase space construction of region in lower right of Figure 9.16, but reflected for ease of visualization. Points of emphasis here are (1) the hatched area is the *primary phase field* for the formation of YBCO-1-2-3, and (2) starting from a stoichiometric 1-2-3 mixture (dashed lines) leads to other phases and *not* to the 1-2-3 phase. Also, when the ambient gas is oxygen and not air, and when the pressure is not one atmosphere, some differences occur. (See Fig. S103j in Ref. 6; originally from Ref. 7.)

that locale. Figure 9.17 is that plot, where in order to ease visualization a reflection from Figure 9.16 has been introduced.

There is one very important piece visible in Figure 9.17, shown by hatching: it is a thin little slice in the foreground, known as the *primary phase field* for forming 1-2-3. Its composition is not that of the 1-2-3 phase, but it is the region of composition from which 1-2-3 forms directly from the liquid upon cooling. When a liquid composition lies in the primary phase field for 1-2-3, the first crystals to precipitate will be 1-2-3. In Figure 9.12, composition L_1 lies in the primary phase field for solid phase α, and not for solid phase β.

The narrowness of that primary phase field for 1-2-3 is the leading impediment to making copious amounts of YBCO superconductor single crystals. Any liquid having the stoichiometric ratios for 1-2-3 lies in the primary phase field for 2-1-1, the nonsuperconducting green phase. Upon cooling, 2-1-1 is what first comes out. The primary phase field for 2-1-1 is much bigger than for 1-2-3. Figure 9.18 is another top-down view like Figure 9.16, but with the primary phase fields for 2-1-1 and 1-2-3 overlaid on it. Evidently, to make 1-2-3 from the liquid state, it is necessary to start out very copper-rich. Crystals of 1-2-3, once formed, ought to be removed from the liquid; or else they will eventually be surrounded by cooled, solid oxides of copper and barium, which are insulators.

A definition introduced earlier enters here. In Figure 9.10, the compound A_2B is a *line compound* because it has exactly that one composition. Similarly, $Y_1Ba_2Cu_3O_x$ is a line compound. Upon cooling a liquid with initial composition in the primary phase field for 1-2-3, the solid phase that forms is the line compound $Y_1Ba_2Cu_3O_x$, which is now in equilibrium with the remaining liquid.

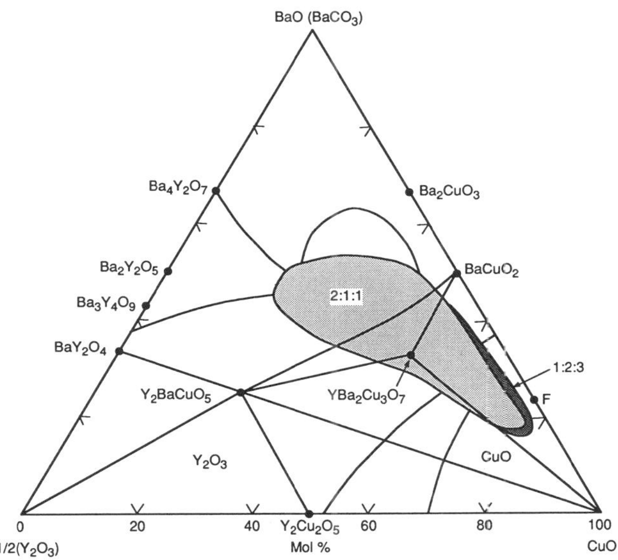

Figure 9.18. Top-down view based on Figure 9.17, with the *primary phase fields* for the 2-1-1 and 1-2-3 phases superimposed on the drawing. From any initial composition within the medium-gray area, 2-1-1 phase is the first to form upon cooling; from within the dark gray area, 1-2-3 forms instead. The point *F* on the barium–copper border is close, but not equal to, the binary eutectic composition near 70% CuO, and is used in Figure 9.19. (Courtesy of R. S. Roth, National Institute of Standards & Technology.)

A pseudo-binary diagram cutting Figure 9.18 along the line connecting 1-2-3 to the point marked *F* on the Ba-Cu border illustrates this transition, and emphasizes just how precarious is the formation of 1-2-3 from the melt. Figure 9.19 presents that cut. For most choices of the liquid composition, cooling leads to the formation of 2-1-1 from the liquid.[7] Only when the composition lies in the narrow shaded band near the right can 1-2-3 form directly from the liquid, and then only in a narrow temperature range near 960°C.

The lever rule tells us that very little 1-2-3 will form along this short liquidus line: for any composition in the shaded band, the lever arm across to 1-2-3 is much much longer than the arm back to the liquidus, so the fraction turning into 1-2-3 must be small. Having formed a little 1-2-3, the cooling liquid evolves to a different composition and begins to produce other unwanted materials.

Fortunately, there are other paths to the formation of 1-2-3, and therein lies much of the art practiced since 1987. In any other portion of Figure 9.19, cooling first produces 2-1-1, then {2-1-1 + 1-2-3}, then {1-2-3 + barium cuprate}, etc. The trouble with such a route is that it is very hard to get rid of the other materials later on.

One trick is to heat a 1-2-3 mixture into the liquid state, and then quickly quench to some nonequilibrium intermediate state. Although not yet superconducting, at least the right atoms are in proximity, and therefore the necessary reactions have a better chance of completing. The techniques known as *melt processing, melt-quench-melt-growth,* and similar names involve partial or complete melting of the starting materials. They are all variants upon the very difficult task of finding a pathway to the 1-2-3 state. To enhance flux-pinning, sometimes the object is to leave tiny grains of 2-1-1 within much larger grains of 1-2-3.

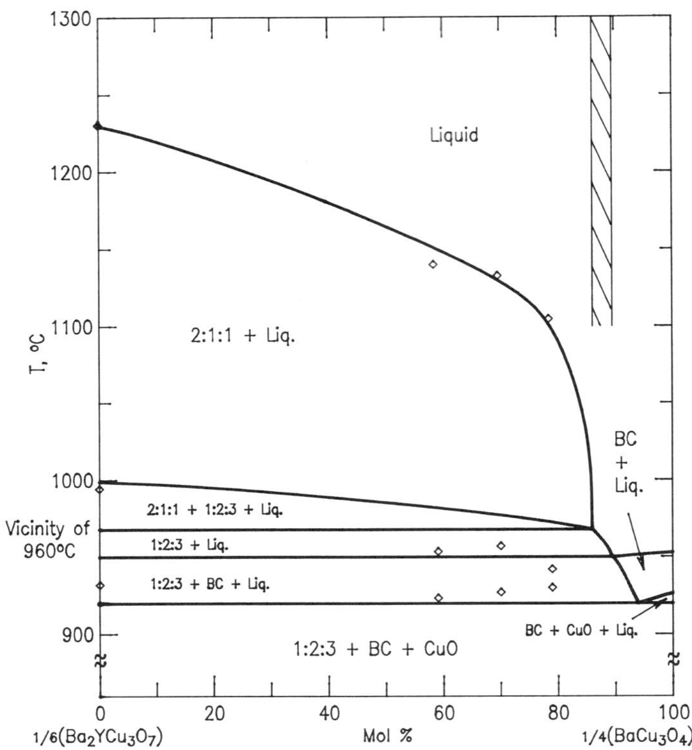

Figure 9.19.   Pseudo-binary slice through the YBCO phase diagram, connecting the 1-2-3 point with point *F* of Figure 9.18. This choice brings out the very large zone of compositions from which 2-1-1 forms, and the very narrow zone (shaded vertical bar) from which 1-2-3 comes out of solution. The notation BC means "barium cuprate". (Figure S103f in Ref. 6; originally from Ref. 7.)

Forming 2-1-1 and then trying to convert it to 1-2-3 by adding copper oxide (moving along a connecting line in Figure 9.18) has also been tried.

In Chapter 15, the pathways to making HTSC powders will be discussed more fully. But it is important here to realize that we are still learning about these very complex phase diagrams. For example, a few years into this research, it was found that the primary phase fields denoted by shading in Figure 9.18 are not correct. In reality, the primary phase field of Y_2O_3 extends well beyond the indicated boundaries (i.e., well into the light shaded region of Figure 9.18). This means that Y_2O_3 will form first as the mixture is cooled. Indeed, the 1-2-3 composition itself has solid Y_2O_3 existing as the primary phase from above 1230°C to approximately 1475°C. Therefore, melt processing must be taken to about 1475°C to avoid extensive production of Y_2O_3 crystals in the melt. None of this was understood at first,[8] and a lot of early work was confused because of it.

On the other hand, if one comes up from below in temperature with the right stoichiometry for 1-2-3, and never quite enters the melting region, but instead heats for a long time in the hope of carrying a solid-state reaction to completion, 1-2-3 can form with little of these nuisance phases appearing. The trouble is that the very slow solid-state reaction may not go fully to completion, and during the time of the process various impurities can enter the material (e.g., platinum from the crucible), eventually to interfere with the desirable properties.

Again, tricks have been used with some success. For instance, the 1-2-3 phase does not begin to form in air below 750°C, as it needs some Cu^+ ions in its structure. It can be formed much more easily at low oxygen pressures and temperatures near 600°C. The result is $Y_1Ba_2Cu_3O_6$, which can then be oxidized at low temperature to $Y_1Ba_2Cu_3O_7$. Remember, the fully oxidized phase is always metastable because the reduced phase *must* be formed first.

No one is very happy with any process to form YBCO. All the paths that produce good crystals are very slow, although the product carries high superconducting current. All the paths that are practical for manufacturing lead to a product that cannot carry much current. In a major review paper in 1991, AT&T Bell Labs researcher Sungho Jin commented[9] on the state of affairs:

> While remarkable progress has been made in materials science and all aspects of processing techniques for wire fabrication, weak-link removal and flux-pinning enhancement, further processing innovations are needed in order to accelerate progress toward major bulk applications.

What is meant by "further processing innovations"? Finding a means to evade the difficulties presented by the very uncooperative phase diagram for YBCO, and steering a course through them with a process that is manufacturably fast.

9.5. FOUR-COMPONENT PHASE DIAGRAMS

It doesn't get easier with BSCCO and TBCCO. There are five elements in these compounds, but as in the YBCO case, oxygen is suppressed by basing the diagram on the oxides of each metal. With four such metals, the triangle graduates into a tetrahedron, of which Figure 9.20 is a representative sample. As before, each vertex means 100% of that component, and the opposite face contains 0% of it. Thus, in Figure 9.20, dealing with BSCCO, the composition denoted 8:2:5:0 contains no copper. The 80 K superconducting phase $Bi_2Sr_2Ca_1Cu_2O_x$ (= 2212) appears somewhere inside the tetrahedron.

Just as we constructed pseudo-binary diagrams to simplify three-dimensional figures, it is convenient to construct triangles (pseudo-ternary diagrams) to save having to visualize looking obliquely through a tetrahedron. For example, if the amounts of strontium and calcium are always equal, then those two can be combined into one single vertex, and the dimensionality of the drawing reduced. We have done this in Figure 9.21(a), where now the top vertex is (SrO + CaO)/2, and the other two vertices correspond to oxides of bismuth and copper. This plane cuts the tetrahedron of Figure 9.20 almost perpendicular to that page. This plane bisects the line connecting strontium with calcium, and goes through the dashed line at the rear of the tetrahedron. The compound BSCCO-2223 lies in this plane, because Sr and Ca contribute equally.

An alternate slice is shown in Figure 9.21(b): all that has changed is the ratio of strontium to calcium, now 2:1. Again the plane contains the rear dashed line, but now it trisects the Sr-Ca line. It contains the phase BSCCO-2212, which is shown interior to the tetrahedron of Figure 9.20. As before, there is risk of misunderstanding here, because certain variables are being suppressed. Analysts must use great care when dealing with such limiting projections. For instance, the BSCCO-2212 phase may never occur in equilibrium at this

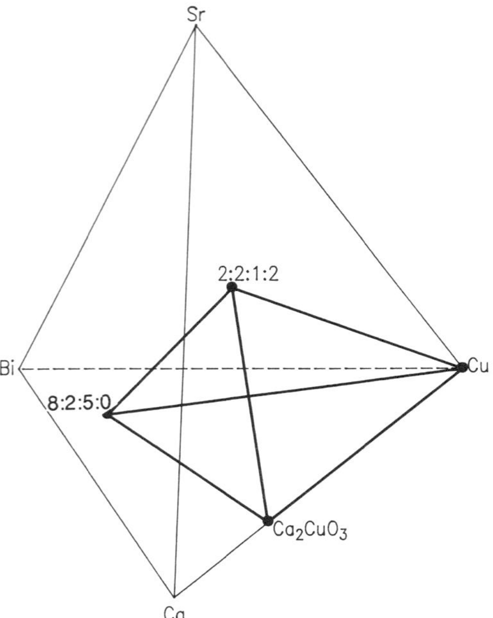

Figure 9.20. Tetrahedral phase diagram for BSCCO, highlighting just a few of the many phases that occur. In analogy with Figure 9.10, each vertex corresponds to 100% of the oxide containing the element marked. [From B. Hong *et al., J. Amer. Ceram. Soc.* **73**, 1965 (1990).]

composition,[8] and it has been suggested the proportions $Bi_4Sr_3Ca_3Cu_4$ may be closer to the equilibrium composition of this phase.

Cutting a line anywhere across the triangles of Figure 9.21 and introducing temperature as a variable makes a pseudo-binary diagram, akin to Figures 9.14 or 9.19. Such a construction brings out that Bi_2O_3 alone melts at a relatively low temperature compared to SrO, CaO, or CuO. The "liquid" region of Figure 9.21 near the bismuth-rich corner testifies to this point, as well. Figure 9.21(a) contains the highest-T_c phase of BSCCO, the 2223 phase. A region of liquid from which it forms is shown, together with certain other phases. However, in the real world it is extremely difficult to form BSCCO-2223 directly.

There are many experimental difficulties blockading these routes. For one thing, fine powders must be used, and they must be very carefully mixed. If the powders are not homogeneously mixed, the local composition at any point may be unrepresentative, and so the final local phase assemblage will be dictated by the local starting composition rather than the global average. Very long times would be needed to allow solid-state diffusion to attain equilibrium.

Whenever a local composition is near the composition of a eutectic, liquid may form at a temperature hundreds of degrees below the temperature at which liquid would form for the overall composition. With regard to achieving equilibrium, liquid formation is a two-edged sword: further reactions may occur more quickly in the presence of a liquid, hastening the homogenization of the composition; or the liquid may drain through the powder before the reactions are complete, if it has high fluidity and the reactions with the solid phases are slow.

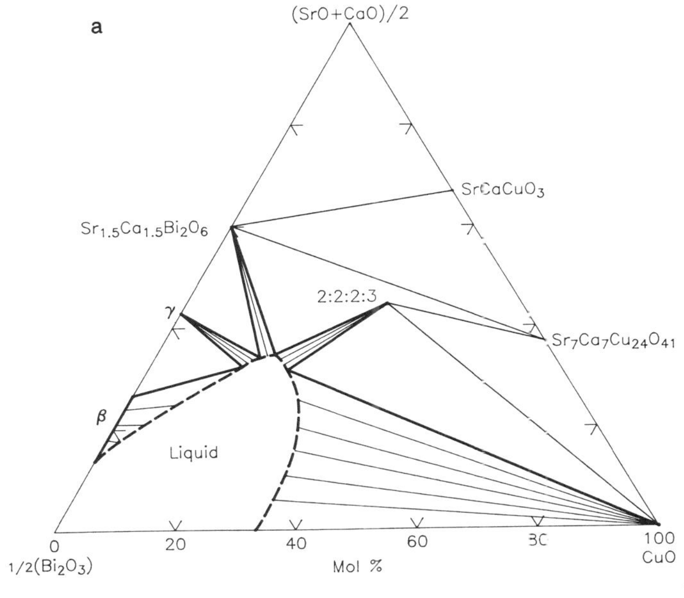

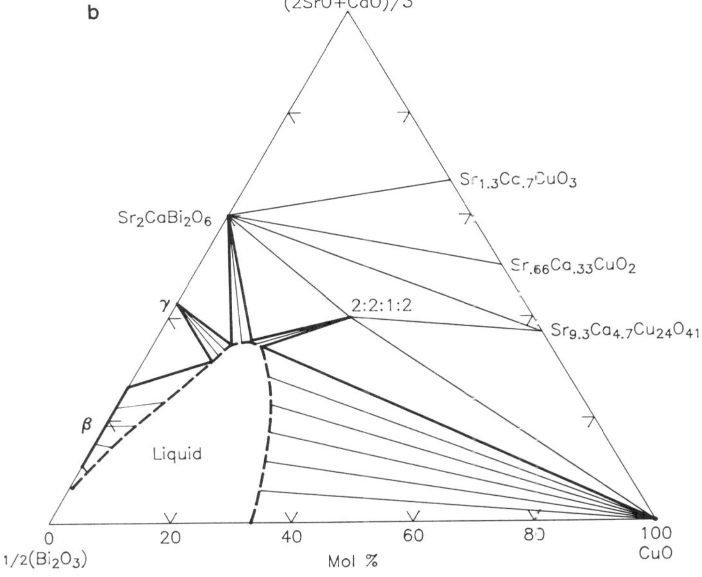

Figure 9.21. Pseudo-ternary triangular diagrams pertaining to BSCCO at 850°C. Here the strontium and calcium components are held fixed relative to each other, and are condensed into one vertex. The other two vertices are oxides of bismuth and copper (a) The Sr:Ca ratio is 1:1 everywhere in a slice through the tetrahedron of Figure 9.20. The compound BSCCO-2223 lies in this projection, along with a large bismuth-rich liquid region, and (b) The Sr:Ca ratio is 2:1 everywhere in this slice. The corresponding diagram differs little, but the compound BSCCO-2212 appears in this projection. [Figs. S224b and S224d in Ref 6; originally from P. Majewski *et al., Proc. Dtsch. Ges. Mater. Mtg. HTSC Mater. Aspects*, Garmisch-Partenkirchen, Germany (May 1990).]

Unfortunately, there has been little success in trying to make BSCCO-2223. It is basically a kinetics problem rather than an equilibrium problem. In these superconductors, various phases form at different rates, so opportune timing of process steps can be exploited to advantage. The phase diagram can sometimes teach us how to evade equilibrium or guide us on an indirect path to equilibrium. Rapid quenching works very well in making certain alloys and metallic glasses. This same notion has been tried with YBCO, but certain reactions go so fast that it doesn't work. Fortunately, the kinetics of BSCCO are slightly slower, and it is possible to quench into a glassy state. This is an enormous advantage, because that state provides a new starting point from which further processing can be done.

Researchers have found that by partially substituting lead for bismuth, the phase $(Bi,Pb)_2Sr_2Ca_2Cu_3O_{10}$ can be made. So, we are really dealing with a six-component system. There are not yet any phase diagrams of this system, and, indeed, no one knows whether the lead all sits on bismuth sites or precipitates out as $CaPbO_2$. In Chapter 15, we revisit this topic and consider the possible role of $CaPbO_2$ as a liquid flux that facilitates conversion of (Bi,Pb)SCCO-2212 into (Bi,Pb)SCCO-2223.

The possibility of exploiting kinetics makes it even more important that reports include details on how samples are made. The risk of having spurious phases is always present, and probably accounts for most of the erroneous input to the literature during the first few years of high-temperature superconductivity.

The thallium compounds pose even more problems, and hence relatively little is known about their phase diagrams. Thallium oxide (Tl_2O_3) is volatile (and poisonous!), and so it is hard to know, while a reaction is proceeding, just how much thallium is really present. In metallurgy and most ceramics, the state variable of vapor pressure is safely ignored when constructing phase diagrams; for thallium superconductors, that is an invalid assumption. TBCCO phase diagrams by different investigators can look quite different, depending on their pressure and temperature conditions.

To map any part of the TBCCO phase diagram, it is necessary to control the thallium pressure with an external thallium buffer. Even then, the sensitivity to slight variations in temperature is great. Thallium oxide starts to volatilize at 700–800°C, which is right where the interesting phase formation takes place. The experimental constraints on temperature and composition are very severe,[10] and it will be a while before dependable phase diagrams are available for TBCCO.

9.6. SUMMARY

This chapter has covered the subject of phase diagrams, beginning from simple familiar examples and progressing to the HTSCs. Diagrams of this type are limited to representing equilibrium mixtures. On the one hand, this limitation allows a systematic understanding of the major relationships between components; on the other hand, it conceals the existence of a rich variety of nonequilibrium compounds such as stainless steel, glasses, and many ceramic composites. Researchers trying to optimize HTSCs for making practical wire exploit nonequilibrium configurations routinely; thus, the study of phase diagrams can only serve to convey part of the information about these materials.

Water is a good example of a one-component system, and its phase diagram is reviewed here as a vehicle for introducing concepts and terminology about phase diagrams in general. Tin–lead solder is a typical example of a two-component system; such concepts as the liquidus and solidus lines, and the eutectic point, are introduced here. As each new chemical constituent is added, the degree of freedom of the phase space increases, and therefore the complexity increases as well. The phase rule brings some order to the problem: $F = C - P + 2$. Using this rule, it is possible to know how many different phases of a mixture can be in equilibrium with one another at any point on the phase diagram.

Three-component systems are difficult to visualize on paper because the phase diagrams are at least three-dimensional. In general, what had been a point in two dimensions becomes a line; a line becomes a surface; a surface becomes a volume. The practice of slicing the phase diagrams with isothermals is used to explain how certain solid phases can freeze out of a quite different liquid. Pseudo-binary phase diagrams are slices through such phase diagrams that promote interpretation in terms of two-component analogs; these are convenient and simpler, but risk misunderstanding when a complex possibility is hidden this way.

YBCO has a complicated phase diagram, and the pathway to forming the 1-2-3 phase is intricate because of the many alternative phases that can form. The phase diagrams commonly used presume the base components are the oxides of Ba, Cu, and Y; but the role of oxygen is not nearly so fixed and concise. In reality, the phase diagram changes depending on whether the YBCO is being made in air or in oxygen of various pressures. Appreciating the intricacy of the phase diagram helps to understand why researchers put so much emphasis on the details of the processes by which YBCO is made.

When yet another component is added, the phase diagrams for BSCCO and TBCCO can best be termed *challenging* since they are based on a tetrahedron with additional spatial dimensions of temperature and pressure. By combining strontium and calcium in a fixed ratio, these drawings can be reduced to pseudo-ternary diagrams; this simplifies the task of representing the myriad phases. However, experimenters are finding it advantageous to partially substitute lead onto the bismuth or thallium sites and mix barium with strontium on their sites. Thus, the real phase diagram is getting further out of hand, so that pseudo-ternary projections become less useful. The preparation of these compounds is largely empirical; researchers eschew the term *scattergun* in favor of the more refined *Edisonian*. The outcome is that phase diagrams are helpful for explaining what happens after the fact.

In the most advanced cases of interest for HTSC—that of the thallium-based superconductors—very little is yet known about their phase diagrams. The great volatility of Tl_2O_3 in the range 700–800°C makes it difficult to conduct measurements to establish the phase diagram of the compound.

REFERENCES

1. F. W. Sears, M. W. Zemansky, and H. D. Young, *University Physics*, 7th ed. (Addison-Wesley, Reading, MA: 1987).
2. C. Kittel and H. Kroemer, *Thermal Physics*, 2nd ed. (Freeman, San Francisco: 1980).
3. L. S. Darken and R. W. Gurry, *Physical Chemistry of Metals* (McGraw-Hill, New York: 1953).
4. J. D. Verhoeven, *Fundamentals of Physical Metallurgy*, Ch. 9 (Wiley, New York: 1975).
5. F. A. Hummel, *Phase Equilibria in Ceramic Systems* (Marcel Dekker, New York: 1984).
6. J. D. Whitler and R. S. Roth, *Phase Diagrams for High-T_c Superconductors* (American Ceramic Society and NIST: 1991).
7. M. Maeda *et al., Jpn. J. Appl. Phys.* **28**, 1417 (1989).
8. R. S. Roth, NIST, private communication. By scrutinizing Ref. 6, the interested reader can trace these details. (e.g., for the primary phase field of Y_2O_3, see Figure S-082c in Ref. 6.)
9. S. Jin and J. E. Graebner, *Materials Sci. Eng.* **B7**, 243 (1991).
10. S. Freiman, National Institute of Standards and Technology, private communication.

10

Effects of Doping

The simple copper oxide perovskites are insulators, not metals. It is only when more complex crystal structures are created (by substituting for certain atoms in the unit cell) that these materials behave as metals and hence possibly as superconductors. This is why the discovery of HTSCs was so startling. No one expected to find such a property among these compounds. As we saw in Chapter 6, the key characteristic of any superconductor is an energy gap exactly at the Fermi level, which means that Cooper pairs of electrons will not scatter off of lattice sites and break. "No scattering" means electrons propagate with no resistance, which *is* superconductivity. The transition temperature depends strongly on the electronic density of states at the Fermi level. In turn, that parameter is strongly affected by doping the initial ceramic with other atoms of a different valence, to provide extra electrons (or too few electrons, called holes), which are then available to participate in the superconducting mechanism.

The practice of doping has a long history in the field of semiconductors, and chemists have developed great skills in modifying the properties of what would ordinarily be an insulator (e.g., pure germanium or silicon) to allow practical electronic devices to be made. It is not surprising to see similar efforts applied to the HTSCs. This chapter is predominantly devoted to the topic of doping.

Chemists have tried a tremendous array of substitutions, seeking to improve mechanical, magnetic, or current-carrying properties of these materials. The complexities of multicomponent phase diagrams, cited in Chapter 9, prevents some attempted compounds from forming at all. Of those that can be made, either in chemical equilibrium or via a narrow path of kinetics, success is not guaranteed because it is hard to change the electronic properties precisely at the Fermi level. Moreover, the Fermi level is not one single number, but rather, there is a *Fermi surface* in *energy-space*, which is badly distorted from a simple spherical shape by the anisotropy present in the HTSC compounds. Consequently, there is considerable empiricism and guesswork present in any particular choice of chemical substitution in the HTSCs.

It is not easy to make sense of this bewildering array of substitutions. It is impossible to be perfectly up to date with reports of the latest clever idea; but it is possible to explain the motivation for why certain substitutions are attempted, and what outcomes might plausibly be expected. However, the reader must remember that it was exactly the notion of what might be expected that prevented other scientists from carrying out the tests that led Bednorz and Muller to the Nobel Prize. These are extremely complex materials—multicomponent, anisotropic, and so forth—and it is easy for an important change to escape detection.

187

10.1. STRUCTURAL DEFECTS

There are many different types of defects possible in any crystal, including vacancies and interstitials, distortions of normal lattice lengths, changes of crystal-axis directions, etc. For the HTSCs, there is particular importance associated with irregularities in the oxygen at certain sites within the crystal structure.[1]

10.1.1. CuO₂ Planes as Conduction Layers

In HTSCs, the supercurrent flows in a sandwich formed by two CuO_2 planes, which are separated by a single atom such as yttrium.[2] These are termed the *conduction layers*. Ideally, the oxygen atoms in these layers are best left undisturbed. Indeed, if the CuO_2 planes are seriously disrupted, superconductivity is destroyed. On the other hand, the remainder of the unit cell serves as a *charge reservoir*, and the primary role of a defect (or substituted atom) in any HTSC is to alter the supply of electrons in otherwise-filled CuO_2 planes. By this means, discrepancies anywhere in the unit cell can cause changes in the electronic state of the material. In all the HTSCs, the CuO_2 planes determine the size of the unit cell.

The mercury compounds have shown that perfectly flat CuO_2 planes produce the highest T_c values. In YBCO, the CuO_2 planes are distorted from a flat geometry into corrugated planes. (This is due to the different Coulomb attraction of the adjacent Ba and Y atoms that reside on either side of a CuO_2 plane.) Superconductivity in all the HTSCs depends delicately on the details of the layer structure, and that in turn depends on the distortions produced by lattice defects or by various dopant atoms.

10.1.2. Oxygen Vacancies

In Chapter 9, the phase diagrams for all of the HTSCs were based on the oxides of the metal constituents, e.g., the corners of the YBCO triangle were CuO, BaO, and Y_2O_3. This eliminated oxygen as a component, and reduced all phase diagrams by one dimension. However, in real-life chemistry, it is not assured that the oxygen will always appear in exact stoichiometric proportions. For the HTSCs, the electronic role of oxygen is often crucial. Some of the most interesting properties are determined by the nonstandard oxygen content of these compounds.

Normally, an oxygen atom takes on two electrons from another atom; if it is absent, then two more electrons are free to go elsewhere in the crystal. This is how vacancies affect the charge balance in a crystal. The shorthand YBCO refers to $Y_1Ba_2Cu_3O_{7-\delta}$, where the subscript δ denotes the deficiency in oxygen atoms from the normal number (7) that corresponds to a complete crystal lattice. Similar oxygen deficiencies are observed in the lanthanum, bismuth, and thallium HTSC compounds. In most HTSCs, superconductivity is optimized when $\delta \sim 0.15$. In YBCO, superconductivity disappears entirely when $\delta > 0.50$. Generally speaking, there is a fairly narrow range of oxygen deficiencies required to achieve superconductivity in the copper oxides.

The first thing that oxygen vacancies do is change the number of free carriers available in the crystal lattice, which in turn adjusts the Fermi level slightly. The density of states at the Fermi level (N_0) is a key parameter of superconductors: the transition temperature T_c depends on N_0 as $\exp(-1/N_0 V)$, where V is the Cooper pairing potential. Small changes in N_0 caused by oxygen vacancies can translate into substantial changes in T_c. (This is a well-known phenomenon, long since exploited in niobium and its alloys to maximize T_c.) It doesn't take much to change N_0 appreciably. There are about 3.5×10^{13} vacancies/cm² in a double-layer compound such as BSCCO, which corresponds to having about 1% of the oxygen atoms missing.

Explaining the role of oxygen vacancies depends upon the *charge transfer model*, by which the electrons normally in the CuO_2 planes are transferred to sites elsewhere in the unit cell.[3] In YBCO, the oxygen deficiency shows up especially in the CuO chains, making their formula $CuO_{1-\delta}$, while the CuO_2 layers remain chemically complete. Referring back to Chapter 8 and Figure 8.4, some of the O1's are missing.

A key experimental quantity in this model is the oxidation state of the copper atoms in the CuO_2 planes. Any deviation from 2.0 indicates that charge transfer is occurring. Because of the mild geometric distortion in the unit cell, the copper–oxygen bonds are stretched slightly in this plane. By measuring the bond lengths around that copper atom, a *bond valence sum* can be calculated, and this gives the *oxidation state*.[4] As the number of oxygen vacancies δ varies, both this sum and the superconducting transition temperature T_c vary in precisely the same pattern, as shown[5] in Figure 10.1. This presents a very convincing argument for the charge transfer model.

In TBCCO, there is competition between several different means of causing charge transfer: missing metal atoms, extra metal atoms, missing oxygens, extra oxygens, and so forth.[1] BSCCO has all these conditions, plus the size mismatch between the normal dimensions of the CuO_2 planes and the other planes in the stack is so large that atoms occasionally wind up in the wrong layer.[6] The presence of many slightly different but similar structures all in one crystal makes it extremely difficult to interpret experimental data. Nevertheless, despite all the complexity, the average oxidation state for copper is 2.21, which tends to confirm the charge transfer model.

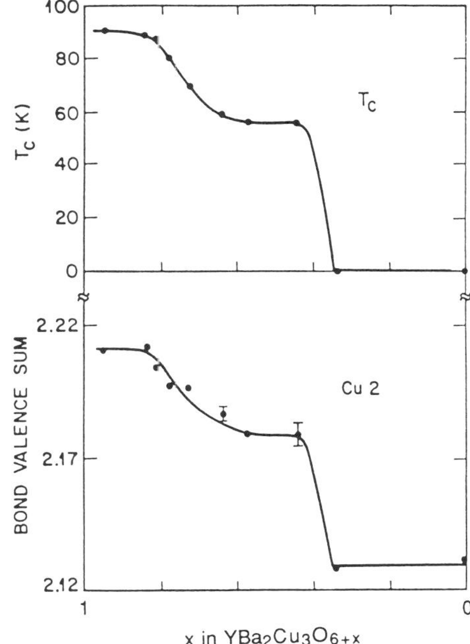

Figure 10.1. Evidence for the charge transfer model. Top: T_c vs. oxygen content in YBCO. The horizontal scale is x in the formula $YBa_2Cu_3O_{6+x}$, so $x = 1 - \delta$, where δ is the number of oxygen vacancies. Bottom: Bond valence sum around the copper atom in the CuO_2 layer. The similar behavior confirms that T_c is proportional to the oxidation state of the copper atoms in the conduction layer (from Ref. 5).

Finally, it should be noted that an *excess* of oxygen atoms also acts as a dopant, because it increases the number of sites where electrons can reside and this amounts to creating hole-carriers in the crystal. The compound La_2CuO_4 is normally an insulator, but when extra oxygen is forced in, changing it to $La_2CuO_{4+\delta}$, it becomes a superconductor.[7] In Section 10.5 we return to consider doping lanthanum copper oxide with strontium, which also changes the number of hole-carriers in the lattice.

The point is that a small number of oxygen vacancies offer a way to make small changes in the carrier concentration. What is less clear is the possible role of an ordered relationship between the structure and the position of the vacancies. Oxygen vacancies are equivalent to substitution of metal atoms in terms of their effect on the availability of charge carriers, because either one affects the charge reservoir layers. In the following section, we review the way that atoms of different valences change the electronic state of the HTSCs.

10.2. VALENCE ELECTRONS AND CHARGE BALANCE

For a material to be called a metal,[8] it must have electrons free to move about the lattice, which can only happen if there are available unoccupied states in one of the energy bands of the solid. As mentioned in Chapter 6, an alternate way to say this is that the Fermi level must fall within a band of available states. An insulator has the Fermi level fall between bands. Starting from an insulator, the way to make a metal is to add or subtract electrons until the number of filled states rises to enter the next band or falls to the preceding (lower) band. The way to do this is by doping.

The experiment of Bednorz and Muller[9] involved doping the parent compound La_2CuO_4 with barium, to see how its properties would change. The outcome was (to understate it) a pleasant surprise. The element lanthanum has three outer electrons, for a valence of +3. Barium or strontium have a valence of +2, so when strontium partially replaces lanthanum in $La_{2-y}Sr_yCuO_4$, the number of electrons drops. The substituted compound becomes a "metal" when the level of filled states declines and leaves some states in a band unoccupied, thus giving freedom of movement to the electrons.

Yttrium, with a valence of +3, would seem to indicate an *increase* in the number of electrons when substituted for barium (in $BaCuO_3$). However, the incompatibility between the valence of yttrium (+3) and oxygen (–2) precludes the formation of an yttrium oxide plane to replace the barium oxide plane. The result is to drop the number of oxygens from 9 to 7, leaving $Y_1Ba_2Cu_3O_7$ with two fewer electrons per unit cell than in $BaCuO_3$, an insulator. Oxygen vacancies revise this to $O_{7-\delta}$, and create more hole carriers.

All of the rare earths and yttrium produce just about the same distortion in YBCO, and they all change the electron density by about the same amount. It makes almost no difference to the superconducting properties which rare earth is chosen. The first ceramic superconductor above 77 K was YBCO, but more generally, this entire class of compounds is written as $RE_1Ba_2Cu_3O_{7-\delta}$, where the notation RE is shorthand for most of the rare earths. Among other things, it is possible to *partially* substitute RE for Y.

Because T_c does not change significantly when substitutions are made, we can draw an important conclusion about HTSCs: It is extremely unlikely that the f-shell electrons in larger atoms play any role in the HTSCs. The reason we know this is because each rare earth has a different number of f-shell electrons, but they all have the same valence state as lanthanum or yttrium (+3). This reinforces the assertions that the Y atom (or RE) only serves as a spacer; and the CuO_2 planes are superconducting without regard to the details of that spacer.

The compound YBCO is still a brittle ceramic even though electronically it has enough free carriers to be classed as a metal. The change in electronic properties from a small amount of doping does not change its mechanical properties. This remains true when yttrium is replaced by a rare-earth element. In TBCCO and BSCCO, the partial substitution of lead for thallium or bismuth does not change the structure, and changes the electronics only a little; but it does change both the phase diagram and the kinetics, which makes the chemistry of preparation easier. This point will reappear in Chapter 15.

10.3. HOLES VS. ELECTRONS

So far, no mention has been made of whether the supercurrent is carried by holes or electrons. It doesn't matter. As we know from long experience with semiconductors, within a solid the wavefunction of a missing electron is equivalent to that of an extra electron. The same has proved true in superconductivity. The earliest HTSCs were hole carriers, but by 1989 reports appeared of perovskite superconductors in which the charge carriers were electrons. These n-type HTSCs were not given great public attention, because their transition temperatures were not impressively high ($T_c \approx 20$ K at first, later somewhat higher), and therefore they had no obvious superiority. However, these n-type superconductors could be important for clarifying certain theoretical questions about how the ceramic superconductors behave.

In the conventional HTSC LaSrCuO, atoms of lower valence are substituted into the lattice in place of higher-valence atoms. This causes charges to depart from the copper oxide planes, leaving behind holes that carry the supercurrent. The compound La_2CuO_4 is an insulator, but when some strontium (valence = +2) is substituted for lanthanum (+3), there results $La_{2-x}Sr_xCuO_4$, which is superconducting below $T_c = 38$ K. In a fully parallel manner, the compound $Nd_{2-x}Ce_xCuO_4$ is a superconductor with *electrons* carrying the current, while the parent compound Nd_2CuO_4 is not.

Many of the n-type superconductors contain the element cerium, and are similar in structure to the other 2-1-4 compounds: for example, $Sm_{1.85}Ce_{0.15}CuO_4$ is the analog of $La_{1.85}Sr_{0.15}CuO_4$; Sm_2CuO_4 is an antiferromagnetic, just like its lanthanum counterpart.

The more complex thallium compounds can also be made into electron carriers.[10] A variant upon the 1212 configuration is $TlSr_2Ca_{1-x}A_xCu_2O_{7+\delta}$ (where A denotes a rare earth element). This material is "tunable" from p-type to n-type by varying x.

The HTSC materials are so unusual that theoretical explanations of their superconductivity contain a lot of leeway for innovation and speculation. After a rocky beginning, better samples and better data weeded out most of the more exotic theories. Still, many theoretical questions remain open, and widely variant ideas compete on approximately equal footing. When theory is required to explain both n-type and p-type superconductivity, this constraint limits the plausible range of explanations. Perhaps most important, the n-type ceramic oxides provide useful clues to the relationship between magnetism and superconductivity.

10.4. MAGNETISM AND SUPERCONDUCTIVITY

One very unusual characteristic of the HTSCs is the juxtaposition of magnetism with superconductivity. One consequence of the discovery of the copper oxide HTSCs has been to force physicists to re-examine the role of electron spins (responsible for para- and ferromagnetism) in the pairing interaction (responsible for superconductivity). Therefore, the study of simultaneous magnetism with superconductivity is directly relevant to understanding the mechanism in HTSCs.

The details are far from resolved, but at least there is agreement that there must be a common underlying relation between superconductivity and antiferromagnetism in the copper oxides. Early on (1987) it was recognized that La_2CuO_4 is an antiferromagnet, while $La_{2-x}Sr_xCuO_4$ is a superconductor. Subsequently, the electron-carrier HTSCs have shown similar properties: Nd_2CuO_4 is antiferromagnetic, while $Nd_{2-x}Ce_xCuO_4$ is superconducting.

10.4.1. Alignment of Spins

The connection between superconductivity and magnetism has fascinated both theorists[11] and experimenters for years. The study of heavy-fermion superconductivity is motivated by questions about the importance of magnetic-moment coupling to the mechanism of superconductivity. The pairing interactions in both cases have certain similarities.

The *Heisenberg exchange interaction* between electrons[12] favors parallel spin alignment (written ↑↑), which is responsible for creating a magnetic moment as in a ferromagnet. There is another term, due to magnetic scattering, that favors antiparallel alignment (↑↓) and demands a negative interaction between ↑ spins and ↓ spins. This term is believed responsible for all heavy-fermion phenomena, which occur predominantly in rare earths and actinides. In magnetic materials, there is a Curie temperature T_c below which ferromagnetic materials have adjacent spins align ↑↑, and there is a Neel temperature T_n below which antiferromagnetic materials have adjacent spins oppositely aligned ↑↓.

In superconductors, Cooper pairs are formed from electrons having spin up and momentum $+k$ with electrons having spin down and momentum $-k$. Within the Abrikosov–Gorkov theory[13] (an expansion of BCS to type II superconductors), we find that $+k$↑ electrons scatter the same as $-k$↓ electrons when they strike a nonmagnetic impurity. When they scatter the same, they stay paired. But if a magnetic impurity is inserted, radically different scattering from $+k$↑ and $-k$↓ electrons takes place. This leads to depairing and causes T_c to drop rapidly.

Other magnetic interactions that are influential in superconductors include the *RKKY effect*, which inhibits superconductivity in dilute magnetic alloys[14,15]; for example, adding 1% manganese to zinc will completely suppress superconductivity, as will ppm levels of iron in molybdenum. The point is that a little magnetism normally impairs superconductivity.

The fact that HTSCs survive despite this condition suggests that the pairing mechanism somehow actually takes advantage of the different scattering from magnetic moments. But that is quite incompatible with Abrikosov–Gorkov theory. The issue goes right to the heart of the mechanism of superconductivity: What causes the pairing to take place in HTSCs? How is that pairing changed by the presence of nearby magnetic moments? Resolving this puzzle would be a major step forward, but it certainly hasn't been done yet.

10.4.2. Magnetic Moments and Sublattices

The exact role of magnetic moments in the mechanism of superconductivity is not well understood. Indeed, to assert that HTSC is conventional superconductivity would not be just an oversimplification; it would be a misrepresentation. The data is sufficiently ambiguous that no individual theory accounts for all of it.

It is agreed that, in HTSCs, superconductivity is localized in the copper oxide planes. Separately, it is very probably true that localized magnetic moments, such as those associated with the rare earths, will align antiferromagnetically, and this in turn will tend to polarize the copper spins toward aligning antiferromagnetically as well. If that polarization seriously affects the copper lattice, superconductivity will be disrupted in the material. Therefore, we must ask about the interaction between these two phenomena.

First, it is necessary to grasp the concept of a *sublattice*. The electron density of states is not uniform throughout these materials (as it would be in a simple metal), but has sizable variations across the unit cell; it is fair to say the electrons are not everywhere, but occupy a sublattice. It is a matter of *band-structure calculations*, not readily verified by experiment, to determine where the maxima and minima are in the density of states. Meanwhile, the rare-earth magnetic moments are localized at those atoms, and they too form a sublattice. The two different sublattices are *interlocking* in real space, but they are not necessarily *interacting*.

If the magnetic moments are located at nulls in the electron density of states, superconducting electron pairing goes on, oblivious to the presence of the magnetic moments. Pair-breaking scattering by magnetic moments simply does not take place. Reality, of course, is messier; when a lattice of rare-earth magnetic moments is *almost* isolated from the electronic state, then they have a very weak depairing influence. The result is that the rare-earth magnetic moments have insufficient influence to destroy superconductivity. This scenario has been used in the past to explain the simultaneous magnetism and superconductivity in the Chevrel compounds and in the rare-earth rhodium borides.

There are major difficulties in applying this to HTSCs, the foremost of which is the inaccessibility of experimental verification of the model, which depends so heavily on calculated electron densities. Another is that an extended lattice has many different possible antiferromagnetic orderings, such as described by the Neel model, the Hubbard model, and others. No one knows which model is appropriate to any given HTSC compound. For example, the Hubbard model may apply to the copper oxide lattice, which can have *itinerant magnetic moments*; but the localized (rare-earth) moments may not follow the Hubbard model. The question is still wide open.

10.4.3. Praseodymium

Magnetic ordering of rare earth ions definitely affects superconductivity. It is instructive to go into detail on one example of how researchers have studied the influence of magnetic moments in HTSCs.

Ever since the early experiments of 1987, the role of the rare-earth element praseodymium in the 1-2-3 compounds has been very puzzling. Praseodymium has a large magnetic moment; and although other rare-earth elements form superconductors with the formula $REBa_2Cu_3O_7$, Pr does not. A research team at the University of California at San Diego (UCSD), led by M. Brian Maple, studied the effects of changing the concentration of Pr substituted into YBCO. In general, the variable of interest is the number of electrons or holes in the copper oxide layers of this class of materials. One step toward maximizing the transition temperature T_c is to finely tune the carrier concentration by doping the lattice with selected elements. The San Diego group studied[16,17] the effects of doping carefully controlled amounts of praseodymium (valence = +4) and calcium (+2) onto the yttrium site (+3) of YBCO.

Maple's team made a collection of samples, all of which had the oxygen content carefully controlled to 6.95 ± 0.2, and had Pr and Ca fractions up to 15%. They measured the T_c values of all these samples, and then fit the data empirically to the form

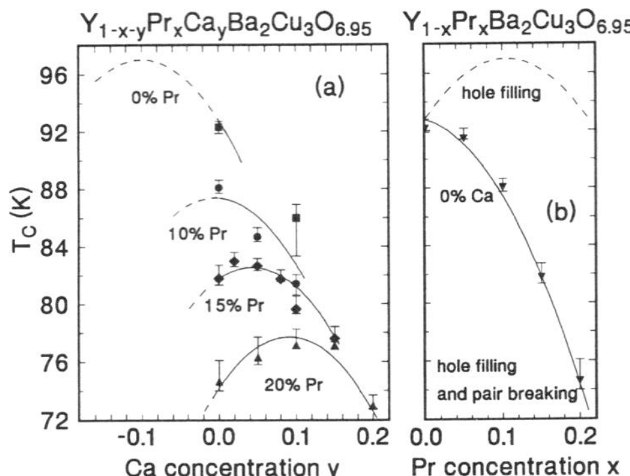

Figure 10.2. (a) T_c vs. calcium concentration y for four concentrations of praseodymium x. Data points indicate the 50% resistivity values of T_c, while the vertical bars are the "10% to 90%" transition widths. Curves formed by solid and dashed lines represent equation (10.5). The negative Ca concentrations are hypothetical and would correspond to tetravalent substitution. (b) T_c vs. Pr concentration x. The dashed line describes pure hole-filling and the solid line represents equation (10.5) with $y = 0$, which includes both hole-filling and pair breaking.

$$T_c (x,y) = 97 \text{ K} - (425 \text{ K})(0.1 - 0.95x + y)^2 - (96.5 \text{ K})x \tag{10.1}$$

where x = praseodymium fraction and y = calcium fraction. The result is shown in Figure 10.2. Note that when $x = 0 = y$, we have standard YBCO and $T_c = 92$ K as usual.

To interpret this data, the UCSD group attributed the quadratic term to changing the electron concentration, and the linear term to magnetic pair breaking. On that basis, the effect of praseodymium is twofold: first, it dopes the CuO$_2$ planes with electrons, countervailing the role of calcium. Second, it has a magnetic-moment interaction with the conduction electrons which sharply inhibits superconductivity. Indeed, PrBa$_2$Cu$_3$O$_7$ is the only nonsuperconductor among the rare-earth series.

The issue is still controversial, because of other data that argues against pair breaking. In the *n*-type 1-2-3 superconductor with neodymium replacing yttrium, the same kind of investigation as this with various amounts of praseodymium being added in the form Nd$_{1-x}$Pr$_x$Ba$_2$Cu$_3$O$_{7-\delta}$ found[18] that superconductivity was present only for $x < 0.25$. For the compound Sm$_{2-x}$Ce$_x$CuO$_4$, neutron diffraction studies led Sumarlin *et al.*[19] to the conclusion that if pair-breaking were important, these electron-doped materials would not be superconducting at all, regardless of the effects of copper magnetic moments.

The essential point to be understood is this: Finding antiferromagnetism and superconductivity together in the copper oxides comes as a surprise. En route to understanding this, the entire relationship between magnetism and superconductivity must be re-examined; and to do this, a progression of well-chosen experiments must build on one another's results.

10.5. SUBSTITUTION ON THE "A" AND "B" SITES

The great majority of substitutions carried out in the HTSCs are done on the "A" sites of the perovskite lattice. Comparatively little substitution has been done for copper, which is commonly the element on the "B" site. Of course, the HTSC BaKBiO$_3$ can be called a complete substitution by bismuth for copper, but here we are usually talking about partial substitution, or doping, on certain sites. Since the prevailing theory holds that the CuO_2 sandwiches carry the supercurrent, and the role of the other atoms is merely to change the availability of electrons in those sandwiches, then logically most attention and experimentation should go to modifying the "A" site. That in fact has happened.

10.5.1. Substitutions in 2-1-4 Compounds

Pure La_2CuO_4 was recognized long ago as an antiferromagnet. By doping it with strontium, it converts to a superconductor, but excessive Sr again causes superconductivity to vanish. The dependence of T_c upon Sr concentration is sketched in Figure 10.3. (In the antiferromagnetic regime, what is plotted is the Neel temperature T_n of antiferromagnetic ordering.)

A close-up of the T_c data in the superconducting range appears in Figure 10.4. This plot is very typical of doping studies in the HTSCs. It is instructive to compare this with the shape of Figure 10.1 for $YBa_2Cu_3O_{7-\delta}$. (Actually, the mirror-image of Figure 10.1 should be imagined.) Both plots show a steep rise in T_c, then a plateau, and a broad maximum. Figure 10.4 falls off with excess Sr doping because oxygen vacancies start to form in the CuO_2 planes.[3] T_c can usually be maximized by carefully scanning the range of possible doping values. Occasionally a compound cannot be made for obscure reasons of chemistry, but even that obstacle can often be overcome by employing elevated pressure or very special time–temperature profiles during processing.

In La_2CuO_4, the same change in T_c can be achieved either by increasing the number of oxygen vacancies or by doping with strontium. Sr has valence $+2$ compared to La's $+3$, giving a net change of -1 for each substitution: one Sr atom introduces one hole. Meanwhile, O has valence -2, so one O vacancy introduces two holes. The two kinds of doping are electronically equivalent, and confirm the charge-transfer model.

10.5.2. Substitutions in 1-2-3 Compounds

The most familiar substitution in the 1-2-3 series is to replace yttrium with almost any rare-earth element. Very little difference is seen. This has already been discussed, especially in connection with the determination of the effects of the magnetic moment of praesodymium

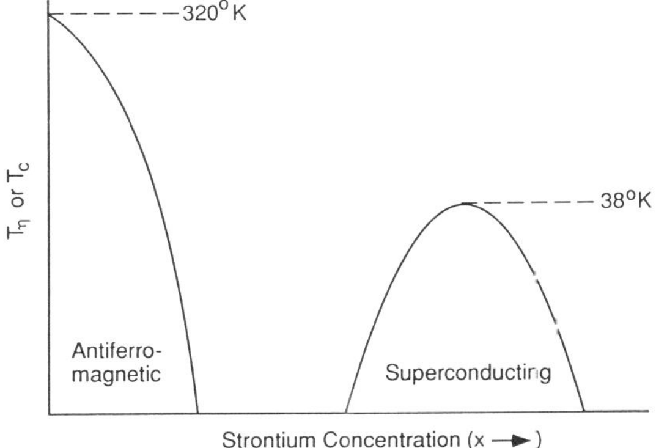

Figure 10.3. Effects of doping La_2CuO_4 with strontium: $La_{2-x}Sr_xCuO_4$. Vertical axis is either Neel temperature of antiferromagnet or transition temperature of superconductor; horizontal axis is strontium concentration x. (Schematic diagram.)

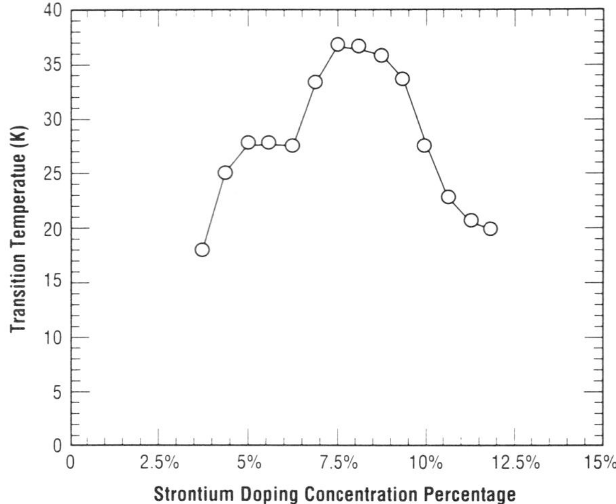

Figure 10.4. T_c vs. strontium concentration. Actual data of T_c vs. doping percentage in $La_{2-x}Sr_xCuO_4$. The region of rising T_c should be compared with the mirror image of Figure 10.1; the shape of the plateau is similar in the two cases, indicating that Sr doping is akin to oxygen-vacancy "doping".

dopants.[16] The lesson learned from such doping is that the details of the spacer between the CuO_2 planes is irrelevant, and this has been used to rule out a role for the f-shell electrons in HTSC. The motivation for doing most substitutions is to investigate one question or another pertaining to the mechanism of superconductivity. By putting in an entirely different atom, one can force changes in the interactions between electrons. The hard part is to understand the relation between the changes being made and the properties of the superconductor.

In YBCO, the element lanthanum (valence = +3) can be substituted on the barium (+2) planes, resulting in the compound $Y_1Ba_{2-x}La_xCu_3O_{7-\delta}$. By introducing electrons, this reduces the hole-carrier density; with only 3% of Ba substituted by La, the observed T_c *rises* to a maximum[20] of 94.4 K. Meanwhile, the normal-state resistivity ρ goes up much more; at 100 K, $\rho = 0.4 + 6.13\,x$, where x is the amount of La doping. Even at the 3% level ($x = 0.06$) where T_c is optimized, ρ is nearly doubled. Clearly, small amounts of doping can make a big difference. In this particular case, the upper critical field H_{c2} is believed to increase by 50%. In the more general case, substitutions with different valences can be expected to change the electronic state of the material dramatically, even if T_c does not change tremendously.

The compound $La_{1.5}Ba_{1.5}Cu_3O_{7-\delta}$ is one extreme case of substitution for Y; this is semiconducting instead of superconducting. The gradual substitution of calcium into this material, via the formula $La_{1.5-x}Ba_{1.5-x}Ca_{2x}Cu_3O_{7-\delta}$, leads eventually to $LaBaCaCu_3O_{7-\delta}$, and along the way[21] superconductivity is restored for $x > 0.2$. This change is reminiscent of the way LaSrCuO changed from an insulator to a superconductor with increasing Sr doping. The governing factor seems to be how many holes are put into the CuO_2 planes by the valences of La (+3), Ba (+2), and Ca (+2). It is always necessary to advise caution for such a conclusion, because the very complex chemistry of formation of these compounds risks the possibility that the sample is really a mixture of two phases, and all the superconducting data comes from one phase. For this reason, any new announcement is met by several other researchers immediately trying to duplicate it.

It is possible to make the analog of YBCO using strontium[22]: $YSr_2Cu_3O_{7-\delta}$ requires very high pressure to synthesize. By attending to the difference between the two different kinds of copper sites, this can be written in a form akin to the 1212 phase of the thallium superconductors: $Cu_1Sr_2Y_1Cu_2O_{7-\delta}$. The La-doped version of this, $Cu_1Sr_{2-x}La_xY_1Cu_2O_{7-\delta}$, has been synthesized[23] and has $T_c \approx 60$ K. Again, substituting La (+3) for Sr (+2) changes the electron density. The remarkable thing is that these compounds are tetragonal (lattice $a = b$) and are still superconducting, despite the early observation that YBCO had to be orthorhombic to be a superconductor.

A great variety of such substitutions have been made, involving many dopants, and the outcomes usually have T_c below 70 K. Thus, there is no special reason for wiremakers to be interested in these compounds, although they may be very helpful for investigating the physics of layered superconductors. However, it could happen at any time that one certain substitution will result in a HTSC with T_c well above 77 K. Then the interesting question will become whether the material can be manufactured into wire. The considerable empiricism associated with this entire field, and the recurrence of surprises, keeps this hope alive.

10.5.3. Substitutions in Bismuth and Thallium Compounds

BSCCO is shorthand for BiSrCaCuO, and as described in Chapter 8 the particular structures of greatest interest are BSCCO-1212 and BSCCO-2223. The most familiar substitution in the BSCCO series is to replace bismuth with lead. This has relatively minor effects on the superconducting properties, but it strongly affects the kinetics and phase equilibrium of the mixture. The result is that $(Pb,Bi)_2Sr_2Ca_2Cu_3O_{10}$ is much easier to make than BSCCO-2223 without lead. Somewhat in analogy with the rare-earth substitution in YBCO, the partial replacement of Bi with Pb makes little difference electronically, and thus changes T_c only very modestly. For purposes of understanding the mechanism of superconductivity within these compounds, attention goes to substitutions onto the other layers in the sandwich.

In both the thallium and bismuth HTSCs, the strategy of doping is driven by the conjectures of each investigator about what will produce desired properties. The number of substitutions possible on those unit cells, together with the complexity of their phase diagrams, invites widespread testing of hypotheses by empirical means. In the thallium series, this involves an expensive commitment of laboratory resources; but it is not as demanding to experiment with the bismuth series. The result is that new compounds are announced frequently. At any point in time, it is only possible to present a snapshot of selected compounds, to illustrate the variety of possibilities.

The familiar series of thallium superconductors are made of TBCCO = thallium, barium, calcium, copper, and oxygen; the transition temperatures of the various phases are generally above 100 K. Less well known is the series based on Tl-Sr-Ca-Cu-O, with strontium playing the equivalent role of barium. Straight TSCCO appears not to superconduct at all down to 4 K. Earlier, Sheng *et al.*[24] showed that the substitution of the trivalent yttrium for the divalent calcium brings an extra electron onto that layer in the unit cell, leading to $T_c \approx 80$ K. Owing to lower T_c's, these compounds are of less interest than the TBCCO group. The question naturally arises whether further substitution can elevate T_c still further.

The goal of most substitutions is to change the number of carriers (holes, usually) by doping the various layers of the unit cell. Just as in the 1-2-3 series, the carrier concentration

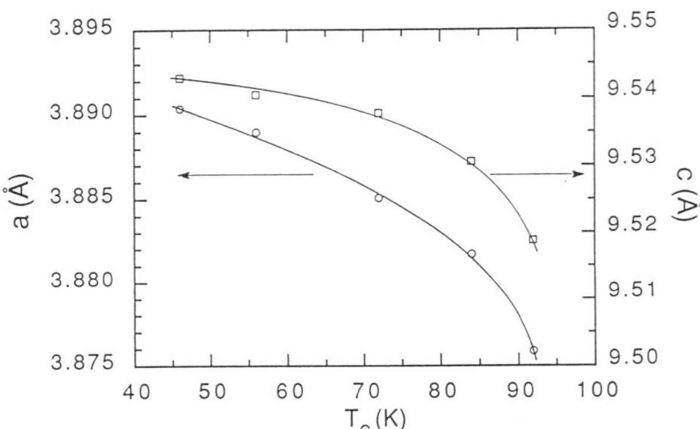

Figure 10.5. Lattice parameters *a* and *c* vs. T_c (midpoint, as measured by AC susceptibility) for $HgBa_2CuO_{4+\delta}$ samples with different oxygen contents (from Ref. 28).

can be manipulated to optimize T_c by doping with elements of different valence. In this way, Liu and co-workers at the University of Cambridge, England, achieved $T_c = 110$ K in TSCCO by combining the lead–bismuth substitution for thallium with a substitution of yttrium for calcium.[25] Liu went on to investigate a new family of elements.[26] In these compounds, calcium is partially substituted by yttrium, and thallium is partially substituted by any of M = Ti, Zr, V, Nb, Hf, or Ta. The chemical formula is typically

$$(Tl_{1-x}M_x)Sr_2(Ca_{1-h}Y_h)Cu_2O_{7-\delta}$$

with various choices of x and h between 0 and 1, and δ around 0.1. They obtained T_c values scattered between 40 and 100 K. What is interesting is not the T_c values, but the success in utilizing elements of high valence to manipulate the carrier concentration. Here, the effect of substitution is to diminish the overdoping originally present in Tl-Sr-Ca-Cu-O. There is no end of possibilities with so many elements to work with.

10.5.4. Mercury Compounds

The first[27] mercury-based HTSC is $HgBa_2CuO_{4+\delta}$. As in the case of $La_2CuO_{4+\delta}$, the primary doping mechanism is an interstitial oxygen. The maximum $T_c \approx 95$ K when $\delta \approx 0.06$, but drops to 59 K when $\delta \approx 0.01$. The changing oxygen content changes the unit cell a and c dimensions, and the associated change in T_c correlates with these dimensions as shown[28] in Figure 10.5. Experimenters have also found that about 8% copper sometimes substitutes on the mercury site, but no systematic pattern of changes in T_c has been established yet. The new sister compounds with two or three CuO_2 layers that have $T_c = 133$ K also contain a slight excess of oxygen.[29] With time, data from samples with varying oxygen content will be gathered, and the relationship to T_c quantified. It is plausible, but not assured, that the effect of oxygen interstitials and vacancies will be similar to that found in previous HTSCs.

10.6. FLUX PINNING BY VACANCIES

In Section 10.1 we described how the oxygen vacancies affect the carrier concentration. A second and very important role of the oxygen vacancies is to provide pinning sites for magnetic flux lines. Disruptions in the crystal structure are always potential pinning sites; for example, twin boundaries in YBCO contribute to pinning the flux lines. A vacancy rate of only 1% is very hard to detect experimentally, but this is enough to provide pinning sites.

It is an essential goal of wire manufacturers to achieve good flux pinning. However, flux pinning is the subject of Chapter 14, so it would be premature here to go into detail about the experimental methods and observations that link oxygen vacancies to pinning. Suffice it to say that oxygen vacancies contribute to pinning; for BSCCO, it appears that oxygen vacancies are the foremost mechanism of pinning magnetic flux lines. This has one very important consequence: the quest for high-current wire made of BSCCO will have to follow paths that do not seriously change the oxygen-vacancy rate, and this will place restrictions on various candidate methods of heat treating and annealing.

10.7. EXPERIMENTAL DIFFICULTIES

In a rapidly moving new field of research, the urgency to produce an explanation for every piece of data is fraught with risk. The HTSCs are no exception. Poor sample quality or adverse experimental conditions lie at the root of many of the rumors and speculative theories that dominated the news for the first few years. In this section we review some of these difficulties.

10.7.1. Ordering Scale Length

It is entirely possible that the local ordering of defects affects superconductivity in the copper oxides, but it is impossible to be sure. Ordinarily, x-ray diffraction methods are used to determine the structure of a compound, but this is only reliable if the unit cell is repeated multiple times without change. Neutron profile refinement techniques are the only way to determine the oxygen positions and concentrations. Thus, if oxygen defects take on some kind of local ordering on a length scale too short to be seen this way, the exact role of the oxygen defects is obscured.

10.7.2. Sample Size Limitations

Certain experiments require exceptionally clean and large crystals of the HTSCs. As one example of this, consider inelastic neutron scattering experiments, which measures the spin or lattice dynamics. Geometric considerations demand large crystals. (Fortunately, magnetic diffraction can be done with samples as small as 2 mg.) As we saw in Chapter 9, the lever rule applied to the YBCO phase diagram ensures that only tiny single crystals come out of solution readily. Either the cold-neutron facility at NIST or the IPNS (Intense Pulsed Neutron Source) at Argonne could be used to probe the ordering of oxygen deficiencies, but it would require centimeter-size crystals. Moreover, those crystals would need to have their oxygen content precisely controlled, and oxygen does not go into large crystals uniformly. Consequently, investigations of this topic remain on the back burner pending better crystals.

10.7.3. Chemical Stability

Many early samples of HTSCs changed properties when exposed to moist air, and a great deal of effort went into achieving reliably reproducible and stable samples for experiments. The mainstream HTSCs now have overcome the stability problem, but many of their substitutional variants have not. Almost every attempt to dope fluorine onto oxygen sites has struggled with the problem of chemical stability. In 1988, rumors of $T_c > 160$ K arose from some fluorine-doping attempts. In 1993, the Siberian work[27] that found $T_c = 117$ K in a thallium–lead compound leaves unanswered a number of questions about structure and stability.

10.7.4. Surface Conditions

Another experimental problem is that the HTSCs have surface properties that are not necessarily representative of the interior material, owing to absorption or desorption of oxygen from the surface layer of atoms. This led to confusion early in HTSC research, and there was considerable debate over whether or not the HTSC materials even contained a Fermi surface. Data from early studies consistently indicated nonmetallic electronic levels, which was contrary to expectations. Because of those experiments, many exotic theories were proposed to explain the apparent contradiction.

Eventually, experiments that recognized and avoided this misleading surface condition cleared up the confusion regarding the metallic behavior of the HTSCs. Very clean photoemission experiments were conducted by a team of researchers from Los Alamos, Argonne, Sandia, and Iowa State. (Photoemission experiments study only the surface.) They began by reducing the temperature of a 1-2-3 superconductor (erbium barium copper oxide) to 20 K, and then cleaving the crystal, thus exposing a fresh surface in a vacuum. At so low a temperature, desorption of oxygen to the surroundings was extremely slow. This enabled them to conduct photoemission measurements that clearly indicated fully metallic behavior of the 1-2-3 material. At higher temperatures, including especially 77 K, the surface layer changes into an insulator. This is what had caused prior photoemission measurements to be distorted by the presence of an insulating layer.

Gradual experimental progress like this is commonplace in physics research. The importance of this clean experimental evidence is to constrain theories to treat the HTSCs as metals, instead of as insulators with localized electronic states.

10.7.5. Confining the Theory

Throughout the history of superconductivity, discoveries have been made by experiment, and theory has followed later to explain the observations. In HTSC, the early experiments were so surprising that many exotic theories were put forth. Unfortunately, few theorists realized both just how bad the early samples were and how unreliable the data was. Within a few years, samples improved and experiments placed much tighter bounds on possible theories. Nevertheless, because of the complexity of the HTSC materials, even very good experiments seldom allow only one explanation.

10.7.6. Special Thallium Problems

Thallium-based superconductors are attractive for several reasons, but many researchers have been hesitant to explore that class of HTSCs because of certain problems peculiar to thallium:

1.   Since the discovery of superconducting compounds containing thallium, there has been much written about the toxicity of thallium. Thallium has been used as a rat poison, has a high partial vapor pressure, and is water soluble and thus capable of being absorbed through the skin. However, thallium is no worse than many other compounds routinely used in chemistry laboratories. Precautions must be taken to ensure processing in a chemically safe environment. This includes use of a fume hood or, better yet, a glove box, protective

clothing, a face mask, and gloves. Thallium is not as bad as arsine, which is used throughout the electronics industry to produce gallium arsenide.

It is worth mentioning that barium also is toxic. It enters the human liver and stays there, breaking down essential enzymes. Barium also combines with water to form barium hydroxide, a very caustic material. Barium rarely receives the safety precautions that it warrants.

2.   Another complication relates to thallium's high partial pressure, which results in thorough contamination of any piece of apparatus used to process thallium. Many laboratories are reluctant to dedicate equipment solely for the purpose of handling thallium. As mentioned in Chapter 9, this volatility has prevented scientists from determining the phase diagram of the thallium compounds.

3.   Thallium HTSCs occur in numerous phases, each with different numbers of layers of copper oxide. The repeat-pattern of layers can change capriciously within a crystal. In cases where change occurs several times, the same material will give different x-ray or neutron-scattering results depending on how the crystal is cleaved prior to examination, thus promoting controversy among workers in the field. Fortunately, $T_c > 77$ K for nearly all phases, but this convenience could deter researchers from attending to other problems associated with the crystal structure.

10.8. SUMMARY

In HTSCs, the copper oxide layers are the conduction layers, while the other intercalated planes comprise the charge reservoir layers. The charge state of the conducting copper oxide layers in HTSCs can be modified either by introducing defects in the structure or by substituting elements of a different valence at specific lattice sites. As long as the conduction layers remain intact, the modifications will appear only in the charge reservoir.

For the case of defects, Jorgensen[3] summarizes the behavior well:

> If defects form in the charge reservoir layer, they can function as a doping mechanism that creates carriers and gives rise to superconducting behavior in a material that may normally be insulating. Defect ordering can enhance or suppress this charge transfer. Conversely, defects associated with the CuO_2 planes in the critical conduction layers can destroy superconductivity. These important concepts are already serving as a guide in the optimization of new superconducting compounds.

For the case of chemical doping, the range of possible substitutions in the HTSCs is enormous, and only a fraction of them have been explored. In this chapter we have presented some examples, but have tried to concentrate on explaining why researchers preferentially try particular kinds of doping. The motivation is to manipulate the interacting electrons so as to learn more about the metallic state of the ceramic oxides. Many variations stick close to the tried-and-true HTSCs, but occasionally a very different compound produces a remarkably high T_c or other parameter.

At the start of the HTSC field, there was a rush to report ever higher T_c values to the newspapers. However, today the goal of doping research is not to raise T_c through empiricism, but to establish an experimental foundation on which a theory can be built to explain the mechanism of HTSC. It is reasonable to think that when the interaction mechanism is well understood, it may be easier to select the optimum path to practical applications.

There are very significant theoretical difficulties, predominantly due to the irregular location of defects in the crystal structure of the HTSCs. Moreover, the relationship between spin pairing, antiferromagnetism, and superconductivity is acknowledged, but is certainly not well understood.

On the experimental side, the difficulty of knowing whether or not a given sample is truly representative of the state allegedly being studied is still great, although many experimenters have overcome their early deficiencies. To the extent that the exotic theories sought to explain data that was faulty in the first place, they have faded away; but when better samples produce better data, there will still be plenty of novel opportunities to explain it.

REFERENCES

1. J. D. Jorgensen *et al.*, *Supercond. Sci. Technol.* **4**, S11-18 (1991).
2. I. K. Schuller and J. D. Jorgensen, *MRS Bulletin*, pp. 27–30 (January 1989).
3. J. D. Jorgensen, *Physics Today* **44** (6), 34 (1991).
4. J. D. Jorgensen *et al.*, *Phys. Rev. B* **36**, 3608 (1987).
5. R. J. Cava *et al.*, *Physica C* **165**, 419 (1990).
6. V. Petricek *et al.*, *Phys. Rev. B* **42**, 387 (1990).
7. J. D. Jorgensen *et al.*, *Phys. Rev. B* **38**, 11337 (1988).
8. N. F. Mott and H. Jones, *Theory of the Properties of Metals and Alloys* (Dover, New York: 1958).
9. J. G. Bednorz and K. Muller, *Z. Phys. B* **64**, 189 (1986).
10. For a thorough review of this entire field, see M. Parathaman and A. M. Hermann, "Thallium-Based HTSC Oxides: A Summary," in *Organic Conductors: Fundamentals and Applications*, edited by J. P. Farges (Marcel Dekker, New York: 1993).
11. J. S. Griffith, *Theory of Transition-Metal Ions* (Cambridge Univ. Press: 1961).
12. C. Kittel, *Quantum Theory of Solids* (Wiley, New York: 1963).
13. A. A. Abrikosov *et al.*, *Methods of Quantum Field Theory in Statistical Mechanics* (Prentice-Hall, Englewood Cliffs, NJ: 1963).
14. M. A. Ruderman and C. Kittel, *Phys. Rev.* **96**, 99 (1954).
15. K. Yosida, *Phys. Rev.* **106**, 893 (1957).
16. J. J. Neumeier *et al.*, *Phys. Rev. Lett.* **63**, 2516 (1989).
17. J. T. Markert *et al.*, *Phys. Rev. Lett.* **64**, 80 (1990).
18. H. Jhans *et al.*, *Physica C* **207**, 247 (1993).
19. I. W. Sumarlin *et al.*, *Phys. Rev. Lett.* **68**, 2228 (1992).
20. J. J. Neumeier, *Applied Phys. Lett.* **61**, 1852 (1992).
21. R. A. Gunasekaran *et al.*, *Physica C* **208**, 143 (1993).
22. B. Okai, *Jpn. J. Appl. Phys.* **29**, L2180 (1990).
23. B. Dabrowski *et al.*, *Physica C* **208**, 183 (1993).
24. Z. Z. Sheng *et al.*, *Phys. Rev. B* **39**, 2918 (1989).
25. R. S. Liu *et al.*, *Physica C* **159**, 385 (1989).
26. R. S. Liu *et al.*, *Appl. Phys. Lett.* **57**, 2492 (1988).
27. S. N. Putalin *et al.*, *Nature* **362**, 226 (1993).
28. J. L. Wagner *et al.*, *Physica C* **210**, 447 (1993).
29. A. Schilling *et al.*, *Nature* **363**, 56 (1993).

11

Mechanical Properties

This chapter looks at the HTSCs from the viewpoint of their mechanical properties—the physical characteristics needed to make wire and useful devices out of these materials. Later chapters go into detail on other electrical problems such as weak-link behavior and flux pinning, but here attention is focused on their serious mechanical limitations.

Unquestionably, the biggest disappointment in the early history of HTSCs has been the inability to make wire. Most superconducting applications of interest to electric utilities require superconductors in the form of wire or tape. Useful superconductors need to be strong, flexible, ductile, and able to carry large current densities. Hence, the extreme brittleness of ceramics is the leading obstacle to the practical implementation of HTSCs.

Wire is commonly manufactured by repeatedly drawing (tensile force) or extruding (compressive force) the material; this requires that material be ductile. For use in magnets or motors, the wire is wound in the form of coils; this requires flexibility. The wire is subject to various forces. During cool-down, the wire may be strained due to thermal contractions. During each cycle the cells in a SMES device are subject to strong cyclic magnet forces. In motors and generators, the coils are subject to large rotational forces; this requires suitable fatigue strength. Furthermore, economic considerations dictate large current densities for most utility devices.

11.1. DEFINITIONS

Such properties of materials as tensile strength, ductility, fracture toughness and fatigue strength are associated with how materials respond to stress.[1] These mechanical properties of solids are best described with reference to a stress/strain diagram, such as Figure 11.1.

Stress is defined as force per unit area applied to the material; *strain* is the resulting deformation (expressed in percent) of the material. In general, materials deform elastically when they are stressed a small amount, and upon relaxation of the stress materials return to their initial shape. Elastic deformation is fully reversible. The slope of the stress/strain curve in the elastic region represents the *stiffness* of the material, and is given by *Young's modulus of elasticity*.

When the stress becomes too great, any material will change its shape, and no longer return to the material's original configuration upon relaxation of the stress. This behavior is called *plastic deformation*, and the bend (or "knee") of the stress/strain curve is the *yield point*.

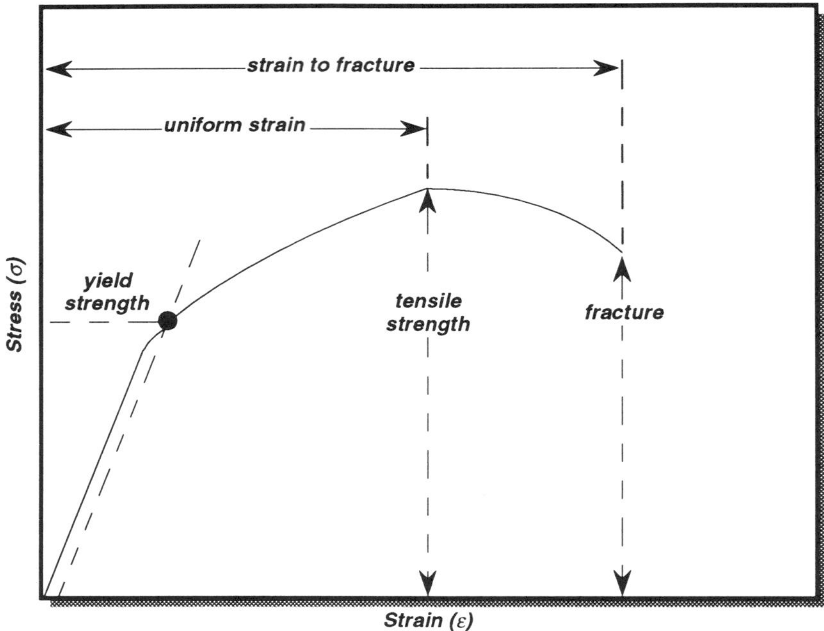

Figure 11.1. Standard stress/strain curve.

Still greater stress will cause the material to break. The stress at which something breaks (when stressed in tension) is called the *tensile strength*; the stress at which it breaks (when stressed sideways) is called the *shear strength*. The surface stress at which it breaks (when bent) is called the *flexural strength*. Isotropic materials, such as most metals, have shear strength about half the tensile strength, but layered materials (e.g., mica) have very different strengths in different directions. Similarly, the strain at which the material breaks is the *fracture elongation or strain limit*.

The total energy per unit area required to break an object defines its *toughness*. Toughness does not necessarily imply stiffness: something that is very easily stretched, even though easily deformed, can require enormous energy before actual breaking occurs, in which case the material correctly would be called tough.

Fatigue is the cumulative damage to a material from repeated stress (strain) applications (cycles), none of which exceed the ultimate tensile strength of the material. The number of cycles required to produce failure decreases as the stress or strain level per cycle increases. The *fatigue strength*, or *fatigue limit*, is defined as the amplitude of cyclic stress that will produce failure in a specified number of cycles, typically 10^7 cycles.

Fatigue measures the response of the material to time-varying stress or strain. On the other hand, *creep* measures the time-dependent deformation under constant load or stress; and *stress rupture* measures the time required for fracture to occur. In general, creep processes become significant only for temperatures in excess of one-half of the melting temperature of the material on the absolute scale. Thus creep is not likely to be important for HTSCs, particularly at liquid nitrogen temperatures.

Table 11.1. Mechanical Properties of Conductors

	Material				
Property	Cu	NbTi	Nb$_3$Sn	YBaCuO	YBaCuO (with Ag)
Stiffness: Young elastic modulus (GPa)[a]	130	80	160	80	100
Tensile strength (GPa) or Flexural strength (GPa)	0.3	1.1	1.3	0.06	0.09
Yield point (%)	0.2	NA	0.2	Does not yield just breaks	No yield, just breaks
Fracture elongation (%)	>5	1.0	0.6	~0.1	Not yet measured

[a]1 GPa = 10^9 N/m^2.

How does breaking take place? In Section 11.3, the subject of fracture mechanics is covered in more detail. Here we give only a brief summary. Tiny cracks begin to grow; propagation releases a certain amount of strain energy. However, it requires some energy to form the crack, because two new surfaces are created by the crack; and that requires energy input. When a crack exceeds a certain *critical length*, it becomes energetically advantageous for the crack to propagate further, the process diverges, and the material breaks.

The *reciprocal critical length* is closely related to the *brittleness* of a material. When the critical length is very short, even a tiny crack readily propagates and causes breaking, so the material is termed brittle. (Ceramics contain small cracks, near the critical size.) Any attempt to flex or deform a ceramic will induce the cracks to grow. This is of no consequence for ordinary copper wire because the cracks remain small and so the material conducts electricity well. However, with a brittle ceramic like YBaCuO the cracks propagate easily and the material breaks. The mechanical characteristics of the HTSCs are often far worse than what we customarily expect of electrical wire.

The core of the problem for HTSCs is that brittle materials such as YBCO rarely exhibit a yield point and subsequent inelastic behavior. They tend to be elastic right up to fracture. They also exhibit very little ductility. Finally, they are greatly weakened by the introduction of surface cracks. For example, cracks 1 μm (1 μm = 10^{-6}m) in length lead to fracture of ordinary glass. The customary parameter used to characterize resistance to crack propagation is the *fracture toughness*, which carries the unusual units MPa (m)$^{1/2}$. Fracture toughness is a property of the material itself, and is not dependent on flaw size.

Table 11.1 gives representative values for Young's modulus, ultimate tensile strength or flexural strength, yield point, and elongationat fracture for several common materials. Brittle materials, having no mechanism for relieving strain energy, seem to explode at fracture. Brittle materials (such as glass, diamond, Nb$_3$Sn, and YBCO) exhibit very little elongation, but can have very high strength (diamond and Nb$_3$Sn).

11.2. MICROSCOPIC PERSPECTIVE

An understanding of why materials behave as they do requires an examination at the microscopic level.[2] In most solid materials, atoms are bound together in a regular crystal

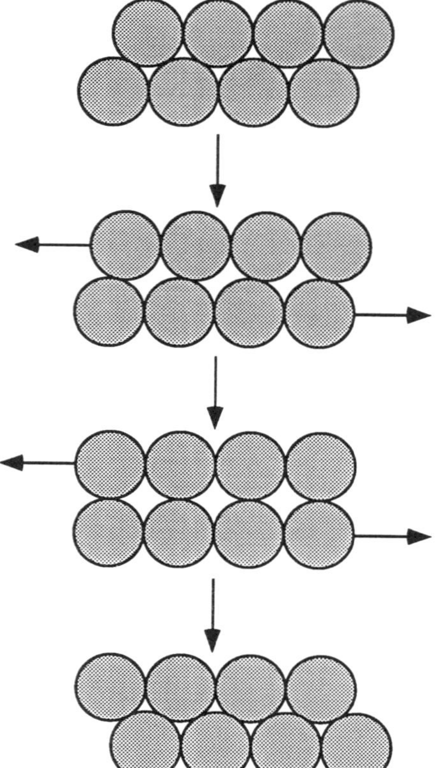

Figure 11.2. Conceptual drawing of shearing mechanism in a solid. At the top is the initial relaxed position. Second shows about 15° angular displacement, at which the resistance to slipping is a maximum. Third shows about 30° angular displacement, at which there is no more resistance to slip. At the bottom is a new rest position, with total motion through 60°. (After Ref. 2.)

lattice. As a crystal is subjected to an external tensile force, the bonds between the atoms in the crystal stretch and exert an initial restoring force proportional to the displacement. As the external force is relaxed, the atoms return to their original positions. The stiffer the bonds, the smaller the displacement (strain) for a given external force and the larger Young's modulus. This explains the stress/strain curve in the elastic region of Figure 11.1.

As the external force is increased the atoms are displaced further and further until a maximum resistance to slip occurs (see Figure 11.2). This is the yield point. A further increase in force causes the atoms to move over one position in the crystal lattice. A relaxation of the force will not result in a return of the atoms to their original positions; a permanent set has been introduced. If the material consisted of a perfect crystal, fracture would occur at the yield point since all bonds would be broken at the same time and there would be nothing to prevent complete fracture of the crystal. In fact, all macroscopic samples contain defects; that is, the constituent atoms are not all located at points in the regular crystal lattice. Movement occurs preferentially at defect locations.

The hardness and brittleness of ceramics, such as HTSCs, stems from the nature of their interatomic bonding. This can be clarified by comparing them with metals. The ductility of a metal is a consequence of nondirectional interactions; the roughly spherical interaction of metallic ions is due to the collective sea of conduction electrons, and the packing of the ions is essentially governed by geometry. Deformation occurs by the sliding of one plane of ions over another as described above. Slipping is easy because the associated distortion to the bond angles has little influence on the metallic crystal's energy.

However, the interatomic bonds in a ceramic are partially covalent, which results in a significant directionality associated with these bonds. This directionality reduces the number of directions available for slippage. Furthermore, the unit cell in ceramics is bigger than metal (many atoms versus single atoms), which means that slippage occurs over greater distances to repeat the unit cell. Thus, the slipping motion of one plane over another is normally unfavorable, and the ceramic cannot easily be deformed; ceramics are hard.

Materials ultimately fail under a tensile load when their interatomic bonds are broken. If this occurs with little or no prior deformation, the substance is brittle. Ductile behavior occurs when atomic motion, most commonly dislocation motion, is favorable. Because such motion is difficult in ceramics, they are brittle.

If crystalline materials *could* be fabricated defect free, they would not exhibit inelastic or plastic behavior. Ductility is associated with the ability to yield to stress without breaking. As an external macroscopic tensile force is applied to a material, an internal microscopic shear stress develops. The shear stress is maximum for planes oriented 45 degrees relative to the direction of stress. This causes parallel planes of atoms to shift in relation to one another, usually along the plane of maximum shear stress (see Figure 11.3). At the micro level, the planes slide (first and third diagrams). At the macro level, necking down occurs (second and fourth diagrams).

11.3. FRACTURE MECHANICS

To understand how a solid fractures, it is first necessary to appreciate the difference between the two states (before and after) of the object. Note in particular that there are *two* surfaces after fracturing where there had been none before. Just as liquids have a surface tension that tends to keep the volume compact, so also solids have a surface energy; it costs energy to form a new surface. This is the key to the modern theory of fracture mechanics. In this section, we follow Gordon's[2] presentation very closely.

To estimate the theoretical tensile strength of solids σ, we ask what condition will be fulfilled just at the breaking point, where a crack opens up. There is strain energy needed to separate two adjacent layers of atoms in the material. Let that equal twice the surface energy of the crack surface. Assuming that the two layers of atoms are initially d meters apart (typically 2 Å, 2×10^{-10} m), then the maximum theoretical stress that a material can withstand is:

$$\sigma_{max} = 2\sqrt{\gamma E/d} \qquad (11.1)$$

where γ is the surface energy of the material and E is the elastic modulus, or Young's modulus. This is derived on the assumption that the interatomic bonds obey Hooke's law all the way to breaking, which is not quite true. A more quantitative and precise derivation would eliminate the factor of 2.

It doesn't matter, however, because real materials virtually never attain their theoretical maximum strength. Real materials have so many imperfections, dislocations and cracks in

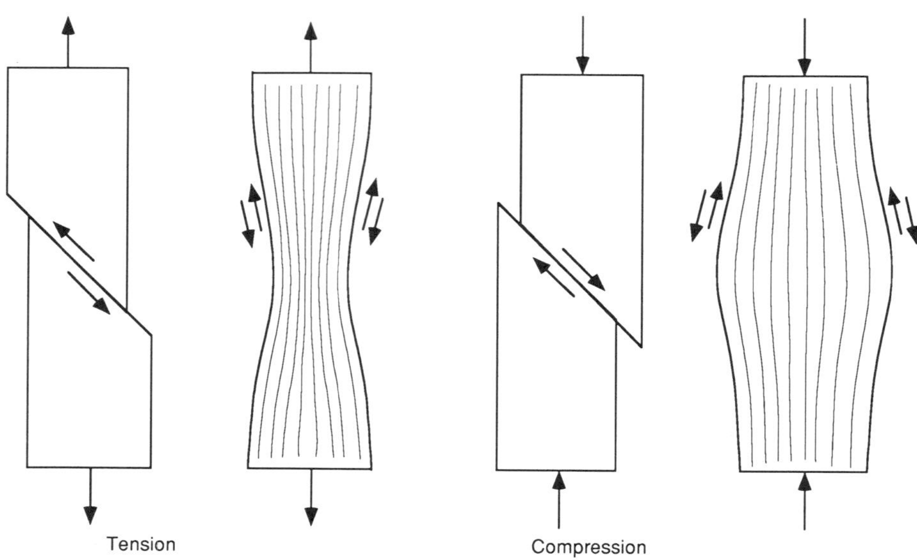

Figure 11.3. Plastic shearing of a ductile material takes place along planes that are at a 45° angle to the applied force. On the left are drawings pertaining to tension; on the right, to compression. The microscopic motion of planes in response to a tensile force is shown at far left. Second from left is the resulting macroscopic change in shape of the test piece, known as necking down. Third is the microscopic motion under compression. At right is the buckling of the test piece. (After Ref. 2.)

their structure that the tensile strength is typically 2 orders of magnitude below the theoretical limit.

So far, the sample was assumed to have uniform cross section. Local changes in geometry of a stressed material can lead to concentrations of stress. An easy way to see this is to plot *stress trajectories*, which are paths along which the stress is transmitted from one molecule to another. Figure 11.4 illustrates this concept. If a simple bar is stressed along one axis, the stress trajectories will be evenly spaced parallel lines. However, if the bar is necked down, the stress will be higher in the neck because the stress trajectories will be crowded together there. If a sharp notch (such as a crack) is added, the local stress at the tip of the crack will be greatly increased, because the stress trajectories will concentrate near the tip.

Mathematical solution for the concentration of the stress for an arbitrary crack is very difficult. However, if the crack is elliptical in shape with length $2L$ and a tip radius r, then a solution is possible and the stress near the tip will be increased by a factor:

$$1 + 2\sqrt{(L/r)} = \text{tip stress/average stress} \tag{11.2}$$

If r is equal to the intermolecular spacing of about 2 Å (2×10^{-10} m) and the crack length is about a micron (10^{-6} m), then the stress-concentration factor is about 100 and increases as the crack gets longer.

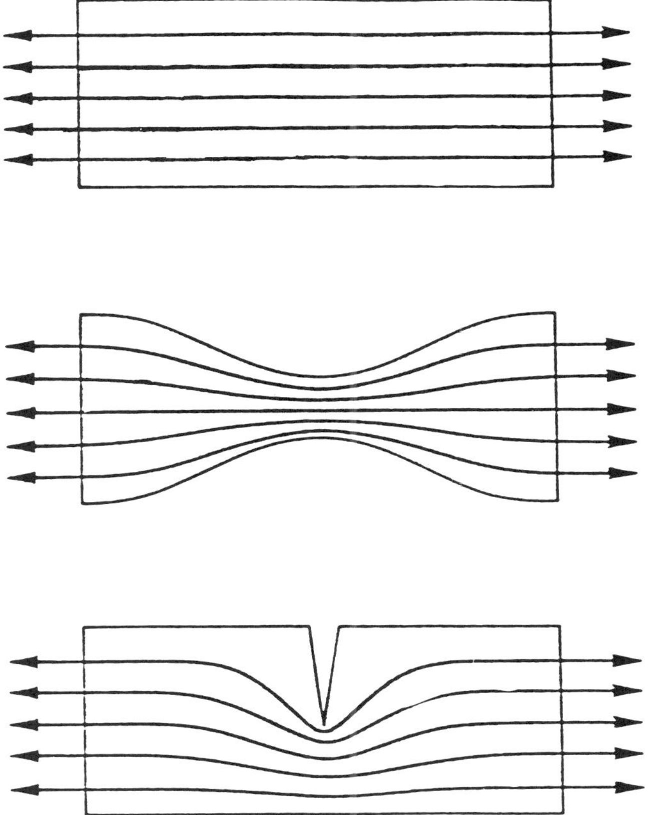

Figure 11.4. Stress trajectories. Top: uniform material subject to axial tension; middle: with a smooth constriction; bottom: around a sharp constriction or notch (From Ref. 2. Reprinted with permission of W. H. Freeman and Company.)

The left side of Figure 11.5 show an unstressed sample. Next, imagine applying a tensile stress σ along the axis. The strain energy per unit volume throughout the material becomes $\sigma^2/2E$. In the middle of Figure 11.5, consider the case where the material is held fixed at both ends: in that case, mechanical energy can neither enter nor leave the system, but the stress is still uniform. Now introduce a crack of length L (right of Figure 11.5) The material immediately adjacent to the crack (heavily shaded triangles) is free to relax, releasing strain energy, which will only promote further cracking.

When the local stress at any point within the sample is sufficient to separate neighboring molecules, then the separation will spread right across the sample and result in fracture. This argument is a necessary condition for fracture, but it is not sufficient. It ignores the role of energy. For crack growth to continue, the energy released by the crack formation must exceed the energy needed to form the crack surface.

The interplay of these various factors leads to fracture parameters that are very different in going from ductile to brittle materials. Table 11.2 illustrates that the work of fracture can

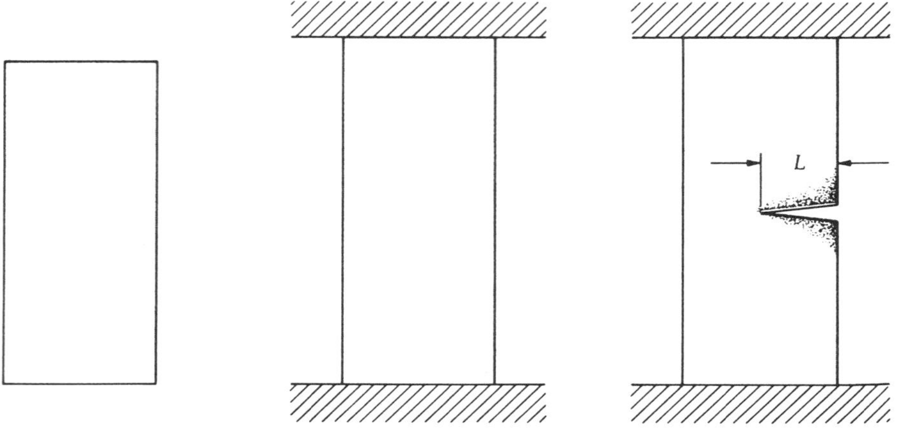

Figure 11.5. Strain-induced cracking. Left: unstrained material; center: material strained and rigidly clamped; right: clamped material cracks. The shaded areas around the crack relax and release strain energy, which is then available for crack propagation. (From Ref. 2. Reprinted with permission of W. H. Freeman and Company.)

be more than 100,000 times the free surface energy in ductile materials and is much smaller in brittle materials such as glass and high-strength steel.

If the release of strain energy exceeds the work required for fracture, the material will fracture. Figure 11.6 presents the energy balance for this process. The amount of strain energy released by a crack of length L goes up roughly parabolically with L. Meanwhile, the work of fracture (W) is linear in L. Thus, for small crack sizes, extension of the crack consumes energy. Beyond point Z in Figure 11.6, called the *Griffith critical crack length L_g*, the crack becomes energetically unstable. Every tiny extension of the crack releases energy, which guarantees that the crack will propagate faster still; the result is fracture. Denoting the tensile strength by σ and the elastic modulus by E, the Griffith critical crack length is given by

$$L_g = 2\ WE/(\pi\sigma^2) \tag{11.3}$$

Table 11.2 contains calculated Griffith critical crack lengths for some common materials. From this table, it is evident why glass is so hard to work with; its work of fracture is very low, resulting in a very small L_g. Microcracks of one micron in length can lead to fracture in glass. L_g has not been directly measured for HTSCs, but it is calculable.

This aspect of fracture mechanics has important applications in construction practices: Although high tensile steel is very strong (2.5 times mild steel) it has an L_g almost 2 orders of magnitude smaller than mild steel, making it much more subject to fracture. For small structures (less than a meter), the longest crack is unlikely to be more than several millimeters, and high tensile steel is suitable. For larger structures, a more ductile material such as mild steel is more suitable.

All materials contain defects and small (micro) cracks. As the material is subjected to stress these cracks grow until L_g is exceeded and fracture occurs. *Fracture toughness* is a measure of how subject to fracture a material is. It has units of Pa-$\sqrt{m}$ and is related to the

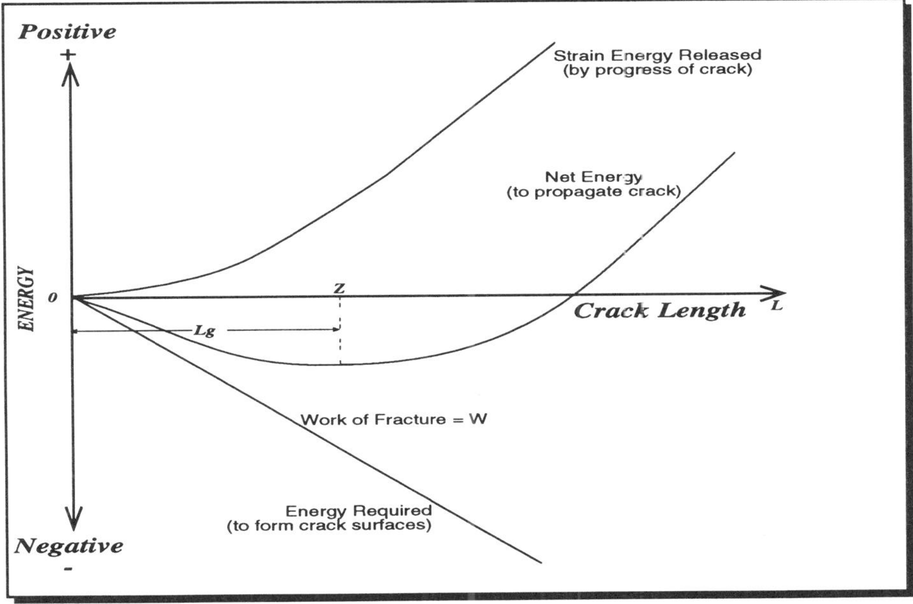

Figure 11.6. Energy balance for crack propagation. Cracks of length less than L_g are stable with respect to further crack growth (crack growth consumes energy). Cracks of length greater than L_g are unstable (crack growth releases energy, resulting in catastrophic growth leading to fracture).

stress concentration factor given in Equation (11.2). If we generalize Equation (11.2) to consider an arbitrary crack-tip with radius d and multiply by the external stress σ, then Equation (11.2) can be rewritten as

$$\sigma + K/\sqrt{4d} = \text{tip stress} \tag{11.4}$$

K is called the stress intensity factor and has the same dimensions as fracture toughness. K is proportional to σ and the square root of the crack length L. However, unlike yield strength

Table 11.2. Calculation[a] of Griffith Critical Crack Length[b]

Material	Surface energy, J/m^2	Work of fracture, J/m^2	Ratio	Tensile strength, MPa	Young's modulus, GPa	Critical crack length (m)
Mild steel	2.0	500,000	250,000	400	200	NA
High tensile steel	2.0	10,000	5,000	1,000	200	0.1
Al alloys	0.9	100,000	110,000	400	90	NA
Glass	0.5	1–10	2–20	170	70	0.4–4.0×10^{-6}

[a]Data columns 1, 2, and 4 are taken from Ref. 2.
[b]Critical crack length calculated for a working stress equal to 0.2 of the tensile strength.
NA = For materials that deform plastically, the calculated critical crack length is meaningless.

Table 11.3. ·Fracture Toughness for HTSCs and Some Common
Materials

Material	Fracture toughness, MPa $\sqrt{m}$	Fracture stress, MPa
Steel	50	350
Al Alloy	50	350
Glass	1	170
YBCO	1.8	191
YBCO/15% Ag	3.1	225
NbTi	NA[a]	1,000
Nb$_3$Sn	NA[b]	1,300

[a]NbTi usually comes in the form of a Cu, Cu/Ni, NbTi composite wire. This composite has a fracture toughness comparable to steel.

[b]Nb$_3$Sn usually comes in the form of a Cu, NB$_3$Sn, steel composite tape. This composite has a fracture toughness higher than YBCO and much lower than NbTi. More importantly, electrical performance decreases rapidly with increasing strain and sets the limit for tolerable strain levels.

and ultimate strength (which are reproducible material properties for specimens of identical composition and microstructure), the value of K varies with the depth of the precrack (L) relative to the width (W) and thickness (B) of the test sample. As the ratio L/W and L/B is decreased, K decreases to a lower limiting value. This limiting value is the fracture toughness K_{Ic} of the material, and is a material constant with respect to crack geometry.

Fracture toughness also controls the rate at which microcracks within a sample grow, and this rate can be calculated; but here it suffices to note that in ceramics the propagation speed is too fast to permit recovery or evasive action to be taken.

Fracture toughness varies with temperature, strain rate, and microstructure. Table 11.3 compares fracture toughness and tensile strength for several common materials, LTSCs, and HTSCs. The fracture toughness of YBCO without silver is comparable to that of glass and 50 times lower than more conventional metallic materials.

If a material is made cleverly, it is not mandatory that a crack lead to fracture. Should a crack, once propagating, encounter a different material with a lower energy-release factor, the propagation will halt and the sample will not break. The strategy used in fabricating most composite materials is to juxtapose regions having different properties in order to capture the best properties of each material. Addition of silver improves the fracture toughness and strength of YBCO. However, the strength of such composites is only comparable to window glass, and about a factor of 10 lower than conventional metallic material or LTSCs.

11.4. MEASUREMENT METHODS

A standard stress-strain curve such as Figure 11.1 is produced by stretching (straining) a sample at a controlled rate and measuring the force (stress) required to deform it. The strain is measured by observing the extension between gage marks.

The point where the stress/strain curve departs from linearity is termed the *yield point*. In practice, this point is difficult to determine because the deviation from linearity is gradual. A common technique to overcome this is to determine the stress required to produce a prescribed inelastic (i.e., irreversible) strain, typically 0.2%. Upon relaxation, a linear curve (the dashed line in Figure 11.1) which is displaced from but parallel to the original curve results. The intersection of this line with the original curve gives the *offset yield point.*

Hardness testing provides a relatively inexpensive and fast method for determining the resistance of a sample to deformation. In a hardness test, a standardized load is applied to a standardized indenter, and the dimensions of the indent are measured. The units for hardness are kg/m^2. Two common tests are the Brinell and Rockwell tests. In the Brinell test, a hardened steel ball is forced into a surface, by a force appropriate to the hardness of the material being tested, for a standard time, typically 30 secs. The Brinell hardness number is defined as the ratio of the applied force to the area of contact between the ball and the sample after the ball has been removed. The Rockwell hardness test consists of applying a minor load to an indenter which establishes a reference point and then a major load. The Rockwell hardness number is the amount by which the indent has been deepened on a reversed, linear scale (soft materials have a lower number).

The value of maximum strength (the tensile strength) is determined by pulling on opposite ends of a test piece until it breaks, and noting the stress needed to achieve breakage. By also measuring the total elongation of the sample up to the point of fracture, the tensile test provides a measure of the sample's ductility, the capacity of the material to deform by extension. The elongation to the point of necking down is distributed uniformly along the length of the sample and is proportional to the length of the sample. This region is referred to as the region of uniform strain. Beyond this point, elongation is confined to the region of necking down and is independent of sample length but proportional to the sample cross-sectional area. Thus, the total elongation-to-fracture, and hence the measure of ductility, depends on the length of the sample. For this reason, a standard must be specified. The American standard ratio of gauge length to diameter for a cylindrical test specimen is 2 to 0.505. The shape of the cross section has no effect up to ratios of width to thickness of 5 to 1. Thus, either cylindrical or rectangular samples can be used. In practice, the percentage reduction in cross-sectional area at the point of fracture provides a better measure of ductility than total elongation.

The resistance of the sample to crack propagation is quite different from the relationship between stress and strain during deformation. To be quantitative, a measurement of fracture toughness is needed. Like elongation, this measurement depends on the sample's geometry. Therefore, several ASTM standards have been developed to specify sample geometry and test conditions.

Basically, three types of tests are used to measure fracture toughness: (1) a notch tensile test; (2) a notched-bar impact test; and (3) a crack-opening displacement test. In the first test, a tensile stress test is applied to a standard bar with a standard notch. Stress is applied in a direction perpendicular to the plane containing the notch. The stress at which the sample fractures is measured. In the second test, a standard bar containing a standard notch is struck by a pendulum in a direction parallel to the plane containing the notch. The work required to fracture the sample is measured. In the third case, a standard-notched bar is subjected to forces which flex the bar in a direction so as to open the notch. Figure 11.7 illustrates this latter case.[3] The maximum crack opening at time of fracture is measured. In all three cases, the measured quantity can be converted into the fracture toughness.

When dealing with ceramics, the scatter in fracture toughness measurements from one laboratory to another is considerable. NIST organized a round-robin test to measure fracture toughness in Si_3N_4 and similar ceramics,[4] and got results scattered in a band 100% wide.

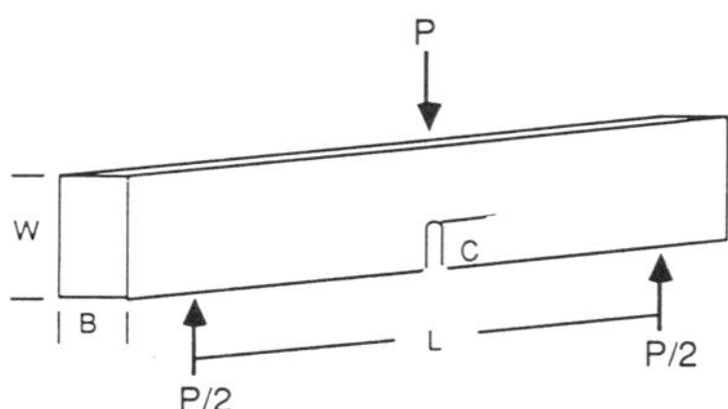

Figure 11.7. Schematic of single-edge notched beam for K_{Ic} testing. (From Ref. 3.)

Thus, fracture toughness results quoted by diverse researchers for HTSCs are considered in agreement if they are within 50% of one another.

11.5. MECHANICAL PROPERTIES OF HTSCs

The most elusive goal in this field is to achieve good mechanical strength simultaneously with good electrical properties. In this section we review the progress made toward that goal.

A comparison from LTSC is helpful in seeing which way this field is headed. Niobium-titanium superconductors (in the form of multifilamentary composites) are the overwhelming choice for low-temperature applications of superconductors. These composites are flexible, strong, and can carry large current densities. Although Nb_3Sn has the potential to carry larger current densities in higher magnetic fields and at higher temperature, its brittleness has restricted its use to those special-purpose applications which can not be satisfied by NbTi superconductors. The lack of ductility in Nb_3Sn has prevented the manufacturing of wire using conventional extrusion or drawing techniques. However, a special bronze process has been developed to produce Nb_3Sn composite wires and tapes, consisting of a thin layer of Nb_3Sn embedded in a sandwich of copper (which acts as a stabilizer) and steel (which provides mechanical support). These products are strong and somewhat flexible; but are much less flexible than conventional NbTi composites.

Several points are worth mentioning here, to underline the difficulty of working with brittle materials. First, no one has yet made Nb_3Sn consistently or profitably, despite its existence for over 30 years. Second, the strain limit of Nb_3Sn is about 0.6%, whereas the strain limit of YBaCuO is estimated to be about 0.1%. This means that the "brittleness problem" of the HTSCs is likely to be six times worse than that of Nb_3Sn. The data in Table 11.3 compares certain parameters.

11.5.1. YBCO

Most of the measured numbers on the mechanical properties of YBCO come from one of two laboratories: Argonne National Lab and the Texas Center for Superconductivity at the University of Houston (TCSUH). Early on, Argonne reported[5] fracture toughness values ranging between 0.8 and 1.0 MPa$\sqrt{m}$. Similar studies led by Salama at TCSUH[6] got up to 2.0 MPa$\sqrt{m}$. Ever since, it has been recognized that a way must be found to strengthen YBCO, or else it will never fulfil its potential.

For purposes of making wire, the most relevant parameter is the *strain to fracture*, a relative measure of fracture toughness. That is 0.1% in plain YBCO but rises to 0.4% with the addition of silver. (Fracture toughness is a useful way to characterize bulk materials, but the highly aligned grains in melt-textured material are very anisotropic, which makes strain to fracture the figure of merit.)

Several techniques, developed to improve the mechanical properties of ceramics, might be applicable to YBCO. For example, one such process is *transformation toughening*. Here, particles of a material that undergoes a *displacive transformation* are dispersed in a matrix of material that does not undergo the same transformation. During the cooling cycle, the dispersed particles undergo a transformation accompanied by a volume change. The surrounding matrix material is either cracked or locally stressed by this volume change, resulting in a significant increase in fracture toughness. This technique has been used to produce a new class of steels which have the strength of high-strength steels while maintaining the ductility of low-strength steels. The transformation occurs not during cooling but upon exposure to strain.

Another technique is *second phase reinforcement*. In this technique, fine particles having a different rate of thermal expansion from the base material are introduced throughout the structure. One possible outcome is that, upon cooling, a large number of microcracks are introduced into the microstructure. The series of microcracks prevent formation of one big crack by redistributing and absorbing stress. This probably will decrease the strength of the material but can increase the strain tolerance of the entire structure. ZrO_2 is believed to behave in YBCO this way.

11.5.2. Alloying with Silver

The foremost means of improving mechanical properties is through alloying. It is tempting to try to improve ductility by alloying with copper or aluminum. In the case of the HTSCs, this doesn't work, because the copper reacts chemically with YBCO (or BSCCO) to destroy superconductivity. Only the expensive noble metals of gold, silver, and platinum are suitable; and silver is the least expensive of these. Thus, silver is the metal of choice for strengthening the HTSCs.

The effect of adding silver to YBCO has been examined extensively. The use of silver can have several potential benefits in YBCO. First, silver is a good electrical conductor; this promotes the conduction of current (although not supercurrent) across grain boundaries. Second, silver substitutes for the extraneous material that tends to accumulate at grain boundaries. Both of these properties should help carry current. Addition of silver up to 15 or 20 percent volume actually improves the current density in bars of YBCO. However, at higher concentrations, the current density falls off rapidly. For these concentrations, the structure of YBCO probably reverts to tetragonal which no longer is a superconductor.

With regard to the mechanical properties of YBCO, silver puts the YBCO matrix in compression. Addition of either silver oxide or silver improves sample density, hardness, Young's modulus and flexural strength. Flexural strength increases by almost 80 percent for both silver and silver oxide while fracture toughness increases by approximately 50 percent. The flexural strength and fracture toughness of silver/YBCO alloys are better than window glass by about a factor of three.

Why does the addition of silver improve mechanical properties so much? First of all, silver is a very ductile metal (see Table 11.1). It is also a very soft material with the strength of a metal. For low concentrations of silver, the thermal mismatch between YBCO and silver that occurs during processing causes the superconductor to be placed in compression. Brittle materials tend to be stronger in compression than extension. In addition, cracks in many materials have been observed to propagate along grain boundaries within the material. In YBCO, the silver tends to accumulate along grain boundaries. As a crack begins to propagate, it will run into a region of silver. Because silver is soft and ductile, it easily deforms inelastically. This has two effects: First, the inelastic deformation at the crack tip tends to increase the characteristic dimension of the crack tip, thereby reducing the stress concentration factor. Second, the inelastic deformation absorbs strain energy being released by the crack propagation. These two effects can greatly reduce the propensity of the crack to propagate. In short, the silver acts to release the strain energy before it can reach catastrophic proportions and lead to fracture.

In the special case of HTSCs, YBCO-123 is chosen for its superconductivity, YBCO-211 is added to create flux-pinning sites, and silver is added to create mechanical flexibility and resistance to crack propagation. The fracture toughness of typical YBCO samples, as measured at TCSUH,[7] are compared with other familiar ceramics in Figure 11.8. Evidently, YBCO fractures comparatively easily, although silver is distinctly helpful. Similar numerical results have also been achieved at Argonne.[8]

Improving mechanical properties is not achieved without costs, however. Generally speaking, adding more silver helps the mechanical problem, but the critical current J_c degrades when the silver content is large. Figure 11.9 offers a clever way of visualizing the trade-off by displaying the fracture toughness alongside the logarithm of critical current. At the bottom, ordinary sintered YBCO-123 is mechanically weak and has poor J_c (only 200 A/cm^2). With 20% silver but without melt texturing (second from bottom),

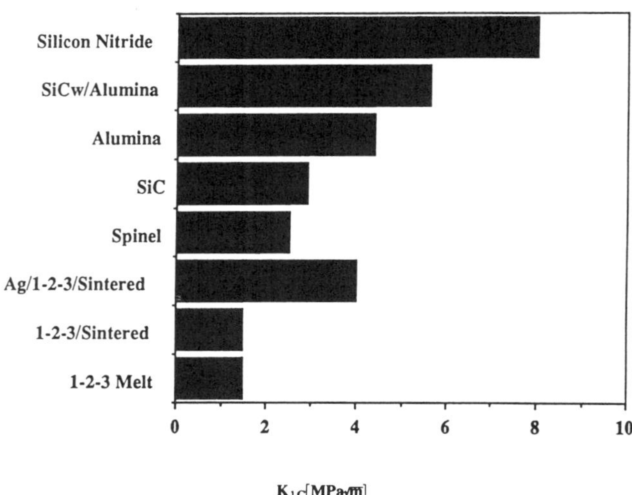

Figure 11.8. Fracture toughness of selected materials for comparison with YBCO and YBCO/Ag. (From private communication, K. W. White, University of Houston, Houston TX 77204-4792.)

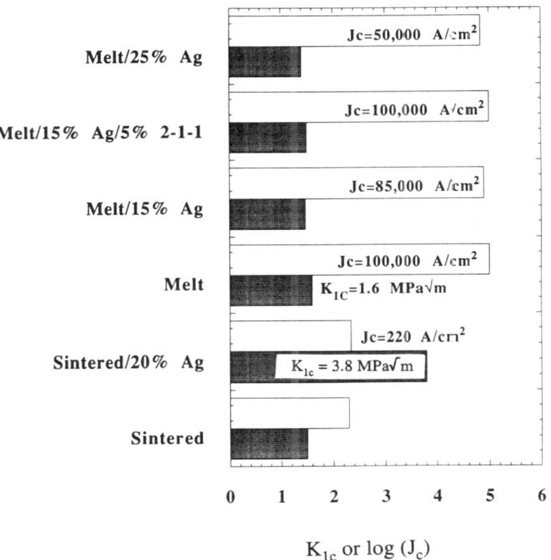

Figure 11.9.   Fracture toughness and current density (0 field) of monolithic YBCO and composites. (From private communication, K. W. White, University of Houston, Houston TX 77204-4792.)

the fracture toughness K_{1c} rises to 3.8 MPa$\sqrt{m}$; but J_c remains down at 220 A/cm². Melt-texturing alone (third from bottom) brings J_c up to 100,000 A/cm², but does not significantly improve K_{1c}. With 15% silver (fourth), both properties diminish slightly. Adding YBCO-211 (fifth) boosts J_c only a little. Going up to 25% silver (top) diminishes both K_{1c} and J_c. Obviously the desirable but elusive outcome is to obtain $K_{1c} > 4$ combined with $J_c > 10,000$ A/cm².

Figure 11.10 is a photomicrograph of a welded junction with a silver-YBCO composite. Made by American Superconductor, Inc., this material is approximately 70% silver. The fact that a welded junction can be made at all is a significant step forward. Using a high percentage of silver yields desirable mechanical properties, but J_c falls off as more of the available cross section is made of silver. Once connectivity of YBCO grains is lost, J_c vanishes. Research toward striking the right compromise continues.

11.5.3. BSCCO

In addition to work on YBCO, there is effort toward improving the mechanical properties of bismuth-system superconductors. Wire made of BSCCO enclosed in a silver tube is covered in detail in Chapter 16; but it is appropriate here to mention certain mechanical results:

Researchers at Kobe Steel[9] have studied uniaxial stress in silver-sheathed BiPbSrCaCuO. Goto and Maruyama (Nagoya Institute of Technology)[10] measured tensile strength = 55 MPa and elongation (at fracture) of 1.1% in filaments of BiPbSrCaCuO. Mechanical tests were run by Sumitomo[11] on BSCCO/Ag-sheathed wires, including thermal cycling, repeated bending, and strain tolerance. They found that cycling between room

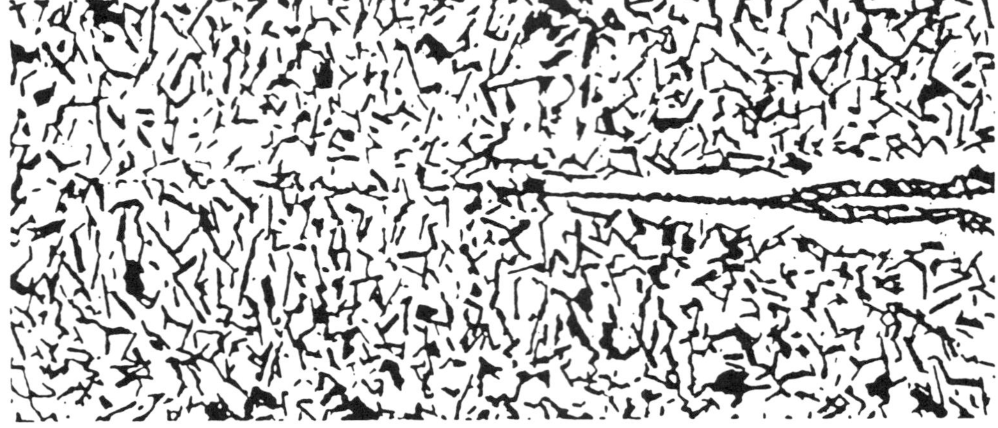

Figure 11.10.   Welded junction of YBCO. (From American Superconductor Corp.)

temperature and 77 K did not reduce J_c significantly: after 100 such cycles of a composite conductor, J_c remained at 97% of its original value.

However, cycling to elevated temperatures was indeed harmful: Vacuumschmelze[12] reported that repeated soldering seriously degrades J_c in BSCCO-2212/Ag wires. Evidently, wires are sensitive to the thermal shock of cooling, and soldering causes cracks due to internal temperature variations.

11.5.4. Strength of Multifibers

One approach to achieving the best of both worlds is to combine many fibers into one conductor wire. Multiconductor wires retard crack propagation and provide quench protection, but they must match the thermal expansion coefficients of several materials.

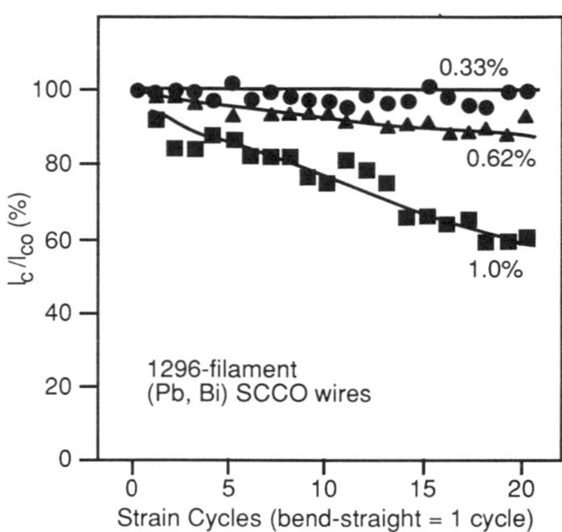

Figure 11.11.   Effect of repeated bending on current capacity in BSCCO wire. Labels on individual curves are percent strain applied to the fibers within the wire. (From Ref. 11.)

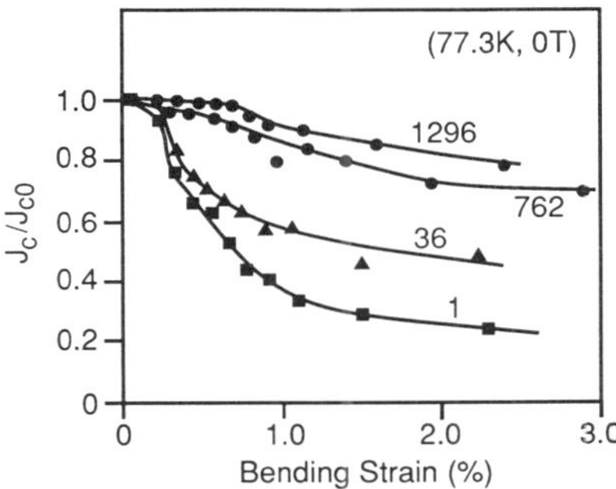

Figure 11.12. Comparison of strain tolerance behavior in BSCCO, for different numbers of filaments in a wire. (From Ref. 13.)

Repeated bending also reduces current. The Sumitomo 1,296-core BSCCO wire was put through 20 cycles of bending and straightening,[11] with results depicted in Figure 11.11. When the strain is only 0.33%, no current deterioration is seen; but at 1% strain, perhaps as many as 40% of the individual filaments have broken. This information is relevant because when a magnet is made, bending is done only once; but for a cable, several bending events occur throughout the total process of making and installing it.

Strain tolerance is definitely better in multifilament conductors. More data from Sumitomo[13] appears in Figure 11.12, to illustrate the point that a single strand degrades above 0.3%, while a 1,296-filament wire retains good J_c out to 1% strain. This is believed to be due to the effect of multifilaments in retarding crack propagation.

11.6. NOVEL WAYS TO IMPROVE STRENGTH

Beyond alloying with silver, there are ideas to improve strength that utilize different means. Here are a few examples:

- Argonne found[3] that it is possible to use zirconium oxide (ZrO_2) to add strength, provided the small ZrO_2 grains could first be coated with YBCO-211 phase to prevent it from chemically reacting with the YBCO-123 phase. In this way, fracture toughness rose to over 4 MPa$\sqrt{m}$.
- Similarly, a group in Nagoya Japan found[14] that a silver/YBCO alloy can be reinforced with zirconium metal, gaining about a factor of 5 in mechanical strength compared to conventional sintered YBCO. Specifically, the flexural strength of this new alloy is 280 MPa, when the silver content is 33% (by volume) and there is 80% as much zirconium as yttrium. What happens is that $BaZrO_3$ and CuO form and fill in the voids that might otherwise be present, thus adding strength to the material. The silver adds a bit of plasticity to the overall alloy. Meanwhile, T_c stays above 77 K.

    This high level of flexural strength is close to that found in commercial MgO and Al_2O_3, which implies it is machinable and suitable for bearings or other applications requiring mechanical durability. However, this high percentage of silver diminishes the total current flowing (the effective J_c), which limits its usefulness for making wire.

- The bismuth compounds also allow for some unusual processing techniques. Throughout the field of materials science, very small and extremely thin crystals of substances (known as whiskers) usually have properties that are close to the ideal limit. Anticipating similar behavior, Matsubara *et al.* (Osaka) studied whiskers of BSCCO.[15] Their J_c values were encouraging, being comparable to filamentary conductors.

- Babcock & Wilcox (in collaboration with Ames Laboratory) developed a way to produce flexible, superconducting composite wires consisting of microfilamentary BSCCO fibers embedded in a lead matrix. The process yielded needlelike strands of BSCCO, 1–50 μm in diameter. These were spun to produce a mass similar to steel wool or cotton candy. However, the elusive goal remains: to produce a well-connected, superconducting structure that carries large current densities.

11.7. COMPARISON TO FIBER OPTICS

Since the maximum strain experienced by a wire of thickness d bent with a radius of curvature R is proportional to d/R, a common technique for improving flexibility is to make the diameter of the wire smaller. Fiber optics cables are an example of this technique. Fiber optics cables consist of micron-sized filaments made from glass under a carefully controlled process. Although ordinary glass is quite brittle, fiber optic cables are strong and flexible. These fibers are made by melting the glass and drawing the melt into filaments under carefully controlled conditions that minimize flaw size. This is followed by either an ion exchange process or quench process to place the surface of each filament in compression. The ion exchange process involves heating the glass and exposing it to cations that are larger than those initially in the glass. The larger ions replace the smaller ions near the surface but not the interior. When the glass subsequently cools, the larger ions near the surface force the surface into compression. The quench approach is similar. The glass is heated above its softening temperature and then cooled rapidly. The surface solidifies first and then is pulled into compression as the interior cools more slowly.

A number of cheerful predictions have been made about YBCO fibers based on the history of fiber optics. Unfortunately, the constraints imposed by brittleness on thin fibers of YBCO are very strict. A calculation by Johnson[16] makes the comparison between SiO_2 fibers and YBCO fibers, and shows the severity of the problem.

To fabricate a coil from YBCO, either it must be wound into its final shape before sintering, or the wire diameter must be so small that the superconducting (sintered) YBCO wire can be wound around a certain radius. Johnson considers the latter case, in which the presintered wire is to be bent, and asks what fiber diameter D_f is necessary to allow a certain bend diameter D_{mb}. The two are related through the elastic properties of the YBCO, specifically the fracture toughness K_{1c}. We have

$$\frac{D_f}{D_{mb}} = \frac{S}{E} = \frac{K_{1c}\sqrt{a}}{E} \tag{11.5}$$

where E is the elastic modulus, S is the flexural strength, and a is a typical flaw size.

For YBCO, Johnson[16] uses a value $E = 200$ GPa, whereas Argonne has measured[5] $E = 75 - 120$ GPa (depending on silver content), but that is not a crucial distinction. The fracture toughness of YBCO is $K_{1c} = 1.3$ MPa-m$^{1/2}$, and guessing a typical flaw size near 10^{-4} yields

Table 11.4. Estimation of Minimum Bend Diameters for Fibers

Material	Flexural strength	Elastic modulus	Ratio of fiber diam. to bend diam.
SiO_2	7.0 GPa	70 GPa	0.1
YBCO	0.1 GPa	90 GPa	0.001

$S = 100$ MPa $= 0.1$ GPa. It follows at once that the diameter ratio is about 10^{-3} for YBCO, as shown in Table 11.4. Optical fibers made of SiO_2, by contrast are made in compression, so that their strength is far greater, near 7 GPa. The comparison in the table is striking: fiber optics can be bent 100 times as much as YBCO. This calculation is indicative of just how far YBCO has to go to be a practical material. Clearly, if YBCO fibers can someday be made in compression the way optical fibers are, this will greatly enhance the flexibility of YBCO.

11.8. SUMMARY

For a conductor to have practical applications, it must withstand appreciable stresses. Normal use may require thermal cycling (between ambient and operating temperatures), repeated strain (during fabrication or operation), and consistent behavior over long lengths. Consequently, it is just as important to investigate the mechanical properties of the HTSCs as to study the critical current.

This chapter introduced the terminology pertaining to strength of materials, including the terms stress, strain, Young's modulus, hardness, etc. These are reviewed in standard mechanical engineering textbooks.[1] A brief description of how some of these parameters are measured is included.

Fracture toughness is one of several parameters of mechanical strength, along with more familiar indicators. The fracture toughness is a measure of how well a material resists crack propagation. The symbol used is K_{1c}, and the units are MPa $\sqrt{m}$, that is, mega-Pascal-root-meter. This very unusual unit originates from the nature of what is actually being measured: It takes energy to overcome binding and open up a small crack; but once that is done, some energy is released. If the released energy exceeds the required input energy, it will be energetically favorable for the crack to propagate, and if that proceeds unchecked, the material will fracture.

The HTSCs are brittle ceramics, and are difficult to make into wire as the word is commonly understood. To overcome this, various alloying techniques have been studied, and it is known that alloying with silver can increase the mechanical strength. Separately, melt texturing can improve critical current J_c. The goal is to get *both*, and this has so far been elusive. Making HTSC wires and tapes is still in the early stages of development.

Most research efforts to date have been devoted to understanding the physics and chemistry of HTSCs, or to raising the value of J_c. This emphasis has resulted in little direct effort being devoted to understanding and improving mechanical properties. Because both the electrical and mechanical properties of HTSCs depend on the microstructure of the material, it is plausible to hope that improvement of electrical performance may contribute to improvement in mechanical characteristics.

Will the various alloying techniques ultimately be successful for HTSCs or will new techniques be needed? We do not know. What is needed is to first understand how to manipulate the microstructure so as to improve pinning (and consequently current density). Once acceptable J_c values are achieved, attention can shift to improving mechanical properties, hopefully without reducing current density.

REFERENCES

1. S. H. Crandall *et al.*, *Introduction to the Mechanics of Solids* (McGraw-Hill, New York: 1972).
2. J. E. Gordon, *The Science of Structures and Materials*, Ch. 4, Scientific American Library series (W. H. Freeman, New York: 1988).
3. K. Goretta *et al.*, "Toughened YBCO/ZrO2 Composites," in *High-Temperature Superconducting Compounds II*, edited by S. H. Whang *et al.* (The Minerals, Metals & Materials Society: 1990).
4. G. D. Quinn *et al.*, *J. Res. NIST* **97**, 579 (1992).
5. J. P. Singh *et al.*, *Appl. Phys. Lett.* **53**, 237 (1988).
6. K. Salama, private communication.
7. D. Lee & K. Salama, *Jpn. J. Appl. Phys. Lett.* **29**, 2017 (1990).
8. J. P. Singh *et al.*, *J. Mat. Res. Soc.* **7**, 2324 (1992).
9. S. Hayashi *et al.*, "Effect of Uniaxial Stress on the Electromagnetic Properties of BiPbSrCaCuO Superconductors," paper PWB-34, ISS 89 Conference (ISTEC, Tokyo, November 1989).
10. T. Goto and T. Maruyama, "HTSC BiPbSrCaCuO and TlBaCaCuO Filaments Produced by the Suspension Spinning Method," paper PWB-1, ISS 89 Conference (ISTEC, Tokyo, November 1989).
11. H. Mukai, Third ISS Conference (Sendai, Japan, November 1990).
12. J. Tenbrink *et al.*, *IEEE Trans. Magn.* **MAG-27**, 1239 (1991).
13. K. Sato *et al.*, *IEEE Trans. Magn.* **MAG-27**, 1231 (1991).
14. T. Oka *et al.*, *Jpn. J. Appl. Phys.* **29**, 1924 (1990).
15. I. Matsubara *et al.*, *Appl. Phys. Lett.* **57**, 2490 (1990).
16. D. W. Johnson, Jr. "Bulk Processing Methods for High T_c Ceramic Superconductors," International Forum on Fine Ceramics '90, Nagoya, Japan.

12

Theory of HTSCs

In Chapters 2 and 6 we explained some of the theoretical aspects of superconductivity applied to conventional superconductors. This chapter deals with applications of known theories to HTSCs. HTSCs are so unusual, so unexpected, that there is still plenty of controversy concerning these materials. In the future, as better experimental data appears, the statements made here may be superseded and revised.

Since 1957, the BCS theory of superconductivity[1] has guided research and development efforts in this field. Indeed, as late as 1986 it was believed (on the basis of BCS, as then understood) that superconductivity could not occur above 30 K. The discovery of HTSCs received the Nobel Prize because it experimentally destroyed that belief. Subsequently, as experiment outran theory, debate ensued as to whether BCS theory was relevant to the HTSCs, and a variety of new theories were proposed.

With improved understanding of the properties of the HTSCs, it has gradually become clear that the BCS theory still gives the correct explanation for superconductivity, but modifications are needed to account for the extreme anisotropy (indeed, near two-dimensionality) of the ceramic oxide superconductors. None of the exotic theories answer as many questions as does the modified BCS theory. Thus, confidence has been restored in BCS theory, despite the upheaval of 1987. This has important consequences for the applied research now being pursued, the goal of which is to produce practical wire and superconducting devices.

This chapter tries to present enough background information to enable the reader to appreciate the on-going theoretical efforts toward understanding HTSCs. It relies heavily on the concepts and terminology introduced in Chapter 6. We begin by reviewing the normal state and point out how the Fermi surface differs in HTSCs from ordinary metals. Next, we draw attention to some characteristics of layered superconductors. In the pairing interaction within the HTSCs, the atomic wave functions overlap in an unusual way, owing to the crystal structure of these materials. We also present some tunneling data, to show the difference from conventional superconductors.

After this tour of exceptional features in the HTSCs, we go on to present the broad outlines of one particular theory—that of Kresin and Wolf. It needs to be stressed that this is no guarantee of a single right theory; rather, it is an illustration of just how challenging it is to create *any* theory of HTSCs. There are so many unusual features, so much conflicting data, that a comprehensive understanding of HTSCs (comparable to BCS for LTSCs) is still many years away.

223

12.1. THE NORMAL-STATE FERMI SURFACE

Because superconductivity is caused by an instability in the normal state, it is first necessary to have some model for the normal state before trying to understand the superconducting state. For any metal, common normal-state measurements such as resistivity, tunnelling conductance, NMR, etc., provide data that are only compatible with certain theoretical models, and thus the possible choices among models are restricted. For compounds as complex as the HTSCs, this limitation is very important.

12.1.1. Momentum Space

The simplest case of all is that of a metal with one single numerical value for the Fermi energy E_F. At absolute zero, all electronic states below that energy level are occupied, and all above are unoccupied. At any finite temperature, there is a slight tail in the Fermi distribution function, which governs the occupancy of the available energy levels. Repeating from Chapter 6, that function is

$$\frac{1}{\exp[(E - E_f)/k_B T] + 1} \tag{12.1}$$

At any temperature of interest, this function drops suddenly from 1 to 0 very near $E = E_F$. The difference in a superconductor is that there is an energy gap exactly at the Fermi level, so the tail is forced to extend slightly further on either side of E_F. Figure 12.1 compares the two cases.

In the three dimensions of momentum space, this kind of uniform filling of low-energy levels amounts to filling up a sphere, because $E = k^2/2m$ (and $k^2 = k_x^2 + k_y^2 + k_z^2$). Figure 12.2 is a sketch of this, where the shaded area represents the partially filled states near the Fermi level. That boundary between filled and unfilled states is termed the Fermi surface. As simple as this surface is in Figure 12.2, it is still fuzzy and diffuse at any finite temperature (although the diffuseness is exaggerated here for emphasis).

The occupied states below the Fermi level comprise the Fermi sea. The many electrons interact weakly with one another, in a disordered way. Continuing the analogy of terminology, these electrons are said to form a Fermi liquid. The normal state is a conventional Fermi liquid.

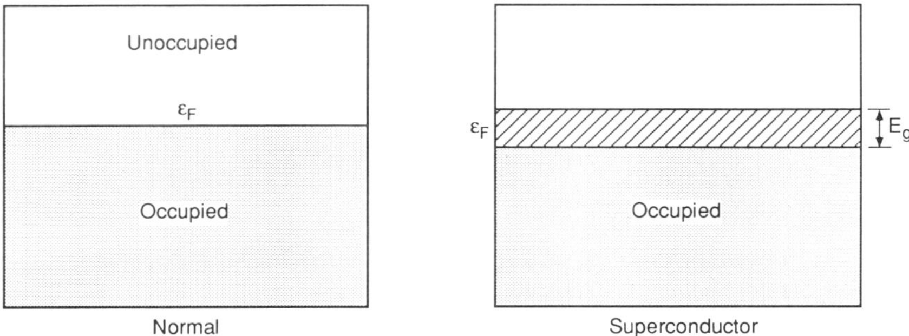

Figure 12.1.   Comparison of the nature of the Fermi level in a normal (left) and a superconducting (right) metal. A key feature of BCS theory is that the energy gap is centered right at the Fermi level. [After C. Kittel, *Introduction to Solid State Physics*, 6th Ed. (Wiley, New York: 1986). Reprinted by permission of John Wiley & Sons, Inc.]

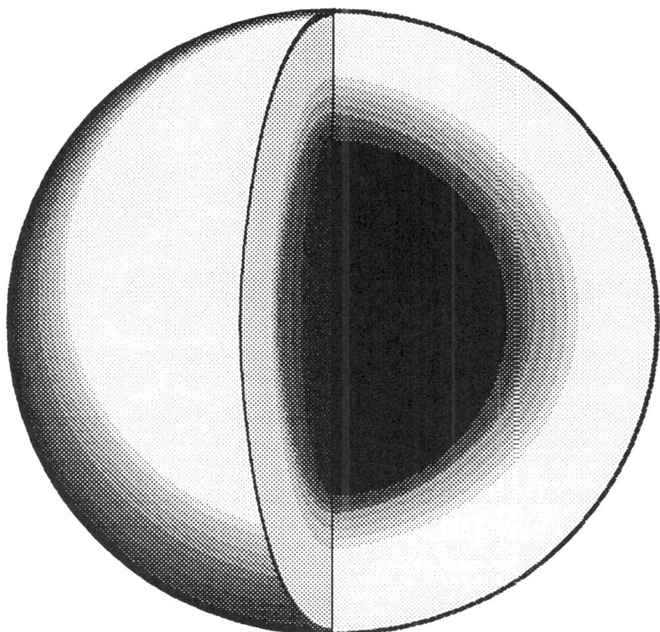

Figure 12.2. Sketch of the Fermi surface in a conventional simple metal. The x-, y-, and z- axes correspond to dimensions in momentum space: k_x, k_y, and k_z. The radial distance outward (k^2) is increasing energy. The black inner region denotes fully occupied states in the metal; the light outer region denotes empty states. The transition from totally filled to totally empty takes place over a narrow band of energy states centered on the Fermi level E_F.

As soon as a material becomes anisotropic, the simplicity of the Fermi surface goes away. To begin with, the Fermi surface becomes a prolate ellipsoid if the effective mass in one direction is much larger than in the other two. Then the Fermi level of energy can be different along different axes k_x, k_y, k_z. Depending on the direction within momentum space, the sequential filling of levels may be in an entirely different energy band. We say that different bands cross the Fermi energy at different momentum vectors $\mathbf{k}$.

As shown in Chapter 8, the HTSCs are extremely anisotropic materials and so, even in the normal state, the HTSCs have a radically altered Fermi surface. Figure 12.3[2,3] is the result of a band-structure calculation[4] for YBCO, and bears no resemblance to a sphere. The Fermi surface is electronlike in some places, and holelike in others. Some parts of it are due to the CuO_2 planes and some due to the CuO chains. Furthermore, a number of conditions mentioned in Chapter 10 were omitted from the band-structure calculations: magnetic fluctuations, charge fluctuations, and undulations in the CuO_2 planes. When this kind of complexity is combined with the experimental limitations associated with imperfect samples, it is little wonder that any theory must be considered tentative.

There are alternatives to the Fermi liquid theory of HTSCs: these include *bipolarons*, in which the charge carriers bind into pairs to form bosonic (not fermionic) excitations[5]; *resonance valence bond* coupling[6]; a *marginal Fermi liquid*[7]; and *anyons*.[8] However, these are dwindling in attractiveness as experimental data improves.

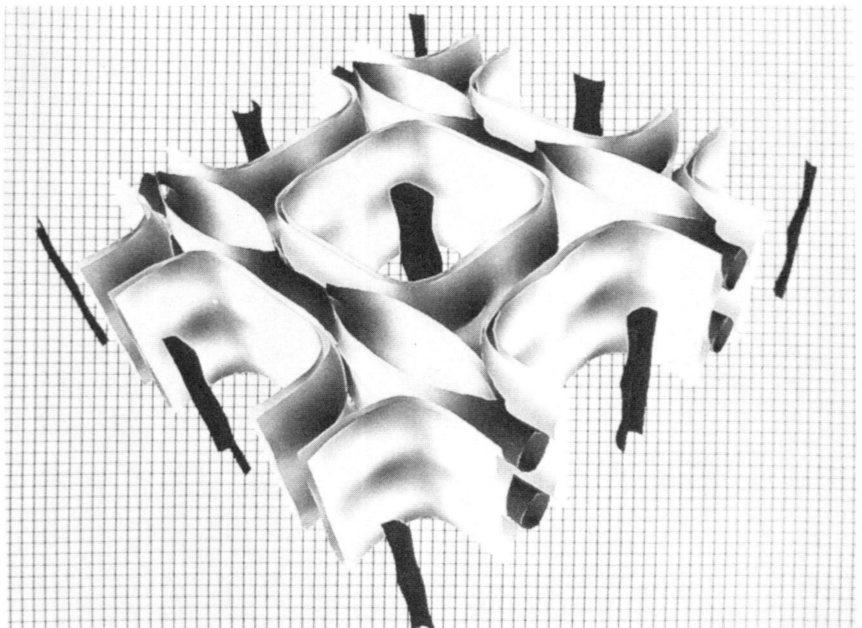

Figure 12.3. Three-dimensional picture of the calculated Fermi surfaces of $YBa_2Cu_3O_7$, slightly broadened and extended periodically. This is derived from band-structure calculations reported by R. E. Cohen of the Carnegie Institute of Washington.

12.1.2. Experimental Results

Experimental data about the Fermi surface has helped to clarify the picture. It is instructive to consider briefly the experimental aspects of determining the shape of the Fermi surface. There are basically three ways to measure a Fermi surface: Angle-Resolved PhotoElectron Spectroscopy (ARPES); Angular Correlation Annihilation Radiation (ACAR); and deHaas–vanAlphen resonance (dHvA). All three have certain advantages, and certain problems as well. For example, dHvA sees electrons orbiting the Fermi surface and determines their elapsed orbiting time through a resonance with an applied magnetic field. But when samples contain voids and impurities, scattering events interrupt the electron trajectories, degrade the data, and leave the interpretation uncertain.

Photoemission preferentially sees the surface of a sample, so if there is surface contamination, the results can be misleading. Los Alamos used photoemission data to deduce the presence of a Fermi surface at $T < 50$ K. Since (a) this technique must be performed in a vacuum, (b) it is very sensitive to surface effects, and (c) $Y_1Ba_2Cu_3O_{7-x}$ loses oxygen rapidly to a vacuum at temperatures above 50 K, experiments performed above 50 K showed no Fermi surface. Subsequently, experiments performed at Argonne provided definitive evidence that $Y_1Ba_2Cu_3O_{7-x}$ has a Fermi surface. The Argonne experiments used positron annihilation data together with complicated computer computations to deduce the existence of a Fermi surface at temperatures bracketing T_c.

Early in the history of HTSC, a number of exotic theories were proposed to explain superconductivity via a non-BCS mechanism. Some of those were committed to a no-Fermi-surface picture of the HTSCs, and with the Fermi surface now firmly established, they are no longer viable theories. However, the fact that they were proposed at all is testimony to the considerable experimental difficulties that the HTSCs encountered.

12.2. MACROSCOPIC THEORIES

The BCS theory is a *microscopic* theory, because it deals with individual electron pairs linked by phonons, and considers their interactions. Of greater practical interest, however, are those *phenomenological* theories that account for *macroscopic* quantities such as electrical current. The original London theory,[9] last mentioned in Chapter 2, is one example. The most useful phenomenological model is the Ginsburg–Landau[10] theory, which was applied to type II superconductors (including the HTSCs) by Abrikosov.[11] It was shown by Gor'kov[12] that this theory is a limiting form of BCS, and thus today it is called the GLAG (Ginsburg–Landau–Abrikosov-Gor'kov) theory.

It would take us well beyond our scope to present the details of the GLAG theory, but it is important to note that it emphasizes the importance of an order parameter, commonly denoted by ψ, which is directly proportional to the gap parameter Δ. The local density of superconducting charge carriers is $n_s = |\psi(x)|^2$, thus giving ψ a significance similar to that of the wavefunction of quantum mechanics.[13] In fact, the differential equation governing ψ is reminiscent of Schroedinger's equation, but with a nonlinear term included. Because ψ varies with position x, this model allows treatment of charge densities that vary spatially. This is particularly useful for type II superconductors where the flux vortices have normal cores and the charge density varies over a coherence length ξ. Since ξ is only of the order of the lattice constant in HTSCs, it is mandatory to use GLAG for such materials.

In order for its equations to be soluble, the GLAG model must have a small order parameter, which means the model is really only good near T_c or H_{c2}. However, in fact it still works well at values far from T_c or H_{c2}, giving qualitatively correct answers. Needless to say, anisotropy adds another degree of complexity, and the solution of the anisotropic GL equations are the subject of more advanced texts.[14]

One simple way to model HTSCs is to adapt the London model by adding anisotropy. This involves replacing the mass of the charge carriers by an effective mass tensor, with different mass values for the several directions.[15] This model has had some success, especially with YBCO, but since the London model ignores the normal vortex cores, it is limited in how much it can do. To go further, a more advanced model that allows both normal and superconducting regions is needed.

BSCCO and TBCCO are even more anisotropic than YBCO. The properties in the c-direction are so different from those in the a-b plane that it is more accurate to regard them as two-dimensional superconductors. In fact, the most appropriate model is that corresponding to a layered superconductor; in this case, a series of CuO_2-plane sandwiches layered one upon another, with intercalated planes of insulating metal oxides separating them. The communication between successive layers is entirely due to tunnelling.[16]

Fortunately, there is a theory that treats this case. Developed by Lawrence and Doniach,[17] this model revises the conventional GL equations, which contain a gradient operator. The gradient operator is restricted to only the x- and y-directions, and perpendicular to the layers (the z-direction) there is a discrete set of equations (one for each layer). Current in the x-y planes is governed by the usual GL equations, but perpendicular current is restricted to tunnelling. In the regime near H_{c1}, one must treat discrete coupled nonlinear equations, but near H_{c2}, the equations can be linearized and solved.

The Lawrence–Doniach model has built into it a crossover from three-dimensional to two-dimensional behavior, which depends on the effective mass ratio, the coherence length, and the spacing between planes. The model was originally devised to treat transition-metal dichalcogenides,[18] but later it was applied to organic layered superconductors (see Chapter 27). Because of the similarities to such compounds, computer models of the LD model applicable to HTSCs are being developed.

12.3. INTERACTING ELECTRONS

The nearly two-dimensional structure of the HTSCs means these materials are far different from conventional LTSCs. One very important consideration is this: What kinds of electrons (or holes) engage in the pairing mechanism that causes superconductivity? In a simple LTSC, the answer is trivial: the free electrons in the conduction band. In the HTSCs, where superconductivity occurs primarily in the CuO_2 planes, this is not necessarily so. The interaction may be between very specific copper and oxygen electrons.

In order to appreciate the new features that are involved here, it is first necessary to provide some background about interactions between electrons:

12.3.1. Overlapping Wave Functions

In quantum mechanics, whenever two electrons interact, they do so by having their wave functions overlap. Denoting the two wave functions by ψ_1 and ψ_2 and the interaction potential by the letter H, the interaction depends upon the *overlap integral*

$$\int_{-\infty}^{+\infty} \psi_1 H \psi_2 \, d^3x \tag{12.2}$$

where d^3x denotes an element of volume; the integral is carried out over all space. Obviously, a change in the wavefunctions will change the numerical value of the integral, and hence the interaction.

A totally free electron traversing a solid is not confined anywhere, and thus its wavefunction ψ is just a plane wave, usually written $\psi = \exp[i\mathbf{k}\cdot\mathbf{x}]$, where $\mathbf{k}$ denotes the momentum vector and $\mathbf{x}$ the position vector. The wavefunctions of different electrons differ only in their frequency of oscillation, which is related to the momentum $\mathbf{k}$. Incidentally, the classical billiard ball picture of electrons is useless for describing the interaction taking place in superconductivity.

By contrast, electrons bound to atoms fill up available energy levels consecutively, starting with the lowest energy state, and following the *shell model* familiar from introductory chemistry. Any electron orbiting an atom occupies one of many different *orbitals*, that is, its wave function must be of a very specific type. Recalling that angular momentum is quantized, when an electron has zero angular momentum, it has quantum numbers l and m = 0, and it occupies the *s*-shell of the atom. It has an *s-wave* orbital, which is spherically symmetric. If it orbits with one unit of angular momentum, it is called the *p-wave*; the energy of this state is slightly higher, and the wavefunction is no longer spherically symmetric. An electron with two units of angular momentum is called a *d-wave*, three units is the *f-wave*, and so forth. Each wavefunction ψ is the product of a radial function (which depends on the energy) and a function of angle. The Schroedinger equation yields solutions for the angular part which are *spherical harmonics*, and these have very specific shapes.[13] In general, these are written

$$Y_{lm} = \frac{1}{\sqrt{2\pi}} \exp{(im\,\phi)}\Theta_{lm}(\theta) \tag{12.3}$$

where the Θ functions are the *associated Legendre polynomials*. The angle θ is defined with respect to the z-axis of a coordinate system that aligns with the crystallographic plane in which that atom lies. The value ϕ is an azimuthal angle within that plane. For the case $m = 0$, the ϕ-dependence vanishes, and the first three functions become

$$Y_{00} = \frac{1}{\sqrt{4\pi}}$$

$$Y_{10} = \frac{\sqrt{3}}{\sqrt{4\pi}} \cos\theta \tag{12.4a}$$

and

$$Y_{20} = \frac{\sqrt{5}}{\sqrt{16\pi}}(3\cos^2\theta - 1)$$

Also of interest is the case $l = 2$, $m = -1$, for which we have

$$Y_{2,-1} = \frac{\sqrt{15}}{\sqrt{8\pi}} \cos\theta \sin\theta \exp{(-i\phi)} \tag{12.4b}$$

Figure 12.4 shows polar plots of these four functions;[13] they are the angular part of the s-wave, p-wave, and d-wave orbitals. These differ because of the amount of angular momentum associated with each orbital. The simplest of these is a spherically symmetric wave function known as the s-wave; the angular momentum is zero, and the energy is low. The important thing to notice is that with increasing angular momentum, the wave function becomes less uniform and more directional. Thus for the s-wave, the electron has equal probability of being found in any direction; but even for the p-wave, the wave function is localized to the left and right of the drawing, which means the electron is likely to be found at either the left or right, and not likely at the top or bottom.

From Figure 12.4, it is clear that the s-wave orbital is independent of how θ is defined. On the other hand, Y_{20} goes through zero at $\theta \approx 55°$, $125°$, $235°$, and $305°$; this d-wave is quite sensitive to orientation. Similarly, $Y_{2,-1}$ has a four-leaf clover shape with zeros at $0°$, $90°$, $180°$, and $270°$. Thus, if the plane containing an atom is tipped with respect to nearby atoms, the overlap of its d-wave orbital with the orbitals of nearby electrons will be far different from that ordinarily associated with flat planes. This condition presents a very difficult numerical computation problem to begin with. When there are irregularities in any real crystal, the problem worsens.

What happens when the outer electrons of two adjacent atoms interact? It doesn't matter how the atoms are oriented with respect to one another if the electrons are in s-wave orbitals. The integral of Equation (12.2) will come out the same. But if one electron is in a p-wave orbital, then any slight change in the orientation of one atom will change the overlap between the two wave functions, and therefore change the net interaction as well. When a d-wave is involved, the sensitivity to change in orientation becomes greater, because of the narrower

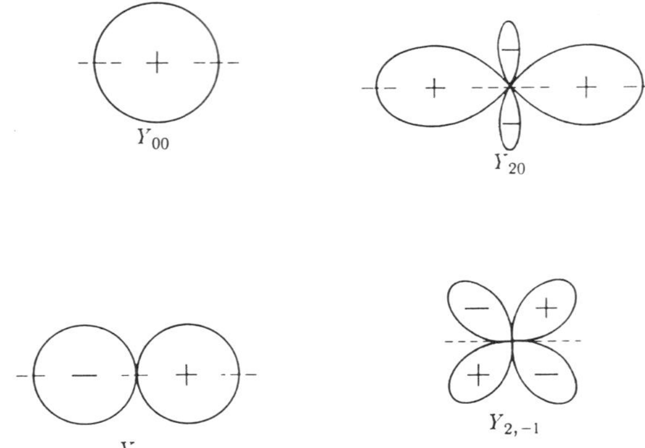

Figure 12.4. Polar plots of associated Legendre polynomials, the angular part of wave functions ψ for atomic orbitals: *s*-wave, *p*-wave, and *d*-wave, corresponding to $l = 0, 1, 2$, respectively. In these drawings, $\phi = 0$, so there is no azimuthal dependence displayed for these functions (although for three of them, the quantum number $m = 0$). The magnitude of the function at any angle θ is given by the radial distance from the origin; the sign is indicated as "+" or "–" within each lobe. As angular momentum levels (values of l) increase, overlap integrals between these become more sensitive to orientation and alignment (from Ref. 13).

lobes of the corresponding wave functions. If the overlap is between a *p*-wave and a *d*-wave electron, the complexity grows. The bottom line is this: relatively minor deviations from perfect alignment can cause dramatic changes in the overlap of neighboring electrons.

12.3.2. Application to HTSCs

It is also possible for an electron to be in a hybrid orbital, that is, a combination of two (or more) wave functions. A typical representation might be

$$\psi_1 = \frac{1}{\sqrt{2}} \left\{ \psi(4s) + \psi(3d) \right\} \tag{12.5}$$

where $\psi(4s)$ denotes the wave function for the $4s$ state of an atom, and so on. Obviously, this increases the complexity of any overlap integral calculation, and it also increases the flexibility of theoretical constructions that purport to account for observed results.

Hybridization in HTSCs means that in the CuO_2 planes, the single-particle state consists of overlapping wave functions of copper $3d$-shell electrons with oxygen $2p$-shell electrons. The case of $Y_{2,-1}$ is of particular interest for HTSCs, because experimental evidence suggests that this is the wave function shape that dominates the copper–oxygen linkage in the CuO_2 planes.

Questions involving wave functions do not yield easily to resolution by experiment. For example, x-ray absorption measurements indicate whether the charges go into the conduction band or into an orbital on one of the atoms. On the other hand, in the case of hybrid orbitals (e.g., copper $4s$ state combined with copper $3d$ and oxygen $2p$), the shape of the orbitals leads to anisotropy in normal conduction: conduction in the CuO_2 planes is favored. This undercuts the conventional interpretation of the x-ray measurements.

Meanwhile, the paired state of electrons that gives superconductivity is normally discussed in momentum space, rather than *xyz* space. It is known that the pairing state is not isotropic. However, that doesn't absolutely rule out an *s*-wave interaction, because anisotropy is so important. To see this, recall that with a lattice constant $a < 4$Å, YBCO has coherence lengths of about 3 Å along the *c*-axis and 16 Å along *a,b*. Now suppose the wave functions are spheres, but in the metric of the anisotropic lattice with $c >> a,b$. In that case they become ellipsoids, more resembling *d*-waves than *s*-waves. Adding further confusion is the fact that the energy gap is anisotropic, and may go to zero in some directions of momentum space.

There is considerable controversy surrounding these issues. Here, it suffices to note that better experimental data is needed before confidence can be placed in any of several competing theories.

12.4. THE DENSITY OF STATES IN HTSCs

In Chapter 6 we explained why the density of states is so important a parameter in superconductors, and how tunnelling experiments are used to determine it. Here, we consider tunnelling for the HTSCs, and present some typical density-of-states data.

12.4.1. Experimental Tunnelling Results

One nuisance that occurs in HTSCs is thermal smearing. En route to having the measured curve of *dI/dV* represent the density of states, factors of the form of Equation (12.1) appear in the combination

$$\frac{d}{dV}[f(E) - f(E + eV)] \tag{12.6}$$

which, conveniently, is a delta function at $T = 0$ K. Unfortunately, at temperatures well above 4 K, that is no longer true, and the raw data must be corrected.

Limitations of sample quality must be taken into account when dealing with the HTSCs. Beyond the effects of anisotropy in a perfect crystal, real samples have other imperfections such as mismatched grains, interrupted layers, and other factors which clutter the experimental data. To appreciate the resulting discrepancies, it is first necessary to recall that in going from one normal metal to another, the current–voltage characteristic would simply be linear. For the HTSCs, samples are not necessarily 100% in the superconducting phase, but have connected normal-state pathways available to carry current. Thus, there is a background conductance which must be added to the superconducting *I-V* characteristic. The combination produces a roughly parabolic-shaped curve with peaks at the gap edges, somewhat like Figure 12.5.

Most of the available YBCO tunneling data is from thin-film junctions and appears as in Figure 12.6. It is entirely possible that these unusual features are simply artifacts of poor samples. In any case, the linear-conductance term is prominent in anisotropic HTSCs, especially when tunneling is in the *c*-direction; for *a-b*-plane tunneling, the parabolic shape is much neater. Researchers generally speak of gaplike behavior anytime they see even a vague dip in the conductance near $V = 0$, but anisotropy of the gap makes for very complicated analysis of data.

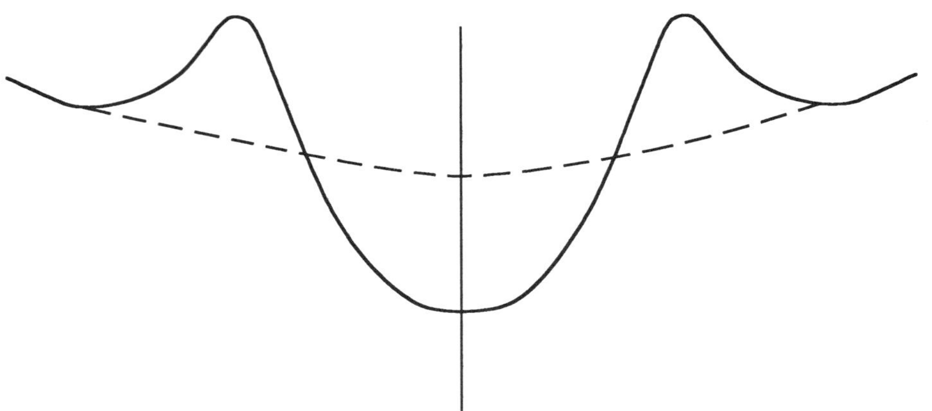

Figure 12.5. Typical tunnelling data for a HTSC in which the peaks and gap ride atop a broad parabolic conductance pattern.

The best way to remedy this problem is to study a HTSC without anisotropy. Fortunately, the compound $BaKBiO_3$ ($T_c = 30$ K) is isotropic (see Chapter 8). Good tunneling data has been obtained for it by Zasadzinski and co-workers,[19] who find that high-energy phonons (≈ 60 meV) are involved in its superconductivity.

It is also possible to determine the phonon density of states by means of Inelastic Neutron Scattering (INS) measurements. Figure 12.7 shows[20] phonon density-of-states data for three variations of $BaKBiO_3$ at 15 K. The data is symmetric about the zero-voltage point, so only positive energy data is displayed. Looking first at $Ba_{0.6}K_{0.4}BiO_3$ (curve *a*), the most striking feature is the presence of two peaks in the phonon density of states, one around 30 meV and the other out at 60 meV. These are due to oxygen vibrations, not the metals. This spectrum

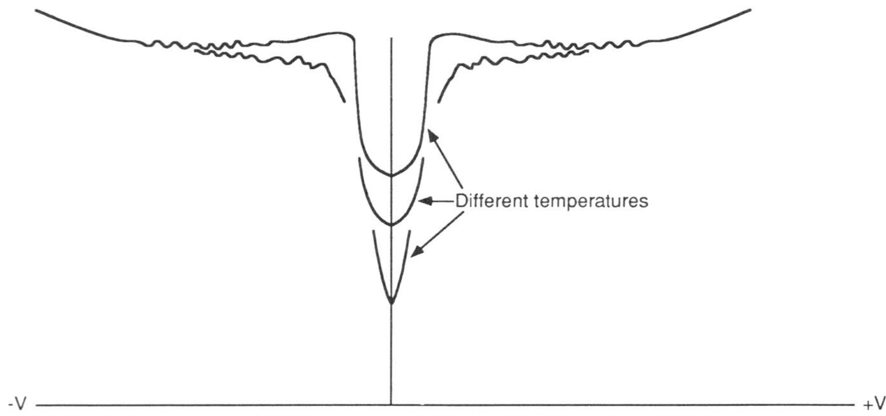

Figure 12.6. Typical tunnelling data for YBCO, in which a very narrow gaplike behavior is observed.

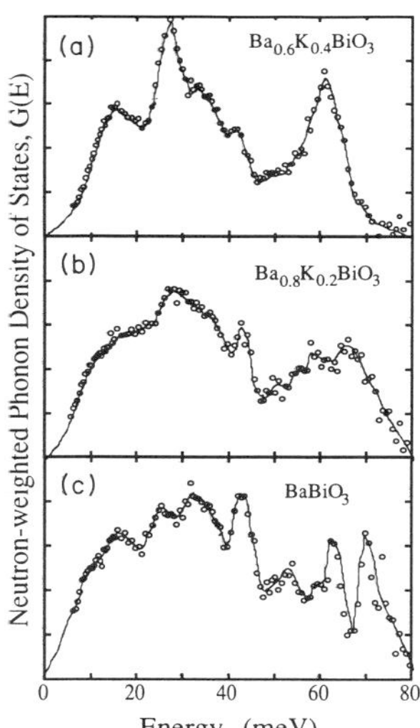

Figure 12.7. Experimental neutron-weighted phonon density of states at 15 K for (a) superconducting Ba$_{0.6}$K$_{0.4}$BiO$_3$, (b) insulating Ba$_{0.8}$K$_{0.2}$BiO$_3$, and (c) parent material BaBiO$_3$ (from Ref. 20).

is substantially different from the simple BCS model presented in Chapter 6, where the density of states fell off monotonically from a singularity at the gap edge. What this means is that in Ba$_{0.6}$K$_{0.4}$BiO$_3$, both a low-frequency and a high-frequency phonon are capable of mediating the superconducting interaction.

Comparison with other amounts of potassium doping is enlightening. Compound (a) has $T_c = 30$ K; compounds (b) and (c) are not superconductors at any temperature. It is evident that there is a qualitative difference here: The superconductor (a) has three phonon bands, centered around 15, 30, and 60 meV, and the phonon spectrum cuts off at about 70 meV. The two insulators [(b) and (c)] have broader spectra which peak at higher energies, and the peaks and valleys are not as pronounced as in (a). When phonons shift to lower energies, we say the phonon mode *softens*. For Ba$_{0.6}$K$_{0.4}$BiO$_3$, this reduction in phonon energy is probably due[20] to screening by holes introduced when K (valence = +1) replaces Ba (+2). It is reasonable to conclude that superconductivity here is due to coupling between electrons and the 30 meV and 60 meV oxygen phonons.

12.4.2. Importance of High-Frequency Phonons

A peak in the density-of-states at a high-energy signals that an adjustment is needed to the standard BCS theory, which considered only low-energy (acoustic) phonons. In BCS, certain assumptions were introduced to simplify the mathematics, which resulted in the familiar equation

$$k_B T_c / \hbar \omega_c = 1.14 \exp\left(-1/N_0 V\right) \tag{12.7}$$

This indicates that the ratio of T_c to the phonon cutoff frequency ω_c is dependent upon $N_0 V$ (or its modern surrogate λ in generalizations of BCS). The biggest question about the HTSCs, one that prompts many excursions into exotic theories, is: why is T_c so high? For any numerical choice of λ of order one, the simplest way to get T_c from around 10 K to around 100 K is to increase the cutoff frequency ω_c.

This is precisely what happens if electrons couple to optical phonons as well as to acoustic phonons. In Section 12.5, we consider one such model. The optical phonons can mediate the Cooper pairing, thus raising the cutoff frequency ω_c and raising T_c as well. Tunneling data for $Ba_{0.6}K_{0.4}BiO_3$ indicate that it is a conventional BCS superconductor, but with optical phonons in a prominent role.

Density-of-states data has also been obtained[21] for NdCeCuO, a 2-1-4 compound which seems to be a BCS superconductor and is only mildly anisotropic. NdCeCuO has $T_c \approx 22$ K and $\lambda \approx 1$, but it would require $\lambda \approx 2.5$ to drive T_c up to 90 K. However, that does not mean that comparable λ values are required for the other HTSCs. YBCO, BSCCO, and TBCCO have not reached the level of experimental confirmation (of BCS) obtained for NdCeCuO, but the significant point is this: the clear demonstration of one HTSC that follows BCS is sufficient to argue against the need to introduce exotic theories to account for superconductivity here.

12.5. A TWO-BAND, TWO-GAP THEORY

In this section we describe one theory of HTSCs, which is successful in accounting for a substantial fraction of the available data. By no means is it the final word, but it is a good start. In particular, it stays very close to BCS in its major features.

The HTSCs differ from conventional LTSCs in a number of ways, but perhaps the most significant differences originate in the dramatic difference between the coherence lengths ξ in the two classes. LTSC coherence lengths are of the order of microns in size, comparable at least to a grain size in a metal. The electron mean free path l is typically much smaller than ξ, and most type II materials are termed "dirty." In HTSCs, coherence lengths are typically angstroms; $l > \xi$ and the materials are effectively clean. Moreover, the anisotropy of the crystal structure makes $\xi \sim 15 - 20$ Å in the a or b direction, but only ~ 3 Å in the c-direction. (Indeed, in the very anisotropic thallium compounds, ξ can be far smaller, perhaps 0.02 Å.) The most immediate consequence of this is to isolate the superconductivity in one unit cell from that in its neighbor (in the c-direction), creating two-dimensional superconductivity confined to individual copper oxide planes. The pairs are strongly hindered from hopping in the c-direction.

The reason the BCS theory was so successful is that it identified the essential interaction, took its simplest form, and then worked out the consequences, which agreed well with experiments. (Actually, BCS was not really rigorous; only 9 years later was that quality obtained.[22]) However, BCS works because the fundamental process is pairing of quasiparticles near the Fermi surface, and the BCS reduced interaction reproduces such a pairing. For simple, weak-coupling, isotropic, conventional superconductors, that is enough to give a very accurate theory.

The objective here is quite similar—to use the simplest possible model that still gives adequate agreement with experiments. There is no guarantee that any one theory will stand the test of time, but it serves to illustrate the way in which a theory is constructed.

12.5.1. Kresin–Wolf Theory

Kresin and Wolf[23,24] analyzed the consequences of anisotropy in the energy gaps of the HTSCs, mainly YBCO. Their theory features two bands and two energy gaps, and takes account of the quasi-two-dimensional nature of these compounds. The extremely complex Fermi surface calculated in Figure 12.3 is idealized and simplified[25] by a combination of a cylinder adjacent to two flat planes, as shown in Figure 12.8.

It must be remembered that the Fermi surface exists in *momentum space*, which is not the same as *position space* (they are Fourier transforms of each other); we speak of the *lattice* and the *reciprocal lattice*. Normally, regions in one space do not correspond to specific points in the other space. However, in the HTSCs, the planes and chain in position space are responsible for very specific portions of the Fermi surface (in momentum space). This characteristic is a very significant distinction between LTSCs and HTSCs.

In general, in any superconductor each electron-band is characterized by its own energy gap. This applies to both conventional and cuprate superconductors. Differences in density of states, pairing interactions, etc., lead to distinct values of the gap. When the various bands overlap, multiple energy gaps come into play. The multiplicity is smeared out in a conventional superconductor where $l \ll \xi$, and so the distinct gaps are not seen experimentally. Therefore, a simple one-gap model generally suffices to account for the observed properties of those materials. The customary presentation of the BCS model uses only one gap parameter Δ.

However, for the HTSCs (especially YBCO) the experimental evidence for two *gaps* is too strong to ignore. Moreover, the band structure calculations that yield Fermi surfaces such as Figure 12.3 stand on very solid ground, and experiments point to a minimum of two *bands*.

Staying within the structure of BCS theory, Kresin and Wolf postulate that the CuO_2 planes are intrinsically superconducting, while the CuO chains are intrinsically normal, and superconductivity must be *induced* in the chains.[26] Some electrons exchange phonons and

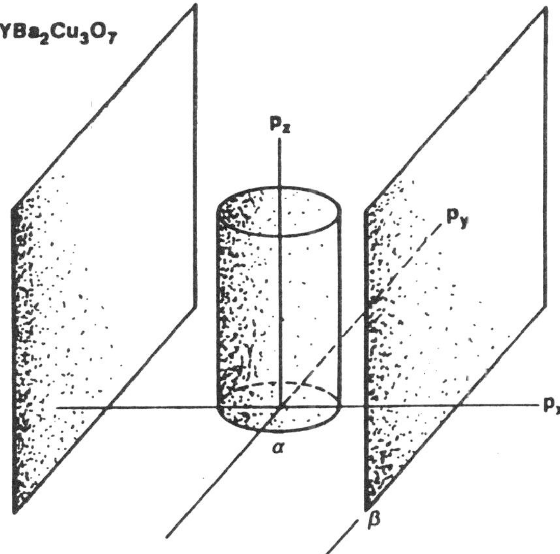

Figure 12.8. Idealized Fermi surface for YBCO in the Kresin–Wolf two-band, two-gap model. This simple combination of planes and a cylinder replaces the more advanced Fermi surface of Figure 12.3.

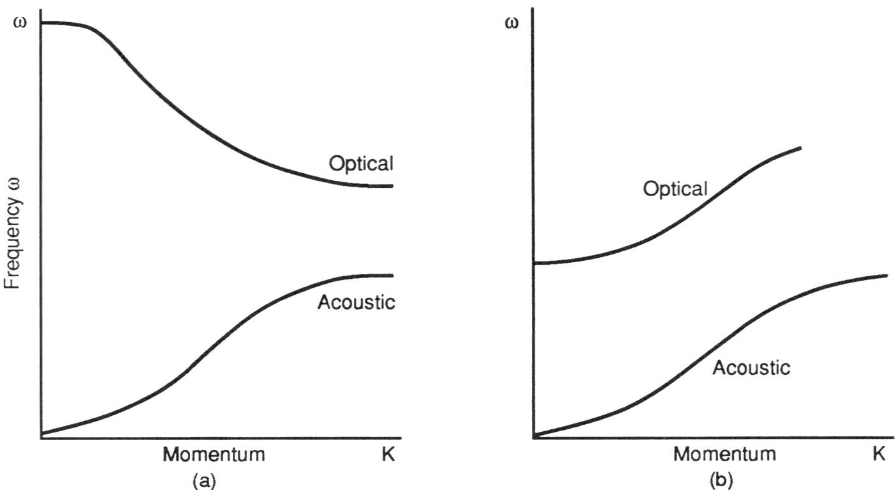

Figure 12.9.   (a) Branches of the phonon spectrum in a conventional solid. (b) Unusual characteristic associated with a two-dimensional material, in which the optical branch has a totally different shape. This feature is utilized in the Kresin–Wolf model of the HTSCs.

form Cooper pairs in the usual manner. However, other electrons in one band give off a phonon and are excited into the second band, where pairing takes place. There are three different values of λ ($= N_0 V$) corresponding to the three possible band-and-gap configurations. Fortunately, if one λ is much larger than the others, then some simplification occurs.

Next, Kresin and Wolf introduce the idea that, in HTSCs, the strong coupling arises from very unusual lattice phonons. In conventional LTSCs, all the action appears below 20 meV, that is, in the conventional acoustic-phonon branch. Not so for the HTSCs! There are optical phonons of a most exceptional kind. Figure 12.9a shows the standard Brillouin-zone drawing ($\omega - k$ drawing) containing acoustic and optical branches of phonons. This is the same as Figure 6.1. In contrast, Figure 12.9b shows an alternate kind of optical branch, which has a shape similar to the acoustic branch, but is elevated to about 15 meV at $k = 0$, rising to 25 or 30 meV near the zone boundary. This peculiar optical branch is characteristic of two-dimensional structures. It is called a *soft optical mode.*

These special high-energy phonons couple strongly to the electrons. In rigorous BCS theory,[27] the trick is to maintain a high $\lambda \equiv N_0 V$ in the presence of energetic phonons, and this is exactly what happens here. Thus, it is a combination of factors that makes the layered cuprates high-temperature superconductors.

When the coupling strength is no longer weak, the BCS formula [Equation (12.7)] for T_c must be revised. Kresin[28] has shown that a general expression covering the full range of λ values is

$$k_B T_c = \frac{0.25\overline{\Omega}}{\sqrt{\exp\,[2/\lambda] - 1}} \tag{12.8}$$

where $0.25\,\overline{\Omega}$ is a generalization of the cutoff frequency term $1.14\,\hbar\omega_c$ to the case of a phonon spectrum with structure. When $\lambda = N_0 V$ is small, this reduces to Equation (12.7), the original BCS formula.

However, in the Kresin–Wolf model, λ is a mixture of the coupling constants corresponding to superconductivity in the CuO_2 planes, the interband interaction, and the proximity effect. For YBCO, the first of these values is about 3, and the others range from 0.5 to 0.9. Under such circumstances, the familiar BCS relation $2\Delta/k_B T_c = 3.53$ no longer holds; it must be modified by a correction factor for strong coupling. One of the gap energies comes out greater than the BCS prediction, and one is less. Utilizing experimental data for YBCO, Kresin and Wolf assign one value $2\Delta_1 = 6\,k_B T_c$, and the other $2\Delta_2 \sim 2.5\,k_B T_c$.

It bears mentioning that these ideas are not universally accepted. No one is positive even that phonons are the mechanism of HTSC, let alone *optical* phonons. Still, this analysis helps to explain the relative importance of the planes and the chains in YBCO, which had been so mysterious in the early days of HTSC. Seen through the eyes of this model, the planes exhibit strong-coupling superconductivity, while the chains have weak interband coupling.[29] Using the cylinder-plus-planes Fermi surface (Figure 12.8), the cylinder corresponds to the Fermi surface for the CuO_2 planes, and the planar Fermi surface corresponds to the chains.

The Kresin–Wolf analysis represents a step forward in our understanding of YBCO. The interesting question, still very much open, is whether similar principles hold for the bismuth and thallium superconductors, which are still more anisotropic than YBCO. Extending the same general concepts to the Bi and Tl compounds, their Fermi surfaces may have two cylinders, thus averting the question of chains entirely. The BiO layers (or the TlO layers) each form a conducting subsystem, but data on these materials is nowhere near as complete as for YBCO, so predictions are not easy to constrain. More detailed investigations are required of BSCCO and TBCCO before any theory can be tested.

12.6. COMPARISON WITH DATA

Many experiments demand either a two-band or a two-gap explanation. The evidence for two gaps is based on tunneling,[30] nuclear magnetic resonance,[31] IR reflection,[32] and penetration depth,[33] among others. Each experiment contributes to an overall mosaic that constrains the theory. For example, rotating the sample in an IR reflection experiment points to two gaps, one for the chains and one for the planes.

Kresin and Wolf summarize the evidence[26] as follows:

The temperature dependence of the Knight shift[34] and NMR relaxation time[31,34] for the plane and chain copper atoms was different and is the most direct evidence for two gaps, one associated with the planes and the other with the chains.[29] Both the real and imaginary parts of the surface impedance are described by the sum of two contributions with different energy gaps.[35] Raman measurements on crystals of $YBa_2Cu_4O_8$ (124) also show the presence of two gaps.[36]

How wide are the gaps? In YBCO the value of $2\Delta/k_B T_c$ comes out anywhere from 2.8 to 6 (for both bands), indicating that there are definitely two gaps; typically $\Delta_1 \sim 2.2$ BCS while $\Delta_1 \sim 0.6$ BCS. A single-gap BCS temperature dependence cannot account for the data.

Among other things, there is an internal consistency within the theory of Kresin and Wolf related to the smallness of ξ: The two-band model allows a prediction of the Fermi energy: $E_F = 200$ meV, which agrees well with experiment. When the Fermi level E_F is this small, the Fermi velocity V_F is likewise small, and the coherence length becomes extremely small: $\xi \propto V_F/T_c$. The coherence length in the planes is thus calculated to be $\xi_{pl} \approx 10$ Å, and in the chains it is $\xi_{ch} \approx 25$ Å. The effective mass along the chains is ≈ 25 m_e, which reinforces the notion of extremely sluggish motion along that direction.

It is plausible that the cylindrical Fermi surface corresponds to hole carriers, and the planar Fermi surface corresponds to electron carriers (obviously in separate bands). Evidence for this comes from thermopower data, where small variations in oxygen content cause a huge difference. It is the case that with negative (electron) carriers, thermopower is negative, and positive with positive carriers. The conduction in the chains dominates when 7 oxygens are present; but at 6.98 oxygens, thermopower starts to change,[37] and when oxygen content is down to 6.88, a positive value for thermopower occurs.

Additional evidence for two bands comes from normal-state resistivity data.[38] In untwinned single crystals of YBCO-123, the conductivity in different directions proved to be different, and this was attributed to the planes and chains, each having its own electron band.

One rather dramatic success of this two-band model is in explaining the behavior observed when praseodymium is partially substituted for yttrium in YBCO. Chapter 10 presents the data showing how T_c declines under this substitution;[39] superconductivity vanishes by the time the substituted fraction of Pr reaches 0.5. The physics of Pr substitution is very complex, so conclusions are only tentative. According to Kresin and Wolf, magnetic impurities drastically affect the value of the induced energy gap—the less important gap—but until there is a sizable change in the number of carriers (due to valence differences), the critical temperature does not change much.

It is reasonable to hope that other data on YBCO (and on the bismuth and thallium superconductors) might yield to analysis along these same lines. Nevertheless, in a field containing as many surprises as HTSC, any theory can be overthrown suddenly by contradictory data; or a more comprehensive theory might arise that explains even more data. Consequently, the only safe strategy is to take this (or any alternate theory) as one example of how this kind of research progresses. The final word is not in yet.

12.7. UNIVERSAL CURVES

One of the real strengths of the classical BCS theory is that it gives scaling laws that reduce a great deal of data from many superconductors to a single universal curve. The best example of this is Figure 6.4, which presents the reduced energy gap, $\delta = \Delta(T)/\Delta(0)$, as a function of the reduced temperature, $t = T/T_c$. Given measurements of T_c and of Δ at $T \approx 0$ for any elemental superconductor, the experimental data at intermediate temperatures will stay very close to the curve presented in Figure 6.4. There are similar laws for other quantities, including the reduced magnetic field, $h = H_c(T)/H_c(0)$.

In the case of the HTSCs, there are not as many superconductors, so data is not as plentiful as for the LTSCs. Furthermore, the theory is not as well-developed as BCS for the LTSCs. Nevertheless, a start has been made toward finding universal trends among these compounds. Schneider and Keller[40] have identified a relation between the transition temperature T_c and the zero-temperature condensate density $n(0)$. Denoting the maximum T_c value by T_c^m, and the maximum condensate density by $n_m(0)$, it is easy to define the reduced quantities $\overline{T}_c = T_c/T_c^m$ and $<n> = n(0)/n_m(0)$. To obtain a value for $n(0)$, we measure the muon

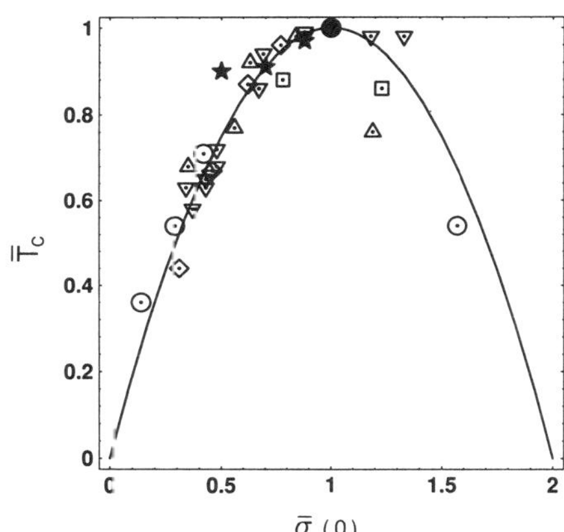

Figure 12.10. $T_c = T_c/T_c^m$ vs. $\overline{\sigma}(0) = \sigma(0)/\sigma_m(0)$ for various HTSCs. The solid line is the simple parabola $T_c = 2\overline{\sigma}(1 - \overline{\sigma}/2)$. Data points include YBCO-123, (Y,Pr)BaCuO, BSCCO-2223, BiSr(Ca,Y)CuO, LaSrCuO, and Chevrel phase compounds.

spin relaxation rate $\sigma(0)$, which is proportional to $n(0)$; and construct $\overline{\sigma}(0)$ in the same way by reference to its maximum. The data for several HTSCs then falls along a simple parabolic curve as shown in Figure 12.10.

These same researchers also applied this kind of scaling to data on the isotope effect and pressure effect, both of which shift T_c by small amounts. Based on fewer data points, the degree of fit is less convincing, but still points toward universal trends for these effects. As additional data becomes available, refinements are expected for this type of analysis.

12.8. SUMMARY

This chapter has introduced some of the key theoretical topics of interest concerning HTSCs. Substantial changes from conventional models are required by the extreme anisotropy of their crystal structures. For example, there *is* a Fermi surface in the HTSCs, but it is more complicated than the Fermi surface of ordinary metals.

One unique consequence of anisotropy is that the charges that engage in superconducting pairing may not be the usual free electrons (or holes), but might be hybrids of d-wave orbitals on the copper atoms with oxygen p-orbitals. Also, there could be anisotropy in the pairing interaction. There is room here for diverse explanations, and experimental results are ambiguous, so the controversy will continue for some time.

At this writing, there is no clear microscopic theory for the HTSCs, comparable to BCS for the LTSCs. However, every indication is that the BCS model is still applicable to these superconductors, but with modifications. Among the many phenomenological models put forth so far, three types are likely to survive: (a) the anisotropic London model; (b) the anisotropic Ginsburg–Landau model; and (c) Lawrence–Doniach models. The London model has the weakness that it ignores the normal vortex cores. GL includes the vortex cores, but only LD treats the layers. Hence, the future lies in the direction of LD.

The density of states is a pivotal parameter of superconductivity, because there is a gap in the density of states located right at the Fermi level. The gap is an essential characteristic of a superconductor. Tunneling experiments investigate the energy gap and the surrounding density of states. As a result of contemporary experiments, it is now entirely plausible to argue that the BCS theory still holds for the HTSCs, and no exotic theories are necessary to explain it. However, the extreme anisotropy of the HTSCs requires that some modifications be added to the simple BCS theory, notably with regard to the energy gap and the phonons that carry the interaction.

The two-gap, two-band model of Kresin and Wolf is cited here, because it modifies BCS in a comparatively simple way. It assigns one energy gap to superconductivity in the CuO_2 planes and another to the chains, and including optical phonons (as well as acoustic phonons) in the coupling mechanism of superconductivity. This works well for YBCO, where pairing is stronger in the planes than in the chains. This clears up several mysteries.

The perplexing question is still "How applicable is BCS?" Weak-coupling BCS predicts that $2\Delta/k_BT_c = 3.53$, but experimentally, $2\Delta/k_BT_c \neq 3.53$ for HTSCs. However, this does not necessarily imply that the coupling is strong. In a remarkably simple paper designed to "cast doubt on two superstitions," Abrikosov[41] has shown that large anisotropy can produce the same change in $2\Delta/k_BT_c$. This sort of thing keeps happening as our understanding of the HTSCs improves: the isotope-effect exponent need not be 1/2, the gap need not be 3.53 k_BT_c, and so forth. The customary numerical outcomes of BCS are not reproduced when BCS is applied to these layered, anisotropic compounds.

The theory of the HTSCs needs all the attributes of BCS theory, including a normal state just like any other metal. This conclusion is important for establishing the future direction of both theory and experiment: it says to keep using BCS as in years past, but be careful to include the effects of anisotropy when trying to explain experimental observations which may at first seem anomalous.

REFERENCES

1. J. Bardeen, L. N. Cooper, and J. R. Schrieffer, *Phys. Rev.* **108**, 1175 (1957).
2. W. E. Pickett *et al., Science* **255**, 46 (1992).
3. W. E. Pickett *et al., Phys. Rev. B* **42**, 8764 (1990).
4. R. O. Jones and O. Gunnarsson, *Rev. Mod. Phys.* **61**, 689 (1989).
5. R. Micnas *et al., Rev. Mod. Phys.* **62**, 113 (1990).
6. P. W. Anderson, *Science* **235**, 1196 (1987).
7. C. M. Varma *et al., Phys. Rev. Lett.* **63**, 1996 (1989).
8. R. B. Laughlin, *Science* **242**, 525 (1988).
9. F. and H. London, *Proc. Royal Society (London)* **A149**, 71 (1935).
10. V. L. Ginsburg and L. D. Landau, *Zh. Eksp. Teor. Fiz.* **20**, 1064 (1950).
11. A. A. Abrikosov, *Soviet Phys. JETP* **5**, 1174 (1957).
12. L. P. Gor'kov, *Soviet Phys. JETP* **9**, 1364 (1959).
13. R. H. Dicke and J. P. Wittke, *Introduction to Quantum Mechanics* (Addison-Wesley, Reading, MA: 1960).
14. R. A. Klemm, *Layered Superconductors* (Oxford Univ. Press, New York: in press).
15. V. G. Kogan, *Phys. Rev. B* **24**, 1572 (1981).
16. R. A. Klemm, A. Luther, and M. R. Beasley, *Phys. Rev. B* **12**, 877 (1975).
17. W. E. Lawrence and S. Doniach, *Proc. 12th Intl. Conf. Low-Temperature Physics* (Kyoto, 1972), pp. 361–362.
18. S. R. Gamble, F. J. diSalvo, R. A. Klemm, and T. H. Geballe, *Science* **168**, 568 (1970).
19. J. F. Zasadzinski *et al., Physica C* **158**, 519 (1989).
20. C. K. Loong *et al., Physical Review B* **45**, 8052 (1992).
21. Q. Huang *et al., Nature* **347**, 369 (1990).
22. J. R. Schrieffer *et al., Phys. Rev.* **148**, 263 (1966).
23. V. Z. Kresin and S. A. Wolf, *Physica C* **169**, 476 (1990).
24. V. Z. Kresin & S. A. Wolf, *Phys. Rev. B* **43**, 2691 (1990).
25. V. Z. Kresin & S. A. Wolf, *Phys. Rev. B* **41**, 4278 (1990).

26. V. Z. Kresin & S. A. Wolf, *Phys. Rev. B* **46**, 6458 (1992).
27. A. Abrikosov, L. P. Gor'kov, and I. Dzyaloshinskii, *Methods of Quantum Field Theory in Statistical Physics* (Prentice-Hall, Englewood Cliffs, NJ: 1963).
28. V. Z. Kresin, *Phys. Lett.* **122**, 434 (1987).
29. V. Z. Kresin, S. A. Wolf, and G. Deutscher, *J. Phys. C* **191**, 9 (1992).
30. M. Gurevitch *et al., Phys. Rev. Lett.* **63**, 1008 (1989).
31. W. Warren *et al., Phys. Rev. Lett.* **59**, 1860 (1987).
32. Schlesinger *et al., Phys. Rev. Lett.* **65**, 801 (1990).
33. S. Anlage *et al., Phys. Rev. B* **44**, 9164 (1991).
34. S. Barnett *et al., Phys. Rev. B* **41**, 6283 (1990).
35. H. Piel *et al., Physica C* **153–155**, 1604 (1988).
36. E. Heyen *et al., Phys. Rev. B* **43**, 12958 (1991).
37. J. L. Cohn *et al., Phys. Rev. Lett.* **66**, 1098 (1991).
38. J. L. Cohn, Naval Research Laboratory, unpublished data.
39. J. J. Neumeier *et al., Phys. Rev. Lett.* **63**, 2516 (1989).
40. T. Schneider and H. Keller, *Phys. Rev. Lett.* **69**, 3374 (1992).
41. A. A. Abrikosov, *Physica C* **214**, 107 (1993).

13

Weak Links

The problem of low critical current J_c is the primary obstacle to the use of HTSCs. Unless a solution is found, HTSCs will have far fewer applications than originally expected. There is hope for such a solution, because thin films of the HTSCs have high J_c values. However, the current carrying capacity in bulk HTSCs is much lower than in thin films. The leading candidate for an explanation is the weak link phenomenon in HTSCs.

Weak link behavior is a term used to describe what happens to a supercurrent as it crosses a nonsuperconductor region such as a grain boundary. Classically, a supercurrent could not cross such a barrier, but quantum mechanics permits the supercurrent to tunnel through the barrier. In doing so, the supercurrent experiences an exponential attenuation which is proportional to the barrier thickness, inversely proportional to the coherence length; and attenuation increases with increasing applied magnetic field. Since the coherence length ξ in HTSCs is very low, typically 3–30 Å, attenuation can be quite large even for a barrier only a few atoms thick. Weak link behavior is responsible for the steep fall off in transport current, particularly in the presence of applied magnetic fields, in bulk samples.

In this chapter, we first review tunneling and describe the properties of a *Josephson junction*.[1] This leads naturally into a discussion of SQUIDs, which exploit weak link behavior to make very sensitive measurements of magnetic fields, as discussed in Chapter 5. However, in the HTSCs, most grain boundary are Josephson junctions, and the extreme sensitivity to magnetic fields becomes a major nuisance. The elimination of weak links is a major goal of researchers concerned with making HTSC wire.

There are three subproblems that must be solved in order to increase J_c: these relate to grain boundaries, grain alignment, and flux pinning. The problem of flux pinning is deferred to the next chapter, while grain boundaries and grain alignment are covered here. By purifying the original material carefully, grain boundaries are kept relatively clean, and consequently they act as very thin barriers, less of an obstacle to current. By aligning grains carefully, a large contact surface between grains is assured, which may allow more total current to flow despite a low critical current *density*. Also, the great difference between coherence lengths in the c-direction and in the a-b plane means that current flows much easier in the a-b plane. Unless the grains are well aligned for the full length of a wire, inevitably the total current flowing will be severely limited. This creates a major difficulty for wire manufacturers.

13.1. JOSEPHSON JUNCTIONS

13.1.1. Tunneling

One of the most dramatic departures from classical mechanics that quantum mechanics shows is the phenomenon of tunneling, by which particles (such as electrons) penetrate insulating barriers. This cannot happen on the classical billiard ball model of particles, but when particles are represented by wave functions it is possible for the wave function to exist within the barrier and for the particle to be found (with finite probability) outside the barrier on the other side.[2] The particle has tunneled through the barrier. This topic was introduced in Chapter 6. Radioactive decay, in which alpha particles leave the nucleus of an atom, was explained by tunneling.

Superconductors also exhibit a phenomenon whereby a superconducting current (supercurrent) can tunnel through a barrier and continue flowing in another superconductor beyond the barrier. Experimentally, this is often done with thin films wherein the insulating barrier is a nonsuperconductor that is evaporated on top of a superconducting film and another superconducting layer is evaporated on top of that. Two types are recognized: SIS (superconductor–insulator–superconductor) and SNS (superconductor–normal–superconductor).

In either case the superconducting current is attenuated by the barrier by an amount which is readily calculable. That attenuation depends heavily on the coherence length ξ in the superconductor. Figure 13.1 (which is very similar to Figure 6.6) illustrates how the wave function changes. While in the superconductor, the wave function representing the superconducting electrons varies sinusoidally as exp $(i\mathbf{k}\cdot\mathbf{x})$, but once at the barrier, the wave function falls off as exp $(-x/\xi)$ inside the insulator. Upon entering the other superconductor beyond the insulator the supercurrent resumes its sinusoidal waveform and is no longer attenuated. The *phase* of the wave function is also changed after passing through the barrier, and that phase difference is crucially important.

13.1.2. The Josephson Effect

In order to explain how tunneling affects the current flowing in HTSCs, it is first necessary to explain how a Josephson junction works. In any superconductor, each grain boundary is an insulator (between superconducting grains), and thus a macroscopic superconductor is an interconnected network of countless Josephson junctions. It is customary to call these *weak links* because the limit on the total current is set by the resistance across the grain boundary barriers.

The primary characteristic of interest here is that at any superconducting weak link, current flows under *zero* voltage up to a certain point, as shown in Figure 13.2. This is the *dc Josephson effect*. The magnitude of that limiting current J_0 depends on the barrier

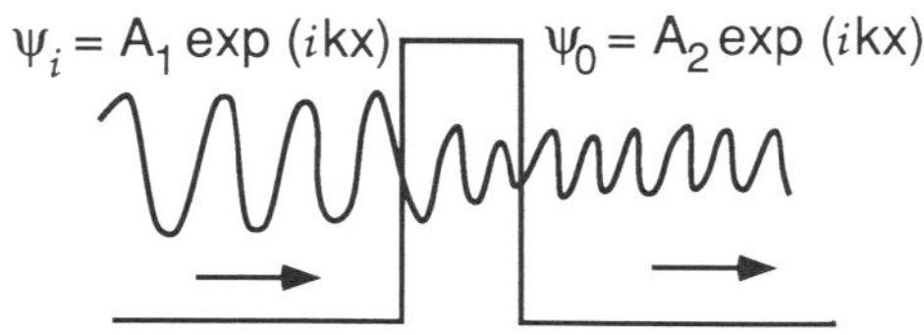

Figure 13.1.   Tunneling through a barrier by a quantum-mechanical wave function.

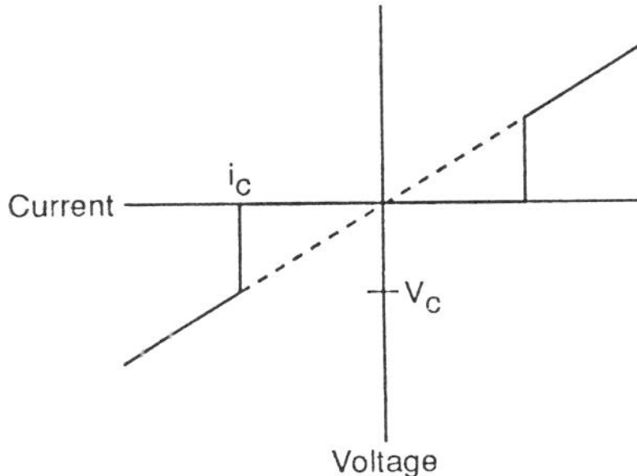

Figure 13.2. Current–voltage characteristic of a Josephson junction. DC current flows without any applied voltage up to a critical current i_c.

thickness compared to the coherence length. Thus, the limiting current through any one weak link is diminished when the grain boundary thickness is comparable to, or greater than, the coherence length. In a very clear derivation, Kittel[3] shows that the Josephson current is given by $J_0\sin(\theta)$, where θ is the phase difference of the wave functions across the junction boundary. Notice that current can be positive or negative depending on that phase. Such a dependence of a macroscopic observable (current) on a quantum-mechanical phase factor is most unusual. It is one of the unique features of superconductivity that make it a "macroscopic quantum phenomenon."

As we saw in Chapter 12, the density of superconducting electrons varies in space as $|\Psi|^2$ (Ψ = the Ginzburg–Landau function). This falls off with increasing temperature, and hence the current near a weak link is smaller at 77 K than at 20 K or lower.

Referring to Figure 13.2, to push more current than i_c through the junction requires applying a voltage. Moreover, applying that voltage causes the phase difference θ to change over time, as $d(\theta)/dt = 4\pi eV/h$. Now since the phase varies in time, the Josephson current will oscillate with a frequency proportional to the voltage. This is called the *ac Josephson effect*, which can be exploited to determine a very precise value[4] of the ratio e/h.

Quantitative accounts of Josephson junction behavior customarily make use of the resistively shunted junction (RSJ) model. In this model, a capacitor and resistor are assumed to be in a parallel circuit with the Josephson junction. The capacitor and resistor reflect the conventional electrical properties of the junction. Using this RSJ model, the phase θ can be related to the external driving current I according to

$$\frac{h}{4\pi e}\left(C\frac{d^2\theta}{dt^2} + G\frac{d\theta}{dt}\right) + I_0 \sin\theta = I(t) \tag{13.1}$$

where C is the capacitance of the junction and $G = 1/R$ is the passive conductivity.

When $I < I_c$, the system is superconducting, θ = constant, and $V = 0$: the dc Josephson effect. When $I \gg I_c$, the system asymptotically approaches Ohm's law, $V = IR$. In between, the behavior is very interesting and depends on the numerical values of C, G, and I_0. Just as with any harmonic oscillator, if G is large the system is overdamped, but for small G it is underdamped, and at exactly $I = I_c$ the system is unstable.[5] The underdamped case is particularly interesting, as shown in Figure 13.3: as I increases, the I-V characteristic is similar to Figure 13.2, but as I decreases through I_c there is a hysteresislike effect as the

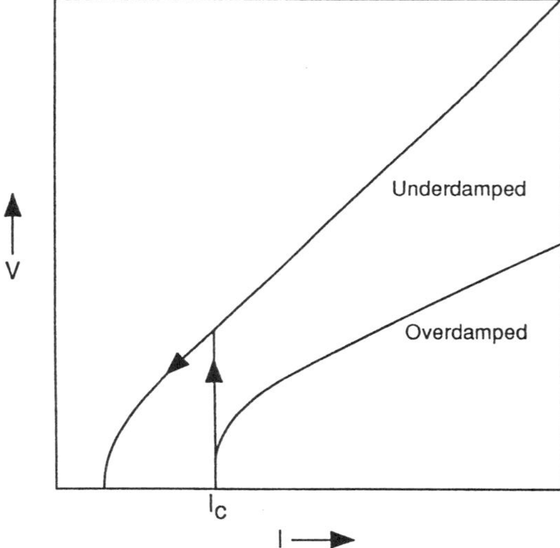

Figure 13.3. Current–voltage plot for a resistively shunted junction, illustrating overdamped and underdamped behavior.

voltage remains finite down to a lower current. This is an interesting curiosity for a single junction, but has important practical consequences for multiple junctions.

13.2. SQUIDs

The DC current in a Josephson junction is extremely sensitive to applied magnetic fields. The reason is that whenever a magnetic flux line lies between two Josephson junctions a phase shift will be introduced, and this changes the value of the maximum current in the DC Josephson effect. This condition is known as *quantum interference*, and the devices that exploit it are called *s*uperconducting *qu*antum *i*nterference *d*evices, or SQUIDs.

Consider two Josephson junctions in parallel. They form a current loop and their circuit diagram is that of Figure 5.3. When a magnetic field is present, integrating around the loop produces a phase change proportional to the magnetic flux Φ: $\delta\theta = \Phi/\Phi_0$, where $\Phi_0 = h/2e$ is the flux quantum. It is convenient to write down the current in one junction as $J_0 \sin(\theta - \delta\theta/2)$ and the other as $J_0 \sin(\theta + \delta\theta/2)$. Using trigonometry, the total current through both adds up to

$$J_{\text{total}} = J_0 \sin\theta \cos\frac{\Phi}{2\Phi_0} \qquad (13.2)$$

So in addition to the $J_0 \sin\theta$ dependence of one Josephson junction, for two in parallel there is also a variation in total current due to the enclosed magnetic flux Φ.

Because any real SQUID must have finite dimensions, some refinement to this analysis is necessary. For a pair of identical weak-link junctions, it can be shown[2,4] that the variation of current with increasing magnetic field B falls off according to a curve similar to the Fraunhofer diffraction pattern, i.e., $(\sin x)/x$, with $x = \pi B/B_0$, where $B_0 = \Phi_0/dL$, d is the

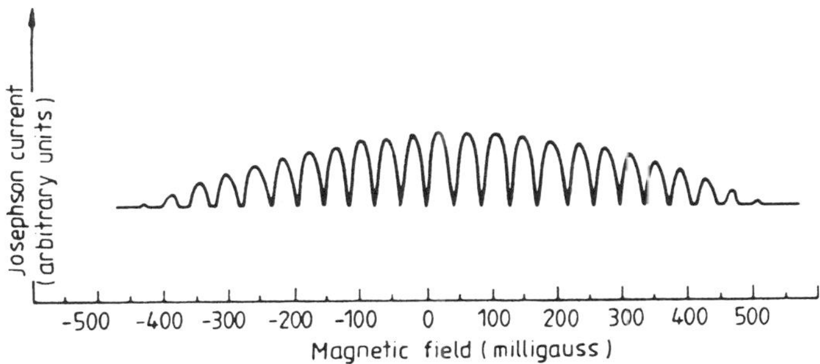

Figure 13.4. Plot of Josephson current amplitude (J_0) vs. magnetic field for a two-junction configuration. Experimental data fits the Fraunhofer diffraction formula quite well.

effective junction thickness, and L is the linear dimension of the junction. This pattern[6] appears in Figure 13.4.

SQUIDs have tremendous sensitivity because the critical current in the device oscillates with period Φ_0 in applied magnetic flux, and $\Phi_0 = 2.0678 \times 10^{-15}$ Webers. That is, the difference of a single flux quantum is enough to change the current in a SQUID from maximum to minimum and back again. When the area enclosed by the superconducting loop is large, a single flux line within the loop may correspond to a very small change in magnetic field. Just as an optical interferometer can be used to measure tiny changes in distance, this quantum interferometer enables the measurement of tiny magnetic fields.

Laboratory measurements using SQUIDs have demonstrated that they are as accurate as theory predicts. Commercial SQUID technology has been carried to a high level through competition to produce devices that are both sensitive and rugged. Practical LTSC SQUIDs have reached within a factor of 100 of their ultimate theoretical sensitivity, which is an exceptional accomplishment. HTSC SQUIDs are only now beginning to be developed commercially. We will return to this topic in Chapter 25.

13.3. GRAIN BOUNDARIES

In superconductors, grain boundaries present barriers to the passage of current and act as weak links. Structural misalignment, dirt or other impurity phases, and deviations from normal chemistry (e.g., oxygen vacancies) are just some of the reasons for this. This section presents an overview of a few key topics bearing on grain boundaries in the HTSCs.

13.3.1. Networks of Junctions

A well-known problem from electrical engineering textbooks is that of finding the Thevenin equivalent resistance of a network of resistors. The current can pass through many different paths in crossing the network, each of which is a combination of parallel and serial connections. When all the resistors are of equal value, the calculation is tractable; but with a distribution of resistances, many random paths carry various currents, and only average approximations may be made.

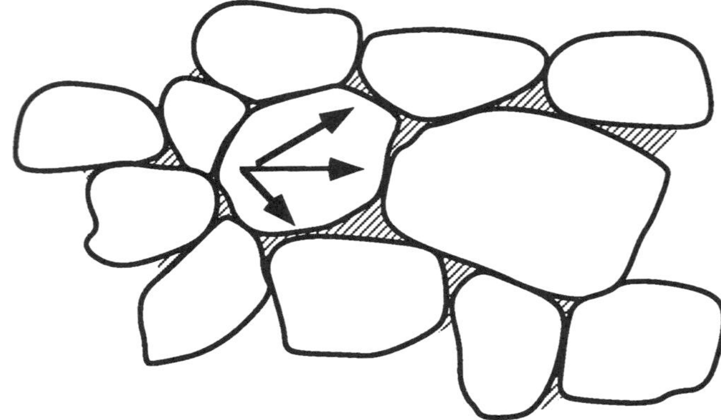

Figure 13.5.   Typical grains in a bulk superconductor. Current can flow along many paths.

Figure 13.5 is a sketch of typical grains in a superconductor. Obviously, the current can take many different paths through the grain boundaries. Thus, a bulk superconductor is a multiple Josephson junction array. The weak-link problem in HTSCs is topographically similar to the resistor network, with weak links causing attenuation of current similar to the Thevenin equivalent resistance.

Of greatest interest for our purposes is the behavior of a weak link array in a magnetic field. Experience with HTSCs shows a very steep fall-off of critical current in relatively minor magnetic fields. For a larger array of junctions all having the same separation between them, the Fraunhofer pattern for multislit diffraction holds. For grains oriented randomly in a magnetic field, the Airy diffraction pattern is more representative than the Fraunhofer pattern, so we have

$$\frac{J_c(B)}{J_c(0)} = \left| \frac{J_1(\pi B/B_0)}{\pi B/2B_0} \right| \tag{13.3}$$

where J_1 denotes the first-order Bessel function.

However, real HTSCs are composed of grains and boundaries of many different sizes. Their effect is to superimpose many curves of the form of Equation (13.3), each having different B_0 values. The result is to smear out the J vs. B curves. Peterson and Ekin[7] have modeled this behavior and gotten good agreement with experimental observations. One important caution needs to be mentioned: The presumed validity of this weak-link model is not assured, because the basic *I-V* characteristic of HTSC weak links may not be the same[8] as in LTSCs.

For realistic grain structure, the multitude of weak links will have various limiting currents, and the question of total current passing through the network will not have a simple solution. Obviously, the total current will be generally related to such parameters as grain size, dirtiness of the boundaries, and so on, but no one has done the full calculation necessary to determine the overall weak-link limited current in the complex pattern of intermixed crystals that constitutes a real sample of bulk superconductor.

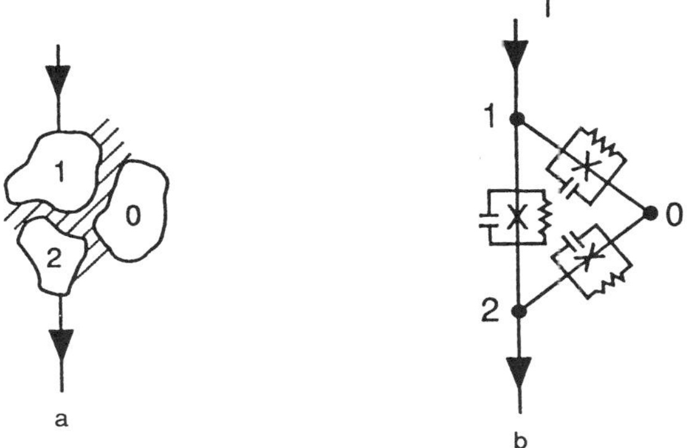

Figure 13.6.   (a) Three Josephson junction configuration with (b) equivalent circuit diagram.

Figure 13.6 presents a simple three-junction configuration with the equivalent circuit displayed as well.[9] The dynamics of this circuit is nonlinear, even when the three junctions are all identical. Recalling that $V = (h/4\pi e)\ d(\theta)/dt$, the frequency of oscillation shows period doubling and subsequent chaotic behavior. This has been observed experimentally in arrays of hundreds of junctions. It is not surprising that three coupled nonlinear equations [akin to Equation (13.1)] lead to chaotic behavior. Here, it suffices to underline the point that coupled weak-link behavior in superconductors is anything but simple.

13.3.2.   Differences Between Low-T_c and High-T_c Superconductors

Weak-link behavior has never been a problem in LTSCs, despite the fact that impurities collect at grain boundaries, and so on. The major reason for the difference between LTSC and HTSC behavior is found in the different coherence lengths ξ in the two classes of material. In niobium compounds, the thickness of the impurity layer at the grain boundaries is negligible compared to the coherence length ξ, and hence the attenuation is likewise negligible.[10] In HTSCs, however, ξ is typically 3–30 Å, which is comparable to the grain boundary thickness. Barrier penetration is therefore much more difficult in HTSCs, and hence the attenuation is finite. This corresponds to a finite resistivity at the boundary.

At each weak link boundary, Cooper pairs of superconducting electrons are broken. However, HTSCs cannot afford to have pairs break as readily as can LTSCs. A typical LTSC grain has a J_c near 10^8 A/cm^2, so when a defect knocks this down by a factor of 100, it's not a problem. By contrast, in HTSCs where J_c is much smaller, defects are more harmful. Basically, even in the normal state, HTSCs have far fewer electrons per cm^3, which causes a low density of Cooper pairs. They cannot lose many pairs before superconductivity is impaired. Weak-link behavior is a condition that was present in principle all along, but only in the HTSCs has the numerical magnitude made it a problem.

If HTSCs are to conduct large currents, three desirable paths immediately present themselves: (1) increase the coherence length; (2) decrease the number of links; or (3) decrease the boundary thickness. The first of these is not easy: the only control we have over the coherence length comes via alignment, because ξ in the a and b directions is about 10 times ξ in the c direction—30 Å vs. 3 Å. The approach is to ensure that the c-axis does not couple a grain to its neighbors in the direction of the current. This means aligning grains carefully, something usually done best in epitaxial thin films. The second approach involves

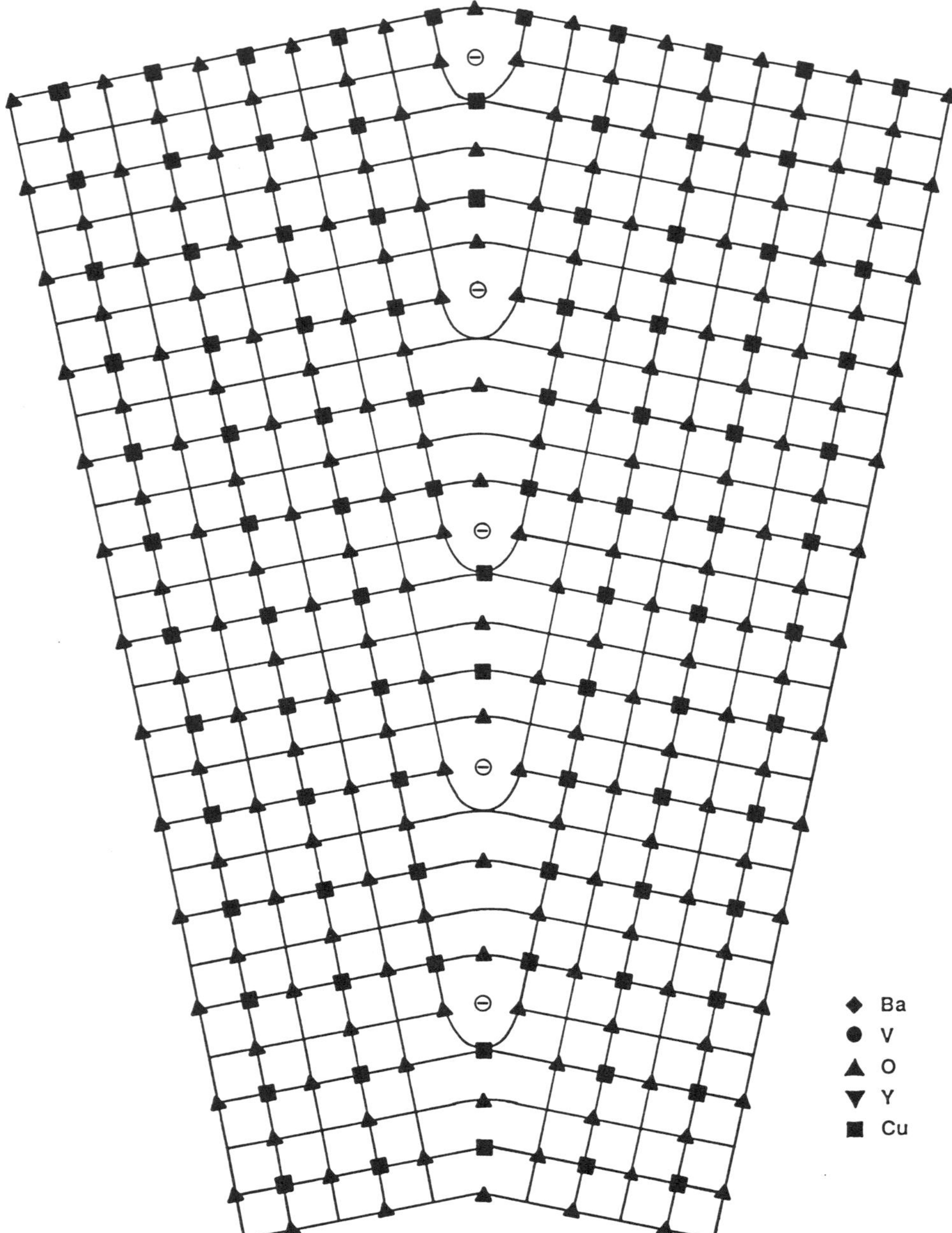

Figure 13.7. Atomic lattice positions in a partially coalesced tilt boundary. The angle between grains is 22.6°. (From Ref. 11.)

making single crystals that are extremely long in the direction in which current is to flow. The third approach is to make high-quality grain boundaries containing few impurities. Here, very careful control of the chemistry and processing conditions are required.

13.3.3. Factors Inhibiting Currents

An individual grain boundary is a place where the ordinary lattice structure is interrupted. In the simplest case, there is only a misalignment of atoms. In dirtier grain boundaries, there may be impurities present, there may be changes in the charge density, or deficiencies of oxygen, and so on. To understand the attenuation that a supercurrent experiences, it is necessary to examine the details of the boundary.

When two adjacent single crystals come together, their boundary looks like that shown[11] in Figure 13.7. (Triangles, circles, and squares suggest various atoms, such as Cu, O, etc.) The irregular spacing right at the boundary means that there will be dislocations, and some regions will be in hydrostatic compression while others are in tension. Any void regions act as very formidable tunneling barriers, transmitting almost zero current. As the angle of a grain boundary increases, voids occur more frequently. To the extent that impurities are present, it only makes the situation worse.

Of the many reasons why grain boundaries obstruct current, this geometrical mismatch one is merely the easiest to visualize. Other equally important contributors are the variation in charge density near a grain boundary (which lowers T_c) and the presence of an unknown thickness of some impurity phase. Early YBCO samples were degraded by $BaCO_3$ (a remnant of the original processing), and all successful processing schemes take care to drive off carbon for exactly this reason. Still, an impurity layer a few nanometers thick makes the grain boundary a resistive junction instead of a tunnel junction.

On the experimental side, there is ongoing interest in grain boundaries made artificially on thin films. This is the most controlled way to study the problem because the type of boundary can be tailormade, the twist or tilt* can be adjusted, and the properties of the boundary can be compared with those of the neighboring grain to isolate the boundary effects. Geometrically, there are many possible facet-planes which can separate two grains of the same relative orientation, some with good structural match and others with poor match. Thus, J_c across the boundary depends on the particular facet.

On the computational side, Jagannadham and Narayan have derived formulas and presented graphs[11] that model the changes in J_c across a grain boundary as a function of temperature, boundary dimensions, and grain alignment. The observed roll-off in J_c with increasing misalignment fits Narayan's model well.

13.4. EXPERIMENTAL OBSERVATIONS

It is very difficult to properly model the current through an array of irregular grain boundaries, and so measurements are indispensable to understanding weak-link behavior in HTSCs. The majority of experiments have been done on YBCO and suffice to establish the contemporary picture of how weak links affect current. BSCCO data essentially confirms the same picture.

Twist is when the rotation axis is in the plane of the boundary. *Tilt* refers to a rotation axis not in the plane of the boundary.

13.4.1. Visualizing Weak Links

At AT&T Bell Laboratories, an imaging technique (LTSEM = low-temperature scanning electron microscopy) was used[12] to see weak links. The process begins by etching a narrow channel (perhaps 2 μm across) onto a HTSC film. A current is run through this channel and the voltage is monitored. The current is set high enough to be near J_c to begin with. Then an electron beam is turned on, to dump extra energy into a single spot. The beam is scanned across the entire region of the channel. If a weak link is present, it will show up as a relatively high voltage, represented as an intensity of light or a spectrum of color. This is shown in Figure 13.8. By keeping track of the electron beam position in the usual way for scanning instruments, a map of the entire channel can be made that shows the weak link quite clearly. Some surprisingly small defects showed up by this technique. Of course, it is a surface measurement, and therefore cannot see deep within a bulk sample. Still, it is quite useful with thin films, and many weak-link experiments are done using films anyway.

13.4.2. Early Experiments

As mentioned above, a true bulk sample presents a bewildering array of grains through which the current may pass. It is better to study a small number of grain boundaries, and this can be done by carefully contriving the shape of the sample. In general, the transport critical current J_c is the parameter of interest, which is determined by measuring both the current through and the voltage across a known cross-sectional area of HTSC material. This is treated

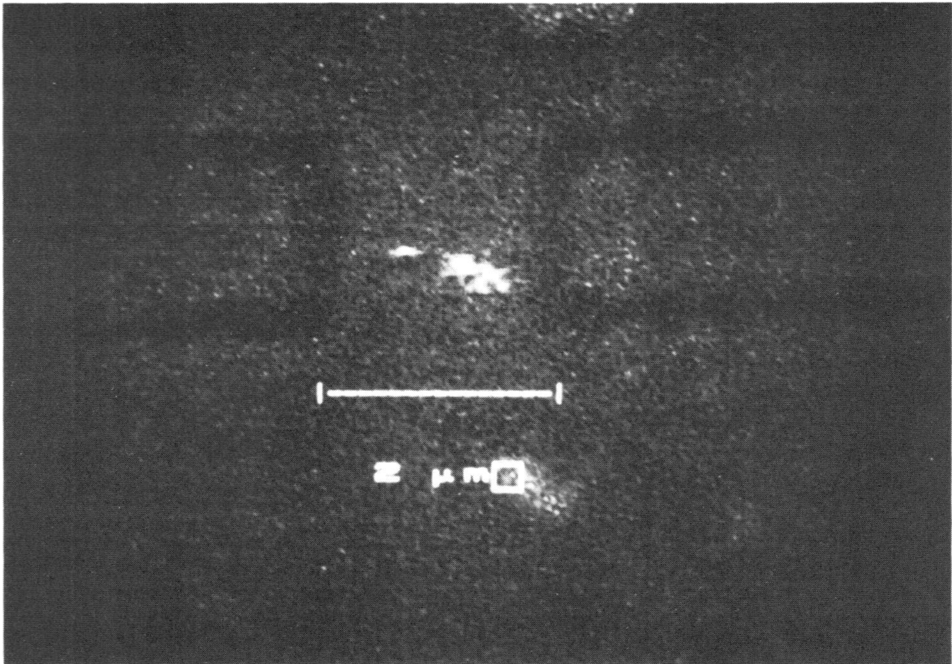

Figure 13.8.   Photograph of a weak link, made by the electron beam scanning technique. The sample sits on a low-temperature stage inside an electron microscope. Current flows from bottom to top through a narrow channel delineated by dark shadows. The bright spots show areas of voltage drop caused by local heating induced by the moving electron beam. (Courtesy of P. Gammel, AT&T Bell Laboratories.)

more fully in Appendix A, where Figure A.4 shows a thin film of YBCO with current and voltage contacts soldered on, and an enlargement of the central section. A microbridge has been etched in the middle, thus controlling the geometry precisely. When the voltage reaches some prespecified level (typically 1 or 10 μV/cm), the measured current is *defined* as the critical current. The advantage of this microbridge geometry is that one or a few grain boundaries block the path of the current, and by eliminating multiple current paths, analysis of the data is simplified.

In 1988, researchers at IBM developed techniques to produce large-grained samples with well-defined grain boundary geometries. Each grain boundary is a Josephson junction, with a voltage drop of about 8 meV ($= J_cR$). Taken together, they reduced J_c by a factor of 100 or 1000. They used these samples together with patterning techniques to measure current transport across grain boundaries.[13–15] They found that the RSJ model works well to describe the behavior; in particular, J_c falls off very rapidly in a magnetic field, which is characteristic of Josephson junctions.

Subsequently, the IBM team extended their studies to include variations in the tilt angle (*c*-axis relative to substrate plane) and alternative processing techniques (laser ablation versus evaporation). Because the dependence of J_c on magnetic field is the same for various types of misalignment (rotation, tilt, twist), they concluded that depression of J_c across grain boundaries is *not* due to an anisotropic pairing mechanism. Because films prepared by evaporation at substrate temperatures of 900°C show the same properties as films prepared by laser ablation at substrate $T \approx 700$°C, they concluded that an interaction between the film and the substrate does not reduce J_c. They observed Josephson junction behavior at the grain boundaries but could not distinguish between superconductor–insulator–superconductor (SIS) and superconductor–normal metal–superconductor (SNS) behavior. Additional experiments showed that certain other things are not responsible for the observed fall-off in J_c: anisotropy, pair symmetry breaking, second phases (impurities), nonequilibrium composition or related kinetic effects. Twinning has, at most, a small effect.

Numerous corroborating experiments from many laboratories showed essentially the same thing: clearly, it is the grain boundaries that limit J_c. The central question concerning multigrain (bulk) HTSCs then became, Is this *intrinsic*? That is, are we stuck with it, or can we get around it?

13.4.3. Dependence upon Angle of Alignment

Alignment is important because the HTSC crystals are anisotropic,[16] with J_c larger in the *a-b* crystalline plane and smaller in the crystalline *c*-direction. At first, the literature was filled with conflicting results and interpretations, until the importance of grain alignment was realized. Data from samples with randomly oriented grains was often confusing. Misalignment between grains at their boundaries causes degradation of J_c, particularly in finite magnetic fields. Also, even when the grains are well aligned with one another, rotating the applied magnetic field affects J_c tremendously. Ekin *et al.* measured[17] the way J_c falls off in YBCO as the magnetic field is rotated relative to the *a-b* plane. As shown in Figure 13.9, the greater the value of field, the worse is the fall-off.

Fortunately, grain alignment does not have to be perfect. Scrutiny of Figure 13.9 reveals that when B is roughly perpendicular to the *c*-axis ($\pm$ 30°), J_c stays reasonably high. This means that it is not necessary to achieve perfect alignment of axes, which in turn relaxes the constraints on manufacturing wire. Ekin states[17]: "This unexpectedly wide shoulder region

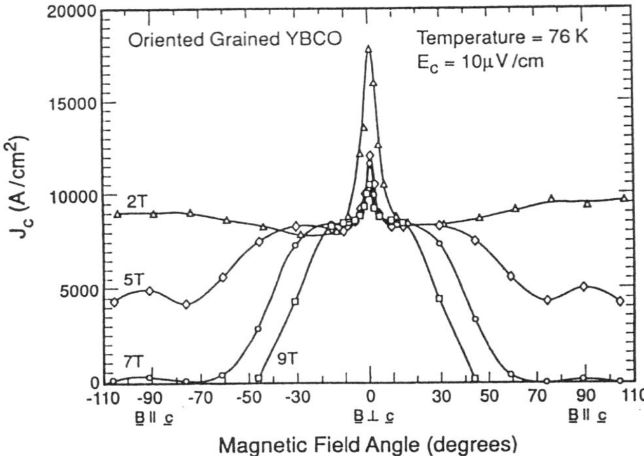

Figure 13.9. Transport critical current density J_c as a function of the angle between magnetic field and the *a-b* plane for bulk oriented-grained $YBa_2Cu_3O_7$ at LN_2 temperature (from Ref. 17).

is important from the standpoint of enabling practical design of high-field superconducting magnets at high temperatures."

BSCCO has a much more[18] anisotropic unit cell than YBCO, which means that it conducts much better along the copper oxide planes than perpendicular to them. Therefore, when grains are misaligned, the impact of anisotropy is more severe. For example, in BSCCO, a small misalignment between crystalline axis and applied magnetic field can dramatically change the J_c value. Figure 13.10 displays the ratio of J_c when a magnetic field is applied perpendicular to the crystalline *c*-axis to J_c when the field is parallel to the *c*-axis, for various applied magnetic fields, as a function of temperature.[19] At stronger fields and higher temperatures (of interest for most applications), the ratio is high, which in turn means that good crystal alignment is more critical. Generally, if grains are aligned within 5°, this problem does not come up.[20]

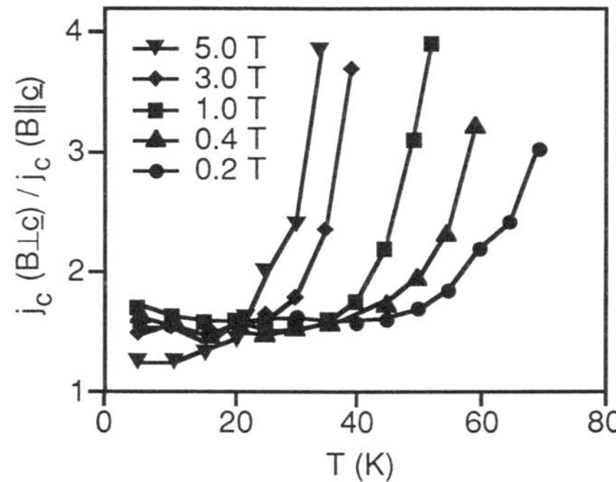

Figure 13.10. Ratio of J_c for different magnetic field orientations as a function of temperature for BSCCO. At higher temperatures, J_c depends much more on the orientation of the magnetic field direction relative to the axis of a single crystal. (From Ref. 19.)

13.5. OPTIMIZING CURRENT ACROSS GRAIN BOUNDARIES

The weak-link problem can be mitigated through proper grain alignment, which is the subject of this section.

13.5.1. The Brick Wall Model

Even before HTSCs came along, Gray and co-workers at Argonne[21] experimented with niobium nitride superconductors. Niobium nitride exhibited Josephson junction behavior (weak links) across grain boundaries. The group produced samples having a needlelike, oriented structure, which did not exhibit weak-link behavior. The orientation of the needlelike grains results in a large surface-area overlap between grains. Although at each point, the grain boundary acts as a Josephson junction and reduces the current density, the large area of overlap allows a large total current to be transmitted across the grain boundary.

Essentially the same explanation applies to HTSCs. Single crystals in well-aligned material (especially BSCCO) form a brick wall structure.[22] The several bricks are weak-linked one to another, and the path of current through a wire is a network of branches across weak links. This is sketched in Figure 13.11. Although the critical current J_c is low at each weak link, the large area of face-to-face contact compensates for this, and a moderate net current flows down the wire. The total current is the sum over the network:

$$I_c = \sum_i J_c{}^i A^i \tag{13.4}$$

where the index i runs over all connected grains. Although the grains are too irregular to actually compute each A^i, the idea is quite clear from Figure 13.11: the current I can weave through as many grains as necessary while the current *density* J never exceeds J_c. The *effective* J_c observed for the whole wire is greater than that at any individual weak-link junction.

13.5.2. Melt Texturing

One method of preparing bulk samples of HTSCs is *melt texturing*, which is described in more detail in Chapter 16. Briefly, the components are raised in temperature into the liquid state, and then cooled so as to produce extremely long, thin single crystals adjacent to one another and aligned in the direction of current flow. On the basis of the brick wall model, it is plausible to suggest that melt texturing ought to improve J_c and reduce weak-link behavior in HTSCs in three ways: (1) undesirable impurity phases are mitigated, so the grain

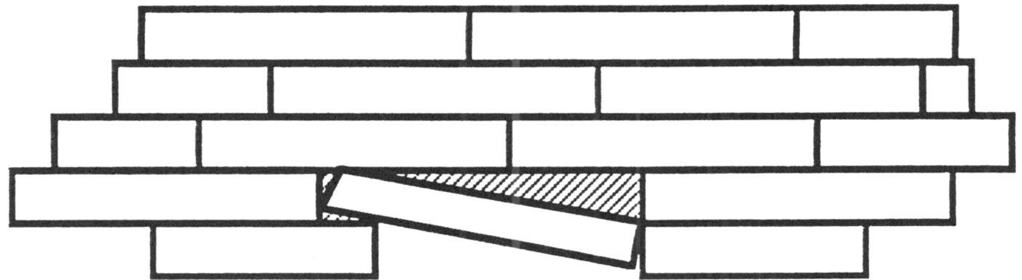

Figure 13.11.   The brick wall configuration typical of well-aligned grains of HTSCs.

boundaries are cleaner; (2) the number of grain boundaries is reduced; and (3) the effect of each grain boundary is reduced due to the large contact area between grain boundaries.

Sungho Jin *et al.* at AT&T Bell Labs[23] were the first to achieve $J_c > 10^4$ A/cm^2 in YBCO by this means. Photos of his melt-textured material showed individual grains about a centimeter long and a few microns across. Unfortunately, Jin's results could not be duplicated because his samples were composed of a small, finite number of weak links (as contrasted to a huge array of weak links in more conventional bulk samples). The particular choice of sites to attach the lead wires fortuitously affected the number of weak links in the path of the supercurrent in melt-textured samples: the difference between three weak links and twelve might easily depend on minor details of sample fabrication that would ordinarily escape attention but could greatly affect the apparent J_c.

Subsequently, Salama *et al.* at the University of Houston[24] and Murakami *et al.* at Nippon Steel[25] obtained even better results via melt texturing. Melt texturing seemed to produce samples immune from weak-link effects, and so great effort went into developing this technique. Eventually, melt texturing became a standard practice.

However, it is a very slow process, requiring about a week to make a one-meter sample of YBCO. Furthermore, in a series of melt-textured grains long enough to be called a wire the number of weak-link junctions would inevitably become very large. That would cause total critical current to decline to the low values typical of bulk YBCO. Therefore, melt texturing of YBCO is directed mainly toward applications other than wire.

Melt texturing works better for BSCCO. Randomly oriented BSCCO is quite a poor conductor, but BSCCO is also a micaceous compound (i.e., like the mineral mica), and shears along certain planes. This property leads to good grain alignment in BSCCO when it is rolled mechanically. When BSCCO is subsequently melt textured, still further alignment takes place, and so melt texturing is a helpful step in wiremaking.

13.5.3. Platelets

Looking at a YBCO sample, one sees stacked, parallel platelets with the *a-b* plane in the platelets and the *c*-axis perpendicular. The obvious conclusion is that this is indeed a brick wall geometry. Moreover, current within the *a-b* plane seems to fall off rapidly in modest magnetic fields. However, current in the *c*-direction, although lower to begin with, does not fall off as rapidly as might be expected. This puzzle was considered of secondary interest as wiremakers strove to obtain alignment along the *a-b* plane.

A group at Oak Ridge National Laboratory[26] analyzed some melt-processed YBCO and challenged the weak-link brick wall concept. They used convergent beam electron diffraction (CBED) to observe that there is no orientation change between adjacent platelets. This surprising outcome indicates that the platelets are actually portions of a common single crystal. In the *a-b* plane, portions of adjacent crystals still abut with grain boundaries acting as weak links; but in the *c*-direction, the current follows a continuous path across many platelets.

Why are there platelets? They result from the pattern of crystal growth, together with the entrapment of impurity species that are being driven ahead of the grain-growth front. Figure 13.12 presents the model proposed by the Oak Ridge group. Here, a domain grows as a single crystal and the growth in the *a-b* plane is rapid, lateral ledge growth.[26,27] Meanwhile, growth along the *c*-axis is comparatively slow. As growth proceeds upward (Figure 13.12(a)), rejected liquid phase material (BaCuO$_2$, copper, etc.) gets trapped and prevents the structure from filling in and closing. The eventual shape is that of Figure

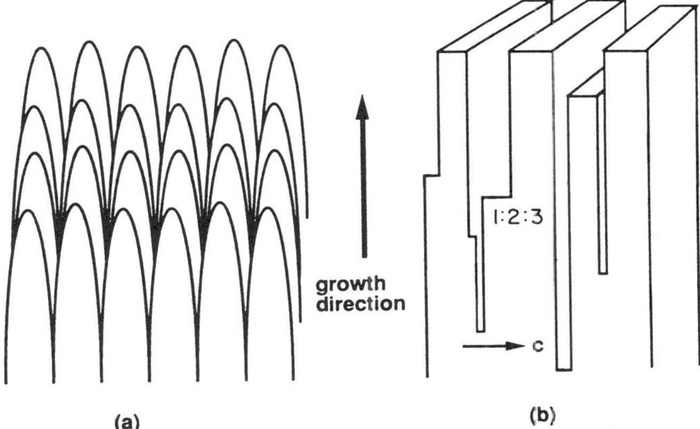

Figure 13.12.   (a) Typical growth front during cellular solidification. (b) Growth front during solidification of melt-textured YBCO. (From Ref. 26.)

13.12(b). They consider this to be a two-dimensional analog of a traditional cellular solidification process. The gaps simply don't heal well as growth proceeds, and thus platelets remain in the final solid. A number of photomicrographs[27] verify this model in real samples of YBCO.

Thus, instead of having a brick wall, each domain has a mazelike pattern, as sketched in Figure 13.13. Current flowing in the c-direction may have to wend its way through a winding path, but since the material is superconducting the length of the path doesn't matter. The cross-sectional area of the narrowest connection in the c-direction limits the current flow; but that area is anybody's guess.

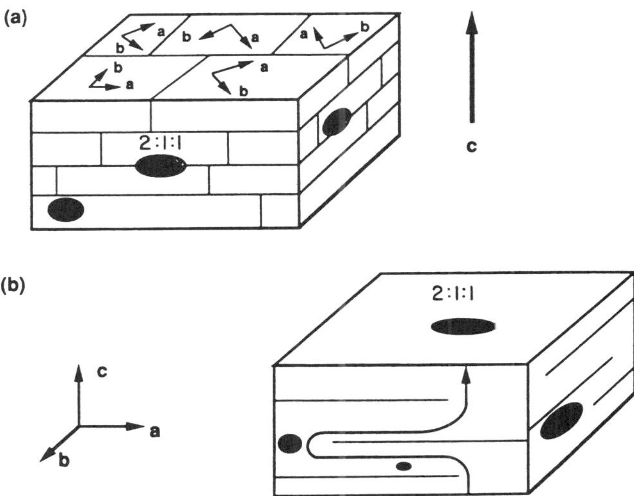

Figure 13.13.   (a) Schematic of a brick wall microstructure. (b) Schematic proposed microstructure of YBCO(123) domains. Current flow along the c-axis is solely through single-crystalline material within a domain of melt-textured YBCO. (From Ref. 26.)

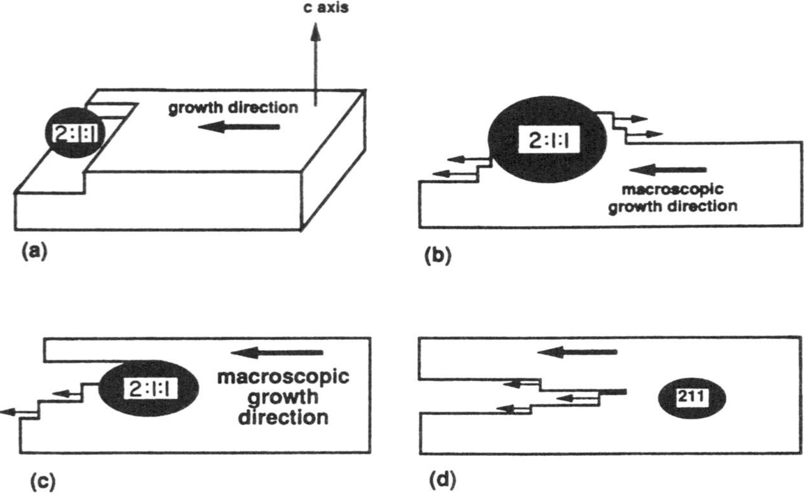

Figure 13.14. Schematic of possible role of YBCO(211) particles on the growth of melt-textured YBCO. (a) YBCO platelet abuts a growing platelet. (b) Heterogeneous nucleation occurs at the junction between platelet and the YBCO(211) liquid. (c) Once sufficient growth along the c-axis occurs such that the 211 particle can be bypassed, rapid lateral growth will occur. (d) This process results in a gap, which may not heal completely. (From Ref. 26.)

Photomicrographs also reveal that a lot of the gaps terminate very near to a 2-1-1 particle. The Oak Ridge group suggests that as growth proceeds, these 2-1-1 inclusions may be the cause of platelet formation, as indicated by Figure 13.14. As crystal growth approaches a 2-1-1 barrier and propagates around it, a gap can readily form on the far side which will not heal as growth continues.

Current propagation along the c-axis is not itself of great interest, but the discovery of this form of crystal growth is very relevant. In the next chapter, we shall consider ways to enhance flux pinning in HTSCs, and the topic of crystal growth around impurities will be quite important.

13.6. SUMMARY

Weak-link behavior is best understood by imagining the HTSC as an interconnected network of Josephson junctions. The individual grains superconduct, but the grain boundaries do not. Therefore, it is necessary for the quantum-mechanical wave functions of the electrons to tunnel through the grain boundaries, which is strongly akin to the tunneling that occurs in a Josephson junction. The very short coherence lengths ξ in HTSCs makes the grain boundaries much more formidable obstacles than they are in LTSCs. In NbTi, ξ is many orders of magnitude greater than the grain boundary thickness,[10] and the attenuation at each barrier is negligible. In the HTSCs, ξ is typically less than 3 nm—in the same range as the grain boundary thickness. Furthermore, these layered cuprates exhibit properties vastly different in the crystalline c-direction from those in the a-b plane. With $\xi_c \approx 0.1\ \xi_a$ (or ξ_b), a change in alignment suddenly makes a barrier ten times higher.

To focus on wiremaking, we must consider grain structure, because kilometer-length single crystals are impractical. When the grain boundaries are occupied by a nonsuperconducting phase (such as CuO or YBCO-211), the barriers are high; but if the interspersed material is a good conductor such as silver, the resistance is low. Clean grain boundaries enhance J_c by diminishing the barrier thickness that the supercurrent must tunnel through. By contrast, impurities (e.g., $BaCO_3$) make J_c worse, as does improper chemistry (e.g., variations in the oxygen concentration), disturbed crystal structure, and microcracking.

Total current is the sum over diverse paths. Calculations of the expected current are hard to compare with experimental data because real samples seldom have ideal grain patterns. Still, it is clear that no matter what path the current follows, careful alignment of grains is an essential part of making good wire. Because of its micaceousness, BSCCO tends to achieve grain alignment during the drawing and rolling steps in the wire manufacturing process. Consequently, finite progress has been made toward making BSCCO wire, although its performance at 77 K is intrinsically limited. Wiremaking is discussed in Chapter 16. YBCO is not micaceous and does not preferentially align grains one way or another. Meanwhile, work with TBCCO is at an early stage. Its alignment behavior is not readily predicted.

REFERENCES

1. B. D. Josephson, *Phys. Rev. Lett.* **1**, 251 (1962).
2. E. Merzbacher, *Quantum Mechanics* (Wiley, New York: 1961).
3. C. Kittel, *Introduction to Solid-State Physics*, 6th ed., Ch. 12 (Wiley, New York: 1986).
4. W. H. Parker *et al., Phys. Rev. Lett.* **18**, 287 (1967).
5. D. R. Tilley and J. Tilley, *Superfluidity and Superconductivity*, 2nd ed. (Adam Hilger, Bristol: 1986).
6. R. C. Jaklevic *et al., Phys. Rev. A* **140**, A1628 (1965).
7. R. L. Peterson and J. W. Ekin, *Physica C* **157**, 325 (1989).
8. S. S. Bungre *et al., Nature* **341**, 725 (1989).
9. S. Doniach, "Granular Superconductors and Josephson Junction Arrays," in *Percolation, Localization, and Superconductivity*, ed. by A. M. Goldman and S. A. Wolf pp. 401–429 (Plenum Press, New York: 1984).
10. M. Tinkham, *Introduction to Superconductivity* (Kreiger Publ. Co., Malabar, FL: 1980).
11. K. Jagannadham and J. Narayan, "Structure and Properties of Grain Boundaries in HTSCs," in *Superconductivity and Its Applications*, AIP Conference Proceedings #273, 6th NYSIS Conference, Buffalo, NY pp. 37–49 (September 1992).
12. D. Monroe *et al., Appl. Phys. Lett.* **53**, 1210 (1988).
13. P. Chaudhari *et al., Phys. Rev. Lett.* **60**, 1653 (1988).
14. D. Dimos *et al., Phys. Rev. Lett.* **61**, 219 (1988).
15. J. Mannhart *et al., Phys. Rev. Lett.* **61**, 2476 (1988).
16. S. Hu *et al., Phys. Rev. B* **43**, 2878 (1991).
17. J. W. Ekin *et al., Appl. Phys. Lett.* **59**, 360 (1991).
18. D. E. Farrell *et al., Phys. Rev. Lett.* **63**, 782 (1989).
19. H. Krauth *et al., Proc. Third Int. Symp. on Superconductivity* (Sendai, Japan, November 6–9, 1990).
20. S. Jin and J. E. Graebner, *Materials. Sci. Eng.* **B7**, 243 (1991).
21. H. L. Ho *et al., Ultramicroscopy* **22**, 297 (1987).
22. A. P. Malozemoff, in *High-Temperature Superconducting Compounds II*, ed. by S. H. Whang *et al.*, p. 3 (TMS Publications, Warrendale PA, 1990).
23. S. Jin *et al., Phys. Rev. B* **37**, 7850 (1988).
24. K. Salama *et al., Appl. Phys. Lett.* **54**, 2352 (1989).
25. M. Murakami *et al., Jpn. J. Appl. Phys.* **28**, L1125 (1989).
26. K. B. Alexander *et al., Phys. Rev. B* **45**, 5622 (1992).
27. A. Goyal *et al., Physica C* **210**, 197 (1993).

III

CARRYING ELECTRICITY

14

Flux Pinning

This chapter resumes where Chapter 2 left off. The basic concepts of current flow and magnetism in superconductors still hold, but with substantial modifications for the HTSCs. Foremost among these are the effects of higher temperatures, which allow the lines of magnetic flux to move within the material. Flux line motion is a means of dissipating energy within a superconductor. When current is flowing, flux line motion requires a voltage to sustain that current, and hence it acts as a surrogate resistance—defeating the purpose of superconductivity. Since the flux lines are lines of magnetic field H, when a current J flows there is a Lorentz force $J \times H$ perpendicular to both which forces the flux lines to move. In LTSCs this is a minor effect that occurs very near T_c, and practical devices are operated far enough away from this point that the problem is avoided. However, it is a very significant obstacle to practical applications at 77 K.

Furthermore, the highly anisotropic nature of the HTSCs, with superconductivity taking place in the CuO_2 planes, undermines certain conventional models, including the notion that the flux lines are like miniature telephone poles. Consequently, the dynamics of flux line motion is quite different in the copper oxides from that expected for LTSCs. The model known as *thermally activated flux flow* (TAFF), valid for LTSCs, does not adequately describe the HTSCs. A new *vortex glass* model is more appropriate. This brings with it a number of implications about what can be done to make wire that will carry high currents.

This chapter begins by explaining a number of new concepts, including *irreversibility, flux motion, giant flux creep, flux flow*, and *flux lattice melting*. We then introduce the vortex glass model. The effects of anisotropy are covered next, including Josephson junction effects, *pancake vortices*, and the crossover from two-dimensional to three-dimensional behavior. We describe various methods of enhancing flux pinning. The consequences of flux motion for making practical wire are also presented.

14.1. THE IRREVERSIBILITY LINE

In Chapter 2, we described the behavior of an ideal type II superconductor and showed how its magnetization varies in Figure 2.9. Section 2.10 then mentioned realistic type II superconductors and described how flux is pinned and even trapped in a superconductor, leaving residual magnetization. The result is that the magnetic induction B is finite even when the applied magnetic field H is zero, a condition that mimics a permanent magnet until the sample is warmed up and superconductivity is lost.

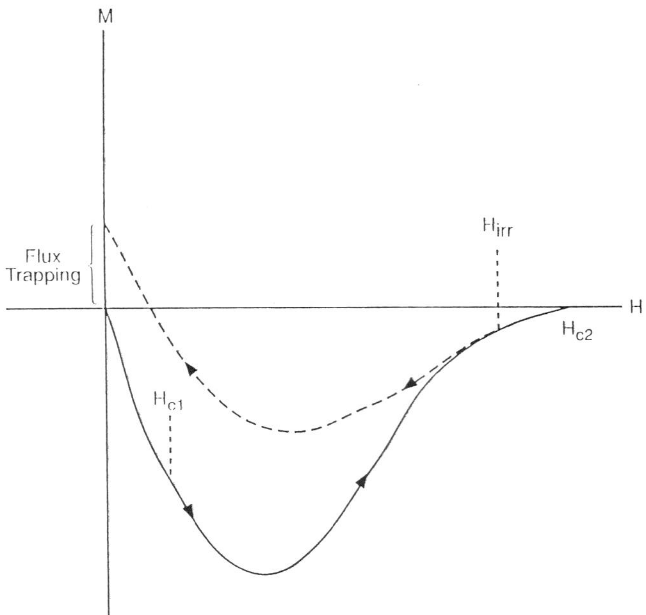

Figure 14.1. A typical HTSC has a magnetization curve that at first retraces the forward curve as the applied field **H** is reduced below H_{c2}. The magnetization is reversible above H_{irr}. As **H** reduces still more, flux pinning occurs and **M** becomes irreversible at lower fields. Only with flux pinning can a finite current flow.

Stationary, nonmoving flux lines are not harmful to superconductivity. The dissipation of energy occurs by two different mechanisms: conventional electrical resistance, or changing magnetic flux. In superconductors, the first is absent, but Maxwell's equation, curl **E** = $-\partial\mathbf{B}/\partial t$, continues to hold. In hard superconductors, hysteresis dissipates energy and a voltage is needed to sustain the current. The hysteresis loop in the **M–H** plane determines the energy loss per cycle for low-frequency applications. For DC applications, $\partial\mathbf{B}/\partial t$ is zero, flux is pinned, and there is no loss.

In a superconductor, the field at which the **M** vs. **H** curve is no longer double-valued is known[1] as the *irreversibility field* H_{irr}. In NbTi or Nb$_3$Sn, this field is extremely close to H_{c2}, and there is no important distinction between them. By contrast, in a HTSC, thermal activation is much greater, which leads to easier flux line motion near H_{c2}. Hence, a new phenomenon takes place, as illustrated in Figure 14.1: After reaching H_{c2}, when H is reduced, flux lines are at first free to move, and so the trajectory of M retraces its path. The superconductor is "soft." There comes a point when flux pinning becomes stronger, **B** declines slower than does **H**, and **M** deviates from the increasing-H curve. The superconductor changes to "hard." This value of H_{irr} in HTSCs is appreciably different from H_{c2}, although still quite far above H_{c1}. Typical numerical values at 77 K are $H_{c1} \simeq 0.05$ T, $H_{irr} \simeq 10$ T, and $H_{c2} \simeq 20$ T.

For any given choice of temperature, there is a particular **M** vs. **H** diagram, and with it a value of H_{irr}. These values can be assembled into an irreversibility line as a function of T. Figure 14.2 shows some typical H_{irr} vs. T data, for certain forms of the HgBaCuO compounds.[2]

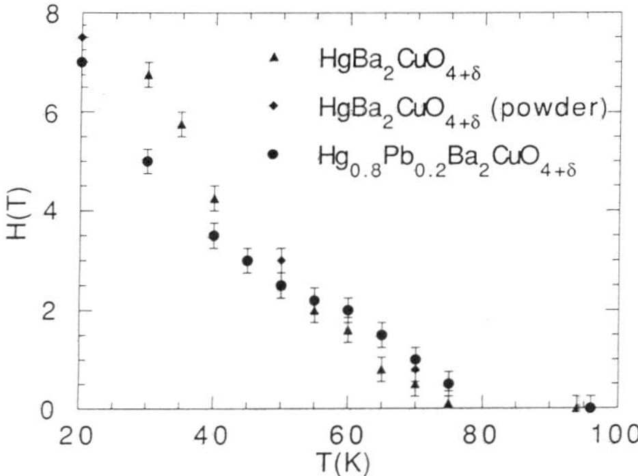

Figure 14.2. Experimental data showing the irreversibility line for various compounds of the superconducting series HgBaCuO. (From Ref. 2.)

14.2. BASIC CONCEPTS OF FLUX PINNING

The flow of current in a superconductor is limited by several factors. Above a certain *critical current density* J_c the material becomes normal and has a high resistance. J_c itself is not even one simple number. A very important difference between LTSCs and HTSCs is that the HTSCs are not really continuous materials. Rather, they are small crystals (grains) within which current flows well, but impurities at the grain boundaries act as insulators. Consequently, current passes only via Josephson junction effects known as weak-link behavior. It is important to distinguish between *intra*granular (J_{cg}) and *inter*granular (J_{cj}) currents. Moreover, at higher temperatures thermal activation of flux motion becomes significant.

14.2.1. Pinning Strength U_0

When a modest number of flux lines penetrate a type II superconductor, they begin to feel one another's presence when separated by about one penetration depth λ. Their mutual repulsion forms the flux lines into a regular geometric pattern[3] called the *Abrikosov lattice* because of its similarity to a crystal lattice formed by interacting atoms. In a conventional LTSC the pattern is triangular (see Figure 14.3). Figure 2.7 is a computer drawn image showing how neutrons are scattered by individual vortex lines. In HTSCs, the extreme anisotropy of the crystal lattice results in certain changes in the flux lattice: the vortices are elliptical, not circular, due to differences between coherence lengths in the a,b directions and in the c direction; and the vortices are spaced much farther apart along one axis than another.[4]

Pinning is characterized by an *energy well* of depth U_0 and *interwell distance d*. Both U_0 and d are somewhat imperfectly defined[5]: U_0 is related to the thermodynamic critical field H_c and the coherence length ξ, whereas d is the hopping distance through which a flux line moves when a jump occurs. These terms are taken over from LTSC, where flux lattice behavior only needed to be understood in a qualitative way. For HTSC, greater precision for these concepts is needed.

The pinning strength U_0 is normally determined via a technique pioneered by Beasley *et al.*[6] It relies on measuring the variation of pinning strength U with the parameter $B\nabla B$, and extrapolating from the slope of such curves to estimate U_0, the pinning strength when $B = 0$. Some typical U_0 data for HTSCs obtained from measurements near $B = 0.1$ T are:

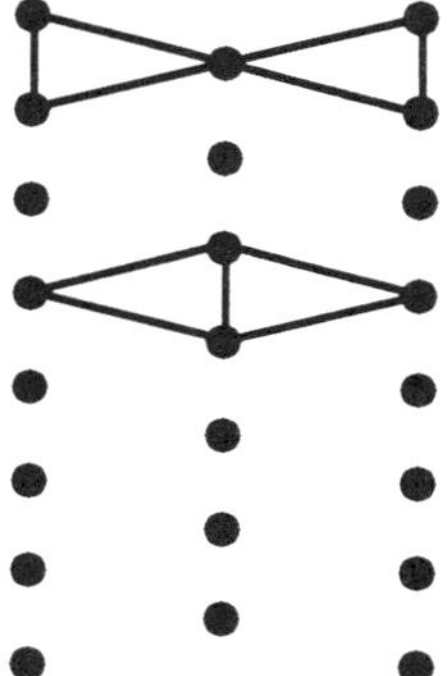

Figure 14.3. Quantized lines of magnetic flux form a regular pattern known as the Abrikosov lattice.

$Ba_{0.6}K_{0.4}BiO_3$:	0.08 eV
$YBa_2Cu_3O_{7-x}$:	0.18 eV
$Tl_2Ca_2Ba_2Cu_3O_x$:	0.33 eV

For HTSCs, it is better to measure U_0 via transport measurements rather than via magnetic measurements. Transport measurements, especially the widening of the transition, give accurate values for U_0.

For analysis of LTSCs, it was sufficient to employ a single value for U_0. In reality, however, not every pinning well has exactly the same depth. Therefore, the constant value U_0 should be replaced[7] by a probability distribution function $f(U_0)$. Average values of U_0 range from below 0.1 eV up to about 0.4 eV in carefully prepared melt-textured samples, but relatively little has been measured concerning the distribution of U_0 values.

14.2.2. The E vs. J Curve

Sometimes the resistivity of a superconductor is plotted as a function of current. This is misleading because the resistivity[8] is really only the slope of the curve of electric field E vs. current density J. It is more informative to work with E vs. J plots, recognizing that when the relationship is nonlinear the resistivity is no longer constant.

The use of linear graph paper is also misleading. When the transition to superconductivity is presented on linear paper, it creates the illusion that the resistance is true zero below T_c. In reality, the resistance never reaches zero in these materials. When plotted on semilog paper, the resistive behavior below T_c is emphasized. In Figure 14.4 we see that BSCCO has an onset transition at $T_c = 80$ K, but in a 1 T magnetic field the resistivity only drops below that of copper at 35 K. In general, to be useful as a superconductor the resistivity must be far enough below copper to compensate for the extra cost of refrigeration fluid to carry away the heat dissipated. This implies a factor of 10 below copper at 77 K and a factor of 1000 at 4 K. On this basis, for the HTSCs represented by Figure 14.4 we would say that YBCO is useful, but BSCCO is not.

At zero temperature, the variation of E with J is simple. The solid line in Figure 14.5 shows that below J_c, $E = 0$, because the material is fully superconducting and the flux vortices

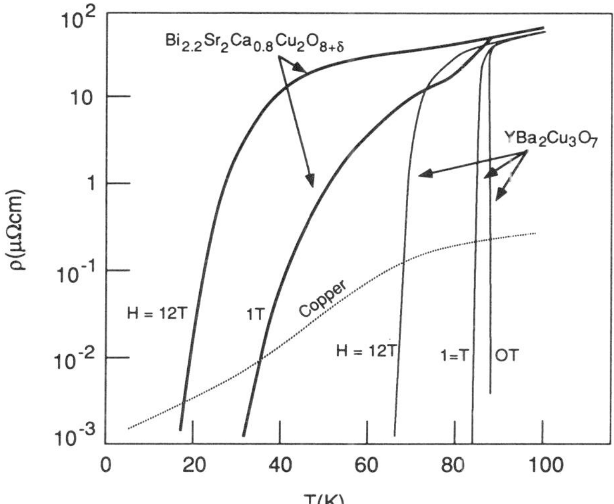

Figure 14.4. Electrical resistivity of HTSCs in various magnetic fields. The resistivity remains above that of copper at temperatures well below the nominal T_c.

are pinned and immobile. Above J_c, flux flow begins, and E rises proportional to $(J - J_c)$; the proportionality constant is termed the *flux flow resistivity* ρ_f. As the applied magnetic field increases, J_c decreases and the slope of the E vs. J plot in the flux flow regime increases.[9] Ultimately, in a very high field, the material becomes normal and J_c is driven to zero, so the slope is just the normal resistivity ρ_n.

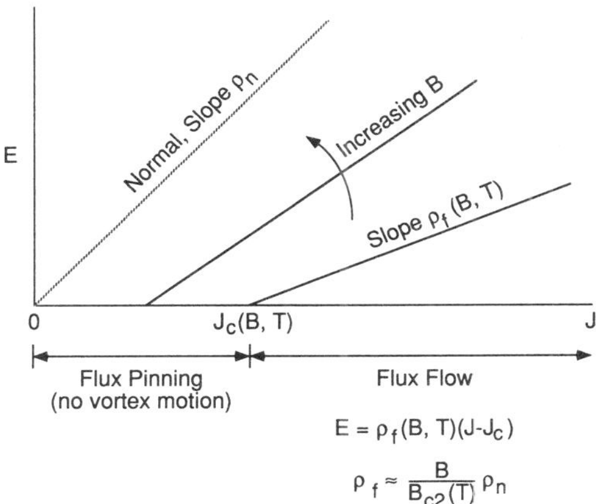

Figure 14.5. Plot of electric field as a function of current (E vs. J plot) at absolute zero. Below J_c all flux vortices are pinned; above J_c the flux flows with energy dissipated as $E = \rho_f(J - J_c)$, where $\rho_f = \rho_n[H/H_{c2}(T)]$.

14.3. THERMAL ACTIVATION

This section appears here mainly because of its historical importance in the evolution of our understanding of flux pinning. In later sections of this chapter, we shall explain why very little of this is applicable to the HTSCs, whose anisotropy and other unusual properties cause their flux motion to differ from this conventional model. Nevertheless, because nearly everyone active in the development of better flux pinning models began from this starting point, it needs to be presented.

At 4 K, the zero-temperature picture of E vs. J almost holds, except for a bit of curvature (instead of a sharp break) right near J_c. That small effect is due to thermally activated motion of flux vortices, known as flux creep. At still higher temperatures, the effect of thermal activation becomes much more pronounced. (Up to 18 K it is still small, and hence little attention has been given to thermal activation in NbTi or Nb$_3$Sn.)

However, at 77 K it is mandatory to take thermal activation into account. By defining a current parameter J_{th} (proportional to k_BT), it is convenient to identify regions in the E vs. J plot corresponding to thermally activated resistance, flux creep, and flux flow. This is shown in Figure 14.6. The electric field varies as

$$E = E_0(T) \sinh (J/J_{th}) \tag{14.1}$$

for currents below J_c. Above J_c full-scale flux flow sets in and the E vs. J relationship changes to a different form. Again, E is linear in $(J - J_c)$ with a proportionality constant that depends on the applied magnetic field and the normal state resistance

$$\rho_f = \rho_n (H/H_{c2}) \tag{14.2}$$

Referring to Figure 14.6, we define J_{co} as the J_c value that would have obtained in the absence of flux creep. The prominent excursion away from the tiny curvature of E vs. J associated

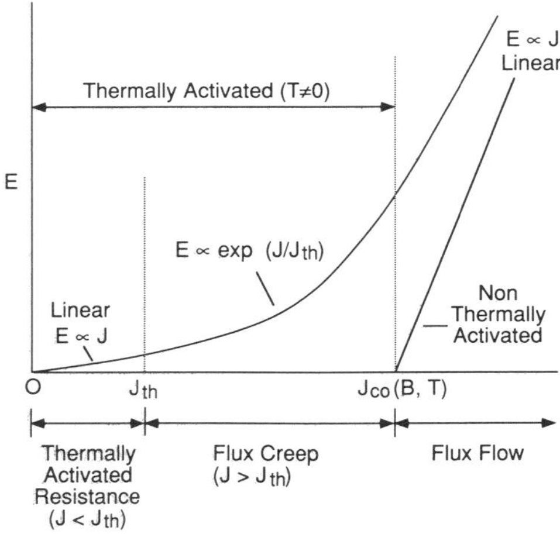

Figure 14.6. E vs. J plot at finite temperature, showing effect of thermally activated flux motion. This giant flux creep causes energy dissipation (equivalent to resistance) at currents well below J_c.

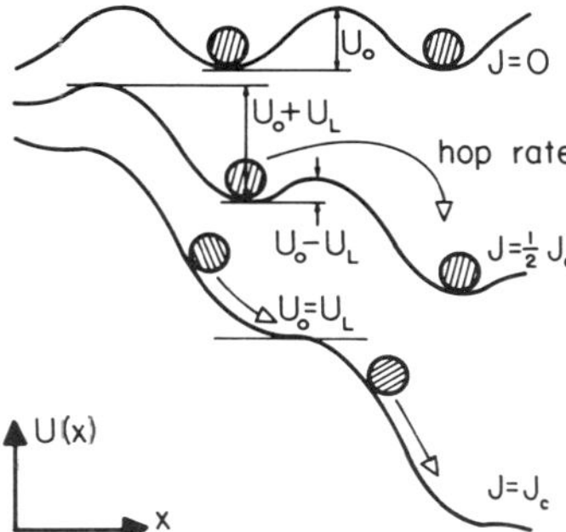

Figure 14.7. Mechanism of flux flow. The presence of current in a magnetic field generates a Lorentz force which tilts the staircase, allowing flux lines to hop out of their pinning wells more easily (from Ref. 10).

with 4 K justifies the term *giant flux creep* to describe the behavior of HTSCs. The linear region at very low currents has no more significance than that $\sinh(x) \approx x$ for small x.

We lump these effects together under the label TAFF. In either flux creep or flux flow, a flux line or a flux bundle is imagined to hop over the pinning energy barrier. Figure 14.7 illustrates[10] how, under the influence of a Lorentz force (proportional to J), a series of wells evolves into a staircase with progressively less opposition to flux motion. On this model, the energy dissipated (the resistance) is given by an Arrhenius-type equation,

$$\rho(T) = \rho_0 \exp(-U_0/k_B T) \tag{14.3}$$

The observational data is simply resistance R as a function of temperature T, but since that varies over several orders of magnitude it is best to plot $\log(R)$. Moreover, by plotting it against $1/T$ instead of T itself, one obtains an Arrhenius plot, which is most useful for studying activation energies: the slope of the data gives the activation energy U_0. However, U_0 itself is a function of temperature and magnetic field, and $U_0 \to 0$ as $T \to T_c$; so the exponential relationship is by no means pure. For HTSCs, several researchers[11] found experimentally that the pinning strength U_0 falls off as $(1 - T/T_c)^{3/2}/H$.

For classical type II superconductors, the Anderson–Kim model[12] of flux creep relates the measured critical current density J_c to the pinning strength and voltage criterion

$$J_c = J_{co} \left[1 - (k_B T/U_0) \ln (Bd\Omega/E) \right] \tag{14.4}$$

where J_{co} is J_c at zero temperature, Ω is the attempt frequency of a flux-hopping event, d is the hopping distance, B is the magnetic induction, and E is the voltage criterion (usually the smallest experimentally discernible electric field) whose units are volts/meter. Parameters d and Ω depend on temperature and magnetic field, so solutions of Equation (14.4) must be carried out numerically. This much is clear: For any voltage-per-meter criterion E, there will be a magnetic field or temperature for which J first gives a discernible voltage drop across the length of the superconducting specimen.

The role of the discrepancy between J_c values measured in different ways (discussed in Appendix A) becomes apparent once the effect of flux creep is understood. In contrast to the low-temperature behavior of Figure 14.5, HTSCs will have E vs. J curves like Figure 14.8. If one uses a SQUID magnetometer (with high sensitivity) to define the criterion at which superconductivity disappears, a low J_c value will result. A less sensitive criterion upon E (more typical of transport measurements) will give a correspondingly higher value of J_c. For

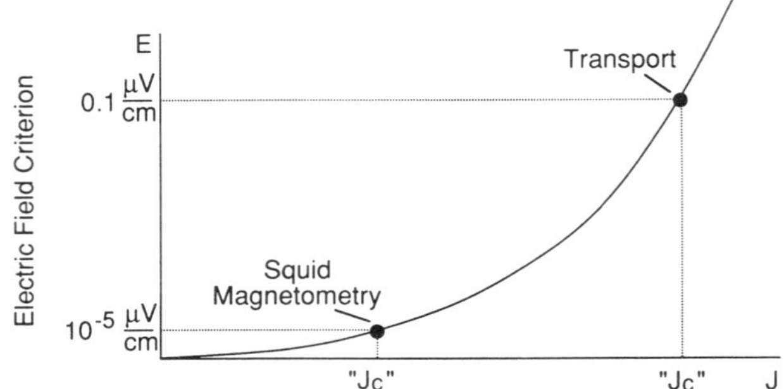

Figure 14.8. *E* vs. *J* plot at finite temperature, marked to show that different choices for the voltage criterion will lead to different measured values for J_c. For SQUID magnetometry, $E \approx 10^{-5}$ μV/cm, while for transport measurements, 0.1 μV/cm is a typical value.

this reason, it is mandatory that *some* voltage criterion be set as a standard. Appendix A goes into more detail on this point.

It should be noted that thermally activated flux motion is always present at 77 K, and thus the effective resistance *never* goes to zero. This is quite unlike the situation at 4 K, where flux is so well pinned that current decays only on geological time scales. The relevant question at 77 K is, how much better is this superconductor than copper?

In order to understand what has to be achieved for practical 77 K superconductors, it is useful to plot *E* vs. *J* on log–log paper. Figure 14.9 is a sketch that conveys the expected behavior. With this kind of graph, it is easy to recognize regions of thermally activated resistance, flux creep, and flux flow. Construction lines on the graph show the target current capacity of 10^5 A/cm^2, as well as the resistivity of copper and the magnet designer's criterion of 10^{-14} Ωm, the resistivity needed for application of these materials to large-scale superconducting magnets. This graph establishes goals to be reached en route to building a useful magnet.

14.4. IRREVERSIBILITY AND FLUX CREEP

It was found experimentally[1,13] that $H_{\text{irr}} \propto (1 - T/T_c)^{3/2}$. The phenomenon of thermally activated flux motion likewise presents the condition $(1 - T/T_c)^{3/2}/H \propto U_0$. The proportionality constant in this relation is within a factor of 2 of that in the expression for H_{irr}, which led Yeshuran and Malozemoff[13] to assert that H_{irr} is identical with the H that makes J_c zero. They coined the term *giant flux creep* to describe this identification. Under this interpretation, below H_{irr} flux pinning takes place and a finite critical current flows, whereas above H_{irr} there is no flux pinning and J_c is zero or very small. In the regime above H_{irr} but below H_{c2}, energy is dissipated by the freely moving unpinned flux lines. Therefore, in order to have a persistent transport current it is necessary to stay below H_{irr}, not just below H_{c2}.

This analysis brings unity to the picture of magnetic behavior in HTSCs. The giant flux creep model, which emphasizes weak pinning forces together with temperatures (77 K) much higher than for LTSCs, explains both the irreversibility line and the sharp fall-off in J_c with the magnetic field. The fact that H_{irr} is significantly less than H_{c2} is an important difference between HTSCs and LTSCs. The flux pinning and flux creep model is conventional, but the numerical values of the HTSC parameters are much greater than the LTSC parameters.

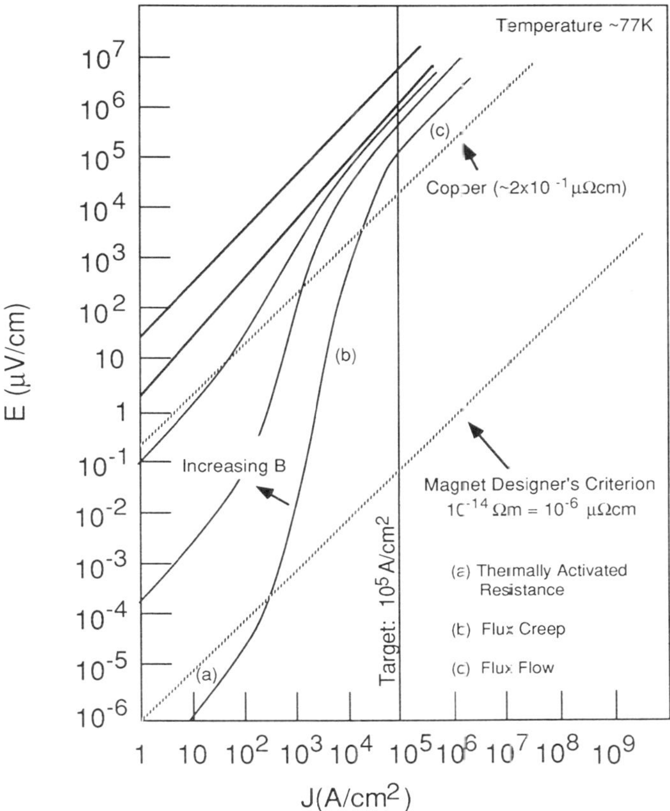

Figure 14.9. Log–log plot of *E* vs. *J* displaying family of curves akin to Figures 14.6 and 14.8 (for various values of applied field). Dotted lines draw attention to the resistivity of copper and the resistivity customarily demanded for LTSC magnets. The acceptable current level of 10^5 A/cm^2 is also emphasized.

What is the role of flux creep at high current density? Maley and co-workers[14] analyzed this question using typical values for U_0 together with the *E* vs. *J* behavior predicted by the long-established Anderson–Kim[12] model,

$$E = LB\,\Omega_0 \exp(-U_0/k_B T)\,\sinh\,(JBV_c L/k_B T) \qquad (14.5)$$

where Ω_0 is the hopping attempt frequency, L is the width of the pinning well, and V_c is the volume of the flux lattice surrounding that well. Basically, $\mathbf{J} \times \mathbf{B}$ is a force, V_c measures the number of flux lines that move together, and L is the distance they move—the product $JBV_c L$ has units of energy.[15] Inverting the *E* vs. *J* formula yields

$$J_{cf} = \frac{k_B T}{BV_c L}\,\sinh^{-1}\left[\frac{E\exp(U_0/k_B T)}{LB\,\Omega_0}\right] \qquad (14.6)$$

McHenry *et al.*[14] inserted explicit numerical values into these formulas. To estimate how J_c changes with T, they set a criterion of $E = 10^{-5}$ V/m and considered the case $B = 10$ T.

They assumed $U_0 = U_{00} (1 - T/T_c)^{3/2}$, $T_c = 93$ K, and plausible values[16] for the other factors. Their results are shown in Figure 14.10 for three particular choices of U_0. The remarkable feature is that J_c drops by 11 orders of magnitude between $T = 0$ and 93 K. Since an acceptable J_c value is around 10^5 A/cm^2, it is clear that for this set of values the temperature will have to be below 50 K for J_c to be satisfactory.

Figure 14.10 makes it clear that anything that can be done to run U_0 up toward 1 eV would greatly improve current carrying capacity at 77 K. If pinning strengths cannot be increased and if the calculation is correct, then HTSCs will only conduct appreciable currents below 40 or 50 K—well below 77 K. The importance of flux pinning and irreversibility is so great at 77 K that a material must have *all three* of T_c, H_{irr}, and U_0 large to be a practical high-T_c superconductor.

The TAFF model was carried further in 1988 by Tinkham,[17] who found that at still higher T_c and higher magnetic fields, the giant flux creep problem gets worse. For plausible values of certain parameters, the resistance would never be zero in a 30 T field, no matter how high T_c were to become. Alternately, if $T_c = 400$ K, then zero resistance at room temperature would require that the magnetic field be kept below 10 T. This analysis seriously truncated the optimism for room-temperature superconductors.

With hindsight, we can see that the TAFF model contributed significantly at the early stages of understanding how HTSCs carry current. For the first time, a phenomenon (flux creep) emerged that was present in principle all along, but hidden by its extremely small numerical significance at 4 K. Only later would models be developed that obviated the TAFF model by revising the fundamental picture of the flux lines within HTSCs.

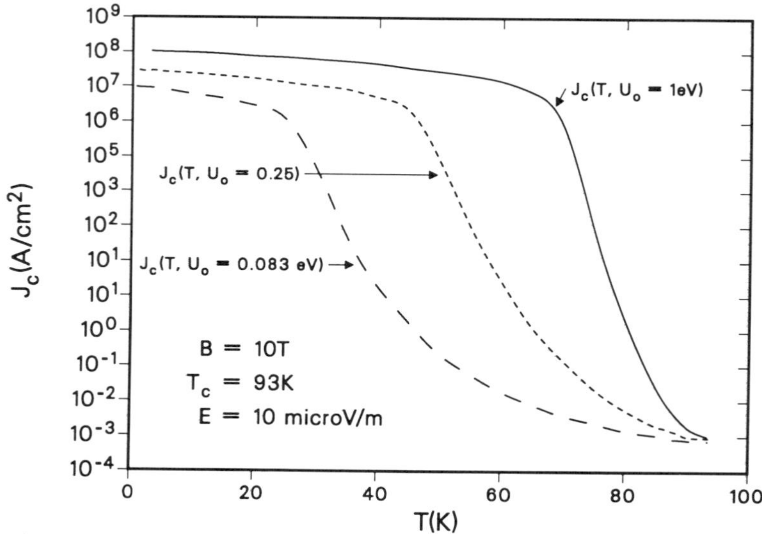

Figure 14.10. Calculated values of J_c vs. T for HTSC with $T_c = 93$ K, in 10 T field, under different choices for the flux pinning strength U_0.

14.5. FLUX LATTICE MELTING

Experiments have made it clear that in HTSCs the flux lines do not behave in the same way as in LTSCs. The first question is whether flux lines move as a group or as individual lines, and the second question concerns the behavior of those individual lines.

14.5.1. Flux Lattice Dynamics

One model has the flux lines arranging themselves into a glassy state (a vortex glass).[18] In analogy with ordinary glass there is no regular lattice, but the flux lines remain rigidly positioned relative to one another. At high enough temperatures this glassy state can melt, in which case the flux lines move about independently.[19] If the transition temperature T_c is lower than the melting temperature T_m this is of no consequence because superconductivity vanishes anyway above T_c. In HTSCs, for the first time it is possible that T_m may be substantially lower than T_c.

The distinction between a glass and a melted lattice depends on the viscosity. In general, melting of any lattice is defined by the Lindeman criterion, by which the average random excursion of a point on the lattice (an atom) exceeds the lattice spacing. This should not be confused with motion of the entire flux lattice relative to the crystal. For a lattice of flux lines, the lattice spacing decreases with increasing field as more flux lines penetrate the material. Therefore, sufficiently large excursions occur at either high temperatures or high magnetic fields. In particular, at a fixed temperature (say 77 K), increasing the magnetic field can cause a flux lattice to melt.[20]

With an operating temperature of 77 K for the HTSCs, it is not of engineering importance to ask how they behave above 85 K. However, studying the range above the irreversibility temperature and field gives insight into the flux-pinning mechanism. In the particular case of YBCO, Farrell et al.[21] used a delicate torsion balance to measure dissipated energy in crystals of YBCO, scanning the temperature across the irreversibility line. (The magnetic field was aligned along the c-axis.) With crystals containing twin boundaries no dissipation was observed. However, untwinned crystals showed dissipation so strong that it could only be explained in terms of flux line motion that would correspond to melting of the flux lattice. In the untwinned crystals, according to Farrell, the Abrikosov lattice of flux lines could contain dislocations, and the motion of these dislocations sets off the melting of the flux lattice. Whenever twin boundaries are present the dislocations are prevented from moving very far.

Farrell's data (shown in Figure 14.11) is best fit by a flux lattice melting theory, in which $H = H_0 (1 - T/T_c)^\beta$, with $\beta = 2$. By contrast, a thermal-activation model would suggest an exponent of $\beta = 4/3$ or $3/2$ instead of 2—quite inconsistent with the data. It is fair to conclude that defect-free single crystals of YBCO are subject to flux lattice melting, which certainly limits their current-carrying capacity.

The presence of flux lattice melting in HTSCs prompted another look at LTSCs. It was found[22] that in NbTi and Nb$_3$Sn, a previously unobserved region of reversible flux line motion exists, and is about 1 K wide. (This was always presumed to be so narrow as to be indistinguishable from the upper critical field curve.) The data is best explained by the flux lattice melting model,[23] as contrasted with the thermally activated giant flux creep model.[24] This indicates that when the ratio $\kappa = \lambda/\xi$ is very large (as is the case for Nb$_3$Sn and the copper oxide superconductors), the standard understanding of flux line motion (TAFF) no longer applies.

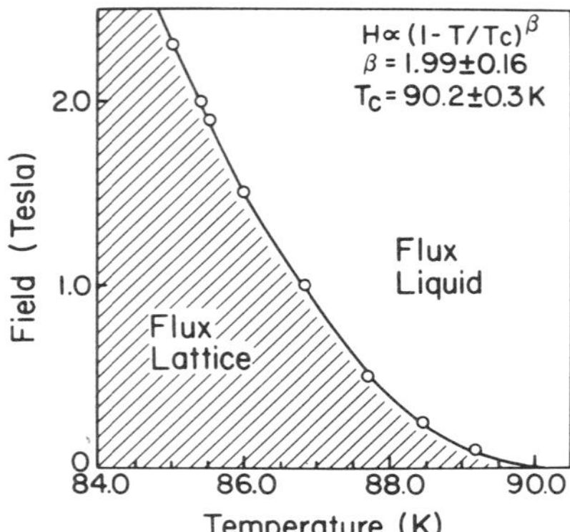

Figure 14.11. Relationship between the temperature of flux lattice melting and the field. The solid curve through the data is the equation in the figure. (From Ref. 21.)

14.5.2. Variation of U_0

In Section 14.3, it was mentioned that the slope of the Arrhenius plot gives the pinning strength U_0, but that U_0 falls off as T approaches T_c. However, there is a more important variation of U_0 with temperature or field, which we now consider.

In the simplest model, the pinning energy U_0 is given by

$$U_0 = 0.02 \, n_1 n_2 \frac{H_c^2 \, \Phi_0}{8\pi B} \xi^m \tag{14.7}$$

Here, n_1 is the number of flux lines bundled together that move as a unit when they move, and n_2 is a measure of the length of a flux line—the number of multiples of ξ that a line occupies. Typically, in canonical type II superconductors n_1 and n_2 are both around 10, which makes the numerical prefactor of order unity. The temperature dependence is embedded in H_c and ξ. The exponent m lies between 1/2 and 3, depending on geometrical considerations.

If the rigidity of the flux lattice is lost, then the *shear strength* of the lattice vanishes and pinning one vortex does nothing to restrain the movement of other flux lines. Melting of a previously rigid flux lattice causes n_1 to drop from 10 or more to 1, causing a precipitous drop in U_0. When melting occurs, much more flux flow takes place.

Actually, this simple model is too simple. In reality, U_0 is inversely proportional to current density J because the motion of partial flux lines involves a type of nucleation effect. As mentioned immediately above in Section 14.5.1, the Abrikosov lattice of flux lines may contain dislocations. But what can a dislocation be in a magnetic field line?

Imagine an isolated vortex line that would like to move sideways under a Lorentz force $\mathbf{J} \times \mathbf{B}$, but where only one finite piece is actually able to move due to stronger pinning elsewhere. (This condition never arose in the Anderson–Kim model, where flux lines moved in their entirety.) Next, remember that the Maxwell equation *div* $\mathbf{B} = 0$ demands that lines of magnetic field must be continuous. Thus, the only way to represent a dislocation is as the superposition of the original vortex line plus an adjacent vortex loop in the opposite direction. Figure 14.12 shows this configuration.

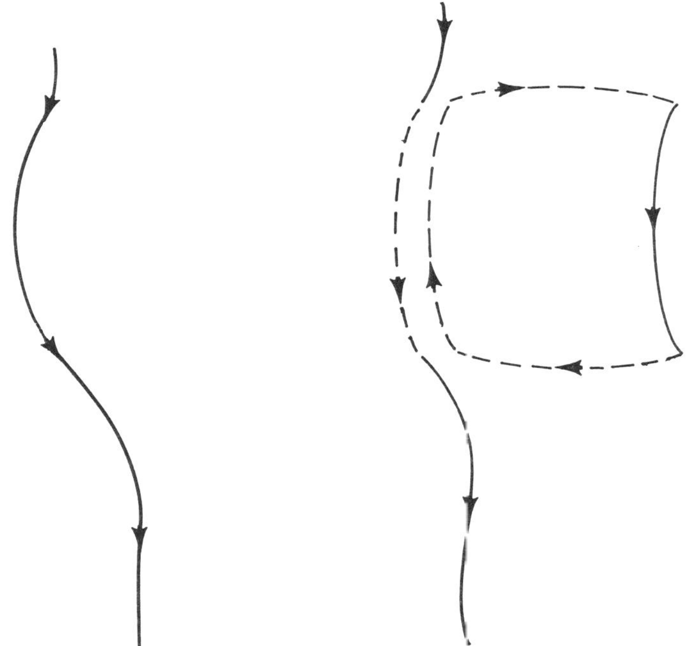

Figure 14.12. Sketch of a single vortex line and the model used to characterize it when one segment of the line jumps to a new position.

To create a vortex loop, some nucleation energy is required. The amount of energy depends on the radius of the loop (in other words, how far the segment of flux line jumped): $U_1 = \alpha r$. A loop of very small radius is not stable; the dislocation would merely hop right back to its original position. Large loops cost too much energy to form. Meanwhile, the system also *gains* some energy when this loop is forming: $U_2 = \Phi_0 J\pi r^2$. The sum of these two competing terms has a maximum, known as U_0, at some critical loop size r_0. This is quite reminiscent of the discussion of crack formation in Chapter 11, and Figure 11.6 (if inverted) could equally well be a sketch of these two effects. Notice that as J increases, r_0 gets smaller.

The net result is that the activation energy $U_0 \sim (J_c/J)^\mu$, where $\mu \approx 1$ in a clean superconductor with a line defect, as sketched here. In the vortex glass model of Daniel Fisher *et al.*,[25] it turns out that $\mu \approx 0.2$. Referring back to Equation (14.7) above, the J dependence comes in via the dependence on n_2. The central point is that where hopping of partial flux lines is possible, the activation energy is a function of current, whereas in conventional TAFF theory U_0 is constant.

14.6. VORTEX GLASS MODEL

Whenever the flux lines (vortices) form a *lattice*, they interact with one another and travel together; hence, pinning a *few* lines is sufficient to prevent *all* of them from moving. Should that lattice melt, so that blocking one vortex no longer influences adjacent vortices,

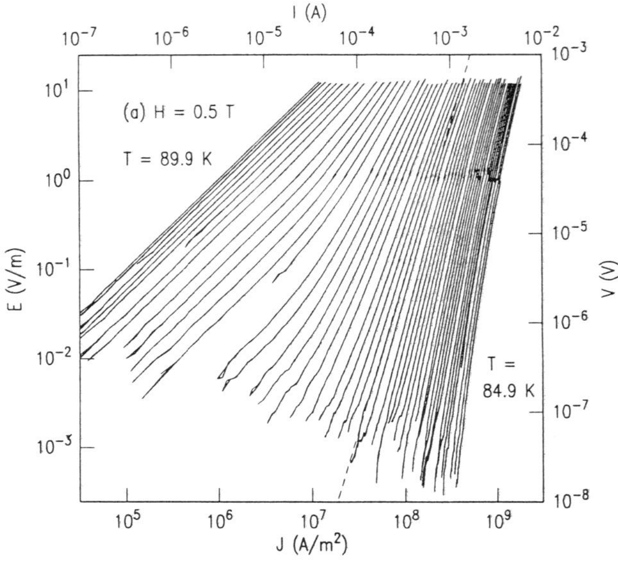

Figure 14.13. Experimental *I-V* curves of Koch *et al.*[26] at 50 different temperatures (84.9–89.9 K) in a field of $H = 0.5$ T. The dashed line defines T_g.

considerable energy dissipation would take place and the advantage of superconductivity would be lost. The vortex-glass model[25] offers an alternative to this either/or condition: it does not require an ordered lattice, but still keeps the vortices related to one another. It is applicable over a considerable temperature range.

The early simple models of flux motion were based on quite limited data, often from poor samples. With better experiments, it became clear that there is actually a more subtle shape to the *I-V* curve, or the ρ-*T* curve: separate regions of flux *creep* and flux *flow* can be identified. A set of 50 *I-V* curves obtained by Koch *et al.*[26] at various temperatures in a field of 0.5 T are shown in Figure 14.13. Comparable data was gathered at many different magnetic field values.

In fact, it is equally possible to sit at one temperature and step the magnetic field to generate a very similar plot. Shibutani *et al.*[20] obtained the data of Figure 14.14 in this way. Whereas Figure 14.13 determines a value of T_g in fixed H, Figure 14.14 finds H_g at a fixed temperature.

Matthew Fisher developed a scaling relation[18] by which a great deal of this data in the *I-V* plane collapses onto two simple curves. The first curve is a *flux creep* line that goes as $E \propto \sinh (J/J_c)$ at low enough J; the resistance is finite everywhere. The second curve, corresponding to the vortices being frozen into a glassy state, has E going to zero at finite J. That means zero resistance and true superconductivity, thus no flux creep. Figure 14.15 shows these two lines for Koch's data; the axes are scalings of I and V, in which the scaling factors are functions of temperature.

Looking back at Figure 14.7 it is clear that when U_0 is fixed there will always be *some* hopping over the barrier. But with a glass model, the barrier height can approach infinity, in which case there will be no dissipation.

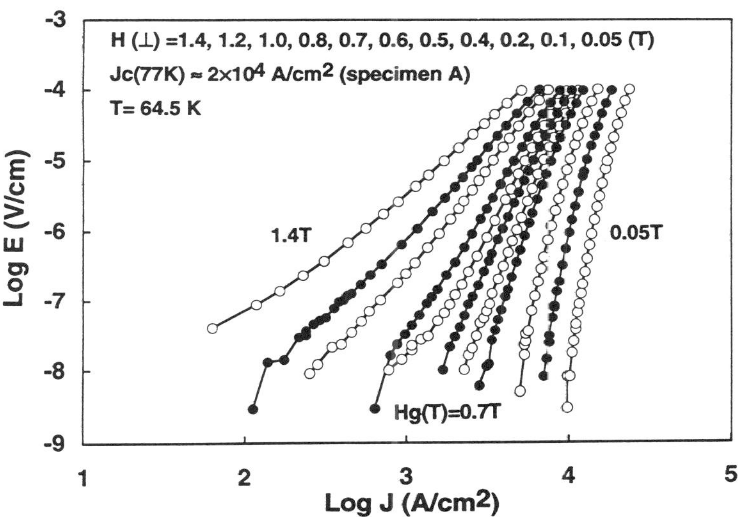

Figure 14.14. Experimental *I-V* curves of Shibutani *et al.*[20] at 11 different field values (0.05–1.4 T) at $T = 64.5$ K. At this temperature the glass transition field $H_g \approx 0.7$ T.

The split between the two curves of Figure 14.15 is a second-order phase transition which varies as a function of pinning strength. (Whenever there is a second-order phase transition, scaling is expected.) In this model, the transition to a glassy state takes place at a temperature T_g, and the two universal curves lie to either side of that temperature. The theory asymptotically reverts to ohmic behavior or to flux creep behavior at certain extremes. The plot of the *H-T* plane (Figure 14.16) shows T_g as a function of H looking just like a critical-field curve.

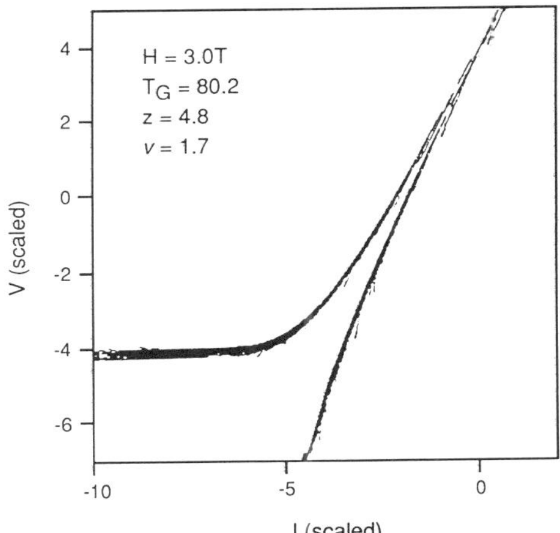

Figure 14.15. Voltage vs. current relationship for HTSCs. Each axis is scaled via Fisher's vortex glass model.[18] A great deal of data collapses onto two curves—the upper one showing flux motion, the lower one with flux lines frozen.

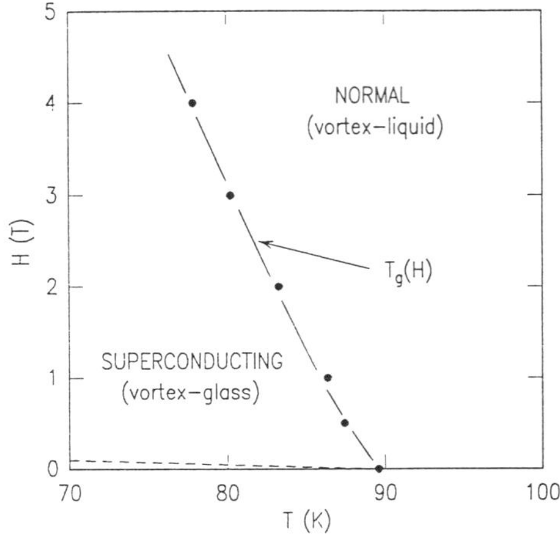

Figure 14.16. Magnetic field vs. the temperature at which a vortex glass transition occurs. The resemblance to a conventional critical-field curve is striking.

Koch *et al.*'s data[26] is best explained on the basis of random pinning sites tugging on the vortices, bending and distorting them somewhat, thus leaving the vortices frozen in a glassy state; that is, no regular Abrikosov lattice, just amorphous positioning of lines, but rigid nonetheless. The vortex glass model[18] may be compared to Nelson's flux-lattice melting[19] via the analogy of comparing partially cooked spaghetti to fully cooked spaghetti. In a glass, the vortex lines wiggle, but there are no big deviations in vortex separation. The vortex lines deviate partially, eventually becoming hung up on pinning sites of strength 1 eV or so, after which the whole block has to move together. By contrast, the flux lattice melting theory has no pinning in it and vortex lines easily become entangled.

The cooperative role of neighboring vortices is the principal difference between the vortex glass model and various melting models; it requires a collective behavior of many flux lines, which is not melting at all. The definitions of these characteristic temperatures are summarized in Table 14.1.

Table 14.1. Parameters of the Flux Lattice.

Quantity	Definition	LTSC	HTSC
$H_{irr}(T)$	The field at which flux vortices are no longer pinned at temperature T. Above H_{irr} behavior is reversible.	Only occurs very near H_{c2}.	Occurs well below H_{c2}.
$T_{melt}(H)$	Temperature at which Abrikosov lattice melts and each pinning site pins only a single flux vortex. Varies with applied field.	Lattice does not melt except possibly at H_{c2}.	Lattice may melt well below 77 K (especially for TBCCO and BSCCO).
$T_{glass}(H)$	Temperature at which a rigid but amorphous pattern (not an Abrikosov lattice) of flux vortices forms. Varies with applied field.	Not applicable. Always forms Abrikosov lattice.	Occurs below $T_{melt}(H)$.

14.7. ANISOTROPY EFFECTS

Not surprisingly, the two-dimensionality of the CuO_2 layers has a profound effect on flux pinning, just as it does on most other properties of the HTSCs. Because the superconductivity takes place primarily between conducting planes, the flux vortices are likewise confined. This means that the vortices are no longer miniature telephone poles, as in conventional superconductors, but are similar to stacks of pancakes, that is, segmented in the direction perpendicular to the CuO_2 planes. If the penetrating magnetic field is not aligned with the c-axis, then the stack corresponding to each flux vortex will be slanted. This changes the nature of flux-line motion tremendously in HTSCs, with the result that the simple models (e.g., TAFF) must be completely reconsidered.

14.7.1. Josephson Junctions

As has been discussed in earlier chapters, especially Chapters 8 and 13, the linkages along the crystalline c-axis are predominantly through Josephson junctions. It is plausible to argue that this will influence flux pinning as well. In particular, grain boundaries are a well-known example of Josephson junctions. In conventional (LTSC) superconductors, the coherence length ξ is large enough that the grain boundary poses a very minor barrier to the passage of a supercurrent. However, in HTSCs (where ξ is about equal to the lattice constant) the wave function of the superconducting pair is severely attenuated by the Josephson junction at a grain boundary, and hence the current drops off substantially in a bulk sample made up of many grains.

To investigate this experimentally, Gray and his research team at Argonne National Laboratory[27,28] studied flux line motion in thin films of both YBCO and TBCCO. They also studied LTSC films[29] of niobium nitride (NbN). Their data is explainable using a Josephson junction model. They reason that within a single crystal or thin film dislocations or other crystal imperfections create Josephson junctions. The current flowing along a CuO_2 plane comes to one of these obstacles and is forced to switch to a different plane. That change is accomplished by a quantum-mechanical tunneling through adjacent (nonsuperconducting) planes of BaO, CaO, TlO, or whatever—those insulating planes create a Josephson junction between adjacent CuO_2 planes. Overall, the superconducting current must hop through a series of such Josephson junctions in traversing the sample, even when the sample is a single crystal, a well-aligned epitaxial thin film, or is otherwise nominally perfect.

Their Josephson junction model predicts a simple temperature-independent relation between U_0 and the critical current J_c, whereas the Anderson–Kim[12] flux-creep model predicts $J_c = (1 - T/T_c) U_0$. This distinction allows the competing models to be compared, and the experimental data agrees better with the Josephson model. This further undermines the TAFF model, which relies entirely on the Lorentz force as its driving mechanism.

14.7.2. Crossover Between Dimensionality

In the extremely anisotropic BSCCO and TBCCO materials, when the magnetic field is applied along the c-axis, the flux vortices have become pancakes of magnetic flux localized in the copper oxide multilayer regions. The coupling of pancakes from layer to layer is determined by Josephson tunneling. Therefore, the weaker Josephson coupling of BSCCO permits much greater thermally activated flux motion, which in turn implies finite resistance at very modest applied magnetic fields. This constitutes the broadening of the resistive transition in a magnetic field, recognized[30] in HTSCs since 1987.

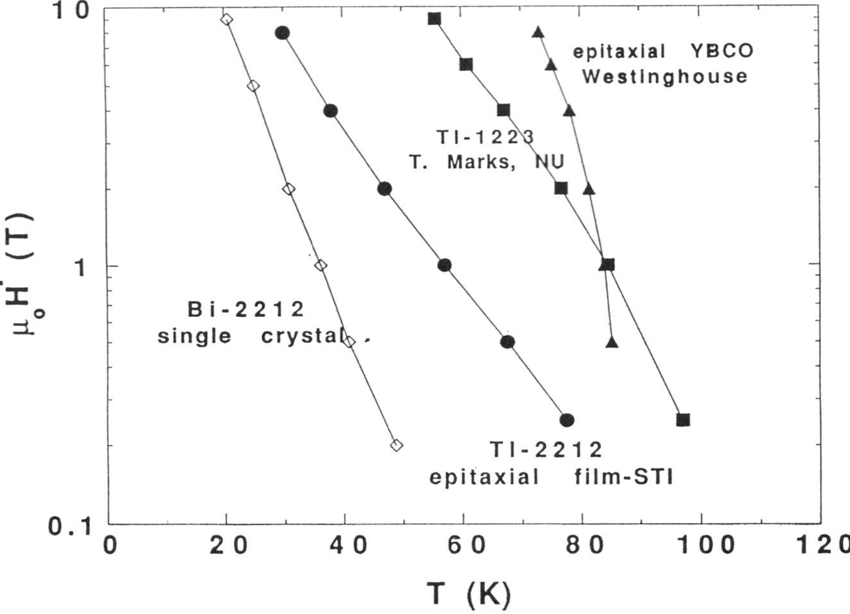

Figure 14.17. Crossover fields ($\parallel$ c-axis) for dissipation due to thermally activated flux motion in various HTSCs.

Kim *et al.*[31] characterize this broadening via a crossover field $H^*(T)$, which is related to the irreversibility line in the magnetic phase diagram for the HTSCs. Figure 14.17 shows some typical data for $H^*(T)$. Essentially, there is a combination of pinning forces both within and perpendicular to the *a-b*-plane, whose effectiveness diminishes with increasing temperature.

Each pancake vortex is pinned locally at its position within the *ab*-plane, but that pinning is weak and the associated horizontal potential well is shallow. Each vortex is also pinned to the ones above and below it (in the next multilayer of CuO_2 planes), and that vertical potential well can be substantially deeper. In order for a flux vortex to move it must acquire an energy $k_B T$ sufficient to escape from both potential wells. That condition yields the values of $H^*(T)$ displayed in Figure 14.17. The combination of effects is shown in Figure 14.18 where the thin solid curve corresponds to vertical pinning, the dashed curve to horizontal pinning, and the thick solid curve to their combined effect. Clearly, the relative sizes of both wells depends on T and H.

In Chapter 8 (Table 8.2) we enumerated the lattice parameters relevant to the layered structure of BSCCO and TBCCO. The key parameter is the distance d_i between consecutive CuO_2 multilayer pairs. That distance is an intrinsic property of the unit cell, which cannot be modified by insertion of pinning sites. The claim put forth by Kim *et al.* is that the vertical pinning depends on d_i. On the other hand, the horizontal pinning can be enhanced by adding pinning sites; the trouble is, to be effective every CuO_2 multilayer needs a large number of pinning sites, and such a vast collection of sites could only be achieved at the cost of greatly diminishing T_c with all those impurities. Pinning of the vertical kind, on the other hand,

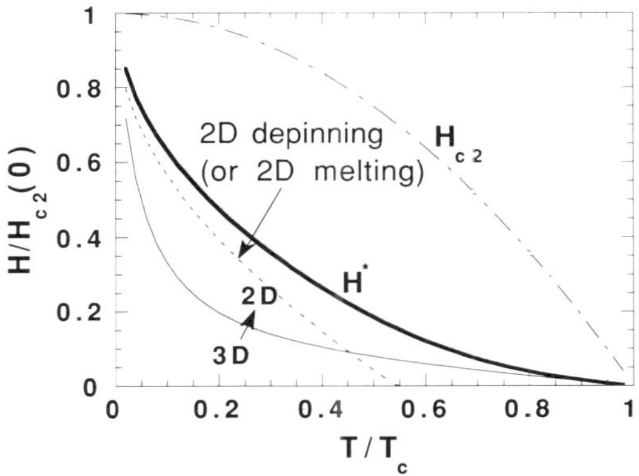

Figure 14.18. Characteristic fields for thermally activated three- and two-dimensional crossover (solid), flux depinning (dashed), and combined effect of both (heavy solid line). The temperature dependence of $H_{c2}(T)/H_{c2}(0) = [1 - (T/T_c)^2]$ is also shown (dot-dash line).

depends on Josephson tunneling and will be strong if the vertical barriers are thin, and weak if they are thick.

In the case illustrated by Figure 14.18, vertical pinning is the dominant contribution to $H^*(T)$ above $T/T_c > 0.5$. This means that the spacing between CuO_2 multilayers is a dominant parameter throughout the temperature regime that includes 77 K. The data displayed in Figure 14.17 confirms this, because the Tl-1223 phase (with $d_i = 9.59$ Å) has a much larger H^* than the Tl-2212 phase (with $d_i = 11.55$ Å) or the Bi-2212 phase (with $d_i = 12.29$ Å).

Initial experimental verification of the Kim–Gray model came from General Electric[32] and from Japan,[33] where TBCCO was studied. Both groups' results confirm that the irreversibility line for TBCCO-1223 is substantially shifted toward higher temperatures and fields.

The conclusion drawn by Kim *et al.* is that the thallium compounds will postpone flux lattice melting better than BSCCO in a magnetic field at 77K, because of their relative vertical spacings between the superconducting layers. As this interpretation gains experimental confirmation, attention is shifting away from BSCCO and toward TBCCO.

One caution is in order: This model does not apply to ordinary YBCO, which is less anisotropic to begin with and contains copper oxide chains that short-circuit the Josephson tunneling between layers; whereas tunneling is a key feature of this model. However, experiments have been performed on specially contrived YBCO samples that mimic the layer separation inherent in BSCCO and TBCCO. A group at the University of Maryland[34] deposited thin films of alternating layers of pure YBCO with praseodymium-substituted $(Y_{1-x}Pr_x)BCO$. From Chapter 10 we recall that small amounts of Pr lower T_c in YBCO dramatically. The superlattice thus formed by the alternating layers was thoroughly superconducting below $T_c^x \approx 50$ K, but contained insulating layers above 50 K. The observed relationship between J_c and H confirmed that pinning switched from YBCO-like to BSCCO-like as T_c^x was crossed, and the CuO_2 double-planes were being decoupled. This says that the difference between pinning in YBCO and BSCCO is in the interplanar coupling. The experimental results demonstrate unambiguously the important effect of dimensionality on flux pinning in HTSCs.

14.8. CREATING STRONG PINNING SITES

The art of making NbTi wire includes a lot of clever engineering to introduce lattice defects and thus increase the number of pinning sites, resulting in wire that carries more current. A similar effort has been in progress for the HTSCs, but with less promising results. Much of the early work was frustrated by a lack of understanding of the nature of layered superconductivity.

Certainly, pinning is a collective phenomenon; in YBCO, for example, oxygen vacancies in the lattice are the primary source of pinning.[35] For typical experimental magnetic fields, when 1% of the oxygens are absent (i.e., $\delta = 0.07$ in $YBa_2Cu_3O_{7-\delta}$), the spacing between flux lines is sufficiently large[36] that each flux vortex is pinned by 7 or 8 vacancies.

Twin boundaries in YBCO definitely improve pinning. As mentioned above in Section 14.5.1, twin-free single crystals of YBCO suffer from flux-lattice melting. However, specific details of the effect of twin boundaries are lacking. Proper alignment of both grains and applied magnetic field is crucial. At Argonne, Kwok *et al.*[37] distinguished between intrinsic pinning and twin-boundary pinning through a series of measurements between 88 and 91 K. When $H = 1$ T or more, they found it necessary to align the magnetic field very precisely along the *a-b* plane in order to achieve zero resistance. This shows that intrinsic pinning occurs when the flux lines are captured between the CuO_2 planes.

At 77 K, questions about the irreversibility line vanish, but the role of pinning is still very important. Other forms of defect production are described in Chapter 15. For example, to enhance flux pinning, tiny precipitates of YBCO-211 are embedded within melt-textured grains of YBCO-123.

In contrast to the experimental configuration most often used (B parallel to the *c*-axis), when the magnetic field and current both lie in the *a-b* plane, the lines of flux are continuous, pinning is easier, and flux lattice melting can be avoided. Ekin *et al.*[38] placed a melt-textured sample of YBCO in magnetic fields as high as 30 T and observed that J_c still remained fairly high, above 1000 A/cm². Their data above 8 T appears here in Figure 14.19. This shows that the flux lines are pinned quite well in between the CuO_2 planes, and flux lattice melting is not occurring.

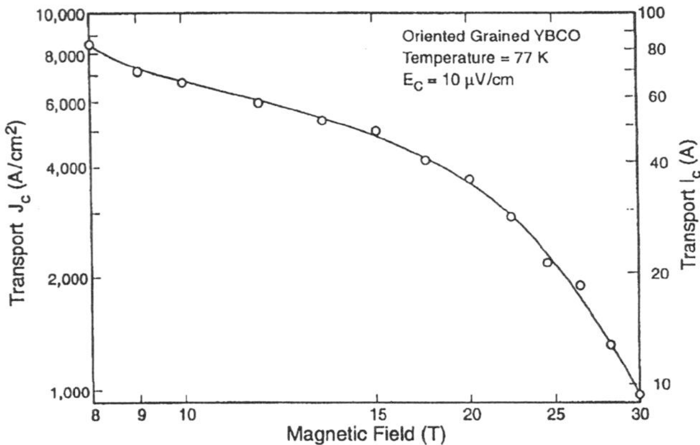

Figure 14.19. Transport critical current density J_c vs. magnetic field for bulk oriented-grained $YBa_2Cu_3O_7$ at LN_2 temperature with **B** perpendicular to the *c*-axis. (From Ref. 38.)

There are many good ways to create pinning sites and thus evade the condition of flux lattice melting. In one series of measurements, Civale *et al.*[39] contrived a very special type of defect in YBCO. They irradiated YBCO with heavy ions (tin, 580 MeV) in order to produce damage tracks that serve as containers for flux lines. The irradiation dose corresponded to the expected number of flux lines for the particular choices of magnetic fields 1, 3, and 5 T. Thus there was one columnar defect with which to confine each vortex; the columns were 15 μm long and about 50 Å in diameter (about two coherence lengths). As soon as the magnetic field was aligned with these tracks, flux pinning was greatly enhanced; J_c values remain above 10^5 A/cm^2 out to several Tesla. Moreover, the irreversibility line moves upward considerably, demonstrating the presence of strong pinning.

14.9. IMPLICATIONS FOR CONDUCTING CURRENT

The topic of flux pinning was very controversial for several years. Despite many proposed models, the experiments were never decisive, and researchers called for more and better data. In the intervening years, this has been obtained.

14.9.1. YBCO

To make a long story short, the case for a glass model is now very strong.[40] In a review article in *Science* magazine, Bishop *et al.* at AT&T Bell Labs explained how this came about.[41] Basically, what was first shown in thin films by IBM[25,26] was confirmed and strengthened by others[42] and ultimately established for crystals by Bell Labs.[40] A SQUID picovoltmeter developed by Gammel added almost 6 orders of magnitude sensitivity to their previous measurements, and hence they obtained definitive data on both YBCO and BSCCO in the intermediate-temperature range, which eliminated the thermal-activation flux flow (TAFF) explanation. The Bell Lab group stated:

> Our measurements of the temperature dependence of both R, the linear response resistivity, and the onset of nonlinear response, strongly constrain theoretical fitting parameters and have allowed us to rule out the class of models that attempt to explain the dynamics in these systems as due solely to thermally activated hopping over barriers.

Two figures summarize the observations well: Figure 14.13 is the data of Koch *et al.*[26] The faintly visible dashed line is used to define the vortex glass transition temperature T_g, where the curvature of these E vs. J traces switches from positive to negative. Figure 14.20 is from Gammel *et al.*[40] and shows the stunning contrast between the TAFF model (horizontal dashed line) and the vortex glass model (solid line) which follows the data reasonably well. It is remarkable that this data was taken on the very same sample of YBCO that was once used[30] to advance the TAFF model. In the range of higher resistivity (above $R/R_n > 0.2$), the data agree. What is different is the greater sensitivity at much lower resistivity obtained by new experiments,[40] which confirms the vortex glass model and eliminates TAFF.

Further confirmation, with a slight twist, comes from Kwok *et al.*[43] who studied flux lattice melting in single crystals of YBCO, with and without twin boundaries. Twin boundaries pin flux lines, so by carefully rotating the magnetic field direction relative to the twin-boundary plane they were able to adjust the pinning strength and thus turn the melting transition on or off. With **H** along the *c*-axis, the twin boundaries act as regions of *correlated disorder*, which gives rise to a *Bose glass transition*[44]—a relative of the vortex glass, but still far from TAFF.

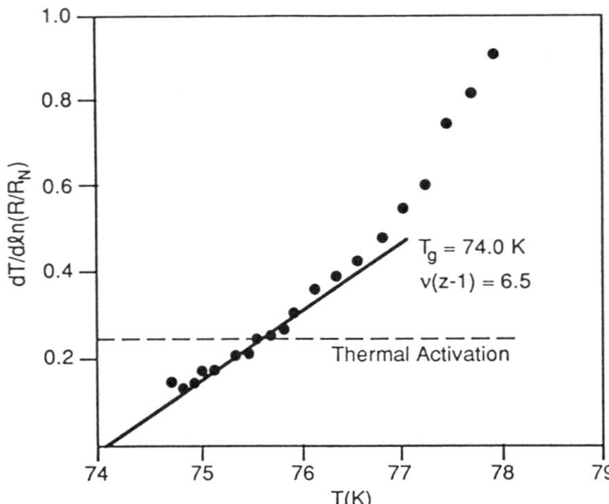

Figure 14.20. The inverse logarithmic derivative of the resistivity should vanish in any model with a phase transition. This data (dots) suggests that the value $T_g = 74$ K in a magnetic field of 6 T, using the vortex glass model (solid line) for extrapolation. The dashed line is the prediction of the TAFF model. (Adapted from Ref. 40.)

On the theoretical side, computations by Sasik & Stroud[45] of the melting curve of YBCO utilize a *superfluid density tensor*, which distinguishes the *c*-axis from the *ab*-plane, and obtain excellent agreement with experiments by Safer *et al.*[46] Flux lattice melting is definitely confirmed.

What are the consequences of trusting the vortex glass model for HTSCs? First of all, after the glass has melted it is necessary to pin every single flux vortex individually, in order to prevent dissipative movement. This amounts to a huge penalty imposed on the integrity of the crystal structure. This is an intrinsic limitation of the material. The practical consequence is that a material will carry very little current above it's glass transition temperature T_g, where the flux lattice begins to move freely, dissipating energy as it does.

That temperature T_g is a function of magnetic field. For example, for YBCO the Bell Lab data[40] gives the curve shown in Figure 14.21. Note that at 77 K, 4 T is the highest field in which one can expect to use YBCO. The curve can only be shifted a very modest amount by clever engineering of defects because flux lattice melting is an intrinsic property of the material. On the other hand, if higher fields are mandatory, then a partial vacuum would be required to lower the LN_2 bath temperature. This imposes significant design constraints: Any apparatus design that incorporates YBCO wire must either hold the expected magnetic field below 4 T, or plans to absorb the economic cost of adding vacuum equipment to the device.

14.9.2. BSCCO

The behavior of BSCCO at intermediate temperatures, and indeed the hope that BSCCO will prove useful at 77 K, is directly linked to our understanding of flux motion. When dealing with YBCO, the proximity of the melting temperature T_m to the transition temperature T_c obscured the distinct characteristics of melting because the pinning strength U_0 was varying with temperature anyway near T_c. In the case of BSCCO, $T_c > 77$ K, but flux lattice melting

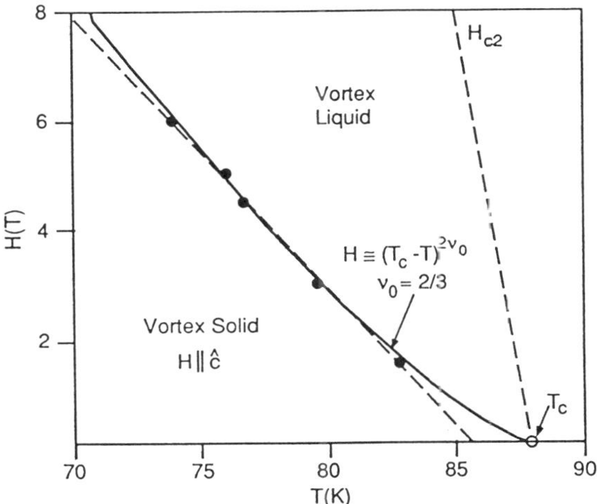

Figure 14.21. The phase diagram in *H–T* space for the vortex lattice. T_c is taken from the zero-field resistive transition. The solid line is a 4/3-power law fit to the data; the dashed line is a linear fit.

takes place below 30 K. The separation makes it possible to identify the formation of a vortex glass at low temperatures, as well as the crossover from three- to two-dimensional behavior at intermediate temperatures. This represents an important step forward in understanding BSCCO and serves to clarify the severe flux pinning problem in BSCCO at 77 K.

Experiments at AT&T Bell Labs[47] in magnetic fields from 0 to 6 T gives strong encouragement to the vortex glass model for BSCCO. The data is shown in Figure 14.22. The horizontal axis is temperature itself (not 1/*T*), and the vertical axis gives the slope of the

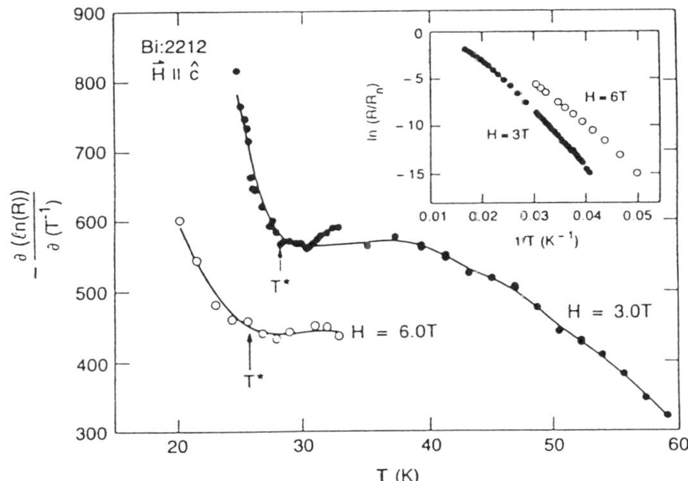

Figure 14.22. Temperature dependence of the activation energy U_0 for 3- and 6-tesla magnetic fields applied $\parallel$ *c*-axis. $U_0(T)$ is defined as the local slope in the Arrhenius plots, shown in the inset for the same magnetic fields. T^* denotes the beginning of the divergence of U_0. The solid lines are guides to the eye. (From Ref. 47.)

Arrhenius plot which is clearly not a single constant value U_0. (However, the inset displays at a glance how easy it would be to look at an Arrhenius plot and *assume* a constant slope through all the data.) What stands out clearly here is that there is a qualitative change in the resistivity behavior below 30 K. The Bell Labs team interprets this as a crossover point T^*, above which the superconducting behavior is predominantly two-dimensional and below which it becomes three-dimensional. That is, at high temperatures there is little correlation between copper oxide planes, and the lines of magnetic flux have become dismembered into weakly correlated two-dimensional pancake vortices. Such vortices move rather easily under thermal activation. By contrast, below T^* there is much more correlation between planes, the system crosses over to three-dimensional behavior, and the flux lines are once again the miniature telephone poles of the familiar model.

At a still lower temperature T_g, a firm and immobile vortex glass forms. Figure 14.23 presents the magnetic phase diagram showing both T^* and T_g as a function of the applied field. When the field is relatively low, there is no important distinction between the two. However, above 2 T there is a spread of several degrees K. In the range between T_g and T^* the vortex glass of flux lines is undergoing a phase transition. Now, historically, phase transitions occur at a single unique temperature, or at least in a narrow critical range. Here, the critical range is very wide, the phase transition is by no means sharp, and we cannot expect to see the traditional characteristics of a first-order phase transition except below about 1 T. This point bears on the question of disorder in the flux lattice, and is important for understanding why the vortex glass model was not accepted sooner.

The result of all this is to validate the vortex glass model, and consequently to say that flux motion in BSCCO is qualitatively different above 30 K: two-dimensional pancake vortices move rather freely. To pin each one would require an extremely large number of lattice defects. To circumvent this limitation, it would be necessary to obviate the vortex

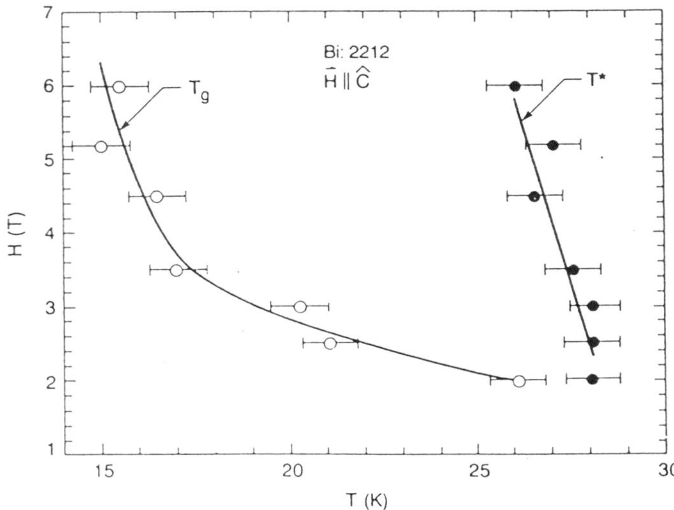

Figure 14.23. Magnetic phase diagram, where T_g represents the glass transition. T^* marks the crossover to a low-temperature regime dominated by three-dimensional cooperative effects. Typical error bars are shown, while the solid lines are guides for the eye. (From Ref. 47.)

glass model, which would require modifying the lattice very dramatically (such as by making the BiO planes noninsulating). One can say this is only a model, but it is a very successful model so far. Therefore, it is reasonable to conclude that BSCCO will be useful as a high-field superconductor only below 30 K.

14.9.3. Thallium

The compounds BSCCO-2223 and TBCCO-2223 both suffer from two-dimensionality, because of the wide separation between superconducting planes caused by the two insulating layers of BiO or TlO. However, TBCCO-1223 has only a single TlO layer, and as such is more three-dimensional, with the result that flux lines do not move as easily at 77 K, and J_c values are much higher.[31]

Confirming data for a variant upon the thallium-1223 compound comes from Liu *et al.*,[48] who investigated lead-substituted TSCCO of the form $(Tl_{0.5}Pb_{0.5})Sr_2Ca_2Cu_3O_9$ and derived a J_c value of 124,000 A/cm^2 at 1 T and 77 K. Although their result comes from magnetic hysteresis data (not from direct current measurements), it is still supportive of the general behavior found in the single-TlO-layered compounds. Liu *et al.* observe that this performance is comparable to YBCO and is far superior to TBCCO-2223 at 77 K. They also measured specific heat near T_c and from those observations concluded that three-dimensional behavior is taking place. This in turn supports the assertion that the Josephson coupling between conducting CuO_2 planes is strong, which then provides far better flux pinning than occurs in a two-dimensional situation. Once again, the conclusion to be drawn is that flux pinning at 77 K requires three-dimensionality. It appears that the thallium-series is the most promising place to find that.

The thallium compounds seem to have a brighter future than BSCCO, based on the vortex glass model. As mentioned in Section 14.7, the spacing between planes in the unit cell[31] dramatically affects motion of flux lines. The difference is most clear in the case of TBCCO-1223 which performs better at high temperatures than TBCCO-2223 (and consequently BSCCO-2223). Interest in making wire from TBCCO has increased greatly.

14.10. SUMMARY

In this chapter, we have discussed the behavior of magnetic flux lines in HTSCs. The notion of an irreversibility line was introduced and then linked to the pinning strength U_0, which is the depth of the potential well of the pinning sites. Because the coherence length is only angstroms in HTSCs, the individual flux vortices are much tinier than in conventional type II superconductors, and hence are much harder to pin. Typical pinning strengths U_0 in HTSCs are only 0.2 eV.

The thermally activated flux flow (TAFF) model was carried over from LTSC. This model holds that flux flow depends upon $\exp(-U_0/k_BT)$. However, at 77 K the factor $U_0/k_BT \approx 30$ for YBCO, compared to 3000 for NbTi at 4.2 K. The factor $\exp(-U_0/k_BT)$ in the E vs. J relation drastically changes the behavior of HTSCs compared to LTSCs. The concepts of flux creep, flux flow, and flux lattice melting all come into play. In the TAFF model, lines of magnetic flux would stay pinned until thermal energy became great enough to allow them to hop over barriers. From this came the idea that the way to achieve high currents at 77 K was simply to construct deeper and deeper pinning wells, so that very few flux lines would have enough thermal energy to hop high barriers. Seen in this way, the pinning problem was one of engineering deeper wells through specifically tailored lattice defects or other means.

Unfortunately, reality was different. The TAFF model has several deficiencies, most of them related to the pinning strength U_0. Its tilted staircase picture is tantamount to using $U_0 = (1 - J/J_c)^{1.0}$, and experimental data for HTSCs refutes that. Moreover, the anisotropy of the HTSCs causes a number of anomalies not seen in LTSCs. The vortex lines are no longer miniature telephone poles, but are more like stacks of pancakes. A crossover between two-dimensional and three-dimensional behavior occurs at an intermediate temperature.

All this is highly relevant to the matter of carrying high currents in substantial magnetic fields at 77 K. The key question is whether flux line motion is an intrinsic or extrinsic property of a HTSC material. If it is intrinsic, then nothing can be done to avoid sacrificing current density in finite magnetic fields; if it is extrinsic, then clever engineering involving the creation of microscopic pinning sites will eventually achieve a high-amperage conductor at 77 K.

Early data was not sensitive enough to resolve the choice between TAFF and competing models. When such experiments were performed, the picture clarified considerably, to the detriment of the TAFF model. Our best understanding now is that at high temperatures (or magnetic fields!) the flux lattice melts, but at a lower temperature T_g it freezes into a vortex glass state which dissipates no energy, and true superconductivity occurs. This resolves the intrinsic vs. extrinsic controversy: flux lattice melting is an intrinsic property of the material. Deeper pinning wells (higher U_0) cannot change this.

In YBCO, the best available data indicates that the vortex glass temperature T_g ranges downward from 90 K (in zero field) to about 70 K (in a field of 8 T). In particular, $T_g = 77$ K when $H = 4$ T, which implies that YBCO cooled by liquid nitrogen will not perform well in fields above 4 T. That is a major restriction upon the use of YBCO.

For BSCCO, the situation is much worse: flux lattice melting takes place around 30 K. There is considerable effort directed toward flux-pinning enhancement; for example, irradiating BSCCO samples with high-energy ions strives to create columns of damage into which flux vortices will fall and remain firmly trapped. It is reasonable to expect this approach to find some modest success in the 30 K range, but not to extend the operating range of BSCCO out to 77 K. Only when the magnetic field is very small (as in transmission lines) can BSCCO be considered a candidate for 77 K applications.

TBCCO-1223 appears more promising than BSCCO because it has relatively thin spacing between the pairs of conducting CuO_2 planes. However, it lacks the easy-slip property (micaceousness) of BSCCO and is expected to be much harder to make into wire. Hope is strong that one of the newest discoveries, such as $HgBa_2CuO_4$, will have an optimum combination of qualities.

By no means is flux pinning completely understood. Between the Josephson junction model, the vortex glass model, and the Bose glass model, there is still plenty of room for controversy. It is entirely possible that some new model will someday replace them all, just as the TAFF model was replaced.

REFERENCES

1. K. A. Muller, M. Takashige, and J. G. Bednorz, *Phys. Rev. Lett.* **58**, 1143 (1987).
2. U. Welp *et al.*, *Appl. Phys. Lett.* **63**, 693 (1993).

3.   A. A. Abrikosov *et al., Methods of Quantum Field Theory in Statistical Mechanics* (Prentice-Hall, Englewood Cliffs, NJ: 1963).

4.   G. J. Dolan *et al., Phys. Rev. Lett.* **62**, 2184 (1989).

5.   M. Tinkham, *Introduction to Superconductivity* (Kreiger Publ. Co., Malabar, Florida: 1980).

6.   M. Beasley *et al., Phys. Rev.* **181**, 682 (1969).

7.   C. W. Hagen and R. Griessen, *Phys. Rev. Lett.* **62**, 2857 (1989).

8.   J. C. Slater and N. H. Frank, *Electromagnetism* (McGraw-Hill, New York: 1947).

9.   M. Wilson, *Superconducting Magnets* (Oxford University Press: 1983).

10.   J. Lohle *et al., J. Applied Phys.* **72**, 1030 (1992).

11.   M. Suenaga, private communication.

12.   P. W. Anderson and Y. B. Kim, *Rev. Mod. Phys.* **36**, 39 (1964).

13.   Y. Yeshuran and A. P. Malozemoff, *Phys. Rev. Lett.* **60**, 2202 (1988).

14.   M. E. McHenry, M. P. Maley *et al., Phys. Rev. B* **39**, 4784, 7339 (1989).

15.   A. D. C. Grassie, *The Superconducting State*, pp. 94–96 (Sussex Univ. Press: 1975).

16. $\Omega_0 = 10^{11}$ Hz, $L = 10^{-8}$ m, $V_c = 10^{-24}$ m^3.

17.   M. Tinkham, *Phys. Rev. Lett.* **61**, 1658 (1988).

18.   M. P. A. Fisher, *Phys. Rev. Lett.* **62**, 1415 (1989).

19.   D. Nelson and S. Seung, *Phys. Rev. B* **39**, 9153 (1989).

20.   K. Shibutani *et al., Appl. Phys. Lett.* **63**, 3513 (1993).

21.   D. E. Farrell *et al., Phys. Rev. Lett.* **67**, 1165 (1991).

22.   M. Suenaga *et al., Phys. Rev. Lett.* **66**, 1777 (1991).

23.   A. Houghton *et al., Phys. Rev. B* **40**, 6763 (1989).

24.   R. Griessen, *Phys. Rev. Lett.* **64**, 1674 (1990).

25.   D. S. Fisher *et. al., Phys. Rev. B* **43**, 130 (1991).

26.   R. H. Koch *et al., Phys. Rev. Lett.* **63**, 1511 (1989).

27.   D. H. Kim *et al., Phys. Rev. B* **41**, 11642 (1990).

28.   D. H. Kim *et al., Phys. Rev. B* **42**, 6249 (1990).

29.   K. E. Gray *et al., Physica C* **174**, 340 (1991).

30.   T. T. M. Palstra *et al., Phys. Rev. Lett.* **61**, 1662 (1988).

31.   D. H. Kim *et al., Physica C* **177**, 431 (1991).

32.   J. E. Tkaczyk, Paper presented at MRS Meeting, Boston, MA (December 1991).

33.   T. Doi *et al., Physica C* **183**, 67 (1991).

34.   Q. Li *et al., Phys. Rev. Lett.* **69**, 2713 (1992).

35.   E. M. Chudnovsky, *Phys. Rev. Lett.* **65**, 3060 (1990).

36.   P. H. Kes and C. J. van der Beek, in *Proc. ICMC-90 Conf. on HTSC Materials*, Garmisch-Partenkirchen, Germany, May 1990.

37.   W. K. Kwok *et al., Phys. Rev. Lett.* **67**, 390 (1991).

38.   J. W. Ekin *et al., Appl. Phys. Lett.* **59**, 360 (1991).

39.   L. Civale *et al., Phys. Rev. Lett.* **67**, 648 (1991).

40.   P. L. Gammel *et al., Phys. Rev. Lett.* **66**, 953 (1991).

41.   D. J. Bishop *et al., Science* **255**, 165 (1992).

42.   R. G. Beck *et al., Phys. Rev. Lett.* **68**, 1594 (1992).

43.   W. K. Kwok *et al., Phys. Rev. Lett.* **69**, 3370 (1992).

44.   D. R. Nelson and V. M. Vinokur, *Phys. Rev. Lett.* **68**, 2398 (1992).

45.   R. Sasik and D. Stroud, *Phys. Rev. Lett.* **72**, 2462 (1994).

46.   H. Safar *et al., Phys. Rev. Lett.* **70**, 3800 (1993).

47.   H. Safar *et al., Phys. Rev. Lett.* **68**, 2672 (1992).

48.   R. S. Liu *et al., Appl. Phys. Lett.* **60**, 1019 (1992).

15

Processing Methods

The HTSCs are very complex materials to begin with, and when partial substitution of particular elements is added they become even more of a challenge to comprehend. To say it is an art form to make good-quality HTSC powder is an appropriate term: HTSC powder of any specified composition can be made by many different routes, and the characteristics of the powder often reflect the preparation techniques. Contaminants (NO_3, CO_3, etc.), size distribution, porosity, and other factors act as wild cards in the process, and can lead to wide variations in properties. In the early stages of HTSC research, a considerable amount of empiricism was required. With the evolution of more thorough and precise preparation techniques, good-quality HTSC powders can now be made routinely.

This chapter explains the basic features common to all processing techniques, and describes how researchers go about making HTSC materials. In Chapter 16 we will discuss making wire, another art form closely allied with that of making the material itself.

It is both impossible and unwise to attempt to describe the one correct way to make any of the HTSCs. The state of the art changes too rapidly to pin down the latest news, and even the compounds of greatest interest keep changing. Therefore, this chapter is instead devoted to introducing the reader to the *techniques* used, and to explaining the types of problems encountered by researchers in the field.

As we saw in Chapter 9, at equilibrium the components of a HTSC chemically combine to form certain specific phases that depend on the starting composition and temperature, but not on the pathway taken by the reaction. However, in reality the pathway matters very much: the *kinetics* of chemical reactions strongly influence the outcome of a process. For the copper oxide ceramics, when the component materials are elevated to temperatures around 1000°C, typical reaction times are measured in minutes or hours, and so the kinetics can be manipulated to advantage. Doing so leads to a final product which may not be in equilibrium at all, but which has optimum superconducting properties. Usually this means the ability to carry high current in a substantial magnetic field.

We begin by reviewing standard production methods and then examine how nonequilibrium chemistry demands a more sophisticated approach. Two key experimental techniques are explained en route, and these are then employed in an example of HTSC production being monitored in real time. The three major HTSCs each have different preferred production methods: the two-powder process for BSCCO, melt-texturing for YBCO, and a nonstoichiometric method for TBCCO. Additional means of enhancing the superconducting properties are briefly noted. Perhaps most important, the door to future discoveries is opened a little wider by the experience gained with the copper oxides.

291

15.1. KINETICS AND THERMODYNAMICS

The most important thought to bring forward from Chapter 9 is that the pathways to producing good HTSCs are very intricate and depend on deviations from equilibrium chemistry. Anybody can make "plain old" YBCO, but as soon as certain desirable properties are sought (especially, high current-carrying capacity), a major step upward in sophistication is required. To show why this is so, first we review the simple production methods which adequately served ceramists for several centuries.

15.1.1. The Standard Process: Calcining

The easiest path to making HTSCs is colloquially known as the "shake and bake" method. The proper stoichiometric ratios of all component are mixed together and heated in a furnace until the superconducting phase forms. For example, this pathway to YBCO-123 starts with a half part Y_2O_3, two parts $BaCO_3$, and three parts CuO, which is known as the *stoichiometric composition*. Ideally, on a phase diagram such as Figure 9.18, the ratio of Y:Ba:Cu is correct in the first place and does not change as the mixture is heated. The trick is to keep the temperature low enough so that YBCO-123 forms, but other phases do not. If the temperature gets too high, YBCO-211 is formed instead, and that material is nonsuperconducting.

This "easy" path generally produces material with very poor superconducting properties. That is so because the *primary phase field* for YBCO-123 begins at a different composition, as discussed in Chapter 9. The stoichiometric composition leads instead to the production of YBCO-211. If any liquid phase forms, segregation results. Attempting to produce the YBCO-123 phase by coming up from lower temperatures is virtually guaranteed to produce a sample with many undesirable phases tagging along. Exactly that was experienced during 1987–1989; the YBCO produced in this way suffered from very impure grain boundaries, which caused a severe weak-link effect and led to unsatisfactory J_c values.

Perhaps the single most troublesome impurity is carbon, which collects at the grain boundaries and kills J_c. In making YBCO, the precursor chemicals are heated (termed *calcining*) to drive the reaction to form YBCO-123. One of the precursor chemicals is barium carbonate ($BaCO_3$). Failing to remove all the carbon (in the form of CO_2) leaves impurity phases in YBCO. The first step upward in sophistication is to remove the carbon without hurting the other properties of the material.

Through a lengthy experimental program,[1] Argonne found that the best method of calcining is to hold the YBCO at 890°C for 24 hr in a vacuum, repeating this process twice. By exhausting CO_2 gas from the furnace chamber, most of the impurity phases that troubled earlier samples of YBCO could be eliminated.

Once good phase-pure powder is in hand, it is possible to obtain a good bulk sample of YBCO via a sintering step, followed by annealing. Sintering is an art that balances the amount of oxygen in the furnace atmosphere with the time and temperature. Densification takes place during sintering, and the final volume shrinks a cumulative 15–25%. Separately, thermal expansion causes volume changes, and care must be taken to prevent cracking.

For BSCCO, the same principles hold, but the optimum calcining temperature is around 850°C. For BSCCO, the existence of two different superconducting phases complicates things: the dominant phase depends on the sintering profile. Furthermore, J_c can be increased by additional heat treatment, as described below in Section 15.4.

The mercury-based superconductor $HgBa_2CuO_4$ starts with powders of HgO and Ba_2CuO_3 in the stoichiometric ratio, which are then placed in an evacuated quartz tube and heated for 8 hr at 845°C. A small amount of elemental mercury remains in the quartz tube afterward.[2] The two-layer versions of this compound (containing calcium and known as Hg-2212, etc.) are prepared by variations of this process. Sophisticated production scenarios of the type employed for YBCO and BSCCO have not yet been developed for HgBaCaCuO.

The thallium compounds are prepared by a different route. In order to delay introducing the volatile and toxic thallium, the nonthallium elements are first combined (at 900°C) to form *precursor oxides* which are subsequently reacted with thallium oxide to obtain the superconducting phase. TBCCO requires sintering in an overpressure of oxygen, to prevent the formation of a low-melting-point thallium oxide and excessive volatilization of thallium.

The final step is annealing, which takes place at a much lower temperature. The purpose of annealing the ceramic superconductors is to restore lattice oxygen which had been lost during the elevated-temperature sintering. For YBCO, we saw in Chapter 10 that the oxygen content must be carefully tuned to be slightly below 7. To bring the sintered material to this state, annealing at 450°C for 2–10 hr in air is about right, although some very large samples (diameter >1/2 in.) require 100 hr. For TBCCO, annealing at 750°C is better. For BSCCO, annealing may be unnecessary because little oxygen is lost in the sintering step.

Despite the employment of many clever techniques in preparing bulk HTSCs, the J_c values obtained (typically $< 10^5$ A/cm^2) are still far below those of thin films (typically 10^7), which are deposited by such techniques as laser ablation, chemical vapor deposition (CVD), and others. Without going into those processes, it suffices to note that the material thus formed is quite free of impurities, has good grain alignment, and has clean grain boundaries—all of which are important to minimizing weak-link effects. The values of J_c obtained in thin films sets a benchmark toward which bulk samples are pushed.

15.1.2. Effects of Slow Reaction Rates[†]

The study of the *rate* at which chemical reactions proceed is the field of *kinetics*. In Chapter 9, the idea that *kinetics* are important was mentioned but not developed; that chapter stressed *equilibrium*. For ceramics in general, solid-state reactions are slow, so kinetics can make a difference. For the HTSCs, the exploitation of kinetics has achieved a number of advantages which would never occur in equilibrium.

One classic example of nonequilibrium is the formation of a solid solution. Returning to Figure 9.7, the cooling along either vertical line L_1 or L_2 could be used to illustrate this situation. Figure 15.1 is a close-up of the right-hand side of that drawing and emphasizes what happens during cooling of a liquid of composition X_2 to form a single phase (β) over a fairly wide temperature range.

On cooling from above T_1, the first solid to form has composition β_1. On further cooling, *at equilibrium*, the compositions of the liquid *and the solid* adjust incrementally so that both are homogeneous throughout. That is, the solid particles that first formed are not only growing but also *continuously changing in composition*, as more A diffuses from the liquid into the solid particles to raise their A content uniformly to the solidus value for any particular temperature.

[†]This section was contributed primarily by Mark De Guire of Case Western Reserve University.

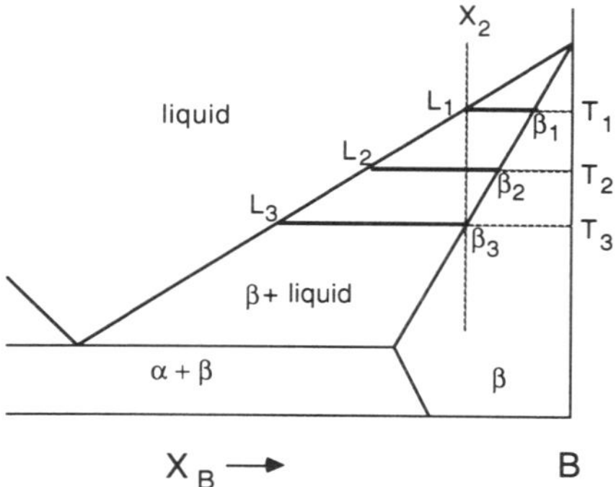

Figure 15.1. Close-up of the β-rich region of a two-component phase diagram. As a liquid of starting composition X_2 cools, phase $β_1$ first begins to precipitate out at temperature T_1, and this material is very rich in component B. Further cooling produces additional β-phase material of compositions $β_2$ and $β_3$, while the remaining liquid grows richer in component A. (Drawing courtesy of M. R. DeGuirre)

This is illustrated by the upper tier of drawings in Figure 15.2. The composition of the liquid is also continuously changing, following the liquidus curve from L_1 to L_2 to L_3. Because compositional homogenization of the β particles involves solid-state diffusion, very slow cooling is required to achieve it.

Under the faster cooling that is usually experienced, local equilibrium at the liquid–solid interface is maintained, but global equilibrium throughout the particle is not. The situation is illustrated by the lower tier in Figure 15.2: as the β particles grow, the new layers of β have the composition dictated by the phase diagram; but because they constitute a low-diffusivity layer encasing the first-formed β, their very presence makes it more difficult for the β underneath to equilibrate with the liquid and achieve the new equilibrium composition. The resulting β particles develop composition gradients from inside to out.

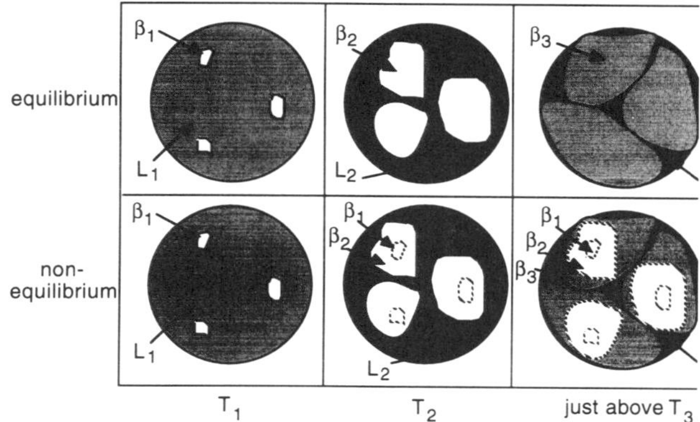

Figure 15.2. The solidification process corresponding to Figure 15.1. Under equilibrium conditions (top tier), the composition of the solid particles changes from $β_1$ to $β_2$ to $β_3$ as cooling proceeds and more and more solid forms. This requires solid-state diffusion within the solidified grains, and is extremely slow. Under realistic conditions (bottom tier), the process is not truly in equilibrium, and successive outer layers of solidified β-phase have different compositions. (Drawing courtesy of M. R. DeGuirre)

A second classic case of nonequilibrium is similar to this one, except it involves incongruently melting compounds. Refer back to Figure 9.9 and imagine cooling a liquid of composition 66.6% A, 33.3% B through the *peritectic* reaction

$$A + L \rightarrow A_2B$$

The first A_2B forms at the interface between the existing A particles and the liquid. Equilibrium would require all of the solid phase to be used up in reacting with the liquid to form A_2B. However, it often happens that before the A particles are completely consumed, they become isolated from further reaction with the liquid by a coating of the solid reaction product A_2B. The reaction at the liquid–solid interface proceeds as if the remaining A were not there, locked out of the action by the A_2B layer. Whereas at equilibrium the solidification would have stopped at the peritectic temperature with the formation of 100% A_2B, under the nonequilibrium situation A persists metastably inside its A_2B crust. Furthermore, liquid is still present at temperatures below the peritectic. The reaction at the interface will continue along the $\{A_2B + L/L\}$ liquidus curve right down to the eutectic, ending with the eutectic reaction to form solid B and A_2B. Instead of single-phase A_2B, the final mixture is A_2B, A, and B—three solid phases, in violation of the phase rule.[†]

Applying this description to HTSC, if we replace A_2B with YBCO-123, A with 211, and B with $BaCuO_2$ + CuO, we see what can happen during solidification of the YBCO-123 composition. In fact, the melt-quench–melt-growth process exploits this phenomenon in a couple of ways, thereby making lemonade out of lemons. The retained 211 particles seem to enhance oxygen diffusion into the interior of the 123 grains, and improve flux pinning as well.

One problem that cannot be ignored is the nonuniform mixing of starting powders. On the crudest level, if the powders are not homogeneously mixed, the local composition at any point might not be representative of the overall composition, and the final local phase assemblage will be dictated by the local starting composition rather than the global average composition. Very long times would be required for solid-state diffusion to bring the resulting heterogeneities to equilibrium.

If the local composition is near the composition of a eutectic, liquid may form at that eutectic temperature. This can be hundreds of degrees below the temperature at which liquid would first form for the overall composition. With regard to achieving equilibrium, liquid formation is a two-edged sword: further reactions may occur more quickly in the presence of a liquid, hastening the homogenization of the composition; or severe *in*homogenieties could result, if the liquid drains through the powder before the reactions are complete; e.g., if it has high fluidity and the reactions with the solid phases are slow.

Once again, the phase diagrams of Chapter 9 provide insight. Re-examining Figure 9.17 for the YBCO system, we see that there is a low-melting eutectic (900°C) along the

---

[†]A nonequilibrium system is a thermodynamic outlaw; that is, the phases present will not possess the minimum free energy for the given composition and conditions; and there may be more phases present than the phase rule allows. However, no system (at equilibrium or not) will ever spontaneously increase its free energy. Even at nonequilibrium, a system always tends toward lower free energy, even though it may be kinetically prevented (as in the present case) from achieving the configuration with the lowest free energy. Call it a thermodynamic Code of the West.

BaCuO$_2$-CuO edge. In the BSCCO system, pictured in Figures 9.20 and 9.21, low melting points dominate the Bi-Cu edge of the tetrahedron, with the lowest believed to be the eutectics in the Bi$_2$O$_3$-CuO binary; adding Pb introduces several more low-temperature eutectics. Obviously, good mixing of the starting materials is imperative to attaining homogeneous superconductors in these systems.

15.2. MEASUREMENT OF PROCESSED MATERIALS

One of the questions that arises in the course of employing these complex processes is "How do we know what we have?" After all, there are so many starting components, and so many intermediate way stations en route to the superconducting phases, there is plenty of room for error.

There are two principal experimental techniques used to investigate these materials: x-ray diffraction (XRD), and differential thermal analysis (DTA).

15.2.1. X-Ray Diffraction

The wavelength of X rays is a few angstroms, which is about equal to the distance between atoms in a crystal lattice. Consequently, X rays scatter very well off the arrayed atoms in a solid structure. In fact, from the point of view of a beam of incoming X rays, the atoms aligned in their repeating patterns of a crystal lattice appear to be a diffraction grating, and so the outgoing beam of X rays form a diffraction pattern.

In optics, when light enters an interferometer the pattern detected at the output (the *interferogram*) is related to the spectrum of the incoming light through a Fourier transform. Therefore it is possible to recover the spectrum computationally from a measured interferogram. This is the principle on which most infrared spectrometers work today. It is called Fourier spectroscopy.

In an analogous way, a diffraction pattern is related to the Fourier transform of the incoming light and grating that produced it. If the incoming beam is monochromatic, then the details of the grating can be recovered by an inverse Fourier transform from the measured diffraction pattern. This technique is used with X rays to recover information about the crystal structure: starting from the measured x-ray diffraction (XRD) pattern, in principle, a Fourier transform can recover the crystal structure. In practice, crystallographers have a general idea of the structure beforehand, and calculate various XRD patterns until they match the experimental data.

Just as in an optical diffraction pattern, X rays interfere destructively except at points where special conditions are met.[3] Imagine the crystal lattice as a series of planes spaced by d apart, and let the incoming X rays have wavelength λ. The scattering off one plane will only interfere constructively with that from adjacent planes at those special angles θ that meet the *Bragg condition*:

$$2d \sin \theta = n\lambda$$

This geometry is illustrated in Figure 15.3. A crystal contains parallel planes along three different axes, so the actual measured values of θ vary with the alignment between the crystal planes and the incoming beam of X rays. A plane perpendicular to the crystallographic a-axis is called the [100] plane; one perpendicular to the b-axis is the [010] plane; and so forth. For

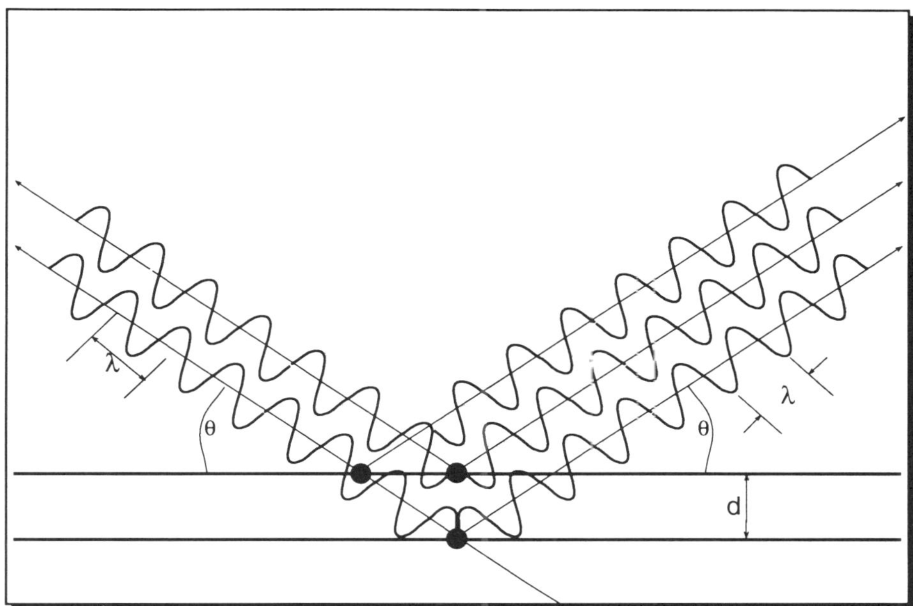

Figure 15.3. Sketch of interfering waves to illustrate the Bragg condition. For scattering of X rays of wavelength λ off planes separated by distance d, the outgoing wave will interfere constructively only at selected angles θ that satisfy the Bragg condition.

complicated crystals like the HTSCs, there are many choices of planes that can contribute to the x-ray diffraction pattern.

Figure 15.4 is a standard x-ray diffraction pattern. The highest peaks are due to the strongest x-ray scatterers. The scanning motion goes through consecutive angles from 0 to $+90°$ as it passes over the pattern, so the independent variable is the angle. The angle at which each peak occurs is determined by the Bragg condition. The custom of plotting 2θ is related to the definition of reciprocal lattice vectors.

An unknown sample is analyzed as follows: First, chemical analysis is used to determine what elements are present in order to limit the number of possibilities. Then, knowing ahead of time that certain compounds form distinct crystal types, whenever the characteristic diffraction pattern of that crystal appears it is a good bet that the corresponding compound is present. If several different crystals are present within the same sample, the total diffraction pattern will be the superposition of each one's pattern; this can get complicated in a hurry.

When there is no particular symmetry to the material under study, the diffraction pattern will also lack structure. A disordered material like glass produces such patterns. It is very easy to tell glass apart from a crystal lattice via x-ray diffraction, and this is quite helpful to understanding the HTSCs.

In a bulk sample, there may be many grains, all of different alignments, and thus it is impossible to be sure that the incoming beam is aligned with one particular axis. The pattern emerging will then have peaks corresponding to the {100} crystallographic direction, the {110} direction, {111}, {200}, {220}, {311}, {420}, and so on. Even relatively simple substances might therefore show complicated x-ray patterns.

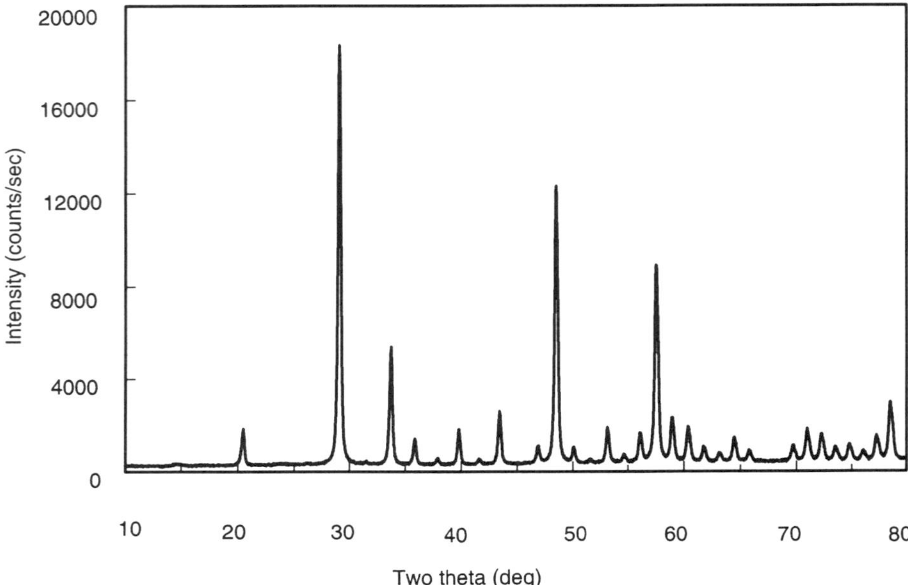

Figure 15.4. Typical x-ray diffraction pattern for a two-element crystal, Y_2O_3, showing different intensities produced by different reflecting planes and atoms. (From Ref. 7.)

The XRD pattern of the 94 K superconductor $HgBa_2CuO_{4+\delta}$ is presented[4] in Figure 15.5, with the peaks labelled according to their corresponding crystalline directions. This labelling activity is called indexing, and is a very important step in utilizing XRDs to identify unknown compounds. In principle, knowing the crystal structure allows one to calculate how the various planes will reflect X rays, so it is not hard to select the proper label for each peak. In practice, the experimental XRD from a pure single crystal is used to help determine the crystal structure. When you can account for all peaks in the XRD, you know you have the crystal structure right.

Once a pure substance has been indexed, the information is catalogued and put on-line, where computer programs can call up its particular pattern to compare with experimental data from an unknown sample. Indexing has been done for an extremely large number of compounds, so that almost any crystalline material can be identified through XRD. Only fairly general supplementary knowledge is required; for instance, in dealing with the materials that make up YBCO, you know that TiO_2 is not present.

The HTSCs have x-ray patterns that are quite complicated, not only due to their several different planes of CuO_2, BaO, etc., but because of extraneous unwanted phases such as $BaCO_3$ that exist within the sample. The x-ray patterns of all these appear juxtaposed in the measurement and have to be sorted out. This is where indexing pays off, especially when aided by computer algorithms. Fortunately, YBCO-123 has a different pattern from YBCO-211 and other YBCO phases, so the skilled XRD analyst can use the data to determine the presence of several different phases. Figure 15.6 is from a YBCO sample containing several impurities[5]; each peak is labelled with the phase that produced it.

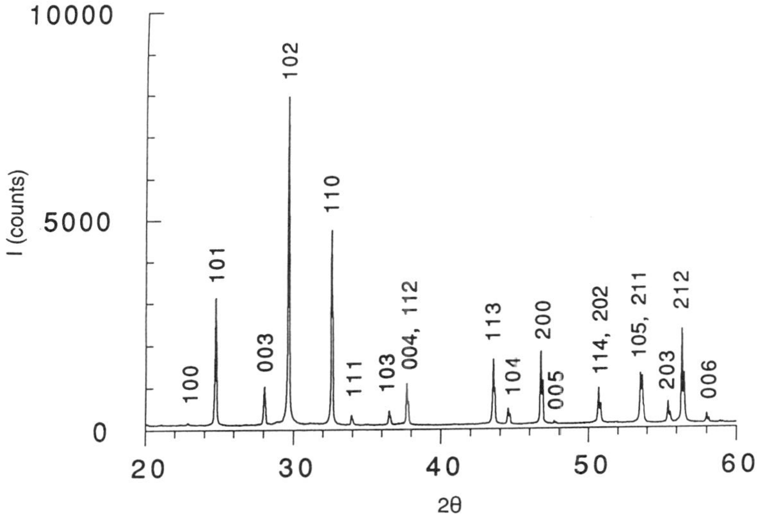

Figure 15.5.   XRD of the high-T_c compound $HgBa_2CuO_{4+\delta}$, with each peak labelled according to the crystallographic direction that produced it. (From Ref. 4.)

Recognizing the possibility of extraneous material being present, analysts usually make rather tentative interpretations of these patterns. Additional information from other sources is needed to positively identify exactly what a sample contains.

15.2.2. Differential Thermal Analysis

Differential thermal analysis (DTA) relies on the fact that heat is evolved or absorbed during a phase change. By observing temperature changes due to this transfer of heat, one can surmise that a phase change is taking place.

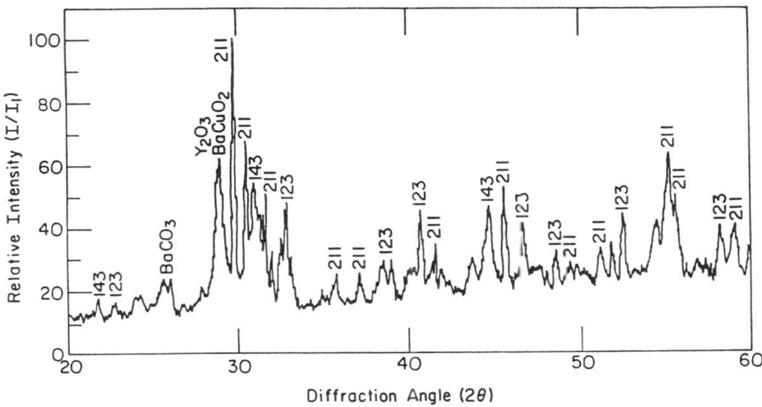

Figure 15.6.   XRD of YBCO, containing several phases including 123 and 211. Each peak is labelled according to the phase that produced it. (From Ref. 5.)

Imagine starting with an ultracold block of ice and heating it with a constant heat input; also, the temperature must be monitored throughout the process. As time passes, the temperature will increase. The rate of change of temperature will be constant for a while, but will suddenly drop to zero at the melting point because the temperature remains at 0°C until the entire heat of fusion is supplied, i.e., until the ice melts. Once that is done, the temperature will begin to rise again, and its rate of change will take on a new value as the liquid phase is heated. At the boiling point, once again the rate of temperature change will drop to zero for a while, and eventually it will jump back up to a high value as the gas phase is heated.

Now suppose the exact same heating process is applied to a fairly boring solid in which no phase changes take place, such as a block of copper. Across the same temperature range, there will be no deviations in the rate of change of temperature during the time of heating. If the difference between the copper temperature and the water temperature is recorded, it will undergo excursions at both the phase changes of water. A plot of that difference as a function of the reference temperature (in this case, the copper temperature) is a DTA plot. This simplest case of a DTA plot is shown in Figure 15.7: it consists of three separate constant plateaus separated by two notches.

Recalling the discussion of phase equilibrium in Chapter 9, extend this thinking to a two-component system, such as that pictured in Figures 9.7 and 9.8. In the (unlikely) event of having the composition exactly equal the eutectic composition, the rate of temperature change would drop to zero at the eutectic temperature until the whole system turned to liquid. But for any other composition, reaching that temperature would only be the start of liquid formation, which would continue across a range of temperatures until reaching the liquidus

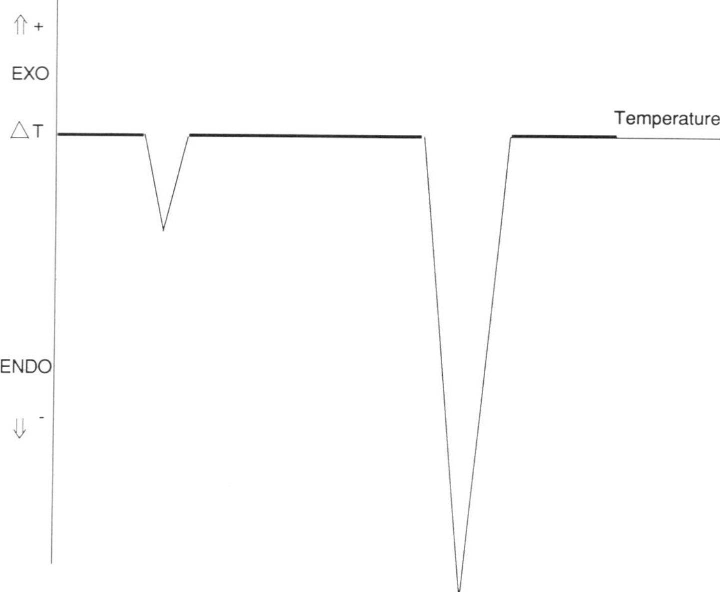

Figure 15.7. Extremely simple DTA for water, a one-component system, showing notches at both phase transitions.

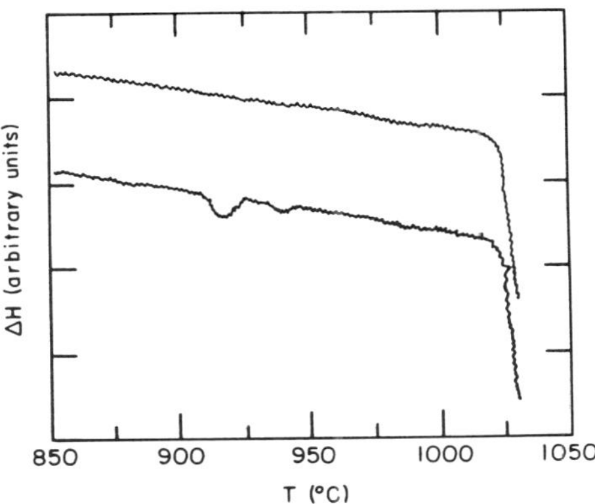

Figure 15.8.   DTA for YBCO. The upper line is for pure YBCO-123; the lower line (offset for better visibility) is YBCO containing impurities. In either case, the steep drop at the right is melting. The shallow endothermic notches around 920–940°C are associated with phases of $BaCuO_2$ and CuO. (From Ref. 6.)

line. Because liquid continuously forms as the temperature still rises, the rate of change of temperature cannot fall to zero. Rather it will take on a range of values as the mixture passes through its various combinations of phases. The details of the DTA plot are dictated by both the heat of fusion (for phase α or phase β) and the relative proportions of liquid and solid, which are in turn determined by the overall composition of the two-component system. Only when these factors are known can a DTA curve be forecast correctly.

In fact, for an unknown mixture a DTA plot can be used to derive information about the composition, the heat of fusion, and any phase transitions that occur in the system. Observing a drop in the DTA between 800°C and 900°C indicates that there must be a two-phase region there. The depth and breadth of the notch in the DTA can be converted into information about the shape of the liquidus and solidus lines.

When it comes to the HTSCs, a typical DTA will contain several such notches. Figure 15.8 shows both a pure and an impure sample of YBCO.[6] Ceramists interpret these graphs in the light of what they already know about certain phases, but often the DTA holds surprises about previously unsuspected phases or multiphase regions. As the number of components increases, so does the complexity of the phase diagram. Any data taken during an excursion through the phase diagram (which the DTA is) certainly may be equally complex.

15.2.3. Complementary Information

The XRD tells you what the particular phases are; the DTA tells you when and where (in temperature) a reaction is taking place, and whether it is exothermic or endothermic. To reach an understanding of the process of formation of the various HTSCs requires both kinds of data. Combining the information from x-ray diffraction with that from DTA, it is possible to determine in what temperature range the various phases form, and what their constitutents are.

This is how phase diagrams are produced. Any of the very complex figures in the back of Chapter 9 require many careful experiments and precise analysis of the XRDs and DTAs before the complete phase diagram is determined. Needless to say, as the number of components increases, so does the complexity. The partial substitution of lead in the bismuth and thallium compounds greatly enriches the DTAs and XRDs.

15.3. REAL TIME MONITORING

It is important to find out at what temperatures various phases form or disappear. The usual way to do this is to heat a mixture into its liquid state, cool it slowly to some intermediate elevated temperature where partial crystallization has taken place, and then suddenly quench the sample to room temperature. It is hoped this will capture the phase distribution of the material at the elevated temperature. The problem with this is twofold: some other changes might take place during quenching, and different samples must be used for each selected temperature.[7]

XRD measurements can be done at any temperature, including those encountered during formation of the HTSCs. This is called HTXRD (high-temperature XRD). By observing the x-ray patterns at many temperatures, it is possible to watch certain characteristic intensities rise and fall, corresponding to the formation or disappearance of certain phases in the mixture.

One specific example of this comes from the task of making Bi-2212. It is already known that a solid-state reaction is slow and gives poor phase purity. The discussion of Section 9.5, together with Figures 9.20 and 9.21, indicated how tricky it is to make the bismuth compounds. One approach is the glass-ceramic method, in which the mixture is first melted above 1000°C (to achieve good mixing), then quenched into a glass state. Subsequently the glass is reheated to a temperature at which devitrification takes place and surprisingly clean Bi-2212 forms. Along the way, extraneous phases such as CaO participate in the process, but ultimately vanish. The mechanism of all this is by no means obvious, but HTXRD has been used to unravel it.[8]

The DTA of the glass powder (having the stoichiometric composition of Bi-2212) appears in Figure 15.9. It indicates that Bi-2201 forms via an exothermic reaction at around 475°C, while the desired Bi-2212 itself forms above 800°C. Melting takes place above 900°C; indeed, there are two stages of melting. However, this figure doesn't really show the formation of Bi-2212; that had to be derived from the x-ray data.

The corresponding XRDs are combined in Figure 15.10. Certain peaks (solid dots) indicate Bi-2212, and these begin to rise above 800°C. Other peaks (open circles) correspond

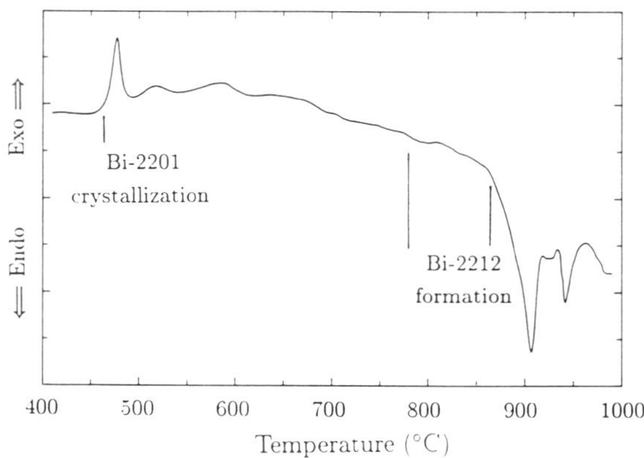

Figure 15.9.   DTA for the process leading to Bi-2212, involving devitrification from the glass . The phases formed were identified by comparison to HTXRD data. (From Ref. 8.)

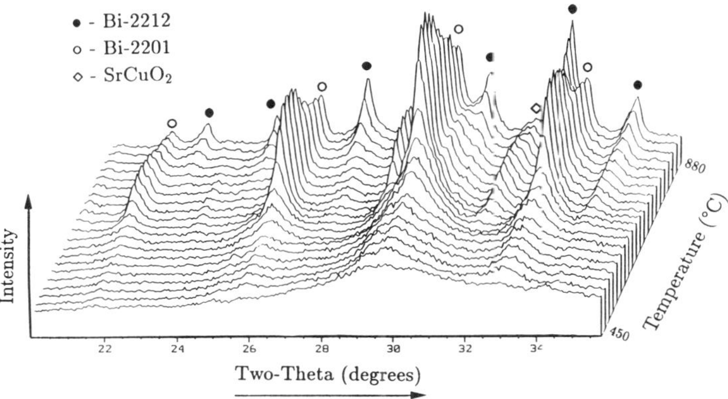

Figure 15.10. HTXRD profiles for the Bi-2212 system at consecutively higher temperatures between 450 and 880°C. The formation and disappearance of Bi-2201 is apparent; Bi-2212 begins to form above 800°C and is the dominant phase by 880°C. (From Ref. 8.)

to the calcium-free phase (Bi-2201), which first begins to form below 500°C, maximizes around 800°C, and then diminishes. What is happening above 800°C is that the Bi-2201 is transforming to Bi-2212 by picking up Ca and Cu. This is confirmed by noticing that the small peak ($2\theta \approx 32°$, open diamond) due to $SrCuO_2$ and Ca_2CuO_3 diminishes above 800°C. The significant point is this: Bi-2212 is *not* forming from the glassy phase; rather, Bi-2201 is reacting with other phases present to form Bi-2212. The reaction is slow, because it all takes place inside a glass and is therefore diffusion-limited.

The ability to discern reaction chemistry in solids is one of the great benefits of XRD, and doing it in real time as the sample heats up is particularly helpful. This same technique has been used to explain how Bi-2212 breaks down above the melting temperature.[8]

Still other investigators have used XRD and DTA to understand the way that lead enhances Bi-2223 formation.[9] The difficulty of forming pure Bi-2223 is well known, but a modest amount of lead enhances[10] the formation of (Bi,Pb)-2223. Figure 15.11 is a DTA for a glass devitrification process very similar to the one that produced Figure 15.9, but with composition $Bi_{1.84}Pb_{0.34}Sr_{1.91}Ca_{2.03}Cu_{3.06}O_y$ instead of the standard 2223 stoichiometry. The formation and decomposition of Ca_2PbO_4 are noted on the figure, because this compound plays a pivotal role. At about 800–820°C, Ca_2PbO_4 decomposes to CaO and PbO, which acts essentially as a liquid solder flux that transports CuO and CaO to the 2223 phase as it grows. Indeed, the 2223 phase forms just below 840°C, but at 842°C it starts to degrade into the 2201 phase plus other compounds. The stable range of temperatures is very narrow, which explains why Bi-2223 has been so difficult to make.

The point to be made here is that it is the combination of XRD and DTA that enable researchers to figure out the pathways to formation of the HTSCs.

15.4. BSCCO: THE TWO-POWDER PROCESS

Considering the narrow temperature range for formation of the BSCCO compounds, the tendency to melt or transform into other constituents, and the uncertain role of interme-

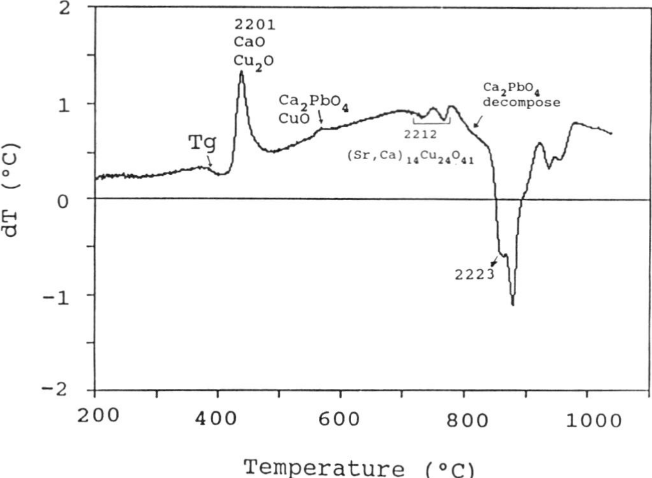

Figure 15.11. DTA of lead–BSCCO system, leading to the formation of 2223 phase at 840°C, enhanced by presence of Ca_2PbO_4 at intermediate temperatures. (T_g denotes the crystallization temperature.) (From Ref. 9.)

diate species, it is little wonder that phase-pure BSCCO is very elusive. The problem is well-stated by Dorris *et al.*[11]:

> Bismuth-2223 is usually prepared by combining all constituents in a single step and then calcining to produce a mixture of distinct phases. Although calcination conditions can be varied to produce a dominant phase, numerous phases can be present in the mixture, such as 2201, 2212, 2223, . . . [etc.] Lead-bearing compounds are also often present, because lead must be added to stabilize 2223. Processes involving such a large number of phases, any one of which might significantly affect the formation of 2223, are difficult to control on a microscopic level. As a result, 2223 produced by such a single-powder process often contains significant amounts of relatively large (5–10 μm) second phases. Also, formation of 2223 by such a process is slow, typically requiring heat-treatment times of several hundred hours to obtain materials with ≈ 10 vol. % nonsuperconducting phases.

The Argonne research team[11] has made BSCCO-2223 via a two-powder process, in which pure (Bi,Pb)SCCO-2212 is combined with $(Sr,Ca)CuO_2$ to reach a final state of pure (Bi,Pb)SCCO-2223. The group tried various fractional combinations of Sr and Ca (i.e., $Sr_{2-x}Ca_{1+x}$), but found that standard 2212 (= Sr_2Ca_1) gave the purest 2223.

Table 15.1 presents the steps in the two-powder process. The individual powders, here termed intermediate precursors, were prepared via conventional calcining. The bismuth/lead fractions of 1.8 and 0.4 were present in the 2212 phase and did not change during the final step. The two powders were blended into a mixture having the right stoichiometry (2223), but not yet chemically reacted, and then were put inside a silver tube. (This is a way of making wire, to be discussed further in Chapter 16.) Only after drawing and pressing was the final heat treatment carried out to reach the 2223 phase, and the proximity of the silver was relevant to that reaction.

The optimum choice of heat-treating temperature is guided by examining the DTAs of the intermediate precursors, as well as that of the final product, shown in Figure 15.12. No matter what choice is made for *x*—the Sr/Ca fraction—all the DTAs show approximately the same behavior, with melting occurring around 860°C. The optimum processing tempera- ture turns out to be ≈ 820°C, which corresponds to the onset of partial melting. At that

Table 15.1 Two-Powder Process

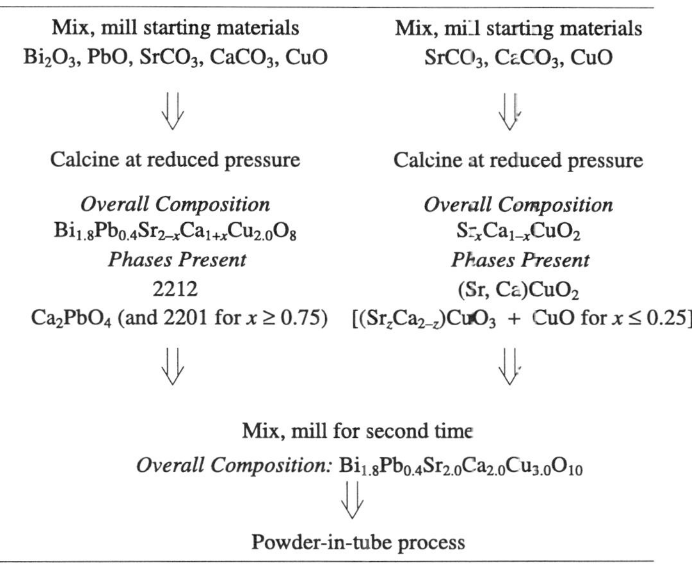

Mix, mill starting materials	Mix, mill starting materials
Bi_2O_3, PbO, $SrCO_3$, $CaCO_3$, CuO	$SrCO_3$, $CaCO_3$, CuO
⇓	⇓
Calcine at reduced pressure	Calcine at reduced pressure
Overall Composition	*Overall Composition*
$Bi_{1.8}Pb_{0.4}Sr_{2-x}Ca_{1+x}Cu_{2.0}O_8$	$Sr_xCa_{1-x}CuO_2$
Phases Present	*Phases Present*
2212	$(Sr, Ca)CuO_2$
Ca_2PbO_4 (and 2201 for $x \geq 0.75$)	$[(Sr_zCa_{2-z})CuO_3 + CuO$ for $x \leq 0.25]$
⇓	⇓

Mix, mill for second time

Overall Composition: $Bi_{1.8}Pb_{0.4}Sr_{2.0}Ca_{2.0}Cu_{3.0}O_{10}$

⇓

Powder-in-tube process

temperature, highly pure 2223 forms in only 50 hr, which is much faster than most production methods for 2223. At lower temperatures the kinetics are too slow, and at higher temperatures the 2223 decomposes, as mentioned above in Section 15.3.

Longer annealing times seems to increase J_c appreciably, particularly in the $x = 0$ case (starting powder 2212); after 350 hr, $J_c \approx 20,000$ A/cm^2 in zero field. Dorris *et al.* state[11]: "The high purity and good reproducibility obtained with this process might greatly facilitate the fabrication of long lengths of superconductor with high J_c."

What has been accomplished here is a means of producing relatively pure BSCCO-2223 in a reasonable time frame. Unfortunately, the magnetic behavior is no better than samples produced by slower means. BSCCO still suffers from flux-lattice melting, which makes J_c low at 77 K in finite magnetic fields.

15.5. MELT PROCESSING IN YBCO

One widely studied new processing scheme for the HTSCs has been melt processing and its variants. The primary reason for this is because it mitigates the weak-link problem at grain boundaries, and thus greatly increases J_c in materials that would otherwise be rather poor conductors. A second advantage is that it can improve flux pinning. While the technique has been used most heavily with YBCO, it has also been shown to improve the properties of BSCCO.

An excellent review of this entire field has been published by Salama *et al.*,[12] and the reader seeking more detail will find it there. Briefly, the foremost problem with YBCO is that the peritectic transformation, in which Y_2BaCuO_5 (211) reacts with the liquid to form $YBa_2Cu_3O_{7-x}$ (123), is terribly slow. The problem was mentioned in Chapter 9 and expanded in Section 15.1 above. This creates a demand for clever new processing techniques to circumvent the difficulties that this slowness brings.

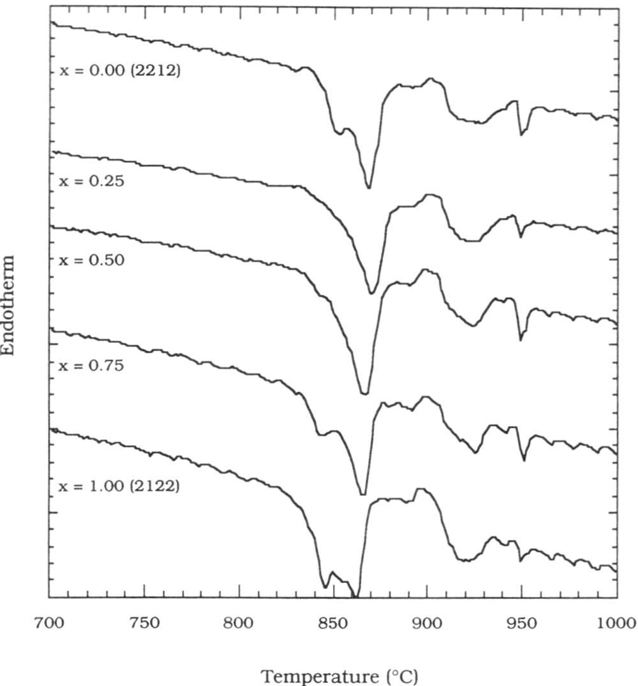

Figure 15.12. DTAs for mixtures made by combining $Bi_{1.8}Pb_{0.4}Sr_{2-x}Ca_{1+x}Cu_2O_8$ and $Sr_xCa_{1-x}CuO_2$ for like values of x.

15.5.1. Static Techniques

Figure 15.13 defines the three major variants of melt processing, which are distinguishable by their time–temperature profiles.

The earliest of these techniques, *melt-textured growth*, was used by Jin *et al.*[13] in 1988 to raise J_c in YBCO from ≈ 100 A/cm² to 17,000 A/cm². The starting material is sintered YBCO-123 with the proper stoichiometry. The process is to partially melt it and then slowly

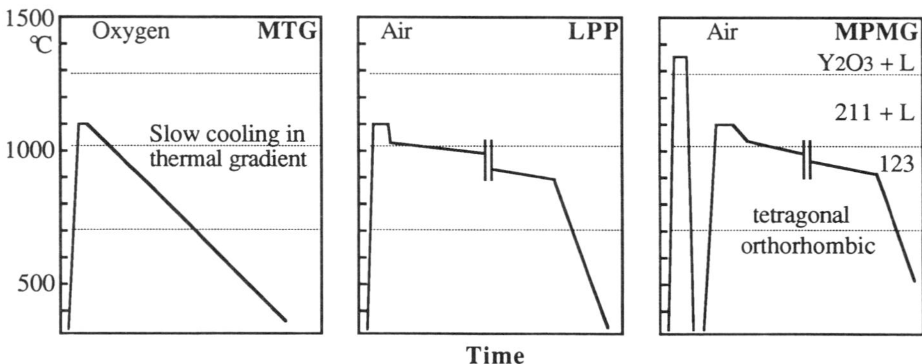

Figure 15.13. Three varieties of melt-processing for YBCO. The evolution of the thermal profiles from left to right led to better values of J_c. MTG = Melt-Textured Growth; LPP = Liquid-Phase Processing; MPMG = Melt Partial-Melt Growth.

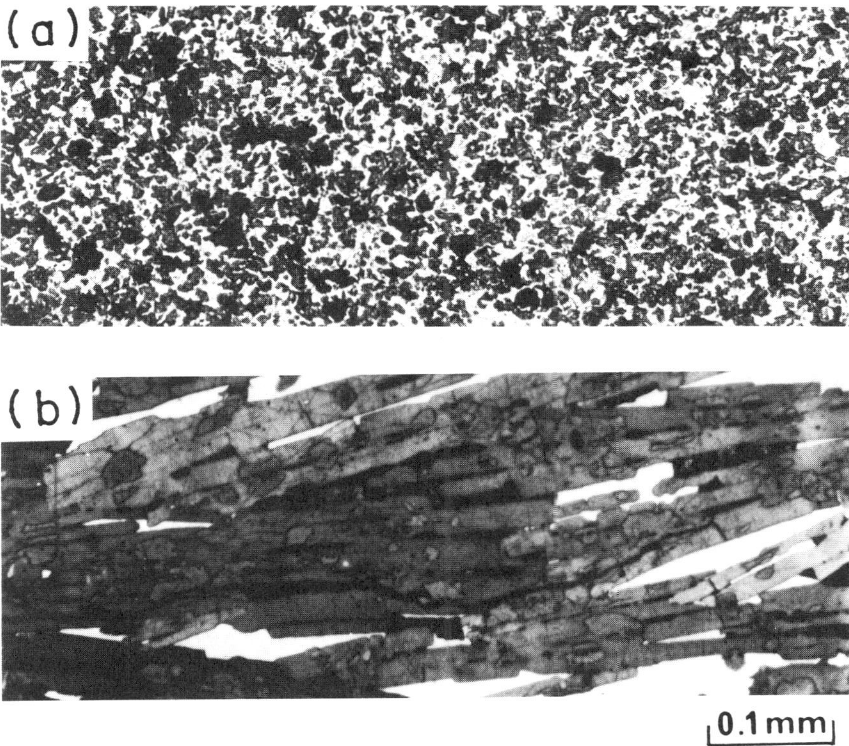

Figure 15.14. Examples of YBCO samples before and after melt texturing (a) A conventional sintered sample characterized by small-scale grains, dirty grain boundaries, and low current density. (b) The results of melt texturing: this sample now has long parallel grains, clean grain boundaries, and improvement of J_c by a factor of 40. (Photos courtesy of S. Jin, AT&T Bell Laboratories.)

cool it in a thermal gradient of 20–50 °C/cm. Figure 15.14 shows before-and-after photographs of this material. The outcome was a collection of well-aligned, needlelike grains in which the weak-link effect was diminished simply by having large contact area at grain boundaries. The advantage of this was described in Section 13.5.

An improvement upon this, shown in the center of Figure 15.13, was the *liquid-phase processing* method of Salama *et al.*[14] Here, the sample is held around 1100°C (above the peritectic temperature) for 10–20 min, which ensures that the YBCO-123 has all turned to YBCO-211 plus liquid. After sudden cooling to 1025°C, the material is ready to return through the peritectic transition, which begins at a slightly lower temperature. A long, slow cooling follows, at about 1–2 °C/hr, until $T = 925$°C is reached—in other words, 50–100 hr. This had the effect of producing large crystals of YBCO-123, in a stacked plate microstructure. Furthermore, J_c was even higher, around 68,000 A/cm^2.

The next step forward was the *melt partial-melt growth* (MPMG) method of Murakami *et al.*[15] at Nippon Steel. Here, the starting material is yttrium-rich YBCO; it is raised to a very high temperature, about 1450°C. The liquid is then quenched to room temperature, forming a precursor material containing finely dispersed particles of Y_2O_3. The main quenched material is mostly frozen phases of copper oxide and barium copper oxide. Next this precursor is ground into powder, reheated to 1100°C and held for 20 min, during which time the Y_2O_3 reacts with the copper and barium to form YBCO-211. The 211 phase remains as finely dispersed as was the Y_2O_3 in the precursor. Third, the material is cooled slowly through the peritectic transition, at which point the change to YBCO-123 occurs. Additionally, the final cooling to room temperature is done in flowing oxygen and a temperature gradient, thus promoting directional growth of the 123 phase.

The resulting 123 contained many finely dispersed remnants of 211 phase. The surprisingly favorable outcome was that this material conducted current well, with $J_c = 1500$ A/cm^2 at 1 T and 1000 A/cm^2 at 10 T.[†] The Nippon Steel group attributed this good performance to the well-connected path for supercurrents and the paucity of grain boundaries associated with large grains.

This is a good example of how serendipity advances research. Initially, Murakami *et al.* were simply trying to achieve good connectivity of the 123 phase. The presence of finely dispersed particles of 211 was not originally intended. Nevertheless, it turned out that J_c held up in higher magnetic fields, apparently due to the flux pinning action of the dispersed particles. Because of these experiments, many others began working to find ways to include pinning sites within the YBCO-123 grains. Today, this strategy is well established.

There are a number of other variations on melt processing. They all have in common a final oxygen anneal at about 450°C (not shown in Figure 15.13), which is required to make the product superconducting. Moreover, no melt processing technique gets around the limitation that the peritectic reaction is slow. This means that the cooling rate must also be slow, which in turn severely limits the applicability of this technique to wiremaking.

15.5.2. Long Lengths

Efforts directed toward making long lengths of melt-processed YBCO have had limited success. The general idea is to move a sample continuously through a furnace, in which the temperature profile assures that each point along the moving sample sequentially experiences the temperature cycle of melt-processing. Using a horizontal configuration, the sample has to be contained when in the liquid state. In a vertical-tube furnace, zone-melting is possible without losing the sample.[16] However, the requirement for very slow cooling through the peritectic transition remains, so the sample must move through the temperature profile slowly, and the production rate is only a few millimeters per hour.

A separate problem is that because of anisotropy, any misalignment of grains will cause J_c to fall off, as described in Chapter 13. This translates into a requirement for good alignment of the *a-b* planes over the full sample length. Should the *a-b* planes not be aligned straight down the wire, the transport current will flow at an angle to those planes, resulting in lower J_c. To make matters worse, for a sample moving through a furnace profile, temperature gradients outward from the cooling sample tend to promote directional growth with the *a-b* planes, *not* along the wire axis. The perversity of mother nature strikes again.

In the *partial melt–growth* process, Shi *et al.*[17] produced multicrystalline bars of YBCO up to 50 mm long, with well-aligned grains and very little other phases in the material. Like other melt-growth processes, their method begins by heating the sample into the melting range (1150°C) for 10 min, then quenching it to 1050°C, so that only a partial melt takes place (a liquid–solid mixture occurs). The next step involves slow cooling over a long period: a thermal gradient of 2°C/mm is used to drop from 1050°C to 800°C by gradually withdrawing the bar from the furnace hot zone over a period of 10 days. A later oxygen annealing step at 450°C requires 5 more days. There are several novel features to this

---

[†]Determined via transport measurements.

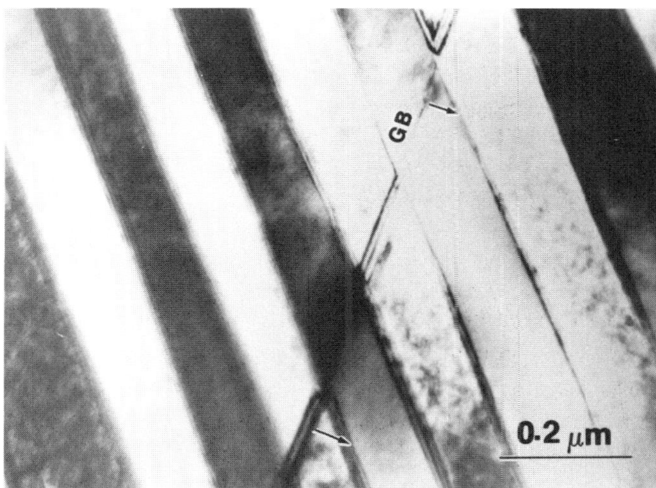

Figure 15.15. TEM photograph showing the small-angle grain boundary in a sample processed by partial melt growth. Despite crossing a grain boundary, there is no interruption in the twinning pattern. (From Ref. 17.)

approach. First, the furnace is vertical (i.e., the temperature gradient is axially aligned with gravity). Second, the bars of YBCO are coated with silver. It seems that the vertical furnace promotes good alignment of the adjacent grains, and the silver (which melts at 961°C) enhances the partial-melting process and lessens the formation of YBCO-211 phase. Figure 15.15 shows two adjacent grains that are so well aligned that the twin boundaries run smoothly through the grain boundary. Figure 15.16 compares YBCO samples prepared by the old zone-melting process (in which a small region is completely liquified) and by this process. The absence of undesirable phases is evident.

This process is not a practical manufacturing technique for making wire; it is still very slow. However, it shows that the same basic process that earlier reached high J_c values in zero field can be used to obtain high J_c in considerably higher fields (≈ 2 T). Moreover, grains can be aligned well enough to sustain high J_c over many consecutive grain boundaries. Both of these are important steps toward the eventual development of YBCO wire.

Suenaga[18] developed a zone-refined method of melt-texturing in which a specimen moves through a quartz tube, past an intense lamp, after which 211 reacts with liquid phase to form 123, which is then zone-refined. The result is improved J_c and fewer microcracks. A statistical study of their grain-boundary misorientation indicates that most interface angles are held below 10°. Refinements to processes such as this may ultimately lead to acceptable J_c in long lengths of YBCO.

Figure 15.17 presents the general overall state of progress toward making YBCO that has high J_c in high magnetic fields. The benchmark is set by thin-film results, and melt-processed YBCO is still far below that, although it has surpassed the magic numbers of 10^4 A/cm^2 in a 1 T field. Of course, it is still far better than ordinary sintered YBCO. If it were possible to make wire reliably out of melt-processed YBCO, such material would be adequate for a number of electrical applications.

Still, it must be kept in mind that no demonstration has yet been made of high and uniform J_c over long lengths of melt-processed YBCO.

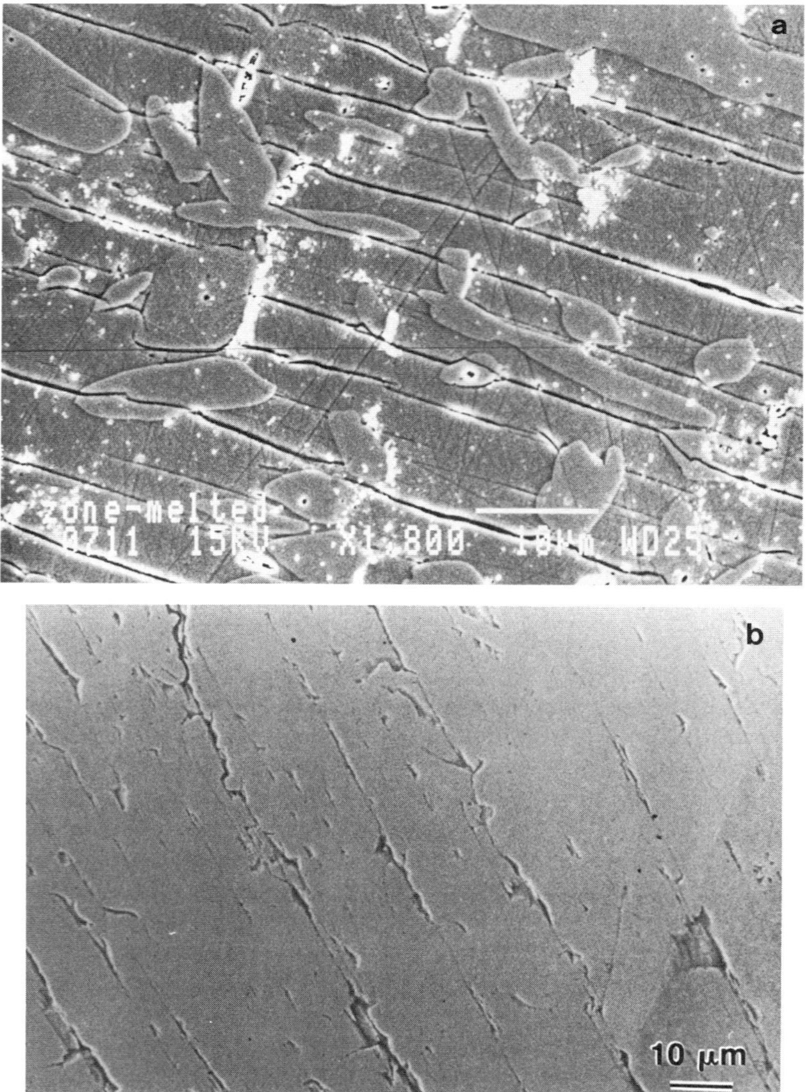

Figure 15.16. Scanning Electron M photographs showing well-textured grains of samples processed by (a) zone melting and (b) partial melt growth. The zone-melting method has a much less precise temperature-vs.-time regimen. Second phases are nearly absent in the sample processed by partial melt growth. (From Ref. 17.)

15.6. VOLATILITY AND THALLIUM COMPOUNDS

Interest continues to build in the thallium system as prospects for using BSCCO at 77 K decline. Despite their high T_c values, many experimenters have eschewed working with thallium because it is both toxic and volatile.

The volatility problem is very severe: a significant fraction of the thallium gets up and leaves during processing at typical temperatures. To see this, observe Figure 15.18. This is

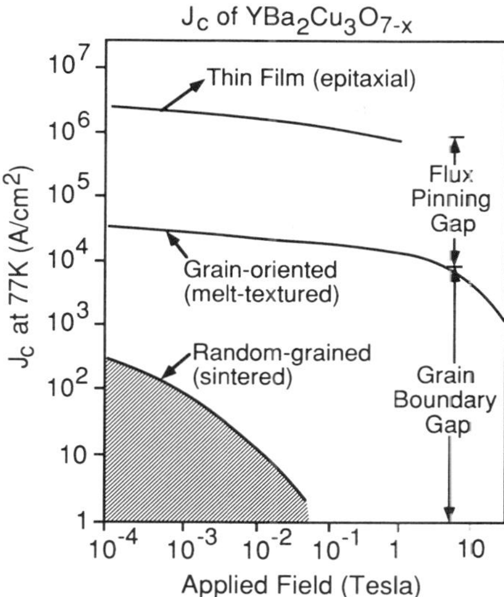

Figure 15.17. Improvement of J_c at 77 K in magnetic fields due to melt texturing of YBCO. Although still far below thin-film performance, the properties are vastly improved over conventional sintered YBCO.

a pair of DTAs for $Tl_{0.5}Pb_{0.5}Sr_{1.6}Ba_{0.4}Ca_2Cu_3O_x$, one variant of the thallium system. The major difference is that the upper one was heated in air, the lower one in pure oxygen. Each chart also contains a new line, the TGA (*t*hermogravimetric *a*nalysis), which is the upper curve on either graph. It is essentially a scale showing mass gain (rising) or loss (falling) as temperature increases. Mass diminishes slowly below 900°C. In the air-heated case (upper), the significant dip in the DTA near 910°C indicates an endothermic reaction, and the TGA contains a slight kink at that point. (This information does not resolve whether it is oxygen or thallium that is leaving the material.) Meanwhile, the sample heated in pure oxygen (lower) shows very little activity below 940°C, when the major phase change (melting) begins. Clearly, the choice of oxygen atmosphere makes a tremendous difference.

What this means is that the pressure is no longer a variable that can be dismissed from consideration in the phase diagram; indeed, a 5-component phase diagram that includes oxygen as a variable really ought to be used. This is why good phase diagrams for the thallium system have not yet been produced. Furthermore, the vapor pressure of thallium is intimately related to the duration of the experiment, which in turn is tied to kinetics. To make matters worse, because a layer of thallium oxide coats the inside of the apparatus being used, it is necessary to dedicate certain equipment to thallium compounds and no other, which greatly increases the cost of research.

Goretta *et al.* at Argonne investigated[19] various pathways to making the thallium series of HTSCs. It was believed originally that although all the thallium compounds have $T_c >$ 100 K, the best one was TBCCO-2223, with $T_c = 125$ K. After a number of attempts to make 2223, Goretta found that the composition actually achieved is by no means stoichiometric, but is instead quite deficient in thallium due to its volatility. Favorable T_c and J_c values resulted nonetheless.

Starting with nominal 2223 composition produces {1.25, 1.74, 1.7, 3.3} in 1 atm oxygen, or {1.33, 2.0, 1.8, 3.1} in 3 atm oxygen. In other words, starting with 2223 materials,

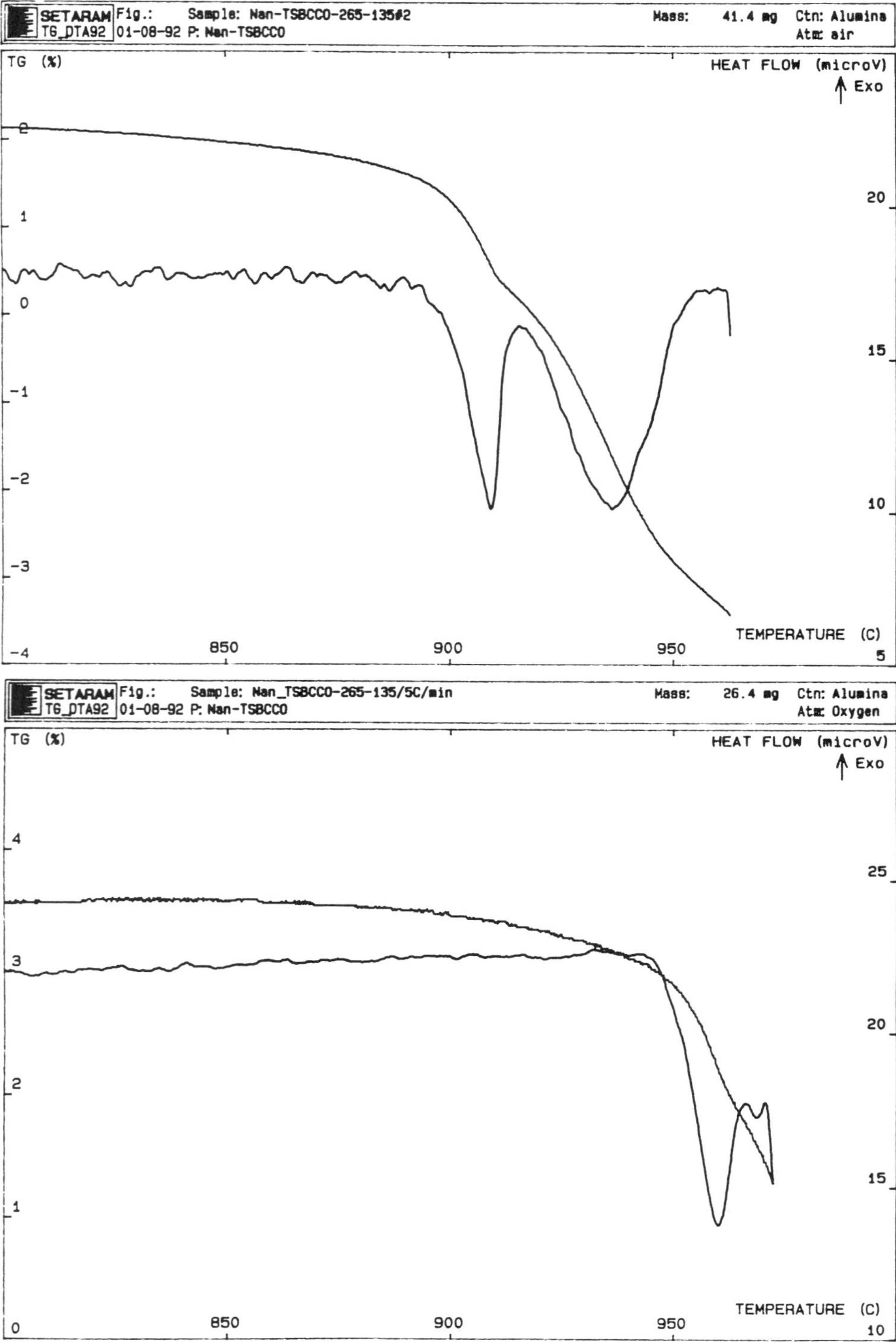

Figure 15.18.   DTAs for $Tl_{0.5}Pb_{0.5}Sr_{1.6}Ba_{0.4}Ca_2Cu_3O_x$ heated (a) in air and (b) in pure O_2. The upper curve in each chart which falls off monotonically to the right is a TGA (thermogravimetric analysis). As explained in the text, the temperatures at which phase changes occur are dramatically affected by the choice of oxygen pressure.

the synthesis drives the composition toward 1223, and the volatile thallium simply leaves—it contaminates apparatus and poses a health risk. If one begins with excess thallium, 2212 is formed, but with lots of extra compositions that aren't superconducting at all: $\rho(T)$ measurements show large resistive tails in such samples.

Happily, TBCCO-1223 is likely to be the best superconductor in the series anyway. As mentioned in Chapter 8, the single-thallium layer compound has a lower value for d_i, the distance between consecutive pairs of CuO_2 layers in the unit cell. This leads to better flux pinning, and the outlook is optimistic for use of TBCCO-1223 in high magnetic fields. Since $T_c > 110$ K, operating in a bath at 77 K leaves sufficient margin below T_c to allow the value of J_c to remain large. The slight temperature advantage of TBCCO-2223 ($T_c \approx 125$ K) is more than offset by its weaker flux pinning, with the result that 1223 is likely to be the better choice at 77 K.

The first task is to make substantial quantities of powder of TBCCO-1223. A team effort to investigate various pathways to this goal is underway, involving researchers at Argonne, Los Alamos, and Sandia, as well as several others in the private sector.

As described in Section 15.4 above, the best processing route for BSCCO often begins with precursor intermediate phases which are then further reacted to reach the desired final state. The same philosophy applies to experiments with the thallium system, but it is much harder because of the high vapor pressure of thallium. Therefore, a variety of approaches have been tried:

Sandia deposits a thin film of precursor metals and heats that in a crucible to form TBCCO.[20] To compensate for the volatilization of thallium, the atmosphere is made thallium rich by including a pellet that boils off gas-phase Tl_2O_3. (This is known as thallation from the vapor phase.) A number of tricks have been tried to reach the 1223 phase. Argonne has adjusted the ratio of thallium (near 1.0), and has found that starting with a few percent extra leads eventually to a slight deficiency of thallium in an otherwise 1223 phase. Unfortunately, TBCCO is hard to densify, and when Argonne tried sinter-forging,[21] as soon as some liquid formed the thallium started to leave.

Silver seems to play a vital role in forming 1223. In a process developed by General Electric,[22] it lowers the melting point and the processing temperature, and enhances liquid phase formation. With silver absent, the process simply doesn't work. This catalytic action by silver is not fully understood, but GE finds that silver does not react with the final 1223 phase.

Putting in lead helps reach the 1223 phase. In the Republic of China (Taiwan), Huang *et al.*[23] investigated mixtures in which Pb or Bi partially substituted for Tl: $(Tl_{0.64}Pb_{0.2}Bi_{0.16})Sr_{1.6}Ca_{2.4}Cu_3O_y$. They found that by varying the Ca/Sr ratio (up to 2.4/1.6), they could more readily form a compound with three copper oxide planes, i.e., the 1223 phase. A comparatively short time is needed for this to work, and the fraction of 1223 is very high. These compounds also have T_c near 120 K. Lead may act as a barium "getter", forming $BaPbO_2$. Once formed, $BaPbO_2$ is an excellent flux-pinner. Hitachi observed wide hysteresis loops in lead-substituted TBCCO.

More important than the processing details, this discussion shows how researchers grapple with the diverse possibilities presented by the chemistry of the thallium system. Many different laboratories are contributing pieces to the puzzle, which it is hoped will speed the arrival of useful thallium HTSCs.

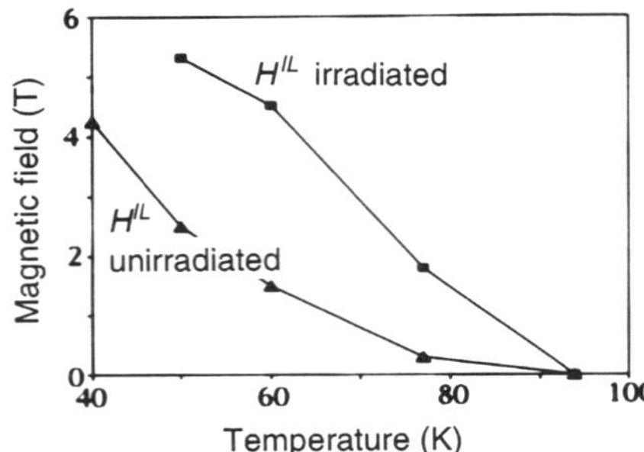

Figure 15.19. Effect of neutron irradiation upon the irreversibility line (H^{IL}) of $HgBa_2CuO_{4+\delta}$. Triangles show performance before irradiation; squares are after irradiation. The data for H^{IL} (unirradiated) are the same as the triangles in Figure 14.2.

15.7. POSTPROCESSING: IRRADIATION

The goal of enhancing J_c in the HTSCs has stimulated a wide variety of research avenues: postprocessing steps involve special annealing, lattice-defect creation or removal, and so on. It is not possible to enumerate them all. One single example of a creative approach to improving flux pinning will suffice to illustrate the kind of things being tried.

In LTSCs, flux pinning is primarily due to crystal imperfections. By increasing the number of such imperfections (e.g., by cold-working the metal), the critical current can be raised[24] in LTSCs. Applying this principle to HTSCs suggests that by introducing crystal defects in a well-controlled fashion, flux pinning should improve and J_c should increase.

As discussed in Section 14.8, it is possible to create columnar defects in reasonably thin YBCO samples via ion bombardment[25]; the channels thus formed tend to contain the flux lines. In bulk samples, more general defects can be created with either electron beams or neutron beams.[26] Atomic displacement of oxygen atoms makes pinning sites. Displacing an oxygen requires 20 eV energy transfer to the oxygen, which sets the threshold for damage at 120 KeV for electrons.[†] Consequently, not everyone is equipped to pursue this avenue.

The availability of excellent equipment invites new experiments. The Los Alamos Meson Physics Facility (LAMPF) has been used to fire a beam of 800 MeV (0.8 GeV) protons into BSCCO-2212. The advantage to high-energy protons is that they penetrate to a depth of $\approx 1/2$ meter, whereas heavy ions would stop a few microns below the surface. Thus, a silver sheath is no obstacle to protons. These energetic protons cause bismuth to fission, sending Xe or Kr nuclei crashing through the material with energies near 100 MeV, and forming columnar defects deep within the material.[27] Although the columns lie in totally random directions, they are still effective in pinning flux, perhaps because of the extreme anisotropy of BSCCO. After bombardment, the irreversibility line moves upward by about 20 K, a remarkable improvement.

Neutrons are also effective in displacing oxygen, but they tend to activate the other elements (i.e., make them radioactive), which limits their appeal. In the case of $HgBa_2CuO_{4+\delta}$, irradiation with neutrons substantially enhances J_c, because of better flux pinning by the defects produced. The enhancement is 1 to 2 orders of magnitude, which is much larger than in other oxide superconductors.[27] Figure 15.19 shows how the irreversibility line (H^{IL}) moves upward after neutron bombardment. At 77 K, $H^{IL} = 1.8$ T, which

---

[†]Because electrons are very light compared to oxygen atoms, only a small amount of energy can be transferred from the electron to the oxygen atom in a single collision.

is competitive with YBCO; before irradiation, $H^{IL} \approx 0.2$ T at 77 K. This result is important because it suggests that $HgBa_2CuO_{4+\delta}$ is not intrinsically limited to low J_c, unlike BSCCO. In turn, this raises the possibility that $HgBa_2CuO_{4+\delta}$ might eventually become a practical material for bulk applications at 77 K.

There is considerably more work to be done before a definitive understanding of radiation damage can be given. Experience with LTSC materials showed that there are better ways (mechanical and thermal) to increase pinning than by radiation; the same may well be true for HTSCs. Nevertheless, the fact that J_c can be changed via selective radiation damage offers one more path for investigating the properties of the HTSCs.

15.8. SUMMARY

This chapter reviewed the principal means of making HTSCs. The detailed chemistry of such processes require careful attention to kinetics because equilibrium chemistry is simply inadequate. Most of the reaction paths contain one or more steps involving solid-state diffusion, which is notoriously slow. Accordingly, nonequilibrium configurations are commonplace results among the HTSCs.

The experimental techniques of x-ray diffraction (XRD) and differential thermal analysis (DTA) are used to determine which phases form at what temperatures in the HTSCs. This information is essential for picking the best pathway to producing a material with optimum properties. Thus, the judgement that a certain variant of a process is a success (say, in producing YBCO-123 with YBCO-211 grains inside it) is made on the basis of combining *both* these measurements.

By making optimum use of such techniques, it has been possible to discover several clever routes through the phase diagram to produce the various HTSCs. One excellent example of this is the two-powder process for making BSCCO, in which Bi-2212 powder is mixed with cuprates of strontium and calcium, and packed into a tube. After forming the tube into wire and heat-treating it, the resulting interior material is Bi-2223. It is extremely difficult to make pure Bi-2223 directly from a starting composition of the several oxides, so this is a distinct improvement. Unfortunately, all forms of BSCCO suffer from flux lattice melting in the vicinity of 30 K, and so BSCCO cannot be used in high magnetic fields at 77 K.

It is comparatively easy to make plain old YBCO, but it is quite another story to make YBCO-123 that will carry plenty of current in a magnetic field. The melt-texturing technique and its variants work well for making YBCO crystals with high J_c. By using complex heat-treating cycles that contain steps of melting, quenching, partial melting, and so on, it is possible to attain almost 10^5 A/cm^2 in low fields at 77 K. Unfortunately, all such processes are extremely slow (≈ 1 mm/hr). Even in the best of cases, the crystals obtained by these methods are about 1 cm long.

Chapter 16 considers ways to scale-up processes so as to make wire. In the case of melt-textured YBCO, the likely outcome would be a long sequence of crystals containing myriad grain boundaries, each of which attenuates the superconducting current. Under such circumstances, J_c would fall off considerably, especially in a magnetic field. The cause of this fall-off is twofold: grain boundaries contain impurities or other phases (nonsuperconducting) such as YBCO-211, which act as barriers to the supercurrent; and consecutive grains are aligned at different angles, thus diverting the current from its smooth flow within the sandwich of the copper oxide planes. Consequently, melt-texturing is not given much chance to become a useful production method for wire.

TBCCO is still in its infancy. Researchers are exploring a number of techniques to make it, and are focusing primarily on Tl-1223, a phase with $T_c > 100$ K and good flux-pinning qualities. So little is certain about the phase diagram of the thallium system that researchers are following predominantly empirical paths, with guidance drawn from the analogies with the bismuth system. At present, the goal is to establish a reliable way to produce Tl-1223 powder; consideration of long lengths will come later.

In the category of thin films, progress has been greater. Deposition techniques taken over from the semiconductor industry have made it possible to control the layering of the HTSCs with great precision, and this has allowed the exploration of many different fabrication strategies. As a consequence, thin-film applications of HTSCs are more advanced than bulk applications such as wire. Nevertheless, if superconductors are ever to serve in power-handling equipment, it will be necessary to persevere toward manufacturing long lengths of large cross-section HTSCs within reasonable time and cost constraints.

REFERENCES

1. R. B. Poeppel *et al.*, "High-Temperature Superconductors," in *Engineered Materials Handbook, Vol. IV: Ceramics and Glasses*, pp. 1156–1160 (ASM International: 1992).
2. J. L. Wagner *et al.*, *Physica C* **210**, 447 (1993).
3. L. H. Van Vlack, *Materials Science for Engineers* (Addison-Wesley, Reading, MA: 1970).
4. U. Welp *et al.*, *Appl. Phys. Lett.* **63**, 693 (1993).
5. J. P. Singh *et al.*, *J. Materials Res.* **7**, 2324 (1992).
6. K. G. Goretta *et al.*, *Materials Lett.* **7**, 161 (1988).
7. J. Blendell, NIST, private communication.
8. S. T. Misture *et al.*, in *Superconductivity and its Applications*, pp. 582–592, 6th NYSIS Conference, Buffalo NY 1992, H. S. Kwok *et al.*, eds., AIP Conference Proceedings #273 (AIP, New York: 1993).
9. W. Wong-Ng *et al.*, *Bull. Amer. Ceramic Soc.* **71**, 1261 (1992).
10. K. H. Loon and H. B. Lee, *J. Materials Sci.* **26**, 5101 (1991).
11. S. E. Dorris *et al.*, *Physica C* **212**, 66 (1993).
12. K. Salama *et al.*, "Melt-Processing and Properties of YBCO," Chapter 4 in *Processing and Properties of HTSCs*, Vol. I, edited by S. Jin *et al.*, (World Scientific Publ. Co.: 1993).
13. S. Jin *et al.*, *Appl. Phys. Lett.* **52**, 2074 (1988).
14. K. Salama *et al.*, *Appl. Phys. Lett.* **54**, 2352 (1989).
15. M. Murakami *et al.*, *Jpn. J. Appl. Phys.* **28**, L1125 (1989); *Modern Phys. Lett. B* **4**, 163 (1990).
16. P. J. McGinn *et al.*, *Appl. Phys. Lett.* **57**, 1455 (1990).
17. D. Shi *et al.*, *Appl. Phys. Lett.* **57**, 2606 (1990).
18. M. Suenaga, Brookhaven National Laboratory, private communication.
19. K. C. Goretta *et al.*, *Supercon. Sci. Techn.* **5**, 534 (1992).
20. D. S. Ginley, "Thallium Films for Microelectronic Applications," in *Thallium-Based High-Temperature Superconductors*, edited by A. M. Hermann and M. V. Yakhmi (Marcel Dekker, New York: 1993).
21. K. C. Goretta *et al.*, in *High-Temperature Superconducting Compounds II*, pp. 263–274, edited by S. H. Whang *et al.* (Minerals, Metals & Materials Society: 1990).
22. J. A. deLuca *et al.*, *Physica C* **205**, 21 (1993).
23. Y. T. Huang *et al.*, *Appl. Phys. Lett.* **57**, 2354 (1990).
24. E. W. Collings, *Applied Superconductivity, Metallurgy, and Physics of Titanium Alloys* (2 vol.) (Plenum Press, New York: 1986).
25. L. Civale *et al.*, *Phys. Rev. Lett.* **67**, 648 (1991).
26. M. A. Kirk and H. W. Weber, "Electron Microscopy Investigations of Irradiation Defects in the High-T_c Superconductor YBa$_2$Cu$_3$O$_{7-x}$," in *Studies of High-Temperature Superconductors*, edited by A. Narlikar, vol. 10, p. 243 (Nova Scientific Publ.: 1992).
27. L. Krusin-Elbaum *et al.*, *Appl. Phys. Lett.* **64**, 3331 (1994).
28. J. Schwartz *et al.*, "Large J_c in Neutron-Irradiated Polycrystalline HgBaCuO," *Phys. Rev. B* **48**, 9932 (1993).

16

Wire

Thomas P. Sheahen and Alan M. Wolsky*

Making **wire** is the primary goal of applied HTSC research today. To say the least, making wire out of ceramic superconductors is difficult. If it were easy, there would be all sorts of devices in operation today. But the obstacles discussed in prior chapters —brittleness, weak link behavior, flux lattice melting—all conspire to make the attainment of "good" wire extremely difficult. Fortunately, there are thin-film applications of YBCO to microwave-cavity wall-coatings; there are levitation demonstrations that may become practical bearings; but the holy grail of a 77 K wire that will carry plenty of current in high magnetic fields is still elusive.

In this chapter, we explain the methods used to attempt to make useful HTSC wire and review the progress that has been made so far. The *powder-in-tube* method has been the most successful, especially for BSCCO, but the desirable traits of BSCCO seem confined to relatively low temperatures. Nevertheless, up to about 30 K, BSCCO is an excellent material. YBCO and TBCCO seem more suitable for 77 K applications, but their problems with weak links and brittleness are severe. There is always optimistic hope for every new material that is discovered, and $HgBa_2CuO_{4+\delta}$ held that distinction briefly.

Throughout this chapter we cite numerical accomplishments by various researchers in wire performance, i.e., in maximizing J_c at various temperatures and magnetic fields. However, these are not interesting in themselves. For one thing, the best numerical values will change with progress in the field; for another, we do not judge what is the best route to making wire. Rather, the object here is to leave the reader with an understanding of the problems associated with making HTSC wire and the motivations that lead researchers to try new strategies. Then, as future progress unfolds, it may be possible for the reader to judge which news is likely to lead to improved performance.

16.1. THE CHALLENGE

Imagine picking up a delicate ceramic flower vase and announcing, "I'm going to turn this into a spool of wire!" The reaction would be variously laughter, raised eyebrows, and perhaps alarm from the vase owner. No one would give you much chance for success, and you'd have to have a mighty good incentive to undertake a serious attempt to make wire. This is the situation we find ourselves in today. The incentive is the attraction of a 77 K conductor having no resistance.

*Argonne National Laboratory.

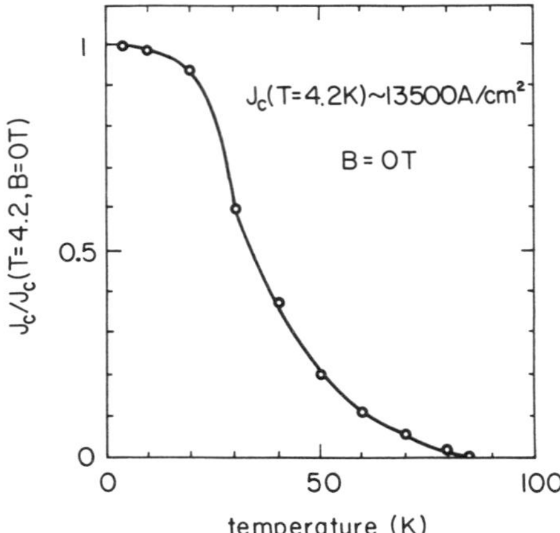

Figure 16.1.   The decline of J_c with increasing temperature, showing the effect of *flux lattice melting* in BSCCO.

The words ceramic and wire are nearly mutually exclusive. Some laboratory progress has been made toward making wire, but a great deal of clever engineering will be needed to achieve a commercially viable product. The success of fibre optics offers hope, but remember that glass has neither crystal structure nor grain orientation. HTSCs, by contrast, not only are solid crystals, but, due to anisotropy, it matters very much *which* orientation the crystals have. For large currents to flow down the wire, the axis of the wire must coincide with the favorable direction for current flow through the HTSC material. This adds a considerable measure of extra difficulty to the wiremaking task.

Chapter 11 dealt with the mechanical aspects of HTSCs, and most of the data there was for YBCO. The other HTSCs also share the brittleness of copper oxide ceramics, but one HTSC (BSCCO) is micaceous, that is, like mica. It shears[1] easily along certain planes, which means that it deforms in a very predictable way; the grains can be aligned during wire fabrication. This property makes it easier to form BSCCO into wire than any other HTSC. Consequently, the only successes so far in wiremaking are for BSCCO.

On the other hand, BSCCO is not very useful at 77 K or even at intermediate temperatures due to flux lattice melting, which was discussed in Chapter 14. For BSCCO-2212, Figure 16.1 shows how J_c drops off above 20 K in a zero applied magnetic field.[2] In a finite field, the performance is even worse; the *irreversibility line* in the *T-H* plane defines a magnetic field and temperature, above which flux lines move freely and the superconductor dissipates energy as current flows. Here we are dealing with intrinsic properties of BSCCO. Clever engineering of pinning sites can move the irreversibility line around a bit, but cannot make the major changes needed to provide high J_c in sizable fields at 77 K.

Hence, there is still a need to find ways to make wire out of materials that carry large currents ($J_c > 10^4$ A/cm^2) at 77 K.

16.2.   YBCO: EARLY ATTEMPTS

In this section, we review some of the ways that have been tried for making wire (or tape) out of HTSCs. Basically, these are a combination of standard wiremaking techniques (drawing, swaging, extruding)[3] with heat treatments adapted to the HTSC ceramics. Most of these techniques were first tried on YBCO and later extended to BSCCO.

a. *Powder.* No wire can be any better than its basic constituents, and so the initial task is to make good HTSC powder. Chapter 15 covered the ways in which HTSC powders are made today. In the initial attempts at wiremaking,[4] inferior powders led to inferior wire, mainly because of impurities collecting at the grain boundaries.[5] Today, wiremakers begin with good-quality powder, and this has eliminated many puzzling and conflicting results.

b. *Shape and Form.* To make wire, the powder must be formed into a proper shape. The principal ways of doing so include (a) using a binder material to congeal the powder; (b) attaching the HTSC powder to a central core; (c) packing powder inside of a tube; and (d) depositing HTSC on a substrate to make a film. This section gives examples of methods (a) and (b); Section 16.3 describes method (c); and Section 16.4 covers (d). It deserves mention that the optimization of the various temperatures, cycle times, and depths of coatings are the result of painstaking engineering over a long period of time. The essence of competition in this field lies in the details of each process.

c. *Wind and React Methods.* The initial attempts to make YBCO wire exploited the strategy of making green-state YBCO, winding it into a coil or other desired shape, and then heat-treating it to form the superconducting YBCO-123 state. This entire class of operations was unsuccessful, because the YBCO thus produced had dirty grain boundaries and poor alignment of grains, both of which degraded J_c unacceptably.

To work directly with YBCO-123 itself, a number of clever tricks were tried to give it a proper wire shape. One example of this method will suffice to illustrate the general idea: AT&T Bell Laboratories[6] used a thick film technique with a silver substrate. Figure 16.2 illustrates the process: starting chemicals were formed into YBCO powder, which was ground to small size, mixed with an organic binder, and placed in a vat. A thin silver fiber was drawn through the mixture, and a film of YBCO adhered to it. This was wound into a coil and taken to a furnace to sinter the YBCO. The final product was a superconducting coil of YBCO with a silver inner core for strength.

Many different wire manufacturers tried increasingly sophisticated variants on this method. Figure 16.3 shows the Pacific Superconductors' process.[7] Here, the inner core was

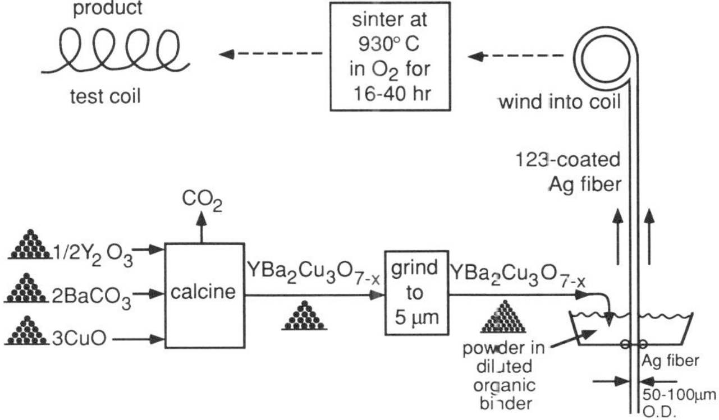

Figure 16.2. Schematic drawing of YBCO wire-production process developed by AT&T Bell Laboratories.

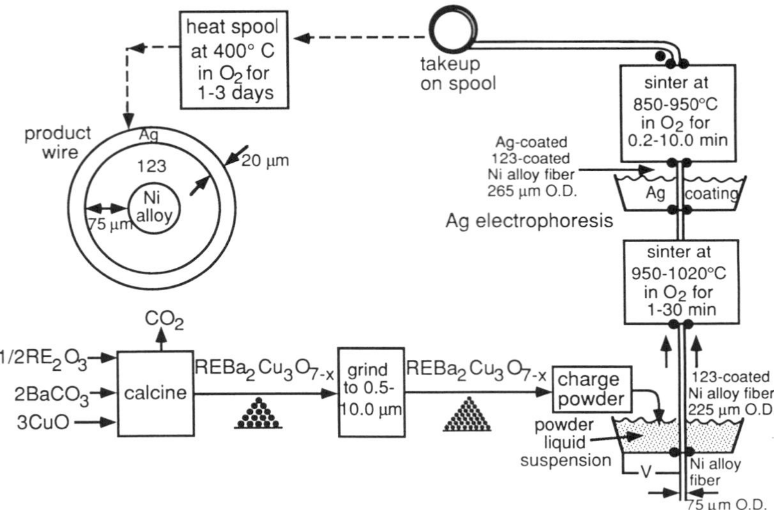

Figure 16.3. Schematic drawing of YBCO wire-production process developed by Pacific Superconductors, using a nickel wire core and an outer silver cladding. For BSCCO, they have coated the core with *two* layers of BSCCO and silver.

a nickel wire, which was coated with YBCO, followed by a silver cladding on the outside. The core was stronger this way, but the current density still was too low.

d. *Silver.* The role of silver deserves some explanation. For any ceramic, a metal sheath provides several services: During fabrication, it defines the shape and protects the HTSC from abrasion. During handling and use it also protects from chemical attack (including moist air!); it facilitates electrical contact; and it conducts heat away.

Experiments at Argonne National Laboratory showed that silver is the preferred cladding medium for YBCO wires. Being a good conductor, silver provides a relatively low-resistivity link between disjointed grains. Figure 16.4 shows that for reasonable numerical parameters, the resistivity along a wire made of YBCO/Ag composite might be 1/1000 of the resistivity of silver alone. At 77 K, that is quite low. Moreover, silver can be mixed with YBCO without hurting its superconducting properties (up to a point), and mechanical properties improve[8,9] because silver increases *fracture toughness* (see Chapter 11).

Copper cladding has the disadvantage of attacking the YBCO and removing oxygen (to form CuO), but silver does not; indeed, Ag_2O decomposes into silver and oxygen at about

Figure 16.4. Idealized sketch of adjacent grains of HTSC, with intervening gaps occupied by silver. This choice of reasonable geometric parameters reduces the effective length of the silver path by a factor of 1000.

200°C. When silver gathers at the grain boundaries, it offers a diffusion-pathway for oxygen to reach the interior. Of all metals, silver has the highest oxygen diffusion rate (even more than gold), and hence has minimum impact upon the oxygen stoichiometry of the HTSCs.

e. *Multifilaments.* Some experimenters went directly to a multifilamentary approach. In America, the Babcock-Wilcox corporation, in conjunction with Iowa State's Ames Laboratory, made a cotton-candy type of YBCO.[10] The hope was that by having only very fine filaments, the brittleness problem would be overcome and the ultralong grains (i.e., the short fibers) would connect to form a network that would carry high current. With hindsight, we see that grain-to-grain contact limited this material.

In Minsk, Belarus, a group at the Institute of Solid-State and Semiconductor Physics under the leadership of Prof. Vladimir Novikov[11] used a fine organic fiber coated with YBCO, leading to multifilamentary composite conductors. This very clever technique illustrates the degree of creativity that has gone into YBCO wiremaking: They started with cellulose fibers and oxidized them to form carboxyl-cellulose, which were then woven into a yarnlike unit. Next an ion-exchange reaction was used to make salts of Y, Ba, and Cu with carboxyl-cellulose penetrating each fiber. An yttrium-rich region formed in the interior, and a copper-rich region on the outer edge, with barium salts present as well. This radial concentration gradient was very important for the next step. (Recall that Y-Ba-Cu will not form a metal alloy at the stoichiometric ratio.) When heating began, a diffusion-reaction process took place, such that the eutectic zone propagated both inward and outward in the fiber. At the same time, the organic core was burned off, but the shape of each fiber remained throughout. At the conclusion, fibers of YBCO of a uniform phase[12] were left behind. This multifilamentary form of YBCO was a modest success: at 20 K in 0 T, $J_c \approx 10^4$ A/cm^2, but this fell off to $< 10^3$ at 77 K. Only short samples (a few centimeters) were made in this way.

The point here is not to endorse this one method, but to illustrate that tremendous efforts have gone into YBCO wire. The catalog of techniques is long, and the amount of innovative thinking is impressive.

f. *Weak Links.* From all these methods, whether single-strand or multifilament, the lesson learned was that grain alignment and grain boundaries are the problem. The weak-link effect in YBCO is very severe, and the only way to overcome it has been through melt-texturing in its various forms. But melt-texturing is terribly slow, typically < 1 cm/hr, and it would take too long to make practical lengths of YBCO by such a process.[13] Thus, there is at present no satisfactory method for making YBCO wire.

16.3. POWDER-IN-TUBE METHOD

The goal of every wiremaker is to optimize performance, meaning to obtain the highest possible critical current density J_c under various operating conditions. The most popular method of wire manufacture used by many groups (e.g., Intermagnetics General (IGC), American Superconductor (ASC), Sumitomo (SEI), and Vacuumschmelze) is to begin by packing HTSC powder into silver tubes.[14] The tube gives shape to the HTSC and strength to withstand the forces during elongation. When the silver tube is subsequently processed into fine wire, the inner core is HTSC. Figure 1.2 is a schematic drawing of this process. There are several versions of this same basic process.

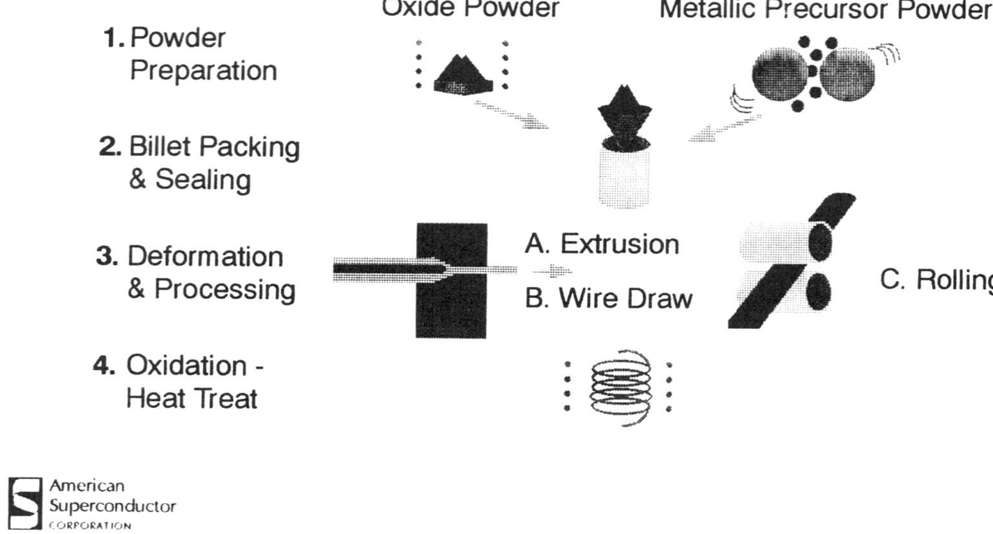

Figure 16.5.   Schematic showing the general OPIT process used to produce BSCCO tapes.

The sheathing tube is virtually always silver, because silver does not react with oxygen. For powder-in-tube processing of YBCO, an early patent by Jin *et al.* of AT&T Bell Laboratories[15] recognizes this point: ". . . at least the portion of the cladding that is in contact with the oxide powder is substantially inert with respect to oxygen and the oxide under the conditions of heat treatment." For BSCCO, Maeda and his colleagues at NRIM in Japan first demonstrated[16] that the mechanical processing of the silver tube actually improves grain orientation[17] and thus enhances J_c. One version, used by ASC[18] and known as the "Oxide Powder in Tube" process (OPIT), appears in Figure 16.5.

16.3.1. Process Parameters

There are several variables associated with each step in the process, and these are tabulated[19] in Table 16.1, the "Powder-in-Tube Parameter Matrix."

As a single example of how these parameters influence one another, consider the seemingly innocuous item "powder size distribution—grinding" in Table 16.1. The packing density of HTSC powder affects subsequent steps to the final wire. Hand packing gives density of about 30%, which would lead to major voids in a sintered material. When the wire is to be extruded, greater initial packing density is required. For that case, *cold isostatic pressing* results in about 70% density. The maximum packing fraction for equal spheres is 74%, so Vacuumschmelze[20] suggested mixing two different size powders to increase the density. On the other hand, most powders are not spheres. Furthermore, BSCCO wire-making is often done by a drawing and rolling process instead of extrusion. In that case, it is desirable to start with *low* packing density, because when the powder is densely packed initially, drawing problems result.[21] After drawing and rolling, the BSCCO density ≈90%. Clearly, very delicate trade-offs must be made during wire manufacturing.

Table 16.1.   Powder-in-Tube
Parameter Matrix

Powder preparation
    Initial powder composition
    Silver oxide additions
    Powder size distribution (grinding)
    Powder heat treatment
Billet geometry
    Shape
    Silver/superconductor ratio
Drawing parameters
Rolling parameters
    Roll size
    Velocity
    Reduction ratios
Heat treatment parameters
    Temperature
    Time at temperature
    Ramp time to and from temperature
Unixial pressing
    Pressure
Hot Isostatic Pressing
    Pressure
    Temperature
    Time at temperature

Every line in Table 16.1 has a similar story. The difference between success and failure is in the details of processing, and these details are precisely the point of competing processes. We are manipulating nonequilibrium processes here; that is, the dynamics of the process are still ongoing, and the final (equilibrium) state has not been reached at the time of stopping. Therefore, small differences in temperature settings or time durations can make enormous differences in properties of the wire.

16.3.2. Heat Treating

The choice of heat-treatment parameters lies at the heart of each company's process; again, a single example is illustrative. Vacuumschmelze, whose business is to make NbTi and Nb_3Sn wire, developed processes[20] for making silver-sheathed BSCCO-2212 wires and (Bi,Pb)SCCO-2223 tapes. Both processes begin by mixing high-purity oxides and carbonates together in the appropriate concentrations. The mixed powder is subjected to a three-step calcination treatment with intermediate grinding between steps to remove the carbon due to the input of $SrCO_3$. Because of the low melting point of Bi_2O_3 (817°C), the first step is performed at 800°C. Subsequent steps are performed at temperatures up to 850°C.

BSCCO-2212 wire[22] is made by filling a silver tube 8 mm in diameter with a wall thickness of 1 mm. The tube is drawn to produce a wire 1 mm in diameter. An 8-to-1 reduction in diameter implies a 64-to-1 elongation; the engineering trick is to reduce the diameter still further, while retaining continuity of the HTSC enclosed within the tube. The 64-to-1 number has been surpassed: using this process, wires up to 300 m in length were produced by 1992, and 1000 m by 1994.

Heat treating follows the wire-drawing step in this process. A two-step anneal process is used. First, the sample is heated above 900°C to produce a partial melt. This is followed by a long-term anneal of approximately 100 hr at 840°C. The resulting samples are textured, with cylindrical symmetry.

The Vacuumschmelze procedure for making the 2223 *tape* is somewhat more complicated.[23] First, the BSCCO powder is *cold isostatically pressed* (CIP) into a silver tube 6 mm in diameter with a wall thickness of 2 mm. The tube is drawn down to produce a wire 1.3 mm in diameter. This wire is cold-rolled to produce a tape 0.1 mm thick and 2.5 mm wide. The tapes are annealed at 835°C for approximately 100 hr, after which they are subjected to a uniaxial press at 0.5 GPa. This annealing cycle is repeated three times. As a result of this thermomechanical treatment, the final tapes are highly textured.

The point here is not in the details themselves, but only that details are terribly important. Every manufacturer has different variations. In addition to Vacuumschmelze, ASC, IGC, and SEI are all very competitive in this field.

16.3.3. Deformation

Reducing the diameter is an essential step in producing single-strand conductor. Drawing, swaging, extrusion, and rolling are examples of such mechanical-deformation processes, and each gives different characteristics to the final product. For example, extrusion was

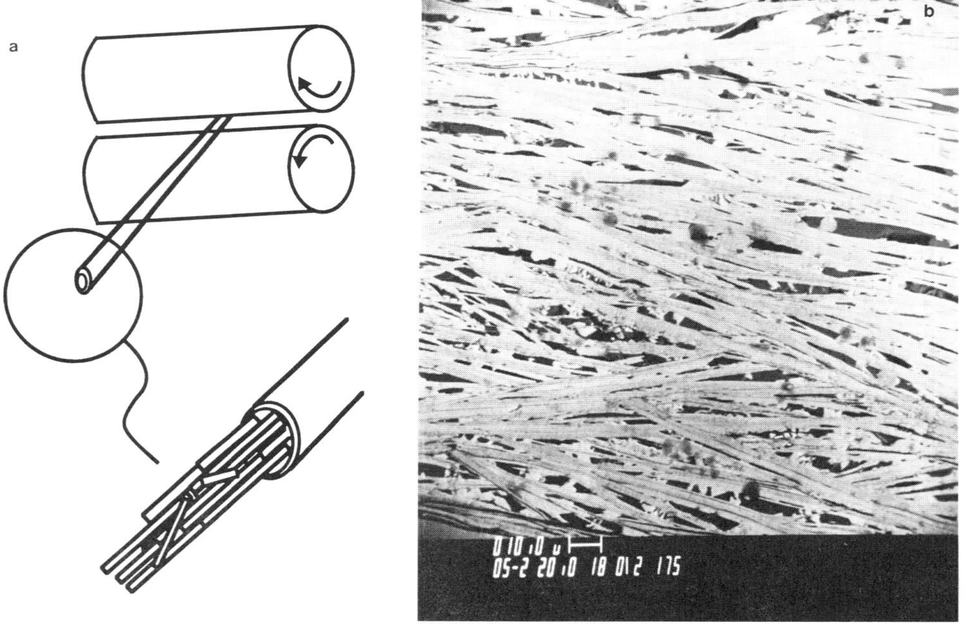

Figure 16.6. The rolling process tends to produce well-aligned grains of the micaceous BSCCO. (a) sketch of concept; (b) photomicrograph after a single roll/heat cycle. (Photo courtesy of K. Goretta, Argonne National Laboratory.)

explored by Sandia National Laboratories, who found that a wire could be produced in a single step (as contrasted to multiple passes through a set of dies with intermediate heat treatments). This resulted in very dense cores (with higher J_c). Rolling, on the other hand, is easier to control precisely and produces a long flat conductor, or tape.

The success achieved with BSCCO is due largely to its micaceous nature: BSCCO grains align well during processing. YBCO, by contrast, is much more troublesome and does not align well during powder-in-tube processing. Hence, its performance is poor. In the case of BSCCO-2223 (which is highly anisotropic at the unit cell level), rolling helps to orient the grains by moving the anisotropic grains along their slip planes. Figure 16.6 illustrates this behavior: the BSCCO grains are rather well aligned.

Los Alamos National Laboratory has a slightly different method of processing[24]: rather than just pour powder into a tube and then pack it, they make a rod of HTSC and machine it to the inside diameter of the silver tube. This has the effect of starting the process off with 80% packing density. The rolling–pressing step is designed to achieve texturing in the final tape or wire. Wire core-thicknesses of 10 μm have been achieved. In a typical finished piece, the cross-section is one-third BSCCO and two-thirds silver. This shows the sophistication that has evolved over several years of improvements.

16.3.4. Resistivity of Silver

Silver itself is a very good conductor, so when the interior HTSC material is not perfectly superconducting, current can be shared between it and the silver surroundings. In fact, the resistivity of pure silver at 20 K is only about 4×10^{-11} Ω-meters. This means that if the voltage criterion defining superconductivity is set at 1 μV/cm (a common choice), a current density of 100 A/cm^2 could pass entirely through the silver without exceeding the voltage criterion. On that basis, a silver tube filled with sand or styrofoam would be called "superconducting." If the HTSC material were only equal in resistivity to silver, it could pass 250 A/cm^2 at 20 K without exceeding 1 μV/cm. Because of this, any value of J_c below 1000 A/cm^2 merits very little attention, and data is seldom graphed down to such low J_c values. Manufacturers of long wires replace the voltage criterion with a more stringent condition for resistance of the full wire, usually 10^{-11} Ω-cm. Since J_c values typically are 20% lower with 10^{-11} Ω-cm than with 1 μV/cm, it is essential to specify the voltage criterion in any report of J_c.

16.3.5. Chemical Variations

There are so many minor variations possible that it is impossible to guess at which are the best ones, but it deserves emphasis that details of chemistry are as important as thermal and mechanical details.

As we saw in Chapter 15, precursor powders are used to reach the final state of each HTSC. With powder-in-tube processing containing some steps in the > 800°C range, it is possible to select a mix of powders for packing that will lead to optimum material after reacting inside the tube. For example, the fraction of calcium and copper can be varied in the initial powder, and this will affect the properties of the final wire. Intermagnetics General has explored[25] the many options here: Figure 16.7 shows how J_c differs in samples of BSCCO prepared from different starting powders: the optimum processing temperature decreases when more calcium and copper are added, because the melting temperature is lowered. As a result of changing the time–temperature profile, the maximum J_c also changes.

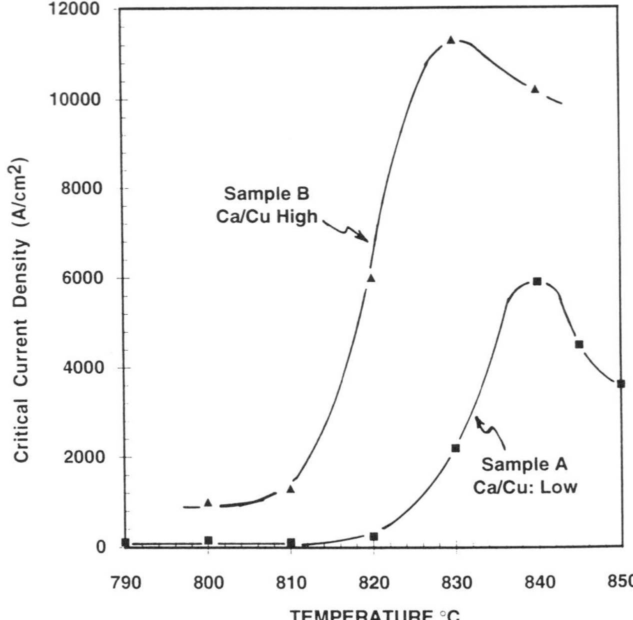

Figure 16.7. Critical current density as a function of the processing temperature, for (Bi,Pb)SCCO powders of two different starting compositions. Sample A yields $Ca_{2.2}Cu_{3.0}$, while sample B gives $Ca_{2.5}Cu_{3.3}$. (From Ref. 25.)

Most chemistry attention goes to the BSCCO material itself, but the details of the silver tubing are important too, because the tubing and the powder react during processing. Vacuumschmelze used a sheath made of silver–nickel–magnesium (Consil[®] 995) and found that the nickel promoted more uniform grain size.[26] Los Alamos found that Consil[®] 995 improved the densification of the BSCCO inside. Moreover, it retained strength after annealing. Intermagnetics General used a silver sheath containing Al_2O_3, and found that the yield stress just about doubled.[19] This in turn permits using a thinner wall sheath, which means a higher percentage of superconductor.

16.4. DIRECT TAPE METHODS

The powder-in-tube process results in a long thin conductor, essentially a tape made of silver with HTSC inside it. There is no fundamental reason why the silver has to be on the outside; it is equally possible to manufacture ribbon by coating silver with HTSC.

16.4.1. HTSC Coating Methods

One of the first, and still most successful, methods of coating silver is the dip coating process pioneered by Maeda's research group at NRIM.[27] This method is very reminiscent of Figure 1.2 or 16.2: a silver wire passes through a slurry of BSCCO with a binder; after the wire dries, it is wound into a coil, and then heat treated. The main problem with this method is that the composite becomes very heavy, and the silver tape sags.

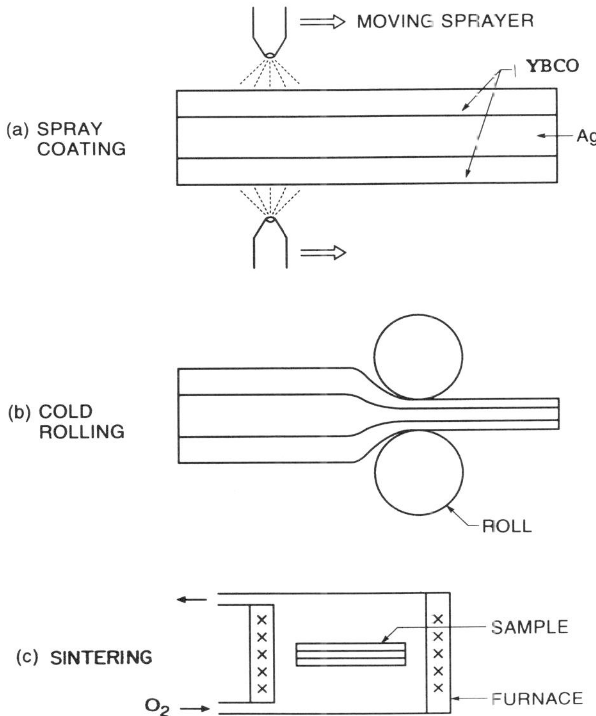

Figure 16.8. Typical method of making ribbon with YBCO coated onto silver. (From Ref. 29.)

The electrophoresis technique tried by Pacific Superconductors for YBCO[7] and pictured in Figure 16.3 has been more successful[28] for BSCCO. Electrophoresis has the advantage that you don't have to pack a tube with powder; moreover, there is no binder to bake out. A 100-μm thickness can be deposited in 5 seconds. In practice, Pacific Superconductors begins with a 250 μm silver core, electrophoretically deposits 100 μm of BSCCO, then 100 μm more of silver, then 100 μm of BSCCO again, and finally another 100 μm of silver, for an overall diameter of about 1 mm. A further attraction is that the step of drawing is eliminated—all that is necessary is to *roll* the wire. Because it is not being pulled through a die, much less silver is required, thus enhancing the cross-sectional percentage of BSCCO. At 77 K in zero field (using a 1 μV/cm criterion), a 115-meter sample of (Bi,Pb)SCCO-2223 coated wire gave $J_c = 8100$ A/cm^2 in the BSCCO, or 1000 A/cm^2 for the overall geometry.

Figure 16.8 is a sketch of another typical process.[29] Starting with a substrate that is either silver or a metal coated with silver, a moving sprayer deposits the HTSC on it. After a rolling operation, the sample is sintered (indeed, partially melt-processed) in a furnace to reach the end product.

A multilayer variation of this has been developed at SUNY-Buffalo.[30] As shown in Figure 16.9, this involves several layers of silver tape forming a sandwich, with intervening thick films of BSCCO-2212 applied via a *doctor blade* technique. Figure 16.9 also shows the furnace geometry by which a partial melt-texturing step is included in the process. The end product is a three-layer BSCCO tape.

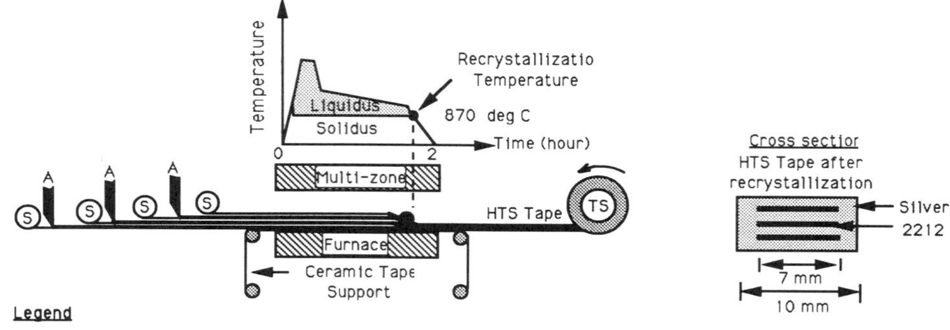

Figure 16.9. Schematic of the continuous BSCCO-2212/Ag composite tape processing, using the Thick Film/Partial Melt Texturing technique. (From Ref. 30.)

It is also possible to make a *jelly roll* conductor, by first placing a layer of HTSC on some normal metal and then rolling up this laminate, as shown[31] in Figure 16.10. After further mechanical processing, the result is a long, thin multilayer conductor.

Thin films regularly exhibit much higher J_c values than do bulk samples, but for practical applications it is I_c that counts as much as J_c: total amps, not just current density. To make high-amperage conductors out of thin films, attention must be given to the substrate because the film itself will be only microns thick.[32] It must be strong, flexible, thin, inexpensive, and have a lattice structure that helps orient the HTSC thin film deposited on it (or at least be compatible with an *orienting buffer*[†]). There is always a considerable amount of cross-sectional area associated with the substrate and stabilizer that coats the final HTSC thin

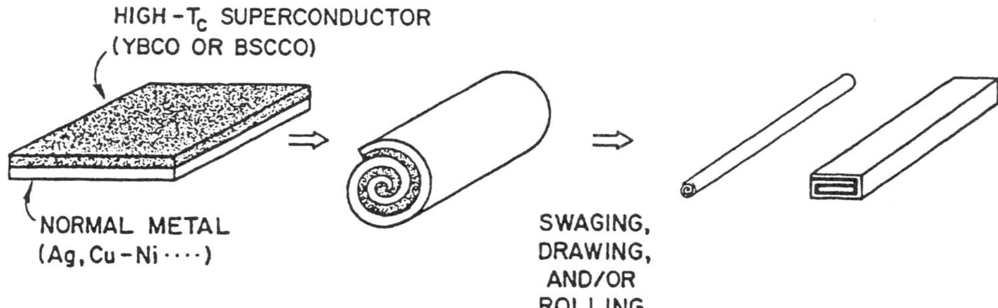

Figure 16.10. Basic steps in the Jelly-Roll process. (From Ref. 31.)

[†]An orienting buffer is an epitaxial layer whose lattice constants are such that when HTSC is deposited on it, the lattice structure of the HTSC matches sufficiently well to provide a good film.

film, and this increase in area penalizes the effective current density (or engineering J_c), leaving it substantially lower than J_c in the film itself. Nevertheless, when J_c is two orders of magnitude higher in films than in sintered material, a penalty in cross-sectional area of a factor of 100 is acceptable.

16.4.2. Metallic Precursor Methods

Other methods of making ribbon are also in use: American Superconductor (ASC) experimented with *melt-spinning* to make a 100-μm thick metallic precursor ribbon. They call the metallic precursor composite Bisc-Ag.[33] After depositing it on the spinning wheel, the metal is oxidized to form platelets of BSCCO. These platelets are randomly aligned, so the next thing to do is improve the orientation, by pressing and sintering the ribbon; the alignment is then much better. Following these treatment steps, the silver is still dispersed throughout the material. This ribbon is 70 or 80% silver by weight, the remainder being platelets of (Pb,Bi) material of 2223 composition. ASC finds $J_c = 3,000-5,000$ A/cm^2 in zero field; their hope is to get to 30,000 or 50,000 A/cm^2. The advantage is that this ribbon is very long.

The National Renewable Energy Laboratory (NREL) makes TBCCO as follows[34]: They begin by electrodepositing precursor films on a silver tape. Next, a two-zone process holds a container of Tl_2O_3 at $\approx 730°C$ while the precursor sample is at 840–860°C, in flowing oxygen, for about 30 minutes. After cooling, the sample is given a post-growth oxygenation, in which its temperature ramps up to 600°C over 4 hours, stays at 600°C for 4 hours, and then ramps down to 200°C over 4 more hours. Subsequent XRDs show that the sample develops highly phase-pure, c-axis oriented TBCCO-1223. The two-zone annealing process uniformly melts the layer. The onset of superconductivity is at 113 K, and zero resistance occurred at 106 K. Results are encouraging: at 77 K and zero field, $J_c = 68,000$ A/cm^2; in a field of 5 T (oriented in the *ab* place and parallel to the current), $J_c = 8200$ A/cm^2 at 77 K.

16.5. MONOFILAMENT WIRES

Modern LTSC wire (made of NbTi or Nb$_3$Sn) is an engineering marvel, with thousands of micron-sized filaments embedded in a matrix of copper.[35] Figures 2.11 and 17.11 show their cross-sections. But first it was necessary to make a single strand of NbTi; later, the importance of *flux jumping* and *stability* provided the incentive to go on to multifilament wire. For the HTSCs, only BSCCO has moved to the multifilament stage; the others are still working on single strands.

16.5.1. YBCO

The biggest disappointment in the entire field of HTSCs has been the difficulty[36] of making wire out of YBCO. YBCO is now in practical use as a coating for microwave cavities, in SQUIDs, and in several other thin-film applications, but it has eluded all attempts to turn it into wire, as the word is commonly used. Only short (< 1 m) samples of melt-textured YBCO have shown the kind of $J_c(H)$ needed for practical applications at 77 K. Meanwhile, at 4.2 K, BSCCO performs better than YBCO anyway, so there is no motivation to endure the brittleness problem associated with YBCO.

There definitely has been progress in melt-texturing,[13] as recounted in Chapter 15. Figure 15.17 sums up the state of affairs for YBCO. There are 15 cm samples of melt-textured

YBCO fibers that carry 4×10^4 A/cm^2 at 1 T. But the usefulness of melt-texturing for long lengths remains questionable. The rate of melt-texturing small filaments is up to 1 cm/hr, compared to 1 mm/hr in 1990, and still further increases in speed are a worldwide goal.

Are there fundamental limitations on the speed of certain metallurgical processes that will prevent further gains here? No one knows how to speed up the melt-texturing process, and there is substantial evidence that it cannot be speeded up. A quote that appears in section 9.4 above is worth recalling: in summarizing the YBCO situation, Jin and Graebner[36] stated that "further processing innovations are needed in order to accelerate progress toward major bulk applications."

"Further processing innovations" means some entirely new and creative approach, as exceptional as melt-processing was in 1988. Otherwise, there will not be any high-current wire made from YBCO.

One example of such an innovation comes from Los Alamos,[37] where a thick film of YBCO was applied to a substrate with carefully aligned grain boundaries. The technique of Ion Beam Assisted Deposition (IBAD) was used to reduce high-angle grain boundaries while depositing Yttrium-Stabilized Zirconia on a nickel tape. The resulting YSZ/Ni substrate led to well-aligned YBCO, which carried $J_c = 800,000$ A/cm^2 in zero field at 75 K, and retained J_c above 30,000 A/cm^2 in H = 1 tesla at 75 K. This process might be scalable to produce commercial lengths.

The goal of all hopeful HTSC wire manufacturers is to develop a practical production method for wire carrying 10^4 A/cm^2 at 5 T. Finding an alternate way to achieve high grain alignment with clean grain boundaries in YBCO might achieve that goal without resorting to melt-texturing.

16.5.2. BSCCO

BSCCO has a complicated pattern of properties, owing to its great anisotropy. At 4.2 K, BSCCO is an excellent superconductor,[38] and if inexpensive, durable wire can be made from it, BSCCO will promptly become a favorite replacement for NbTi or Nb$_3$Sn: it has a far higher critical magnetic field at 4 K (not yet measured, but believed to exceed 100 T). The J_c of BSCCO declines very slowly with increasing magnetic field at 4.2 K, as shown in Figure 16.11. In high magnetic fields BSCCO is still hanging in there, long after NbTi and Nb$_3$Sn have given up. BSCCO is very likely to gain a prominent role in high-field low-temperature superconductivity.[39]

The performance of BSCCO at 20 K is nearly as good as at 4 K. Figure 16.12 makes it obvious that some amount of melt-texturing in the production process is very helpful. Clearly, it is an engineering goal to optimize both J_c performance and wire production speed. Intermagnetics General used powder-in-tube methods (with a two-step sintering process[40]) to make a tape conductor of BSCCO-2223, and measured J_c in fields up to 20 T. Data from these experiments[41] appears in Figure 16.13. At 4.2 K, BSCCO continues to hold $J_c > 50,000$ A/cm^2, even up to a 20 T field. At 20 K, the numerical values are slightly lower, but the trend line is the same.

Unfortunately, the phenomenon of flux lattice melting sets in at slightly higher temperatures in strong magnetic fields, and J_c drops rapidly. Figure 16.13 shows that even at 27 K, J_c falls off sharply. This presents a very serious problem for BSCCO. Unless the pinning strength can be greatly increased somehow, the eventual uses of BSCCO will be restricted either to low magnetic fields or low temperatures.

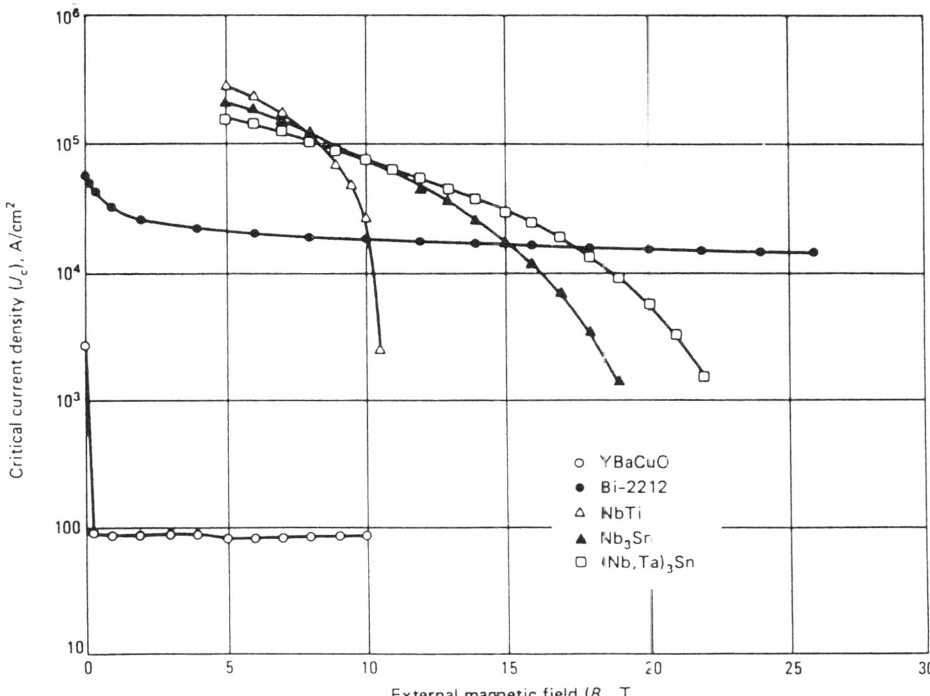

Figure 16.11.   Plot of J_c vs. applied magnetic field at 4.2 K, to compare two PIT HTSC wires (BSCCO-2212/Ag, and YBCO) with three conventional multifilament wires. J_c data is for the *superconductor* cross-section, not the cross section of the composite wire. [From K. Heine *et al., Appl. Phys. Lett.* **55**, 2441 (1989).]

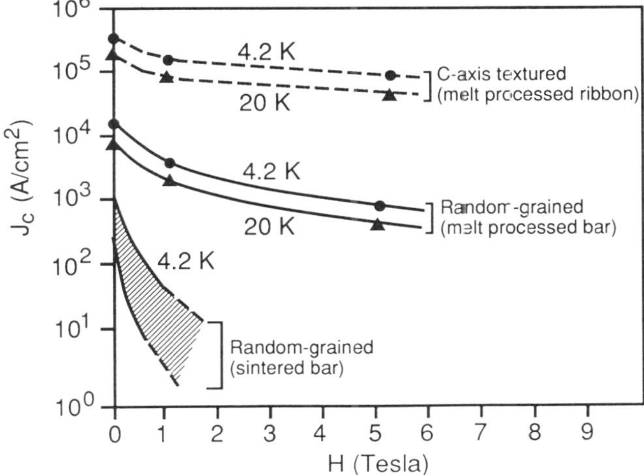

Figure 16.12.   Decline of J_c with increasing magnetic field in BSCCO at low temperatures. Results at 20 K are reminiscent of those at 4.2 K. Melt-processing of BSCCO creates a substantial advantage. [From S. Jin *et al., Appl. Phys. Lett.* **59**, 366 (1991).]

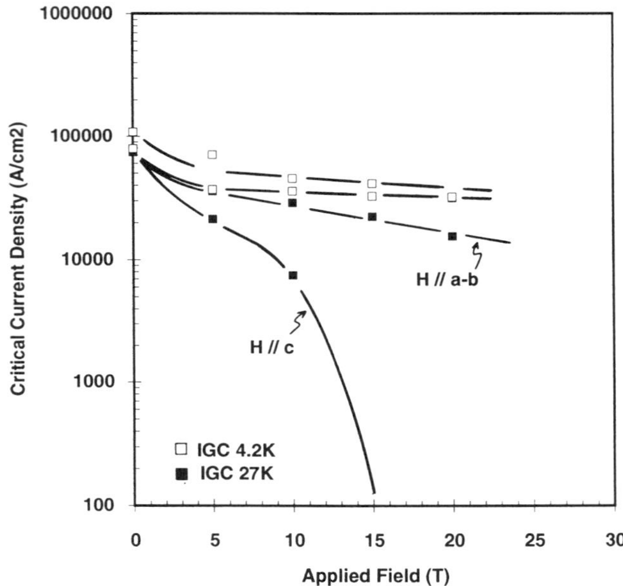

Figure 16.13. Decline of J_c with increasing magnetic field for BSCCO. At 27 K, the roll-off depends strongly on the orientation of the magnetic field. (From Ref. 41.)

The dramatic difference between $H \parallel c$ and $H \parallel ab$ is attributable to flux line motion. With $\mathbf{J}$ and $\mathbf{H}$ both in the ab-plane, the Lorentz force lies along the c-axis; but it is hard for flux lines to move along c; this is one type of *intrinsic pinning*. On the other hand, with $\mathbf{H}$ along c, the Lorentz force lies in the ab-plane, and flux lines can move easily between the CuO_2 planes. As shown by the two different 27 K lines in Figure 16.13, there is very little restraint upon flux line motion in the ab plane ($\mathbf{H} \parallel c$), even at 27 K. The best way to pin c-aligned flux lines is with *columnar defects*, described in Section 14.8. These can significantly increase the fields tolerated in BSCCO tapes.[42]

At still higher temperatures, J_c in BSCCO deteriorates rapidly due to flux lattice melting. Figure 16.14 presents the full picture up to 90 K for both orientations of applied magnetic field. Of greatest interest is the data at 77 K, where J_c is dropping like a rock. Only for fields below 1 T can BSCCO even be considered any more. A close-up of this low-field region at 77 K appears in the log–log plot of Figure 16.15. BSCCO-2223 tape holds up much better[20] than either BSCCO-2212 wire or YBCO.

There are some applications in zero or weak fields (notably transmission lines), so effort is continuing toward making BSCCO conductors for 77 K use. Table 16.2 presents typical data of BSCCO tapes for various temperature and magnetic field conditions; back in 1990, this was "champion" data. Most of this data comes either from Vacuumschmelze in Germany, or from Japan, where Sumitomo Electric Industries is a leading developer[43] of bismuth tapes. While there is little difference between BSCCO-2223 and BSCCO-2212 at low temperatures, at 77 K the 2223-phase is obviously preferable. However, as was discussed in Chapter 15, this phase is extremely difficult to make, unless the bismuth is partially substituted by lead. Consequently, most engineering effort is devoted to the lead-bismuth variety of 2223.

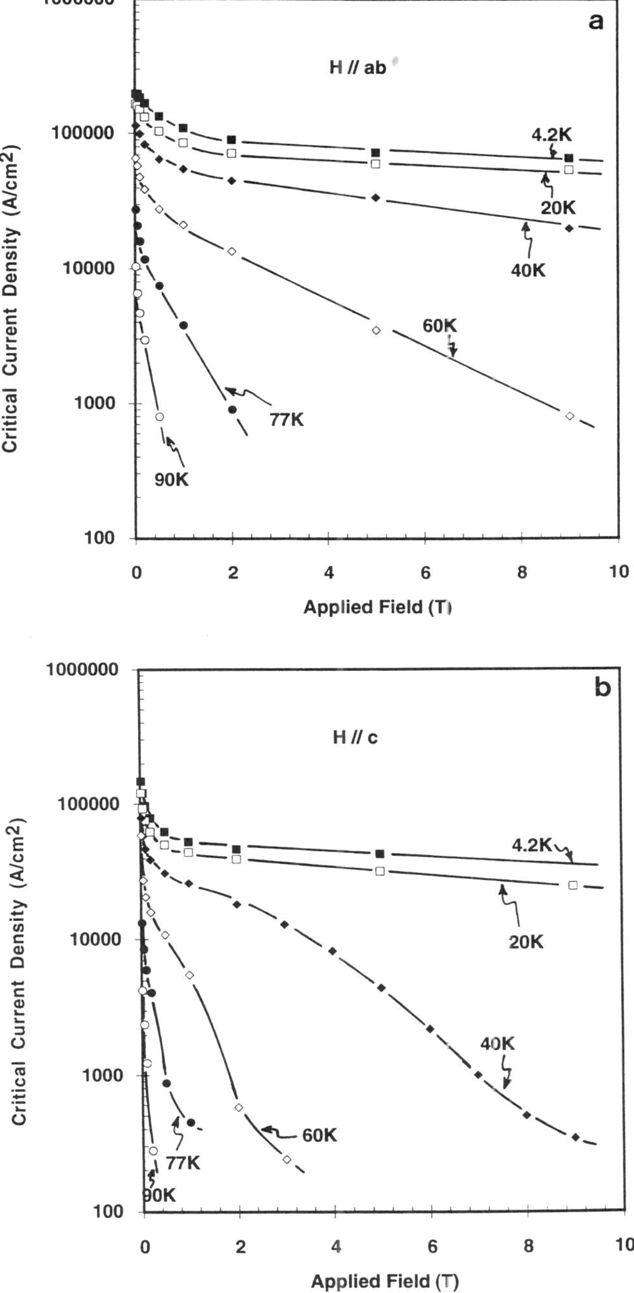

Figure 16.14. Decline of J_c with increasing magnetic field for BSCCO, over a wide temperature range. a) Magnetic field H ∥ ab plane; b) magnetic field H ∥ c axis. Flux lattice melting effects are more severe with H ∥ c. (From Ref. 25.)

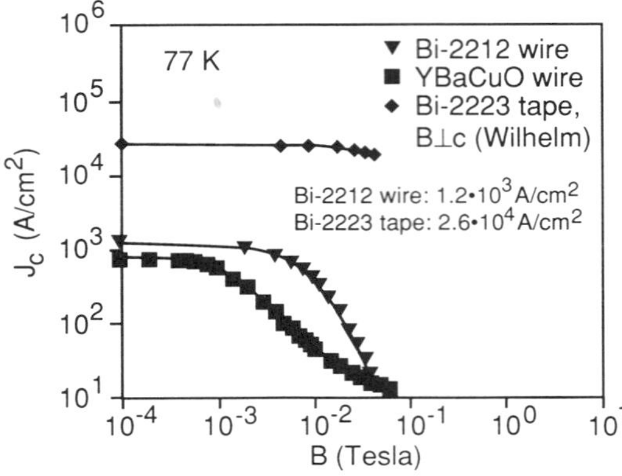

Figure 16.15. For 77 K, log-log plot of J_c vs. magnetic field, giving a close-up of the low-field region. BSCCO-2212 wire is compared with BSCCO-2223 tape and with YBCO. [From Ref. 20, © 1991 IEEE.]

When producing long lengths of wire, it is essential not to overlook the importance of aligning the highly anisotropic BSCCO crystals. All the data in Table 16.2 had a magnetic field in the plane of the tape, and perpendicular to the current. When the magnetic field is no longer aligned with the *ab*-plane, J_c deteriorates faster in moderate magnetic fields. Thus, any misorientation of crystals during wire manufacture carries a severe performance penalty.

What is the bottom line? At 77 K, forget BSCCO, except in a zero field. However, all is not lost. For BSCCO-2212, flux lattice melting has not yet set in at 20 K; similarly, BSCCO-2223 is still useful up to about 35 K. Therefore, in high magnetic fields, BSCCO holds great promise in the intermediate temperature range. As we shall see in Chapter 24, advances in cryogenic refrigeration systems are rapidly making that range accessible.

16.5.3. TBCCO

How will thallium compounds perform when made into wire? Any effort to commercialize TBCCO wire is still well downstream, but the research is focused primarily upon TBCCO-1223. As discussed in Chapters 8 and 14, the closer spacing associated with the single-thallium-oxide layer of 1223 enhances flux-pinning and thus improves hope for high J_c at 77 K. On the other hand, TBCCO grains do not easily align the way BSCCO grains do, so it may be much harder to make wire.

Table 16.2. Performance [J_c (A/cm²)] of Bismuth Tapes[a]

Material	2212	2223
Temperature K		
4.2	2.0×10^5 ($H = 30.0$ T)	1.0×10^5 ($H = 23.0$ T)
20.0	6.0×10^4 ($H = 12.0$ T)	5.5×10^4 ($H = 20.0$ T)
77.0	3.5×10^4 ($H = 0$ T)	4.7×10^4 ($H = 0$ T)
	3.5×10^3 ($H = 0.1$ T)	3.1×10^4 ($H = 0.1$ T)
		1.1×10^4 ($H = 1.0$ T)

[a]Maximum reported J_c A/cm² with H in the tape plane and $H \perp J$.

A. *Powder-in-Tube Processing.* A group at Los Alamos, headed by Dean Peterson, used the powder-in-tube (PIT) method to make tapes[44] of TBCCO-1223. Their initial 1223 powder suffered from various deficiencies, but it was still useful to see how well the powder-in-tube method would work for TBCCO. The object was to simultaneously overcome both the flux pinning problem and the grain boundary problem.

The deformation process used by Los Alamos began with a tube of 6.35 mm outer diameter and ended with 1.0 mm o.d. Next it was rolled into a tape of final thickness 0.24 mm. Then various sections were cut and subjected to assorted heat-treatment cycles, sintering, and additional pressing. Selected sections were used for magnetic susceptibility measurements and some for transport measurements. The microstructure of some pieces of the tape were examined. As expected, TBCCO does not have the favorable grain alignment associated with PIT BSCCO.

Initially, *texturing* and *densification* were the major problem. The volume fraction of the superconductor was about 50%, and although the onset of superconductivity was above 100 K in all cases, only half the tape or wire was superconducting at 77 K. Figure 16.16 shows their direct transport current measurements,[44] as of 1992. Obviously, this is disappointing data, because J_c drops precipitously in less than 0.1 T magnetic field, to about 1000 A/cm^2. This is a clear indication of weak-link behavior. But this was expected; the same thing was seen in early attempts at YBCO and BSCCO.

Texturing is the more difficult problem; densification can usually be achieved through process modifications. The Los Alamos group went on to improve both their initial powder processing and subsequent thermomechanical processing. By 1994, the outlook was sufficiently promising to enter into a CRADA with Intermagnetics General, hoping to scale up the process to commercial lengths.

Looking closely at Figure 16.16, the J_c values holds up rather well in the lower temperature range, showing acceptable decreases out to 5 T. Even at 50 K, the data is not bad. What this shows is that flux pinning is nowhere near as bad a problem as it is for BSCCO, which fades badly even by 2 T at 35 K. Figure 16.17 compares BSCCO-2223 with TBCCO-1223 at 35 K. The difference is compatible with the structural features

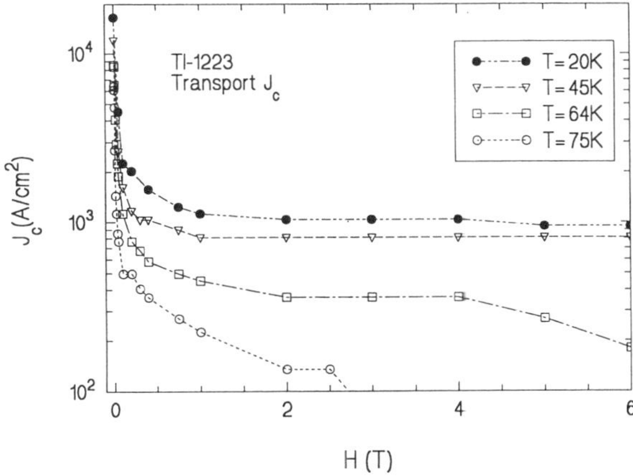

Figure 16.16. Transport J_c vs. magnetic field at selected temperatures for a rolled and pressed TBCCO-1223/Ag tape, with the field oriented perpendicular to the tape plane. [From Ref. 44, © 1993 IEEE.]

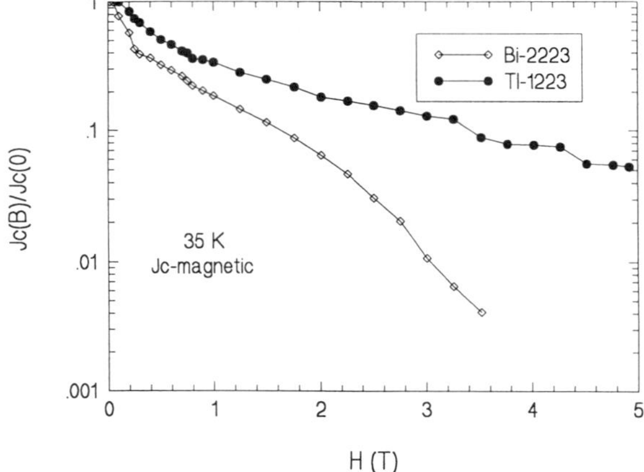

Figure 16.17.   Magnetization J_c vs. magnetic field at 35 K for rolled and pressed tapes of TBCCO-1223/Ag and BSCCO-2223/Ag. Each J_c is normalized to J_c at zero field. The magnetic field is oriented perpendicular to the tape plane.

discussed in Chapters 8 and 14 that relates flux lattice melting to spacing between insulating planes.[45] The overall conclusion to be drawn is that TBCCO-1223 is a fully viable candidate for making useful HTSC wire.

B. *Thick Film Methods.* The National Renewable Energy Laboratory (NREL) has applied several kinds of thick-film fabrication techniques[46] to TBCCO. The method described above in Section 16.4.2 yielded a tape[34] with the properties shown in Figure 16.18. (Recall that the

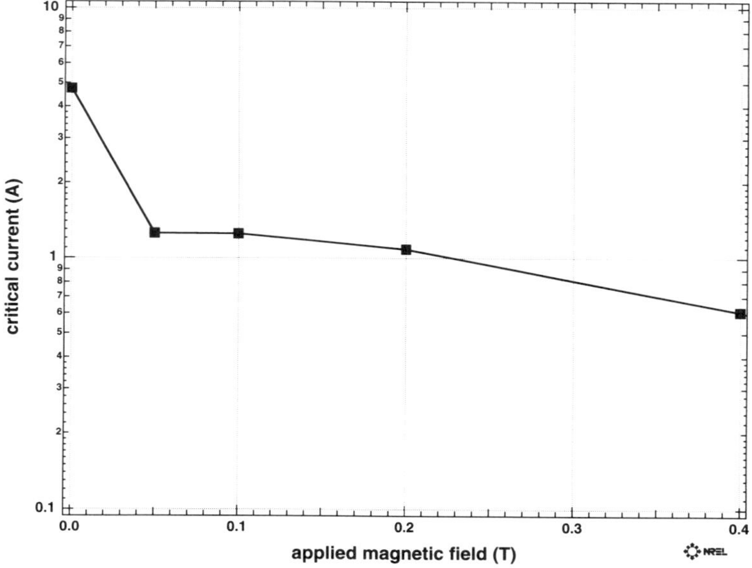

Figure 16.18.   NREL data showing I_c vs. magnetic field for a thick film of TBCCO-1223 prepared by their metallic precursor method.

irregular cross-section prevented conversion of I_c data to J_c.) At 76 K and zero field, the total current was 5 A. This dropped sharply by a factor of 3 in 0.05 T, but then a relatively flat region began and extended out to at least 0.4 T. This behavior in a magnetic field indicates that flux lattice melting is not occurring.

General Electric, in collaboration with NREL, used a spray pyrolysis technique[47] to make thick films of TBCCO-1223, which gave much better results than earlier TBCCO-2223 material. Here, the J_c values were scattered between 10,000 and 100,000 A/cm^2; for example, J_c = 56,000 A/cm^2 for TBCCO-1223 on a SrTiO$_3$ substrate, but on a silver foil it drops to 8,000. The most noteworthy number observed was a critical current in 1 T field at 77 K of 10,000 A/cm^2. By contrast, earlier TBCCO-2223 data showed a sudden drop-off around 1 T.

16.5.4. Effect of Irregularities

Clearly, there have been successes on the long road toward making wire. Still, it must be remembered that long wires do not perform as well as short segments, and most data is taken on short segments. Along any one wire, various segments will have different values of J_c, which will be distributed about some mean value. Actual wire frequency distributions of J_c are shown[20,23] in Figure 16.19. Unfortunately, in a wire these are all in series, and so the low values dominate the wire behavior. Very often, experimenters only report their best values for short segments. What is needed eventually is for the *worst* J_c value in a long segment to be rather high. One way to overcome this limitation is by offering multiple paths to the current, so as to gain statistical safety in numbers. That is the topic of the next section.

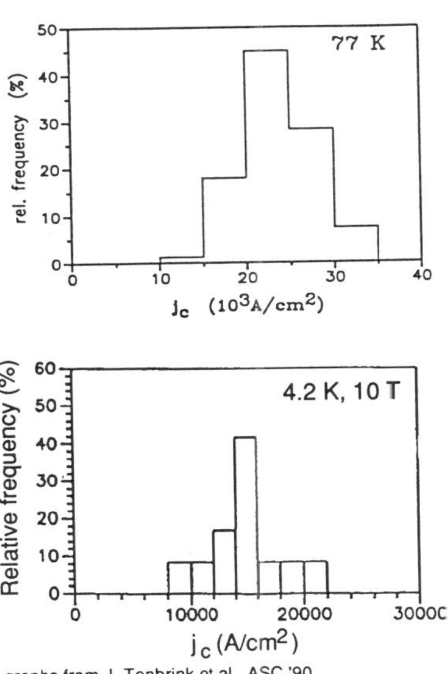

Figure 16.19. Histograms of the J_c values obtained at various places along BSCCO wires, showing distribution of local J_c values. Top: (Bi,Pb)SCCO-2223 tapes at 77 K and 0 T, from Ref. 23. Bottom: BSCCO-2212 at 4.2 K and 10 T. (From Ref. 20; © 1991, 1993 IEEE.)

16.6. MULTIFILAMENT WIRE

Following the experience of type II superconducting wire manufacturers,[35] there are ongoing efforts to make multifilamentary HTSC wire as well. The reasons for doing so are:

- Sharing the load, so there is negligible loss if one strand breaks.
- Adjacent filaments do not necessarily break when one single filament breaks.
- The *matrix* in multi-conductor wires retards crack propagation.

Multiconductor wires have distinct advantages, but they must match the thermal expansion coefficients of several materials. Also, the matrix surrounding the HTSC filaments must be a conductor (for stability and quench protection, as will be discussed in Chapter 17), but it must be a *poor* conductor so as to generate very small eddy currents.

A practical definition of a *composite* is this: two or more materials in a product that have beneficial properties relative to either material in bulk form. It is certainly valid to refer to silver-sheathed HTSCs as composite wires.

16.6.1. Powder-in-Tube Approach

One method of making multifilament HTSC wire invokes a multiple-stage process of reducing powder-in-tube samples of BSCCO. Sumitomo Electric Industries[48] first demonstrated this technique. HTSC powder in a single silver tube is one filament; after extrusion or rolling, these can be bundled together to make a multi-filament wire. A group of 19 single powder-in-tube BSCCO wires, once extruded, are bundled together and that unit is extruded. Then in turn, a group of these units are bundled together and the extrusion step repeated

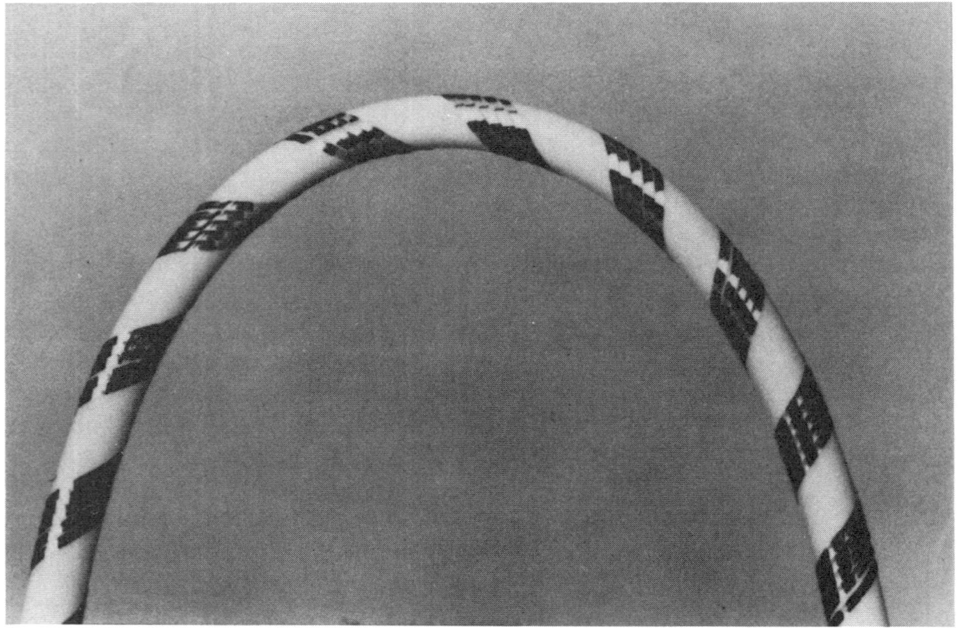

Figure 16.20.   Photograph of flexible cable made of BSCCO by Sumitomo Electric Industries.

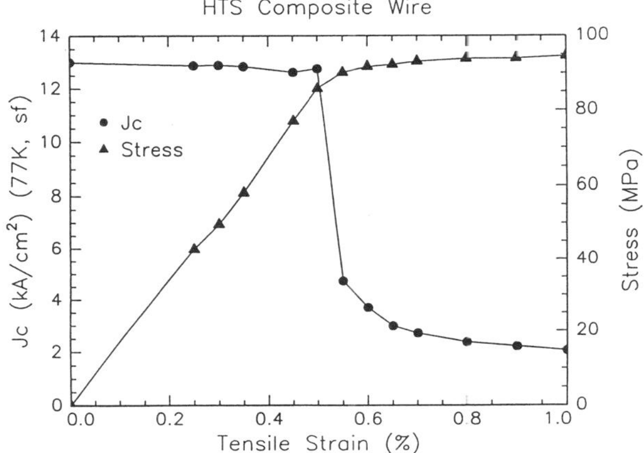

Figure 16.21. American Superconductor data showing stress-strain relation in multi-filament BSCCO wire, along with J_c as a function of strain. A major change takes place at 0.5% strain. [From G. N. Riley Jr. and W. L. Carter, *Bull. Amer. Ceram. Soc.* **72**, 91 (1993).]

again. In this way, Sumitomo constructed a multicore wire having 1,296 individual filaments, or "cores."

Sumitomo has carried its multifilament conductor work still further: four strands of 1,296-core wire have been wound with a 6-cm pitch on a 2-cm diameter Teflon rod, in order to make a short length of a flexible prototype. Figure 16.20 is a photo of this. They have also wound a cable of 22 strands of 1,296-core wire on a 2-cm diameter copper rod.

The advantage here is to gain flexibility, because each core will bend only a fraction of a percent as the whole cable is bent through a few-centimeter radius. The mechanical properties of these Sumitomo multifilamentary cables have been covered in Chapter 11.

Other manufacturers have advanced the state of the art as well. ASC is able to combine either 7 or 19 filaments, embedded in silver, with only a 2% deviation in cross-section after deformation through processing. For one ASC composite wire, the stress-strain relation is shown in Figure 16.21, with J_c superimposed on it. The drop in J_c is unmistakable: yield occurs at a strain of 0.5%, corresponding to a *composite yield stress* of 85 MPa. The BSCCO filaments themselves have a yield stress closer to 300 Mpa, which may be compared with a yield stress of $\approx$ 200 MPa for YBCO (see Chapter 11).

Intermagnetics General (IGC) fabricated the multifilament BSCCO conductors shown in Figure 16.22; samples (a) and (c) have a standard silver sheath, whereas (b) and (d) use silver impregnated with Al_2O_3. There is very little difference in the final cross-sections. However, the stress-strain relations change dramatically: Figure 16.23 shows that the Ag/Al_2O_3-sheathed conductor tolerates approximately twice as much stress.

Electrically, the race continues, and any snapshot will be obsolete by tomorrow. In spring 1994, several major wire manufacturers reported values of J_c above 10,000 A/cm^2 in long multifilament wire, as shown in Table 16.3.

16.6.2. Metallic Precursor Approach

The ease with which multifilament NbTi wire can be manufactured has made it the workhorse of the superconductivity industry. It is everyone's *wish* to have a HTSC wire with

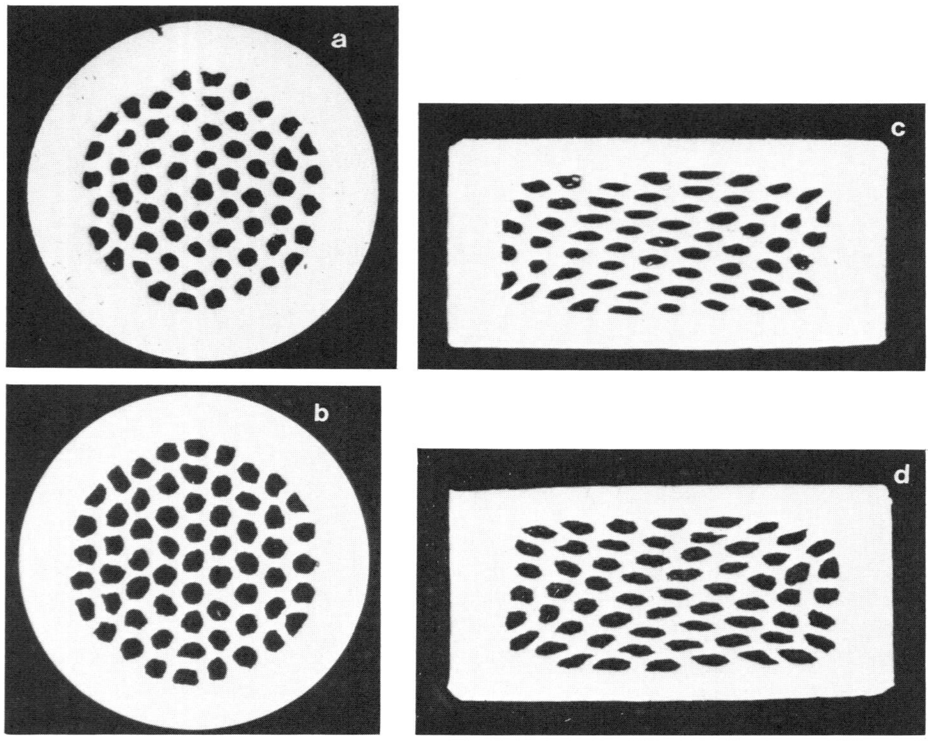

Figure 16.22.   Intermagnetics General multi-filament BSCCO-2212 conductor. (a) Ag sheath at 1.1 mm o.d.; (b) Ag-Al$_2$O$_3$ sheath; (c) Ag sheath pressed to 0.7 mm × 1.4 mm; (d) Ag-Al$_2$O$_3$ sheath. [From L. R. Motowidlo *et al.*, *Applied Superconductivity* **1**, 1503 (1993).]

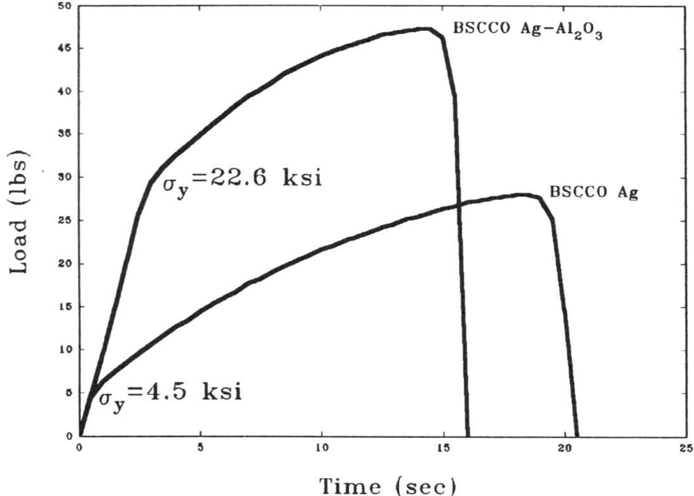

Figure 16.23.   Stress-strain curves for BSCCO multifilament conductors encased in Ag and in Ag-Al$_2$O$_3$. [From Ref. 21.]

Table 16.3. Performance of Multi-filamentary BSCCO-2223 Conductor; Snapshot as of July 1994; All data at 77 K and zero field

Company	Length (m)	Number of filaments	J_c (A/cm^2)
American Superconductor	446	19	13,800
American Superconductor[a]	1160	19	12,700
Intermagnetics General[b]	850	37	10,500
Sumitomo[c]	623	61	15,900
Sumitomo[c]	1080	61	4,020

[a]G. N. Riley, presented at DOE Annual Peer Review, July 19–20, 1994
[b]P. Haldar, Intermagnetics General, private communication
[c]T. Hikata *et al.*, Proc. 1994 International Workshop on Superconductivity, Kyoto, Japan, June 6–9, 1994, pp. 69–72.

the same ease of manufacture. American Superconductor Corp. (ASC) has advanced this avenue with its metallic precursor method, in which all the drawing and rolling steps are carried out on a metal alloy, and only after a full-length multifilament wire exists is the oxygenation step applied to make it superconducting.

To make their metallic superconductor, ASC uses mechanical alloying. Powders of Bi, Pb, Sr, Ca, and Cu are mixed (in the proper proportions to make BSCCO-2223) in a device known as a *ball mill*, which is reminiscent of a paint-shaker, but contains very hard tungsten or ceramic balls to pound the elements together. The final product, a powder of the alloy that is homogeneous on the atomic scale, is packed into a silver can, which is then extruded into a long hexagonal rod. Cut pieces of the rod are stacked into a multirod bundle; in turn, that bundle is drawn down, and slices of that are stacked together for the next round. This process can be repeated several times. In each cycle, the cross-sectional area decreases by a factor of 100, leading to final filaments of the metallic precursor which are typically only 5 μm thick. After that, this silver-BSCCO precursor is rolled into tape. Figure 16.24 shows both the transverse and longitudinal cross-section of a tape containing 9583 filaments.[49]

With all the mechanical deformation successfully completed, it is finally time to convert it to a superconductor. Needless to say, the details of the process are highly proprietary; ASC says only[49]: "This is made possible in part by the remarkably high diffusivity of oxygen through the silver matrix."

BSCCO-2223 is intended primarily for use in low fields at 77 K ; and preliminary results are encouraging: an 85-meter length of conductor containing 351 filaments showed $J_c \approx$ 17,900 A/cm^2 in self-field at 77 K. The key accomplishment is that the cable is moderately flexible, which means it can be wound into coils *after* reacting—thus fulfilling the definition of the react and wind process.

The engineering advances in fabrication of multifilament conductors are impressive and likely to continue. The use of metallic precursors allows extrusion, drawing, and rolling without sacrificing the cross-sectional uniformity of the conductor, thus obtaining the same manufacturing advantages now enjoyed by NbTi multifilament wire. It is reasonable to hope that scale-up will succeed without deterioration of J_c, which gives cause for optimism.

16.7. COILS

The key to making good coils is *flexibility*. This has been the downfall of nearly all attempts to make coils of bulk HTSCs. Multi-filamentary conductors have several advantages: First, when multiple filaments are embedded in silver, crack propagation is inhibited.

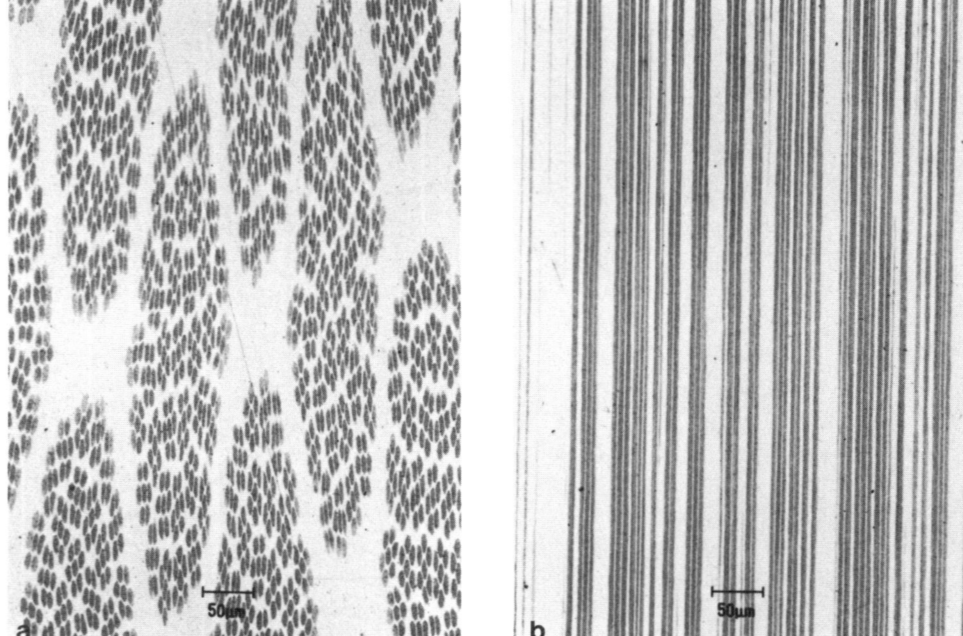

Figure 16.24. Transverse and longitudinal cross-sections of silver clad BSCCO-2223 tape prepared by the multifilament metallic-precursor approach. (Photos courtesy of American Superconductor Corp.)

Second, if one filament does crack, neighboring filaments can carry its share of the current. Third, when a filament only 1 micron in diameter bends on a 10 cm radius, the strain is only 1 part in 10^5. (Section 11.7 treats this in more detail.)

One issue that demands attention for coils constructed from HTSCs is the *hoop stress* created by the Lorentz force. Consider a coil of 5 cm radius, carrying 60 A and producing a field of 20 T: the hoop stress of that configuration is 38 Mpa. This compares with a yield strength of 25 MPa for a typical BSCCO-2212/Ag wire. Vacuumschmelze overcomes the problem by using a sheath made of a silver–nickel–magnesium alloy, with a yield strength of 100–150 MPa.[26] As a side benefit, Vacuumschmelze found that the Ag-Ni-Mg alloy contracts upon annealing, which leaves the BSCCO-2212 in compression, thus reducing the chance of cracking.

Optimism is high for BSCCO-2223/silver composites. Figure 16.25 is a photo of some typical pancake-coils.[19] Dimensions and electrical properties (as of early 1994) are listed in Table 16.4; at 4.2 K, the best coils produce 2.6 T field. The dip-coating method[27] has also led to pancake coils,[50] the best of which also yield 2.6 T field. New experimental techniques, such as *Magneto-Optic Imaging* (which sees flux lines moving at 77 K), are enabling researchers to advance rapidly.

Tapes prepared by the metallic precursor approach[49] have been wound into coils (overall lengths up to 85 m) and tested up to 5 T at various temperatures.[51] The phase YBCO-124 ($T_c = 75$ K) can be made this way,[52] but it still suffers great attrition in $J_c(H)$ due to weak

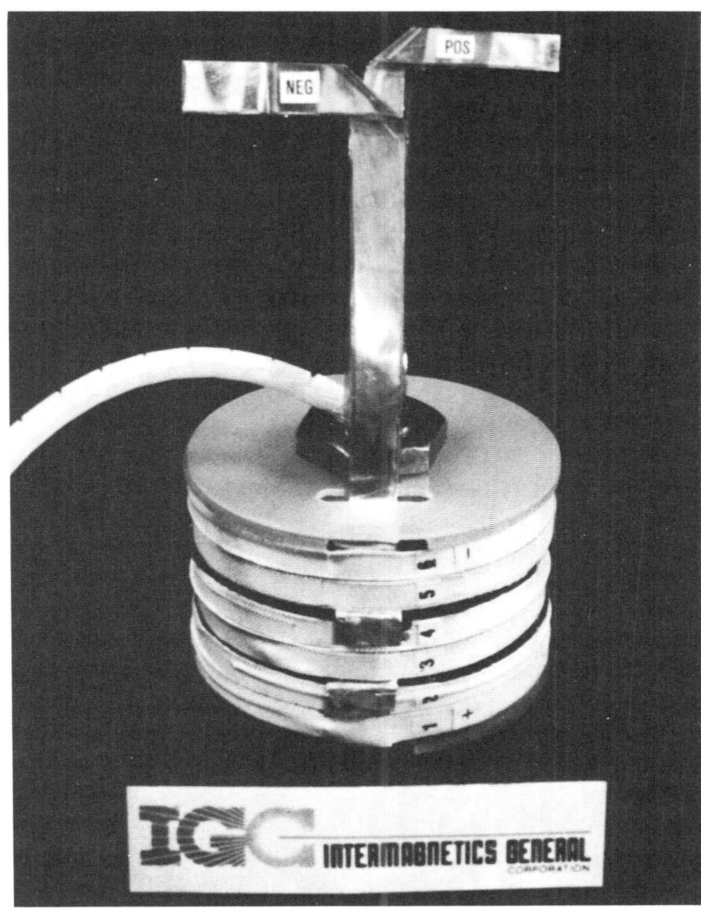

Figure 16.25. Photograph of pancake coils made by Intermagnetics General Corp.

Table 16.4. 1993 Coil Data

		Coil 1 (June 93)	Coil 2 (Aug. 93)	Coil 4 (Oct. 93)
Winding inner diameter (cm)		2.50	2.50	2.50
Winding outer diameter (cm)		7.50	9.65	11.30
Coil height (cm)		5.33	9.84	6.35
Number of Co-wound tapes per pancake		3	5	3
Total length of tape in magnet (m)		153	570	480
Total number of turns in the magnet		330	612	700
Overall winding cross-section (cm^2)		0.0348	0.0532	0.0363
Number of pancake coils		6	10	10
Magnet constant (gauss/amp)		62.9	64.7	111.2
4.2 K	I_c max	170 A	255 A	234 A
	B_0 max	1.01 T	1.65 T	2.60 T
27 K	I_c max	125 A	171 A	160 A
	B_0 max	0.78 T	1.10 T	1.8 T
77 K	I_c max	19 A	41 A	32 A
	B_0 max	0.12 T	0.26 T	0.36 T

links, so it is not very interesting for coils. Tapes made of BSCCO-2223 have a much more optimistic outlook. Figure 16.14(a) suggests that with good alignment of the *ab*-plane, at 50 K the attrition of J_c in a 5 T field would be only about a factor of 3 from the zero-field value. In fact, ASC found a factor of 2.6 fall-off for a coil wound from 35 m of BSCCO-2223 tape made by the metallic precursor approach. Although J_c itself was only 400 A/cm^2, unacceptably low, the implication of this prototype test result is that BSCCO-2223 could conceivably be used up to 50 K. The potential commercial value associated with such coils assures that a strong research effort will continue here.

TBCCO-1223 has good magnetic behavior (sufficient flux pinning at 77 K), so it is a natural candidate for making coils, if it can be made flexible. However, it does not have the easy slip planes of BSCCO; indeed, NREL refers to the "unforgiving mechanical properties" of TBCCO. The trick is to obtain both good compaction and densification (to mitigate the weak link problem) at the same time as good flexibility. This goal is elusive.

One approach is to depart from the simple chemistry of conventional TBCCO. The 7-element compound $Tl_{0.5}Pb_{0.5}Ba_{0.4}Sr_{1.6}Ca_2Cu_3O_{8.5}$ was investigated at SUNY-Buffalo by Ren and Wang,[53] who followed a powder-in-tube procedure leading to silver-sheathed tapes. They succeeded in making a tape 24 m long, with J_c nearly 10^4 A/cm^2 in zero field at 77 K. Subsequently, they wound this into an 89-turn coil, but it was necessary to follow this with a 10-hr anneal at 845°C; so the full procedure is a relative of the wind and react method. What everyone wants is wire that does not require any final heat treating after winding it into the geometry of its application. Nevertheless, this is a step forward, and it suffices for some applications.

16.8. FUTURE DIRECTIONS

For all HTSCs, it is desirable to increase J_c at 77 K, no matter what the magnetic field is—zero field or many tesla. This is the flux pinning problem, and it is especially acute in BSCCO. If pinning strength in BSCCO can be upgraded enough to permit operation at 35 K, additional practical applications will open up as new refrigerators are developed. This has been recognized by the U.S. Department of Energy's HTSC wire research program since 1988, when an operating temperature of 35 K was established.[54]

The future directions for high-amperage conductor research can be categorized as either basic or engineering tasks. The basic issues include the following:

- For YBCO, speed up the melt-texturing process, which is essential for good alignment of grain boundaries.
- For BSCCO, improve pinning strength (required for 77 K operation).
- For TBCCO, align grains in multifilamentary wire.

There are more engineering tasks; a list provided by Westinghouse[55] states them well:

- development of long lengths of multifilamentary wire;
- filament *uniformity* over long lengths;
- heat treatment that minimizes core voids (and core-to-sheath discontinuities);
- measurement of I_c over long lengths at 4.2 and 77 K, in finite fields;
- braiding and coiling to minimize strains on brittle cores;
- materials, methods, and timing of putting on insulation.

Significantly absent from this list is mention of reaching higher J_c. The emphasis on uniformity, long lengths, insulation, and so on shows that the major goal is making magnets. BSCCO has good enough electrical properties at 4.2 K, so the problems of fabricating actual wire have taken center stage.

At the same time, comparatively little attention is being given to such topics as reducing the ratio of silver to HTSC, fabricating low-resistance (i.e., superconducting) joints between HTSC wires, or making practical lengths of conductors from thin films. Each of these issues will have to be met eventually, but they are of secondary concern.

The question of tapes versus wires also sits in the background. Although BSCCO-2223 tapes have better performance than BSCCO-2212 wires, most magnet designers prefer to use wire conductors rather than tape conductors. Wires are essentially one-dimensional, which permits easy winding of coils, especially for magnets requiring complicated coil designs. Tapes are two-dimensional, which complicates winding. Tapes are also susceptible to edge effects. BSCCO tapes will be even more difficult to use because of their anisotropic behavior. On the other hand, a higher packing density can be obtained by using tapes, so they are preferred for some applications.

One new topic that is beginning to receive attention is exploring the economics and engineering feasibility of operating at intermediate temperatures. Advances in refrigeration systems have lifted the either/or restriction of 77 K versus 4 K. Also, every hopeful wire manufacturer is acutely aware of the need to hold down manufacturing costs in any full-scale production process.

Intermagnetics General Corp. (IGC) is America's leading producer of NbTi wire, and they have more experience with Nb_3Sn than anyone else in America as well. Therefore, they are keenly aware of the importance of choosing a process which can be scaled up, and an important part of their planning is to eventually use their existing wire-drawing and rolling facilities. IGC has had some success with finite lengths of BSCCO, and they assert that "scale up to much longer lengths is not expected to pose insurmountable problems."

At a 1992 Wire Development Workshop, researchers with widely different specialties came to very similar conclusions about the status of HTSC wire. Defining "good" as having a resistivity of 10^{-11} Ω-cm with $J_c > 10,000$ A/cm^2 in 1 or 2 T field, they concluded:

- BSCCO-2212 is good up to 20 K.
- BSCCO-2223 is good up to 35 K.
- TBCCO-1223 is good up to 70 K.
- YBCO-123 is good up to 77 K.

It would be incorrect to call YBCO better than TBCCO because of this. The relatively well-studied YBCO faces severe processing difficulties [13,36] By contrast, little is known about processes for making wire out of TBCCO. It would be overly optimistic to presume that the as-yet unexplored TBCCO-1223 will overcome these very same problems.

Meanwhile, comparisons with BSCCO processing may or may not be relevant, because TBCCO lacks the micaceousness of BSCCO. The thallium compounds are less easy to draw out, but fortunately the kinetics are much better: the same conditions can be obtained in thallium after only 3 hr that require 100–150 hr for BSCCO. Also, cracks in TBCCO heal in a short time. Moreover, the toxicity problem of TBCCO goes way down when a little Bi or Pb is substituted for Tl, and that makes it easier to develop the powder-in-tube process. A typical prognosis for wire-development in TBCCO is given in Table 16.5.

Table 16.5. Prospects for Thallium

A workshop hosted by NREL in 1993 gathered researchers from six national labs, seven corporations, and four universities who concluded in part:[a]

- For wire and tape, there is interest in the Tl-1223 compound due to its promising transport properties in technologically useful magnetic fields at 77 K.

- Although silver is known to lower the melting point of the thallium oxide superconductors, its exact role in producing highly textured films with good current-carrying properties requires further investigation. Silver appears to provide a liquid assisted mechanism that influences the reaction kinetics.

- The two-zone annealing process for maintaining the thallium partial pressure, either through equilibrium or transport approaches, is becoming a standard technique among researchers and industry for reacting thallium compounds.

- The thallium powder-in-tube (PIT) process shows poor texturing, exacerbated by the thallium compounds' unforgiving mechanical properties. Some texturing adjacent to the silver interface was observed for Tl-1223 PIT.

---

[a]Reported in U.S. Department of Energy *Superconductivity Bulletin* **2**(2), 6 (June 1993).

Most important of all, TBCCO pins flux better than BSCCO. BSCCO does not seem to be weak-linked to the extent that TBCCO or YBCO are, but it may be easier to solve the weak-link problem for thallium than to solve the flux pinning problem for bismuth.

16.9. SUMMARY

This chapter explains how HTSC wire is made, gives a snapshot of where progress stands, and points toward possible future directions for experimentation.

The combination of high current, field and temperature (in long lengths) has not yet been achieved. In thin films, we find $J_c > 10^6$ A/cm^2, but bulk samples do not reach that level. For YBCO, melt-texturing gives good results in very short lengths; but J_c deteriorates with increasing length, and it may be impossible to speed up the melt-texturing process.

The powder-in-tube (PIT) method is the leading means of producing HTSC wire. Here, powder is first packed into a silver tube, which is then drawn and rolled. (Silver is used because it does not react with the core, and readily permits the passage of oxygen, which must be restored to the structure in a final annealing step.) The rolling step is followed by a heat treatment at elevated temperatures, to cause a partial melt. This partial melting step is very important. The micaceous BSCCO has the unique property that when heated and deformed, it produces aligned crystals in the finished product. Thallium compounds, lacking similar micaceousness, have not been treated as successfully by this method.

At 4.2 K, both BSCCO-2212 wires and -2223 tapes outperform conventional NbTi wires or Nb$_3$Sn tapes in high fields. The first applications of BSCCO is likely to be for generating very high fields at 4 K, because no niobium-based superconductor remains superconducting in fields above 20 T. Thus, BSCCO has an assured niche market, if wire can be made from it economically. At this time, BSCCO-2212 appears easier to process than BSCCO-2223.

On the other hand, above 20 K, the onset of flux lattice melting reduces the performance of BSCCO. As we saw in Chapter 14, flux lattice melting does not happen all at once, so the 2223 phase survives out to above 35 K. However, all BSCCO phases are very nearly ruled out of high-temperature applications. Only in low field does BSCCO (2223 phase) offer any hope at 77 K; as soon as a magnetic field is applied, J_c degrades even further due to flux lattice melting. BSCCO will not be used at 77 K for coils, electric motors, or similar applications.

The net result of all this discouraging experience with YBCO and BSCCO is to focus attention on the thallium compounds, especially on TBCCO-1223. There is optimism (but no guarantee) that clever methods will be found to achieve satisfactory grain alignment in TBCCO, thus overcoming the weak-link problem.

The appearance of the mercury series of HTSC compounds ($HgBa_2CuO_4$, $HgBa_2CaCu_2O_7$, etc.) has generated considerable interest, because they have $T_c \approx 134$ K. If we are lucky, a mercury-based compound may have the flux-pinning characteristics of Tl-1223, and less brittleness. That would qualify it as a commercially important material.[56] Of course, the mercury series may face other problems, such as weak links or grain alignment. Nonetheless, people are starting to think about making wire.

REFERENCES

1. S. H. Crandall *et al.*, *Introduction to the Mechanics of Solids* (McGraw-Hill, New York: 1972).
2. J. Lohle *et al.*, *J. Appl. Phys.* **72**, 1030 (1992).
3. "Forming of Wire" in *Metals Handbook*, 8th ed., vol. 4: *Forming*, pp. 347–352 (American Society of Metals: 1969).
4. R. B. Poeppel *et al.*, in *Chemistry of High Temperature Superconductors*, pp. 261 ff (American Chemical Society: 1987).
5. U. Balachandran *et al.*, *Mater. Lett.* **8**, 454 (1989).
6. S. Jin, *J. Metals* **43** (3), 7–12 (March 1991).
7. L. D. Woolf *et al.*, *Appl. Phys. Lett.* **58**, 534 (1991).
8. J. P. Singh *et al.*, *J. Mater. Res. Soc.* **7**, 2324 (1992).
9. D. Lee and K. Salama, *Jpn. J. Appl. Phys. Lett.* **29**, 2017 (1990).
10. S. E. LeBeau *et al.*, *Appl. Phys. Lett.* **55**, 292 (1989).
11. V. P. Novikov *et al.*, *Superconductivity: Phys., Chem., Tech.* **2**, 178 (1989).
12. V. P. Novikov *et al.*, *Superconductivity: Phys., Chem., Tech.* **4**, (1991).
13. K. Salama *et al.*, "Melt-Processing and Properties of YBCO", Chapter 4 in *Processing and Properties of HTSCs*, Vol I, Ed. by S. Jin et al., (World Scientific Publ. Co.: 1993).
14. S. X. Dou *et al.*, *Superconductivity Sci. Tech.* **3**, 138 (1990); K. Heine *et al.*, *Appl. Phys. Lett.* **55**, 2441 (1989).
15. S. Jin *et al.*, U.S. Patent #4,952,554 (filed April 1987, issued August 1990).
16. H. Sekine *et al.*, *J. Appl. Phys.* **66**, 2762 (1989).
17. H. Kumajura *et al.*, *J. Appl. Phys.* **67**, 3443 (1990).
18. A. P. Malozemoff *et al.*, *Cryogenics* **32**, ICMC Supplement, 478 (1992).
19. L. R. Motowidlo *et al.*, MRS Spring Meeting, San Francisco CA (1994).
20. J. Tenbrink *et al.*, *IEEE Trans. Magn.* **MAG-27**, 1239 (1991).
21. L. R. Motowidlo *et al.*, *Appl. Superconductivity* **1**, 1503 (1993).
22. K. Heine *et al.*, *Cryogenics* **32**, ICMC Supplement, 504 (1992).
23. J. Tenbrink *et al.*, *IEEE Trans. Appl. Superconductivity* **3**, 1123 (1993).
24. D. E. Peterson, private communication.
25. P. Haldar and L. Motowidlo, *J. Metals* **44** (10), 54–58 (October 1992).
26. J. Tenbrink, Vacuumschmelze, presentation at Argonne Nat. Lab. (9 July 1993).
27. K. Togano *et al.*, "Fabrication of BSCCO/Ag composite Superconductors by Dip-Coating process", Advances in Cryogenic Engineering **38**, 1081, ICMC Conference, Huntsville AL, 1991.
28. L. D. Woolf, private communication.
29. S. Jin *et al.*, *Physica C* **198**, 333 (1992).
30. S. Patel *et al.*, presented at 6th annual NYSIS Conference on Superconductivity, Buffalo NY (September 1992).
31. S. Patel *et al.*, in *Superconductivity and Its Applications*, AIP Conference Proceedings #273, pp. 567–574 (Amer. Inst. of Physics: 1993).
32. E. Narumi *et al.*, *IEEE Trans. Magn.* **MAG-27**, 1648 (1991).

33. G. Yurek, American Superconductor Corp., presentation at SC^{3}EP Meeting at NIST, April 1993.

34. R. N. Bhattacharya and R. D. Blaugher, *Physica C* **225**, 269 (1994).

35. E. W. Collings, *Applied Superconductivity, Metallurgy and Physics of Titanium Alloys*, (Plenum Press, New York: 1986).

36. S. Jin and J. E. Graebner, *Materials Science & Engineering* **B7**, 243 (1991).

37. X. D. Wu *et al.*, "High Current YBCO Thick Films on Flexible Nickel Substrates with Textured Buffer Layers" to be published in *Applied Physics Letters* (1994).

38. J. Tenbrink *et al., IEEE Trans. Magn.* **MAG-27**, 1239 (1991).

39. N. Enomoto *et al.*, Furukawa Review #8, 15–20 (Aug. 1990).

40. L. R. Motowidlo *et al., Appl. Phys. Lett.* **59**, 736 (1991).

41. P. Haldar *et al., Advances in Cryogenic Engineering (Materials)* **40**, 313 (Plenum Press, New York: 1994).

42. A. P. Malozemoff *et al., Advances in Cryogenic Engineering (Materials)* **40**, 25 (Plenum Press, New York: 1994).

43. T. Hikata, *Cryogenics* **30**, 924 (1990); H. Mukai, Proc. Third Int'l. Symp. Superconductivity, (Sendai, Japan, November 6–9, 1990).

44. D. E. Peterson *et al., Physica C* **199**, 161 (1992); D. E. Peterson *et al., IEEE Trans. Appl. Superconductivity* **3**, 1219 (1993).

45. D. H. Kim *et al., Physica C* **177**, 431 (1991).

46. R. N. Bhattacharya, P. A. Parilla, and R. D. Blaugher, *Physica C* **211**, 475 (1993).

47. J. A. DeLuca *et al.*, pp. 531–543 in *Superconductivity and Its Applications*, AIP Conference Proceedings #273, (American Institute of Physics, New York: 1993); *Physica C* **205**, 21 (1993).

48. K.-I. Sato *et al., IEEE Trans. Magn.* **MAG-27**, 1231 (1991).

49. A. Otto *et al., IEEE Trans. Appl. Superconductivity* **3**, 915 (1993).

50. J.-I. Shimoyama *et al., Jpn. J. Appl. Phys.* **31**, L 163 (1992).

51. J. W. Lue *et al., Advances in Cryogenic Engineering (Materials)* **40**, 327 (Plenum Press, New York: 1994).

52. L. J. Masur *et al.*, to be published in *Physica C* (1994).

53. Z. F. Ren and J. H. Wang, *Appl. Phys. Lett.* **61**, 1715 (1992).

54. J. G. Daley, U.S. Department of Energy, 1988 document.

55. G. A. Whitlow and J. C. Bowker, Proc. HTS Wire Development Workshop, DoE report 920286, 5–38 (February 1992).

56. A. Umezawa *et al., Nature* **364**, 129 (1993).

17

Protecting Against Damage

Thomas P. Sheahen and Robert F. Giese*

Stability affects the engineering of practical magnets and power transmission lines. This chapter discusses the stability of high-temperature superconductors (HTSCs), which differs greatly from that of low-temperature superconductors (LTSCs) primarily because there is a difference of orders of magnitude between the specific heats of LTSCs at 4.2 K and those of HTSCs at 77 K. This difference has a direct bearing on both the thermal stability of superconducting electric systems and the design of conductors for them.

Accordingly, this chapter begins with a discussion of specific heat. We then go on to show how this affects the three categories of stability: *adiabatic, dynamic,* and *cryogenic* stability. This presentation closely follows that of Martin Wilson's book *Superconducting Magnets.* Finally, we describe some experimental data pertaining to the stability of HTSCs.

17.1. PHYSICS VS. ENGINEERING

There are both physics and engineering reasons to be interested in specific heat. Measuring the specific heat of any solid provides information about the thermal energy within the crystalline structure, and hence about the lattice vibrations (phonons) and the electrons. The specific heat of the electrons is used to find the density of states at the Fermi surface, which is a critically important property of any solid. Furthermore, the specific heat is a property of the bulk material, which supplements other measurements (such as resistivity) that are easily altered by filamentary paths through the material. Perhaps most important, specific heat is a thermodynamic property, and since we believe very strongly in thermodynamics, measured values of specific heat can be used to constrain any new theory that might be proposed to explain superconductivity.

On the engineering side, the design of superconducting wire is heavily constrained by the need to absorb and remove heat, which is produced in the wire when a flux jump occurs. If the amount of heat thus produced is sufficient to drive a small local region into the normal state, the event is termed a *quench.* Until that heat is carried away to the refrigerant bath, the wire is a normal conductor, with resistive dissipation. The specific heat of the superconductor is essential to heat management; it tells how high the temperature will rise due to any given pulse of heat. In a typical application, the wire used in large superconducting magnets for medical MRI units consists of a bundle of filaments of NbTi embedded in a matrix of copper. When one local region becomes normal, current is shunted around the NbTi and

*Argonne National Laboratory.

349

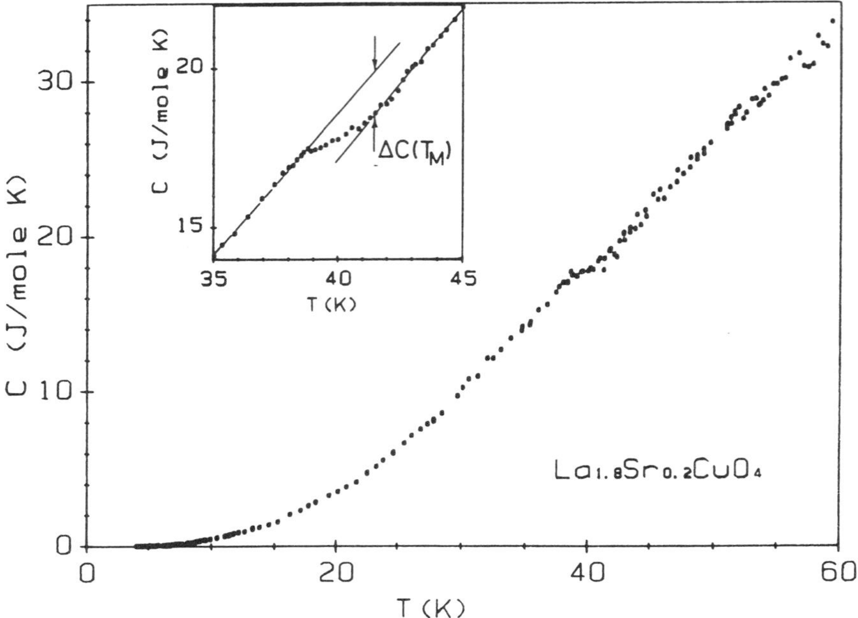

Figure 17.1.   Specific heat of $La_{1.8}Sr_{0.2}CuO_4$ from 0 to 60 K. (From Ref. 1).

passes through the copper matrix until superconductivity is restored. The choice of the number of filaments and the radius of each is an outcome of a stability analysis to prevent thermal runaway. A key factor in that analysis is the specific heat of the combined copper–NbTi matrix. Therefore, any discussion of stability hinges on the specific heat of the material.

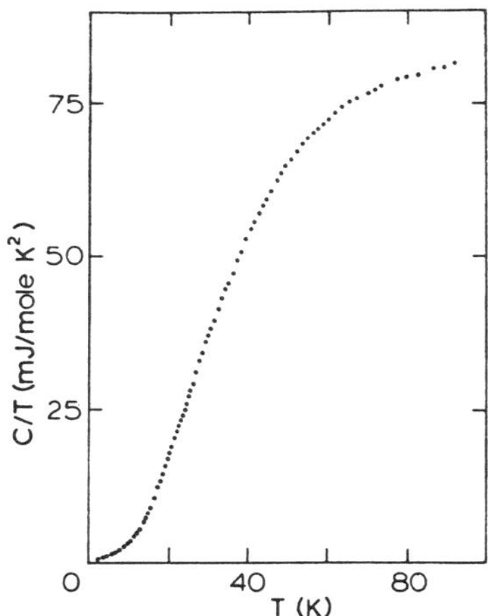

Figure 17.2.   Specific heat of $La_{1.8}Ba_{0.2}CuO_{4-y}$ divided by temperature, from 0 to 90 K. Plotting C_p/T emphasizes the low-temperature region. (From Ref. 2.)

are usually millijoules per mole (K). The specific heat per mass and the volumetric specific heat are related to the specific heat per mole by the molecular weight and density, respectively.

There is an extremely important difference between materials operating at liquid helium temperatures (4.2 K) and liquid nitrogen temperatures (77 K): NbTi, for example, has a specific heat of 0.82 mJ/gram-K at 4.2 K, whereas the specific heat of YBCO at 77 K is much larger, 158 mJ/g-K. This is due to a much greater contribution of lattice vibrations in the 77 K range.

Figure 17.1 shows early measurements[1] of the specific heat of La-Sr-Cu-O, the Bednorz–Muller HTSC with $T_c = 38$ K, over the range 0–60 K. The specific heat at 60 K is several orders of magnitude greater than at 4 K. This is entirely typical of all materials; indeed, it has become customary in this field to plot not specific heat itself (C_p) but specific heat divided by temperature (C_p/T), in order to emphasize the low-temperature features of the specific heat. Figure 17.2 is such a plot[2] for La-Ba-Cu-O; the straight-line segment from 15 to 35 K says that C_p varies as the square of temperature in that range.

17.2. MEASUREMENT OF SPECIFIC HEAT

The customary means of determining the specific heat of a solid is by using a calorimeter. In room-temperature experiments (as in high school physics labs) this might be a large aluminum chamber, but in low-temperature experiments, the calorimeter is a long thin chamber designed to fit inside a dewar. Figure 17.3 is a drawing of a typical low-temperature calorimeter. In operation, everything in Figure 17.3 would be housed inside a vacuum chamber, which is immersed in liquid helium.

The concept of the measurement is quite simple: apply a known quantity of heat ΔQ to the specimen via an electrically resistive wire, and measure the temperature change of the specimen ΔT. The mass m of the specimen is determined separately, and the specific heat C_p is determined through $\Delta Q = mC_p \Delta T$.

There are two complicating factors: First, the heater, the platform and the thermometer all have heat capacities of their own. This is known as the *addenda*, and a separate measurement must be done later with the specimen absent to determine those nuisance values. It is for this reason that nobody does specific heat experiments on HTSC thin films—the mass of the film is so many orders of magnitude less than the mass of the substrate that all the real information would be lost in the noise.

Second, extraneous heat leaks degrade the accuracy of ΔQ measurements. Extremely careful experimental design is necessary to minimize these heat leaks. The walls of the chamber radiate heat to the sample; at 4.2 K this is negligible, and at 77 K it is not serious. Gas molecules in the chamber convect heat to the walls. In the hard vacuum of a 4.2 K cryostat, this too is negligible, but at 77 K any nitrogen or inert gas not evacuated would be troublesome: a low-pressure gas convects heat quickly between walls and sample. Finally, heat is conducted into or out of the specimen through supporting members and electrical connecting wires. To minimize this, the specimen platform is supported by thread, and the smallest possible wires are selected for electrical connections. All of them are connected to a guard ring and every effort is made to match the guard ring's temperature to the specimen's temperature as the experiment progresses. Hence, the guard ring has its own heater and thermometer to facilitate tracking of the temperature.

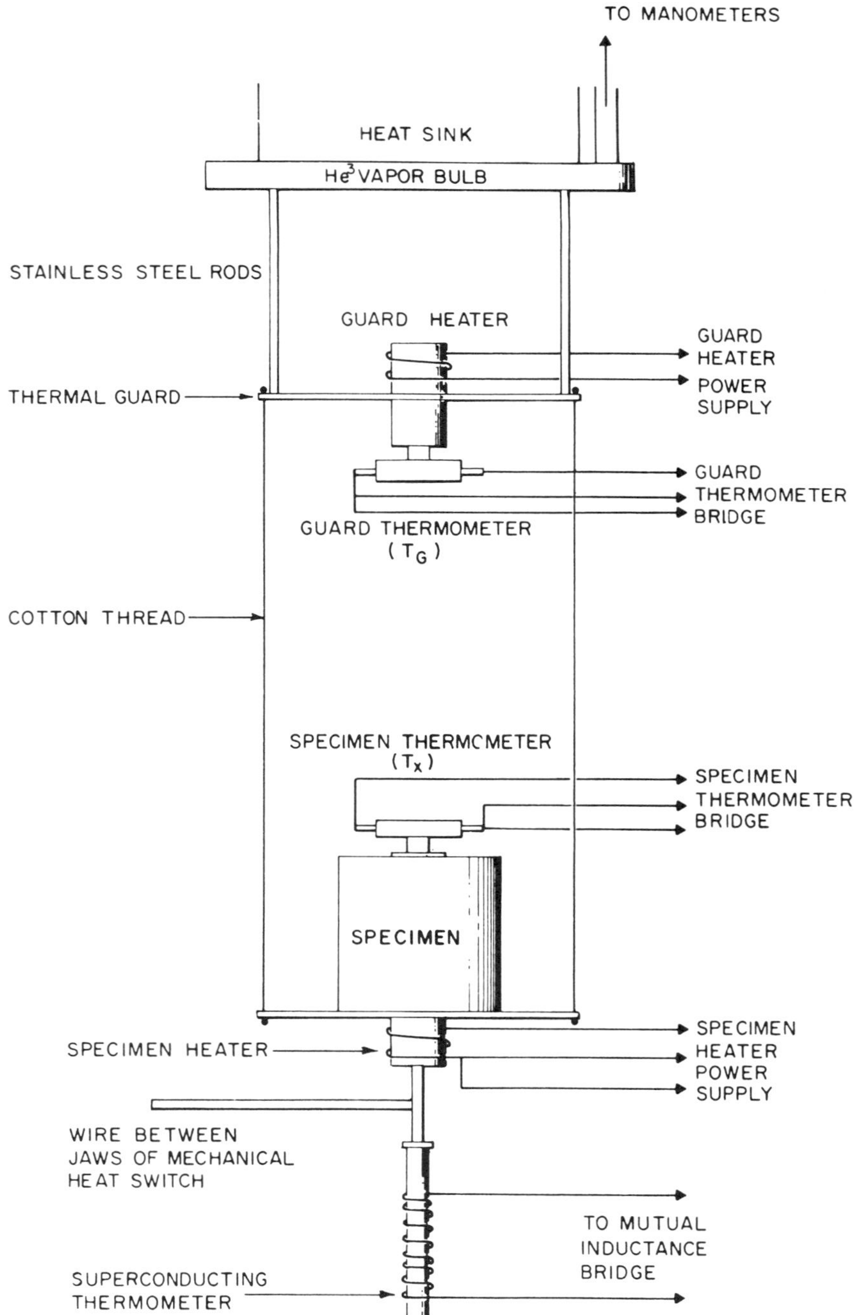

Figure 17.3. Typical calorimeter used in low-temperature specific heat experiments.

In a specific heat experiment, the specimen is initially cooled to the lowest possible temperature by mechanically clamping it to a heat switch at the temperature of the helium bath (often substantially below 4.2 K, achievable by vacuum pumping the helium bath). The same thing could be done with a 77 K nitrogen bath. The thick wire stub at the bottom left of Figure 17.3 serves this purpose. Once the cooling is complete, the mechanical switch is opened and the specimen is free to start warming. If zero current is applied to the heater, then the specimen will warm about 1 K in one day, due to unavoidable heat leaks of less than a microwatt due to antenna-pickup from ambient electromagnetic interference (EMI) by the connecting wires.

To carry out a continuous warming experiment, a current far greater than the heat leak is driven through the heater, and the voltage is measured across the heater. These two determine the ΔQ per unit time. Simultaneous measurements from the specimen thermometer give ΔT per unit time, provided that the specimen itself is massive enough not to sustain any thermal gradient across it. For highest accuracy, the thermometer is customarily a semiconductor or an electrical resistor, either of which has an accurately calibrated relationship between R and T (usually, $\log R \approx 1/T$ near 4.2 K). In a step warming experiment, the current is only on for a short time so that a known heat pulse ΔQ is delivered and the system is given time to reach a new equilibrium at a temperature higher by ΔT.

Low temperature calorimeters seldom are able to reach above ≈ 10 K. The need for data near 100 K has motivated many experimenters to build entirely new calorimeters to operate in that higher range, and high quality data is now achievable. At first, impurity phases and non-superconducting constituents within the samples led to very irregular specific heat results for HTSCs. Today, good samples assure the accuracy of specific heat measurements.

17.3. SPECIFIC HEAT OF SUPERCONDUCTORS

17.3.1. Lattice Specific Heat

The dominant contributor to the specific heat near 77 K is the lattice, whose vibrations (phonons) comprise most of the energy in the solid. The Debye temperature T_D is a characteristic parameter often used to classify the strength of these vibrations; $T_D \approx 200$–300 K, and T_D can be found from other experiments. At very low temperatures, the lattice specific heat starts out as a cubic term, i.e.,

$$C_{pL} = \zeta \, (T/T_D)^3$$

where ζ is a universal numerical factor. For temperatures greater than 10 K, this expression is usually not accurate. However, because this formula is valid below 5 K for nearly all metals, it's had an important influence on the classical way of presenting specific heat data for analysis. Figure 17.4 shows a common way of plotting data: C_p/T as before, but now plotted against T^2. In the representation

$$C_p = \gamma T + \beta T^3$$

the intercept on the y-axis gives γ, and the low-temperature slope of the data gives β, which is related to ζ and T_D above. These parameters are interesting because they shed light on the mechanism of superconductivity. Because this form of graph was an important aid to "eyeball" analysis years ago, it has remained customary to display specific heat data as plots of C_p/T vs. T^2. This custom no longer serves a purpose near 100 K.

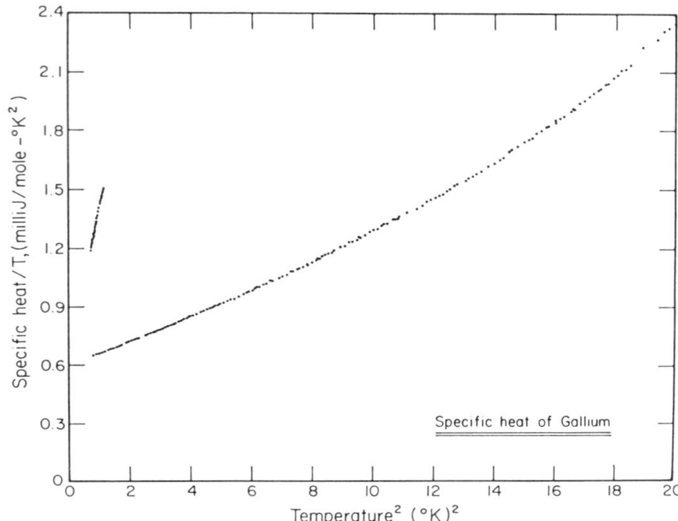

Figure 17.4. Conventional plot of C_p/T vs. T^2 for a type I superconductor. This method of plotting allows visual determination of the lattice term and the electronic term in the specific heat. On graphs of this type, when the horizontal axis reads 100 units, that represents only 10 K in temperature.

17.3.2. Electronic Specific Heat

The other contribution to the specific heat is that of the electrons in the lattice. In the normal state this is usually linearly proportional to the temperature, with a proportionality constant γ first studied by Sommerfeld, who showed that the density of states at the Fermi level (commonly written N_0) is the principal variable determining γ. Hence, a determination of the normal-state specific heat of the electrons is tantamount to a determination of N_0, yielding important information about the electronic structure of the material.

In the superconducting state, the electronic specific heat is quite different in conventional LTSCs. The BCS theory predicts a certain dependence of C_p upon the energy gap Δ: At very low temperatures, where Δ does not change with temperature, the electronic specific heat grows exponentially with T; at temperatures nearing T_c, Δ falls off as shown[3] in Figure 17.5a, so C_p grows less rapidly. Finally, at T_c itself, there is a discontinuity in the electronic specific heat as the superconductor returns to the normal state. Just before the transition, the numerical values within the BCS theory work out to give a specific heat value of

$$C_p(T \lesssim T_c) = 2.43\gamma T_c$$

Since in the normal state just above the transition

$$C_p(T \gtrsim T_c) = 1.00\gamma T_c$$

the size of the discontinuity is simply

$$\Delta C_p = 1.43\,\gamma T_c \tag{17.1}$$

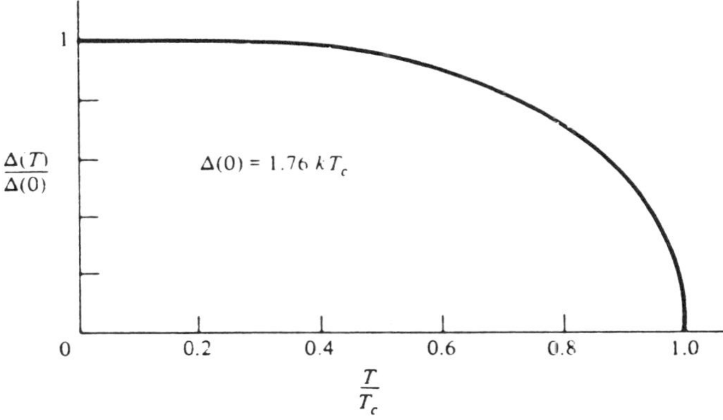

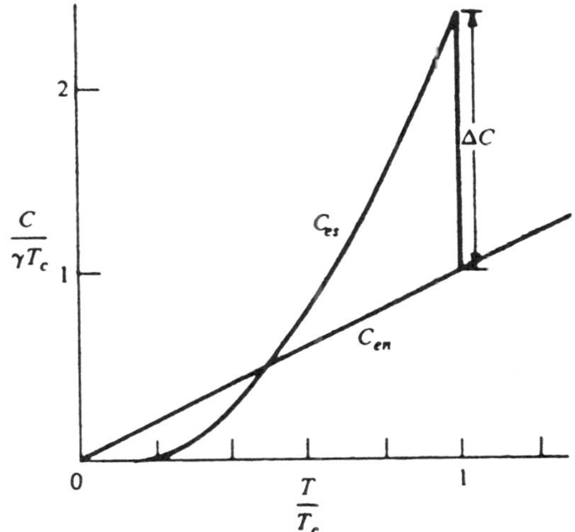

Figure 17.5. (a): Energy gap of a BCS superconductor vs. temperature, with each scale normalized to one, at $\Delta(T = 0)$ and at $T = T_c$. The electronic specific heat in the normal state is simply γT, but in the superconducting state it is a function of the gap. (b): Numerical predictions of the BCS theory for both normal (C_{en}) and superconducting (C_{es}) specific heat. (From Tinkham, Ref. 3.)

$$\Delta C_p = 1.43\,\gamma T_c \qquad\qquad (17.1)$$

according to BCS theory. Figure 17.5b shows this comparison. The fact that most type I superconductors show good agreement with this prediction is a major reason why BCS has been accepted as the theory of superconductivity. Actually, the numerical value 1.43 holds only when the attractive pairing interaction between electrons is very weak; this is known as the *weak-coupling* limit. As the superconducting interaction strength increases, the value increases as well. For intermediate-coupling superconductors, the numerical values of ΔC_p can be explained in terms of an *effective interaction strength*.[4]

17.3.3. High-Temperature Superconductors' Specific Heat

The specific heat jump ΔC_p provides a way to test the relevance of the BCS theory for the HTSCs. If we hypothesize that HTSCs follow the BCS theory, but with stronger coupling, then for $\Delta C_p/\gamma T_c$ we should expect a numerical factor well above 1.43, perhaps 3 to 6. On the other hand, a measured value near 1.43 would indicate that the BCS weak-coupling model holds well beyond its presumed range of validity.

It is quite difficult[5] to make accurate measurements of ΔC_p near 100 K. Figure 17.6 shows[6] the specific heat of YBCO in the 100 K range, and evidently the huge lattice specific heat swamps the relatively small electronic specific heat. Figure 17.7 is similar data[7] plotted

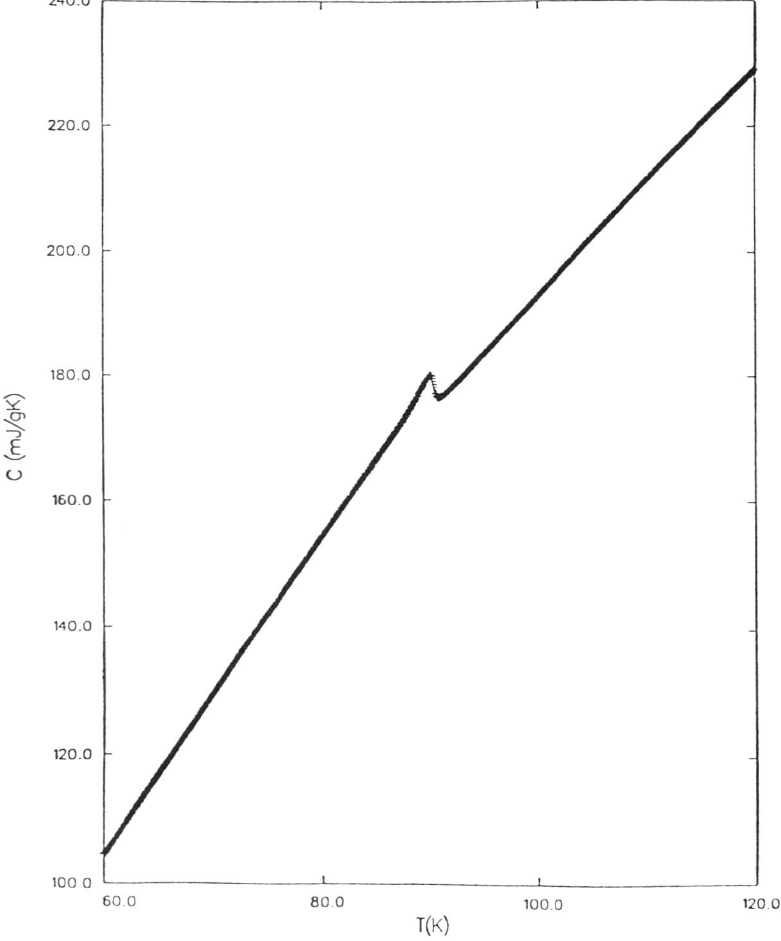

Figure 17.6.   Specific heat of $YBa_2Cu_3O_{7-x}$ between 60 and 120 K. Note that (1) the zero on the vertical axis is suppressed, and (2) the data is plotted on a per gram basis rather than the conventional per mole basis. In either the normal or the superconducting state, the specific heat is dominated by the lattice contribution. (From Ref. 6.)

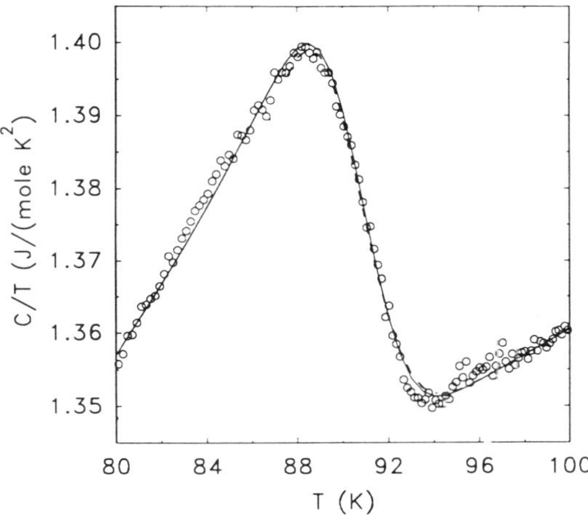

Figure 17.7. Close-up of the 80–100 K range of typical YBCO specific heat data, now plotted as C_p/T. The specific heat jump must be determined by extrapolating the normal-state specific heat down to T_c, despite the curvature of the data. Substantial error creeps in through this analytical process. (From Ref. 7.)

as C_p/T vs. T. After subtracting the lattice contribution, a reasonable estimate for the jump is $\Delta C \simeq 7$ J/mol-K. The rounding of the discontinuity limits the accuracy of determining ΔC_p. Also, it is difficult to isolate the normal electronic specific heat γT from the large phonon contribution; $\gamma = 16 \pm 2$ mJ/mol-K^2 is the best value.[5] Consequently, there are large error bars on any experimental ratio, but the value $\Delta C_p/\gamma T_c = 5$ is representative and is very far from the weak-coupling BCS value of 1.43.

Similar measurements on bismuth and thallium compounds give results that are the same in character, although numerically somewhat different. There is a wealth of other knowledge to be gained (in principle) from accurate specific heat measurements near T_c; fluctuations between normal and superconducting states, energy gap variations, and so on. However, such "physics" measurements, while interesting, are less significant than the practical consequences of the much higher specific heats of the HTSCs.

17.4. SPECIFIC HEAT AND STABILITY

Of greater interest for engineering considerations is the way that the large specific heat of HTSCs affects the stability of superconducting devices. In the past, for LTSCs, stability has been a major worry and a prominent design constraint. For example, all of the practical devices described in Chapters 4 and 5, which use NbTi as their superconductor, were built amid great attention to stability considerations.

By comparison, the investigation of stability for HTSCs can be termed "relaxed." At first, the same cautions applicable to LTSCs worried designers of HTSC equipment. However, after measured numerical values were inserted into the applicable equations, it soon became apparent that at 77 K, the stability problem is far less severe. At intermediate temperatures (about 30 K), stability consideration still demand attention, but they do not drive the design of apparatus.

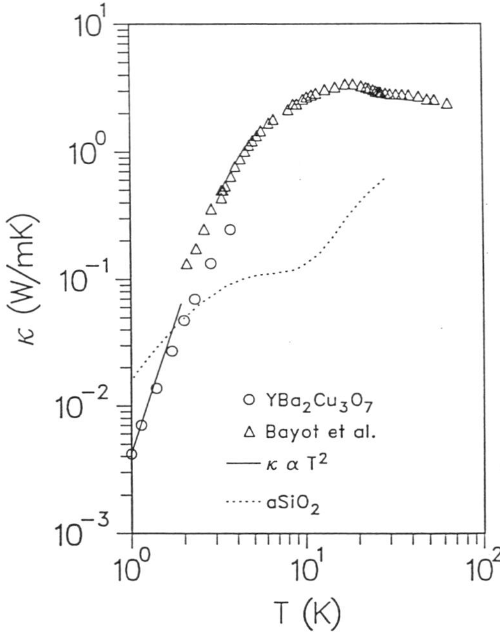

Figure 17.8. Thermal conductivity of YBCO below 300 K. In the vicinity of 77 K, thermal conductivity is slightly higher than at room temperature. Moreover, there is a "kink" in the curve (difficult to see on this scale) at the transition temperature T_c. (From Ref. 8.)

In addition to large specific heat values, the thermal conductivity of HTSCs is an order of magnitude greater at 77 K than at 4 K. Figure 17.8 displays thermal conductivity[8] for YBCO. Just as specific heat influences adiabatic stability, thermal conductivity influences dynamic stability. Nevertheless, both of those are subordinate to cryogenic stability, wherein a system returns to the superconducting state even after it has become normal. This depends on the ability of the system—superconductor *plus* cryogen—to conduct away heat from any incipient hot spot. The numerical values associated with 77 K are much more favorable than those for 4 K, no matter what the material.

If operation at 77 K was assured, stability would no longer be an important topic. However, there is a finite likelihood that the ceramic oxide superconductors will function in commercial devices at 20–35 K. In that case, they will probably be cooled by cold helium gas, which does not transport away heat nearly as well as liquid nitrogen (LN$_2$) at 77 K. Moreover, their specific heat values, near $T \sim 30$ K, while large compared to their 4 K values, are substantially lower than at 77 K. Because of the possible relevance of this intermediate-temperature range, we present the elements of stability analysis in the following sections.

17.5. QUENCHING AND FLUX JUMPING

The entire questions of stability—whether adiabatic, dynamic, or cryogenic—grew out of observing the performance of NbTi magnets bathed in liquid helium (LHe). Early efforts during the 1960s to produce high-field magnets based on LTSCs such as NbTi invariably

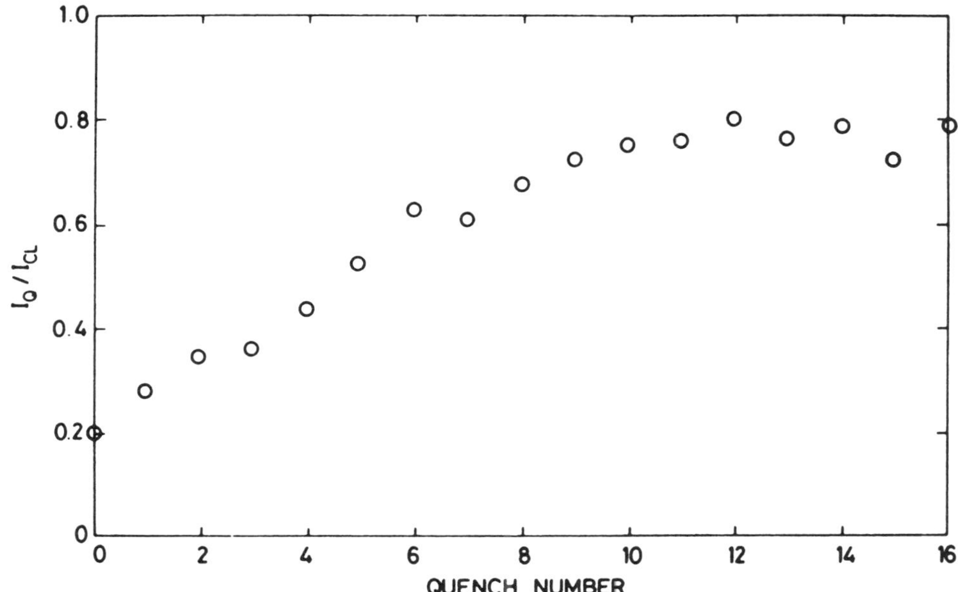

Figure 17.9. The occurrence of *training* in a superconducting magnet. The current required to initiate a quench increases over time as consecutive quenches occur. (From *Superconducting Magnets* by Martin N. Wilson, publ. by Oxford University Press 1983, 1987; © M.N. Wilson 1983.)

led to devices that exhibited performance far short of what was expected. Attempts to carry design currents in the magnet winding caused a loss of superconductivity that resulted in an explosive release of He gas from the coolant. This behavior is termed a *quench*. It accompanies the intense local heating caused by energy dissipation ($I^2 R$) when all or part of the winding switches from the superconducting to the normal state. Since the normal-state resistivity and current density are both large, the amount of heat released is likewise quite large, and the region surrounding the quench point reaches temperatures above the critical temperature T_c. That results in a complete dissipation of the energy stored in the magnet ($1/2\ LI^2$) as heat.

Investigation of this phenomenon led to the conclusion that transient disturbances are the dominant cause of quenching. Two types of disturbances were identified: (1) flux jumping and (2) mechanical disturbances. *Flux jumping* refers to a sudden dissipative rearrangement of magnetic flux within a superconductor. *Mechanical disturbances* refer to physical movement of wires within a coil to relieve magnetic hoop stresses.

It was also noted that the current at which quenching occurred increased with repeated quenching. This behavior, shown in Figure 17.9, is termed *training*. Mechanical disturbances can explain the training behavior exhibited by most superconducting magnets. Repeated magnet quenching causes release of larger and larger magnetic hoop stresses at higher and higher magnetic fields. Finally, as shown in Figure 17.9, magnet performance reaches a plateau after the various parts of the coil have readjusted to relieve all stresses. Any subsequent degradation is due primarily to flux jumping.

Flux jumping is not quite the same thing as giant flux creep[9] or flux lattice melting[10] mentioned in Chapter 14. Those are steady-state conditions that create surrogate resistivity

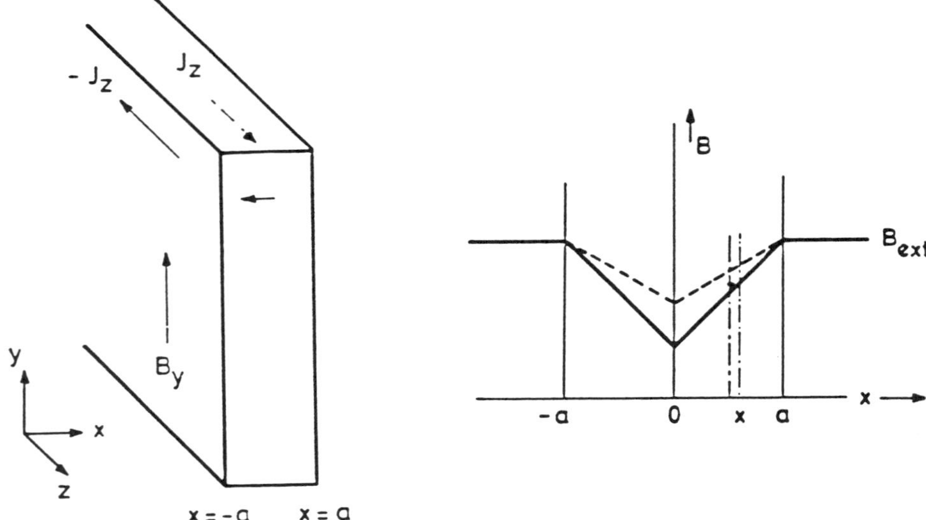

Figure 17.10.   Geometry of a slab penetrated by a magnetic field to illustrate the Bean critical state model.

in HTSCs. Flux jumping, by contrast, is the sudden release of substantial energy when a whole bundle of flux lines move to a new position. Flux jumping is a kind of electromagnetic/thermal instability which affects all high-field superconductors.

The *Bean critical state model*,[11] familiar from Chapter 2, is helpful in understanding flux jumping. Consider Figure 17.10, which shows how the magnetic field varies within a slab-conductor of width $2a$, and recall that at any point the current is either $\pm J_c$ or zero. (The simple slab geometry allows the Maxwell equation $\nabla \times B = \mu_o J$ to yield a linear behavior of B when J is constant.) What is the effect of a small heat pulse on this slab? As the temperature rises, J_c falls. This causes the screening currents to decay to a lower value, permitting the magnetic field to penetrate further into the slab—the dashed line in Figure 17.10(b). The resulting flux motion generates heat, which causes the temperature to rise further, creating a positive feedback loop. If the initial heat pulse is large enough, the temperature of the coil will grow unstably, driving the coil normal and resulting in a quench.

Several techniques have been employed to prevent quenches. The first technique is *adiabatic stability*. This is a passive technique that relies on the heat capacity of the material, either pure superconductor or composite, to absorb the heat pulses and dampen the inherent temperature rises. The next technique is *dynamic stability*. This technique relies on the use of a composite and a coolant (LHe for LTSCs, and LN_2 for HTSCs) to conduct heat away fast enough to prevent a quench. The third technique, *cryostability*, relies on the coolant to absorb the energy of the flux jump and return the conductor to the superconducting state.

17.6. COMPOSITE CONDUCTORS

The term composite refers to the type of superconducting wire. If one had pure NbTi, the risk of a quench would be so high that damage (local burnout) would likely occur very

Table 17.1. Representative Stability Parameters for HTSCs at 77 K

System	Critical temperature, T_c(K)	Normal-state resistivity,[a] $\rho(\mu\Omega cm)$	Volume specific heat, C_p (J/cm^3-K)	Thermal conductivity, κ(W/cm-K)
YBCO	92[b]	570	0.92	0.02
BSCCO	105[b]	< 2000	0.95	0.02
TBCCO	115[b]	< 2000	0.95	0.02
Generic HTSC	92[c]	570	1.0	0.02

[a]Values are very sample dependent.
[b]Zero magnetic field.
[c]In a 6 T magnetic field.

soon. Recognizing this, engineers make superconducting wire in which the NbTi is surrounded by copper. Indeed, the typical cable contains multiple thin filaments of NbTi embedded in a matrix of copper. It is not cheap to make wire this way; the reasons why it is necessary have to do with stability.

Ideally, the copper surroundings *never* carry current—it all flows through the resistanceless NbTi. However, when a local spot heats up and goes normal, the copper carries the current until superconductivity is restored, and during that time the magnet does not burn out. In this way the copper justifies the manufacturing expense of including it in the wire.

What is the right combination of copper and NbTi (or, stated more generally, matrix and superconductor)? That depends on the numerical values of certain properties of the component materials. We call these the stability parameters.

Table 17.1 presents representative values of stability parameters for YBCO, BSCCO, and TBCCO at 77 K. The resistivity ρ measures how rapidly the material generates heat when current is passed through it in the normal state. The thermal conductivity κ measures how easily the material conducts heat. The volumetric heat capacity C_p measures the material's ability to absorb heat. These numbers are difficult to measure and are sample dependent. Therefore, for purposes of analysis we use a generic HTSC, also an entry in Table 17.1.

Table 17.2 displays stability parameters for NbTi and Nb$_3$Sn at LHe temperature and HTSC at LN$_2$ temperature. The NbTi and Nb$_3$Sn values for J_c and T_c are representative for

Table 17.2. Comparison of Stability Parameters for NbTi, Nb$_3$Sn, HTSC and Cu

System	Critical[a] temperature, T_c(K)	Critical[a,b] Current J_c(kA/cm^2)	Normal-state resistivity $\rho(\mu\Omega$-cm)	Volume specific heat, C_p (J/cm^3-K)	Thermal conductivity, κ((W/cm-K)	Coolant type	Temperature of bath T_0(K)
NbTi	6.5	100	70	.005	.001	LHe	4.2
Nb$_3$Sn	13.2	100	10	.001	.0005	LHe	4.2
HTSC	92	100	570	1	.02	LN$_2$	77
Cu	—	—	.038[c]	.001	11	LHe	4.2
Cu	—	—	.25[c]	2	5.5	LN$_2$	77

[a]In a 6 T magnetic field.
[b]Depends upon manufacturer for LTSCs. HTSC value assumed to be equal to LTSC value.
[c]Depends upon Cu purity and magnetic field strength at 4.2 K but not at 77 K.

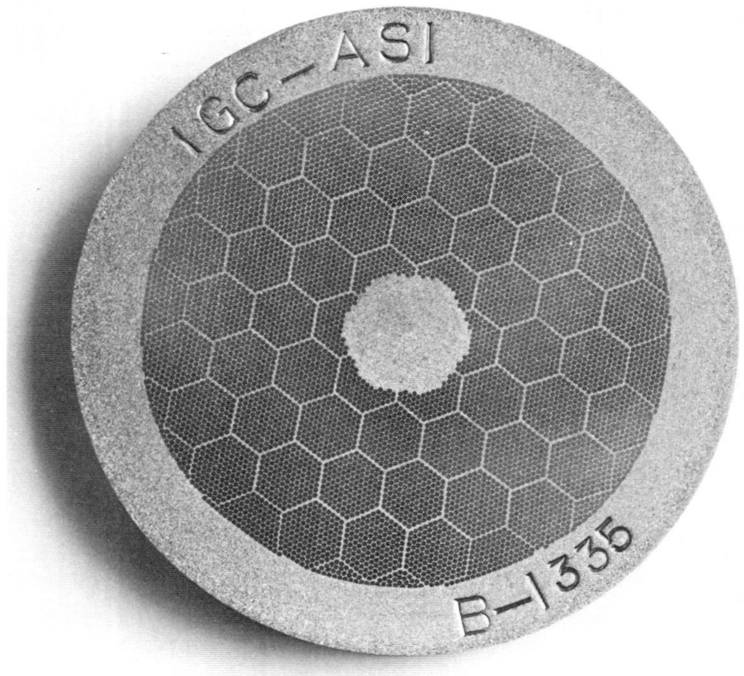

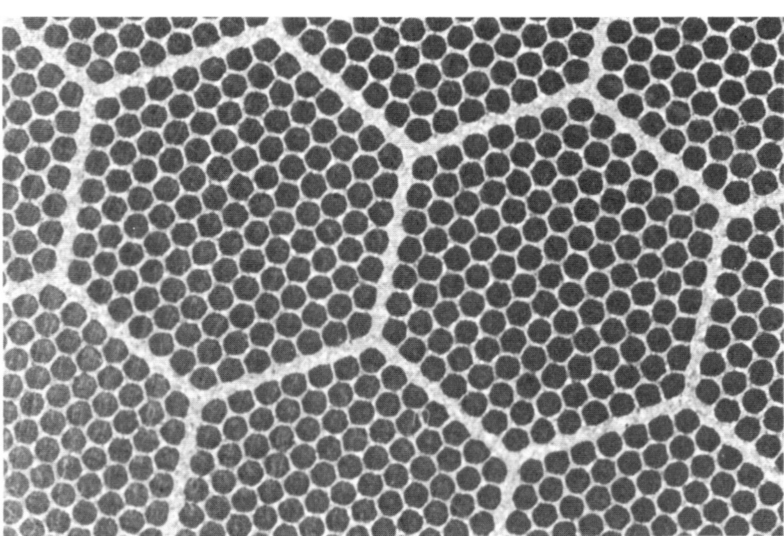

Figure 17.11.   Cross section of a typical superconducting wire made of NbTi filaments embedded in a copper matrix. (Courtesy of Intermagnetics General Corporation.)

operation in a 6 T field. The J_c value for HTSC is assumed to be the same as for the LTSCs. As we shall see, lower J_c values imply greater stability so this is a conservative assumption.

In a composite wire, the superconductor carries the current when things are going well, but the matrix (copper) carries the current during abnormalities. The cost-minded wire designer provides just the right amount of each. Figure 17.11 is a cross section of a typical commercial NbTi/Cu wire.

The penalty for including copper is a loss in the effective cross-sectional area of the wire. Typically, commercial superconducting wire is about 5 parts copper to 2 parts NbTi; the ratio is called the *volume fraction of superconductor* and is customarily denoted by λ, not to be confused with the penetration depth. That ratio is a trade-off between the need for safety and the desire to have the highest possible current density (usually in order to get a high magnetic field).

When safety grossly outweighs current density as a criterion, the ratio λ can be changed considerably. Thus, in the Eriez magnetic separator discussed in Chapter 4, the wire has a ratio of 39 copper to 1 NbTi. That magnet has never quenched, and probably never will. The magnetic field in the bore is only 2 T, but that's all that is needed to saturate an iron mesh.

By comparison, the magnets for major research accelerators are designed to run very close to the margin: $\lambda \approx 0.4$, *and* the current pushed through the NbTi is about 90% of J_c, so the magnetic field is as high as possible. Researchers learn to live with an occasional quench.

17.7. QUENCH PROPAGATION

To appreciate a stabilized magnet, it is necessary to understand how hot spots grow and how they dissipate. To begin with, we inquire into the energy density and absolute size of the energy pulse required to initiate a quench. As energy is deposited in a piece of superconductor, its temperature rises and eventually drives the superconductor normal. If the superconductor were already operating at its critical current density, only an infinitesimal amount of energy would be required. For this reason, coils are usually operated at a fraction of the maximum current density, no more than $0.9\,J_c$. The heat capacity of the material determines how much energy is required to drive the coil normal. As pointed out in Section 17.4, the heat capacity of YBCO at 77 K is roughly a factor of 200 times greater than NbTi at 4 K. Therefore, HTSCs are far more stable.

A numerical example is helpful here. Using the values from Table 7.2, and remembering that at low temperatures C_p varies as T^3, while it is roughly linear at higher temperatures, we calculate that deposition of 1.2 mJ/cm^3 will drive a NbTi wire (no copper stabilizer) operating at $0.9\,J_c$ into the normal state. Due to its much higher heat capacity, a HTSC wire could absorb 1.5 J/cm^3—one thousand times more.

The most significant concern is, what magnitude of energy pulse would be sufficient to initiate a quench? At this point, it is necessary to choose a specific model for the behavior of the superconductor. In his book *Superconducting Magnets*, Wilson selects a model appropriate for NbTi; in the following paragraphs, we follow that approach. However, remember that the HTSCs may not imitate NbTi in their behavior; and the heat transfer environment at 30 K may not resemble that of either 4 K or 77 K.

Consider Figure 17.12, which shows a local hot spot of length l_h in a wire of area A above temperature T_c. We also assume that the temperature returns to the bath temperature T_0 after another length l_h (in cm). In that geometry, heat conduction equals heat generation I^2R, leading to:

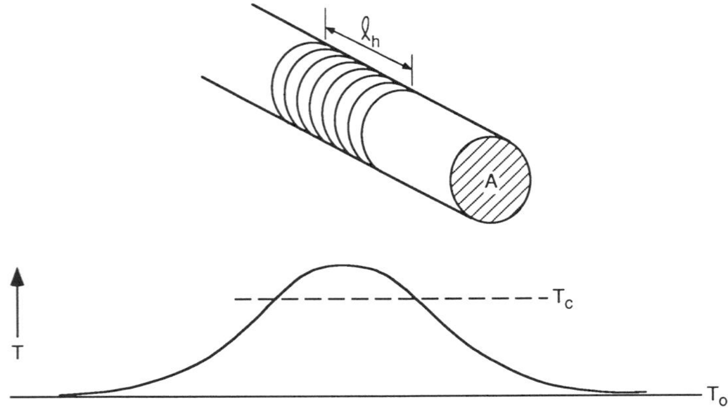

Figure 17.12. Propagation of a small normal zone starting from a single hot spot.

$$2\kappa(T_c - T_0)A/l_h = J_c^2\rho A l_h \tag{17.2}$$

where ρ is normal state resistivity and κ is thermal conductivity. Solving for l_h yields

$$l_h = \sqrt{\frac{2\kappa(T_c - T)}{J_c^2\rho}} \tag{17.3}$$

l_h is called the *minimum propagation zone*, MPZ. A normal zone longer than l_h will grow because generation exceeds cooling, whereas a zone shorter than l_h will collapse with a return to superconductivity.

Using the values in Table 17.2, we calculate an l_h of 0.8 μm, 3.0 μm, and 3.2 μm for NbTi, Nb$_3$Sn, and HTSC, respectively. If we assume a wire 0.3 mm in diameter—such a wire would carry 70 A—then the heat required to initiate a quench, the *minimum quench energy* (MQE), would be 0.001 μJ for NbTi and 3.4 μJ for HTSC. To put this in perspective, 1 μJ deposited in a gram of water would raise its temperature 2×10^{-13} K, a very tiny number. Alternatively, this is equivalent to depositing 1 W of power for 1 μs.

Should a quench occur, another important factor is how fast the quench propagates. If the quench propagates too slowly, hot spots will develop, possibly resulting in permanent conductor damage. On the other hand, rapid quench propagation spreads the heat dissipation throughout the conductor, averting damage.

How fast do quenches propagate? Note that the minimum propagation length l_h is independent of the specific heat C_p, so that l_h is about the same for Nb$_3$Sn and for HTSC. In contrast, the *quench propagation velocity* (QPV) depends inversely on C_p. Therefore, in HTSCs, when heat is generated it does not leave very quickly.

To design a conductor, it is prudent to make conservative guesses about what becomes of the heat. The most conservative assumption is that heat cannot escape from the wire to the external surroundings; that is called the adiabatic limit. For that special case, Wilson finds for the quench propagation velocity:

$$V = \frac{J_c}{C_p}\sqrt{\frac{\rho\kappa}{T_s - T_0}} \tag{17.4}$$

                                                    365

Table 17.3. Comparison of Quench Parameter for LTSCs and HTSCs With and Without Copper Stabilizer

System	Minimum propagation zone, MPZ (μm)	Minimum[a] energy density, (μJ/cm^3)	Minimum[b] quench energy, MQE (μJ)	Quench propagation velocity, QPV (m/s)
NbTi	0.8	0.001	0.001	49
Nb$_3$Sn	3.0	0.001	0.02	33
HTSC	3.2	1.5	3.4	0.1
NbTi with Cu[c]	1200	0.001	1.7	110
Nb$_3$Sn with Cu[c]	2400	0.001	17	230
HTSC with Cu[c]	860	1.9	1100	0.3

[a]Energy density required to drive a wire operating at 0.9 J_c into the normal state.
[b]Assumes a wire diameter of 0.3 mm for pure superconductor and 0.35 mm for composite. Composite wire carries same total current as noncomposite wire.
[c]25 percent Cu by volume.

In this expression, the intermediate temperature value, T_s lies between T_0 and T_c and depends on operating conditions. It is convenient to let T_s be the mean of T_0 and T_c. Using the values in Table 17.2 yields QPVs of approximately 50 m/s for NbTi and 1.2 m/s for HTSC.

Examining Equation (17.3) shows that the tiny MPZs result from a combination of high ρ, low κ, and high J_c. Although lowering J_c would alleviate the problem, high J_c is desirable for most applications. Pure metals such as copper and aluminum have low ρ, particularly at very low temperatures, and high thermal conductivity. This explains why practical LTSC wires are usually made in composite form, containing both superconductor and normal metal.

Table 17.3 summarizes the comparison of MPZ, MQE, and QPV for LTSCs and HTSCs in composite and noncomposite form. In this table, the size of the composite conductor has been increased to carry the same total current. Notice that in noncomposite form NbTi, Nb$_3$Sn, and HTSC all have MPZs of approximately 1 μm, a very small number. Addition of 25% copper by volume dramatically increases MPZs, a factor of approximately 1000 for NbTi and Nb$_3$Sn and a factor of 270 for HTSC. Copper is less beneficial for HTSCs due to the poorer thermal and electrical properties of copper at 77 K relative to 4.2 K.

Due to the large specific heat of HTSCs, the MQE is 2 to 3 orders of magnitude greater than for LTSCs. A HTSC wire containing 25% copper requires almost a million times more energy to initiate a quench than a pure NbTi wire. This means that HTSCs are inherently more stable than LTSCs. However, after a quench has been initiated it is much harder to control in HTSCs because the quench propagation velocities are comparatively small, that is, provided the adiabatic limit is still the appropriate model.

In reality, an actual composite conductor has heat-removal paths other than simply down the wire. In practical LTSC conductors, during a flux jump current flows through the copper and the ohmic heat thus generated is transferred to the coolant. The overall system is designed so that the cooling capacity exceeds heat generation, causing the temperature to fall and superconductivity to be restored. This strategy usually requires a composite with a high ratio of copper or aluminum to superconductor and thus a rather low overall current density. Large magnets with large amounts of stored energy generally employ this technique.

17.8. TYPES OF STABILITY

17.8.1. Adiabatic Stability

Wilson has carried out the derivation of the quench-propagation behavior in the adiabatic limit for certain other special cases, including stability in an external field and in the self-field of a conducting wire. For LTSCs, the designer is forced to go with fine-filament superconductors. On the other hand, HTSCs are stable at sizes 40 times greater than LTSCs, so the design is much more relaxed. HTSCs are capable of carrying very large currents in single strands even without stabilizers.

For example, consider a DC transmission line rated at 1000 MVA. If we provide two cables, each carrying 10 kA and operating at + and −100 kV, each cable could carry the total 1000 MVA load providing 100% redundancy. If each cable consisted of three strands of HTSC superconducting wire with $J_c \approx 100$ kA/cm^2, under customary operating conditions ($J < 0.9\,J_c$), then the radius of each strand would be 0.34 cm, which is well within the adiabatic stability limit for self field; thus, stability is assured. In fact, if J_c were 10 kA/cm^2, the strand radius would have to rise to 1.9 cm, but the stability would be even greater.

17.8.2. Dynamic Stability

In adiabatic stability, the conduction of heat during a flux jump is ignored. Dynamic stability considers the time dependence of heat, current, and magnetic flux. For LTSCs, dynamic stability provides increased stability for composite conductors. However, because of the much greater heat capacities of HTSCs and Cu at 77 K, dynamic stability has little additional impact for HTSCs. Much more information concerning dynamic stability appears in Wilson's book.

17.8.3. Cryogenic Stability

Both adiabatic and dynamic stability are designed to prevent flux jumping. Cryostability allows the magnet to return to standard operation even from a condition where the entire coil winding has gone normal (a quench). In LTSCs, cryostability involves the use of large amounts of normal metal such as copper. Since the resistivity of copper is much less than

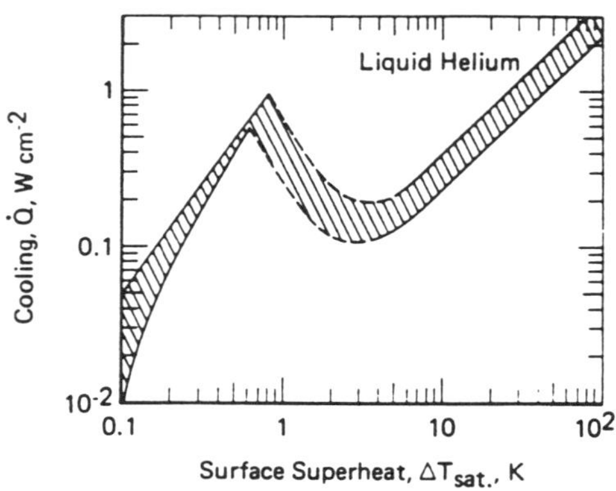

Figure 17.13.   Heat transfer as a function of temperature for liquid helium. Note that as temperature increases by only about one degree, there is a point of sudden discontinuity from nucleate boiling to film boiling.

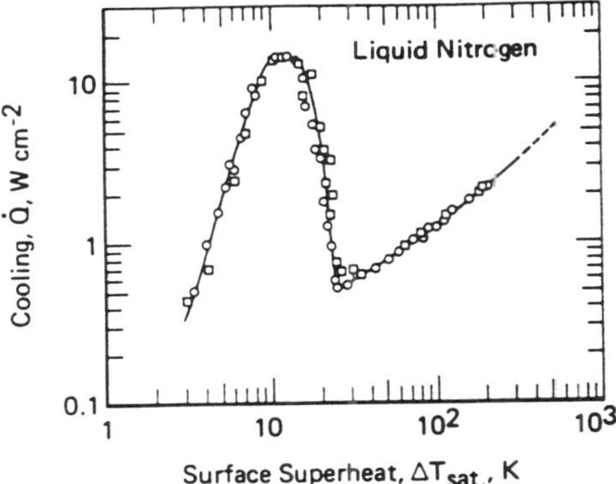

Figure 17.14. Heat transfer as a function of temperature for liquid nitrogen. The range of safety before nucleate boiling stops and film boiling begins is much wider than for liquid helium.

that of the superconductor in the normal state, inclusion of large amounts of copper in the composite conductor greatly reduces ohmic heating while increasing heat transfer to the coolant.

The key to cryostability is heat transfer to the surrounding bath. Figure 17.13 shows $\dot{Q}$ the heat transfer rate of helium as a function of temperature differential ($\Delta T \equiv \Theta$) above the bath temperature. Figure 17.14 is similar, for nitrogen. Both curves show a rapid increase in cooling capacity when temperature first rises above the bath. This regime is termed *nucleate boiling.* As the temperature of an object increases still more, the heat transfer coefficient drops, but eventually rises again, entering the regime of *film boiling.*

There is a tremendous difference here between helium and nitrogen cooling. For helium, even minor warning (to perhaps 5 K) leads to an unstable, instantaneous jump into the *film boiling* regime at about 8 K. A thin veneer of helium gas surrounds the warm object, and therefore it cools only slowly, in a medium whose effective $\dot{Q}$ is below 1/4 W/cm². If the heat-generation rate is faster than that (as in a local quench), the object might warm still further, unless the wire is designed to by cryostable even at the minimum in $\dot{Q}$.

For nitrogen, even when the temperature reaches 87 K ($\Theta = 10$ K above the bath), $\dot{Q}$ is still increasing, and the very effective heat transfer of nucleate boiling takes place. There is no discontinuous jump to film boiling, either; the nucleate boiling continues, with quite large heat transfer values, even up to a 20 K temperature differential.

Film boiling certainly does occur in liquid nitrogen, but at temperatures above 100 K. In fact, the reason you can quickly jab your finger into a dish of LN_2 without harm is because of film boiling. A film of gaseous nitrogen forms around your finger and insulates it briefly so that it does not cool very much below 300 K in less than a second.

In the case of the HTSCs, nucleate boiling is the rule. Cryogenic stability can be represented numerically in terms of the ratio of area to perimeter for a wire, which must be less than a factor involving $\dot{Q}/J_c^2\rho$. (That makes sense, if the I^2R losses are to be carried away to the bath.) Again, the details appear in Wilson's *Superconducting Magnets.* The effect is to place an upper limit on the radius of a wire. Depending on how close to the margin one is willing to operate (i.e., $J \sim 0.9\,J_c$ compared to $J \sim 0.5\,J_c$, etc.), the calculated maximum allowable radii range from 10 μm to 1 mm.

It is interesting to note that the nucleate boiling criterion for the HTSCs yields wire radius values that are comparable (differing by less than a factor of 2) to those obtained using the helium film-boiling criterion for NbTi and Nb$_3$Sn. The thermal properties of various materials and cryogens at either temperature (4 K or 77 K) apparently counterbalance each other in roughly the same way.

17.9. EXPERIMENTAL VERIFICATION OF THE MODEL

When all the provisos of the preceding sections are put together, the model that emerges is nowhere near as simple as the equations we have presented for QPV, and so on. The fact that current is shared between the matrix material and the suddenly normal superconductor has not yet been considered, but in all plausible wire configurations the HTSC will be surrounded by a sheath probably made of silver. Multifilamentary wires have been analyzed for the LTSCs, but the HTSCs will not have the exact same geometry. To move beyond the level of generalities about stability, it is necessary to both carry out computations using a realistic model and verify the predictions experimentally.

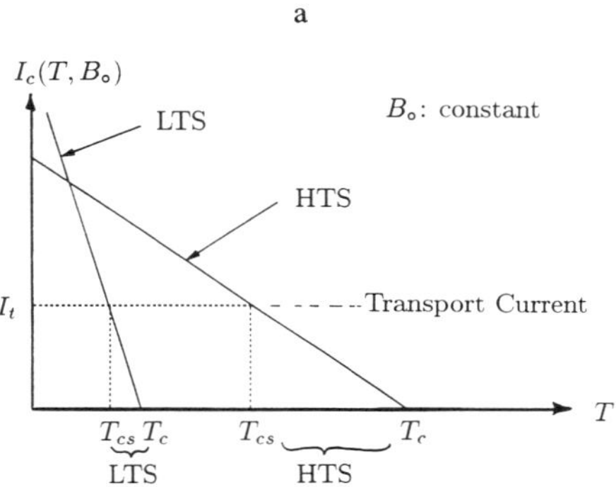

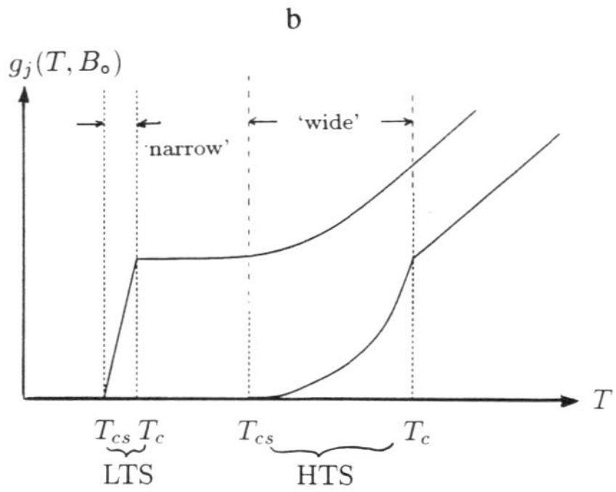

Figure 17.15.   (a) Plot of critical current I_c (T,B_0) vs. temperature for typical low-temperature and high-temperature superconductors, emphasizing region in which current sharing takes place. (b) Plot of heat generation g_j (T,B_0) at a constant value of transport current I_t for the two cases considered in (a). In the HTSC case, g_j at first rises as T^2, because ρ (copper) is proportional to T above 30 K. (From Ref. 12.)

This effort has been advanced by Iwasa of MIT and his students.[12] The first step is to take a simple model of the variation of critical current with temperature, and this is shown in Figure 17.15(a). For either LTSCs or HTSCs, the critical current I_c is taken to decline linearly with rising temperature; but the HTSCs go to zero at a much higher temperature, of course. The analysis is carried out in terms of current I rather than current density J, because current sharing takes place during the transition phase to the normal state.

In LTSCs, the transition phase to the normal state is quite a narrow temperature range, and this is depicted in figure 17.15(a). For any given choice of transport current I_t, there is a temperature T_{cs} below T_c at which some of the superconductor goes normal and current sharing begins. In any real situation, this generates so much heat that the entire wire quenches and boils off liquid helium, often very noisily. The heat generated $g_j(T, B_0)$ is shown in Figure 17.15(b). Unless the current drops enough to allow the helium bath to recover and restore superconductivity, the temperature will continue to rise. From the point of view of practical engineering, the narrow distinction between T_{cs} and T_c can be ignored.

It is quite a different story for HTSCs. As shown in Figure 17.15(a), there is a considerable temperature difference between the start of current sharing and the wire becoming totally normal. Numerical values are 10 to 30 K wide. Current sharing becomes an important parameter to include in the model. The corresponding heat generation [Figure 17.15(b)] is much more gradual. Since the specific heats of both the HTSC itself and the coolant are much higher in this temperature range, it is not necessarily true that the wire will reach its full normal state. The concept of a MPZ must be reconsidered and generalized to include this possibility. Similarly, the quench propagation velocity, which is much smaller for HTSCs than for LTSCs and is the source of burnout risk, must be reconsidered. If a sudden spike in current creates a momentary excursion into the current-sharing temperature regime, the probability of a safe recovery is a complicated function of several factors.

To investigate this, Iwasa built the apparatus[12] shown in Figure 17.16. In a 12 cm long configuration, a heater could be pulsed to momentarily drive the wire into the current-sharing regime, and voltages were measured at selected points downstream to observe how fast the

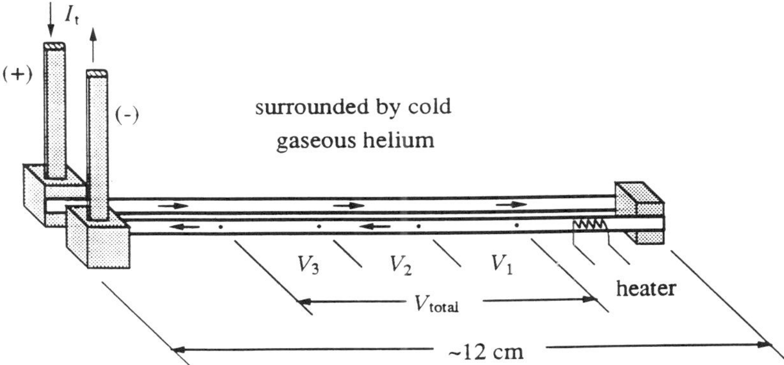

Figure 17.16. Experiment to study normal-zone propagation. With transport current I_t held constant, a pulse of heat is applied to create a small normal zone and its propagation down the wire is monitored by measuring voltages at selected points. After the normal zone reaches a particular voltage tap, there should be little further increase in voltage with time. (From Ref. 12.)

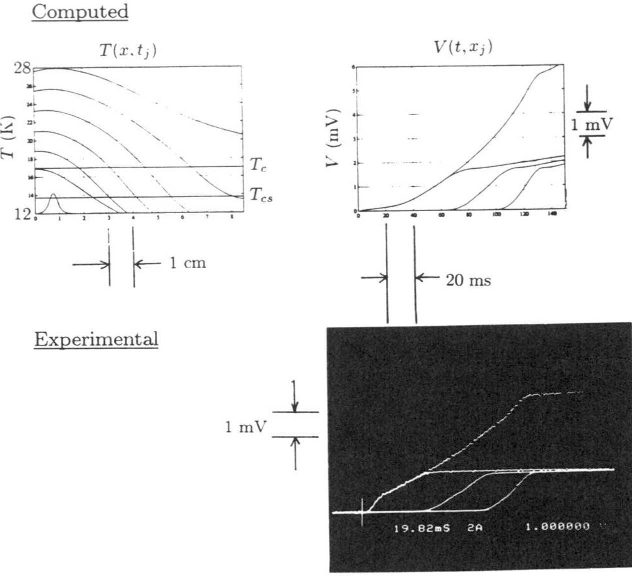

Figure 17.17. Comparison of computed and experimental voltages for the apparatus of Figure 17.16, for the special case of Nb₃Sn tape at 12 K, cooled by helium gas. The simulated temperature variation is shown as a sequence of lines at discrete time intervals, beginning with a pulse at $t = 0$ (lower left) and eventually rising to 28 K. The predicted voltage as a function of time between each of the three voltage taps, as well as the total voltage, is shown on the right. Experimental data (oscilloscope photo) confirms the validity of this model. (Ref. 12)

normal zone propagated. Computations were done to predict the ensuing temperature excursion and the resulting voltages expected at the established points. Measurements then served to check the accuracy of the model.

Figure 17.17 shows the results for a LTSC: a Nb₃Sn tape (made by GE) operating at 12 K with a transport current of 225 A, cooled at 40 mW/cm^2 by helium gas. The agreement between predicted and experimental voltages is very good, and thus shows that the detailed model is satisfactory. It is worth mentioning that the QPV is about 2 m/s and the maximum temperature reached by the tape is 28 K.

When a HTSC was tested, the very slow QPV interfered with the experiment. They used a (Bi,Pb)SCCO-2223 tape (made by Sumitomo) at various combinations of operating temperature, transport current, and external field. In one example ($I_t = 100$ A at 32 K), the predicted QPV was about 4×10^{-3} m/s, and the predicted temperature rose to over 360 K!

Table 17.4. Quench Propagation in BSCCO
at Intermediate Temperatures

T(K)	J_c(A/cm^2)	QPV (mm/sec)
10	41,600	35
20	9,600	0.5
36	6,300	0.08

The experiment did not exactly agree with the model, because of uncertainties involved in modeling the cooling conditions.

Subsequently, with improved apparatus, Iwasa was able to measure the QPV at several intermediate temperatures,[13] as shown in Table 17.4. Evidently, the QPV drops very rapidly as the temperature increases.

What all this shows is that if a quench ever does occur in a HTSC, it is not possible to rely on normal-zone propagation to spread out the heat load and achieve uniform dissipation within the HTSC. Changing to a different HTSC or going to 77 K does not affect this conclusion. Nearly uniform dissipation of the stored electromagnetic energy during a quench is necessary to make superconducting magnets self-protecting. Therefore, protection schemes that do not rely on normal-zone propagation must be devised for HTSC magnets.[12]

17.10. SUMMARY

Devices such as motors and generators, power transmission lines, and SMES require superconducting wires or tapes capable of carrying large currents and—except for transmission lines — in the presence of high magnetic fields. Magnets built during the 1960s using LTSCs were subject to thermal instabilities due in part to rapid and dissipative flux motion (flux jump) within the conductor, which drove the conductor normal and resulted in a tremendous release of heat called a *quench*. The solution to this problem for NbTi was to make composite conductors consisting of micron-sized NbTi filaments embedded in a copper matrix.

A similar analysis applied to HTSCs shows that, because of the much greater heat capacity of HTSCs at 77 K relative to LTSCs at 4 K, HTSC conductors should be adiabatically stable at much larger (up to forty times) filament sizes.

For the purpose of designing stable superconducting wire that will not burn out in a quench, the specific heat and the thermal conductivity are of great importance. Referring back to Figure 17.6, it is obvious that from an engineering point of view, there is no important change in specific heat near 77 K. This is in sharp contrast to LTSCs, whose specific heat varies dramatically near their operating point. Moreover, if driven into the normal state by a sudden local quench, the sharp drop in specific heat of an LTSC will allow a much greater temperature excursion in that local region, promoting the chance of thermal runaway. HTSCs are exempt from this behavior.

A comparison between specific heats of the HTSCs and NbTi (at their respective operating temperatures) illuminates the contribution of specific heat to adiabatic stability. The data of Figure 17.6 show that YBCO has a heat capacity near 200 mJ/g-K at 100 K, and about 150 mJ/g-K at 77 K. By comparison, NbTi has 0.82 mJ/g-K at 4.2 K. The electronic specific heat for NbTi is $C_{el(s)} \stackrel{\sim}{<} 0.8$ mJ/g-K, and for YBCO it is $C_{el(s)} \simeq 9.2$ mJ/g-K. However, it is the lattice specific heat (which is over two orders of magnitude greater in YBCO at 77 K) that dominates the heat capacity. This is what makes the question of adiabatic stability far less worrisome for HTSCs. Had we chosen a bismuth or thallium compound to illustrate this point, the result would have been similar. When a magnetic field is applied, the small electronic specific heat may vary, and T_c will decrease, but the lattice specific heat does not change. This, too, simplifies stability analysis compared to the low-temperature regime.

Another technique to stabilize very large magnets is cryostability. This technique employs large copper/superconductor ratios and large cooling capacities. Cryostability can restore superconductivity, even in the case where the entire coil has gone normal. HTSCs have cryostability conditions similar to LTSCs. The poorer thermal and electrical properties of copper at 77 K relative to 4.2 K are offset by the greater cooling capability of LN_2 relative to LHe.

Compared with LTSC conductors, HTSC conductors have extremely low quench propagation velocities (QPVs), which means that if a quench occurs it will likely burn out the wire unless some new protection method is devised. At the same time, however, HTSCs are far less likely to quench. They should be highly stable against flux jumps, allowing conductors to contain much larger filaments, but, as we shall see in the discussion of AC losses, smaller filament sizes lead to smaller AC losses. Also, smaller filaments are more flexible. It is clear that the engineering trade-offs associated with HTSC designs will be substantially different from those of LTSC designs.

Finally, the remaining unsettled question is, what happens in the intermediate temperature range (20–35 K)? Experimental values of heat capacity, thermal conductivity, and other parameters[13] are only beginning to appear for that temperature region. It is fair to assume that helium gas will be the refrigerant, but at what pressure? Film boiling will be the cooling mechanism. Will that limit the allowable filament size? As superconducting engineers begin designing apparatus for intermediate temperatures, these questions will have to be answered.

BIBLIOGRAPHY

E. W. Collings, *Applied Superconductivity, Metallurgy, and Physics of Titanium Alloys* (2 vol.), Plenum Press, New York (1986).
Y. Iwasa, *Case Studies in Superconduting Magnets*, Plenum Press, New York (1994).
J. E. C. Williams, *Superconductivity and Its Applications*, Pion Ltd., London (1970).
Martin N. Wilson, *Superconducting Magnets*, Oxford University Press, Oxford (1983).

REFERENCES

1. G. Nieva *et al.*, *Phys. Rev. B* **36**, 8780 (1987).
2. Wenger *et al.*, *Phys. Rev. B* **35**, 7213 (1987).
3. M. Tinkham, *Introduction to Superconductivity* (Krieger Publ. Co., Malabar FL: 1980).
4. T. P. Sheahen, *Phys. Rev.* **149,** 370 (1966).
5. N. E. Phillips, R. A. Fisher, and J. E. Gordon, Workshop on Low-Temperature Calorimetry, Tsing Hua Univ., HsinChu, Taiwan (Apr. 10–11, 1992).
6. S. E. Inderhees, Ph.D. thesis, University of Illinois at Champaign-Urbana (1990).
7. J. E. Gordon, R. A. Fisher, and N. E. Phillips, *Phil. Mag. B* **65**, 1389 (1992).
8. J. J. Freeman *et al.*, *Phys. Rev. B* **36**, 8786 (1987).
9. Y. Yeshuran and A. P. Malozemoff, *Phys. Rev. Lett.* **60**, 2202 (1988).
10. A. Houghton *et al.*, *Phys. Rev. B* **40**, 6763 (1989).
11. C. P. Bean, *Phys. Rev. Lett.* **8**, 250 (1962).
12. R. H. Bellis and Y. Iwasa, *Cryogenics* **34**, 129 (1994); Y. Iwasa, "HTSC Magnets," *Proc. International Superconductivity Symposium*, (Springer-Verlag, Tokyo: 1993).
13. Y. Iwasa, presented at DOE Annual Peer Review, July 19–20, 1994.

18

AC Losses

This chapter discusses AC losses and their implications for producing devices based on high-temperature superconductors (HTSCs). In particular, we will consider how AC losses affect the engineering of practical magnets and AC power transmission lines. Magnets are essential components in electric utility devices such as motors, generators, and superconducting magnetic storage (SMES) and are subject to various levels of AC fields. For purposes of this discussion, we will only consider low-frequency (60 Hz) AC losses.

18.1. BACKGROUND

AC losses are an important consideration for many large-scale devices. Efficiency is a critical design consideration, and is already quite high in conventional equipment. For example, the efficiencies of a conventional transformer, power transmission line, and generator[1] are 99.7, 98.3, and 98.6%, respectively.

These high efficiencies leave little room for AC losses in their superconducting analogues. In addition, each watt of AC loss deposited as heat in a superconductor requires many watts of refrigeration power for its removal. Removing one watt of heat deposited at 4 K to room temperature requires 500–1000 W of refrigeration power.[2] By contrast, 10 W of refrigeration power[1] will be required to remove each watt of heat generated in an AC transmission line operating at 77 K. Thus, for the same level of AC losses, the adverse impact on efficiency is much lower for 77 K HTSCs than for conventional superconductors.

During the 1970s, Brookhaven National Laboratory built a prototype transmission line using Nb_3Sn superconducting tape. The experimentally observed AC losses in this transmission line guided theorists to a fuller understanding of the loss mechanism, which in turn indicated the way to design a transmission line to minimize AC losses. That well-established theory still stands for conventional LTSCs. Initial laboratory measurements of AC losses in small bulk samples of HTSCs have been carried out and indicate that the conventional theory requires revision in order to explain the observed data.

Transformers are intrinsically AC devices and may never accommodate the AC losses of superconductors. Even for devices such as SMES and motors/generators which operate under quasi-DC conditions, large AC losses can be developed during transients. The subsequent deposition of AC-loss-generated heat promotes thermal instabilities, quenching, and even conductor damage. As pointed out in Chapter 17 on stability, because of the much higher heat capacity of HTSCs operating at 77 K compared to LTSCs operating at 4 K, HTSC devices should be much more stable.

373

How are AC losses generated? There are three components: *hysteretic* losses, *eddy current* losses, and *self-field* losses. In the following sections, we discuss these loss components. Next we relate these losses to the HTSCs. Following that, we review several methods for measuring AC losses and then describe measurements of AC losses for HTSCs.

The reader seeking more detail will find it in books by Wilson,[2] Carr,[3] Collings,[4] and a report by Clem.[5] Here we paraphrase and summarize Wilson's treatment; Figures 18.1 to 18.10 are taken from there.

18.2. AC LOSS MODEL

In composite superconductors two types of loss mechanisms operate. First, when a superconductor operates within a magnetic field in excess of H_{c1}, flux penetrates the superconductor. Movement of this flux results in dissipation. A time-varying magnetic field causes flux motion (hysteresis) and hence AC losses. This hysteretic dissipation is proportional to frequency f. Second, varying magnetic fields cause eddy currents to flow in the normal-metal portion of the composite, and their resistive dissipation loss goes as f^2. Therefore, losses *per cycle* are plotted versus frequency. The intercept of such a plot determines the hysteretic coefficient of losses (a) while the slope gives the eddy current loss coefficient (b):

$$w/f = a + bf \tag{18.1}$$

where w is the total loss.

Wilson's treatment begins by defining currents and voltage as shown in Figure 18.1. Each current loop has a self-inductance and they have a mutual inductance between them. The energy loss is $\mathbf{E} \cdot \mathbf{J}$ within any volume, so to get the total loss one integrates over an outer surface. Employing Maxwell's equations, it can be shown that the energy loss in one cycle is the area enclosed by the hysteresis loop in the plane of M vs. H. As usual, we take $B = \mu_0 H$.

One simple derivation that is easy to follow is that for a slab geometry. (Very similar results emerge for cylindrical and other geometries, but the mathematics is simplest for slab geometry.) Figure 18.2 presents the definitions to be used. The slab thickness is $2a$.

What happens when a magnetic field is applied, indeed, a time-varying (AC) magnetic field? For a slab made of a conventional low-temperature superconductor, at any instant we

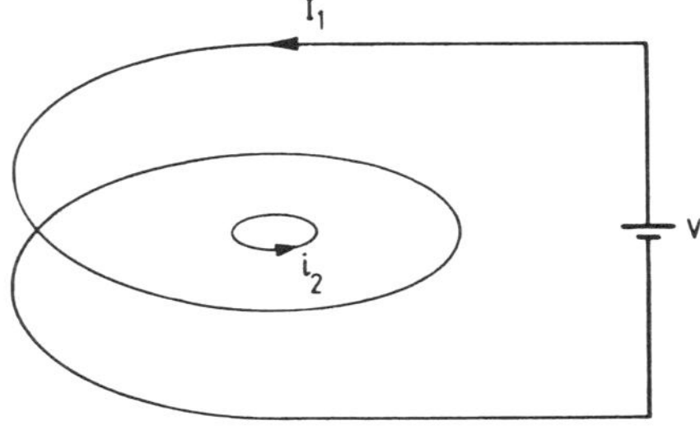

Figure 18.1. Definition of the current and voltage terms used in the text. Figures 18.1–18.9 are all taken from Ref. 2: *Superconducting Magnets* by Martin N. Wilson, publ. by Oxford University Press 1983, 1987; © M.N. Wilson 1983.

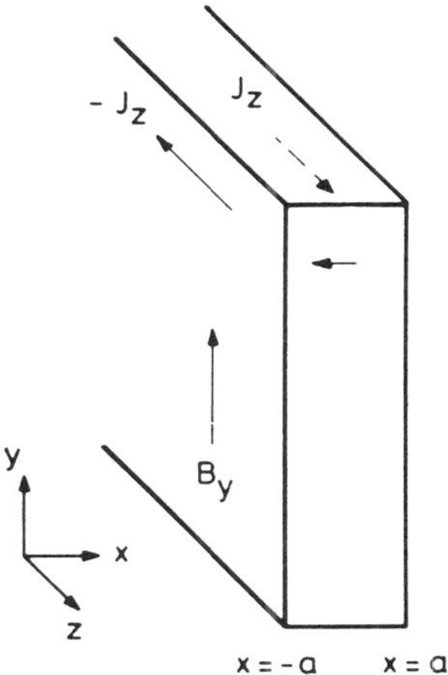

Figure 18.2.   Definition of the symbols used to describe the slab geometry in the text.

can safely assume that the magnetic field inside it, and the current flowing in it, are related by the Bean critical state model, as first presented in Chapter 2. It is an entirely separate question just how good the Bean model is for HTSCs, but, for the moment, assume its validity. On that basis, the magnetic field penetrates the slab partway, as shown in Figure 18.3(a). The current flowing near the outside of the slab is taken to be exactly $\pm J_c$ under the Bean model, and the interior current is zero. Thus, the depth to which current flows is $p = B_m/2\mu_0 J_c$, where B_m is the maximum field value (at the surface).

Next, to model the AC behavior in a very simple way, a negative going field pulse is applied. The field distribution at one-quarter and one-half cycle are shown in Figure 18.3(b) and (c), respectively. Since B_m is the peak-to-peak field strength and since we have assumed B_m does not fully penetrate the slab, we know that the current density has the value $\pm J_c$ and flows in only a portion of the volume of the slab. That condition makes it easy to carry out the volume integral of $\mathbf{E}\cdot\mathbf{J}$, thus obtaining the losses generated per cycle Q. Wilson[2] works through all this to obtain the very important formula

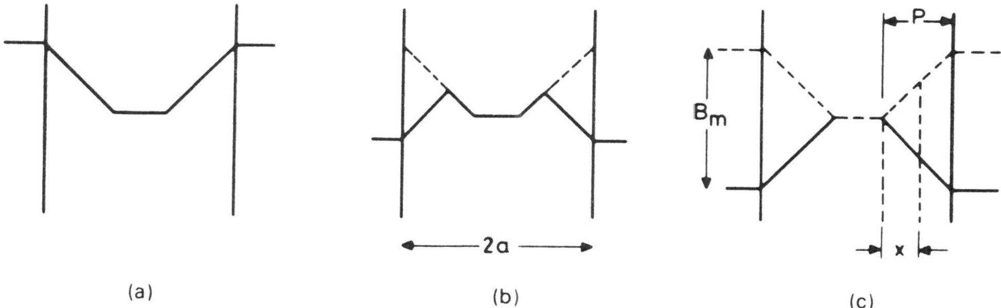

Figure 18.3.   (a) Field pattern within a superconducting slab of thickness $2a$, subjected to a small field change; (b) as the external field is reduced; (c) when the external field reaches minimum value before rising again. B_m is the peak-to-peak excursion of magnetic field; P is the depth to which magnetism penetrates the slab.

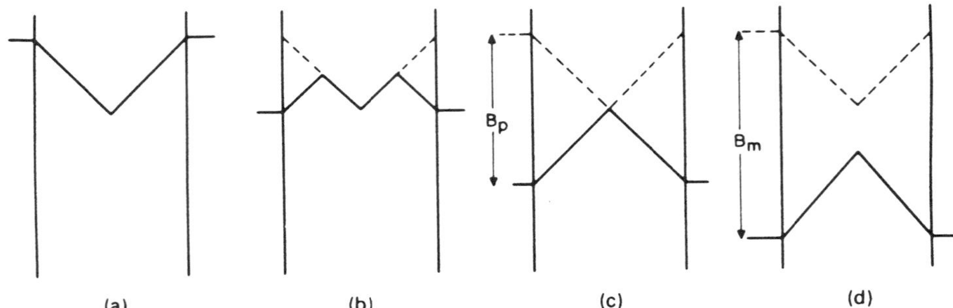

Figure 18.4.   (a) Field pattern within a superconducting slab subjected to large field change; (b) as the field is reduced; (c) when the field change penetrates to center of slab; (d) when the field reaches a minimum value before rising again. B_m is the peak-to-peak excursion of the field, and B_p is the variation in B which just causes full penetration of the slab.

$$Q = \frac{B_m^2}{2\mu_0} \cdot \frac{\beta}{3} \tag{18.2}$$

The dimensionless factor $\beta = B_m/B_p$, where $B_p = 2\mu_0 J_c a$. Actually, B_p corresponds to the field that would exactly allow the magnetic field to just reach the center of the slab. In Equation (18.2), the first factor is the maximum volumetric energy stored by the magnetic field during an oscillation. It is multiplied by the second factor, which is the fraction dissipated. Wilson calls this second factor the *loss factor* $\Gamma(\beta)$.

Figure 18.4 illustrates the case where B_m exceeds B_p. Again starting with the field pattern in (a), a negative going field pulse penetrates to the center of the slab in (c) and continues to its minimum values in (d). The electric field $E = 0$ at the center of the slab, where the screening currents reverse, and a similar volumetric integral over $E \cdot J$ yields for the volumetric loss per cycle Q:

$$Q = \frac{B_m^2}{2\mu_0}\left(\frac{1}{\beta} - \frac{2}{3\beta^2}\right) \qquad \text{for } (\beta \geq 1) \tag{18.3}$$

Similar results can be derived for a cylinder (wire) parallel and perpendicular to the field, with slightly different numerical factors, of course. The case for the field parallel to the cylinder is quite straightforward. However, the case where the field is perpendicular to the cylinder is much more complicated, and that is the case most typical of solenoids. Wilson derives an expression that must be solved numerically, and Figure 18.5 graphically depicts the results for β between 0.01 and 100. In each case, the loss factor is maximum when $\beta \approx 1$. For $\beta \approx 1$ the AC losses are proportional to B_m^2.

To obtain low loss factors, either a very small or a very large β is required. Because B_m is usually determined by the requirements of the application, and because high J_c is desirable for reasons of cost and compactness, only the characteristic size a (usually the radius) is available for controlling AC losses. To obtain a low loss factor, a must either be large or small. Large a means that each conductor element must carry large total current, so thermal stability is reduced. Furthermore, B_m and hence β increases under many transient conditions. If the conductor has been designed to operate in the low β region, an increase in

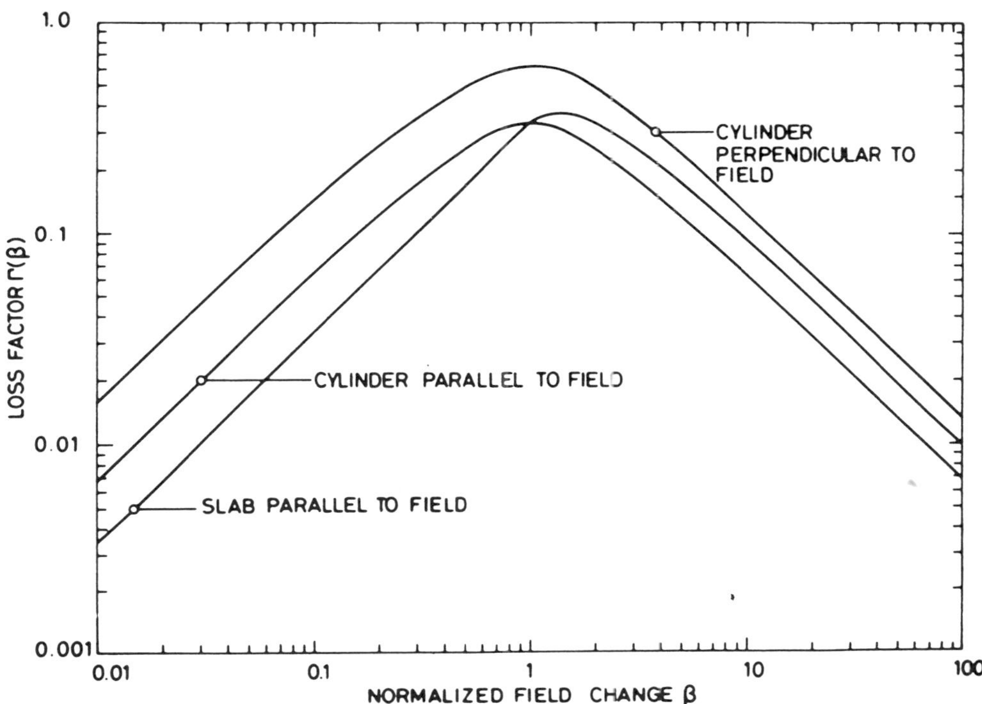

Figure 18.5. Loss factor $\Gamma(\beta)$ for hysteresis loss per cycle in different shapes of superconductor. In the simplest case (small β), the factor is simply $\beta/3$, as in Equation (18.2).

B_m will increase both the stored energy and loss factor, leading to greatly increased AC losses (proportional to B_m^3), which has the potential for thermal runaway. By contrast, if the conductor has been designed to operate in the high β range (small a), an increase in B_m and hence β leads to smaller increases in AC losses (proportional to B_m). This is the reason why most conventional LTSC conductors designed for AC applications employ a multifilamentary design where each superconducting filament is near 1 μm in diameter.

So far we have considered the effects of changing magnetic fields on screening currents as though no transport current were present. But in real devices, there will normally be a sizable transport current J_T, often a high fraction of J_c. It is convenient to denote the transport current in terms of its ratio to J_c, defining $i = J_T/J_c$. Clearly, the addition of a transport current can only make things worse. Wilson[2] considers the case in which the transport currents are held at a constant level during the cycle. The effect of transport currents in a changing magnetic field is to lower the value of B at which current penetrates all the way to the middle of the conductor. This no longer means $\beta = 1$, but now $\beta = 1 - i$. For the case of $\beta < 1 - i$, the derivation leading to Equation (18.2) still holds, that is

$$Q = \frac{B_m^2}{2\mu_0} \frac{\beta}{3} \qquad \text{for } \beta = B_m/2\mu_0 J_c a \leq (1 - i) \tag{18.4a}$$

However, for $\beta > 1 - i$, the new value of volumetric energy loss per cycle Q becomes

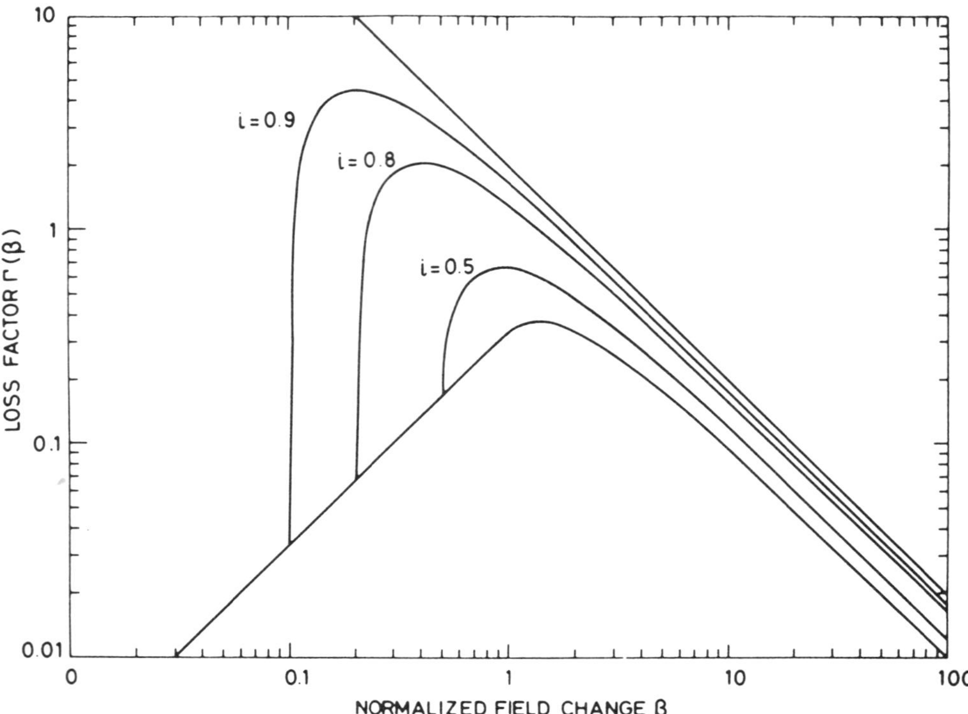

Figure 18.6.   Hysteresis loss in superconducting slab carrying fixed transport current $I_t = iI_c$ and subjected to a changing external field, calculated from Equation (18.4).

$$Q = \frac{B_m^2}{2\mu_0}\left[\frac{1+i^2}{\beta} - \frac{2(1-i^3)}{3\beta^2}\right] \qquad \text{for } \beta \geq (1-i) \qquad (18.4b)$$

Figure 18.6 shows the loss factor as a function of β for various values of i. The effect of transport current is to shift the peak downward to a lower value of β, which also greatly increases the loss factor, particularly for $\beta > (1-i)$ and $\beta \leq 1$. The increase is relatively small for large β, and there is no change for $\beta \leq (1-i)$. Transport currents between half and 90% of J_c are commonplace, and so Figure 18.6 shows it is a serious problem.

This adds one more reason to use multifilamentary cable with small a to get high β. To promote conductor stability and reduce AC losses, most LTSC conductors are manufactured in the form of a multifilamentary composite consisting of a large number of very fine superconducting filaments embedded in a normal metal matrix. A typical conductor can contain 15,000 filaments.

18.3. DESIGNING AGAINST AC LOSSES

The typical application of LTSC cable is not for transmission but in magnets; then the magnetic field is perpendicular to the conductor axis. The mathematics covering that behavior is intricate; either Wilson[2] or Carr[3] treats it adequately. For our purposes it is

Table 18.1. Wire Parameters for Multifilament Composite NbTi AC Cable[a]

Wire diameter (mm)	Filament diameter (μm)	Twist pitch length (mm)	Weight (kg/km)	$J_c{}^b$ (kA/cm^2)	I$^+$ (A)
0.30	1.4	5.0	0.56	55	39
0.20	1.0	2.2	0.24	51	16
0.12	0.6	0.8	0.09	52	5.9

[a]Commercially available wire produced by Alsthom (CCN 14000 LL) contains 14,496 filaments.
[b]1 μV/cm criterion measured at 4.2 K and 5 T. Composite consists of 1/3 NbTi, 2/5 CuNi, and 4/15 Cu.

sufficient to note that changing magnetic fields can couple the filaments together, resulting in greatly increased AC losses due to eddy currents circulating in the normal metal regions of the wire.

The engineering solution that minimizes such losses is to twist the multifilamentary composite; and the twist-pitch length is determined by a certain characteristic length that depends on the product $J_c a$, on the normal state resistivity ρ, and on the time rate of change of field, $\dot{B}$. Wires used for AC applications are engineered carefully to steer the optimum course through conflicting influences of heat transfer, protection, hysteresis losses, and eddy current loss. For example, reducing the pitch length L also reduces Q. However—with standard manufacturing techniques used for LTSCs—as L is reduced, wire strain increases and eventually causes a decrease in J_c and ultimately filament breakage. Table 18.1 presents wire parameters for a multifilamentary composite sold by Alsthom for AC applications.

There are also losses associated with the transport current flowing through a wire. In typical magnet applications, the self-field associated with a single filament is much smaller than the total magnet field. If there are $N = 1,000$ to $10,000$ turns in a solenoid, the self-field is approximately N times smaller than the solenoid field. For a transmission line, the external field is equal to the self-field (i.e., $N = 1$) and is proportional to the transport current. Simple twisting of the filaments does not reduce self-field losses because twisting does not affect the linkage of self-field flux between inner and outer filaments. To avoid this linkage, transposition must be employed. This requires that the wire be braided, such as in Figure 18.7.

Figure 18.8 illustrates the self-field case for reversing and nonreversing oscillations. Again following Wilson,[2] the field distribution can be found using Ampere's law:

$$B(r) = \mu_0 \lambda J_c \left(r - \frac{c^2}{r} \right) \qquad \text{for } c \le r \le a$$

where a is the wire radius, c is the penetration depth, and λ is the volume fraction of superconductor. For the reversing case [(a), (b), and (c) in Figure 18.8] the change in flux can be obtained by integrating $\beta\,(r)$, and then a similar formula appears for the heat loss:

$$Q_R = \frac{B_{ms}^2}{2\mu_0} \left[\frac{2}{\beta} - 1 + \frac{2(1-\beta)}{\beta^2} \ln(1-\beta) \right] \tag{18.5}$$

where $B_{ms} = \beta \mu_0 \lambda J_c a$.

For the nonreversing case [(d), (e), and (f) in Figure 18.8],

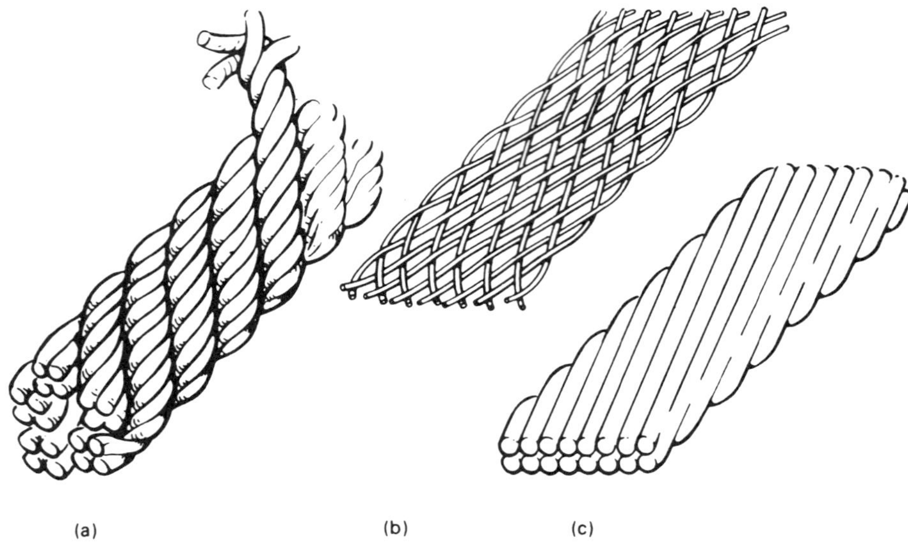

Figure 18.7. Three alternative methods of braiding conductors so as to minimize the effects of self-field on the individual filaments.

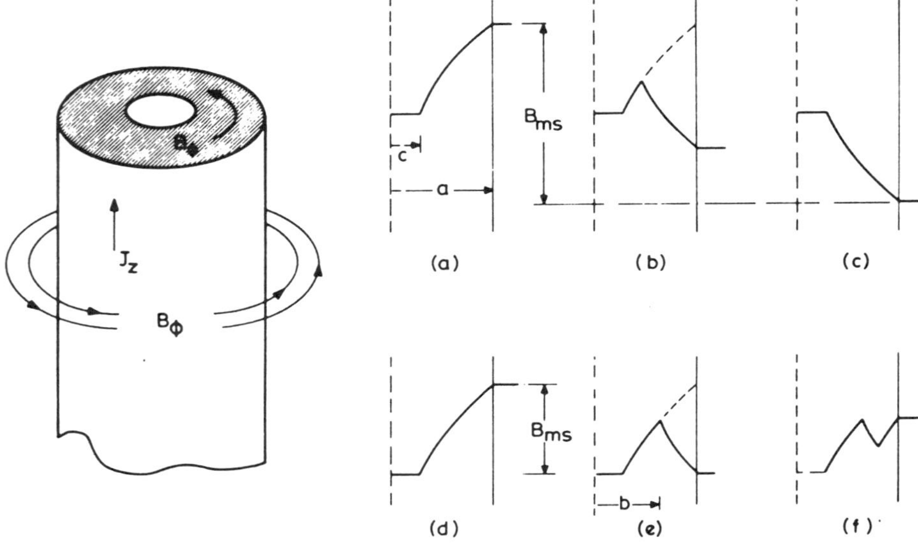

Figure 18.8. Self-field in a superconducting cylinder (or filamentary composite) carrying transport current. As the transport current is reversed, the profile of B within the cylinder changes from (a) through (b) to (c). B is shown for the case where the field does not reverse (unidirectional case) in (d), (e), and (f).

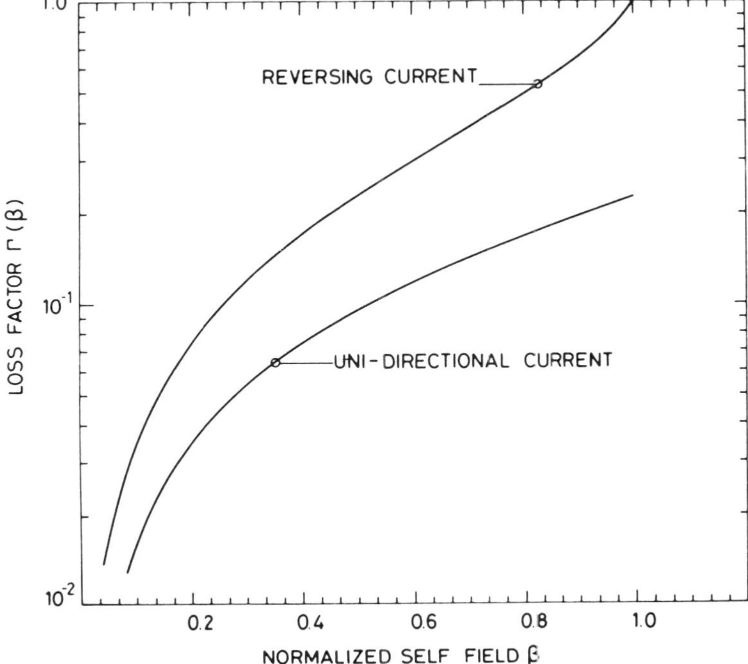

Figure 18.9. Self-field loss factor (per cycle) for the geometry of Figure 18.8. The case of fully reversing current shows greater losses than does the case of unidirectional oscillations from zero.

$$Q_{NR} = \frac{B_{ms}^2}{2\mu_0} \left[\frac{4}{\beta} - 1 + \frac{4(2-\beta)}{\beta^2} \ln\left(\frac{2-\beta}{2}\right) \right] \tag{18.6}$$

Figure 18.9 illustrates the self-field loss factors for the reversing and nonreversing cases. For all β, losses are lower for the nonreversing case. This is because the internal currents penetrate less for the nonreversing than the reversing case. For very small β, $Q_R = 2Q_{NR}$. For $\beta = 1$ (the maximum possible β), $Q_R = 4.4\ Q_{NR}$.

Early in the history of HTSC, these same concepts were brought over from LTSC,[1] but more refined analysis has been done for the HTSCs themselves. However, no one is yet in any position to design HTSC cable for AC-loss minimization. It is hard enough to make wire out of HTSCs at all.

18.4. HTSC THEORY OF AC LOSSES

The intricacies of AC losses are evidently quite specific to the type of superconducting wire composite being used. Thus, it is by no means obvious that the same model will hold for the HTSCs. Because of this possibility, in 1988–1989 EPRI sponsored Prof. John Clem of Iowa State University to make a theoretical study of AC losses in HTSCs. He wrote the report *AC Losses in the New High-Temperature Superconductors.*[6]

18.4.1. Basic Loss Formula

Clem began by modeling the losses in a coaxial transmission line based on the Brookhaven design. In this design, counterflowing current passes through two concentric, superconducting cylinders separated by a dielectric. Because the radius of each cylinder is much larger than its thickness and because the self-field will be parallel to its surface, the slab formula [Equation (18.2)] is a good approximation of the losses within each cylinder. Because the losses are ultimately removed by the refrigerant at the surface of each cylinder, it is customary to express the losses per cycle on a *per unit surface area* rather than on a *per unit volume* basis. To enhance the distinction between these, we use W for the former and Q for the latter.

The magnetic field penetrates from both sides of the slab, but because of the coaxial design, the magnetic field is present only between the cylinders. Thus, the energy loss per cycle per unit surface area, W_a in joules per square meter, is given by

$$W_a = (2a)(\tfrac{1}{2})Q = 2\mu_0 h_0^3/3J_c \tag{18.7}$$

where $2\mu_0 h_0 = B_m$. Specification of the desired power level and operating voltage of the Brookhaven design determines the required transport current, I_T. The critical current density and cylinder thickness determines the required radius of each cylinder, which determines h_0. In practice I_T is chosen to be significantly lower than I_c so as to allow for fault currents and operation under emergency conditions.

In particular, the Brookhaven transmission line had $h_0 = 889$ Oersteds (707 A/cm) and thus $b_0 = 0.0889$ T. (Recall that $b_0 = \mu_0 h_0$.) Brookhaven's conductor was a tape made of stainless steel, tin, niobium and copper, in which two thin regions of diffused Nb$_3$Sn actually carried the supercurrent. Because there are eddy current losses whenever an AC current flows next to a normal metal, the Brookhaven transmission line was designed to minimize eddy currents; neglecting them does not introduce serious error.

18.4.2. J_c Values

There is an important difference between the LTSC and HTSC materials, which was first brought out in Chapter 13. The coherence length ξ is so small in HTSCs that superconductivity does not easily cross grain boundaries the way it does in LTSCs. This means that although each single grain may have a high J_c value (denoted by J_{cg}), the actual transport current that can flow down a wire made of many, many grains is greatly reduced by the presence of barriers (each grain boundary). This so-called weak link behavior makes the carrier a network of Josephson junctions, and the reduced transport current is therefore denoted by J_{cJ}. J_{cJ} is also termed the *inter*granular critical current, whereas J_{cg} denotes the *intra*granular current. In HTSCs, J_{cJ} is typically found to be less than J_{cg} by a factor of 1,000.

Clearly, it is J_{cJ} that counts for transmission of electricity, and thus it is J_{cJ} that appears in the denominator of Equation (18.7). This will make the losses two or three orders of magnitude worse than if J_{cg} could be used. This is one of the principal obstacles today to fabrication of practical HTSC wire. A great deal of research effort is going into finding ways to increase J_{cJ} toward J_{cg}.

18.4.3. Penetration Depth

Clem's report describes some of the details of the mechanism by which magnetic fields partially penetrate the granular HTSC material, and he collects several factors into an effective magnetic permeability which is a function of temperature, μ_{eff} (T). This in turn determines the intergranular penetration depth, which is usually far greater than the intra-granular penetration depth that is characteristic of the HTSC material. Presently Clem obtains the loss formula

$$W_v = 2\mu_0 \mu_{\text{eff}} h_0^3 / 3 J_{cJ} \tag{18.8}$$

which is obviously analogous to the more familiar Equations (18.2) or (18.7). For cases of practical interest, μ_{eff} varies between unity and f_n, the volume fraction of the material in the normal state. Thus, the range of possible losses is delineated.

However, one cannot eliminate AC losses just by driving the normal state fraction toward zero. There will be *intra*granular losses within the superconducting fraction f_s as well. Careful consideration of the relationship between grain size and depth of penetration by the field introduces a factor Ω, $0 < \Omega < 1$, in the form $\mu_{\text{eff}} = f_n + f_s \Omega$; Ω is close to unity in the pertinent cases. Treatment of cylindrical conductors, both solid and hollow, introduces further numerical factors that give different limiting values for the losses when magnetic fields do or do not penetrate the individual grains. When there is negligible penetration (due to low field or high J_{cJ}), then $\mu_{\text{eff}} = f_n$; and when there is complete penetration (due to high field or low J_{cJ}), then $\mu_{\text{eff}} = 1$. The formulas for all cases still contain J_{cJ} in the denominator, and therefore increasing J_{cJ} is of paramount importance.

18.4.4. Surface Barrier

In LTSC materials, a surface barrier forms between superconductor and matrix. It acts to enhance power transmission and reduce losses. Taking an optimistic approach, Clem notes that a similar surface barrier (not yet found experimentally in bulk YBCO) would help in the same way. By analogy with the Nb$_3$Sn case, where the surface barrier field is 14% of the thermodynamic critical field, Clem argues that a surface barrier in YBCO might screen about half the peak field (nominally 0.09 T) from the interior, substantially reducing AC losses. This can best be termed "chancy." Experimental observations of the surface barrier would be most welcome.

18.4.5. Numerical Values

Using the Brookhaven criteria for a transmission line, Clem worked out numerical examples of losses in certain cases. Recall that both Q and W are per cycle; frequency is needed to convert to loss per second. For 60 Hz AC current with $h_0 = 889$ Oe (707 A/cm), corresponding to a cable carrying 500 rms A/cm, the power loss per unit surface area P_a in W/cm^2 is given by

$$P_a = W_a(60) = 177.7 / J_{cJ} \tag{18.9}$$

where J_{cJ} is the *inter*granular current density in A/cm^2.

It can be seen immediately by plugging in typical J_{cJ} values near 1,000 A/cm^2 that the power losses will be fractions of watts, not microwatts as in the Brookhaven case. For example, if $J_{cJ} = 2000$ A/cm^2, $P_a = 90$ mW/cm^2, and if $J_{cJ} = 400$ A/cm^2 (which is what copper carries), $P_a = 440$ mW/cm^2.

Even after noting that a nitrogen-cooled system can afford much greater heat losses than a helium-cooled system, these numbers are still significant. The only way HTSCs can economically carry AC current is to get higher values of J_{cJ}. If thin-film values of J_{cJ} ($\geq 10^6$

A/cm^2) can be achieved someday in a macroscopic conductor, losses would certainly fall to an acceptable range, below 100 μW/cm^2.

Proceeding to more detail, in the HTSCs J_{cJ} is actually dependent on H. And in many samples of YBCO, J_{cJ} falls off rapidly as H increases. Clem did not use any particular model for the dependence of $J_{cJ}(H)$, but calculated a worst case of 200 W/cm^2 loss. As HTSCs with improved flux pinning appear from various laboratories, it may be hoped that the roll-off of J_c with H will soften, so that the losses will eventually not be so great. However, it is still prudent to be suspicious of theoretical predictions until they are verified by measurements.

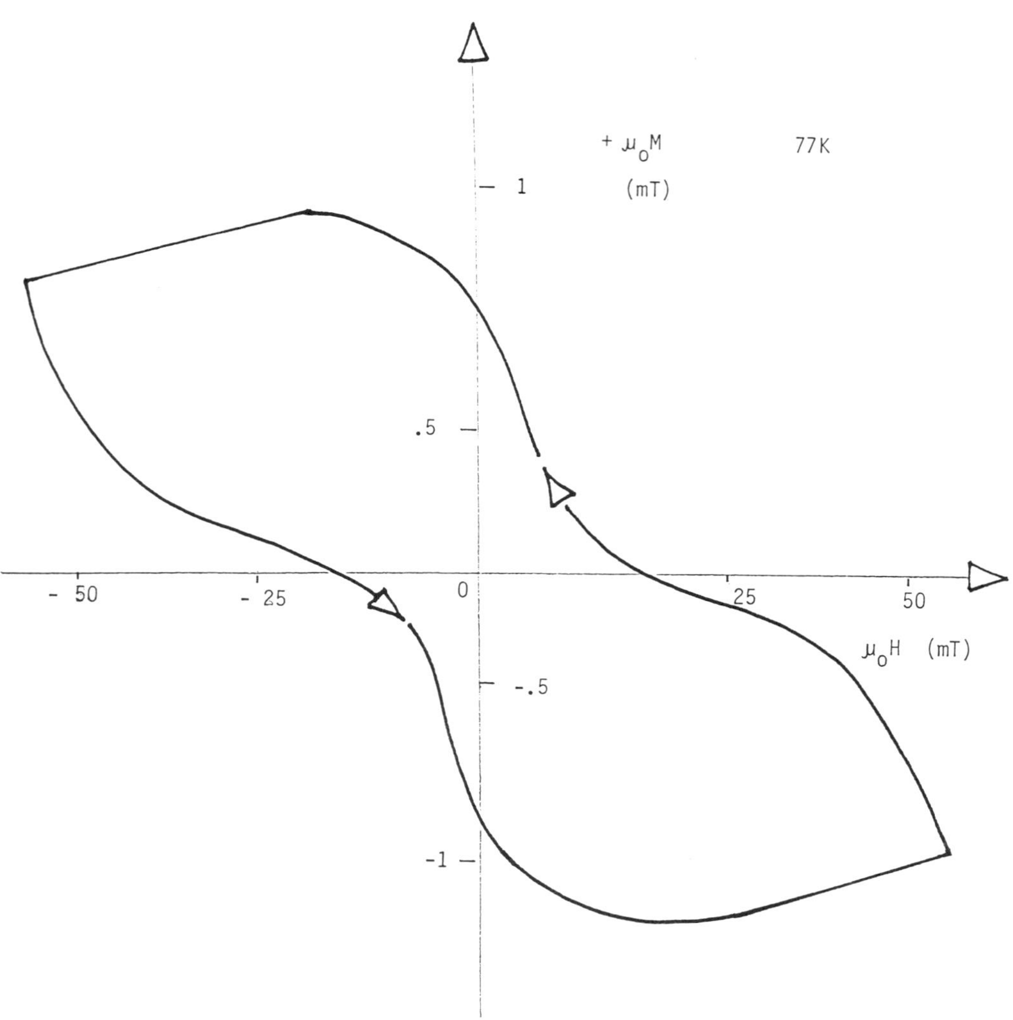

Figure 18.10.   Magnetization plot (*M* vs. *H*) for a typical sample of YBCO. The area enclosed by the curve gives $\oint M \cdot dH$, the AC hysteresis loss in one cycle. (After Ref. 7.)

18.5. MEASURING AC LOSSES

AC losses can either be measured calorimetrically or electronically. However, calorimetric measurements are even more difficult than those for determining specific heat, because of many extraneous heat leaks. The electronic method suffers from having to take the difference between two large numbers, but this can be done with modern equipment and careful attention to avoiding extraneous phase shifts. To be successful, the phase shifter must be accurate to one part in 10^5, because the AC loss is only about one part in 10^3 or 10^4 of the energy flowing in the test circuit. (The rest is stored reversibly in the magnetic field generated by the test coil.) Apparatus of this type, but vintage 1970s, is described by Wilson.[2] The art is steadily improving.

Either the calorimetric or the electronic method studies the superconducting sample in the transport mode. One can also use AC magnetization measurements and determine the losses from the area enclosed by the magnetization versus H plot. This method eliminates the need to make electrical connections to the sample and is therefore very inviting. Figure 18.10 illustrates a magnetization plot[7] for a sample of YBCO at 77 K. The area inside the curve equals the total AC loss for one cycle.

The apparatus for measurements of this type is described by Wilson.[2] The experiment relies on the symmetry of positioning samples within a large superconducting coil inside a cryostat; a trim coil helps balance the circuit. In practice, the results are easily interpretable only when the magnetization is uniform throughout the sample. Application of demagnetizing factors to correct for variations can be quite difficult. This is particularly true for HTSCs where one has to worry about the effects of intergrain and intragrain magnetization currents.

Another technique used for measuring the magnetization properties of HTSCs is to obtain signals while moving the sample in and out of the search coil while keeping the solenoid field constant. This is called a vibrating magnetometer.

18.6. EXPERIMENTAL RESULTS

In the past, two U.S. groups worked extensively on measuring AC losses in LTSCs. At Brookhaven National Laboratory (BNL) in the 1970s,[8] Garber and Suenaga measured AC losses in Nb_3Sn tapes. This work was performed as part of the AC power transmission line project. A team at Westinghouse studied AC losses in multifilament NbTi composites as a part of an effort to develop a superconducting power generator. Numerous experiments performed on LTSC composites validated the qualitative features described above. Quantitatively, theory and experiment agree at roughly the 30% level over a wide range of frequencies, pitch lengths, filament sizes, and applied field strength.

Investigation of AC loss in the HTSCs are less refined. Until genuine wire is made, there is little reason to push such measurements to high accuracy.

18.6.1. (RE)BCO

Since J_c was so low in early samples of YBCO, experimental AC losses turned out very high. Some preliminary results were as follows:

1. In Europe, Ciszek et al.[9] measured AC losses by magnetization in a cylinder of $GdBa_2Cu_3O_7$ at 4.2 K. The results are shown in Figure 18.11. Loss per cycle per unit surface area (W) is plotted as a function of applied field for several frequencies. All the points

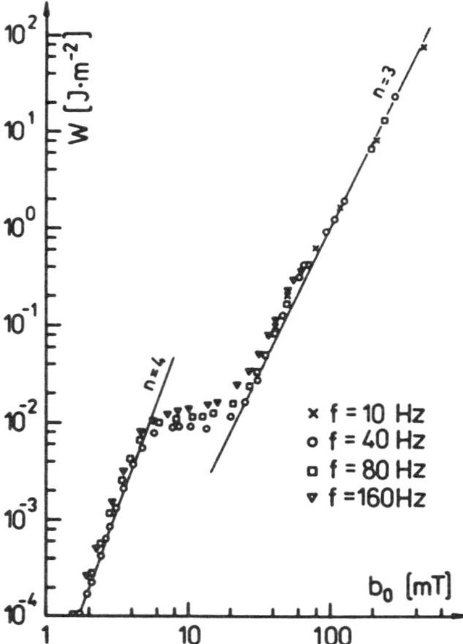

Figure 18.11. AC losses in a HTSC at 4.2 K as a function of applied magnetic field. Although this data was taken at 4 different frequencies, it has been reduced to a common curve by representing the losses on a *per cycle* basis. The lines labelled $n = 4$ and $n = 3$ correspond to different choices of the exponent n in $W \propto B^n$. (From Ciszek, *et al.*[9])

coalescing into a single line indicates that loss per second (in W/cm^2) is linear in frequency. This data is consistent with other AC loss experiments and is characteristic of hysteretic losses.

2. A limited amount of direct transport current data was obtained at Argonne National Laboratory in 1989 by van der Vegt,[10] and is shown in Table 18.2. He tested four coils made of bulk YBCO containing silver, with critical current densities near 100 A/cm^2. The experiments consisted of varying the AC current through each coil and measuring the voltage loss. Several different frequencies were used, but the measured loss data did not depend

Table 18.2. YBCO/Silver Coil Data

	Coil 1	Coil 2[a]	Coil 3b	Coil 4
Wire diameter	0.983 mm	0.962 mm	0.965 mm	0.959 mm
Wire length	1200 mm	1063 mm	1330 mm	1407 mm
Coil diameter	28.2 mm	28.2 mm	28.2 mm	28.2 mm
Number of turns	14	12.5	16	17.5
Silver content	15%	0%	7.5%	7.5%
Critical current	1650 mA	413 mA	750 mA	1080 mA
Critical current density	264 A/cm^2	57 A/cm^2	103 A/cm^2	150 A/cm^2
Magnetic field constant	7.1 G/A	7.3 G/A	8.4 G/A	8.7 G/A
Contact method	Indium	Silver	Silver	Silver

[a]Coil 2 was refired to make the silver contacts: 700°C, 12 hr. The original critical current was 1.43 A (J_c = 197 A/cm^2).
[b]Coil 3 was deformed after the firing and showed overlapping windings.

strongly on frequency, which shows that the losses were mainly hysteretic, not eddy–current losses.

3. The most advanced data comes from Orehotsky and co-workers[11–13] at Brookhaven National Laboratory (BNL), who carried out a series of AC-loss measurements on HTSCs, at both 4 K and 77 K. Their early YBCO samples[11] included two types of sintered specimens: coarse grains, with many impurities at the grain boundaries, and fine grains with relatively clean boundaries. The AC loss performance appears better (less lossy) in the coarse-grain material. Figure 18.12 displays the AC loss *per second* (in W/cm^2) of both grain types as a function of either applied field or induced surface current at 4 K. Figure 18.13 is similar for 77 K.

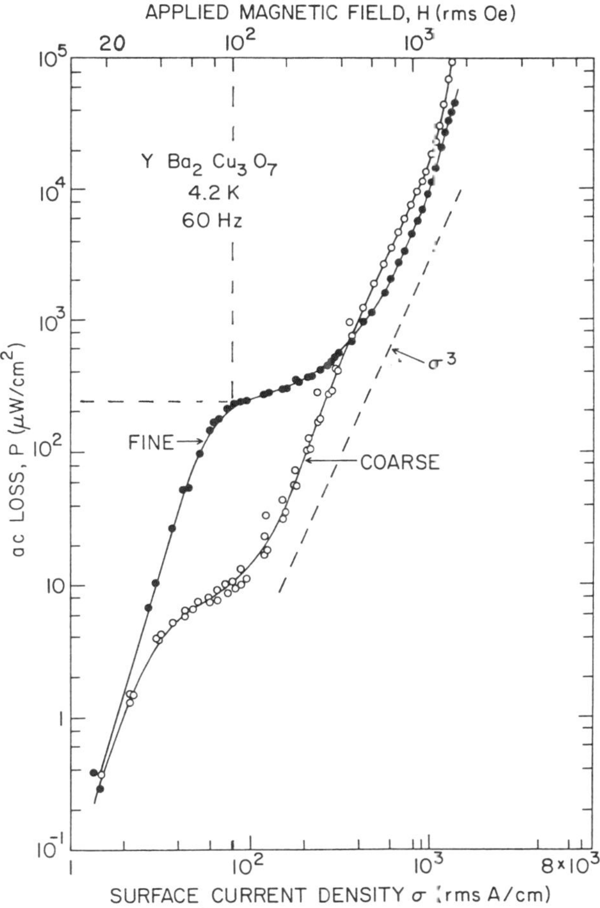

Figure 18.12. AC losses at 4.2 K and 60 Hz. A/cm is a *surface current* per tape width. The samples were sintered-powder $YBa_2Cu_3O_7$ cylinders, measured by the magnetic induction method. Samples made of fine or coarse grains show markedly different behavior. The vertical dashed line at H = 100 Oe shows that fine-grain samples have a loss of 220 μW/cm^2. The slope of a σ^3 line ($\equiv B^3$) is indicated, to make connection with the simplest theory of AC losses. (From Orehotsky *et al.*[11])

Remember that these measurements are magnetization, not transport, and it is very tricky to translate from one to the other. Magnetization measurements only tell what the individual grains are doing, which is not necessarily the same as what transport current might show. This causes some very anomalous results:

- In particular, the fine-grain samples show 77 K losses that are lower than the 4 K losses. The construction lines at 100 Oe on Figures 18.12 and 18.13 draw attention to this fact: losses are 220 μW/cm^2 at 4.2 K but only 110 μW/cm^2 at 77 K. What is happening here is that at the higher temperature, the grain boundaries decouple, and hence the magnetization apparatus sees only *intra*grain current at 77 K instead of *inter*grain current at 4 K. Suenaga of Brookhaven calls this an "apples to oranges" comparison. Failure to take note of this decoupling effect could lead one to believe the performance actually improves at higher temperatures.
- Similarly, the figures show that coarse-grain samples have losses generally lower than fine-grain samples in the low-field range typical of transmission lines. However, recognizing that the transport current is dominated by weak-link effects, and coarse

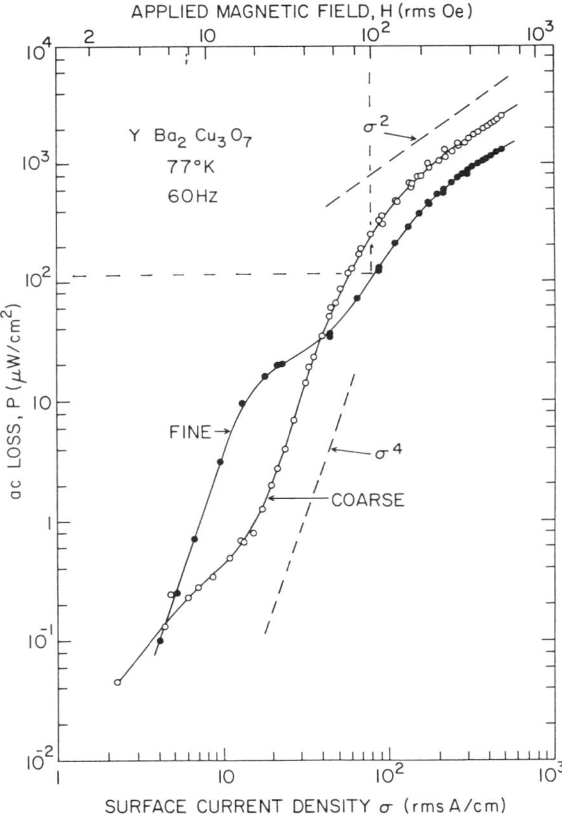

Figure 18.13. AC losses at 77 K and 60 Hz. This data is for the same samples as in Figure 18.12, but at higher temperature. In this case, the dashed line at $H = 100$ Oe shows that fine-grain samples lose only 110 μ W/cm^2.

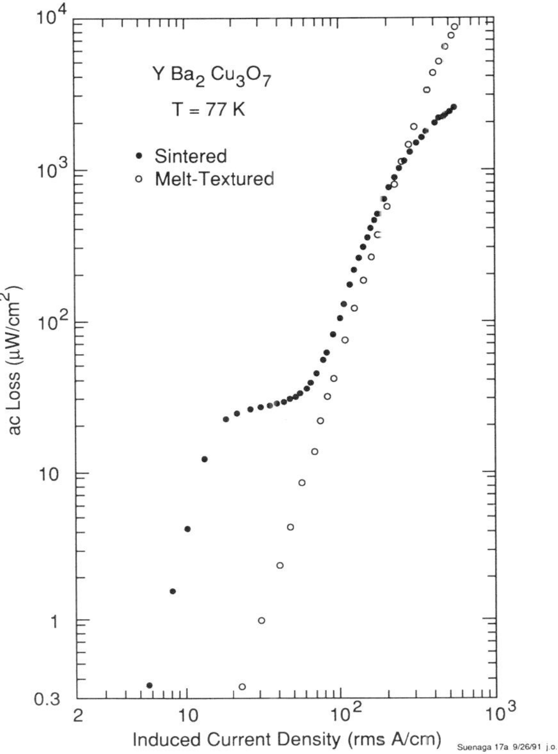

Figure 18.14. AC loss at 77 K and 60 Hz as a function of the rms-induced linear current density σ for a sintered and a melt-textured YBaCuO sample (From Ref. 12.)

grains with dirty boundaries suffer worse weak-link effects, it is unlikely that coarse-grain material will ever be used for transporting current.

Later, the Brookhaven group compared sintered YBCO with melt-textured YBCO,[12] and got the results shown in Figure 18.14. The melt-textured sample does not have the plateau of the sintered sample, which indicates that there are no obvious weak links in the specimen. The power loss varies as B^3 over most of the range.

18.6.2. BSCCO

The Brookhaven team also measured AC losses of BSCCO tapes made by the powder-in-tube method described in Chapter 16. They compared two samples of lead-doped BSCCO-2223, i.e., $Bi_{1.8}Pb_{0.4}Ca_{2.2}Sr_2Cu_3O_x$, one with $J_c = 29,000$ A/cm^2 and the other with $J_c = 9000$ A/cm^2, which are both better than their YBCO samples. They took data at several different temperatures between 4 and 77 K.

At 60 Hz, the results for all temperatures on one sample appear in Figure 18.15; the 4.2 K line lies generally below the higher temperature curves. In the low-field range ($\beta < 1$) the AC loss varies as B_0^3; in the high-field range ($\beta > 1$) it varies as B_0. However, the same uncertainties occur here that impaired the YBCO measurements: it may be that the weak links between grains are affecting the apparent values of J_c. Nevertheless, for a typical value

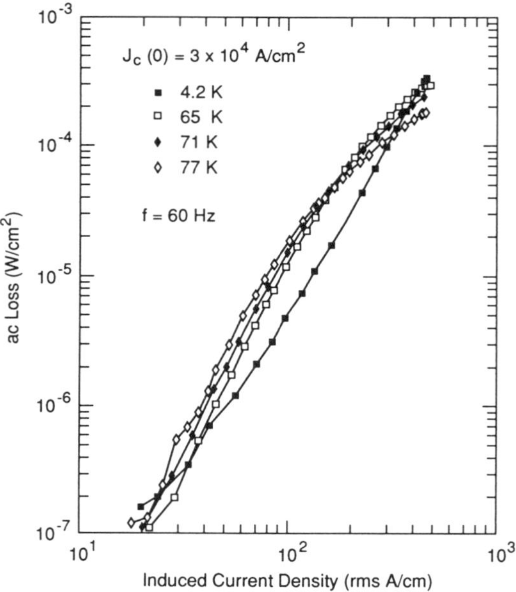

Figure 18.15. AC loss at 60 Hz of a BSCCO-2223 powder-in-tube tape at several different temperatures. (From Ref. 13.)

of applied field (say, surface current of 100 rms A/cm), the BSCCO AC losses are around a factor of 10 lower than the YBCO losses, as may be seen by scrutinizing Figures 18.14 and 18.15.

The mechanism of eddy–current loss was also investigated by varying the frequency of the AC current. It is reasonable to expect the silver sheath around BSCCO tape to offer a path to eddy currents. Figure 18.16 shows BNL's data (for their higher-J_c BSCCO sample)

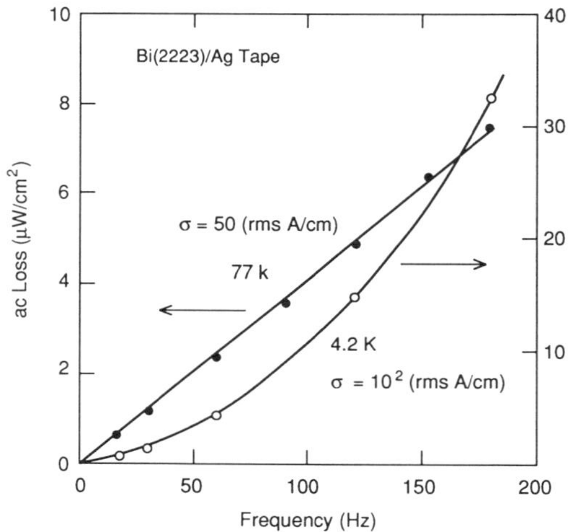

Figure 18.16. AC loss vs. frequency of a BSCCO-2223 powder-in-tube tape having $J_c = 29,000$ A/cm^2 at 4.2 K and 77 K. (From Ref. 12.)

at both 4.2 K and 77 K. The linear slope at 77 K indicates hysteresis loss, whereas the quadratic slope at 4.2 K suggests eddy–current loss, presumably in the silver.

A new apparatus for measuring AC losses has been built by Lanagan[14] of Argonne, and tested on some large (25 cm long) sinter-forged bars of BSCCO having a cross-sectional area of about 0.5 cm^2, and $J_c \approx 1400$ A/cm^2 in zero field at 77 K. The measured AC loss was a very strong function of applied current, but for the special case of I = 100 A at 50 Hz, the losses corresponded to 2 μW/cm^2, which fortuitously falls exactly onto the straight line in Figure 18.16.

18.7. THEORY / EXPERIMENT COMPARISON

All this data affords certain comparisons with the theory developed years ago for LTSCs, and encapsulated above. Using the Wilson model for a cylinder with parallel applied magnetic field, and remembering that β is proportional to magnetic field, we expect

$$W \propto B_o^3 \qquad \text{for } \beta < 1 \tag{a}$$

$$W \propto B_o^2 \qquad \text{for } \beta = 1 \tag{b}$$

$$W \propto B_o \qquad \text{for } \beta > 1 \tag{c}$$

For the Gd-Ba-Cu-O data,[9] the actual behavior illustrated in Figure 18.11 begins with $W = B_o^4$, continues with a region of constant W, and ends with $W \propto B_o^3$. The last region is consistent with hysteretic loss in the low-field region prior to full penetration ($\beta < 1$).

The explanation of the data is similar to Brookhaven's: The sample is presumed to consist of grains with high critical parameters connected by weak links. For low B_o, flux penetrates the weak links, resulting in a strong dependence of critical current on applied magnetic field. In the plateau region, the weak links are driven normal and losses exhibit a frequency dependence similar to normal metals. For high fields, only intragranular screening currents are applicable. These exhibit the characteristic B_o^3 dependence.

The Brookhaven data has also been analyzed. The line having slope = 3 (from Wilson's theory for low β) is drawn on Figure 18.12. At 77 K, Figure 18.14 shows that only the melt-textured material behaves this way. Thus, the agreement is good in the absence of extraneous influences like decoupling. We conclude that the loss mechanism is hysteresis loss.

Eddy–current losses apparently occur in the silver, whether composited with YBCO as in the experiments of van der Vegt[10] or surrounding the BSCCO as in the BNL powder-in-tube tapes.[13] They both observed a mild nonlinear dependence of loss upon frequency [as predicted by Equation (18.1)], especially at low temperatures.

Clem's numerical values for $h_0 = 889$ Oe (707 A/cm) can be compared with the data on the far right of Figure 18.13. If J_{cJ} is taken to be 1000 A/cm^2 for this material, Equation (18.9) predicts 178 mW/cm$^2 \equiv 1.78 \times 10^5$ μW/cm^2, well off the scale of the graph. On this showing, Clem's prediction is too high. On the other hand, it well may be that the BNL experiment effectively used J_{cg} (in magnetic dissipation measurements, rather than transport measurements). If $J_{cg} \sim 100 \, J_{cJ}$ the loss predicted with J_{cg} in the denominator of Equation (18.9) drops to only about 2 mW/cm^2, in agreement with the data.[11]

For low magnetic fields, Clem used unpublished data from Los Alamos to compare with his theory. The agreement was spectacular: 0.9 μW/cm^2 predicted and 1.0 μW/cm^2 measured in 20 Oe, 8.1 μW/cm^2 predicted and 7.9 μW/cm^2 measured in 40 Oe. Clem cautioned that more experiments are needed to confirm this.

In somewhat higher fields where full penetration by the magnetic field has occurred (i.e., the upper right of Figures 18.12 to 18.14), the Bean critical state model[15] predicts that the AC loss should vary linearly (not inversely) with J_c. The BNL group tested a collection of melt-textured YBCO bars[12] and powder-in-tube BSCCO tapes,[13] all having different J_c values, and found AC loss behavior generally in accord with this model. Again, this supports Wilson's theory and Clem's adaptation of it to HTSCs.

Meanwhile, the transport data of van der Vegt[10] is also helpful for vindicating theoretical models. This data is not subject to precise comparisons with Clem's theory, because the grain size, the fraction in the normal state, the fall-off of J_{cJ} with H, and the magnetic field at the position of each coil winding are not known. Nevertheless, the formulas carried over uncritically from Wilson[2] agree with van der Vegt's experiment within about a factor of 2, and there is nothing to suggest that Clem's refinements for HTSCs are inapplicable. It is fair to call this level of agreement adequate. Of course, the losses are unacceptably large, but only because I_c is so small in these coils.

18.8. SUMMARY

In this chapter we have discussed AC losses, which are an important design consideration for many large-scale devices. We began with a discussion of the theory of AC losses as derived from the Bean critical state model. Because calculating AC losses is quite complicated, we broke down the losses into several components: hysteretic, eddy current, and self-field.

In LTSCs, Wilson[2] shows that hysteretic losses are expected to be linear in frequency. For partial field penetration, $B_m < 2\mu_0 J_c a$ ($\beta < 1$), losses are proportional to the field B_m^3 and inversely proportional to the product of the critical current density J_c and the characteristic dimension a. For $\beta \simeq 1$, the losses are proportional to B_m^2, and for $\beta > 1$, the losses are proportional to $B_m J_c a$. For $\beta > 1$, fine subdivision of the superconductor helps reduce loss. This (coupled with stability requirements) generally leads to multifilamentary composite conductors.

Eddy current losses in a composite are due to coupling. To minimize coupling among filaments, filaments are twisted, with a small twist length. With care, eddy current losses at AC frequencies can be kept small relative to hysteretic losses.

Transport currents can double hysteretic losses. Self-field losses are due to varying transport current. In this case, filament twist is not sufficient to prevent coupling. Only transposition is effective in reducing coupling. Self-field loss is proportional to frequency (like hysteretic loss) and decreases with decreasing β (by definition $\beta \leq 1$).

There are a couple of methods for measuring AC losses: calorimetric versus electronic, and transport versus magnetization. Several groups found evidence for hysteretic loss behavior at levels one to two orders of magnitude higher than for LTSCs. They also found evidence for weak-link behavior, which caused intergrain currents to die out at very low fields. Experimental data on AC losses in the HTSCs is quite limited, owing to the use of magnetization measurements, which may not be relevant to actual transport current conditions. The most critical need is for a way to connect the magnetization behavior to the transport behavior.

Still, it is instructive to compare the Brookhaven YBCO results with their earlier work on Nb_3Sn tapes for transmission lines. To compete with Nb_3Sn tapes, losses should be only 20 $\mu W/cm^2$ with 450 A/cm flowing (A/cm is a surface current per tape width). Unfortunately, contemporary measurements of AC loss in YBCO give about 1000 $\mu W/cm^2$. This is discouraging: HTSC losses are much worse than for LTSCs.

The theory by Clem[6] is satisfactory to explain all data on AC losses in YBCO so far. Unfortunately, the weak-link behavior at grain boundaries leads to very large AC losses in ordinary YBCO. At 77 K, the magnitude of AC losses for YBCO (based on estimates of intragrain current densities) was calculated to be 50 to 100 times higher than for LTSCs. The weak link behavior taking place at each grain boundary causes this condition. In addition to being an obstacle to DC conduction, it is now clear that this same limit upon J_{cJ} also causes excessively high AC losses.

Can the AC loss problem be solved? It is one and the same problem of increasing J_c. Wire manufacturers are already making multifilamentary BSCCO wire, but J_c falls below 10^4 A/cm^2 at 77 K. Even if such a cable were able to tolerate low J_{cJ} for total current capacity, it would have large associated AC losses. The achievement of $J_c > 10^6$ A/cm^2 in thin films is a hopeful sign that ways will be found to make conductors with high J_c: either as wires or multilayer thin film tapes.

What does this all mean for the use of HTSCs in AC devices? Development of HTSCs is following a path similar to LTSCs: early efforts concentrated on understanding the behavior (physics and chemistry) of LTSCs. The making of wires came next, then improvement of J_c, and finally reduction of AC losses. HTSC development is still early in this cycle. When high values of J_c are obtained at 77 K, we can expect these wires to exhibit loss characteristics similar to LTSCs, in which case multifilamentary composites should provide acceptable losses.

REFERENCES

1. A. M. Wolsky *et al.*, "Advances in Applied Superconductivity: A Preliminary Evaluation of Goals and Impacts," ANL/CNSV-64, Argonne National Laboratory (1988).
2. M. Wilson, *Superconducting Magnets* (Oxford Univ. Press, London: 1983).
3. W. J. Carr, Jr., *AC Loss and Macroscopic Theory of Superconductors* (Gordon and Breach, New York: 1983).
4. E. W. Collings, *Applied Superconductivity, Metallurgy, and Physics of Titanium Alloys* (Plenum Press, New York: 1986).
5. J. R. Clem, "AC Losses in Type II Superconductors," AMES Laboratory Technical Report ISM280 (1979).
6. J. R. Clem, "AC Losses in the New High-Temperature Superconductors," EPRI Report EL-6277 (1989).
7. J. R. Cave *et al.*, in *Proc. Int'l Conf. on J_c in HTSCs*, Snowmass CO, August 1988.
8. J. D. Bussiere, M. Garber, and M. Suenaga, *IEEE Trans. Magn.* **MAG-19**, 324 (1975).
9. M. Ciszek *et al.*, *Physica C* **152**, 247 (1988).
10. E. van der Vegt, Argonne National Laboratory internal report.
11. J. Orehotsky *et al.*, *J. Appl. Phys.* **67**, 1433 (1990).
12. J. Orehotsky *et al.*, in *Advances in Superconductivity IV*, edited by H. Hayakawa *et al.*, p. 631 (Springer-Verlag, New York: 1992).
13. J. Orehotsky *et al.*, *Appl. Phys. Lett.* **60**, 252 (1992).
14. T. M. Lanagan, Presented at DOE Annual Peer Review, July 19–20, 1994.
15. C. P. Bean, *Rev. Modern Phys.* **36**, 31 (1964).

IV

ELECTRIC POWER APPLICATIONS OF HTSC

19

Transmission Lines

John S. Engelhardt, Donald Von Dollen,[†]
Ralph Samm,[†] and Thomas P. Sheahen*

This chapter examines the potential application of HTSC (high-temperature superconductor) materials to the transmission of electric power.[1] Power *transmission* is loosely defined as the transfer of electric energy from a source to a load over conductors that carry relatively large currents, while being maintained at a high voltage. The power transmitted is the product of voltage and current; the presence of one without the other is of no value. AC transmission is characterized by voltages of 69 kV to 765 kV, whereas DC transmission is not so clearly defined, but generally will be in the range of 100 kV to 600 kV. AC circuits operating at lower voltages are considered *distribution* class circuits.

The textbook power system[2] consists of a *generator* located in a remote area, a *transformer* to raise the voltage and lower the current output of the generator, a *transmission line* or cable to transfer the power to a developed area, and a *substation* to receive the transmitted power and transform it from high voltage to a lower *distribution voltage*. Finally, the power is sent out over many *distribution lines* to loads (customers) in the area. At each customer point, a final transformer will lower the voltage again to the *service level*. This system can be viewed as having two levels; the *bulk power* part of the system, which generates and transports electrical energy, and the *distribution* system which delivers the energy to the customer. These two functions come together at the *utility substation*.

19.1. UNDERGROUND CABLES

Superconducting cables may be used for both functions, but only underground transmission will be considered here. This is because HTSC materials are very unlikely to be cost-competitive for overhead transmission lines. However, for underground transmission, there are scenarios that favor HTSC cable systems over conventional alternatives. This chapter first describes those conditions and then goes on to look at practical HTSC conductor and cable system configurations that might compete with conventional cable technology. As

*Underground Systems, Inc.
[†]EPRI.

societal pressures against overhead transmission lines continue to grow, underground transmission will become more popular and thus HTSC cables may obtain a serious market share.

19.1.1. Conventional Cables

Lack of right-of-way and aesthetic considerations in and around urban centers[3] (as well as river crossings, etc.) sometimes limits the use of aerial transmission lines. In these cases, underground transmission lines are frequently employed. A set of coaxial cables (one for each phase) is enclosed in a pipe which is buried in the ground. Usually, oil-impregnated paper is used as a dielectric. The pipe is filled with dielectric fluid and pressurized to enhance the dielectric's properties and in some cases used as coolant to carry away heat generated within the cable. In addition to the normal ohmic losses, underground transmission lines are also subject to voltage-dependent dielectric losses. Because underground transmission lines use coaxial cables having significant intrinsic capacitance, reactive compensation is sometimes required to restore the desired impedance. Underground transmission lines require enclosure, excavation, and backfill and thus have installation costs that greatly exceed those for overhead power transmission lines.

Studies show that superconducting AC transmission lines can transmit high levels of power with fewer losses and with cost savings of up to 60% relative to conventional underground high-pressure, oil-filled, transmission lines.[4] However, conventional underground technology is improving. The most significant development in the past several decades in underground transmission has been the commercialization of PPP, a laminated paper-polypropylene tape that replaces paper tape as the dielectric of high-pressure fluid-filled (HPFF) cables.[5] Whether or not superconducting transmission lines can compete with PPP cables will depend on the cost and performance that can be achieved in HTSC cables.

Several new technologies were developed for underground cables in the 1960s and 1970s: heavily forced-cooled HPFF pipe-type cables, compressed-gas (SF_6) insulated transmission lines (GITLs), cryoresistive cables operating at liquid hydrogen or liquid nitrogen temperatures, and LTSC cables (cooled by liquid or gaseous helium)[6] based on the A-15 series of metallic superconductors. In the 1970s, several heavily forced-cooled HPFF 345 kV cable systems were built into the New York City area, and a few short GITL lines were installed, mostly as *bus ties* within substations.

Power transmission technology has now reached a mature state. The utility planner has a broad arsenal of tools from which to chose for virtually any type of transmission requirement, using well-established designs with many alternatives to suit particular scenarios. Both 500-kV and 765-kV underground cable designs have been qualified for more than a decade,[7,8] but are not yet required.

19.1.2. Market Size

Before getting into the details of cable design for underground transmission, it is appropriate to mention the size of the market for underground cables. It is a small part of a very big business, and it has excellent growth potential. The bulk of the transmission in the United States (68%) takes place in the 69–161 kV range where the capacities are low (200 MW or less). Superconducting cables have never been thought of as viable for systems of such low capacity.

Considering all transmission in the country that is 69 kV and above, there are about 433,000 circuit miles, of which 2500 circuit miles (0.6%) are underground pipe-type cable systems. Approximately 400 circuit miles (16%) of this underground plant are extra high voltage (EHV) (230 kV or 345 kV) pipe-type cables, which leaves 84% of the underground pipe cables in the low-capacity category, 161 kV or less. New York City is a major user of underground cable, because of the tight space requirements there.

Another place that requires underground cable is Tokyo, which is served by Tokyo Electric Power Co. (TEPCO). Due to space and environmental considerations, all transmission and distribution systems within Tokyo are underground. A series of tunnels and ducts is used to transmit power throughout the city. A typical tunnel carries 2000 MVA and serves two ducts, each carrying 1000 MVA. Tokyo Electric has been experiencing annual growth in electricity demand of 2.5%. This translates to a doubling of demand over the next 30 years. Clearly, it is worth considering HTSCs for such a market.

To meet this increased demand, Tokyo Electric considered three options:

- use conventional oil-cooled cable, build new tunnels and ducts, and upgrade cable voltages;
- replace existing cable with low-temperature superconducting (LTSC) cable with twice the capacity, and use existing tunnels and ducts; and
- replace existing cable with high-temperature superconducting (HTSC) cable with twice the capacity, and use existing tunnels and ducts.

Because of the high cost of tunneling within Tokyo, there is a large cost advantage associated with using the existing tunnels and ducts. To utilize these ducts, the cable diameter must be kept under 150 mm. The conductor size depends on the operating voltage. As the operating voltage increases, more insulation, and hence a larger cable diameter, is required to prevent voltage breakdown. As voltage is decreased, more current must be carried to maintain the power rating. Higher current leads to higher AC losses, hence a greater cooling requirement and a larger cable size.

The results of the 1989 Tokyo Electric design study[9] are as follows: For a liquid helium–cooled system, the smallest possible size is 360 mm, too large to fit within the existing ducts. However, because a HTSC requires only a nitrogen shield, and because the amount of thermal insulation can be less, the smallest possible size for an HTSC cable is 130 mm—small enough to fit inside the existing ducts. (This analysis assumes AC losses are comparable to those for Nb_3Sn.) Under this assumption, the optimum cable voltage is approximately 66 kV and the corresponding J_c requirement is $10^6 A/cm^2$ at 0.1 T. Of course, there is no wire yet that carries $10^6 A/cm^2$ at 77 K, so no experimental test of this design is being done yet. But it serves to illustrate the interplay between the demands of the marketplace and design of specific cable systems.

19.2. CAPACITY LIMITATIONS

When designing a cable to meet specific needs, a number of economic trade-offs must be made. Dominating the list of design criteria is the simple matter of how much power can be pushed through a line.

19.2.1. Thermal Limits

There are several factors that limit the capacity of a transmission circuit. The ultimate limit is the thermal capacity, which is the loading at which the ohmic and dielectric losses cause the conductor or dielectric to reach the maximum temperature that can be physically

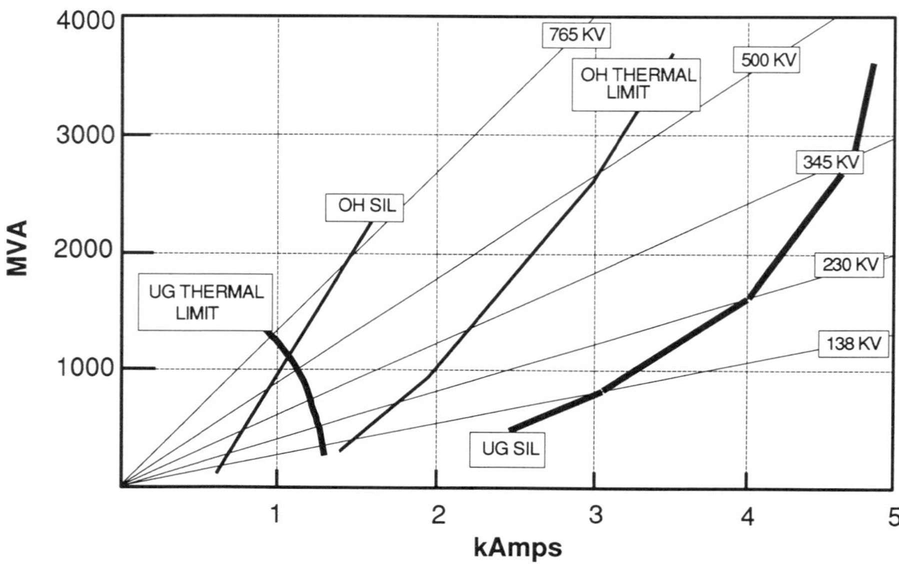

Figure 19.1. Power vs. Current for various transmission lines.

tolerated. These losses must be dissipated to the surroundings, either to the air or to the earth as the case may be.

Figure 19.1 plots the power–current relationship for common voltage levels to illustrate the power capacity behavior for the two different types of transmission: overhead (OH) and underground (UG). Notice that the current carrying capacity of overhead lines increases with voltage, whereas the opposite is true for underground cables. Overhead lines use larger conductors at higher voltages to control corona and radio noise, and at EHV levels each phase will have more than one conductor in order to lower the electric stress in the air at the conductor surface.

The thermal limit of an overhead conductor is usually established by sag; as the conductor heats it expands and the midpoint of each span drops closer to the ground. Sag violation can be hazardous, in addition to lowering the dielectric strength of the circuit. A second concern is aging of the conductor by creep, which is a tendency for slow irreversible extension of the conductor under stress. It is accelerated by hot weather and conductor heating. Short-time excursions to very high temperatures will cause annealing and severe loss of strength.

Conventional underground pipe-type cables are insulated with oil-impregnated paper or PPP which have well-defined temperature limitations based on the thermal decomposition of cellulose. The current produces ohmic losses in the conductor, shield, and pipe, which then cause the operating temperature of the cable's conductor to rise. In addition, dielectric losses occur in the oil-impregnated paper or PPP insulation.

For a HTSC conductor, the thermal limit will be a function of the HTSC characteristics. As discussed in Chapter 18, AC losses vary as H^3/J_c, and the cryogenic system will be designed with limited capacity, so as to control the temperature of the HTSC material. H_{c2}, the upper critical field, in principle sets an upper limit, but this is far above the limit set by J_c. Clearly, the higher the values of J_c, the higher will be the current capacity of the cable.

19.2.2. Stability Limits

Beyond the thermal limits, the size and capacity of a transmission line is driven by other considerations as well. Most important of all is to protect and maintain the stability of the system. In a total electrical system, the various parts (generator, transformer, transmission lines, etc.) are matched to optimize total performance. For example, transformers have serious physical size problems that limit their capabilities to the 750 MW range. (If larger units are desired, they will have to be manufactured on site.) Therefore, superconducting transmission lines with capabilities in the range of other equipment will have more opportunity for application. Eventually, higher-capacity superconducting cables will be needed, but their development will follow smaller systems.

One application for high-capacity transmission lines is the transfer of bulk power between regions. These transfers are motivated by inexpensive generation, which can be prearranged in relatively large blocks. This type of transfer is common for 500 and 765 kV overhead lines that have excess thermal capacity. Asynchronous DC connections are used for similar bulk transfers between frequency-independent power systems, such as that connecting Hydro-Quebec to the United States. Transfers in the range of 1000–3000 MW are possible in this category, and these levels can be expected to grow in the future. These interconnections could be a possible niche market for medium-power AC or DC HTSC cable systems. In urban or environmentally sensitive areas, sub-EHV voltages are required, and hence such cable systems might provide bulk power transfer more economically.

19.2.3. Surge Impedance

A more subtle stability criteria involves the dynamic stability of the power system and its ability to maintain control of the bus voltage at distribution substations. Voltage control considerations dictate loading limits for lines in the 50–200 mile range. A transmission circuit has two characteristic reactances that determine its behavior within, and effect upon, the power system. The first is its *inductive series reactance*. The energy stored in this reactance is proportional to the square of the current in the line. The second is the *shunt capacitive reactance* whose energy is proportional to the square of the voltage on the line. The square root of the ratio of the line's inductance to its capacitance is its characteristic impedance Z_0, also known as its *surge impedance*. Z_0 corresponds to the specific load resistance at which the transmission line neither demands nor contributes reactive energy. When a transmission circuit is loaded with its surge impedance, the power level is known as the *surge impedance load*, or SIL. Looked at another way, at surge impedance loading the series inductive energy exactly balances the shunt capacitive energy. If the load drops below the SIL, the transmission system is capacitive, while above its surge level it is inductive.

Table 19.1 summarizes the relations between surge impedance and load limits for overhead and underground systems. It tabulates the typical surge impedances, SILs and thermal limits, as well as the ratios of thermal rating to SIL.

One can see at a glance that there is a great difference between overhead and underground lines in this category. Overhead lines have Z_0 of 250–300 Ω, which depends on conductor size more than voltage level, but tends to increase with voltage. For conventional pipe-type cables, $20 < Z_0 < 70\ \Omega$, increasing significantly with voltage. The thermal limits of EHV overhead lines tend to be about three times SIL. Conventional underground cables

Table 19.1. Surge Impedance and Load Limits

Voltage (kV)	Z_0 (Ω)		SIL (MW)		Thermal limits (MW)			
	OH	UG	OH	UG	OH	%SIL	UG	%SIL
69	250	24	19	200	200	1050	150	75
115–161	260	27	73	705	350	480	320	45
230	270	33	196	1603	600	306	460	29
345	280	41	425	2900	1400	330	650	22
500	290	50	862	5000	2800	325	930	19
765	300	61	1950	9600	7000	360	1300	13

are thermally limited to a fraction of their SIL. This dramatic difference can be useful when designing a hybrid overhead–underground transmission system.

Clearly, thermal capacity is a minor concern for the bulk of the overhead transmission applications in the United States. Surge impedance characteristics dominate the design criteria for longer high-power transmission lines.[10] Although it is naive to expect HTSC cables to compete economically with overhead lines, they can provide a viable alternative with much lower surge impedance than the overhead option. This may prove to be an enabling technology for some grandiose power supply schemes that aren't viable with conventional transmission technology.

It must be noted, however, that the longest underground AC line in the United States is 25 miles long. The capacitive reactance is not large enough to justify the cost of fixed compensation for underground cables that are less than about 15 miles long. Moreover, many utilities welcome the capacitive reactance to help balance overhead lines and loads, which tend to be inductive.

How can HTSCs change this picture? HTSC cables will have surge impedance characteristics similar to conventional HPFF cable systems. Surprisingly, the fact that the HTSC's ohmic losses are small has little relevance when they are inserted into the power system: The loss component of the series impedance is very small compared to the reactance; and so it is common practice to ignore it in system studies, whether using overhead conductors or conventional underground cables. Dielectric stresses and losses with HTSCs will be similar to conventional systems simply because conventional dielectric materials and design will be used.

The big difference with HTSCs will be a large increase in thermal capacity, limited now by the ability of the superconductors to carry relatively high currents in moderately high fields. The consequence of this one point will be to shift the underground (UG) thermal limit curve in Figure 19.1 to the right, perhaps all the way past the underground SIL curve. This will give the utility the ability to build longer underground circuits with higher capacity and lower compensation costs than conventional cables. These cables will clearly be able to match and exceed an overhead line's thermal capability for short- and medium-length applications. Furthermore, because of their lower surge impedance they will open the door to higher AC transfers per circuit over medium-to-long distances, when underground cables are required and DC is not justified.

19.3. SUPERCONDUCTING TRANSMISSION SYSTEMS

As a preliminary to discussing superconducting cables, it is first necessary to describe conventional cables.

19.3.1. Basic Design Concepts

High-voltage cables must perform two functions: (1) they must have a conductor that can carry a useable current, and (2) they must insulate each conductor from the other phase conductors and ground. Conventional power transmission cables consist of both a conductor assembly surrounded by a dielectric system, which in turn is enclosed in a grounded metallic member that serves as the dielectric shield, and a system enclosure, which holds the dielectric pressure and protects the cable from the earth environment. The pressure medium is a dielectric fluid or nitrogen gas. The conductor material is copper or aluminum strands, depending on the economic trade-offs at the time of manufacture. The dielectric is vacuum-dried oil-impregnated paper or PPP. The enclosure may be a steel pipe holding all three phases, or lead or aluminum sheaths over each phase. The latter is known as a *self-contained cable*, quite common in Europe but not popular in the United States.

Ohmic losses occur in a conventional (nonsuperconducting) conductor when the cable carries current. Ohmic losses also occur in the dielectric shields and enclosure due to induced circulating currents, eddy currents, and hysteresis caused by the AC magnetic fields created by the currents in the three conductors. Dielectric losses occur in the dielectric due to the applied voltage. These losses appear as heat and cause the cable temperature to rise until the heat dissipation to the surroundings balances the heat generated by the losses. Thermal degradation of the cellulose in the dielectric limits the temperature at which the system can operate and thereby limits the power transfer capability of the system.

19.3.2. Superconducting Cable Designs

The motivation for a superconducting cable is to replace the metallic conductor with a superconductor that can carry a larger current with lower ohmic loss (or zero ohmic loss in the DC case). The penalty is the need to keep the superconductor cold. This requires a channel for the flow of the cryogen, plus a cryostat for insulation. Fortunately, the superconductor can support a very large current density, so little material is needed for the conductor. This leaves space for the cryogen channel and cryostat and allows a superconducting cable to be comparable in size to (or smaller than) a conventional cable. This is important because more than half the cost of conventional underground installations comes in digging the trench to contain the system.

Decades ago, development work on LTSC cables considered many alternative design concepts, some of which were truly bizarre. All designs were driven by two factors that may not apply to HTSC-based cable systems. First, the refrigeration energy cost for cooling cables to temperatures below 10 K placed the utmost demands on the minimization of losses. Early LTSC cable designers had to contend with the refrigeration cost of cooling helium, with a penalty of nearly 400 W per watt-of-heat-removed. Second, system capacity targets were set at very high levels in anticipation of continued growth and economies of scale (in vogue between 1965 and 1975). This required cable designs that could not use conventional cable concepts and materials, because of the need to reduce ohmic and dielectric losses to the absolute minimum.

DC cable designs were straightforward, because there is zero ohmic loss and insignificant dielectric dissipation. The loss problem for DC cables was to control the heat in-leak. DC concepts had fewer restraints, and those working to develop LTSC DC cables proposed several different concepts, all of which were viable to some degree.

AC cable designers, however, quickly found they had few choices as to fundamental design. For example, they found that the magnetic fields normal to the surface of the superconductor seriously degraded the superconductor's performance, and losses in the shields and enclosures due to eddy currents and circulating currents would be unacceptable. Their only option was to make the outer shield of the cable also superconducting. In this coaxial form the outer superconductor experiences an induced current equal and opposite to the current in the inner conductor. This confines the magnetic fields to the space between the two and eliminates the driving force for eddy currents and circulating currents in the outer metallic components. The truly coaxial cable has the lowest inductance possible for a given dielectric spacing, which results in the lowest surge impedance obtainable.

The cryogenic aspects of the cable are more complex because the entire cable must be kept cold. The dielectric is immersed in the cryogen, and dielectric losses must be removed by the cooling system. Thus, all AC LTSC cable designs required cryogenic dielectrics and two coaxial superconductors for each phase. Cost and space considerations favored the placement of the three phases in one large cryostat.

19.3.3. Specific Example

With the design constrained in the ways we have discussed, it is not surprising that actual LTSC cables resembled one another. The rigid design developed at Union Carbide,[11] and the flexible design demonstrated at Brookhaven National Laboratory,[12] as well as most cable concepts put forth around the world, had a lot in common. Figure 19.2 shows the Brookhaven flexible cable, and Figure 19.3 is a photograph of the actual Brookhaven prototype transmission line, from which a great deal was learned. The length of this prototype was about 400 ft, long enough to test all the essential features of a transmission system.

The Brookhaven cable assembly used helium gas at 7 K as its coolant. At this temperature, helium is above its critical point, so liquid and vapor are indistinguishable. This is a rather efficient coolant, because (on a per-unit-mass basis), supercritical helium has a greater heat capacity than liquid nitrogen. The 7 K helium flows in one direction down the central core of each of the three conductor phases. As Figure 19.2 shows, all three fit inside a stainless steel pipe, and the returning helium gas is confined within that pipe, although at a higher temperature. The outermost layer of insulation provides the gradient up to ambient temperature.

This choice of operating temperature is one of those engineering trade-offs that optimize cost-effectiveness. To obtain the highest possible critical current, one would operate at 1.8 K, which would involve holding helium at a low pressure and in the superfluid state where vacuum leaks are more likely. Instead, at 7 K the penalty in J_c is not too bad, and the refrigeration cost is substantially lower. Manufacturability of a very long cable was made an essential design feature of this prototype.

The actual superconducting current flows within tapes made of Nb_3Sn layer, which are wrapped as shown in the cutaway drawing in Figure 19.2. Concentric cylinders of a bronze core and a copper stabilizer lie inside the first layer of Nb_3Sn, which is followed by insulating layers out to the next Nb_3Sn layer, after which there is another layer of copper stabilizer, and

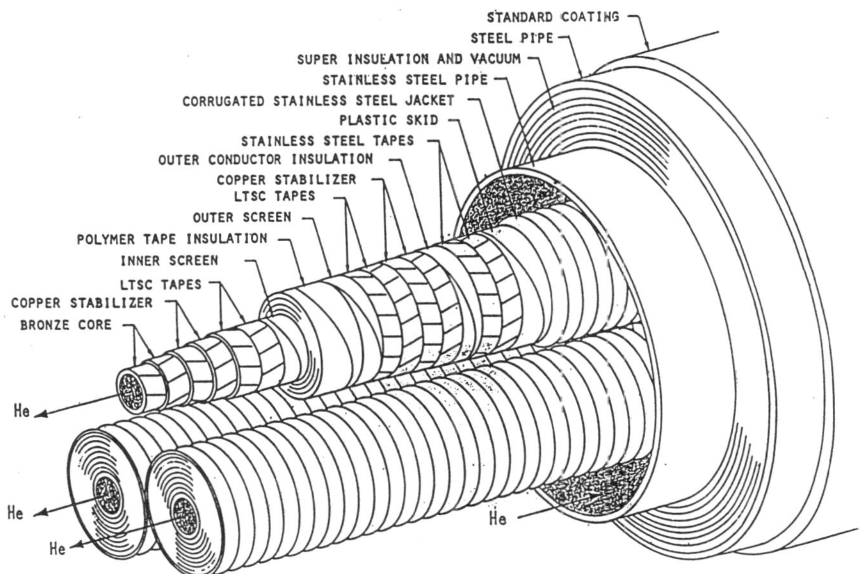

Figure 19.2. Drawing of Brookhaven LTSC cable system.

Figure 19.3. Photo of superconducting transmission line at Brookhaven National Laboratory.

more insulation. Finally, each cable is enclosed by a corrugated stainless steel jacket for strength.

With all these additional layers, the losses are of course greater than those of Nb_3Sn alone,[13] but once again, any real system designed to operate in a utility environment must include provisions for stability and strength under imperfect operating conditions. One practical problem is the need to join segments of superconductor at intervals of about 20 m, and this too increases losses.

In its final configuration, the Brookhaven cable employed helically wound tapes with a pitch angle of 45°. This choice cleverly reduced the annular field (between the two Nb_3Sn layers) to zero, but it allowed other losses to occur because of the current flow patterns in the tapes. As with any major engineering innovation, the Brookhaven transmission line demanded a great number of empirical measurements. However, that data provides guidance in the present day for designers of future superconducting transmission cables.

The summary of Forsyth's review[13] is worth quoting at length, in order to convey the importance of this prototype:

> Despite the many compromises that had to be made in the conductor design, the BNL prototype demonstrated that a large system meeting the predicted technical characteristics could be built and operated with losses that would be economically acceptable to utility companies. It seems likely that the test results obtained with the prototype will lead to ways of reducing the losses. A reduction by a factor of 2 is desirable; further reductions would give designers the option of designing cables carrying three to four times the SIL. Since the work at BNL on Nb_3Sn tape of the mid-1970s, is has become possible to obtain multifilamentary composites with Nb_3Sn filaments less than 2 microns in diameter. These conductors are certainly worthy of evaluation for the next generation of cable designs.

The complexity of trade-offs continues: As was mentioned in Chapter 18, it is desirable to use twisted conductors, but doing so produces magnetic fields in the ϕ and z directions, which then give rise to eddy currents in the normal metal and the stainless steel housing. When eddy current losses exceed hysteresis losses, nothing is gained, and indeed other design trade-offs intended to reduce hysteresis loss become moot. Forsyth[13] has said it best: "the brief discussion above is intended only to demonstrate the scope of the problem."

19.3.4. Application to HTSC Transmission Lines

Can the HTSC materials provide that "next generation?" The experience with LTSC cables suggests that certain features are transferable to HTSC designs. Whereas the coaxial, fully cryogenic cable may be necessary for very high current applications of the future, the lower power levels in use today (and for the near future) open the door to simplification and subsequent cost reductions that take advantage of the smaller refrigeration penalty associated with liquid nitrogen.

The first consequence of a smaller current-carrying requirement is to challenge the need for the superconducting outer shield. Is it possible to revert to conventional shield structures and tolerate the additional losses? The answer is yes; with careful shield optimization the loss can be reduced to an acceptable level.

The next question is, what about magnetic field interactions? The effect of normal fields on the surface of a superconductor is severe, but at lower currents and fields, it too may be tolerable. The lack of phase-to-phase magnetic isolation raises the inductance and surge impedance and reduces the surge impedance load, but at lower power levels a lower SIL is desirable.

The most significant consequence affects the dielectric: if the outer shield is not superconducting, there is no reason for the dielectric to operate at cryogenic temperature. This means that a very simple design put forth for LTSC DC cables in the early 1970s could be applied to HTSC AC cables. This concept placed the conductor inside a small flexible cryostat which was covered with a dielectric operating at ambient temperature.[14]

One lesson learned from LTSC cable systems was that designing and qualifying a new dielectric design for a high-voltage cable system is not a trivial task. Research engineers found that introducing a new dielectric system was far more difficult than qualifying a radically new superconducting conductor assembly. To this day, dielectric concerns can probably be blamed for the lack of acceptance of LTSC concepts.

Among LTSC cables, the only system connected into a utility's operating transmission line was the Kabelmetal[14] system. It used conventional paper impregnated with supercritical helium for its dielectric. The use of a conventional dielectric system operating at ambient temperature would eliminate this formidable hurdle, which in turn would enhance the probability of an early successful system demonstration. Therefore, the first choice for immediate development is to focus on the ambient dielectric cable concept. Fully coaxial, cryogenic dielectric systems may be of interest for higher power levels downstream.

19.4. HTSC DESIGN CONSIDERATIONS

At what temperature should a HTSC transmission line operate? The instant answer is 77 K, but that may not be the optimum. To begin with, there is a certain power required to remove heat generated at a lower temperature. Utilizing the customary temperature ratio for the Carnot efficiency, the power required at 300 K to remove power at temperature T is given by

$$P(T)/P(300) = \{COP\}[T/(300 - T)]$$

where $\{COP\}$ denotes the coefficient of performance, typically 20% or so. At 4.2 K and $\{COP\} = 20\%$, this equation gives a heat removal ratio of about 350:1. By contrast, at 77 K, HTSC cable designers working with liquid nitrogen must pay only about 10 W per watt-of-heat-removed. This luxury permits considerable flexibility in the choices of possible cable configurations.

However, when 77 K is close to T_c for the material, as in the case of YBCO, the variation of J_c with T becomes significant. Since AC losses vary as H^3/J_c, when J_c rolls off at higher temperatures (whether due to flux line motion or any other cause), the losses will be higher at 77 K than at a lower temperature. With J_c assumed to vary as $[1 - (T/T_c)^2]$, Forsyth[13] calculated that the refrigeration power requirement goes through a minimum at $T/T_c = 0.6$, or 55 K for YBCO.

Clearly, with what we know today about flux lattice melting in BSCCO, the optimum temperature using that material is likely to be much lower than 77 K, even for low-field applications such as transmission lines. Still, it is worthwhile to consider how a transmission cable would be designed for operation at 77 K.

19.4.1. Ambient-Temperature Dielectric

Figure 19.4 illustrates an ambient-temperature dielectric, pipe-type cable system built around an HTSC conductor assembly in which the liquid nitrogen channel, the supercon-

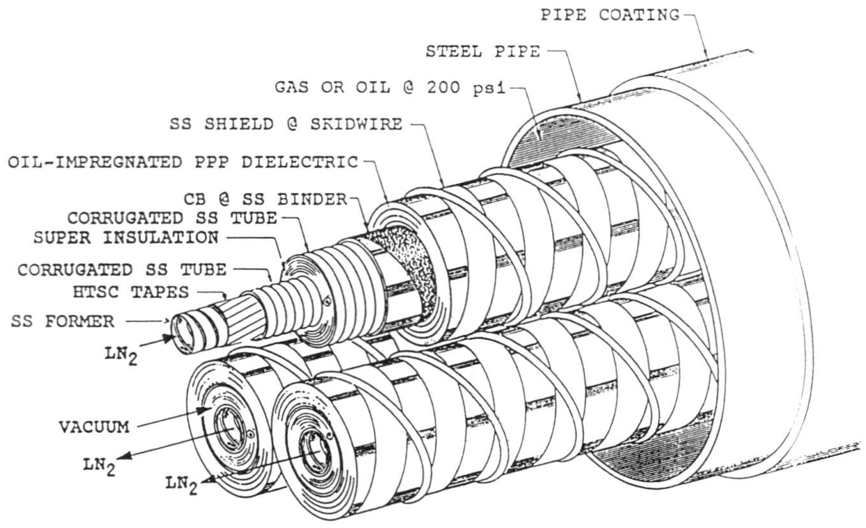

Figure 19.4.   Room-temperature dielectric HTSC cable system.

ductor, and the cryostat are all in one structure. The cable dielectric, its shielding, and the steel pipe enclosure are conventional in every respect. Comparison of this assembly with a standard pipe-type cable reveals that only the conductor has been changed. If the cryostat assembly can be made in the diameter range of normal copper or aluminum conductors (3–6 cm), the physical size and appearance of the cables will be identical to conventional practice and they will be easily assimilated by the cable industry/utility infrastructure.

In comparison to the cryogenic dielectric cable, the ambient-temperature dielectric cable has many advantages beyond the use of an established dielectric system. Only the heat in-leak and the AC and eddy current losses in the superconductor assembly load the refrigerator. Dielectric losses and ohmic losses in the cable shields and enclosure are conducted to the earth; they do not have to be removed by the refrigeration system. This reduces the overall system energy requirements, reduces the size of the refrigeration packages, and lowers the flow requirements, thus extending the distance between cooling plants for a given channel diameter. Thermal contraction phenomena are limited to the former, the HTSC tapes, wires, or coating, and the inner wall of the cryostat.

The remainder of the cable components operate at a temperature slightly above ambient. The mass of material that must be cooled is greatly reduced, which simplifies service procedures and reduces down time, an important consideration for large transmission systems. The cable terminations can be essentially conventional, since the thermal insulation is not exposed to the voltage gradient. Installation does entail one special operation: splicing the conductor assembly. As the Brookhaven experience showed, this source of losses cannot be dismissed lightly.

Manufacturing considerations impact the entire philosophy underlying this ambient-temperature dielectric cable: The key (and only) new component of the system is the conductor assembly, which forms the entire conductor system. Once manufactured, this system can be fully characterized and tested for compliance with its intended service. When its performance has been confirmed, it can enter the conventional cable manufacturing process to be insulated, factory tested for dielectric integrity (at room temperature), shipped, and installed. At all stages in the life of this system, the conductor function and the dielectric function are separated. Voltage testing can be done without cooling the system, and current testing can be done any time, even before installation in the pipe. With this concept, the challenge facing HTSC proponents is to develop only the conductor assembly and cryostat. A new dielectric system is not required.

Perhaps the most important aspect of this entire design is its ability to be retrofitted into existing pipes. This capability presents utility planners with a totally new option for upgrading existing urban facilities.

When installed in a conventional steel pipe with an ambient-temperature dielectric, the limitation of the HTSC cable will be its current-carrying capability. Eddy current and hysteretic losses in the steel will limit the current that each conductor can carry to the 2000–2500 A range, decreasing with increasing pipe size. However, this is still quite good.

The implications of this can be seen by returning to Figure 19.1, which shows power vs. current. A 2500 A capability more than doubles the range of conventional underground circuits and extends the thermal capability beyond that of many single-circuit overhead lines (except at the highest voltage levels). That alone makes such underground cables competitive. If we go one step further and presume that new installations can use a nonmagnetic stainless steel pipe, then it will be possible to achieve current ratings in the 3500 A range. That is close to the SIL for 138 and 230 kV lines. Thus, the ambient-temperature dielectric HTSC cable design could conceivably enable very long underground circuits to operate at or near surge impedance loading with power transfer capabilities in the 1000–1500 MVA range; and that would occur at lower voltages (138 or 230 kV) than overhead lines.

19.4.2. Cryogenic Dielectric

It is by no means certain that HTSC cables will run at 77 K and use LN_2 as their coolant. The growing likelihood that HTSCs will operate at some intermediate temperature, such as at 40 K, makes it imperative to consider a fully cryogenic cable system. The cryogen would most likely be helium gas cooled to the operating temperature.

Figure 19.5 illustrates a preferred assembly for such a system, derived from the earlier LTSC designs. Just as in Figure 19.2, two layers of superconducting tape, helically wound, provide the pathway for current. Again, the cryogen flows one way down the cores and returns in the spaces between the corrugated pipe wall and the three separate conductors. A copper pathway (the stabilizer) is no longer necessary, because of the higher specific heat and greater cryostability of HTSCs at 77 K.

This particular design employs an extruded polyethylene dielectric, which also provides the enclosure for the cryogen channel, and a flexible corrugated inner cryostat wall, which also has the ability to be retrofitted into existing steel pipes. Of course, a conventional rigid cryostat as shown in Figure 19.2 could be used for new installations if a more robust outer covering and skid wires were added to each cable. The current capability of this structure is limited only by the magnetic field and current density characteristics of the HTSC elements.

At this point in time, the cryogenic dielectric of choice may be extruded polyethylene rather than the cryogen-impregnated plastic tapes that were proposed in the 1970s. Extruded dielectrics for transmission cables have evolved into a mature technology, whereas plastic tapes have yet to be successfully demonstrated and adopted by the cable industry. EPRI funded a study of the use of extruded polyethylene at cryogenic temperatures in the early

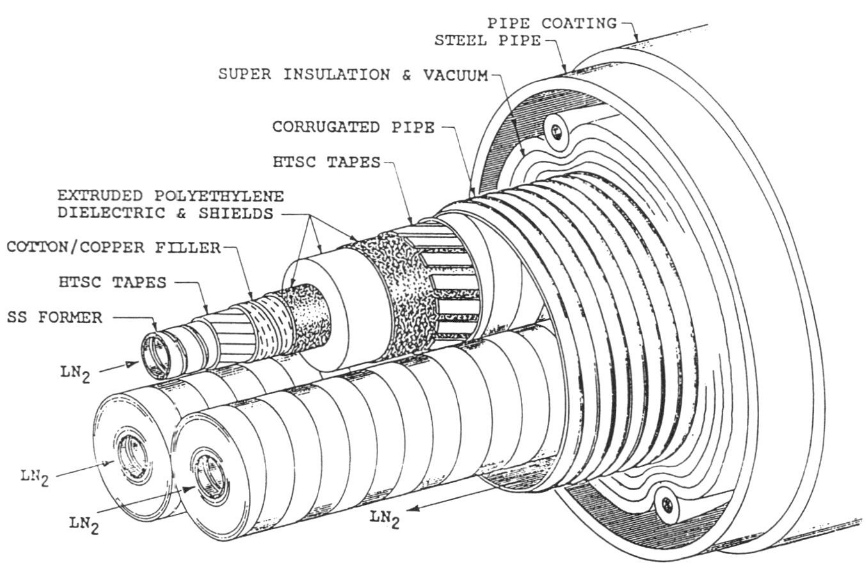

Figure 19.5. Cryogenic dielectric coaxial HTSC cable system.

1980s which demonstrated their viability for operation in liquid nitrogen.[15] More recent work in Japan has confirmed the excellent properties of this material at liquid nitrogen temperatures, and similar HTSC cable designs are being developed.[16] The challenge with extruded polyethylene is to accommodate its thermal contraction when cooled to the operating temperature.

The cooling of extruded polyethylene to 77 K actually improves its dielectric behavior relative to temperature-limited operation in conventional cable applications. This extends the voltage levels that are possible for HTSC systems to the EHV range. Moreover, this enables an underground alternative transmission system to reach the highest power-delivery capability of any known utility applications.

What about operation at 20–30 K? Although extrapolated calculations about the properties of extruded polyethylene may be optimistic, no engineer is going to use it in a system until actual measurements of its dielectric behavior are in hand. The enduring success of the entire electrical utility industry is in part attributable to conservative design and cautious practices. HTSCs are no place to start violating that guideline.

19.5. NEAR-TERM APPLICATIONS FOR HTSC CABLE SYSTEMS

The primary advantages of an HTSC cable over conventional cables is the ability to carry more power per circuit with a lower energy cost per megawatt delivered. Assuming that a single superconducting circuit does not violate reliability criteria, this can produce an economic advantage where space is at a premium, or where more than one conventional underground circuit is required to satisfy the capacity requirement. Two scenarios seem to fall into this category:

The first would take advantage of the room-temperature dielectric concept to retrofit HTSC cables into existing HPFF pipes. By combining the advantage of PPP insulation with an HTSC conductor assembly, an existing pipe's capacity might be increased three to six times. Another opportunity for HTSC cables retrofitted into existing pipes is in the area of inter- and intraregional bulk transfers. The major interconnections are EHV overhead networks, which necessarily skirt urban areas. But the electricity doesn't know that, and would just as soon go straight through the urban center. By replacing one or more underground cables with HTSC cables, a utility could increase their pass-through capability and their local power density without jeopardizing service to the urban area.

The second scenario is the *undergrounding* of sections of overhead lines, typically as they approach populated load centers. This application now favors EHV circuits where conventional cables can't match the thermal capacity of the overhead line. The drawback to the possible application of fully coaxial cryogenic dielectric cable designs for this application is the EHV voltage level. The development and qualification of a cryogenic dielectric capable of EHV levels is many years away. The ambient-temperature dielectric capability is available and qualified now, up to 765 kV using PPP. The missing component is the conductor assembly. Therefore, once a conductor assembly is developed, this scenario could be satisfied by a single circuit HTSC ambient dielectric cable system.

The reduction in energy used (lost) by an HTSC cable is another potential advantage. Unfortunately, this characteristic is not the panacea that has often been claimed. It has merit for circuits that are to be operated at very high load factors (i.e., loaded all the time). An AC HTSC cable can achieve a per-unit energy savings of about 50% when loading conditions are most favorable. A cryogenic cable system requires a baseline amount of energy to be functional, even before it starts to deliver useable power. Although the losses in the transmission circuit are dramatically reduced, the energy consumed by the refrigeration plant eats up much of the savings. Even "lossless" DC HTSC cables will consume considerable energy to balance the heat in-leak. For practical designs of HTSC DC cables, their energy consumption will about equal that of a conventional cable loaded near its thermal limit. Thus, HTSC DC cables are attractive only for loads that are higher than the capability of conventional designs, that is, above 2000 A.

For base-loaded systems of medium length, say 10 to 50 miles in areas hostile to overhead construction, the HTSC cable can enjoy a considerable advantage due to its lower energy costs. The cost of losses for conventional cables when amortized over the life of the system can approach 25% of the installed cost. Therefore, even the 50% reduction in energy consumption per MW delivered gives a sizeable cost advantage to the HTSC cable.

One specific example of a HTSC cable is that designed by Sumitomo Electric Industries. At a joint ISTEC/CSAC conference in May 1992, Tsuneo Nakahara of Sumitomo presented their conceptual design of a 66 kV superconducting cable, which would ideally replace an oil-filled (OF) cable that ran at 275 kV. The characteristics are shown in Table 19.2. Either transmission system carries 4000 MW, but the superconducting system does so with less material and less size. One key feature is that only four parallel cables (instead of eighteen) would be needed to carry the same total power. The biggest advantage of the superconductor is its higher current density, which allows the total power (MW) carried by each cable to be much higher in the superconducting system. Parameters like this motivate cable manufacturers to construct and test prototype superconducting cables. Of course, the crowded conditions faced by Japanese utilities is a stronger motivator than most American utilities confront.

Table 19.2. Conceptual Design of 66 kV SC Cable Replacing
275 kV Oil-Filled Cable[a]

Parameter	SC Cable		OF Cable
Conductor O.D. (mm)	24.1		63.9
Conductor cross-sectional area (mm^2)	45.4		2,500
Heat insulation tube O.D. (mm)	130.0		134
Critical current density[b] (A/cm^2)	10^5	10^6	10^2
Conductor loss[b] (W/m)	30	3.0	43
Electric Energy			
(MW/tunnel)	4,000		4,000
(MW/circuit)	1,000		660
Number of cables (N)	4		18
Number of conductor (N/cable)	3		1
Total number of conductors (N)	12		18

[a]From Sumitomo Electric Industries.
[b]Two different choices of J_c considered for SC cable.

19.6. LONG-RANGE POSSIBILITIES

Given the contemporary low level of growth in the utility sector in the United States and the adequacy of the present transmission industry to meet the needs of utilities over the near term, it is difficult to envision how superconducting cables might impact the future. However, consider the actual numbers. The present demand of North America is 575,000 MW. At present rates this demand will double in about 50 years. That means that the utility industry has to find and build 575,000 MW of power (and the means to transport it to load centers) within this 50 year time frame. Given the realities of the regulatory process, the long delays in obtaining site approvals, public intervention and the like, the magnitude of this task begins to emerge.

Assuming that energy sources and sites for generation can be found, the means for absorbing this additional load into the transmission grid are not obvious. It is unlikely that the public will tolerate even a 50% increase in the density of overhead transmission lines around the country. Thus, transmission considerations are going to impact generation sighting decisions and cause further delay in the process. Larger projects will become mandatory just to stay even with growth.

HTSC cable systems could have a major impact here. Since they will easily perform the same function as the overhead line, and may offer some operational advantages because of their lower surge impedance and higher loading capability, the transmission function may shift to the underground HTSC system. These systems will be environmentally acceptable, will speed the permit process, and may enable an otherwise unacceptable power source option.

The potential for very long, very high power DC lines is also on the horizon, and HTSC cables may be the system of choice because of their potential for very high capacity and low loss. The use of conventional DC cables would consume roughly 10% of the energy they transmit for distances on the order of 1000 miles. HTSC cables could reduce this penalty to the 1–2% range.

19.7. SUMMARY

This chapter has explored the possible applications of HTSCs to underground transmission cables. The principal motivation for using cables is that overhead lines are impermissible in certain crowded circumstances, and underground cables are the only alternative. This condition holds in a small but important fraction of the total mileage of transmission lines. It is reasonable to expect that underground cables will play a greater role in the future, as both crowding and public antipathy to overhead lines increase. Studies from the United States, Japan, and Europe all agree on the increasing importance of underground cables.

Two important characteristics of any transmission system are its *thermal limit* and its *surge impedance*, which directly determine the capacity limit of the system. Overhead and conventional underground lines differ dramatically in these properties, and the design of each is strongly affected by capacity limitation considerations. One of the hopes for HTSC materials is that they will alleviate the design constraints on cables by having capacity limitations that mimic overhead cables.

The first major use of superconducting wire for electrical transmission was the Brookhaven prototype, built to test the engineering aspects of superconducting cables. As in so many applications of new technology, a variety of trade-offs were necessary, and the most difficult obstacles were only understood through experience. A low-temperature dielectric that can handle high voltages is by no means a simple thing, and concerns about dielectrics still pose a major obstacle to LTSC cables. Nevertheless, the lessons learned on that project provide valuable guidance for future uses of superconductivity in transmission cables.

HTSC cables would be similar to LTSC cables in many ways, including the general size, shape, and configuration, the dielectric losses and the eddy–current losses. If both the conductor and the dielectric have to be cooled to 77 K, then the HTSC cable bears a strong resemblance to its LTSC counterpart, albeit with a much less severe thermodynamic penalty for cooling. On the other hand, if only the conductor need be cooled and the dielectric allowed to remain at ambient temperature, then several additional advantages accrue: The cryogen carries away less heat, because only eddy currents, AC losses, and the heat in-leak affect the 77 K level. Dielectric losses, and so on, occur at ambient temperature and are dissipated in the ambient surroundings. Thus, there is a substantial incentive toward the ambient–dielectric configuration.

Obviously, until there is HTSC wire, there will be no HTSC transmission cable. Still, it is easy to see near-term applications for such cables, because of their favorable economics. Retrofitting HTSC cables into existing pipes might increase the capacity by a factor of 3 to 6 (assuming a value of $J_c > 10^5$ A/cm^2 in the HTSC conductor), and this would benefit urban power delivery systems. In the long term, the tremendous growth foreseen in electrical demand and the need to go underground makes the HTSC option very attractive.

REFERENCES

1. J. Engelhardt, D. Von Dollen, and R. Samm, presented at NYSIS Fifth Annual Conference on Superconductivity and Applications, Buffalo NY (September 24–26, 1991).
2. S. Glasstone, *Energy Deskbook*, DOE/IR/051 14-1, U.S. Department of Energy, Technical Information Center, Oak Ridge, TN (1982).
3. "Underground Power Transmission: A Study for the Electric Research Council," by Arthur D. Little, Inc., October 1971, ERC Pub. No. 1-72.

4.   A. M. Wolsky *et al.*, *Advances in Applied Superconductivity: A Preliminary Evaluation of Goals and Impacts*, Argonne National Laboratory Report ANL/CNSV-64 (January 1988).

5.   E. M. Allam *et al.*, "Optimized PPP-Insulated Pipe-Type Cable System for the Commercial Voltage Range," IEEE Paper 86 T&D 569-8 (September 1986).

6.   A. Greenwood and T. Tanaka, *Advanced Power Cable Technology* (CRC Press, Boca Raton, FL: 1983).

7.   J. S. Engelhardt and R. B. Blodgett, "Preliminary Qualification of a HPOF Pipe Cable System for Service at 765 kV," IEEE Trans. on Power Apparatus and Systems, Vol. PAS-94, No. 5 (September/October 1975).

8.   "Development of Low-Loss 765 kV Pipe-Type Cable," EPRI Project RP 7812-1, Final Report EL-2196, EPRI, Palo Alto, CA (January 1982).

9.   T. Hara *et al.*, "Feasibility Study of Compact HTSC Cables by Bean Model," Proc. Second International Symposium on Superconductivity (ISS), Tsukuba Science City, Japan (November 14–17, 1989).

10. H. G. Stoll, *Least-Cost Electric Utility Planning* (Wiley, New York: 1989).

11. "Superconducting Cable System Program (Phase II)," Interim Report, Part I, Union Carbide Corp., EPRI Project RP 7807-1, EPRI (Palo Alto, CA: 1973).

12. E. B. Forsyth, "The Brookhaven Superconducting Underground Power Transmission System," *Electronics & Power* **30**, 383 (May 1984).

13. E. B. Forsyth, *Science* **242**, 391 (1988).

14. P. A. Klaudy and J. Gerhold, "Practical Conclusions from Field Trials of a Superconducting Cable," *IEEE Trans. Mag.*, **MAG-19**, 656 (1983).

15. "Development of Cross-Linked Polyethylene Insulated Cable for Cryogenic Operation," CTL, EPRI Project RP 7892-1, Final Report EL-3907, EPRI (Palo Alto, CA: February 1985).

16. M. Kosaki *et al.*, "Development and Test of Extruded Polyethylene Insulated Superconducting Cable," Proceedings 2nd Intl. Conf. on Properties and Applications of Dielectric Materials (Beijing, China: 1988).

20

Levitation

John R. Hull[*] *and Thomas P. Sheahen*

In addition to zero resistance, the other central property of superconductors is the Meissner effect, by which magnetic fields are driven out of superconductors. This property can be exploited to levitate a magnet. As a consequence, superconductors can be used for noncontacting bearings, which in turn facilitates several useful applications that include the storage of energy in the mechanical rotation of a flywheel.

In this chapter, we develop the central concepts surrounding the use of superconductors in levitation. We first recall the Meissner effect, and then find the forces arising from this repulsion of magnetic fields. Next we introduce the notion of hysteresis in the forces, leading to a graph known as the "force banana". The most familiar practical consequence of such forces is magnetic levitation. The application of this phenomenon to trains is briefly reviewed, with a distinction being made between attractive (electromagnetic) forces and repulsive (electrodynamic) forces. Because this book is chiefly interested in electric power applications of superconductivity, we limit the description of magnetic levitation for trains to a fairly brief section.

We emphasize the aspects of levitation pertaining to bearings and energy storage. After a discussion of bearings that use high-temperature superconductors, we describe the combination of superconductors with permanent magnets in hybrid bearings. After the properties of the bearings are known, it is possible in principle to calculate the efficiency of energy storage in a flywheel. However, success of this storage mechanism depends on engineering details concerning stability, efficiency of energy transfer, and so on; a full-size demonstration unit will be needed to convince utility managers.

20.1. THE MEISSNER EFFECT

Perhaps the most familiar demonstration of HTSCs is that of a small magnet floating in the air above a YBCO block immersed in liquid nitrogen. This surprising, counterintuitive behavior has been used to capture the public imagination and convey the uniqueness of superconductivity. It is due to the expulsion of magnetic fields from a superconductor, which was first discovered by Meissner and Ochsenfeld[1] in 1933. However, no simple visual demonstration was available to the public until high-temperature superconductors came along. Prior to 1987, it required liquid helium to attain temperatures low enough to induce superconductivity and produce levitation. Because liquid helium has to be contained in a dewar, it had long been a matter of some effort with a flashlight peering through narrow

*Argonne National Laboratory.

Figure 20.1.   Small magnet levitated above a block of YBCO in liquid nitrogen.

windows to see anything at all in a liquid helium environment. Therefore, photos of magnets floating above type I superconductors[2] such as aluminum or tin are unknown to the general public. By contrast, today the demonstration of the Meissner effect is easy to perform using HTSCs, and spectators can touch the magnet with a finger, pushing it around to explore the limits of levitation. Passing a dollar bill (or better yet, a $50 bill) between the magnet and the superconductor makes the experience memorable.

The message, of course, is that magnetic forces are strong enough to overcome gravity. The Meissner effect was introduced with Figure 2.1, which shows how the lines of a magnetic field are distorted by being expelled from within the superconductor. When a magnet is placed nearby and its magnetic field is excluded from entering the superconductor, the distortion of field lines results in a force which can lift the magnet against gravity. At equilibrium, the magnet stands a short distance away from the superconductor. It is easy to scale up the experiment to very large proportions. In 1990, Professor Shoji Tanaka stood on

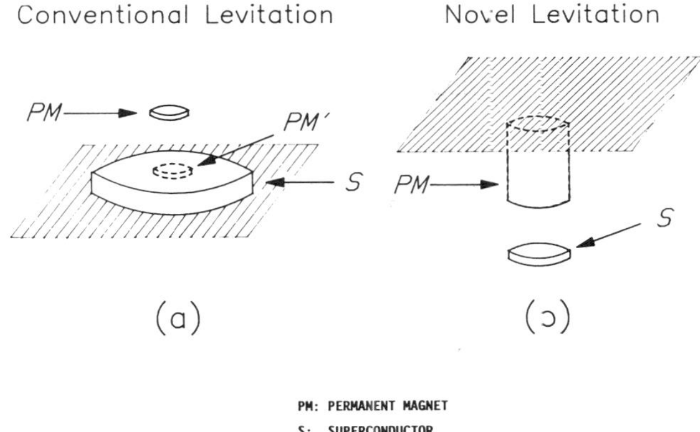

Figure 20.2.   Sketch of interaction between a permanent magnet and a superconductor. Because flux-pinning forces are stronger than gravity, levitation can occur either (a) above or (b) below a sample. (Courtesy of R. D. Shull, NIST.)

a large thin disk which was levitated above an equally large piece of YBCO, thus providing a very memorable HTSC demonstration.

One variation on the demonstration is to let a small magnet sit on top of a warm block of YBCO, and then pour liquid nitrogen around it. Presently, the YBCO cools into the superconducting state, and the magnet suddenly jumps up into the air. Later on, if all the nitrogen is allowed to boil away, the magnet will slowly fall back and settle onto the block of YBCO as it warms toward and then above T_c.

Still another variation is to use two small magnets, one standing off of either end of a rod of TBCCO. During the minute or so it takes TBCCO to warm to $T_c \approx 125$ K after being withdrawn from a 77 K bath of LN_2, the rod can be twirled like a baton, and the magnets follow along with the motion as if held in place by an invisible connection. Figure 20.2 conveys both kinds of levitation. What is actually taking place is a demonstration of the importance of flux pinning: flux lines from each magnet partially penetrate the TBCCO, and the flux lines become pinned while at 77 K. When the TBCCO rod is moved, the only way for the flux lines to remain stationary is for the magnets to move as well, even though they are not physically attached to the TBCCO. The force opposing flux line motion is strong enough to overcome gravity. The magnet attached to the bottom of the rod stands a little further away than the one at the top of the rod; the two exhibit gentle oscillations as the rod is twirled in space. Of course, when the system warms to about T_c, the flux lattice melts, pinning ceases, and the two magnets crash to the floor. Strictly speaking, the levitation phenomena produced by flux pinning is not part of the Meissner effect; however, one often sees it referred to as such in modern writing.

Before going on to examine levitation forces in detail, it is convenient here to explain one other phenomenon often seen in superconductivity demonstrations. When a cylindrical magnet is levitated over a HTSC, under some conditions the magnet will rotate spontaneously and continue to do so indefinitely. A magnet at rest will first begin to oscillate, with the rotational amplitude increasing with time. Finally, the amplitude in one direction becomes large enough that a complete rotation occurs, and the magnet continues rotating in that direction, eventually attaining a maximum rotational frequency of about 1 Hz. These phenomena have been observed by a number of researchers and quantitatively explained by Ma *et al.*[3] While convective instabilities of the boundary layer of air surrounding the magnet seem to enhance the motion, the phenomenon also occurs with NdFeB (neodymium-iron-boron) magnets in vacuum. In this case, the physical cause resides in the temperature dependence of the magnetization, which increases from its value at 300 K as the temperature decreases, peaks at about 160 K, and then decreases again and is nearly constant for temperatures below 100 K. With a temperature gradient across the magnet, the magnetization is larger at the bottom of the magnet. The center of levitation force lies below the center of gravity, a configuration that is mechanically unstable and starts the oscillation.

20.2. THE "FORCE BANANA"

The forces of repulsion between superconductors and magnets are not as simple as first meets the eye. Certainly the force gets stronger as the two come closer. However, due to flux pinning, the levitation force on a magnet over a type II superconductor is different when the magnet approaches the sample than when it is moving away, as shown in Figure 20.3. When the magnet is brought nearer, the lower critical field H_{cl} is reached, and more and more flux penetrates the superconductor. When the magnet is moved away, the repulsive force between the magnet and its image decreases. In addition, the pinned flux lines cause an attractive force that reduces the net repulsive force. This results in a force vs. distance curve with hysteresis, as shown in Figure 20.3.

In Figure 20.3, the repulsive force equals the weight of the magnet, *mg*, at points *A* and *B* or at any point on the line connecting them. Point *B* represents the levitation height for a magnet lowered onto the superconductor. If the magnet is pushed down onto the superconductor and then released at point *C*, it will move to the stable point *A*. Likewise, if the magnet rests on the superconductor when it is cooled, it will rise to point *A*.

Although *any* superconductor can cause levitation, only the type II superconductors (of which the HTSCs are a subset) are of practical use. The combination of flux-pinning forces with the repulsive force of an expelled magnetic field causes magnets and superconductors to exhibit very interesting behavior. In principle, knowledge of the magnetization behavior of the superconductor as the applied field varies should allow one to predict levitation behavior. Because the magnetization is not uniform throughout the superconducting sample, however, many magnetization measurements and a large computational effort would be

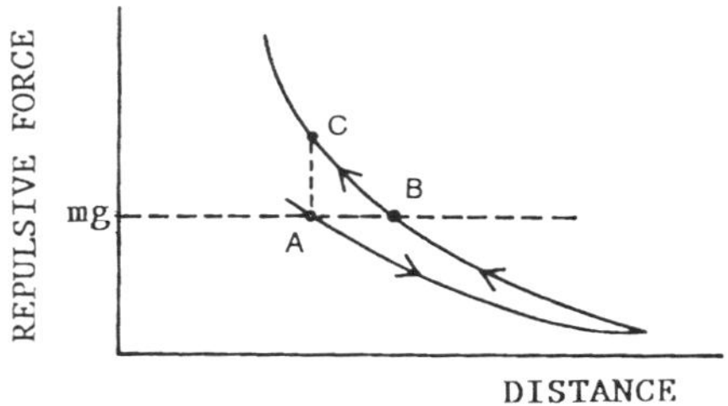

Figure 20.3. Force on a magnet levitated over a type II superconductor. The upper curve represents the force when the magnet approaches, and the lower curve the force for the magnet moving away. A magnet whose weight is *mg* can levitate stably at point *A*, point *B*, or any point on the dashed line between them.

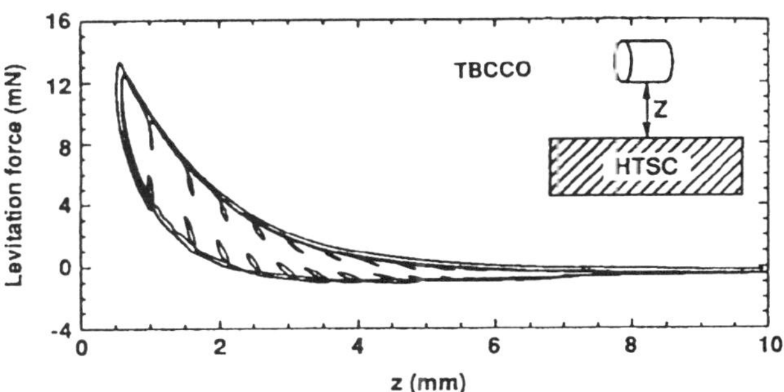

Figure 20.4. Levitation force vs. magnet height for a NdFeB permanent magnet over a TBCCO superconductor, showing one major and several minor hysteresis loops. (From Ref. 4.)

required.[4] In practice, it is more profitable to measure the levitation force as a function of distance, as shown in Figure 20.4.

Of the two stable levitation points adjacent to a YBCO sample, the outer such position corresponds to the usual Meissner-effect repulsion, whereas the inner position corresponds to a flux line being stretched elastically and thus establishing a restoring force.

Figure 20.4 has a shape somewhat like a banana, and researchers have coined the term "force banana" to characterize it. Levitation forces between permanent magnets and HTSCs are quite hysteretic, in some cases even resulting in an attractive force. The minor hysteresis loops in Figure 20.4 were obtained by briefly reversing the ascent/descent motion of the magnet. One useful parameter in the design of magnetic bearings is the dynamic levitation stiffness, given by the slopes of the minor hysteresis loops. The effective magnetic stiffness is notably affected by the amplitude of motion of the magnet[6] and by the size and geometry of both the magnet and the superconductor sample.[7] These concepts will be important in Section 20.5 below.

For the moment, the most important thing to understand from the "force banana" is that when the two components of a bearing interface are moved briefly (relative to each other), they will not necessarily come back to exactly the same position they were in previously. A major excursion can result in a new equilibrium position that differs by several millimeters from the initial equilibrium position. This has to be taken into account at the design stage.

20.3. FORCES ON MOVING MAGNETS

The Meissner effect provides a force to lift a magnet above a superconductor, but whenever a magnet is in motion there is also a drag force. The combination of these two affect the performance, and hence constrain the design for any practical magnetic bearing surface.

The forces on a moving magnet have been described carefully by Rossing and Hull.[8] The key factor is the presence of eddy currents, which are induced by conductors moving past one another. A permanent magnet can be levitated this way above a rotating aluminum disk. Induced eddy currents cause the disk to act as a magnet mirror, the magnet being repelled by its induced image below the conductor. The faster the conducting disk rotates,

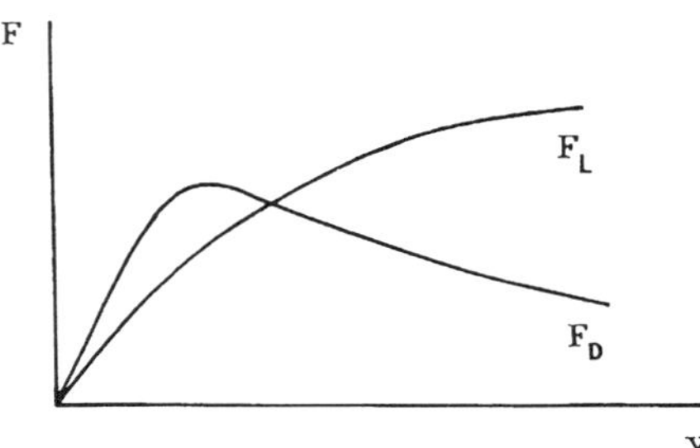

Figure 20.5. Velocity dependence of the lift force F_L and the drag force F_D.

the better the magnetic image it produces, approaching (as a limit) the behavior of a superconductor.

The induced magnetic image was first described by Maxwell[9] in 1872. In his model, when a magnet passes a point on the conducting plane, it induces first a positive image, then a negative image. These images propagate into the plane at a characteristic velocity proportional to the specific resistivity. The lift and drag force on the magnet depend on the ratio of the actual velocity to this characteristic velocity. Except, in the simplest geometries, analytical solutions are not available and numerical calculations are required to determine the magnetic flux lines, and from that follows the lifting force. As the velocity increases, less magnetic flux penetrates the conductor, and the magnetic field due to the induced current makes a greater contribution to the total field. Because of the practical consequences for designing vehicles as well as bearings, both the lift and drag forces need to be calculated quantitatively.

The force on a magnet moving over a nonmagnetic conducting plane can be conveniently resolved into two components: a *lift* force perpendicular to the plane and a *drag* force opposite to the direction of motion. At low velocity, the drag force is proportional to velocity v and considerably greater than the lift force, which is proportional to v^2. As the velocity increases, however, the drag force reaches a maximum (referred to as the *drag peak*) and then decreases as $1/\sqrt{v}$. The lift force, on the other hand, which increases with v^2 at low velocity, overtakes the drag force as velocity increases and approaches an asymptotic value at high velocity, as shown in Figure 20.5. If w is the characteristic velocity of image propagation into the conducting plane, the lift/drag ratio (which is of considerable practical importance), is given by $F_L/F_D = v/w$.

Qualitatively, these forces can be understood by considering magnetic flux diffusion into the conductor. When a magnet moves over a conductor, the field tries to diffuse into the conductor. If the magnet is moving rapidly enough, the field will not penetrate very far into the conductor, and the flux compression between the magnet and the conductor causes a *lift* force. The flux that does penetrate the conductor is dragged along by the moving magnet, and the force required to drag this flux along is equal to the *drag* force.

At high speeds, less of the magnetic flux has time to penetrate the conductor. The lift force resulting from flux compression approaches an asymptotic limit, and the drag force approaches zero at high speed. The lift force on a vertical dipole of moment m moving at velocity v at a height z_0 above a conducting plane can be shown[10] to be

$$F_L = \frac{3\mu_0 m^2}{32\pi z_0^{\,4}} \left(1 - \frac{w}{\sqrt{v^2 + w^2}} \right) \tag{20.1}$$

At high velocity, the lift force approaches the ideal lift from a single image: $3\,\mu_0 m^2/32\pi z_0^4$; at low velocity, the factor in parentheses is approximately equal to $v^2/2w^2$, so the lift force increases as v^2.

The drag force, as already pointed out, is w/v times the lift force, so the drag force is proportional to v at low velocity. According to the *thin-plate model*, which applies here, the drag force should fall off with $1/v$ as the lift force reaches its high-speed limit. However, at high velocity, penetration of the eddy currents and magnetic fields are limited to the skin depth, which is proportional to $v^{-1/2}$. As a first approximation, one might replace plate thickness by skin depth, in which case the drag force takes on a $v^{-1/2}$ dependence at high speed. The transition from thin-plate to skin depth behavior should occur at about 30 m/s in a 1-cm-thick aluminum plate, for example.

When applied to vehicles for high-speed ground transportation, the numerical values change, but the concepts remain applicable.

20.4. MAGNETIC LEVITATION VEHICLES

Undoubtedly, the most well-advertised possible application of superconductivity is that of maglev trains, proposed to run at 500 km/hr (= 300 mph). The public imagination is captured by the image of what amounts to an extremely low-flying aircraft hovering about 10 to 20 cm above a guiderail, whisking passengers silently, swiftly, and comfortably from city to city. This is commonly seen as an idea whose time has come, and high-temperature superconductivity will make it practical and inexpensive to operate.

The reality is that for a major intercity maglev line (say between Washington, D.C., and New York) costing perhaps $10 billion, about $7.5 billion would go for right-of-way acquisition, over $2 billion to build the guideway, and all the rest of the system would cumulatively amount to less than $0.5 billion. In fact, the refrigeration cost distinction between LTSCs and HTSCs is in the noise level of the overall program cost. That cost cannot be dismissed lightly. A $10 billion system requires debt service of nearly $1 billion per year, and it would be necessary to charge a $100 fare to 10 million passengers annually to meet this. Numbers like these can only be justified if the maglev system is seen as a major component of the national air transport network.

Competing with maglev is conventional high-speed transportation (i.e., steel wheels on rails), and proponents of that technology predict speeds above 200 mph soon. There is a very real question as to how much the extra speed is worth to a traveler. (New York to Washington in 45 minutes vs. 75 minutes may not be much of an advantage when the overhead time of getting to the station is factored in.) Nonetheless, if maglev should win out on the basis of economic projections, either LTSCs or HTSCs can be used to advantage. For that reason, designers of maglev are going ahead today, even in the absence of practical HTSC levitation systems. Figure 20.6 portrays a typical maglev system.

20.4.1. History

Magnetic levitation using superconducting magnets was first suggested in 1963 by Powell,[11] soon after the discovery of type II superconductors with their implications for carrying large currents. In 1967, Powell and Danby proposed a system[12] using a less

Figure 20.6. Maglev train being tested at the Emsland (Germany) Test Facility. (Photo courtesy of Transrapid International.)

expensive conducting guideway at room temperature. Later, they conceived the novel idea of a null-flux suspension system that would minimize the drag force and thus require much less propulsion power.[13]

During the late 1960s, groups at the Stanford Research Institute and at Atomic International studied the feasibility of a Mach 10 rocket sled employing magnetic levitation. This principle was later applied to high-speed trains by Coffey *et al.*,[14] among others.[15] In 1972, the group at Stanford Research Institute constructed and demonstrated a vehicle levitated with superconducting magnets over a continuous aluminum guideway 160 m long.

At about the same time, a team from MIT, Raytheon, and United Engineers designed the *magneplane* system in which lightweight cylindrical vehicles propelled by a synchronously traveling magnetic field travel in a curved aluminum trough. One advantage of the curved trough is that the vehicle is free to assume the correct bank angle when negotiating curves, the guideway itself being banked only at approximately the desired angle.[16] The magneplane concept was tested with a $\frac{1}{25}$-scale model system using both permanent magnets and superconducting coils for levitation above a 116-m-long synchronized guideway.[17]

Maglev systems became objects of considerable study in several other countries, most notable Japan, Germany, and the United Kingdom. In the United States, however, virtually all support for maglev research ended about 1975, and very little work has been done from that time until very recently. Research and development continued in Japan and Germany, and full-scale vehicles have been tested in both countries. Plans are being made to construct a system based on the German (*Transrapid*) technology in Orlando, Florida. It will probably be the first public maglev system in the United States.

Two different kinds of maglev are competing in today's prototype phase of development. The difference is in the way the vehicle attaches to the guideway: either attractive levitation forces (electromagnetic suspension) or repulsive forces (electrodynamic levitation). Beginning in the 1970s, research in Germany explored both systems, but more recently only electromagnetic systems are in use.[18] In Japan, both systems were tried and the electrodynamic method was preferred.[19,20]

20.4.2. Electromagnetic Suspension

Electromagnetic systems (EMS) depend on the attractive forces between electromagnets and a ferromagnetic (steel) guideway, as shown in Figure 20.7(a). Because the force of attraction increases with decreasing distance, such systems are inherently unstable and the magnet currents must be carefully controlled to maintain the desired suspension height. Furthermore, the magnet-to-guideway spacing needs to be small (only a few centimeters at most). On the other hand, it is possible to maintain magnetic suspension even when the vehicle is standing still, which is not true for electrodynamic (repulsive force) systems. In the system of Figure 20.7(a), a separate set of electromagnets provides horizontal guidance force, but the levitation magnets, acted on by a moving magnetic field from the guideway, provide the propulsion force. The German Transrapid TR-07 vehicle is designed to carry 200 passengers at a maximum speed of 500 km/hr. The levitation height is 8 mm, and power consumption is estimated to be 43 MW at 400 km/hr.

20.4.3. Electrodynamic Levitation

Here, the principles explained in Section 20.3 above come into play. Electrodynamic systems (EDS), shown in Figure 20.7(b), depend on repulsive forces between moving

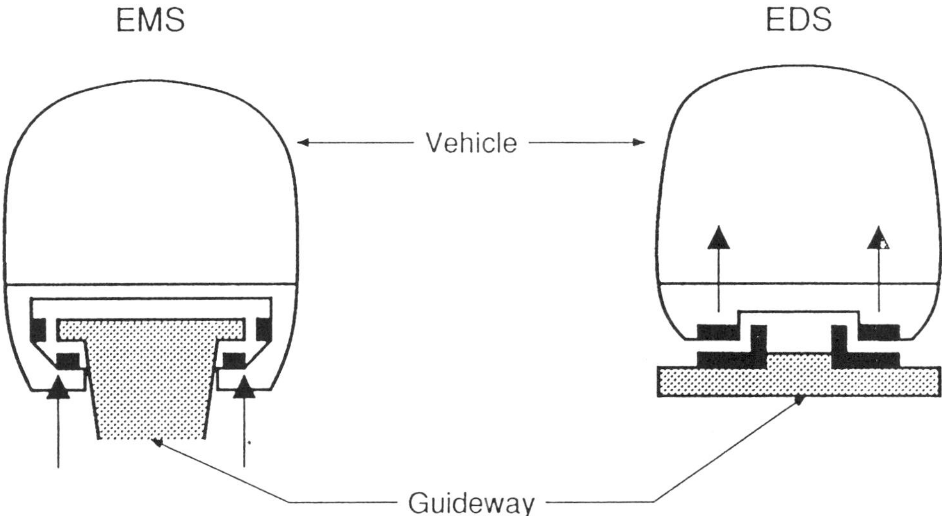

Figure 20.7. Schematic diagram of alternate maglev systems. (a) In the ElectroMagnetic system (EMS), the attractive forces between the magnets on the vehicle and the guideway draw the vehicle upward and lift it slightly above the guideway. Other magnets are used to achieve lateral positioning and stability. (b) In the ElectroDynamic System (EDS), repulsive forces keep the vehicle away from the guideway.

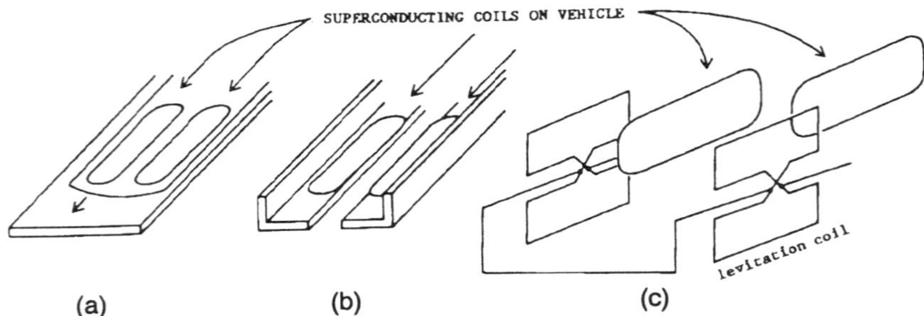

Figure 20.8.   Three different arrangements for electrodynamic (repulsive force) levitation systems: (a) superconducting coils over a flat conducting guideway; (b) superconducting coils with a split L-shape guideway; (c) superconducting coils with short-circuit figure-8 coils connected in a null-flux arrangement. The proposed Japanese maglev system is similar to arrangement (c).

magnets and the eddy currents they induce in a conducting (aluminum) guideway or in conducting loops. The repulsive levitation force is inherently stable with distance, and comparatively large levitation heights (20–30 cm) are attainable by using superconducting magnets. Three different configurations are shown in Figure 20.8: a flat horizontal conductor, a split L-shape conductor, and an array of short-circuit coils on the sidewalls. Each has its own advantages and disadvantages. The proposed Japanese high-speed maglev system uses interconnected figure-eight (null-flux) coils on the sidewalls, as shown in Figure 20.8(c). The null-flux arrangement tends to reduce the magnetic drag force and thus the propulsion power needed.

Clearly, there are reasons for and against each type of system, and which is best may be determined by operating costs and marketplace factors such as passenger comfort. The high-pitched screech of metal against metal at 300 mph may be so unnerving that no efficiency of operation is sufficient to outweigh it. Until real units are built and refined, all such speculations are idle.

20.4.3. Special Considerations for HTSCs

Prototypes of the early 1990s are based entirely on LTSCs or conventional electromagnets; the role of HTSCs, if any, lies in the future. Nevertheless, it is not too early to consider certain design features that are implied by HTSCs. In any future application of HTSCs, it will be necessary to give thought to the mechanical aspects of the superconducting medium. Magnetic forces on superconductors must act through the HTSC itself, and therefore these materials must have substantial mechanical strength. Bonding to steel or aluminum frames will of course be done, but it must be remembered that the forces act first on the HTSC, and are then transmitted to the supporting structure. It is necessary to give considerable attention to the stresses and strains that will take place in various HTSC applications.

For example, in maglev, the train ride may seem very smooth to the passengers because the car is gliding above the rails, but there are repeated pulses of magnetic force acting between the car and the guideway. Each pulse places a stress on the HTSC material. Over a sustained period of operation, if tiny cracks develop and the supercurrent consequently decreases, the pounding will get worse, leading to early failure of HTSC components. Similar scenarios can be imagined for any other HTSC application in which cycling of a component takes place.

One worry associated with HTSCs is the operating cost attributable to AC losses. The smooth ride experienced by passengers is only possible if the undulations in the guideway are compensated by slightly varying the standoff distance between vehicle and guideway, all at 300 mph. That will show up at the superconducting magnets as an AC variation in the magnetic field of the permanent magnets, and AC fluctuations dissipate energy. (The maglev system is designed to run entirely on DC.) Experience at test tracks in Japan and Germany teach us how to design around this problem using niobium-based superconductors; but we have almost no data on AC losses in YBCO or the other HTSCs so far—only preliminary indications that the AC loss problems will be worse for the HTSCs than for the LTSCs. Just as (on a highway) the pavement adjacent to a bridge takes the worst pounding, so this new loss mechanism will tend to exacerbate irregularities and weaknesses at certain points along the entire guideway. It may become cost-effective (indeed necessary) to monitor the shape of the guideway constantly, and repair it often.

20.5. BEARINGS

A frictionless bearing is every mechanical designer's dream, and a bearing in which the two surfaces never make contact is close to that ideal. Magnetic bearings achieve mechanical separation, and when other dissipative factors are minimized, they can be extremely good bearings, allowing rotational speeds not attained any other way.

Magnetic bearings are used in an increasing number of applications. Compared to hydrodynamic or ball bearings, their main advantages are noncontacting surfaces, lower power dissipation, and the potential to achieve significantly higher surface velocities. For stability, conventional magnetic bearings require mechanical support in at least one dimension or, more commonly, active feedback in the electromagnetic circuit. The use of superconducting components in magnetic bearings offers the potential to eliminate the control system and further reduce power dissipation of the system. The main disadvantage of superconductors is the need for a cryogenic refrigerant.

20.5.1. Bearing Principles

Three different types of magnetic bearings are possible:

1. Coil–magnet systems using direct current.
2. Electrodynamic or induced eddy–current devices.
3. Passive Meissner-effect bearings using type II superconductors.

Coil-magnet systems have been successfully used for several years to suspend ultrahigh-speed rotors.[21] The simplest type of magnetic bearing, however, which can be stable without a feedback mechanism, is the third type using type II superconductors. Speeds in excess of 100,000 rpm have been achieved with passive bearings of this type.[22]

A simple magnetic bearing consists of a permanent magnet rotor suspended between two superconductors, as shown in Figure 20.9. Magnetic repulsion forces cause the rotor to be suspended in midair, when it is able to spin freely; the only friction is caused by aerodynamic and magnetic drag. If the rotor tries to drift off center, a restoring force due to

Figure 20.9. A simple magnet bearing consisting of a permanent magnet rotor suspended between two superconductors. (Courtesy of Allied Signal.)

flux pinning restores it. This is known as the *magnetic stiffness* of the bearing and is an important design parameter.

To increase the magnetic stiffness in a practical bearing, a superconducting cylinder that acts as a radial bearing might be added. However, this increases the magnetic drag on the rotor. Superconducting bearings need to be controlled in five directions: up, down, side-to-side, pitch, and yaw. Efforts to increase magnetic stiffness, through clever design and improved materials, are underway at several laboratories. Although a low-stiffness bearing is more forgiving of an out-of-balance rotor, most applications require a larger stiffness than is presently available.

Lift force is not generally considered to be a problem. Actually, magnetic pressure is a better parameter for characterizing bearings, because it does not depend on the area of the magnet of the superconductor. Pressures of 10^5 N/m^2 have been obtained with a permanent magnet and a superconductor, which is probably large enough for magnetic bearings.

20.5.2. Early Experimental Results

Even in 1987, engineers realized the potential for HTSCs to be used in magnetic bearings. YBCO bearings were pushed to high rotational speeds by Professor Frank Moon of Cornell and co-workers. Moon found that YBCO produces better bearings than bismuth or thallium compounds; the ability to carry high current doesn't count where levitation is concerned. Because of increased flux-pinning, the best results were obtained with samples made by the quench melt growth (QMG) process. Such materials showed much better performance than traditional sintered materials.[23] Moon asserts that superconducting bearings will be able to replace many gas bearings, but will never replace ball bearings, because the requisite concentration of forces cannot be achieved with superconducting bearings.

For bearings, the key factors are load capacity, stiffness, and damping. Allied Signal, which built the simple device pictured in Figure 20.9, already has good enough materials for passive bearings and is striving to develop more advanced bearings. The highest speed yet achieved is 500,000 rpm for bearings made from melt-textured YBCO. Load capacities of 5–30 psi are achievable, especially at lower temperatures.

The Fluoramics Corporation also makes bearings out of YBCO, but uses a very different approach. They start with highly porous pellets of YBCO, in which the porosity makes it easy to get the oxygen into the lattice. After grinding, the YBCO powder can be bonded with epoxy, silicone rubber, polyester, and so on, and then molded into a form. Diverse shapes can be made by conventional machine tools. In all cases, the material still levitates a magnet, and the percentage of levitation force is proportional to the percentage of superconductor in the mix. The electrical properties are poor, but for magnetic levitation, this material is excellent. Just as others have found, melt processing seems to improve the performance of Fluoramics' YBCO.

20.5.3. Hybrid Bearings

At the Texas Center for Superconductivity at the University of Houston (TCSUH), advances have been made toward using hybrid bearings made of superconductors and permanent magnets.[24] This concept exploits magnet-to-magnet repulsion to *support* the load and use a superconductor to *stabilize* it. This takes a bit of explaining.

The repulsive force between two like magnets is akin to a ball sitting atop the crown of a hill: any slight perturbation in any direction will start it rolling. Early in the history of magnetism (1842), Earnshaw's theorem[25] showed that a system composed only of static forces with an inverse square law between the system components, such as in a system with permanent magnets and paramagnets, is unstable. Denoting positive constants by α and β, the force between two dipole magnets M_1 and M_2 is $\alpha M_1 M_2$, and the stability parameter is $-\beta M_1 M_2$; the negative sign indicates that it is unstable. Superconductors, by contrast, are very good diamagnets and are therefore able to provide a negative value for one M_i, thus yielding a positive stability parameter.

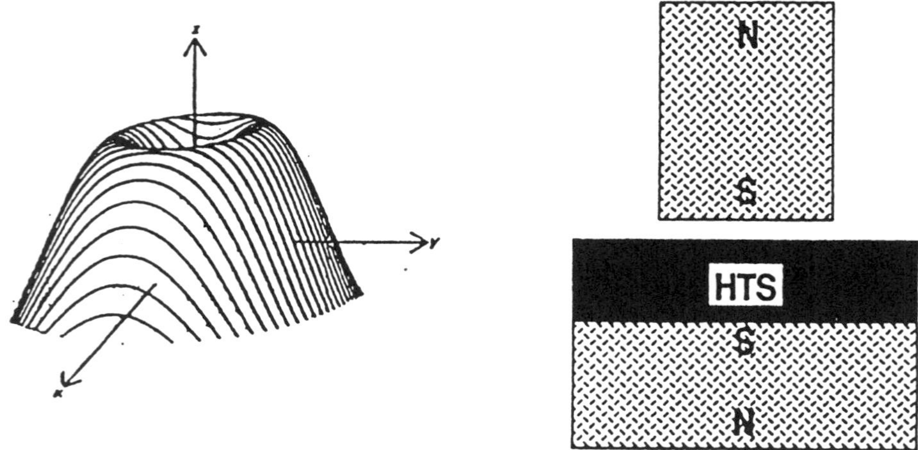

Figure 20.10.   Force potential surface associated with hybrid-superconductor magnetic bearing.

The TCSUH configuration is named the Hybrid Superconductor–Magnet Bearing (HSMB) and is conceptually illustrated in Figure 20.10. The "crown of a hill" potential surface has a dimple in it (left), because a superconductor is placed between the mutually repelling south poles of two permanent magnets (right). This combination provides the needed stability.

In actual practice, several different bearings can be constructed this way. Figure 20.11 shows some of them: At right is a thrust bearing based on magnetic repulsion; in the center is a journal bearing; and at left is a pair of attracting magnets to assist the thrust and journal bearings. The undergraduates at TCSUH built one such device.

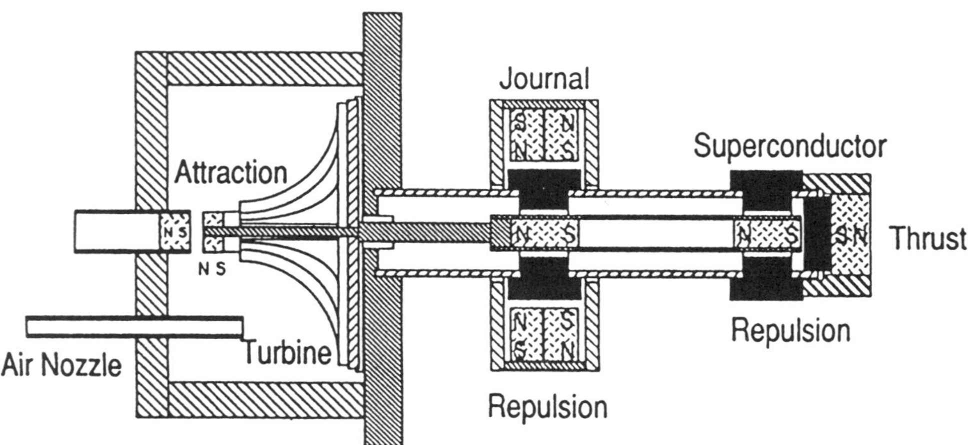

Figure 20.11.   Alternative forms of the hybrid-superconductor bearing.

The next step is to tailor the materials processing methods (for materials used in bearings) toward meeting specific device parameters. This may mean sacrificing some lifting power in order to gain mechanical strength or resistance to damage from vibration. In earlier embodiments, superconducting levitation bearings had limited applications because they simply could not lift the load required in many situations. However, in the TCSUH configuration, the superconductor need only provide the *stability*, whereas the *lifting* is provided by regular magnets. This is very similar to the concept employed since about 1900 in centrifuge bearings in which a set of permanent magnets provides a carefully balanced lift force and a small mechanical pivot bearing provides stability. Using this concept opens the door to a much wider range of practical applications. As in any potential use of HTSCs, the trade-off to be made is between the cost of refrigeration and the benefit of better overall system performance—in this case meaning lower frictional losses in the bearings.

One environment where refrigeration might be free is in space. In a levitation bearing, the rotational dissipation can be reduced by making it passively stable. TCSUH has had some initial success with small laboratory bearings: the TCSUH prototype consumes only 20 mW, compared to 20 W for most typical bearings that might be used by NASA in space. Here again, HTSCs offer an unexpected benefit in an application far from what one would normally expect.

20.6. FLYWHEEL ENERGY STORAGE

One of the foremost applications of superconducting bearings is in energy storage via flywheels. The concept is sketched in Figure 20.12, which is a more detailed version of Figure 1.4. The entire apparatus is housed in a vacuum chamber to eliminate losses due to air drag. Only the HTSC bearing need be cooled to LN_2 temperatures; that bearing acts on a permanent magnet attached to the flywheel, both to suspend it and to keep it stable.

Energy is transferred into and out of a flywheel as follows: A permanent magnet is mounted in the flywheel. Current flows through a coil adjacent to the flywheel, and that repels the magnet, whose motion causes the flywheel to spin. Proper synchronization between the current pulses and the angular position of the rotating magnet is required to accelerate the spinning flywheel. Once the flywheel reaches top speed, the electricity is turned off, and the wheel continues spinning, losing almost no energy to friction or air drag (because of the vacuum). Later, when the circuit is reconnected, the magnet spinning on the flywheel generates electricity in the coil, which can be used to run an external load.

This type of storage is not being done at present. For an electrical utility wishing to store power for 12 hours or so, flywheels have not been considered competitive (compared to batteries, pumped hydro, etc.) because of friction losses in the bearings. However, they could be competitive *if* their bearings were to have very low losses.[26]

How low is low enough? Based on the round-trip efficiency of other means of energy storage, Ken Uherka of Argonne National Laboratory estimated[27] that a flywheel would be superior if it lost only 1% of its energy in 10 hours, a design goal corresponding to $(1/E)dE/dt$ < 0.1% per hour. That means a slow-down rate of $df/dt \approx 10^{-5}$ to 10^{-4} Hz/sec, depending on rotor size. Rapid progress is being made in this area. The lowest value reached by 1993, in a small laboratory model, was 10^{-4} Hz/sec, and this value would be lower if a larger diameter rotor were used. A remaining problem will be to scale up such results to the size of a large industrial flywheel, where added concerns about dynamic stability come into play.

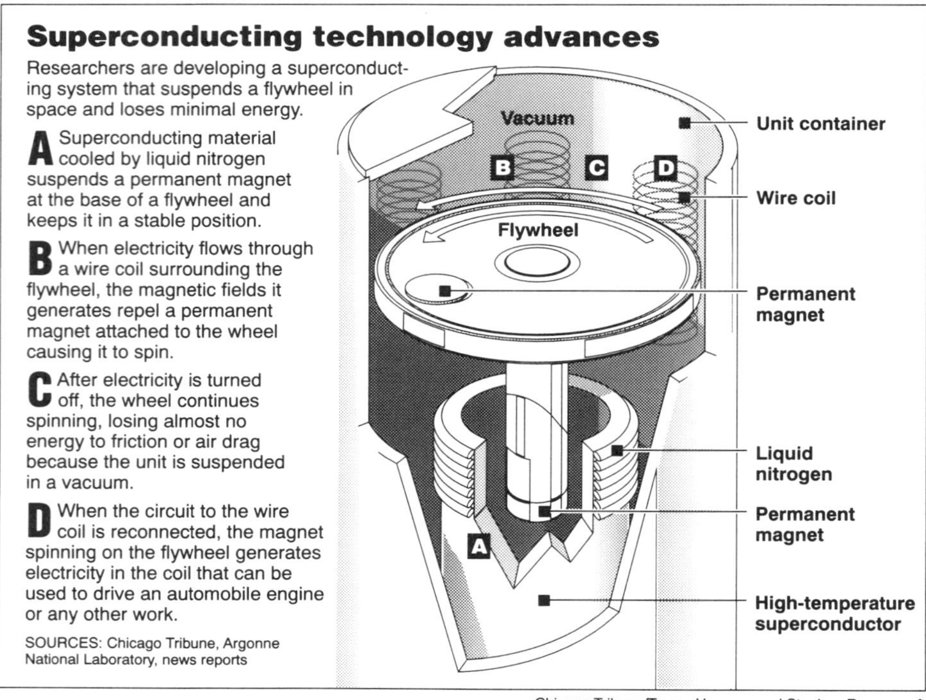

Figure 20.12. Conceptual drawing of flywheel energy storage.

There is intense interest in building a model to demonstrate the feasibility of this approach. Partnerships between industry and research centers are pursuing this goal: There are many engineering strategies being tested; for example, TCSUH is experimenting with a cylinder-tube composite, such that the inside of a bearing is "N," whereas the outside is "S," thus giving a more stable magnet configuration. Also, it has already been found that by pumping a vacuum down to 10^{-6} Torr, air-drag effects are totally eliminated.

20.7. OUTLOOK AND SUMMARY

This chapter has discussed the principles of magnetic levitation and the role that superconductors can play in practical applications of levitation. The most famous of these is for maglev trains, but probably the foremost application in the long run is as a material for magnetic bearings.

There are two distinct types of levitation system being investigated for maglev trains: electromagnetic (attractive magnetic forces) and electrodynamic (repulsive magnetic forces). Each of these have advantages and disadvantages, and engineering research ongoing in Germany and Japan strives to optimize each type of system.

The concept of using superconductors in magnetic bearings originated long before the discovery of ceramic superconductors. Early demonstrations of levitation operated at liquid helium temperatures (4 K), and the difficulty of reaching this temperature was the main argument against using superconductors in this way. Now, with critical temperatures of superconductors as high as 133 K, applications for superconducting bearings are being reexamined. Several applications have been found for superconducting bearings in rotating machinery that can operate at liquid hydrogen temperatures (20 K). Even more have been found in which liquid nitrogen temperatures (77 K) could possibly be used.

To understand and predict the behavior of magnetic bearings that use superconductors, one must know the forces that are generated between the superconductor material and a permanent magnet or a current distribution. Levitation, suspension, restoring, damping, and drag forces need to be determined for a wide variety of geometries, speeds, materials, and so on. While force measurements are perhaps most directly related to bearing behavior, knowledge of material properties is necessary to promote sufficient understanding to choose materials that will optimize the design for any particular application.

Material properties of interest include magnetization, flux-pinning force, and lower and upper critical field. Material characterizations, such as microstructure, chemical composition, grain size, purity, and alignment of grains, are also of interest, especially if they correlate to the force or magnetization measurements. Although present ceramic superconducting materials are satisfactory for some bearing applications, improvement in material properties is expected to increase the number of these applications.

There are also several novel applications for HTSC bearings: replacement of spacecraft bearings that would otherwise require frequent changes is a good example. Moreover, in any device that runs at 20 K or 4 K anyway (miniature cryocoolers for infrared sensors aboard spacecraft and liquid hydrogen turbopumps come to mind), it is smart to employ magnetic bearings, because there is no further penalty for refrigeration.

Current technical issues to be investigated include improved measurements, including standards and common test conditions and better understanding of basic physics.

Finally, for electrical energy storage, HTSC bearings suggest the hope of reaching very high round-trip efficiencies for power stored in a flywheel over 12 hours. Laboratory models have demonstrated very low energy loss rates, but scaling up to a full-size flywheel without introducing new loss mechanisms will be a serious engineering challenge.

REFERENCES

1. W. Meissner and R. Ochsenfeld, *Naturwissen.* **21**, 787 (1933).
2. V. Arkadiev, *Nature* **160**, 330 (1947).
3. K. B. Ma *et al.*, *J. Appl. Phys.* **70**, 3961 (1991).
4. P.-Z. Chang *et al.*, *J. Appl. Phys.* **67**, 4358 (1990).
5. F. C. Moon *et al.*, *Appl. Phys. Lett.* **52**, 1534 (1988).
6. S. A. Basinger *et al.*, *Appl. Phys. Lett.* **57**, 2942 (1990).
7. Y. S. Cha *et al.*, *J. Appl. Phys.* **70**, 6504 (1991).
8. T. D. Rossing and J. R. Hull, *The Physics Teacher* **29**, 552 (1991).
9. J. C. Maxwell, *Proc. Royal Soc. London A* **20**, 160 (1872).
10. J. R. Reitz, *J. Appl. Phys.* **41**, 2067 (1970).
11. J. R. Powell, ASME Railroad Conference, Paper 63-RR-4 (April 23–25, 1963).
12. J. R. Powell and G. R. Danby, *Mech. Eng.* **89**, 30 (1967).
13. J. R. Powell and G. R. Danby, in *Recent Advances in Engineering Science*, edited by A. C. Eringen (Gordon and Breach, New York: 1970).
14. H. T. Coffey *et al.*, *Adv. Cryogenic Tech.* **4**, 275 (1971).
15. C. A. Guderjahn *et al.*, *J. Appl. Phys.* **40**, 2133 (1969).
16. H. H. Kolm and R. D. Thornton, "Electromagnetic Flight," *Sci. Amer.* **229**(4), 17 (1975).
17. H. H. Kolm *et al.*, *Cryogenics* **15**, 377 (1975).

18. W. Menden *et al.*, *Proc. Maglev '89 Intl. Conf.*, pp. 11–18 (IEE, Japan).
19. Y. Kyotani, *IEEE Trans Magn.* **24**, 804 (1988).
20. T. Nagaika and H. Takatsuka, *Proc. Maglev '89 Intl. Conf.*, pp. 29–35 (IEE, Japan).
21. J. W. Beams, *Sci. Amer.* **204**(4), 134 (1961).
22. F. C. Moon & P.-Z. Chang, *Appl. Phys. Lett.* **56**, 391 (1990).
23. J. R. Hull *et al.*, *J. Appl. Phys.* **72**, 2089 (1992).
24. C. K. McMichael *et al.*, *Appl. Phys. Lett.* **60**, 1893 (1992).
25. S. Earnshaw, *Trans. Cambridge Phil. Soc.* **7**, 97 (1842).
26. B. R. Weinberger *et al.*, *Appl. Phys. Lett.* **59**, 1132 (1991).

21

Superconducting Magnetic Energy Storage

Susan M. Schoenung[*] and Thomas P. Sheahen

In Chapter 4, we discussed two kinds of superconducting magnetic energy storage (SMES) units that have actually been used in real power systems. This chapter attends to the possible use of SMES in the future. For present purposes, the relevance of Chapter 4 is that SMES is *not* a futuristic concept; it is real, but needs to be scaled up.

In this chapter we explain certain design parameters for SMES using high-temperature superconductors. Although various specific designs can be imagined readily, the one chosen here[1] serves to illustrate the engineering and economic trade-offs that must be made.

21.1. ECONOMIC MOTIVATION

The desirability of electric energy storage is by now a given, and a number of recent studies[2] have examined the economics associated with various methods of storage. Some are conventional, such as charging and discharging lead-acid batteries; other methods are more innovative. In the storage method known as *pumped hydro*, electricity is generated at night and used to pump water uphill to a basin above a hydroelectric dam; later on, during peak demand hours, the water flows downward through turbines and generates electricity at the time it is needed. In all cases, the figure of merit by which competing methods of storage are evaluated is the *round-trip efficiency*, which means simply the ratio of power delivered upon exit to the power input at the start.

The round-trip efficiently is weighed along with both initial capital cost and annual operating costs to perform a cost/benefit analysis of any particular energy storage pathway. In the case of pumped hydro, for example, Virginia Electric Power has obtained[3] a round-trip efficiency over 80%, but they incurred capital costs in acquiring land and building dams and hydroelectric generators; and, of course, there are finite operating costs of their system. A lifecycle cost analysis incorporates some expected-use profile, and amortizes capital costs over the lifetime of the equipment, so as to arrive at a *net cost per kilowatt* figure. That can then be compared with cost estimates for other forms of storage, and with the option of having no storage at all. Such factors as the estimated future price of coal and natural gas enter into the calculation.

---

[*]W. J. Schafer Associates, Inc.

The options available to a utility are many. Although a blackout is to be avoided through astute advanced planning, gentle reductions in line voltage are not entirely out of the question. Clearly, however, it is better to actually meet the full demand. Doing so may or may not require electricity to be stored. One variation of the no-storage option is to buy power from other utilities to meet peak demand. Not everyone can do that.

A second variation is to get customers to agree to have their power interrupted or limited under circumstances of high demand: if homeowners are offered discounts of \$20/month on the electric bill, many of them will be willing to have the power company shut off their air conditioners in late afternoon, and the outcome at the utility will be to meet the (lower) peak demand without having to build new generating capacity. It is helpful to distinguish between supply-side and demand-side approaches to matching supply and demand.

In any case, storage of electricity has a place in the utility sector. SMES is attractive because it has a round-trip efficiency of over 90% under the right circumstances.

The operating principle of SMES is quite simple: it is a device for efficiently storing energy in the magnetic field associated with a circulating current. An invertor/convertor is used to transform AC power to direct current, which is used to charge a large solenoidal or toroidal magnet. Upon discharge, energy is withdrawn from the magnet and converted to AC power. Figure 21.1 is a schematic diagram of a SMES system. The components include a DC coil, a power conditioning system (PCS) required to convert between DC and AC, and a refrigeration system to hold the superconductor at low temperature. The inverter/converter accounts for about 2–3% energy loss in each direction.

There is a further economic advantage associated with larger SMES units. Denoting the magnetic induction by B, the energy stored in a magnetic field[4] is proportional to B^2. The dimensions of the SMES unit go up only linearly with B, and the refrigeration requirement is proportional to size. Therefore, larger SMES units have the economic advantage of less refrigeration need per stored megawatt.

The most important advantage held by SMES is that it can comply with demands of the form "I want my power NOW!" Alternative storage methods such as pumped hydro, compressed air, and so on, have a substantial delay time associated with the conversion of the power (stored as mechanical energy) back into electricity. Indeed, the limiting case of pumped hydro is to let nature do the pumping and wait for the winter snows to melt in the mountains and thus generate extra electricity in the springtime.

Such a system is in operation today. The Columbia River can produce enough electricity to be able to sell some to Los Angeles, and that power is carried over the *Pacific Intertie*, a 1400-mile high voltage transmission line running north-south through the Oregon and California deserts. During peak-demand periods, power flows southbound; but at night baseload plants near Los Angeles send power northbound.

A major regional electric supply system like this is particularly vulnerable to the unexpected. On January 17, 1994, when an earthquake struck Los Angeles at 4:31 a.m., the "source" was disconnected from the Pacific Intertie and there were power outages in the Pacific Northwest, as far east as Wyoming. A SMES could go a long way towards protecting such as a system. Even a modestly small SMES, which would provide only one or two seconds warning, would mitigate the interruptions associated with a sudden disconnect.

So far, most thinking about SMES for utilities[5] has seen it as a diurnal storage device, charged from baseload power at night and meeting peak loads during the day. Little analysis has gone into the dynamic aspects of electrical power or how SMES might enhance power quality. In reality, there is economic value in better power quality; and just as (today) a factory

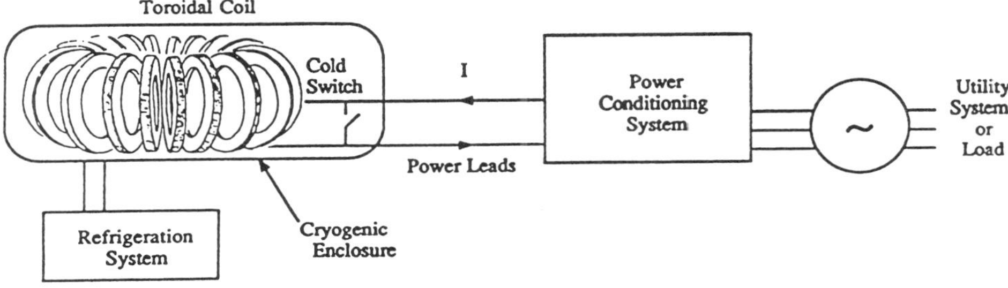

Figure 21.1. Schematic illustration of SMES system. (From Ref. 1, © 1992 IEEE.)

will pay a *demand charge* determined by its peak *amount* of power, in the future it may be feasible to sell extremely *reliable* power at a premium price as well.

21.2. BIG VS. SMALL SMES

There are already some small SMES units in operation, as described in Chapter 4. However, the greatest attention has gone to large-scale storage units, which are of truly massive size. Large-scale SMES devices, 5,000–10,000 MWh, could provide the most economical form of storage. A 5,000-MWh magnet would be in the form of a low-aspect-ratio solenoid, approximately 1,000 m in diameter and 20 m high. It would be buried in a trench in earth or rock to reduce costs, and would use about 1 million liters of liquid helium. The total cost for a 5,000-MWh SMES is about $1–2 billion. Such a device could exploit currently underused, base-load generating capacity to meet up to a 15% load-growth, or to permit early retirement of inefficient peaking and intermediate generating capacity.

The most costly part of a billion-dollar SMES is the mechanical structure required to contain the very large Lorentz forces generated by and on the magnet coils. These costs are the same for a liquid helium based or a liquid nitrogen based SMES, assuming the same mechanical properties of the superconductor and same B. In addition, the refrigeration system is not a likely source of major savings, no matter what operating temperature is selected for large SMES, because the fractional cost is so small. Savings in thermal shield, piping, and refrigerator capital costs totaling roughly 3% ($30 million) are plausible.[6]

On a more modest (but still large) scale, design studies[7] were funded by the Strategic Defense Initiative for a SMES with a capacity of approximately 20 MWh, capable of providing 400 MW of power for 100 sec or 10 MW of power for 2 hr. Each of two SMES designs featured a low aspect ratio solenoid approximately 100 m in diameter buried in earth. This concept, called the engineering test model (ETM), was originally designed for military purposes, but subsequently sought applicability to utility generating systems. The Bonneville Power Administration (BPA) commissioned a cost/benefit study by Pacific Northwest Laboratory[8] of the ETM used as a storage device.

Actually, SMES units could apply to three different categories of the electrical power network. Figure 21.2 portrays the sizes of SMES that would correspond to customer service applications, transmission and distribution, and to power generation. In addition to the ETM, there are a number of smaller SMES units being built in other countries, where interest is high in improving transmission and distribution networks. Several reviews of potential applications for SMES can be found in the literature.[9,10]

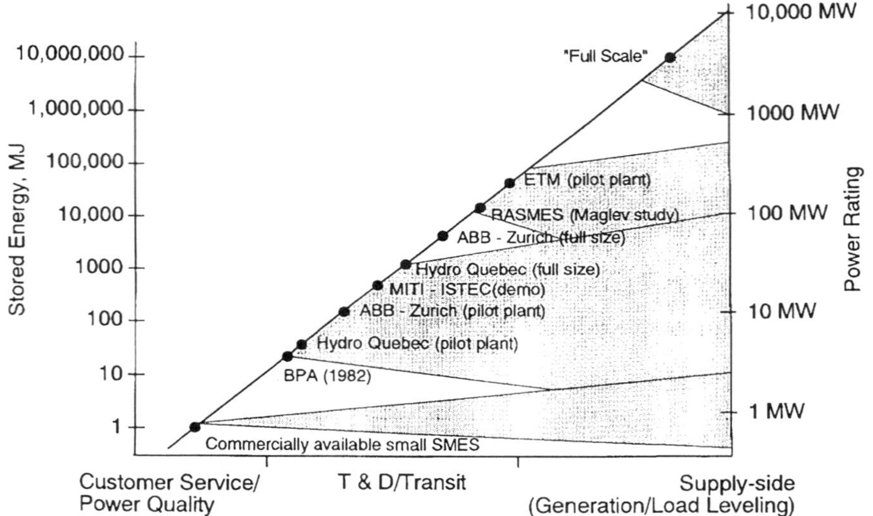

Figure 21.2. SMES units of different sizes, and their scope of applications.

At the low extreme of size is the concept of micro-SMES, referring to the energy storage range near 1 MJ. The device made by Superconductivity, Inc., and described[11] in Chapter 4, can be readily moved about and placed on lines serving single industrial customers with special needs. To justify the cost of a very small SMES, the foremost special need is for very high power quality. Although no small system using liquid helium (or even liquid nitrogen) can be justified at a basic electricity cost of pennies per kWh, the costs associated with inferior power quality[12] are far higher and justify a second look at superconducting technologies. The fact that such units are commercially for sale at around $1 million each justifies the assertion that power quality has great value.

21.3. HTSC SMES CALCULATIONS

The ETM studies were quite detailed, but were based on liquid helium technology and conventional superconductors. Soon after the discovery of HTSCs, their possible application to SMES was considered.[13–15] Subsequently, to consider how HTSC might impact the SMES design, the U.S. Department of Energy (DOE) sponsored a study by W. J. Schafer Associates on HTSC SMES designs.[16] The purpose was to advance beyond qualitative discussions of possible SMES systems to a quantitative description of potentially real configurations. Accordingly, the Schafer team established twelve base case configurations; corresponding to three energy levels (2, 20, and 200 MWh), two geometries (toroid or solenoid), and two kinds of conductors (BSCCO, YBCO).

A SMES system, designed to provide power for a relatively short period of time, is characterized by rapid response and high efficiency. A SMES unit has both a power rating and a storage capacity. In the study by Schoenung *et al.* the energy storage range was 2–200 MWh (7.2–720 GJ) at power levels from 4 to 400 MW. This range of capacities covers utility applications from transit and power stabilization at the small end to spinning reserve and

Table 21.1. Design Parameters Being Used
in Mid-Size SMES Study

Parameter	BSCCO	YBCO
J_c (A/cm^2)	20,000	1,000
I_c (A)	2,000	1,000
B (tesla)	≤ 20	1
f_{sc}	10%	10%
T_{op} (K)	20	77
Strain (%)	0.3	0.3

load leveling at the large end. Also, this range spans the gap between *demonstrated* SMES and *designs* that have been proposed for larger systems.

En route to making an investment decision about building a SMES, managers will make trade-offs among refrigerator costs (and reliability), structural strength, round-trip efficiency, and so on—all of these enter into the cost/benefit calculation. However, at the conceptual design stage, these factors are deferred in favor of technical analysis of the several major components. We follow that pattern here.

21.3.1. HTSC Conductor

In order to limit the analysis to fairly realistic cases, SMES design features were evaluated based on HTSC properties over a range of values. A set of numerical values representative of bismuth-based superconductor (BSCCO) were chosen as the basis for computational modeling. The parameters used in the calculations are listed in Table 21.1, and are drawn from values reported in the literature. SMES configurations using a YBCO conductor were also considered, but because of low J_c and limited operation in an external magnetic field, the results were discouraging for the capacity range considered.

The baseline HTSC parameters were the following: critical current $J_c = 20,000$ A/cm^2 at $T = 20$ K and magnetic field B up to 20 T. As discussed in Chapter 16, these properties have each been demonstrated one at a time,[17–19] and their simultaneous occurrence in a long wire is a near-term target. (The variation of J_c with **B** is very important in a SMES.)

The superconductor was assumed to make up 50% of the conductor volume. In the calculations, the remainder was taken to be copper, even though most samples today are made with silver. The superconductor also needs to be in the form of twisted filaments. It is important to remember that HTSC "wire", as the word is commonly used, is not yet available, but rather is the goal of R&D efforts. Other issues regarding HTSC conductor design for SMES are presented by Stephens.[20]

21.3.2. Mechanical Aspects

Once the properties of the wire are settled, it is possible to move forward to analyze various configurations of the coil itself. For HTSCs, the inferior strain tolerance is a crucial parameter that cannot be ignored or circumvented. Specifically, ceramics cannot carry much tensile load. This condition, along with the effect of thermal contraction upon cooling and the effect of Lorentz forces in a charged coil, seriously constrains the design. The mechanical-engineering aspects of a SMES are as important as the superconductor itself.

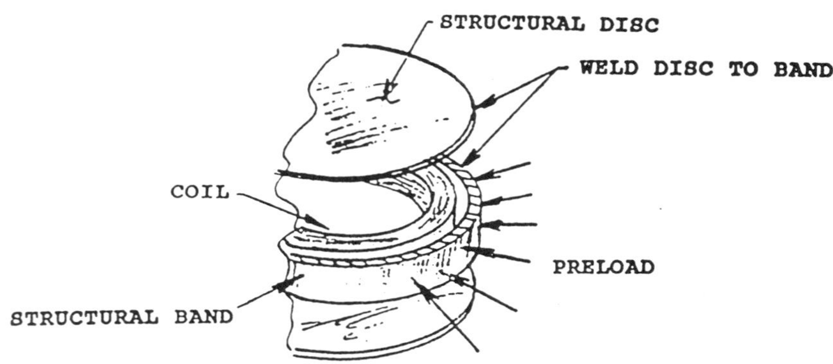

Figure 21.3.   Illustration of SMES module. (From Ref. 1, © 1992 IEEE.)

The strain tolerance is an important input parameter, not because of any electrical effect, but because it determines how much structural material is needed to keep the entire SMES from breaking. In a study[21] of small SMES, the "optimistic" value of 0.3% strain tolerance was selected. For the range 2–200 MWh considered here, a strain tolerance of 0.1% in compression was used as appropriate for HTSC.

The toroidal configuration has low external magnetic field, which is advantageous in locating a small unit near a utility or customer load. The modular approach, illustrated in Figure 21.3, was selected primarily as a way to wind the coil and keep the brittle HTSC under compression at all operating conditions to avoid breakage. The outer hoops provide preload while the disks are an efficient way to carry magnetically-induced Lorentz loads. In fact, both hoops and disks carry a combination of both loads. The modular approach also allows factory fabrication and cost reductions as a result of a learning curve. The reader can find additional assessments of toroidal SMES configurations in Refs. 22, 23, and 24.

Here we focus on the particular case of a cold-supported, modular *torus*. A similar modular *solenoid* was analyzed for comparison. For both configurations, a maximum module diameter of 10 m was assumed. In some toroidal cases, a smaller diameter was optimum. For the coil structure,[16] the toroidal configuration was generally preferred to the solenoidal one.

A cold-supported solenoid was used for the Bonneville SMES and is used in the coils produced by Superconductivity Inc., both described in Chapter 4. However, these coils do not require substantial pre-compression. Larger capacity coils, such as the ETM, do need pre-compression and are designed for warm earth support, because the cost of cold support in larger systems is prohibitive.[25]

In order to enhance comparison with conventional LTSC technology, calculations were also performed using properties of the most common metallic superconductor, NbTi, at 4.2 K. In contrast to BSCCO, J_c in NbTi is a strong function of magnetic field, decreasing with increasing field. A baseline performance of $J_c = 200,000$ A/cm^2 at 5 T was used in the analysis. A tensile strain limit of 0.2% was assumed. The NbTi system has the advantage that precompression of the superconductor is not necessary, resulting in simpler structure design and less mass of structural material.

21.4. UNIQUE FEATURES OF HTSC SMES

The fact that BSCCO at 20 K retains a high J_c even up to 20 T allows us to design a HTSC SMES to operate at much higher magnetic fields than similar LTSC devices. This has the surprising consequence that size (and therefore cost) drops sharply with increasing magnetic field.

Figure 21.4 illustrates this dramatically. For the case of a 20 MWh HTSC toroidal coil, four different parameters all plummet as the magnetic field rises: Q_{sc}, the superconductor required in ampere-meters (A-m); M_{cs}, the mass of the cold support structure in Kg; D_{max}, the outer diameter of the coil in meters; and the refrigerator power requirement in electric watts (W_e). Because of the dramatic reduction in these parameters up to 10 T, a baseline of 10 T for the average field at full charge was chosen for subsequent design considerations. Also selected was a maximum module size of 10 m.

This design included tensile and bending stresses in the support components during preload and operating modes, with the result that the mass of the support structure, M_{cs}, decreases with increasing B in Figure 21.4(a). That would not normally occur in a simple magnet, so this calculation is a caution flag, demanding subsequent detailed analysis. In Figure 21.4(b), the refrigeration power requirement follows the decreasing size curve because most thermal load components (conduction, radiation, and AC losses) decrease with decreasing size and mass of the magnet.

A size comparison with comparable LTSC devices operating at 5 T is shown in Figure 21.5, where module diameter of 10 m was also assumed. The torus maximum diameter D is always smaller for the HTSC magnet because of higher field operation.

Separately, LTSC and HTSC units having solenoidal coils were compared. The solenoid height or length was smaller for HTSC coils, but still much greater than in a toroidal geometry. The modular solenoidal configurations for large (200 MWh) storage capacity may be impractical, to say the least: the calculated height exceeds 100 m. Rather than build a tower 100 m high, such a SMES would have to lie on its side, looking like a gigantic sewer pipe. Earth-supported systems, such as those proposed for the ETM, probably make more sense for stored energy above 20 MWh, if a way can be found to maintain the HTSC in compression, or if the strain tolerance improves to allow operation in tension.

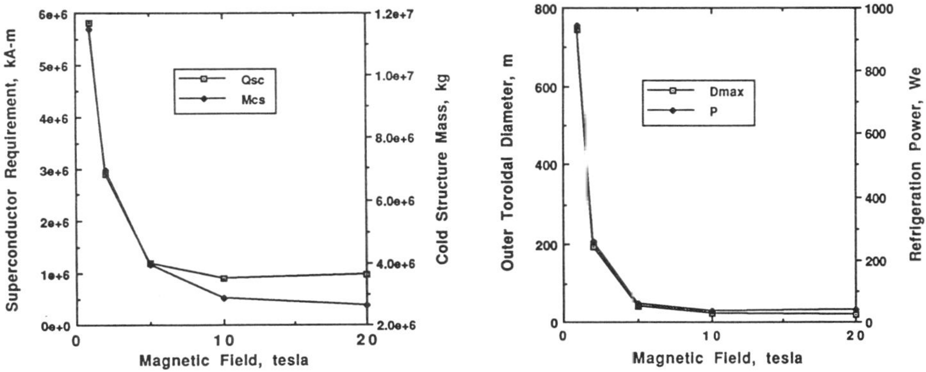

Figure 21.4. (a) Superconductor and cold structure requirements for 20 MWh HTSC torus. (b) Size and refrigeration requirements for 20 MWh HTSC torus. (From Ref. 1, © 1992 IEEE.)

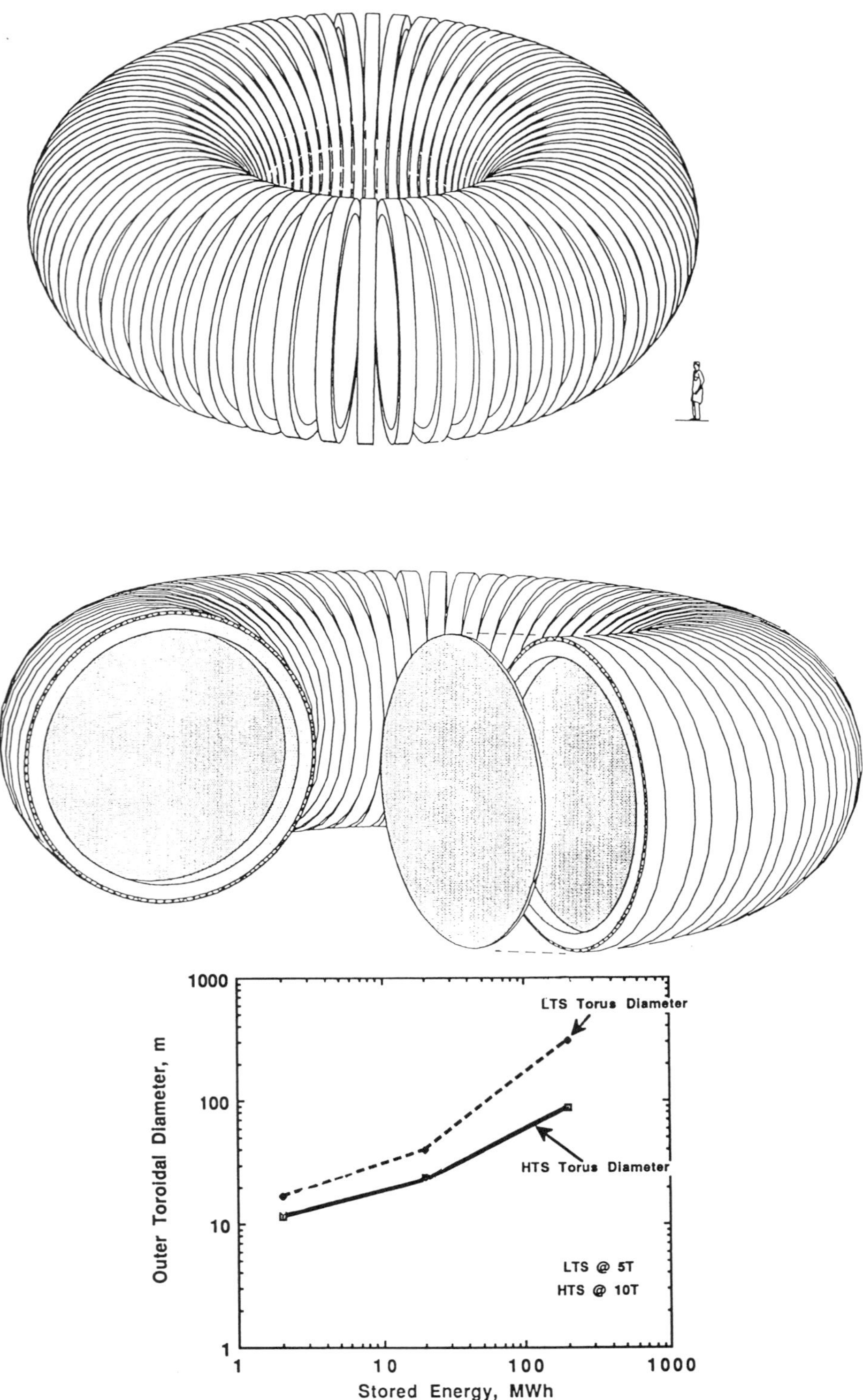

Figure 21.5. Major dimensions of SMES coils. Maximum module diameter < 10 m.

Calculations about specific designs showed that the HTSC must be constrained in compression. In fact, this eliminates from consideration several designs that might otherwise have attractive features. It is necessary to start off with compression when the unit is uncharged, because the Lorentz forces will oppose that compression as the current is increased. It may well be that something like 4, 6, or 8 T is the practical upper limit of magnetic field, even though materials like BSCCO can operate in 20 T fields at low temperatures.

21.5. REFRIGERATION SYSTEM AND ENERGY EFFICIENCY

Under steady-state conditions and in the superconducting state, the coil resistance is negligible. However, the refrigerator requires electric power, and this refrigeration energy must be considered when evaluating the efficiency of SMES as an energy storage device. As described in Chapter 14, although BSCCO has $T_c > 100$ K, flux lattice melting takes place in moderate magnetic fields around 30 K, so a 20 K operating temperature with gaseous helium as coolant was selected here.

The heat loads that must be removed by the cooling system include conduction through the support system, radiation from warmer to colder surfaces, AC losses in the conductor (during charge and discharge), and losses from the cold-to-warm power leads that connect the cold coil to the power conditioning system (PCS) at ambient. Conduction and radiation losses are minimized by proper design of the thermal interfaces. Lead losses can be minimized by good design of the leads, but substantial losses will always occur in normal copper leads when current is flowing. Here, a conductor current of 10,000 A was assumed. Since the lead loss is proportional to the current, it can be reduced by using a lower-current conductor; however, a relatively high-current conductor is desirable to achieve high power discharges with a minimum voltage across the coil, as well as to minimize length of conductor and hence fabrication costs.

AC losses depend on the design of the conductor, the duty cycle of the device, and the power rating. For utility applications, it is plausible to imagine charge and discharge periods of a half-hour and two complete cycles per day. Faster rates or more frequent cycles add proportionally to the AC loss component.

Figure 21.6 indicates the refrigeration requirements (defined as electrical power to operate the refrigeration system) for HTSC and LTSC toroidal coils for the assumed baseline temperatures of 20 K and 4.2 K, respectively. The most significant feature of Figure 21.6 is that as the stored energy increases by a factor of 100, refrigeration cost only goes up by about 20. Also shown for comparison is the refrigeration requirement for a HTSC system operating at 77 K. Although this latter case is in the future, the results show the advantage that would accrue by operating at 77 K. The savings in refrigeration for an HTSC system over an LTSC system range from 60% to 90% for the cases analyzed. Although these savings are significant, they must be judged relative to the overall efficiency and cost of the device.

Figure 21.7 is a new way to plot power consumption data. It shows the daily refrigeration requirement for the assumed operating scenarios as a fraction f of stored energy. One way to define storage efficiency for the system is $\eta = 1 - f$. (The conversion efficiency of the total system must take into account the efficiency of the PCS.) Figure 21.7 shows that improvements in efficiency are not as important for a large plant (where overall efficiencies are good in any case) as they are for a small plant.

One likely near-term application of HTSC in SMES plants is the substitution of HTSC power leads into an LTSC unit. As was discussed in Chapter 20, the electrical connectors

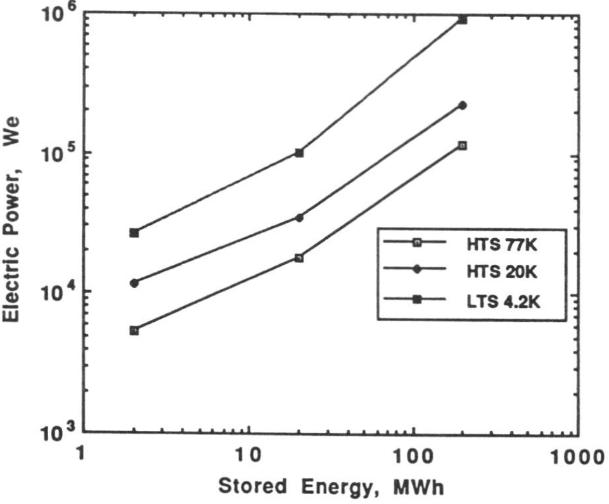

Figure 21.6. Refrigeration power requirement for toroidal coils. (From Ref. 1, © 1992 IEEE.)

between the 77 K plateau and the 4 K bath are normally made of copper, which conducts heat readily. However, they could be made of YBCO, which has very low thermal conductivity but is superconducting over that entire temperature range. The thermal load due to the power leads has been calculated for several different-sized SMES units. If the power leads in either a 4 K or a 20 K system were replaced by HTSC material up to the 77 K plateau, 2/3 of this heat load would vanish. Figure 21.8 displays the fraction of the refrigeration associated with the power leads. For the 2 MWh system, the refrigeration load would be reduced by 25% by utilizing YBCO leads. For units with larger stored energy, the percentage savings

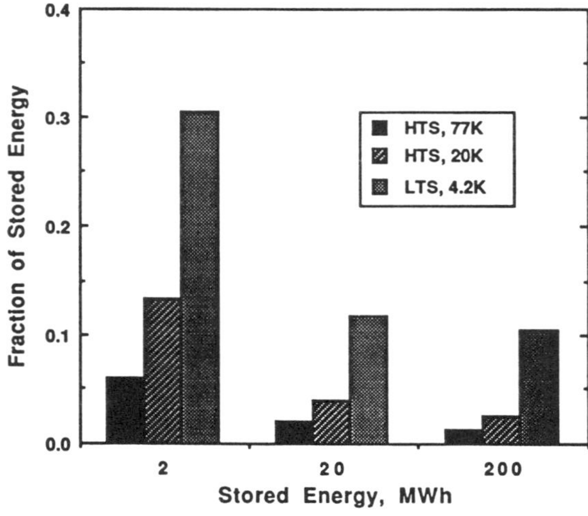

Figure 21.7. Fraction of stored energy required daily for refrigeration. (From Ref. 1, © 1992 IEEE.)

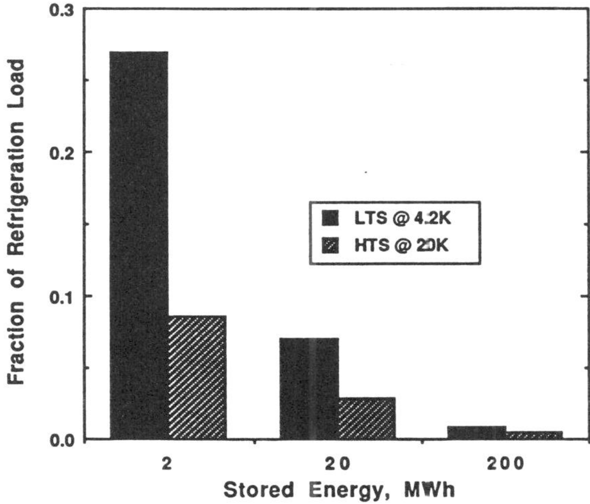

Figure 21.8. Fraction of refrigeration load required for power leads. (From Ref. 1, © 1992 IEEE.)

are less significant, whereas more would be expected at smaller sizes. Advances in HTSC power leads may make this adaptation possible for SMES in the very near future.[26]

21.6. COST OF MAJOR COMPONENTS

There are three major components of a SMES whose costs have to be estimated: conductor (consisting of superconductor and copper stabilizer); cold support (consisting of steel hoops and disks); and refrigerator (at the assumed operating temperatures). These estimates have been made for various SMES units. Given the relative contribution of these components, this allows comparison between the HTSC and conventional LTSC system. The estimates do not include thermal support, vacuum vessel, piping, insulation, cryogens, or any auxiliary systems (such as vacuum, instrumentation, or controls). Although important, these have been shown in other studies[13,27] to be a relatively small part of a large coil cost. The estimates also do not include construction, site preparation, engineering, or the PCS. While these items together can easily cost more than the entire coil (depending on coil and PCS capacity) their contribution to cost should not be significantly different for either the HTSC or LTSC system.

The costs of HTSC materials themselves are presently extremely high, but this is to be expected for a research material. Therefore, to make a comparison applicable to future plausible circumstances, this study assumed that finished HTSC wire costs were equivalent in $/kg to finished NbTi/Cu wire ($100/kg). That may not be a good assumption, and the cost figures affect the calculated outcomes, but any other choice would have been equally weak. Other cost assumptions are indicated in Table 21.2. Refrigerator costs are actually a function of size,[28] but we use average value over the size range. Lower strength, lower cost steel was assumed for the HTSC cases because the module does not support a significant tensile load. Higher strength, higher cost steel was assumed for the LTSC cases where the module strength operates in greater tension.

Table 21.2. Cost Assumptions for SMES Estimates

Component	Unit Cost
HTSC conductor wire	100 $/kg
NbTi conductor wire	100 $/kg
Copper stabilizer	2.2 $/kg
Steel module for HTSC coil (50 ksi)	5 $/kg
Steel module for LTSC coil (60 ksi)	6 $/kg
Refrigerator	
4.2 K	5.0 $/$W_c$
20 K	3.0 $/$W_c$
77 K	2.5 $/$W_c$

The combined costs of conductor, structure, and refrigerator for toroidal coils are shown in Figure 21.9. The same trend is true for solenoidal coils (not shown). For the assumptions made, at any size, HTSC coils are predicted to cost more than LTSC coils by a factor of 2 to 4. This is a disappointing outcome, because one would initially think HTSC coils would be cheaper due to lesser refrigeration requirements.

Why is the HTSC system more expensive? The principal reason lies in the comparative current densities of LTSC and HTSC materials. Although the wire costs the same by weight (by assumption here), if HTSC wire has 1/10 the J_c of LTSC wire, it takes 10 times as much wire to create the same inductance. A breakdown by components is shown in Figure 21.10, which is a logarithmic scale of money! As indicated, the conductor cost dominates the three costs for all HTSC cases and is particularly important at small sizes. This is a direct result of the lower J_c in the HTSC cases. Note that as the SMES size goes up from 2 to 20 to 200 MWh, the LTSC conductor cost also goes up about a factor of 10 at each step. The HTSC conductor cost rises a little slower, but is still by far the costliest item on the chart at each step.

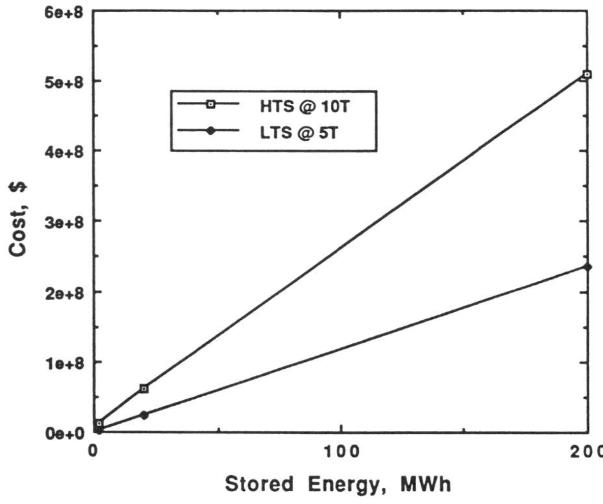

Figure 21.9. Sum of conductor, cold support structure, and refrigerator costs for toroidal coils. (From Ref. 1, © 1992 IEEE.)

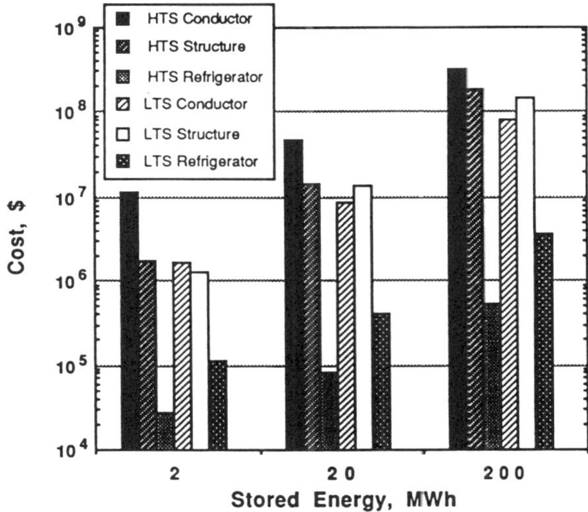

Figure 21.10.   Major component costs for toroidal coils. (From Ref. 1, © 1992 IEEE.)

The structure costs of either LTSC or HTSC SMES go up uniformly (a factor of 10 with each step from 2 to 20 to 200; but even there, the HTSC structure cost is higher, because the strain tolerance of the HTSCs is less than NbTi, which demands more structural materials. Even in the largest cases, the higher conductor cost is not offset by the reduced coil size at higher magnetic field.

It is particularly noteworthy that the refrigerator cost in all cases is so small that there is very little percentage savings associated with reduced refrigeration demands at high temperature. This means that if BSCCO works better at low temperatures, it will certainly be operated there. The advantage that BSCCO offers (high magnetic field) easily outweighs considerations of refrigeration costs. For the case of *very small* SMES, one study[29] indicates that the reduced refrigeration costs of HTSCs will have a more significant positive impact.

21.7. FUTURE OUTLOOK

In order for HTSC SMES to better its competitive standing against LTSC SMES, the following developments are needed:

1. Improve J_c or reduce the cost of HTSC material. Despite other savings, the cost of HTSC materials must be comparable (within a factor of 2) to a conventional superconductor in \$/A-m if HTSC-SMES is to compete on a capital cost basis. Today's HTSC materials are many times more expensive.
2. Design for higher field operation, such as that which may be possible with BSCCO. (However, for small energy systems, some high-field configurations are not possible unless J_c is improved.)
3. Increase the strain range over which the HTSC material can operate so as to decrease the mass of cold structure.

21.8. SUMMARY

This chapter treats the subject of superconducting magnetic energy storage (SMES) from the viewpoint of using HTSCs as the conductor instead of the conventional LTSCs. There are several important differences in design that arise from considering HTSCs. The *a priori* assumption is that with HTSCs allowing the use of LN$_2$ at 77 K, savings would accrue from higher refrigeration efficiencies, greater reliability, and easier acceptance within the utility community. Although all of these things should be true, the significance of the improvements depends on the application and other characteristics of the system.

Conventional SMES designs use metallic superconductors. By comparison, today's HTSC materials have the negative features of lower J_c and greater brittleness than NbTi and Nb$_3$Sn. One must ask just exactly what HTSCs can contribute. The refrigeration cost savings may not suffice to overcome the cost penalties due to structural requirements. Furthermore, significant uncertainties surround AC losses: As mentioned in Chapter 18, initial measurements on both YBCO and BSCCO suggest that J_c must become much greater if AC losses are to be acceptably small.

The variety of applications of SMES depends on the size. Large systems (hundreds to thousands of MWh storage capacity) would be ideal for load leveling; mid-range systems (1–10 MWh can be used for carryover, transient applications such as delivering several hundred megawatts for a few minutes; small systems (1–100 kWh) would be used mostly for stability applications, delivering a few megawatts for several seconds to a minute.

The outlook for very small SMES (using either LTSC or HTSC) is somewhat unique, because the economic motivation is entirely different: power quality considerations drive it. This field is certainly in its infancy. There is not yet any sizable body of experimental data to show how effective a very small SMES will be in regulating power over millisecond time intervals. Utilities are beginning to talk about selling *quality charge* in the sense equivalent to a *demand charge*.

A conceptual design of a modular SMES device based on representative properties of high-temperature superconductors has been analyzed for design and performance features, and major component costs have been estimated. More compact configurations are possible for HTSC SMES if operated at a magnetic field greater than that possible for LTSCs. This leads to appreciable savings. Meanwhile, refrigeration power and energy requirements are reduced at higher operating temperatures. However, refrigeration is a small percentage of the total cost of a SMES, and therefore operating temperature is not an important criterion where SMES is concerned. BSCCO at 20 K is entirely acceptable because of its high magnetic field capability.

Because coil costs are dominated by the cost of superconducting wire, an HTSC SMES unit will be of comparable cost to an LTSC system only if the superconductor material is of comparable cost on the basis of \$/A-m. Superconductor development efforts should focus both on increasing J_c and strain range and on reducing wire manufacturing cost.

REFERENCES AND NOTES

1. Most of this chapter was first presented at the IEEE Power Engineering Society. That paper was co-authored by S. M. Schoenung and W. R. Meier of Schafer Associates, and R. L. Fagaly, M. Heiberger, R. B. Stephens, J. A. Leuer, and R. A. Guzman, all of General Atomics.
2. See, for example, A. M. Wolsky et al., *Advances in Applied Superconductivity: A Preliminary Evaluation of Goals and Impacts*, Argonne National Laboratory Report ANL/CNSV-64 (January 1988).

3. J. Loncoski, Virginia Electric Power Corp., private communication.

4. M. N. Wilson, *Superconducting Magnets* (Oxford Univ. Press, Oxford: 1983).

5. *Superconducting Magnetic Energy Storage,* EPRI Report #EM-3457 (April 1984).

6. W. V. Hassenzahl, "Superconducting Magnetic Energy Storage," *IEEE Trans. Mag.* **MAG-25**(2), 750 (1989).

7. S. M. Schoenung, W. V. Hassenzahl, and P. G. Filios, "U.S. Program to Develop Superconducting Magnetic Energy Storage," *Proceedings 23rd Intersociety Energy Conv. Engineering Conference,* vol. 2 (ASME, New York: 1988).

8. J. G. De Steese, J. E. Dagle, and D. K. Kreid, *Benefit/Cost Comparisons of SMES in System-Specific Application Scenarios,* Presented at the 3rd World Congress on Superconductivity, Munich, Sept. 14–18, 1992.

9. S. M. Schoenung, R. C. Ender, and T. E. Walsh, "Utility Benefits of Superconducting Magnetic Energy Storage," *Proceedings of American Power Conference* **51**, 473 (1989).

10. S. M. Schoenung, *Survey of U.S. SMES Development Programs,* Presented at the 3rd World Congress on Superconductivity, Munich, Sept. 14–18, 1992.

11. C. C. DeWinkel and P. F. Koeppe, "Superconducting Technology Offers Ride-Through Capability for Large Industrial Critical Process Loads," in *Proc. American Power Conference* vol. 2, p. 1252 (Illinois Institute of Technology, Chicago: 1992).

12. J. D. Lamoree, *Analysis of Power Quality Concerns (at Industrial Sites),* Report to Central Hudson Gas & Electric, by Electrotek Concepts, Inc. (August 1992).

13. J. T. Eriksson and J. Kopijarvi, "Economic Potential of Applying High-Temperature Superconductors to Magnetic Energy Storage," *IEEE Trans. Mag.,* **MAG-25**, 1807 (1989).

14. Y. M. Eyssa *et al.,* "The Potential Impact of Developing High T_c Superconductors on Superconductive Magnetic Energy Storage," *Adv. Cryogenic Eng.* **33**, 69 (Plenum Press, New York: 1988).

15. T. Yoshihara *et al.,* "Design Study of SMES System Using High-Temperature Superconductors," *Proceedings 10th Magnetic Technology Conference,* Boston (1987).

16. S. M. Schoenung *et al.,* "Conceptual Design of a SMES Using High-Temperature Superconductors," DOE Report CE34019-1 (1992).

17. H. Sato *et al., Appl. Phys. Lett.* **57**, 1928 (1990).

18. D. P. Hampshire *et al.,* "Temperature-Dependent Critical Current Density of Bi(Pb)-Sr-Ca-Cu-O Tapes in Fields up to 20 Tesla," *Supercond. Sci. Tech.* **3**, 560 (1990).

19. J. Tenbrink *et al.,* "Development of High T_c Superconductor Wires for Magnet Applications," *IEEE Trans. Mag.* **MAG-27**(2), 1239 (1991).

20. R. B. Stephens, "High T_c Wire Design and Power Losses in Cycling SMES Devices," IEEE *Trans. Appl. Superconductivity* **2**, 122 (1993).

21. J. R. Hull *et al.,* "Design and Fabrication Issues for Small-Scale SMES," *Adv. Cryogenic Eng.* **37A**, 369 (1992).

22. Y. M. Eyssa et al., "Design Considerations for High Temperature (High-T_c) Superconducting Magnetic Energy Storage (SMES) Systems," in *Adv. Cryogenic Eng.* **37A**, 387 (1992).

23. J. S. Herring, "Parametric Design Studies of Toroidal Magnetic Energy Storage Units," *Proceedings 25th IECEC* **3**, 409 (1990).

24. R. W. Moses, "Configurational Design of Superconductive Energy Storage Magnets," *Adv. Cryogenic Eng.* **21**, (1976).

25. J. R. Powell and P. Bezler, "A Comparison of Warm- and Cold-Reinforcement Magnet Systems for Tokomak Fusion Reactors," *Proceedings Technology of Controlled Thermonuclear Fusion Experiments and the Engineering Aspects of Fusion Reactors,* pp. 358–383, Austin, Texas, (1973).

26. J. R. Hull, "High-Temperature Superconducting Current Leads for Cryogenic Apparatus," *Cryogenics* **29**, 1116 (1989).

27. S. M. Schoenung, W. R. Meier, and W. V. Hassenzahl, "A Comparison of Large-Scale Toroidal and Solenoidal SMES Systems," *IEEE Trans. Mag.* **MAG-27**, 2324 (1991).

28. M. A. Green and R. Byrns, "Estimating the Cost of Superconducting Magnets and the Refrigerators Needed to Keep Them Cold," in *Adv. Cryogenic Eng.* **37A**, 637 (1992).

29. S. M. Schoenung *et al.,* "Cost Savings and Prospects for Applications of Micro-SMES Using HTSCs," *IEEE Trans. Appl. Superconductivity* **3**, 200 (1993).

22

Electric Motors

*Howard E. Jordan, Rich F. Schiferl[*] and
Thomas P. Sheahen*

Because the greatest single use of electricity is in electric motors, it is of great interest to utilize superconductors so as to capture the highest possible efficiency in an electric motor design. Electric motor-driven applications account for over 50% of the electricity used in the United States. To place this in perspective, U.S. consumers spend $90 billion annually on electricity converted to shaft power by motors, and over $7 billion on new electric motors. Efficient electric motor systems have the potential to reduce industrial electricity consumption by over 240 billion kWh annually by the year 2010. In this chapter, we describe the progress being made toward building an electric motor that runs at 77 K and uses high-temperature superconducting (HTSC) wire.

Electric motors are rather efficient to start with,[1] so a superconducting motor has to do even better in order to offset the cost of refrigeration. There is competition between steadily improving conventional motor designs and new technologies (such as superconductivity).

22.1. CONVENTIONAL MOTORS

Superconducting motors are best understood by comparing them to conventional motor design. A motor and a generator operate on the same principles, although in their practical embodiments they are quite different. A motor converts electricity into the power of a rotating shaft, which in turn can do work. A generator starts with a rotating shaft and produces electricity. Several good textbooks start from Maxwell's equations and establish those principles,[1–4] which will not be repeated here.

In conventional motors, magnetic steel is used to increase the magnetic field produced by the motor coils. This is termed an *iron core* machine However, because iron saturates magnetically at 2.2 T, the maximum field strength in a conventional device is about 2 T.

In general, the power output P_{out} of any rotating machine can be expressed as[5]:

$$P_{out} = C_0 \times V \times B_{gmax} \times J_{bmax}$$

where:

[*]Reliance Electric Company, Cleveland Ohio.

449

$C_0 =$ proportionality constant
$V =$ motor active volume
$B_{gmax} =$ peak air gap radial magnetic flux density
$J_{bmax} =$ peak linear current density at the air gap

If the air gap magnetic field (B_{gmax}) can be increased, the motor size will decrease for the same motor power rating. In conventional motors B_{gmax} is limited by iron core saturation, core loss, and the ability to create a magnetic field with lossy windings. J_{bmax} is limited by the ability to cool the motor coils and by the space available for the current carrying conductors.

22.2. SUPERCONDUCTING MOTORS

The main advantage of using superconductors in electric motors is that they can create an air gap magnetic field without any losses. This advantage must be weighed against the added cost and complexity of having to cool the superconducting windings to cryogenic temperatures. Assuming the superconducting windings are held at 77 K, somewhat below T_c, it is important to remember that the critical current J_c falls off steeply with magnetic field B for any of the HTSC conductors in this temperature range, as described in Chapter 14.

In motors, superconductors are used only in DC windings so as to minimize the necessary cooling costs. Under these conditions only the heat leak from outside the winding cryostat must be compensated by the refrigeration system, because the winding itself (once cooled below its transition temperature) is lossless. Taking all of these attributes of the superconductor into account, an HTSC motor[5] will have the following features:

- DC superconducting windings to produce a large B_{gmax}.
- Air core construction to eliminate the problems of iron core saturation and core loss at the high B_{gmax} levels.
- Normal (copper) AC windings to provide J_{bmax} similar to that of a conventional motor.

The performance advantages of an HTSC motor over that of a conventional motor include the following:

- Higher power density than a conventional motor due to the large B_{gmax} produced by the lossless HTSC winding.
- Higher efficiency than an conventional motor due to the lossless superconducting winding and smaller motor size.

Superconductors can be utilized in any motor type that results in steady-state operation with at least one coil carrying only DC current. A synchronous motor with a superconducting field winding (carrying DC current), and a normal conducting armature winding (carrying AC currents), both fit this description.

There are other machine architectures that were discarded in the past which have been revisited. Lipo[6] described six different architectures that might be used:

Homopolar DC
Synchronous AC
Induction
Induction/synchronous hybrid
Reluctance
Homopolar inductor

The first two machine types, homopolar DC and synchronous AC, have been shown[6] (using low-temperature superconducting materials) to be viable design concepts for the application of superconductivity. Each of the other types was considered at least qualitatively. The conclusion was reached that the homopolar DC and synchronous AC are the best choices. With the design somewhat constrained in this way, it is appropriate to further delimit the design by seeking to optimize the efficiency of the motor.

22.3. EFFICIENCY

In order to discuss what kind of efficiency one expects from an electric motor, it is first necessary to decide what size motor is to be used. The emphasis here is on applying superconducting materials to large motors used in central power generating plants and large-horsepower industrial applications. The selection of large motors was based on two principles: One is that a large motor should be better able to absorb the overhead costs associated with a liquid nitrogen cooling system. The second is that the anticipated space savings and efficiency improvements (which result from machine designs utilizing super-conducting material) will have a significant impact on operating economics. The large motors currently being used in many of these applications take up valuable floor space and are operating a high percentage of the time under load, so motor efficiency improvements are quickly converted to dollar savings.

Typical applications of these motors are for pump and fan drives. Large pump and fan drives are increasingly being served by adjustable speed motor drives, due to the increased

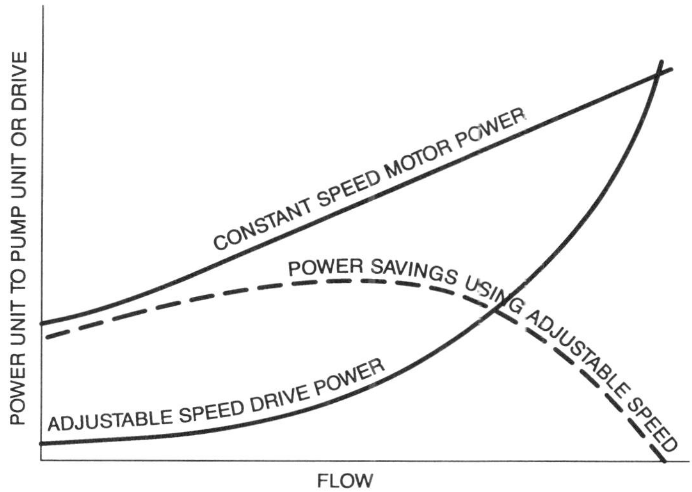

Figure 22.1. Power savings with adjustable-speed pump drive.

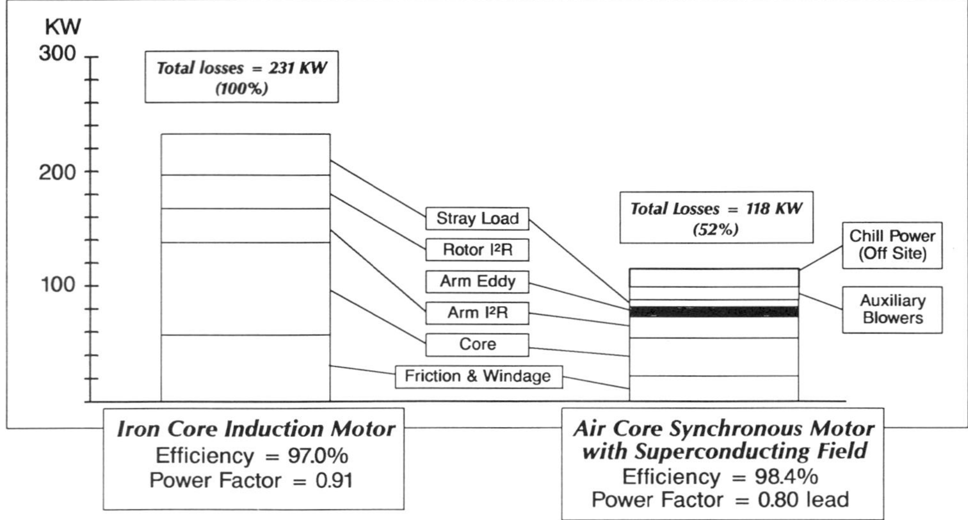

Figure 22.2. Loss comparison of air core (HTSC) and iron-core (conventional) 10,000 hp, 3600 rpm, 60 Hz motors for full-load, rated speed conditions.

system efficiency (compared to throttling) when used for flow control. Figure 22.1 illustrates the power savings that can be achieved by utilizing an adjustable speed pump drive.[1,7] For pump and fan applications the superconducting motor should be designed for adjustable speed use, because this technology is expected to be commonplace when these motors become commercially available. Consequently, the HTSC motor will be started and controlled by an adjustable frequency drive. This means that across-the-line starting[†] is not required.

A conventional large-horsepower motor typically has an efficiency of 97%. Nevertheless, for one specific superconducting motor design (described more fully below in Section 22.5), the efficiency increase vs. a conventional motor is shown in Figure 22.2. The losses in the superconducting motor are only 52% of those in the conventional motor for this 10,000 hp motor design.

The motor designs compared here are rated in thousands of horsepower and are large enough that the cryogenic support system is a fraction of the total motor cost. Another major advantage of these motors is that typically they operate continuously (or nearly so). Loss reductions of greater than 50% can be realized[7,8] by an HTSC motor compared with a large conventional induction motor. Over the operating life of the motor, the resulting cost savings can equal one to two times the initial HTSC motor cost.[8]

The superconducting winding is only one part of this motor. How, then, can it be that the superconducting motor has only 52% of the loss of the equivalent iron core motor? A large part of the loss reduction occurs due to the small motor size. This is an extremely important point. The estimated size of the 10,000 hp superconducting motor is 54% of the volume of the equivalent induction motor. Friction and windage loss, as well as armature I^2R loss, are reduced simply because the motor is smaller. The only increase in loss is the additional armature-winding eddy current loss in the superconducting motor due to the

---

[†]This means that full voltage and full frequency are applied to a motor at rest. This results in high in-rush currents.

increased air gap magnetic field. This is the solid black layer on the right of Figure 22.2. All the other loss contributions decrease in going to the superconducting design.

This comparison is made at full-load conditions. For continuous running full-load applications, the reduced loss of the 10,000 hp superconducting motor represents approximately 1 million kWhr per year saved compared to the conventional iron core motor.

22.4. MOTOR DESIGN PRINCIPLES

The idea of a superconducting electric motor is not new; as soon as superconductors were discovered, it became an obvious goal. However, not until NbTi multifilamentary wire became available could an actual implementation be considered. (Previous superconductors didn't carry enough current to be interesting.) The first superconducting generator designs in the 1970s (using LTSCs, of course) provided guidance for contemporary HTSC motor designs.

At the outset, it is essential to recognize that designing a superconducting motor demands much more than a trivial substitution for copper wire. In conventional motors, a lot of iron and copper are used, and the iron saturates. In a superconducting motor, iron would be eliminated in favor of an air core design, with fields well above the iron saturation limit. This calls for rethinking the entire design.

22.4.1. Initial Design Concepts

In Section 22.2 above, we stated that the homopolar DC and synchronous AC motors are the best candidates. This statement deserves some explaining. Both the homopolar DC and synchronous AC machine types have been constructed using LTSC materials, and the design concepts have been verified by test.[6]

The homopolar DC machine is attractive for using HTSC materials because the coil wound with HTSC materials is stationary, and liquid nitrogen cooling of a stationary coil is a comparatively easy task. This advantage is offset by the low-voltage, high-current power supply requirements imposed by this machine type. High currents must be supplied to the rotating member of the machine, which poses significant sliding current collector problems. Further, the low-voltage, high-current characteristics require an expensive power converter and increase the on-site power cable installation costs. For example, the Fawley homopolar DC machine[9] was rated at 430 V and 5800 A for 3000 hp. The comparable rating for a 3000 hp synchronous motor is three-phase, 4000 V and 350 A, which is much more manageable from existing power distribution systems.

The synchronous AC machine also is well suited to the use of HTSC materials. Advantages of this construction include: (1) easily adapted to air core design; (2) armature is a copper winding designed for a voltage rating appropriate to the horsepower; and (3) increased space available for the armature-winding (due to absence of stator iron teeth) reduces the primary I^2R loss and increases the power density (for a given armature-winding current density). The disadvantages are the difficulties associated with (1) achieving a liquid nitrogen cryostat construction on the rotating member, and (2) designing to withstand the mechanical forces that the HTSC coil experiences.

On balance, the synchronous AC motor concept seemed to have the edge, and therefore further conceptual design followed this path. This does not represent an outright rejection of the homopolar DC motor, but rather exhibits the need to commit to a single design in order to make progress.

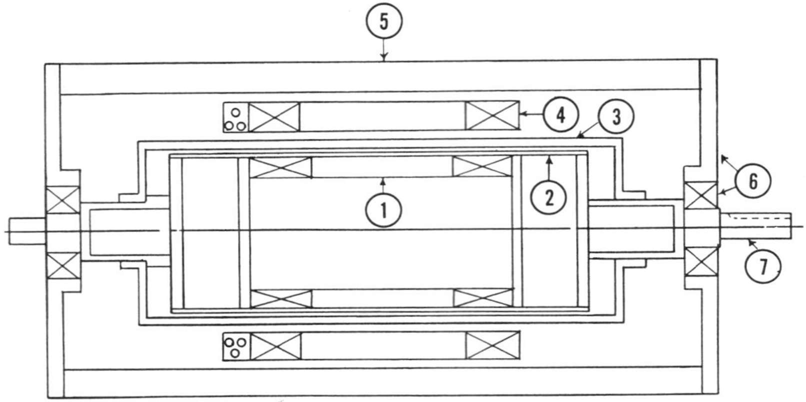

Figure 22.3. General concept of superconducting synchronous motor.

The motor design features a superconducting exciting field winding and a copper armature winding, separated by appropriate flux and thermal shields. There is no magnetic material in the main flux path except for an outer magnetic shield to contain stray flux fields. This construction is referred to as an air core design. Figure 22.3 shows the initial conceptual design.

This design concept presupposes that HTSC material will be available to produce a distributed magnetic field (in the air gap separating the stator and rotor) that is about twice the value found in a conventional machine. To achieve this requires the existence of a HTSC wire that can be wound into a multiturn coil, carrying a current density of 10^5 A/cm^2 in a 5 T magnetic field. The peak field of 5 T in the region of the HTSC conductor diminishes when it reaches the armature conductors. The combination of these electric motor design criteria has become a goal of the Department of Energy's superconductivity technology program. However, researchers and wire manufacturers alike have found this to be a very elusive goal. It is exactly this point that keeps HTSC motors at the design stage instead of at the field-test stage.

22.4.2. Motor Design Challenges

Pending the appearance of satisfactory wire, there are still a number of design problems to be solved. Some of the technical challenges[10] concerned with the motor design include the following.

Magnetic Fields

Large synchronous motors with superconducting field windings (as shown in Figure 22.3) have very little magnetic material in them. Conventional motor design techniques that are based on calculating magnetic fields in a small air gap will not work for HTSC motors. New design techniques based on two-dimensional and three-dimensional magnetic field calculations throughout the entire motor are necessary to model steady-state and transient motor performance. Beyond the unique challenges that the air core geometry presents to motor design, superconducting synchronous motor analysis requires detailed knowledge of the magnetic field distribution in the HTSC winding area. This is because HTSC wire has J_c values that vary with magnetic field. Both steady-state DC as well as transient AC magnetic fields must be determined for all motor operating conditions. This is quite unlike the conventional case where copper wire performance does not depend on magnetic field.

Liquid/Gas Coolant Flow

The combined actions of pressure change, centrifugal force, heat generation, and so on, establish the flow conditions and liquid/gas phase boundaries internal to the rotor cryostat. The fluid mechanics and heat transfer properties of liquid nitrogen in a nonuniform centrifugal force field are especially challenging. This design uses the pool boiling rotor geometry, but an alternative arrangement is to use multichannel two-phase flow. In that case, channel distribution, size, and flow resistance must be adjusted to compensate for radial and axial pressure gradients, so that the liquid/vapor discharge has uniform quality and density. Otherwise, instabilities result that lead to dynamic unbalance problems.

Thermal Analysis

The temperature rise of the field winding must be determined under both steady-state and transient conditions. The analysis must quantify the following:

- Heat influx sources: armature I^2R, field leads, outside ambient, etc.;
- Heat conduction resistances;
- Coolant surfaces and thermal resistances based on the results of the liquid/gas coolant flow study;
- Heat storage, propagation and dissipation during a quench.

Mechanical Analysis

Conventional electric machines operate with fields up to 2 T. The HTSC motor will contain fields up to 5 T. This will increase the stress levels both on the conventional copper winding and the HTSC winding to values well above those experienced due to Lorentz forces in ordinary motors. Stress levels in the superconducting winding are especially critical because these windings are also exposed to other mechanical and thermal forces in addition to the Lorentz forces.

Specifically, the wire sheath material must have mechanical strength to withstand both centrifugal and Lorentz forces, without suffering deformation. Furthermore, the field coil, field support structure and inner flux shield must comprise a prestressed assembly for dimensional stability at rated current, field and speed.

The collection of these considerations dictate the direction of the design concept.

22.5. SPECIFIC DESIGN: 10,000 hp MOTOR

Starting from the general design shown in Figure 22.3, more detailed design studies have been completed for a 10,000 hp, 3600 rpm, 13.8 kV synchronous motor with an HTSC field winding in the geometry shown in Figure 22.4. Scrutiny of Figure 22.4 will reveal that the essential features of Figure 22.3 are embedded in the full design.

This motor has a geometry similar to that of superconducting synchronous generators developed in the past.[11] The motor[10] has an air core (i.e., nonmagnetic) construction so that

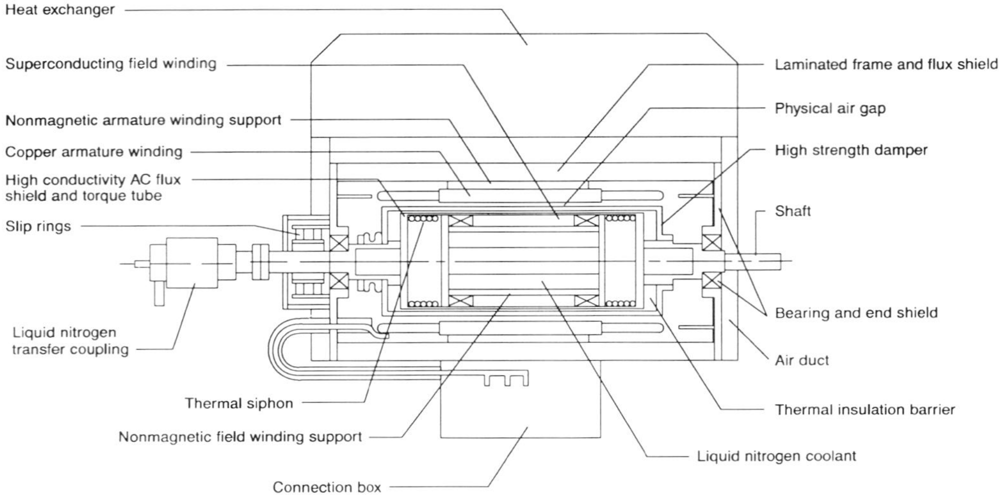

Figure 22.4. Cross-sectional view of a synchronous motor with a superconducting field winding.

the air gap field can be increased without incurring the core loss and saturation problems imposed by a laminated stator and iron core rotor. Only the outer layer, the laminated frame and flux shield, is made of magnetic material. It acts as a flux shunt to prevent the high magnetic fields from escaping from the motor. Inside of the frame is the nonmagnetic and nonconducting support structure for the copper armature-winding.

The normal conducting (copper) armature-winding lies just outside of the air gap. It must be constructed from transposed filaments to reduce eddy current losses. The armature conductors will experience field levels about a factor of 10 greater than those experienced in a conventional motor. In a conventional motor, the conductors are positioned within high permeability teeth; these redirect the flux away from the conductors, so that only the slot-leakage flux actually penetrates the copper. In the superconducting motor the armature conductors see the entire air gap flux density. Therefore, the armature-winding must be carefully designed to minimize eddy current losses.

Under steady-state operation the motor rotor rotates in synchronism with the rotating field created by the three-phase armature currents, and the superconducting field winding experiences only DC magnetic fields. This is how an AC motor can have a HTSC component in it, without incurring intolerable losses. Under load or source transients, however, the rotor will move with respect to the armature-created rotating field and it will experience AC fields. In a conventional synchronous motor, these AC fields induce currents (in damper windings or bars) that create restoring torques to bring the rotor back into synchronism after a disturbance.

In this design, damping is provided by conducting cylinders located outside of and rotating with the field winding. These are designated as the AC flux shield and high-strength damper in Figure 22.4. To prevent AC losses from occurring in the superconductor, the cylinders must also shield the field winding from all AC fields created during transients. A two-layer shielding structure accomplishes the damping and shielding effectively, as follows.

The outer layer is a high-strength material at room temperature (a warm shield) which acts as the damper winding and provides some AC flux shielding. Inside of the outer warm shield is an insulation space that surrounds the rotor cryostat. The inner layer of the rotor damper/shield structure is a high-conductivity cylinder at 77 K. This inner shield provides some damping and, most importantly, acts to shield the superconducting field winding from any AC fields that pass through the outer warm shield. Inside the inner shield is the superconducting field winding on a nonmagnetic support structure.

The motor will be powered by an inverter, to allow adjustable speed operation (because in the future this will be commonplace in this size motor for pump and fan drives). Variable-speed starting will always be used. The presence of the inverter may provide some protection of the superconducting winding during major transients. Large electromagnetic transients will cause fluctuations in the magnetic field and forces on the HTSC winding which can initiate a quench, as discussed in Chapter 17.

22.6. CRYOGENICS

There are two new concerns associated with cooling the motor: the flow of the refrigerant, and its cost.

22.6.1. Coolant Flow

The cooling of the superconducting materials in the field winding of the rotor presents a unique application of momentum and energy transport phenomena. The rotational velocity results in significant radial pressure gradients that affect the flow distribution of the cryogenic fluid. The internal pressure fields can result in significant nonuniformities in the two-phase flow of the coolant. Intending to modify the design to correct for this, Sandia National Laboratories carried out a computational model[12] of the flow process.

Figure 22.5 focuses on the action of the liquid nitrogen coolant in this motor, which is exceptional because of the rotation involved. The field winding is made of HTSC coils. Liquid nitrogen enters the field winding through a rotating inlet header, which feeds axial coolant channels. Nitrogen flow in the channels removes the heat generated due to the small amount of resistive losses of the superconducting winding. The nitrogen flows through these channels and exits into a common outlet header. Both headers are connected to the rest of the coolant loop via central openings on the rotating axis.

A number of basic concerns were identified by examining the unique flow geometry of Figure 22.5. Due to the rotational acceleration, the pressure increases with radius in each of the headers. Because the fluid density is not equal in the two headers (due to the two-phase condition in the exit header), the pressure increase with radius in the two headers is also unequal. This results in nonuniform pressure differentials across the various channels.

Another concern is that conservation of angular momentum requires that the fluid rotating in the outlet header increase its rotational velocity as it is drawn toward the outlet. This will result in a very complex flow because wall shear forces will tend to retard this acceleration. The Sandia model goes into detail on these and other coolant-related factors that influence the design. The point to be understood here is that for a new design of a motor, there are always elements of complexity that must be addressed.

Having constructed a computational model, it was run for the special case of a 5,000 hp motor. The results indicate that there is a large variation in axial coolant flow as a function

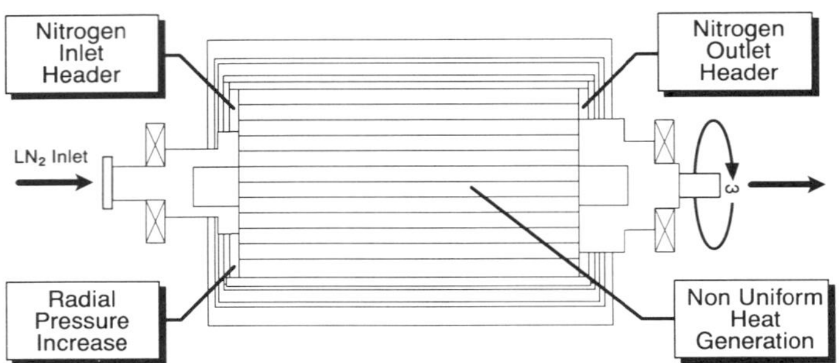

Figure 22.5.   Simplified geometry of the rotor for a 5,000 hp synchronous motor. Liquid nitrogen enters on the left, is distributed radially in the inlet header, then flows axially through the field winding. Liquid and vapor nitrogen are collected in the outlet header and exhausted.

of radius through the field winding. The calculated temperature distribution suggests that a large part of the HTSC field winding coolant is at a temperature above the boiling point of liquid nitrogen at atmospheric pressure. This suggests that HTSC materials must operate well above 77 K for this application.

22.6.2. Cost Considerations

Liquid nitrogen is extremely cheap; about 7¢ per liter in truckload quantities. However, it also doesn't go very far in cooling power. LN_2 has a latent heat of 200 J/g, and a density of 0.808 grams/cm^3. Therefore, its price tag for cooling is 4.33×10^{-7} dollars per Joule, or $1.56/kWh. That is, every time one kWh of electricity is dissipated by boiling off LN_2, it costs $1.56. It may be cheaper to go with a conventional motor of lower efficiency and just buy the extra electricity.

It is possible to construct an indifference curve between paying more for electricity in a conventional motor and paying for LN_2 with a superconducting motor. The final motor choice depends very heavily on just what fraction of the total energy loss goes into the nitrogen pool. If all the losses had to be removed by LN_2, it would never be worth it. However, if only the loss due to heat leaking into an otherwise lossless rotor goes to boil LN_2, then the gain in efficiency might overcome the price tag of liquid nitrogen.

In constructing Figure 22.2, the slice marked "Chill power (off-site)" contains the proper conversion from nitrogen cost to equivalent kW loss. Thus, if electricity sells for 5¢/kWh, then $1.00 worth of LN_2 equates to 20 kWh. The actual heat leak is only a few hundred W, but on this chart the "chill power" is 16 kW (out of total losses of 118 kW). This calculation of an equivalent energy loss is the only way to make an "apples to apples" comparison between motors.

These numbers are for steady-state operation. Transients of any sort increase the "chill power" requirement and lower the efficiency. When the motor changes speed, AC losses dissipate energy, as does the sloshing of LN_2 within the rotor. There are a variety of other parasitic losses that can really only be investigated experimentally.

22.6.3. Lower Temperature Operation

How would these cost factors change if the motor was run with the field coils at a lower temperature, perhaps 20 K? How much more refrigeration cost is tolerable before the advantage of a superconducting motor is lost? Phrased another way, how much heat leak can we tolerate if the motor is to run at 20 K? Based on the same 10,000 hp motor design, we can look at the bar chart of Figure 22.2 and reason as follows: if we can accept a certain amount of refrigeration costs, then what must be the heat leak at the new lower temperature of operation?

The first point to be recognized is that the greater thermal gradients from 300 K to 20 K would increase the heat leak in the first place. Second, figures like "98.4% efficiency" and "52% of conventional losses" already contain a correction factor that converts nitrogen cost to kW. If the coolant were helium gas at a lower temperature (perhaps 20 K), the exact same motor performance would show higher losses and lower efficiency when the corresponding correction factors were applied to convert from coolant price to equivalent kW. Cold helium is far more expensive than LN_2, which is cheap because it is a byproduct of liquid oxygen production.

Utilizing the principles developed in Chapter 3, a refrigerator running at 10% of Carnot efficiency between 300 K and 20 K will have an overall efficiency of 0.7%. Thus the price of electricity gets magnified by a factor of 140 to remove heat at 20 K. If the cost of electricity at the meter is 5¢/kWh, then removing one kWh at 20 K costs $7. Referring once again to the numbers that built up the bar chart of Figure 22.2, a 500 W heat leak would be the equivalent of 70 kW at 20 K (instead of 16 kW at 77 K). The total losses of the superconducting motor would be 172 kW, or 75% of the conventional motor's losses. Operation at 4 K is not cost-effective at all.

Evidently, as operating temperature drops the allowable heat leak drops even faster. Thermal design is absolutely critical to building an economically viable superconducting motor. The design presented above is geared to 77 K and features a thin-walled torque tube to hold the field coils onto the shaft, in order to minimize the heat leak. Axial conduction along the LN_2 manifold, radiation across the flux shield, and conduction through the connecting leads are all roughly equal components of the heat leak. Further reduction of each of these must be a driving factor in design efforts aimed at lower operating temperatures.

22.7. ACTUAL MOTOR CONSTRUCTION

Design studies eventually must be verified by real hardware. In the case of a HTSC motor, the design described above is for a 10,000 hp motor, and wire is not yet available to build such a device. However, that doesn't mean everything has to stop and wait. There are important issues to be understood with respect to instrumentation in a cryogenic environment, cool-down of the components, and so on.

For this purpose, a small DC motor has been constructed and tested. Photographs of that demonstration device appear in Figure 22.6. The motor has a normal conducting armature with the shaft connected to a fan for loading purposes. As shown, the field is a pair of superconducting coils that are visible in Figure 22.6(a). To run the motor, the coils are placed in a liquid nitrogen tank while the motor armature remains above the liquid nitrogen level as shown in Figure 22.6(b).

The limitation at this time is in the size of the coils. The state of the art is advancing steadily, but it is still in a primitive stage. Reliance Electric obtained YBCO coils from Argonne National Laboratory and BSCCO coils from American Superconductor Corp. and Intermagnetics General Corp. The longest coils are those from ASC, and the properties of two such coils are presented[13] in Table 22.1. The oxide powder-in-tube (OPIT) method

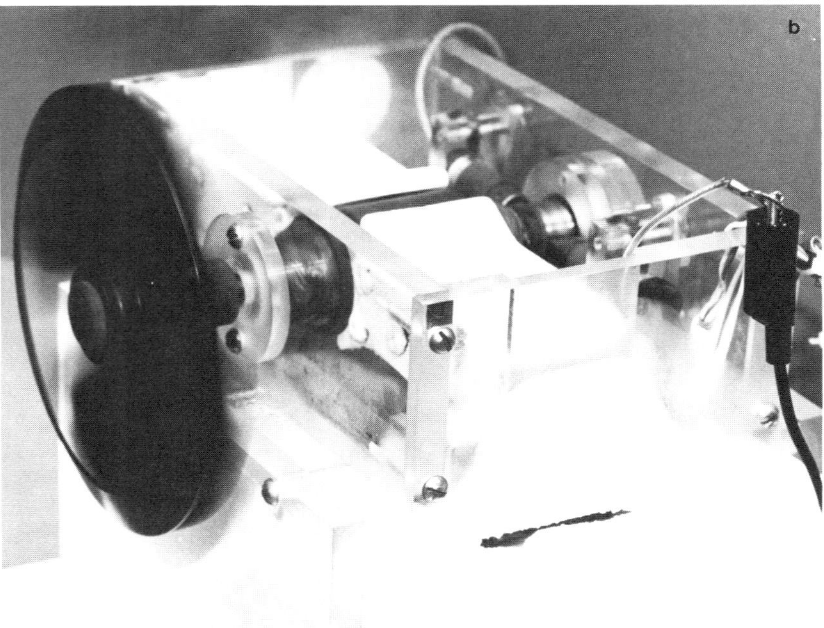

Figure 22.6. DC motor with (a) HTSC field coils without liquid nitrogen tank, showing two HTSC field coils; (b) motor running, supplying power to the fan load.

Table 22.1. Geometry and Electrical Properties of the Coils

Coil	A	B
Inside diameter (mm)	32	32
Outside diameter (mm)	64	65
Coil length (mm)	50	50
Number of turns	326	338
Conductor length (m)	47	50.4
Critical current @ 77 K (amps)	0.92	1.04
Self-field at I_c (gauss)	53	62

NOTE: The critical current is based on an electrical resistivity criteria of 1.95×10^{-9} Ω-cm.

described in Chapter 16 has been used to make 19-filament wires in excess of 1000 m long. Unfortunately, BSCCO carries very little current at 77 K in the kinds of magnetic fields necessary in electric motors. Meanwhile, YBCO continues to be very brittle, the weak-link problem is severe, and lengths over 10 m do not carry enough current to power a large motor.

As a result of all these obstacles, progress has been disappointing. This small test motor has been run repeatedly, but provides only 25 W to the fan load, using the best-performing HTSC field coils. Meanwhile, there are other motor-related efforts going on. A small DC motor using HTSC field coils has been fabricated by Sato *et al.*[14]; it is similar to the motor described here. However, in their study there was no mention of the performance characteristics of the motor other than its ability to turn the armature. A different approach is embodied in a collaboration between TCSUH and Emerson Electric. They have built a small axial air gap motor using permanent field magnets made from melt-textured bulk YBCO. The motor operated[15] at 330 rpm and produced 20 W of power. Because the HTSC components act as permanent magnets, they require no electric connections. Although these prototype motors are too small to be useful, construction of such devices stimulates thinking about how best to take advantage of properties unique to HTSCs.

22.8. FUTURE OUTLOOK

The key to commercially viable HTSC motors, such as the 10,000 hp motor described above, is the ability to create high magnetic fields from the lossless HTSC winding. This translates into the following wire performance specifications for high-horsepower motors:

- Critical current density (J_c) of up to 10^5 A/cm^2
- Critical magnetic field (B_c) of 5 T at 77 K
- Long wire lengths
- High strength (to withstand rotational and $J \times B$ forces on the coil)

The state-of-the-art HTSC wire and coil performance is still a few orders of magnitude below these performance specifications, but progress is being made. Figure 22.7 displays wire performance advances over the first few years for HTSC wire in coil form. As discussed in chapter 16, high values of J_c have now been obtained in straight, uncoiled samples of BSCCO wire. However, in a magnetic field at 77 K, the J_c of BSCCO drops precipitously.

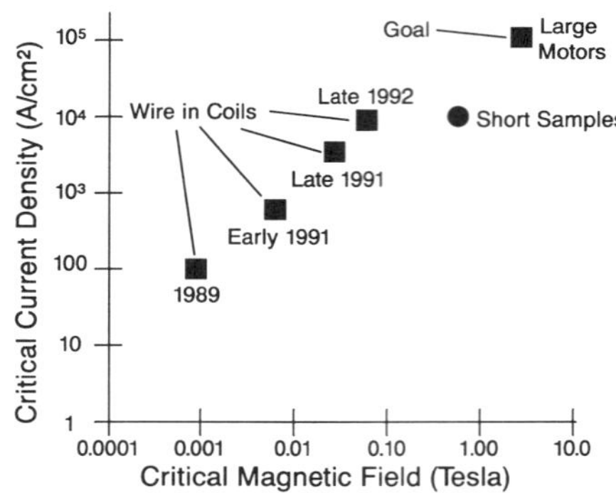

Figure 22.7.   High-temperature superconducting wire performance at 77 K.

BSCCO cannot be used to produce 5 T at 77 K. Therefore, that next order-of-magnitude may be quite difficult to obtain.

One new criterion[16] for measuring progress here is the Linear Current Density (LCD): By reporting the product of the critical current density and the length of the sample, a better indication of progress is presented. For example, if $J_c = 10^4$ A/cm^2 over a 1 meter sample, then LCD = 10^6 A/cm. By this criterion, most melt textured YBCO samples of length ≈ 1 cm have LCDs near 10^4 A/cm. ASC's BSCCO wire (km length) is around 10^8 at 77 K in *zero field*. When major coils are made, having perhaps 10,000 ft in their windings, the LCD measured end-to-end will offer a way to compare them.

In the actual process of making field windings, the superconductor must be fabricated in a complex shape, which will most likely require flexible wires or tapes (wires are better). To date, YBCO and TBCCO have exhibited very brittle behavior. An innovative processing development will be required to permit fabrication of field coils. In addition, the field winding is subject to very large forces due to the self-field, the reaction with the armature winding, and the high rotational speeds. Moreover, the current-density requirements are demanding: approximately 100,000 A/cm^2 in fields of several T.

Further improvement in coil performance are expected, perhaps as efforts to make TBCCO wire prove successful. TBCCO appears to be the best candidate to achieve sufficient J_c in a 5 T field at 77 K. The next step for Reliance Electric will be to construct a 1-hp motor. The implementation of the design presented in Section 22.5 (10,000 hp) is still several years away.

22.9. SUMMARY

The discovery of superconducting materials with $T_c > 77$ K has opened the door for many new applications.[17] One promising opportunity is in superconducting motors, where HTSC coils can be used to create large air gap magnetic fields, thereby increasing the motor power density. Large horsepower, synchronous AC motors have been selected as the size and type most likely to be impacted by HTSCs. Applications for these motors in electric power generating stations are for boiler feed pumps, induced draft, and forced draft fan drives. There are similar pump and fan applications in industry. The ratings for these applications are large enough to economically justify the overhead of the liquid nitrogen cooling system. Also, the loaded, continuous-duty nature of these applications favors taking advantage of the higher efficiency that can be obtained with a motor design using superconducting materials.

The design of a synchronous motor with an HTSC field winding has been described here. This air core motor has peak magnetic field levels exceeding 4 T. Design calculations for a 10,000 hp motor indicate about a 50% reduction in full-load losses and a similar reduction in active motor volume (compared to a conventional iron core motor of the same rating).

Many technical challenges lie ahead in the design and fabrication of air core HTSC motors, not the least of which is the development of HTSC wire that remains superconducting with the high magnetic fields and current density levels that exist in a motor. In parallel with that, and in the expectation that it will come true, work continues on the design of a motor that can effectively utilize the HTSC material. Our projections about the performance and size of a superconducting motor are based on assumed success in the efforts to solve the wire and motor technical problems.

Toward that goal, existing HTSC coils are capable of acting as field coils in a small (25 W) DC motor. The rapid improvement in HTSC wire performance in coil form is encouraging, but it is not yet known whether fundamental limitations exist in the ceramic wire itself. Although BSCCO can be made into wire, its critical current at 77 K may never reach the level required for motors. Nevertheless, the hope is still alive that eventually high-temperature superconductors will be used in large-horsepower motors. The economic advantages of high efficiency and smaller motor size are the main driving forces toward this type of motor.

REFERENCES

1. H. E. Jordan, *Energy-Efficient Electric Motors and Their Application* (Plenum Press, New York: 1994).
2. S. Glasstone, *Energy Deskbook*, U.S. Dept. of Energy Report DOE/IR/05114-1, published by Technical Information Center, Oak Ridge TN (1982).
3. A. E. Fitzgerald *et al.*, *Electrical Machinery*, 4th ed. (McGraw-Hill, New York: 1983).
4. S. A. Nasar, ed., *Handbook of Electric Machines* (McGraw Hill, New York).
5. R. F. Schiferl and J. Stein, "High T_c Superconducting Electric Motors," American Power Conference, Chicago, IL, pp. 1256–1260 (April 13–15, 1992).
6. T. A. Lipo, "The Potential For High-Temperature Superconducting AC and DC Motors," *Electric Machines and Power Systems* **13** (6) (1987).
7. H. E. Jordan, "Feasibility Study of Electric Motors Constructed with High-Temperature Superconducting Materials," *Electric Machines and Power Systems* **16**, 15 (1989).
8. J. S. Edmonds *et al.*, "Application of High-Temperature Superconductivity to Electric Motor Design," *IEEE Trans. Energy Conversion* **7**, 322 (1992).
9. A. D. Appleton, "Motors, Generators and Flux Pumps," *Cryogenics* **9**, 147 (1969).
10. J. D. Edick *et al.*, *IEEE Trans. Appl. Superconductivity* **2**, 189 (1992).
11. J. R. Bumby, *Superconducting Rotating Electrical Machines* (Oxford University Press, London: 1983).
12. R. C. Dykhuizen *et al.*, "Steady-State Cooling of a High-Temperature Superconducting Motor," in *Heat Transfer in Superconducting Equipment*, HTD- vol 229, edited by P. W. Eckels and K. M. Obasih pp. 1–8 (ASME, New York: 1992).
13. C. H. Joshi and R. F. Schiferl, "Design and Fabrication of High-Temperature Superconducting Field Coils for a Demonstration DC Motor," *IEEE Trans. Appl. Superconductivity* **3**, 373 (1993).
14. K. Sato *et al.*, "High J_c Silver-Sheathed Bi-Based Superconducting Wires," *IEEE Trans. Magnetics* **MAG-27**, 1231 (1991).

15. I. G. Chen *et al.*, "Axial Motor with HTSC Permanent Magnet," Applied Superconductivity Conference, Paper LKA-7, Chicago, IL (1992).
16. G. Yurek, American Superconductor Corporation, private communication.
17. T. Schneider and D. Von Dollen, *Energy Applications of High-Temperature Superconductors, A Progress Report,* published by Electric Power Research Institute (Palo Alto, CA: 1992).

23

Fault Current Limiters

Robert F. Giese, [*] *Magne Runde,* [†] *and Thomas P. Sheahen*

This chapter describes one specific application of high-temperature superconductors (HTSCs) to the electrical power industry. Fault current limiters (FCLs) would be an important protective component of any transmission or distribution system, and it is of interest to consider the possible role of HTSC materials in them. Within the purpose and intent of this book, this is a good example, because it shows how HTSCs have both advantages and disadvantages when employed in any particular application.

We begin by explaining the basic function of a FCL and how it would fit into a conservatively designed utility system. Next we go on to examine superconducting FCLs, including the differences associated with HTSCs. The stability and switching conditions are developed, and then design calculations are given for two kinds of devices. The requirement (imposed by any utility) of cost competitiveness is considered, and brief mention is made of other switching applications of HTSCs. Finally, we summarize the key elements of the analysis.

23.1. FAULT CURRENTS

Any electrical circuit is at risk of a short-circuit condition, and some sort of protection is regularly employed. The simple fuse box in older homes and the modern circuit breaker box are examples of this protection. Beyond what might happen in a house, the electrical utility has to worry about short circuits on its transmission and distribution lines, which may be due to lightning strikes, falling trees, or other causes. The current flowing under such conditions can easily be ten times the current normally carried, which means thousands of amperes. Major utility lines carry far more power than a home ever receives, so the equipment used to protect these lines must be much more rugged and efficient.

Whether in the home or in the transmission system, the prescribed response to a short circuit is to open the circuit, interrupting the current before any damage can occur. Circuit breakers on major transmission lines can withstand 63 kA and can open in about 50 msec. But what happens within that margin of 50 msec—which is only three cycles? There is plenty of opportunity for damage to the system if the short-circuit current is large enough. This is where fault current limiters come in: an FCL has a variable-impedance, installed in series

[*]Argonne National Laboratory.
[†]Norwegian Electric Power Research Institute.

with a circuit breaker in a substation. The impedance increases suddenly whenever a short-circuit current occurs. A proper FCL goes into action very quickly and prevents the transient current from becoming excessively large; typically the current excursion is only a factor of 2 above normal. The job of an FCL is to protect a circuit just long enough for the circuit breaker to activate.

One additional requirement upon FCLs is that they return to their low-impedance state promptly and automatically as soon as the fault condition ends. A few cycles is all that is allowed for this restoration to take place.

There are alternatives to a series-wired, or in-line, FCL. An extremely fast switch that shunts the current to an alternate path could provide equivalent protection for the circuit. Devices in which the active element is not itself in the circuit have been considered, at least at the level of test models. The whole idea is to increase the impedance in the line very suddenly, before a fault current can build up. Clearly, a much faster circuit breaker (say, one that operated in less than a millisecond) would make the FCL unnecessary, but no such device has yet been invented. Consequently, there is still a need for FCLs in contemporary electrical power systems. A number of FCL concepts have been proposed,[1] but they all have major cost or performance disadvantages.[2]

23.2. UTILITY CRITERIA

What is meant by "major cost or performance disadvantages"? To answer this, it is necessary to go beyond conceptual designs and look at the actual operating conditions that utilities face.

23.2.1. Cost

The first question to be answered is, "How much does it cost to keep this device on standby?" Just as with a very expensive ballistic missile defense system, the whole idea is never to have to actually use it. In a perfect world, you never have to call the fire department; and, ideally, an FCL is *never* called on to protect an electrical system. Thus, many years go by during which the FCL is inactive. If it causes a slight voltage drop, the cumulative cost of that lost electricity is the cost of operating the FCL. If it must be cooled, the cost of running the refrigerator is the cost of operating the FCL. There is a trade-off to be made between these operating costs and the cost of repairing the damage suffered by an unprotected circuit. To estimate the latter is an exercise in costing out the impact of a low-probability event.

The cost/benefit analysis for FCLs is quite dissimilar in different geographical regions. In remote areas, a single transmission line might be the entire link to the electrical source. In the United States, the electrical grid is so heavily interconnected that the loss of a single transmission line is seldom a disaster (only when coupled with some other problem). Consequently, the United States has very few FCLs in place, and those mostly on an experimental basis.

It is easy to do the arithmetic showing what consumers are willing to pay for protection: consider a remote community of 10,000 people served by a single transmission line. Recall that each person consumes about 1 kW continuously. If FCLs adequately protected that supply, but added one cent per kWh to the cost of electricity, people would soon object to each paying nearly $100 extra per year and would prefer to build an auxiliary power plant in their community, thus eliminating their dependence on remotely transmitted power.

Cognizant of this, when planning new facilities, electrical utilities have generally restricted the allowable cost of FCLs to a few percent of the generating cost.

Whether or not a utility will install an FCL is thus a question of economics, and there is only limited information available about this. Two surveys of American utilities[1,3] indicate that they are not willing to pay more for a FCL than a few times the cost of a circuit breaker. The electrical losses during normal operation should not exceed 25% of the losses in a transformer in the same system.[1] Clearly, any FCL design that consumes lots of power or refrigeration is not going to make it in this cost-competitive environment.

23.2.2. Performance

The performance specifications for FCLs are likewise very demanding. Utility engineers are notoriously conservative and design their systems to survive a worst case condition. The worst case is one where for some reason (a failure of a FCL in an adjacent circuit block?) the current is suddenly very high. Because the highest rating in conventional commercial circuit breakers is 63 kA, utility managers have historically wanted every component in the circuit to withstand such amperage.

To further delineate the issue, it is useful to distinguish two cases of FCLs: those protecting *distribution lines* and *transmission lines*. The easier case is distribution lines, where the voltage is 6–36 kV and the current is 200–2,000 A. For transmission lines, voltages of 100–500 kV are accompanied by a few thousand amp currents. Accordingly, we have constructed two different examples to illustrate the design criteria that prevail in each case. The simplest circuit model is shown in Figure 23.1, and the circuit parameters in the two cases are summarized in Table 23.1.

Under normal (nonfault) conditions, the impedance in the line should be as low as possible to avoid wasting power. Under fault conditions, the impedance should be very high. In typical silicon electronic devices, these two impedances track together; that is, very low normal impedance implies a modest fault impedance. On that basis, it is of interest to ask how low a value of fault impedance will suffice. In the last line of Table 23.1, the maximum fault currents are set to 10 kA and 63 kA for the distribution and transmission systems, respectively. Remembering that there is a finite source impedance even when the load is short-circuited, for an in-line limiting element to handle such currents the minimum impedances would be those given in Table 23.2. It is only a brief further step to show that this does not lead to a viable FCL design.

Whether for a resistive FCL, where the power dissipated is I^2R, or for an inductive FCL, where the magnetic energy stored is $0.5\,LI^2$, the numerical requirements are too extreme. A

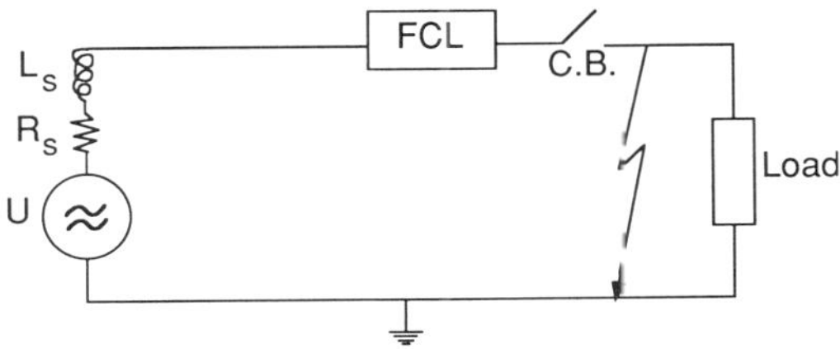

Figure 23.1. Simple idealized circuit containing a fault current limiter (FCL) in series with the source impedance, circuit breaker, and load.

Table 23.1. Parameters for the Distribution and Transmission of
Example Systems (Worst Case Analysis)[a]

Parameter	15 kV System	500 kV System
Source voltage (kV_{RMS}), U	8.7	290
Maximum load current (A_{RMS}), I_n	600	3,000
Source impedance (Ω)	0.43	2.4
X/R ratio	1.7	10
Source resistance (Ω), R_s	0.22	0.24
Source inductance (mH), L_s	1.0	6.3
Unlimited fault current (kA_{RMS})	20	120
Maximum limited fault current (kA_{RMS})	10	63

[a]Power frequency = 60 Hz.

fault current of 63 kA in the 500 kV system requires a resistive FCL to dissipate 240 MJ per power cycle. This value is huge! (If this energy were dissipated in 1 m^3 of water, the temperature would rise by almost 60°C.) Two power cycles can be considered a minimum time of operation for an FCL, so this device must be able to dissipate 480 MJ per fault.

The inductive case is only slightly less hopeless. If an inductive device were used to limit the fault currents to 63 kA in the 500 kV system, about 75 MJ of magnetic energy must be stored in the FCL. An iron core inductor would call for about 75,000 m^3 of iron per phase; this is the size of a two-story warehouse. If the energy is stored in a 1 T field of an *air core reactor*, an air volume of 190 m^3 is required, and the device is still huge (the size of a large room).

23.2.3. The Role of an FCL

The central message of this numerical calculation is that traditional thinking about ways to handle fault currents will not work. The conservative criterion of absorbing 63 kA simply cannot be met by any practical device. Thus, it is necessary to find a new approach to the problem.

The point of an FCL is precisely to prevent having to dissipate such enormous energies. We want something that sits at low impedance nearly all the time, but suddenly increases its impedance when a fault occurs. When an FCL functions correctly, it limits the current to perhaps twice the normal load current, which (on line 2 of Table 23.1) are only 600 and 3,000 amps for the two cases, respectively. Either resistive or inductive types of FCL can be built to handle the corresponding energies.

Table 23.2. FCL Impedances (Inductances)
Required to Accommodate Fault Currents of
10 kA and 63 kA

FCL Type	15 kV System	500 kV System
Resistive Ω	0.56	3.7
Inductive Ω	0.46	2.2
(L, mH)	(1.2)	(5.8)

For the typical 15 kV distribution line, the energies are typically 250 times lower than those in the 500 kV transmission system. Ohmic dissipation of 0.2 to 1 MJ requires some mass and volume, but this is obviously a much simpler task than dealing with gigajoules. Air core reactors in the range of 1–10 mH for 15 kV FCLs are roughly 1 m^3 in volume and are commercially available.

With the problem now put into proper perspective, it is possible to examine what superconductors can contribute to the solution.

23.3. SUPERCONDUCTING FAULT CURRENT LIMITERS

The *a priori* advantage of using a superconductor in a FCL is that the resistance is zero when in the quiescent state, which is nearly all the time. The total cost of running this FCL is the cost of keeping it cold. Obviously this cost is smaller at 77 K than at 4 K. Not surprisingly, a lot of superconducting FCL (SCFCL) conceptual designs are known.[4] However, making a real SCFCL is quite another story.

23.3.1. Operating Principles

In the general case, several different concepts of SCFCLs may be considered.[5] Simplified circuit diagrams for four generically different SCFCLs are shown in Figure 23.2; Figure 23.1 is a special case of these. In all cases, the cryogenic zone is indicated by the broken lines. Common features are that the superconducting element (displayed as a variable resistance) is inserted directly in series with the power circuit to be protected, and they all make use of a superconducting-to-normal transition.

Conceptually, the basic resistive SCFCL, as shown in Figure 23.2(a), is the simplest design. Under quiescent conditions, the superconducting element is in its superconducting state and current passes with virtually no losses. In the event of a fault current, the critical

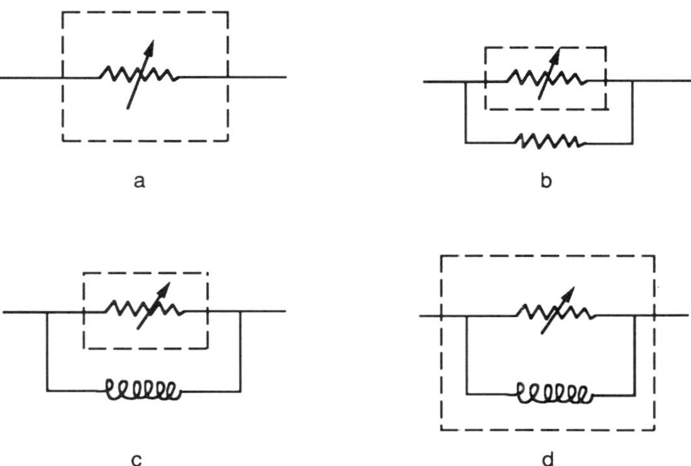

Figure 23.2. Four simple schematics of various resistive and inductive fault current limiters. In all cases, the portion enclosed by dashed lines is to be held at cryogenic temperature.

current density J_c of the superconductor is exceeded and the superconductor transitions to its normal (nonsuperconducting) state; it then becomes a series resistor that limits the current. As we have seen numerically above, a major drawback associated with this strategy is that during a fault, all of the energy is dissipated directly into the superconductor. Consequently, the superconductor must have (or be intimately linked to) a large thermal mass. This arrangement significantly increases the reset time.

The energy dissipated in the superconductor during a fault can be reduced by employing a shunt impedance as the limiting element [Figures 23.2(b)–(d)]. To efficiently commutate the current to the shunt, the normal-state resistance of the superconducting element must be orders of magnitude larger than the shunt. In addition to conventional resistors and air core reactors, a superconducting coil can be used as a limiting impedance, as shown in Figure 23.2(d).

An inherent drawback with all the schemes shown in Figure 23.2 is that the superconducting element is in-line with the power circuit. Because of this, there will be AC losses. Moreover, heat leaking through the current leads and into the cryostat will cause additional thermal losses.

23.3.2. Examples of SCFCLs

Most SCFCLs so far have utilized NbTi multifilamentary wire embedded in a copper matrix. As we saw in Chapter 18, this minimizes AC losses. One example is shown in Figure 23.3, a resistive device built by GEC Alsthom in France. It is intended for use on a distribution system and is rated at 25 kV, with a critical current rating of 330 A (peak). The limiting resistance is 340 Ω. This device has performed well in numerous tests, including some up to 51 kV(rms).

Another FCL test program is that of Toshiba and Tokyo Electric Power Co. (TEPCO); one device is a 6.6 kV-rated FCL carrying 1,000 A and made of NbTi. Figure 23.4 shows its circuitry,[6] together with oscilloscope traces of how it limits current in tests. The way it works is simple: ordinarily current flows through superconducting trigger coils TC1 and TC2, but if a fault occurs and excessive current suddenly appears, these go normal and the current

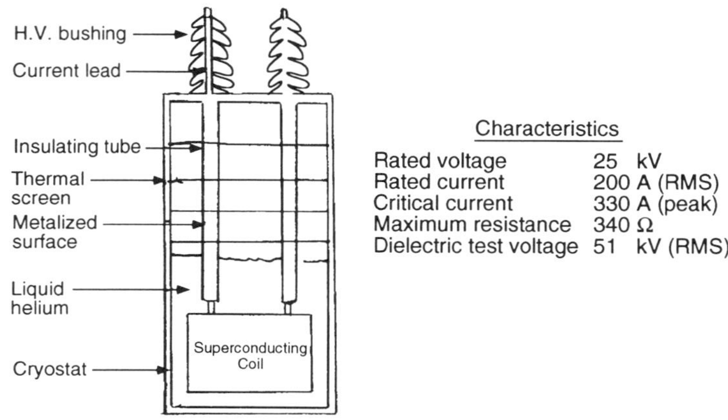

Figure 23.3. Schematic of the GEC Alsthom SCFCL.

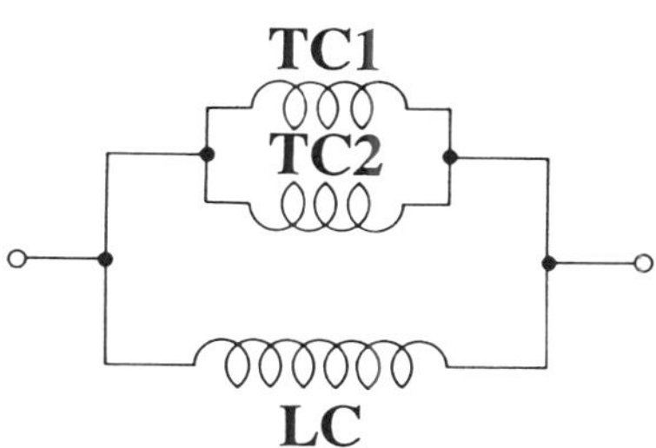

LC: Limiting coil (Superconductor)
TC: Trigger coil (Superconductor)
(Normally, the load current flows
in the trigger coil)
$L_{LC} \gg L_{TC} \approx 0$

(a) Scheme of operation.

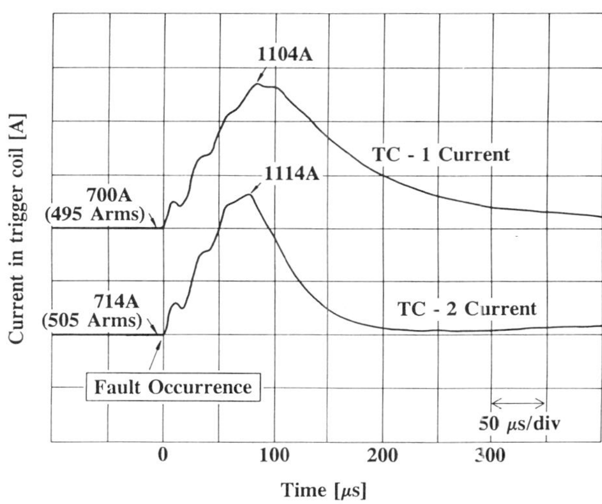

(b) Current suppression

Figure 23.4. New type superconducting fault current limiter from Tokyo Electric Power Co. (a) Principle of operation; (b) experimental data showing current suppression.

must flow through the limiting coil LC. The waveforms show than in only about 80 μsec, the FCL protects the system. The TEPCO configuration employs the principle of Figure 23.2(d). Like most other SCFCLs, this one operates in the low end of the distribution range and cannot begin to serve transmission lines. Still, it is proof that the concept actually can be made to work.

Table 23.3. Representative Values for Thermal Conductivity κ, Heat Capacity c, and Normal-State Resistivity ρ_N for Superconductors, Stabilizers, and Substrate Materials at 4 K and 77 K

Material		κ (W/m K)	c (J/m^3 K)	ρ_N (10^{-8} Ωm)
			Property	
LTSC	(4 K)	0.1	5,000	70
HTSC	(77 K)	2	10^6	100
Cu	(4 K)	3,500	400	0.038
	(77 K)	550	1.9×10^6	0.25
Al$_2$O$_3$	(4 K)	220	20	Insulating
	(77 K)	960	0.3×10^6	Insulating

23.3.3. High-Temperature Superconductors

How will the new HTSC materials affect this picture? We can instantly forecast changes in refrigeration costs, but there will also be different conditions pertaining to AC losses, stability, switching, and transient temperatures. The differences come into focus when numerical calculations for the HTSCs are carried out. To do this, the principles developed in Chapters 17 and 18 are used in examining this potential application.

Material Properties

Table 23.3 summarizes representative material properties for LTSCs (NbTi) and HTSCs. In subsequent calculations, T_c for these materials is taken as 9.5 K for an LTSC and 92 K for an HTSC. The HTSC values are obtained from small samples; the technology is still evolving. The values for copper (a commonly used stabilizer) and sapphire (a possible substrate for superconducting films) are also included in Table 23.3. The table once again shows a point first made in Chapter 17: the most evident difference is that the heat capacity is several orders of magnitude higher at 77 K than at 4 K.

AC Losses

In the concepts of Figure 23.2, the superconductor passes the full load current under quiescent conditions. Hence, the magnitude of the AC losses (i.e., hysteresis losses, and ohmic losses caused by eddy currents if metallic stabilizers are applied) is crucial to the operating costs of the SCFCL. As we have seen in Chapter 18, the electrical losses are primarily determined by the design and operating conditions of the conducting wire, not by the properties of the material itself. Consequently, assuming high-quality materials, we take an optimistic viewpoint and expect the losses to be similar for HTSCs and LTSCs.

Refrigeration Costs

The differences in refrigeration costs depend on (a) the relative thermal loads at 4 K and 77 K, (b) the relative refrigeration efficiency at 4 K and 77 K, and (c) the magnitude of refrigeration costs compared with other SCFCL costs.

Thermal Loads. There are three primary sources of thermal load in an SCFCL: (1) thermal conduction through current leads, residual gas, and structural components; (2) radiation losses; and (3) internally generated heat (AC losses). It has been estimated that the thermal loads at 77 K and 4 K are approximately equal.[7] However, at the higher temperature, the cost of removing the heat is less.

Refrigerator Efficiency. From Chapter 3, recall that a refrigerator's efficiency (relative to the Carnot efficiency) is independent of the lower operating temperature, but efficiency increases with increasing capacity. (In this context, capacity refers to the amount of heat that can be absorbed at the lower temperature and rejected at 300 K.) Thus, provided that the thermal loads are the same, a refrigerator for an SCFCL operating at 77 K would require 24 times less power input than a refrigerator operating at 4.2 K. (The factor 24 is simply the ratio of Carnot efficiencies from 300 K to either temperature.) An intermediate temperature would likewise have some modest efficiency gain.

The cost of refrigeration is expected to dominate the operating cost of an SCFCL. Thus, the savings in SCFCL operating costs by operating at 77 K are expected to be almost as large as the improvement in Carnot efficiency.

Refrigeration Capital Costs. The capital costs of a refrigerator scale approximately as the 0.7 power of the input power.[8] With the same thermal load at 77 K or 4 K, the refrigerator capital cost is reduced by a factor of 10. However, the capital cost of the refrigerator relative to the rest of the SCFCL decreases with increasing complexity of the SCFCL. For a device that employs a high-power shunt, an external triggering unit, and overvoltage protection, the refrigerator cost savings at 77 K relative to 4 K are unlikely to reduce the total cost of the SCFCL by more than 20%. This percentage is based on the cost estimate for a transmission system SCFCL.[4] If it is possible to make a very simple device, the savings become greater, possibly as high as 50% for a distribution system SCFCL. However, because little has been published on the cost of SCFCLs, these percentages must only be considered rough estimates.

23.3.4. A Novel HTSC Device

As we have seen, it is strongly preferable to avoid having the main power pass through the active element of a FCL. To this end, Asea Brown Boveri (ABB) in Switzerland has developed a unique FCL using BSCCO rings to suddenly change the impedance in a power line.

Figure 23.5 is a sketch of the ABB device. It consists of an iron core, a set of superconducting BSCCO rings, and a copper winding in series with the power line. During normal operation, the shielding (screening) currents induced in the superconducting rings isolate the copper winding from the iron core, resulting in a low inductance, almost equal to an air core inductor. Under fault conditions, the screening currents in the BSCCO rings exceed I_c, and the magnetic field due to the copper winding penetrates the superconducting screen. The impedance seen by the primary copper winding (the power line) substantially increases instantly.

This SCFCL has several attractive features, particularly with respect to HTSC conductors. First, the device relies on *intra*granular screening currents (not *inter*granular transport current) and is thus less affected by weak-link behavior. Second, because the superconductor is not in series with the power line (i.e., no current leads), conduction losses are less of a problem. Finally, the device relies on bulk HTSC ring conductors, which are easier to manufacture.

It remains to be seen whether this device can be scaled up. A 20 kVA-rated version employing 8-cm-diam rings was built in 1990, and it occupied one cubic foot. However, for

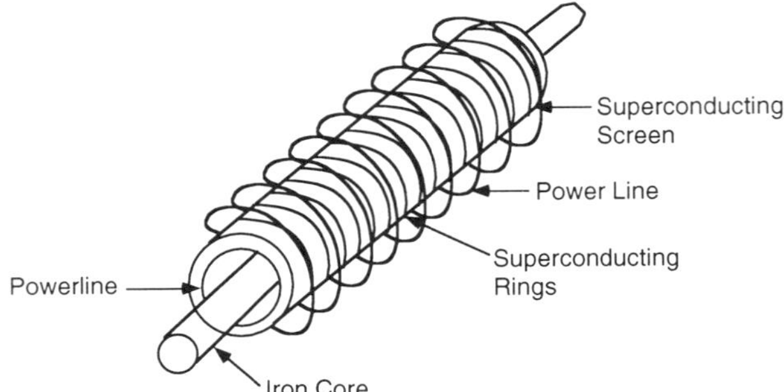

Figure 23.5.   Schematic of the ABB HTSC SCFCL, utilizing BSCCO rings.

the representative distribution system defined earlier in this chapter, the rating must be approximately 800 times that of the 20 kVA device; for the transmission system, the rating must be 100,000 times higher than the 20 kVA SCFCL. ABB is working on scaling up to a device suitable for a generating station.

The biggest difficulty with anything made of BSCCO is that its critical current has fallen off at 77 K, and it is likely to require refrigeration to lower temperatures (20–30 K) to perform satisfactorily. Another disadvantage of this concept is that it requires a lot of iron. The Electrotechnical Laboratory in Japan has proposed another shielding concept using less iron: in addition to a simple coil design, both toroidal and racetrack designs have been proposed. These designs allow the iron core to become saturated, thereby decreasing the amount of iron needed.

23.4. STABILITY AND SWITCHING

This section is really the key to this entire chapter, because in it we show how careful calculation can undercut glib predictions. There is an inherent contradiction in what the SCFCL is being asked to do, and this leads to an impasse. On the one hand, the superconductor should be stable and remain in its superconducting state during ordinary conditions. On the other hand, if a fault occurs, it should switch very rapidly to its nonsuperconducting state. Are these two requirements mutually compatible?

23.4.1. Adiabatic Stability

The topics covered in Chapter 17 are significant here. Wires of superconducting materials are unstable unless they have a small cross section. Recall that slight disturbances of almost any kind (thermal, electrical, mechanical, etc.) will cause a current-carrying superconductor to quench. A common measure of superconductor stability is the adiabatic stability criterion. The maximum conductor radius a_{ad} for which a superconducting wire will not exhibit thermal runaway when passing a current density just below J_c is (according to this criterion)

$$a_{ad} = \sqrt{\frac{c(T_c - T_{op})}{\mu_0 J_c^2}} \tag{23.1}$$

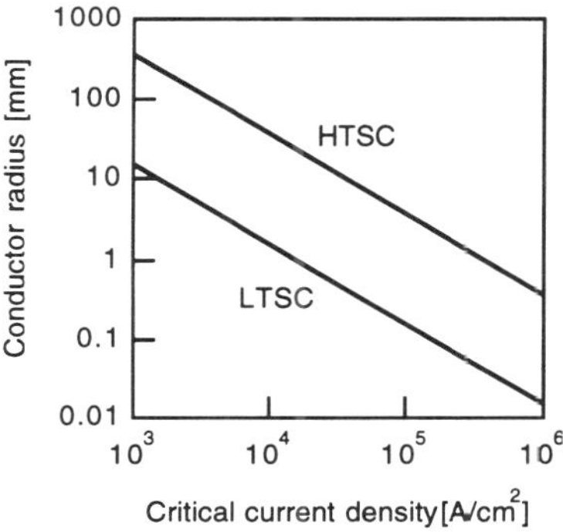

Figure 23.6. Maximum adiabatically stable conductor radius for LTSC at 4 K and HTSC at 77 K.

where T_{op} is the operating temperature (4 K or 77 K) and μ_0 is the permeability of free space. Figure 23.6 shows a_{ad} as a function of J_c for an LTSC and an HTSC with the parameter values given in Table 23.3. HTSC wires are adiabatically stable for diameters 25 times larger than those for LTSC wires. The higher specific heat is the root cause of this happy outcome.

23.4.2. Switching

The required switching time for the superconductor element in an SCFCL depends on the design, but it typically ranges from a few microseconds to a few milliseconds. If the switching is too fast, overvoltages occur across the device, whereas the fault currents can build up to unacceptable levels if the switching is too slow.

For both passively and actively triggered switching the superconductor first enters into the so-called lossy state. In this state, the superconductor is unstable, and the interaction between the current and the magnetic field produces heat, which causes the temperature in the superconductor to rise. However, the superconductor does not reach the full normal-state resistivity ρ_N (and, consequently, the full limiting resistance) until T_c is exceeded. Thus, in the adiabatic case, the switching time τ is given as

$$\tau = \int_{T_{op}}^{T_c} \frac{c}{\rho J^2}\, dT \tag{23.2}$$

where J is the current density and ρ is the superconductor resistivity in the lossy state. An order-of-magnitude estimate of τ can be found by ignoring all temperature dependencies, setting ρ equal to ρ_N, and setting J equal to J_c. Figure 23.7 shows the results of this calculation for a range of J_c by using the parameter values in Table 23.3.

Evidently, switching from the superconducting to the normal state is much slower for HTSCs than for LTSCs; moreover, τ decreases rapidly with increasing J_c. The same large value of specific heat c that improved adiabatic stability is responsible for increasing the switching time. The free lunch escapes again!

There is a much worse phenomenon yet to be considered. Experience with LTSCs has shown that unless active triggering is employed, a superconductor initially goes normal only at one or a few small, weak regions. These nonsuperconducting islands then expand throughout the entire superconductor. As described in Chapter 17, the velocity at which the

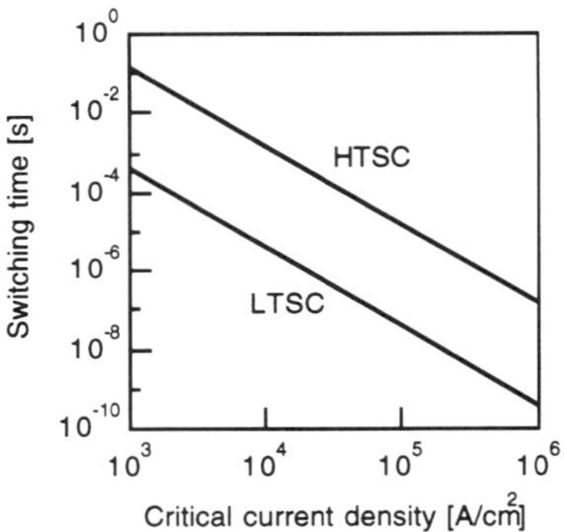

Figure 23.7. Switching times (simplified estimates) for a LTSC and a HTSC.

front of the normal region propagates through the superconductor is called the *quench propagation velocity* v_q. Denoting thermal conductivity by κ and normal-state resistivity by ρ_N, the relationship is

$$v_q = \frac{J}{c} \sqrt{\frac{2\rho_N\kappa}{T_c - T_{op}}} \tag{23.3}$$

which is displayed in Figure 23.8 for an LTSC and an HTSC.

Again, it is the much greater heat capacities of materials at 77 K relative to 4 K that make the difference. Thermal processes are expected to proceed much slower in HTSCs than in LTSCs. v_q is 1.5 orders of magnitude smaller at 77 K, and, as shown above, τ is 2.5 orders of magnitude greater. These conditions will make the transition of an SCFCL from the low-impedance to the high-impedance mode substantially more difficult with an HTSC than with an LTSC. This is the case both for passive and active triggering. However, Figures 23.7 and 23.8 also show that both τ and v_q improve greatly when J_c becomes large. Thus, to obtain sufficiently fast and uniform switching with an HTSC element, $J_c > 10^5$ A/cm^2 may be required.

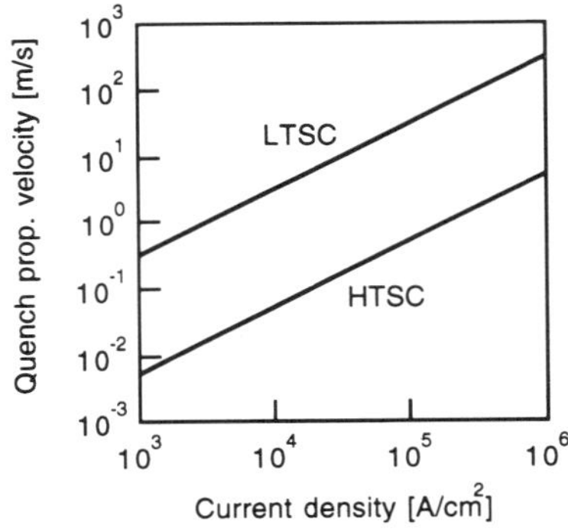

Figure 23.8. Quench propagation velocities in an LTSC and an HTSC.

23.4.3. Temperature Rise During a Fault

When the superconductor eventually has reached the normal state, the temperature will continue to rise as long as current is flowing. The temperature rise is proportional to the reciprocal of the heat capacity. Thus, HTSCs are superior to LTSCs because of their much greater heat capacity. However, this assertion is only true if the temperature rise during a fault is small. As will be discussed in the next section, temperature transients of hundreds of degrees will occur in some schemes, even for HTSCs. With such large energy dissipations, concern for the integrity of the material becomes dominant, and the question of whether the coolant is at 4 K or 77 K is of minor importance.

23.5. CONSIDERATIONS FOR IN-LINE SCFCLs

Now we are ready to assemble all the preceding pieces, to see what must be traded off in a real-world system. From Table 23.1 we take the electrical parameters of the example distribution and transmission systems; from Section 23.3.1 we take the circuit layouts for the SCFCLs; and from Table 23.3 we take the HTSC properties.

23.5.1. Resistive SCFCL

The basic in-line resistive design of Figure 23.2(a) is considered first. The cross-sectional area A of the superconductor is given by

$$I_c = J_c A \tag{23.4}$$

where I_c is the highest current the SCFCL should pass without switching to the high-impedance mode. Typically, $I_c = 5\,I_n$. The normal-state resistance R of a superconductor of length L is

$$R = \frac{\rho_N L}{A} = \frac{\rho_N L J_c}{I_c} \tag{23.5}$$

The volume V of the superconductor can therefore be expressed as

$$V = \frac{I_c^2 R}{J_c^2 \rho_N} \tag{23.6}$$

By assuming a relatively high J_c, the FCL resistance specifications given in Table 23.2 are easily met for reasonable HTSC wire cross sections and lengths. However, if the power dissipation during a fault is also taken into consideration, major problems are encountered. The temperature rise ΔT in time Δt in the superconductor during limiting action is

$$\frac{\Delta T}{\Delta t} = \frac{R I_{\text{lim}}^2}{cV} \tag{23.7}$$

where I_{lim} is the limited fault current given by

$$I_{\text{lim}} = \frac{U}{\sqrt{(R_s + R)^2 + (\omega L_s)^2}} \tag{23.8}$$

where ω is the angular power frequency, $2\pi 60$, and U, R_s, and L_s are given in Table 23.1.

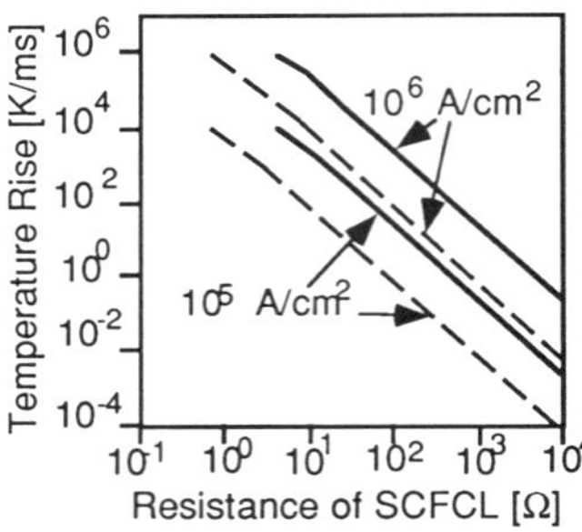

Figure 23.9. Temperature rise per millisecond in the HTSC element of the concept shown in Figure 23.2(a) as a function of the limiting resistance for the example distribution system (broken lines) and transmission system (solid lines).

If at this point we plug in the minimum-resistance values listed in Table 23.2 (0.56 Ω and 2.38 Ω for the distribution and transmission systems, respectively), I_{lim} becomes so high that the sudden temperature rise is catastrophic: between 100 and 10,000 K/msec. *Explosion* would be a better word for what happens than *temperature rise*. The resistive design of Figure 23.2(a) is not technically feasible at all, if the SCFCLs have the minimum required resistances given in Table 23.2. Even with $J_c < 10^4$ A/cm^2, the power dissipation is enough to cause the superconductor to evaporate within the first power cycle (17 msec).

Is there any way around this? The only way to reduce the dissipation (and thereby limit the temperature rise) in the superconductor is to limit the current even further by increasing the SCFCL resistance. Because ρ_N of a HTSC cannot be increased without reducing J_c drastically, this increase in resistance calls for increasing the length of the superconductor. Figure 23.9 shows the rate of temperature rise during a fault as a function of the limiting resistance for two values of J_c.

If the resistance values were those required to limit the current to 10 kA and 63 kA, respectively, the leftmost points on the lines of Figure 23.9 would obtain, and the temperature rise would be explosive (10^4 K/ms, assuming J_c is 10^5 A/cm^2). In order to keep the temperature rise reasonable after two cycles (33 msec), the limiting resistance would have to be about 1 kΩ. In that case, the superconductors must be 3 km and 15 km long, respectively, in our two examples.

Thus, a resistive SCFCL is impractical for transmission lines, and is marginal for distribution lines. This is an awful lot of energy to dissipate in a superconductor.

23.5.2. Inductive SCFCL

It is desirable to avoid dissipation of energy while still limiting the current, typically by applying an inductive shunt instead of a resistive SCFCL [Figure 23.2(c) and (d)]. One consequence of doing so is that the L/R ratio increases, which raises the time constant of the circuit. Therefore, the current commutation to the shunt during a fault becomes slower when employing an inductive shunt. Consequently, the problems related to energy dissipation in the superconducting element in the early stages of a fault are even greater in this design. Most of the potential difficulties related to the switching and resetting of the superconducting element exist also in the inductive schemes of Figure 23.2, because the inductive SCFCLs rely on the same superconducting-to-normal transition.

The TEPCO design (Figure 23.4) compromises by surrendering some response quickness in exchange for a large inductive shunt. This design works well in the low end of the distribution system range, but if scaled up to transmission system parameters the trigger coils would vaporize before the limiting coil came fully on-line.

Because the SCFCL already has a cryostat, an interesting scheme is to use an air core reactor with superconducting windings, as shown in Figure 23.2(d). The reactor remains in its superconducting state during a fault; the current is limited by its impedance. Compared with applying ambient temperature air core reactors [Figure 23.2(c)], this strategy may save real estate (a more compact SCFCL design) because the ampere-turns and the magnetic energy density can be increased substantially. Doubling the magnetic flux density reduces the required reactor volume by a factor of 4, but at the expense of greater structural requirements for balancing the magnetic forces between the conductors. As discussed in Chapter 21 on SMES, the engineering problems related to construction of large high-field (several T) superconducting coils are significant and increase rapidly with the size of the coil. The combined volume and flux density in an SCFCL for our representative transmission system is clearly beyond present technology.

23.6. COST COMPETITION

The superconducting element shown in Figure 23.2 (b)–(d) is essentially only a switch that commutates the fault current to the limiting impedance. Thus, the large energies associated with the limiting action are handled by the shunt, which makes the shunt large and expensive. In a study of a transmission system SCFCL,[4] the shunt resistor was found to account for about half the total SCFCL cost. Consequently, the capital cost of the SCFCL can be reduced significantly if the shunt impedance can be avoided.

To make the shuntless concept of Figure 23.2(a) technically feasible, the fault current must be limited to a small fraction of the nominal load current and the reset time will probably be at least several seconds. For some applications this may be acceptable: the underground transmission cables described in Chapter 19 are one example.

As mentioned in Section 23.1, the in-line FCL is not the only possible way to go. Competing technology to the SCFCL exists, at least at the distribution system level. The feasibility of using gate turn-off thyristors to commutate the fault current to a limiting shunt has been demonstrated.[9] Furthermore, a novel and inexpensive arc-runner scheme for fault-current limiting has been tested.[10]

These and other concepts create cost and reliability criteria that an SCFCL must meet to be competitive. For example, the thyristors that essentially have the same function as the superconductor cost around $40,000 for a three-phase FCL rated for the example distribution system. The losses in this device are about 0.05% of the throughput power; therefore, the refrigeration costs of an SCFCL should not exceed this value. The extent to which these FCL schemes can be scaled up to high-power transmission systems is unclear.

23.7. OTHER SWITCHING APPLICATIONS

Power electronics is often cited as a potential application of HTSCs, because the HTSCs can be switched from a superconducting state to a normal state cleanly and in very short times, $< 10^{-9}$ sec. The HTSCs are also expected to have very low forward voltages, a perennial problem for conventional power electronics. However, it is prudent to remember the amount of current that must be handled in a typical power electronics application, such as an adjustable speed drive.

Competing products include Toshiba's 5,000 V, 5,000 A thyristor, with a switching time of 10 msec. Its disadvantage is that it has several volts forward drop in the off state. (Here is where HTSCs might be useful.) On the other hand, the Power Electronics Applications Center in Oak Ridge, Tennessee, reports that devices now emerging from the laboratory switch 100 A in 1 msec, with a forward drop of only 1 to 1.5 V, while withstanding 500–1000 V. Therefore, the HTSCs will only be an improvement if they can get below 0.2 V forward drop or below 0.1 msec switching time.

23.8. SUMMARY

This chapter analyzed the use of high-temperature superconductors in fault current limiters. The motivation towards FCLs is that as system interconnections and capacities increase, FCLs may be the only solution if circuit breakers with sufficiently high ratings do not exist. Despite real advantages, the actual numerical values of key parameters in FCLs restrict the value of HTSCs. We have brought out the following points:

- Thorough cost/benefit analyses of FCLs are lacking, but surveys among U.S. utilities indicate that the price the utilities are willing to pay for an FCL is, at most, a few times the price of a circuit breaker of corresponding rating.
- The energy management in a FCL is a critical issue. How to safely dissipate the power is the difficult unanswered question. A truly effective FCL limits the current so quickly that it can *never* build up to damaging levels.
- Although using an air-inductor (coil) as the limiting element avoids the problem of energy dissipation during faults, such devices usually require a resistive triggering element. Energy dissipation within this element will still be a problem.
- The refrigerator operating cost may be reduced by a factor of up to 25 and the refrigerator capital cost by a factor of up to 10 for an SCFCL operating at 77 K compared with a device operating at 4 K.
- We *wish* we could have a superconductor with all the properties of NbTi, except for $T_c > 77$ K. Unfortunately, 77 K is accompanied by higher specific heat materials, which takes away some of the advantage of using superconductors.
- The heat capacity of HTSCs, greater by several orders of magnitude than LTSCs, improves the stability of the superconductor. However, it makes switching to the normal state difficult and slower, depending on the critical current density value.
- A J_c value of at least 10^5 A/cm^2 will probably be required to achieve adequate switching speed and an acceptable superconductor volume.
- Any superconducting device along a transmission line requires a very reliable refrigeration system. Also, circuit breakers enjoy a reputation for reliability and performance under field conditions that is hard to displace in the electric utility industry. FCLs must demonstrate comparable reliability before they will be accepted by utilities.

Finally, the advantage to be obtained is that if the FCL performs properly and limits the current to a modest excursion above normal, then many secondary savings will accrue, as circuit breakers and other components no longer need to be as rugged and the size of protective switchgear can be significantly reduced.

REFERENCES

1. *Study of Fault Current Limiting Techniques*, EPRI Report No. EL-6903 (1990).
2. *Fault Current Limiters: Basic Concepts and Associated Technologies*, EPRI EL-6275 (1989).
3. *Effects of Reduced Fault Duration Upon Power-System Components*, EPRI EL-2772 (1982).
4. *Superconducting Fault Current Limiter*, DOE Report No. ET/29063-1 (1982).
5. *Superconducting Fault Current Limiter*, EPRI Report No. EL-329 (1976).
6. T. Mitsui (TEPCO) at the joint ISTEC-CSAC Conference on HTSC, (May 1992); also T. Hara *et al.*, Paper WM 239-4 PWRD, of the IEEE/Power Engr. Soc. Winter Meeting (1992).
7. R. F. Giese and M. Runde, *Fault-Current Limiters*, prepared for the International Energy Agency by Argonne National Laboratory (January 1992).
8. T. R. Strobridge, "Cryogenic Refrigerators: An Updated Study," *NBS Technical Note* (1974).
9. P. G. Slade *et al.*, "The Utility Requirements for a distribution Fault Current Limiter," *IEEE Trans. Power Delivery* **7**, 507–515 (1992).
10. N. Engelman *et al.*, "Field Test Results for a Multishot 12.47 kV Fault Current Limiter," *IEEE Trans. Power Delivery* **6**, 1081–1087 (1991).

V

FUTURE POSSIBILITIES

24

New Refrigerators

As evidence mounts that the high-temperature superconductors (HTSCs) will not be satisfactory at 77 K because of their low J_c values in finite magnetic fields, new attention is being given to the possibility of using HTSCs in the 20–30 K range. To serve this purpose, novel refrigeration techniques are being developed which are aimed at providing cooling to such intermediate temperatures. The goal of 77 K is becoming less of a magic number.

The basic reason to cool only to 20 or 30 K, instead of going all the way down to 4.2 K with liquid helium, is contained in the Carnot-efficiency advantage of the higher temperature. As was enunciated in Chapter 3, the Carnot efficiency is the upper limit of efficiency; all real refrigerators have worse efficiencies. For the efficiency η we have

$$\eta = T_c/(T_h - T_c)$$

where T_h is the temperature of the hot reservoir and T_c is that of the cold reservoir. Evidently, it requires a factor of 5 less energy to get to 20 K instead of 4.2 K. Temperatures of 30 K and 77 K are easier still to attain.

This chapter is devoted to examining some of the refrigerators that reach the intermediate temperatures (in between 4 and 77 K). Refrigerators utilizing hydrogen, helium gas, and neon are reviewed, and a new type of magnetic refrigerator is described as well. Each of these devices has been reduced to practice, but some are not yet standardized, and all of them can be optimized through improvements in efficiency.

24.1. LIQUID HYDROGEN

It has long been known that liquid hydrogen is an excellent cryogen at about 20 K, and NASA has developed the capability to handle great quantities of it safely. Still, it must be remembered that liquid hydrogen is combustible; indeed, NASA uses it as rocket fuel. Where electrical equipment is concerned, the possibility of electrical discharges (i.e., sparks) is so great that liquid hydrogen is never considered as a practical cryogen. The suggestion that the electrical components might be kept separate from the cryogenics is generally greeted with skepticism.

Where we are today is roughly equivalent to the position of sodium-cooled nuclear reactors in the 1950s: the U.S. Navy wouldn't touch them because sodium explodes violently upon contact with water. Water-cooled reactors, first developed for naval vessels, went on to become the mainstay of the nuclear industry. Only in the 1990s has attention returned to

485

sodium-cooled reactors. Similarly, the day may come when liquid hydrogen is used to cool some applications of superconductors, but that day is far over today's horizon.

Nevertheless, it is a useful review to examine the way a hydrogen liquefier works. One typical device is shown[1] schematically in Figure 24.1. Like most refrigerators, the gas is initially at high pressure, and after cooling to close to the boiling point, a sudden expansion through a valve leaves some fraction of the hydrogen liquefied. The input energy is the work

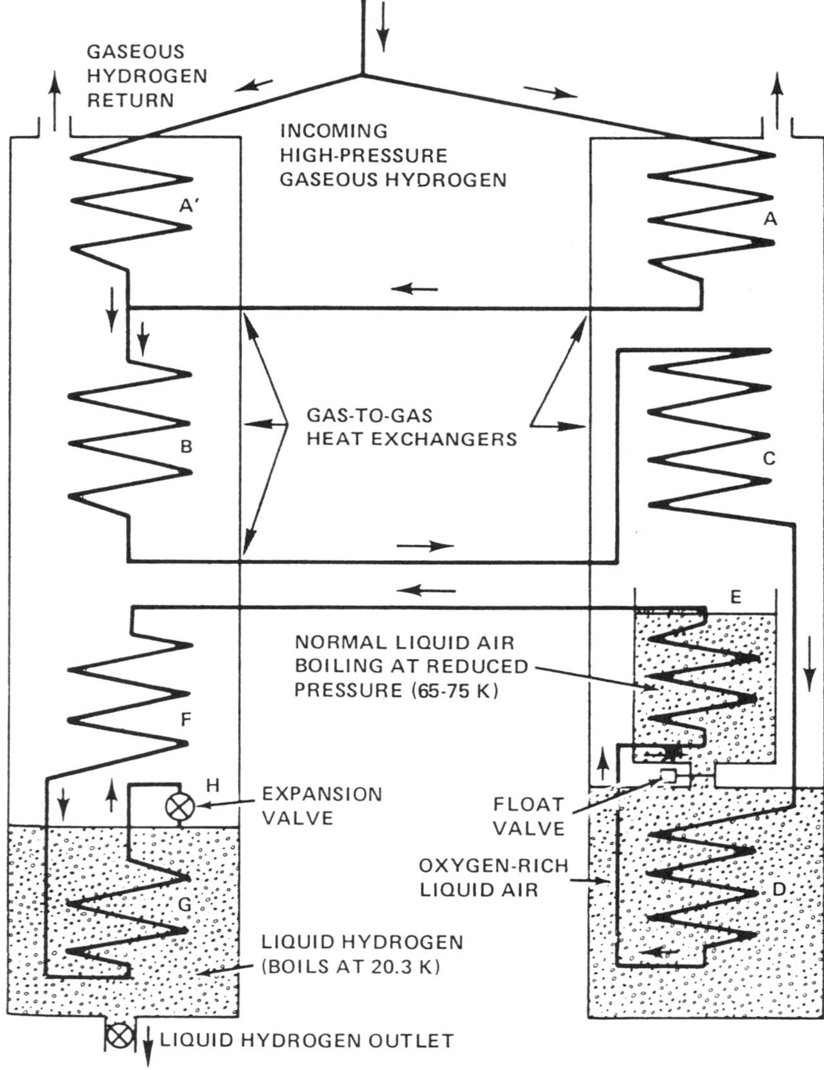

Figure 24.1. Diagram to illustrate the operation of a hydrogen liquefier. Heat is exchanged between counterflowing gases at A, A', B, C, and F; heat is exchanged with liquid air or nitrogen at D and E, and with liquid hydrogen at G. The expansion valve H forms about 20% liquid hydrogen and the rest (80% gas) exchanges heat with incoming gas as it warms. (From Ref. 1.)

of mechanical pumping to high pressure, and liquid air is required to provide intermediate cooling. The heat exchangers do the rest. This refrigerator utilizes counterflowing heat exchangers to optimize the efficiency of the overall system, so that as little energy is wasted as possible.

In actual operation, the system of Figure 24.1 begins at room temperature by pumping hydrogen to a pressure of 125 atm. The left-hand column is the hydrogen chamber, and the right-hand column is the air chamber. (Actually, it contains liquid N_2 in most realizations.) It takes some liquid air to get it started; the configuration illustrated is for steady-state operation in which a reservoir of liquid hydrogen already exists. The pressurized hydrogen is split between heat exchangers A and A', wherein it is cooled by cold nitrogen or hydrogen rising from the liquid bath below. The gas flow stream is recombined, then directed consecutively to heat exchangers B, C, D, E, F, and G. As the drawing shows, B and C exchange heat with cold hydrogen gas and cold nitrogen gas, respectively; D and E are in liquid N_2 and thus reduce the pressurized gas to about 65 K; F utilizes the extremely cold hydrogen gas just above the bath, and G sits in the hydrogen bath. After heat exchanger G, the gas is expanded through valve H, and about 20% of it becomes liquid. The rest (now at 1 atm pressure) rises back up the left-hand column, where it is warmed by exchanging heat with the inward-bound gas at F, B, and A'. Finally, the warm gas is captured by pumps that bring it back up to 125 atm, and it is sent back into the machine. Meanwhile, a small steady resupply of liquid N_2 is required in the right-hand column.

An essential characteristic is that the gas must cool upon expansion at valve H. If this were not so, the machine wouldn't work. This imposes the requirement that the hydrogen must first be cooled below the temperature at which dropping the pressure causes a drop in temperature. This temperature is known as the Joule–Thomson point, and varies from one imperfect gas to another. The Joule–Thomson point for hydrogen is at 205 K, and above that temperature hydrogen warms up as it expands. Thus, for H_2 it is essential that the gas be cooled below 205 K before it reaches valve H. This cooling is accomplished by having liquid N_2 in the right-hand column; it would be impossible to form an initial bath of liquid hydrogen without that precooling. Therefore, this machine is described as a two-stage refrigerator because there needs to be one compressor for H_2 and one for N_2 (or air). The fact that the nitrogen liquefier is at a factory 20 miles away does not change the concept. Other more complex refrigerators count stages in the same way. For example, because the Joule–Thomson point of helium is only 45 K, both a LN_2 bath and a LH_2 bath are required en route, and so we say that a three-stage refrigerator is necessary to liquefy helium.

In modern versions of the hydrogen liquefier, choices of heat exchanger materials improve efficiency and longevity, and the pressure can be more than 125 atm, but the basic principles are exactly the same. This kind of astute engineering translates into considerable cost savings over the total volume of liquid hydrogen needed by NASA. More important for our purposes, this same attention to efficiency may be the controlling factor that determines whether systems using HTSCs are practical or not.

24.2. COLD GASEOUS HELIUM

Cooling helium all the way to its liquid state is a difficult refrigeration task, with poor thermodynamic efficiency at best. Recently, a two-stage Gifford–McMahon[2] refrigerator has been developed by General Electric[3] to cool helium gas to below 10 K. Basically, this is an idea whose time has come: previously, there was no good reason to stop cooling at 10 K, because liquid helium was easier to transport around science laboratories, and anyone wanting to do experiments at 10 or 15 K could heat their apparatus up within the confines of a liquid helium bath. Today, however, the desirability of running a major system continuously at 10 K makes it cost-effective to capture the Carnot advantage associated with cooling only that far. Whenever a superconducting system uses Nb_3Sn wire, it is usually sufficient to cool only to 10 K.

24.2.1. Gifford–McMahon Refrigerators

It is pertinent to recall that Gifford–McMahon refrigerators are *regenerative* coolers, in which an oscillating gas pressure is used to carry heat away from a bed of material.[2] This device consists of a modified air-conditioning compressor and a movable piston (called a *displacer*) which drives helium gas back and forth through the *regenerator*, as shown in Figure 24.2. The lower (colder) section is at the other end of the displacer and regenerator. The motion of the displacer is synchronized with two valves that control a high-pressure inlet and a low-pressure outlet. The pressure varies by a factor of about 2, and typically it cycles at 1 to 2 Hz.

Starting at the load temperature T_L, heat transfers from the load Q_L to cold helium gas. The helium transfers heat back and forth in the regenerator as its pressure varies cyclically through the operation of the valves. Warm helium leaves the regenerator at low pressure and exits to the compressor. Upon recompression, the helium contains both the heat transferred out of the regenerator and heat generated by compression, but this gas is cooled to ambient temperature before returning to the high-pressure inlet valve. Heat Q_R is rejected at temperature T_R in Figure 24.2.

The key to the operation of a Gifford–McMahon cycle is in the timing of the valves and the synchronization of the displacer:

1. With the displacer in its down position, high-pressure gas is allowed to enter the upper volume V_1 and the regenerator. With the inlet valve still open, the "intake stroke" takes

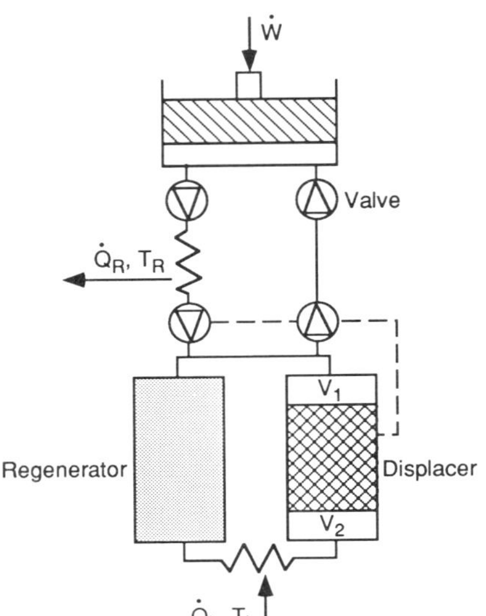

Figure 24.2. Schematic diagram of a Gifford–McMahon refrigerator. Work is done by a compressor at the top. The dashed lines indicate that the displacer position is timed to the position of the valves (see text).

place when the displacer is moved *upward*. In an automobile engine, this would seem like the *compression stroke*, but here what happens is that *additional* gas is drawn through the regenerator to fill the newly created volume V_2 *at the lower end*. This end is quite cold, so more total mass of helium enters the chamber. The contents of volume V_1 are not enough to fill V_2 and keep the pressure high.

2. The inlet value then closes and the exhaust valve opens to bring the system to a lower pressure. The decrease in pressure causes the volume V_2 to expand and drop in temperature, and *this is what provides the cooling*. The *exhaust stroke* follows, in which the displacer moves downward, forcing the cold gas into the regenerator and eventually out the exit valve. The regenerator matrix material is left slightly cooler by this transport, ready for the cycle to begin again. A small mass of helium, now at the higher temperature but at low pressure, enters the newly created volume V_1.

The overall thermodynamic process is to trade work at the compressor for expansion (and cooling) of gas at the lower end of the displacer. The temperature of the exiting gas is slightly higher than that of the incoming gas, and the net refrigeration appears in that difference. The efficiency with which heat transfers to and from the regenerator matrix material affects how much heat is ultimately rejected on each cycle.

In a two-stage Gifford–McMahon refrigerator, there are two displacers driven by a common compressor, and two regenerators in series to handle the gas flow, as indicated in Figure 24.3. This way the first stage runs between about 40 K and 300 K, and the second

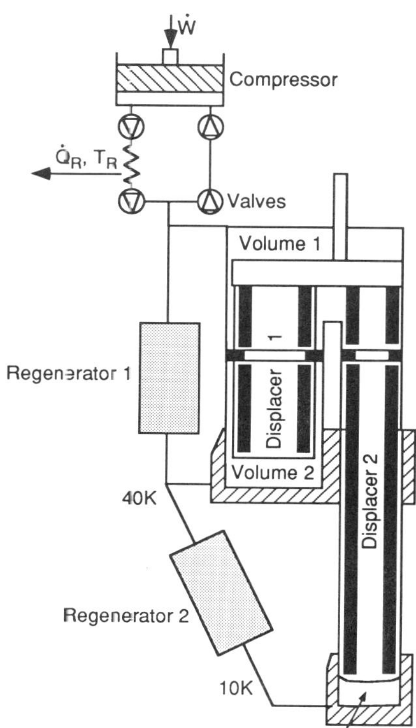

Figure 24.3. Two-stage Gifford–McMahon refrigerator, with separate regenerators but displacers driven by a common synchronizing mechanism. The intermediate thermal shield at 40 K provides much better thermodynamic efficiency than could be realized in a one-stage refrigerator.

stage runs between 10 K and 40 K, which improves thermodynamic efficiency. The load on the first stage is the heat rejected by the second stage.

24.2.2. Specific Implementation

In the GE device, the first stage cools a vacuum jacket to below 40 K, and the second stage brings helium gas down below 10 K, where it contacts a cold head. (This latter device is just an off-the-shelf copper cold finger of the type used in many cryocoolers.) The use of new regenerator materials, such as Erbium-nickel and its derivatives,[4] makes it possible to reach 10 K with only two stages.

The exact temperatures achieved in any particular application are determined by reading a refrigerator load map of the type shown[3] in Figure 24.4. For example, if it can be predicted that the sum of radiation, convection, and so on is 1.5 W at the coldest point and 25 W at the thermal shield, then by plotting that point in Figure 24.3 (using the top and right-hand scales), the temperatures reached by each stage can be predicted (by reading off the bottom and left hand scales): 9.1 K and 36.5 K. Conversely, by measuring the actual temperatures reached in a real device, the measured thermal load can be derived. In one embodiment,[3] GE measured temperatures of 9.5 K at the second stage and 34.6 K at the thermal shield (first stage), which translates into heat loads of 1.9 W and 20 W, respectively. Obviously, the exact operating temperatures depend on the heat load.

With 1 kW electrical power input at room temperature, the cryocooler delivers 30 W of cooling power at the first stage and 2 W of cooling power at the second stage. If that 2 watts

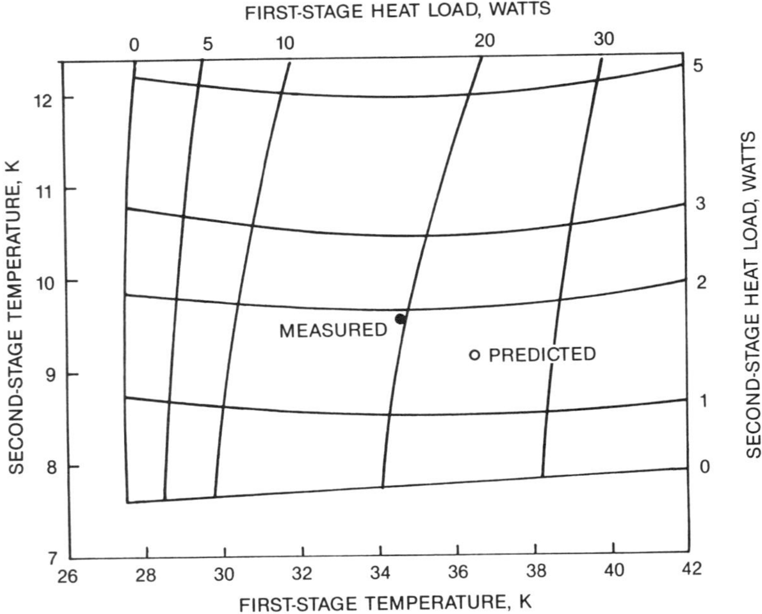

Figure 24.4. Refrigerator load map pertinent to the Gifford–McMahon refrigerator built by General Electric. The basic graph shows how first-stage temperature relates to second-stage temperature, but these variable can be mapped into first-stage heat load and second-stage heat load, as indicated. The predicted and measured points are discussed in the text.

had to be removed via liquid helium cooling, the system would use up 72 liters per day, which is prohibitively expensive. (In fact, conventional designs strive very diligently to get the heat load down to about $\frac{1}{10}$ W, expressly because of the liquid helium expense.) Here the ability to handle 2 W relaxes design constraints enormously.

There are many potential applications of such a refrigerator. For example, General Electric has made MRI systems for many years, and plans to incorporate this technology therein. So far, GE has built an MRI unit that uses Nb_3Sn wire and their two-stage Gifford–McMahon cryocooler. The unit does not even use liquid nitrogen. Although this may seem less than ideal (given the low cost of liquid nitrogen), it must be remembered that not everyone can service liquid nitrogen, and so the higher cost of directly refrigerating helium gas is offset by the lower operating cost associated with not attending to the nitrogen supply. By contrast, a typical MRI unit costs a hospital $20,000 annually for a service contract. Here the cost of helium gas replacement is a fraction of that.

The only role for HTSCs in this GE MRI unit is as current lead-ins from the 40 K heat shield, because HTSCs substantially reduce the heat leak: preliminary results indicate a decline in heat load from 2 W to 0.5 W by using HTSC lead-ins. This would allow the cold end to run at 8 or 9 K, which provides a wider margin away from the point at which the Nb_3Sn would quench—typically 12 K for the kind of currents carried in these systems.

Relaxing the demand in temperature allows still other trade-offs to be made. The specific heat of almost any material rises steeply between 4 and 20 K, which implies that sudden wild excursions in temperature are far less probable in a system that operates at 20 K. In any apparatus, there eventually has to be a piece that sustains the temperature gradient to high temperatures (known as a bushing), and this can be a very temperamental component if it has to reach down to 4 K. In the experience with Brookhaven's transmission line, the cryogenic bushings caused major engineering headaches. For any device in which heat is generated, the difference between 4 and 20 K provides an enormous comfort zone for design engineers.

At the present time, reliability is the greatest limitation of Gifford–McMahon refrigerators. Oil (from the compressor) can find its way to the cold end of the regenerator, where it can freeze out and degrade performance.[3] Oil filters help to some extent. By using an oil-free compressor, this particular problem can be eliminated; but the mean-time-to-failure needs to be around 5 years for adequate reliability, and oil-free compressors are not there yet. Continuing improvements in reliability promise a bright future for Gifford–McMahon cryocoolers.

As applications of BSCCO around 20–30 K are contemplated, it is plausible to imagine that a cryocooler based on a two-stage Gifford–McMahon cycle could be made even more efficient, with concomitant cost-savings during operation. The design parameters are reasonably optimistic, but scaling up a design is always tricky. The large heat loads originating in utility-scale devices may handicap the efficiency and sacrifice cost-effectiveness. It is premature to speculate here on the eventual efficiency achievable.

24.3. LIQUID NEON CRYOSTAT

The boiling point of neon at one atmosphere pressure is 27 K, so a bath of liquid neon would be an excellent cryogen for systems operating around 30 K. Neon has seldom been used to date simply because no one particularly wanted to operate around 30 K. Also, it is

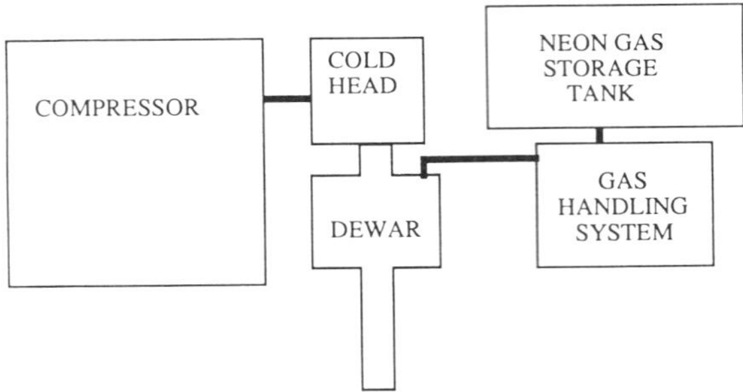

Figure 24.5. Simplified diagram of the liquid neon facility at the Naval Research Laboratory.

very expensive: one liter of liquid neon costs about $100. At that price, even a very small leak in a closed system can become expensive in a hurry.

The Naval Research Laboratory has built a liquid neon cryostat[5] and used it to test some HTSC coils at 27 K. It is shown schematically in Figure 24.5. It was quite simple to construct, and worked the first time it was tried. It takes about one cubic meter of S.T.P. neon gas to make a liter of liquid. In this device, the neon is compressed and pumped to a cold head, which is in contact with a liquid helium reservoir. The pumping flow rate is regulated with a series of valves so as to collect the neon in its liquid state, rather than allow it to solidify at the temperature of liquid helium.

In this device, all the work of cooling is done by the liquid helium bath. If helium leaks out or even boils away, that is no major loss. The cost is in the noise level of the interest on the $100,000 original price. Clearly, there is no real need to use liquid helium at all. A refrigerator like the Gifford–McMahon cryocooler described above could be used to reach a point below 26 K, from whence a cold finger could be used to produce a bath of liquid neon.

What is important here is that NRL has shown that it is easy to work with liquid neon. Clearly, the major obstacle is the cost of neon. NASA Lewis Research Center used neon years ago and reports that the cost of neon is demand-driven. It is not all that scarce, being one of the gases remaining from a liquid–air machine; but since there is so little demand for it, there is little incentive for major producers of oxygen and hydrogen to capture and market it. NASA believes that if the demand were there, the price of neon would drop appreciably. It remains to be seen whether the renewed interest in the 20–30 K temperature range will create a market for neon.

24.4. MAGNETIC REFRIGERATION

Active magnetic regenerators (AMRs) are being developed by the Astronautics Corp. of America (ACA) of Madison, Wisconsin. As mentioned in Chapter 3, the *magneto-caloric effect* allows magnetic salts to be used with adiabatic demagnetization to reduce the temperature of a system below that obtainable by pumping on a liquid cryogen. Historically, that meant starting with liquid helium, vacuum pumping to below 1 K, and then adiabatically demagnetizing to millikelvins. However, what ACA has done is to move the whole enterprise up to the liquid nitrogen range: starting from 70 K (achievable by pumping on LN_2), their

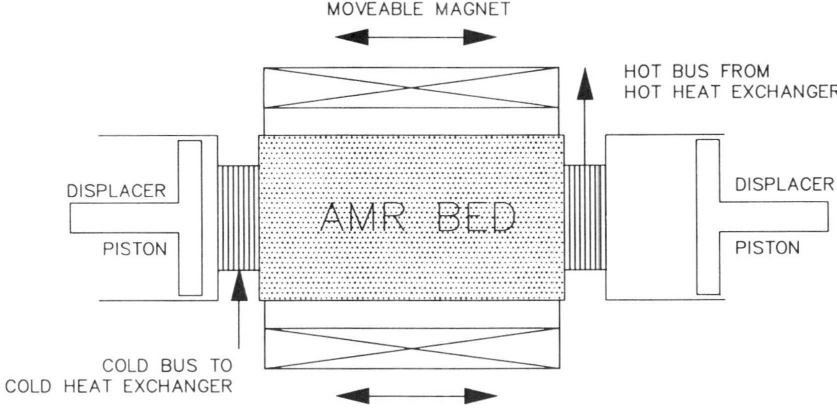

Figure 24.6. Schematic diagram of active magnetic regenerator configuration with regenerator bed surrounded by magnet.

device can lower the temperature to 40 K via magnetic refrigeration.[6] Because the magneto-caloric effect is fully reversible (unlike expansion of a gas), it makes for a very high-efficiency refrigerator.

A schematic diagram of the ACA *active magnetic regenerator* appears in Figure 24.6. First of all, the device is a regenerative cycle, not a recuperative cycle. There is a bed of magnetic material in the middle, and helium gas is going to be pushed through it from a cold (left) side to a warm (right) side. Not shown on the left and right are heat exchangers which constitute the heat load and heat sink, respectively. The left and right pistons are synchronized, moving together to force the helium gas through the bed in one direction or the other. Surrounding the entire magnetic material is a magnet, which can be slid back and forth. The magnet needs to be about 7 T, and would therefore be superconducting.

In the design study carried out by ACA, the cold and hot reservoirs were taken to be 40 K and 70 K, respectively. In the initial state of each cycle, the temperature across the width of the bed of magnetic material (taken to be $GdNi_2$) rises more or less linearly from 40 K to 70 K, as shown in Figure 24.7. When the magnet is moved into position, there will be a slight increase in temperature all along the bed, corresponding to the ΔT associated with magnet-

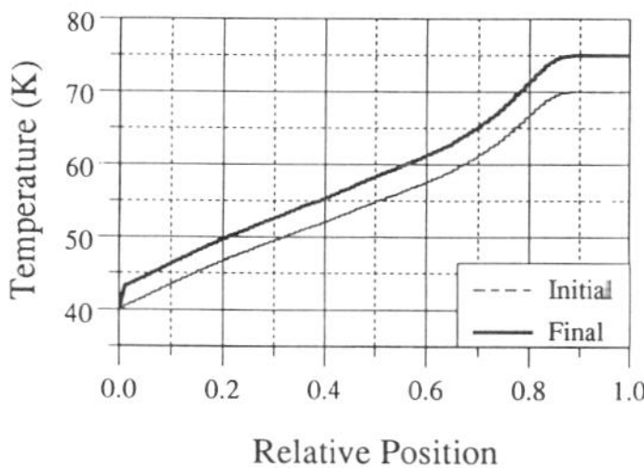

Figure 24.7. Temperature profile across the regenerator bed. When the field is increased, the temperature rises uniformly by a few degrees at each point. This represents step (a) of the cycle of this refrigerator.

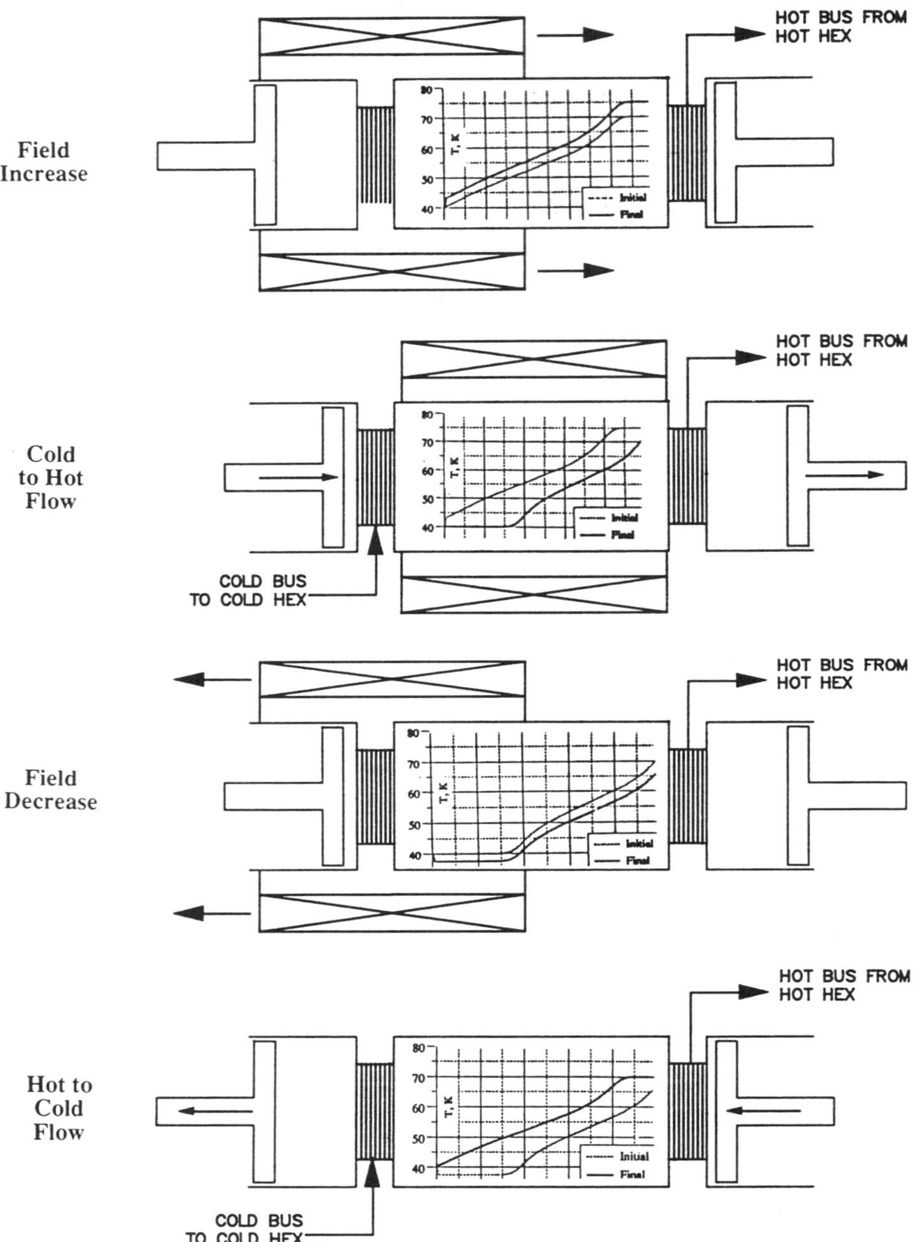

Figure 24.8. The full cycle of the active magnetic regenerator, with temperature profiles for each step superimposed in the center of each drawing. The final temperature profile for each step becomes the initial profile for the next step.

izing the GdNi$_2$. That slightly higher temperature profile, ranging from 43 to 75 K, is also shown in Figure 24.7. Heat will begin to leave the right-hand side because the heat sink is assumed to be held at 70 K.

A better way to visualize all this appears in Figure 24.8. The figure is rather complex and bears some scrutiny. In sections (a)–(d) of this figure a temperature chart (akin to Figure 24.7) is superimposed on a drawing of the apparatus (akin to Figure 24.6) with the magnet and pistons in their consecutive positions. The first step of the cycle, described above, is represented at the top (a) of Figure 24.8. The initial temperature profile of step (b) is the final profile of step (a).

The next step (b) is to drive the pistons rightward, forcing 40 K helium gas into the bed of material. As shown in the central chart of (b), this lowers the temperature uniformly throughout the bed, reducing the right side to 70 K again. By the end of step (b), a fair fraction at the left of the magnetic bed is at 40 K.

In step (c), the magnetic field is taken away by sliding the magnet to the side, and the bed is then cooled by adiabatic demagnetization. Meanwhile, the hot-end heat exchanger exhausts heat from the gas in the chamber at right. Upon completion of the demagnetization step, the bed is prepared to cool that gas because it now has a lower temperature profile.

The cycle is completed in step (d), where the helium transfer gas is displaced from right to left. It warms much of the bed as it passes through, but the gas entering the cold reservoir is at 40 K. The final temperature profile of step (d) is the initial profile for step (a) again.

Astronautics actually built a version of this refrigerator, called a proof-of-principle device, and it behaved generally as predicted. Actually they were able to obtain a low-temperature of 26 K, but that was with no heat load. In that model, a 2-W heat load brought the low-temperature end to 42 K, in line with predictions.

Certain limitations demand recognition here. First and foremost, *work* has to be put into this thermodynamic system someplace. That occurs in sliding the external magnet back and forth with the regenerator bed in different states of magnetization: it takes more force to pull the magnet away than to push it in. This draws attention to the external magnet itself, which generates a 7 T field. Seven tesla means it's a superconducting magnet, and that in turn implies the presence of liquid helium, which means this prototype refrigerator doesn't exactly stand on its own. Also, the device has to operate at 10 to 20 atm pressure, and with moving parts, the leaks around seals mean more headaches. Moreover, the thermal mass of the system is so great that it cools down very slowly.[7] None of these problems are fatal flaws, but they show that a lot of engineering work still lies ahead.

Notice that nowhere does the magnetic material change in temperature from 40 to 70 K. Rather, each portion of the bed changed by only a few degrees during the entire cycle. The presence of a substantial gradient in temperature with position is what makes it possible to span $\Delta T = 30$ K with this device. Alternately, it could be run faster (shorter cycle time), thus offering greater cooling power.

Of course, there is a limit on how wide a temperature range the magnetic refrigerator can span. This is set by the properties of the magnetic salt used in the regenerator, which in the present case is GdNi$_2$. This material is ferromagnetic, with a curie temperature of 75 K. The AMR cannot run above 75 K, because the GdNi$_2$ would warm instead of cooling upon demagnetization, thus defeating the whole concept of the magnetic refrigerator.

Separately, the heat capacity of helium gas is a constant (3R per mole) above about 20 K. This implies that if the AMR is to be fully reversible with a balanced back-and-forth flow, the magnetic refrigerant must have a ΔT proportional to the absolute temperature. Fortu-

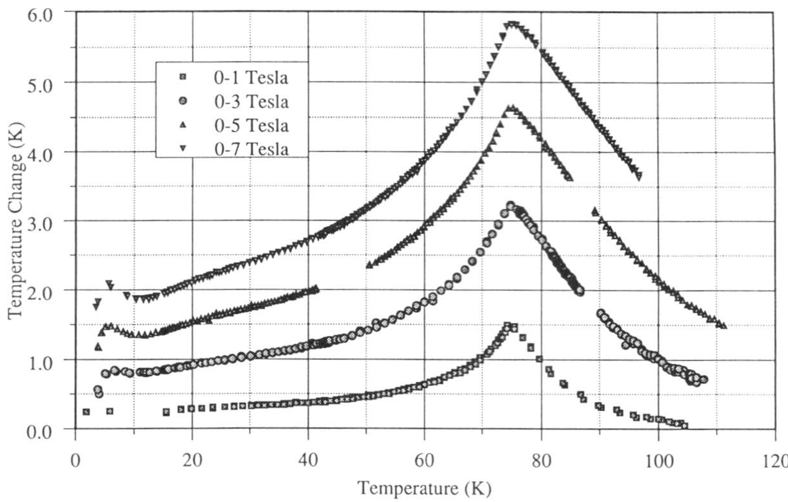

Figure 24.9. Experimental data on the temperature change induced in GdNi$_2$ by various changes in magnetic field. Note that the data between 40 K and 70 K lies reasonably close to a straight line drawn through the origin.

nately, GdNi$_2$ can accommodate this requirement. The heat capacity of a ferromagnet varies as $1/(T - T_{\text{curie}})$, and the temperature rise ΔT associated with magnetizing it varies with the specific heat. Happily, for GdNi$_2$ in the range of 40 to 70 K, this curve (shown in Figure 24.9) bears a reasonable resemblance to a linear increase with temperature. In fact, a line through the origin and the points at 40 K and 70 K is close to the experimental data, so GdNi$_2$ is a good magnetic refrigerant for this device.

The ACA design promises 50 W of cooling power at 40 K, which is appropriate for HTSC motors, generators, and SMES. With the heat sink at 70 K, the modest vacuum required upon a liquid N$_2$ bath does not pose any design difficulty. In a subsequent effort, ACA hopes to build a two-stage hydrogen liquefier that uses magnetic refrigeration, and that would liquefy 1 ton of H$_2$ per day, starting from 77 K. Furthermore, with respect to units that operate near room temperature, ACA argues that for a temperature differential of only $\Delta T = 30$ K, magnetic refrigeration is *more* efficient than Freon-based refrigerators. Typical beneficiaries of such devices would be high-power energy-intensive applications, such as HVAC systems for large buildings.

24.5. SUMMARY

In this chapter, we have previewed some of the coming attractions associated with the new superconductors. Because these materials perform much better at 20–40 K than at 77 K, there is a need to consider cooling to such temperatures without going all the way down to 4.2 K with liquid helium. This attention leads to consideration of several different means of cooling to intermediate temperatures.

The most obvious way to reach 20 K is by creating a bath of liquid hydrogen, which is easily done with a recuperative refrigerator akin to those that liquefy nitrogen or helium. At NASA they have handled large quantities of liquid hydrogen safely for years, but still it *is* a rocket fuel, and if anything goes wrong, it burns explosively. For eventual industrial applications where careful handling cannot be assured, it is necessary to look beyond hydrogen as a means of cooling.

In a liquid helium machine, the final stage (Joule–Thomson expansion) is irreversible and inefficient, so any system that stops short of actually liquefying the helium is assured of being more efficient. For large-scale applications to industry this can tip the economic balance enough to make a high-temperature superconductor economically attractive. Temperatures of 10–20 K can be achieved by cooling helium gas in a regenerative refrigerator, and General Electric has built a Gifford–McMahon refrigerator that does exactly that. Although present configurations have only a few watts cooling capacity, it is plausible to expect that in the future, scaled-up versions of the same general principles will provide much more capacity.

Liquid neon provides a bath at 27 K, which is near the upper useful limit of BSCCO. Neon is terribly expensive to buy in the first place, but if a leak-tight system is constructed carefully, then neon can be a practical coolant. The idea of using cold helium gas and a heat-transfer chamber to liquefy the neon is an obvious means of gaining efficiency while keeping the neon away from mechanical compressors (and possible leaky seals), but no such hybrid refrigeration unit has yet been built.

Recent advances in magnetic materials make it possible to consider a regenerative refrigerator based on adiabatic demagnetization, but running in the 40–70 K range. A proof-of-principle magnetic refrigerator has been constructed to demonstrate this concept. Of course, a 7 T magnet must accompany the device, and if that is a superconducting magnet, then the overall efficiency must be burdened with the cost of cooling that magnet. At this time, it is by no means clear that a magnetic refrigerator can win out economically over a helium gas regenerator, which can reach 40 K with a single stage. Nevertheless, with improvements over time, magnetic refrigeration may gain economic dominance as a means of reaching intermediate temperatures.

The central change that is now taking place is that cryogenics is no longer a simple matter of either liquid nitrogen or liquid helium. The U.S. Navy has a very clear philosophy: the idea of "benchmark" temperatures is old fashioned, and "any system at any temperature is acceptable as long as it meets the reliability and efficiency requirements."[8] Many other users feel the same way. For example, hospitals running MRI systems are oblivious to the presence of LHe *or* LN_2; if a Gifford–McMahon refrigerator cooling BSCCO to 20 K were introduced tomorrow, they wouldn't even notice.

Obviously, there is a large efficiency advantage at 77 K compared to 20 K. But still, there are some circumstances where 20 K operation is warranted. The intermediate-temperature range is of great interest to those striving to make practical devices. The economics associated with refrigeration is an integral part of the overall economic feasibility of any system involving superconductivity. Consequently, we expect the market to drive further improvements in efficiency and reliability of refrigerators in the years ahead.

REFERENCES

1. J. L. Sloop, *Liquid Hydrogen as a Propulsion Fuel, 1945–1959*, NASA special publication SP-4404 (Supt. of Documents, U.S. Government Printing Office, Washington, D.C., 1977).
2. R. A. Ackermann and W. E. Gifford, *J. Eng. Industry*, 273 (1969).

3. K. G. Herd and E. T. Laskaris, "Refrigerated High-T_c Superconducting Devices," in AIP Conference Proceedings #273: 6th NYSIS Conference, Buffalo, NY, edited by H. S. Kwok *et al.*, pp. 462–470 (September 1992).
4. R. Li *et al.*, *Cryogenics* **30**, 521 (1991).
5. T. L. Francavilla and D. U. Gubser, "Measurements of HTSC Wires and Coils: Liquid Neon Facility," in AIP Conference Proceedings #273: 6th NYSIS Conference, Buffalo, NY, edited by H. S. Kwok *et al.*, pp. 455-461 (September 1992).
6. C. B. Zimm and A. J. DeGregoria, "Magnetic Refrigeration: Applications and Enabler for HTSC Magnets," in AIP Conference Proceedings #273: 6th NYSIS Conference, Buffalo, NY, edited by H. S. Kwok *et al.*, pp. 471–480 (September 1992).
7. C. B. Zimm, presented at DOE Annual Peer Review, July 19–20, 1994.
8. M. Superczynski, private communication.

25

Applications to Measurement and Process Control

The theme of these later chapters is to look at what might someday come from high-temperature superconductivity, as contrasted to what is already known about it. Chapter 26 examines those potential applications where the HTSCs make it possible to use extremely high magnetic fields. This chapter is devoted to some potential applications in industrial processes, where the role of HTSCs is in measurement and control. The superconducting quantum interference device (SQUID) is the measurement tool of greatest interest.

Customarily, thoughts of industrial energy conservation begin with large-scale processes,[1] in which great streams of raw materials, energy and capital are combined at a mill to produce some product. In this framework, a new manufacturing process that saves a few percent of the energy used can save millions of barrels of oil equivalent.[2] However, it is seldom noted that the fluctuations inherent in any process line eat up millions of barrels, too. By controlling[3] industrial processes to run at precisely their target specifications, it is possible to reduce waste, increase productivity, make the final product at a more competitive price, and save energy at the same time.

During the 1970s, a variety of energy efficient processes were devised in university research laboratories, but most never became full scale production methods. Sometimes this was because of control problems. Eventually those good ideas should be re-examined to see if their controllability would improve with superconducting sensors. Imaginative engineers might look at other industrial processes, currently in use but far from optimal, and ask whether previously nonexistent sensors might now be possible. Superconductors operating at 77 K open a door to new measurements that may lead to major improvements in control of processes, with corresponding savings in energy.

The basic premise of the field of industrial process control is this: measurement without control cannot increase productivity; control without measurement is impossible. In the current state of this art, the limitations of controllability are set by the accuracy of measurement sensors.

25.1. PRINCIPLES OF SENSORS

What makes a good sensor? Suppose a small change in some hard-to-observe but commercially important variable produces a large change in something else that is easy to observe. The device that converts the former to the latter is a good sensor.

499

The classical example is the Beta-gauge. Here electrons from radioactive decay (β-rays) impinge upon a sheet of steel or paper. The ones that penetrate through the sheet enter an ionization chamber, where a current is generated that is easily detected. The number getting through is proportional to the density and thickness of the sheet (the *basis weight*), which is the commercial parameter by which the product is sold. Gauge designers strive to keep the current linearly proportional to the number of electrons N penetrating the sheet of thickness x, *and* to maximize the logarithmic derivative:

$$\frac{x}{N} \frac{dN}{dx}$$

These two goals are common to all industrial gauges. Maximizing the derivative means selecting some physical interaction that will yield a large change in a readily observable variable due to a small change in what you really want to control.

The quest for superconducting gauges must respect these same criteria. If a gauge is too noisy, the linearity criterion will fail. The requirement of a steep derivative points toward certain relationships that characterize superconductors in general.

25.1.1. Example: Critical Current Sensor

The principles used here are best illustrated by a specific example. Imagine a gauge that takes advantage of the extreme sensitivity of J_c to a changing magnetic field.

Chapters 14–16 describe at length how experimenters have struggled to prevent HTSCs from suffering a steep decline in J_c with applied magnetic field; this is done so as to allow the HTSCs to carry large currents. But consider turning this problem around, and utilizing the steepness of the change in J_c with H. For ordinary sintered YBCO, the relationship is drawn in Figure 25.1. Thus with the system held at a fixed temperature (say 77 K), a tiny change in magnetic field produces a large change in critical current. The logarithmic derivative $(H/J_c)\, dJ_c\, /dH$ is very large, and therein lies the basis for a gauge.

Consider a wire of given diameter which is carrying a current density just below J_c. If the wire goes normal, it will be the equivalent of a diode shutting off, or an electronic gate closing. That in turn can be used to signal a circuit elsewhere to recognize the condition and take appropriate action. When the wire is held in proximity to something that generates a

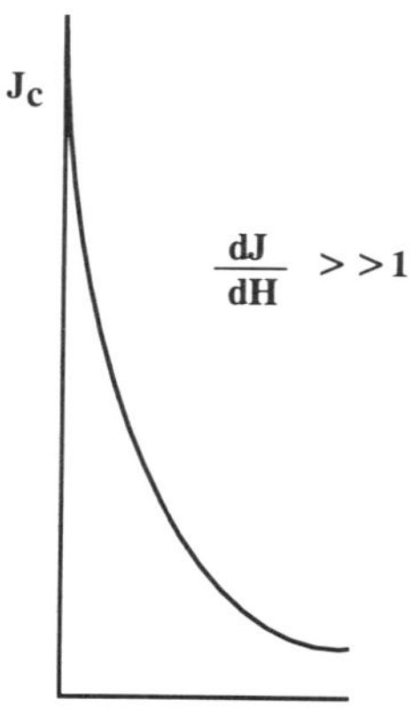

Figure 25.1. Steep decline of critical current with applied magnetic field (plotted on linear graph paper) in a typical sintered HTSC without special preparation.

magnetic field (say, a chemical stream containing a certain percentage of iron), as soon as the magnetic field gets a bit too high, it will drive this wire normal. The value of the magnetic field determines the switch-off point.

Obviously the current in the wire could have a different value, in which case a different magnetic field would be needed to trigger the switch. The curve of Figure 25.1 shows where the relationship is most sensitive. Notice that if the current were swept in time (for example, a sawtooth waveform) at a fairly high frequency, the time at which switching occurs could be used to measure a magnetic field as it varies in time at a lower frequency.

There is a well-known need for detecting time-varying magnetic fields, namely the search for submarines. However, that application is searching for very small magnetic fields, and a Josephson junction device (a SQUID) is a more likely candidate. Here we are thinking of measuring modest magnetic fields with precision, so that the process creating the magnetic field may be controlled.

The gauge possibilities proliferate. If a whole series of wires (having cross-sectional areas that differ by consecutive factors of 2) were used with suitable circuitry to keep track of their on/off conditions, a digital representation of a magnetic field might be obtained. If the magnetic field were simply the field in a solenoid carrying current I, we would have an A-to-D converter for that current.[4]

There is a lot of room for imagination to run here. The basic point is that because the new superconductors have a critical current that drops sharply with magnetic field, the steep dJ_c/dH provides an invitation to devise a gauge. Of course, there must be something relevant to the production process that causes the magnetic field to vary.

25.2. HTSC SQUIDs

As mentioned in Chapter 5, superconductors offer a unique and very sensitive means of detecting small magnetic fields, through the device known as a SQUID. These delicate instruments have found niche applications in laboratories and medical centers, where the value of their measurements outweighs the cost of cooling to 4 K. This chapter looks at the question of possibly using SQUIDs in factories, where the setting is more rugged and the value of the measurement is not as great. Under such circumstances, the SQUIDs to be used are unquestionably HTSC devices, which presumably cost less to operate. With this in mind, it is necessary to ask how such SQUIDs differ from the LTSC SQUIDs described in Chapter 5.

SQUIDs are the most sensitive instruments at the scientist's disposal; not only can they detect a single quantum of magnetic flux, but the resolution of the best SQUIDs allows detection of magnetic field variations corresponding to 1/1000 of a flux quantum. The lower limit of sensitivity in the laboratory is 10^{-14} T, but commercially purchasable LTSC SQUIDs typically detect around 10^{-11} T. Thinking in terms of HTSCs, with good shielding of stray magnetic fields, it is imaginable that fields in the microgauss range (10^{-10} T) might be seen in factories. This information on very weak magnetic fields could in turn be used in a process control system.

25.2.1. Research Results

Progress on HTSC SQUIDs has been very rapid. That subject would fill an entire book. Here we are limited to a few general observations:

SQUIDs are generally made of thin films, and so the research effort breaks down into categories of film properties, junction quality, and substrate effects. Unlike the "bulk wire" applications, here there is no concern with maximizing J_c or wire length. The highest operating temperatures are achieved with SQUIDs made of TBCCO, but since 77 K is the customary operating temperature, there is greater interest in the achievement of low noise.

The biggest difficulty originates in the anisotropy of the HTSCs. Unlike integrated circuits of Si, GaAs, or Nb, HTSC integrated circuit structures must be entirely epitaxial, which means that successive layers must be highly aligned in both the direction of growth and in the plane. This requirement increases the already severe materials problems involved with both multilayer processing (deposition and etching) and with Josephson junctions themselves. Processing steps that are both familiar and automated in semiconductor technology cannot be easily replicated in HTSC technology.

The quality of the SQUID junction is always an issue. Early SQUIDs had a very poor "yield rate," and were very noisy as well, owing to a number of deficiencies in processing methods. Furthermore, they tended to stop working at about 70 K, below the temperature of LN_2. That may have been an artifact of manufacturing edge-junctions, in which the oxygen was depleted around the edges of the film. Anyway, more recently, along with a better understanding of magnetic behavior has come an improvement in the quality of SQUIDs.

As one example of research toward an improved junction, IBM made SQUIDs having a "washer" geometry,[5] as shown in Figure 25.2. They obtained remarkably low noise Figures from such a device. In particular, at 77 K they observed[6] an energy sensitivity as low as 1.5×10^{-30} J/Hz in a YBCO "washer" SQUID, and a $1/f$ noise of 1.2×10^{-28} J/Hz at 10 Hz. The advantage to this geometry is that it tends to promote flux focusing, without increasing noise, thereby enhancing the sensitivity of the SQUID.

The noise in a SQUID comes predominantly from flux creep; indeed, SQUID noise is one way to measure flux creep. The grain boundaries (which themselves are not superconducting) let flux through. In a conventional LTSC, the pinning sites are located on the grain

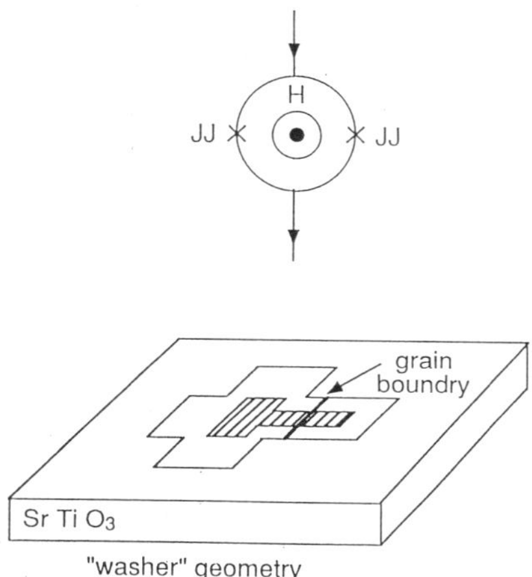

Figure 25.2. (Top) Schematic diagram of two Josephson junctions, with H perpendicular to the plane of the junctions. (Bottom) One configuration of a thin film of YBCO with Josephson junctions at the grain boundary. Vacant area in the middle creates two Josephson junctions.

boundaries. HTSCs allow penetration of the magnetic field through the grain boundaries; even milligauss fields will penetrate. If good quality films without grain boundaries could be made, that wouldn't happen. But in the real world, the vagaries of the grain boundaries prevent similarly-made SQUIDs from being the same. There is a substantial hysteresis effect that creates magnetization, gives apparently different B fields, and different voltage-versus-phase behavior. With experience, important variables become identified, and so improved SQUID performance has come with better control over grain boundaries.

25.2.2. Substrates

Another concern about SQUIDs is the substrate. Every thin film must be deposited on a substrate, and the HTSC materials are no exception. For commercial SQUIDs outside the laboratory, the film-substrate combination must be very rugged.

Popular substrates for YBCO include $SrTiO_3$ (strontium titanate), MgO (magnesium oxide) and $LaAlO_3$ (lanthanum aluminate), among others. The optimum choice of substrates is a compromise between trying to match the interatomic dimensions of the substrate to the film (to create dislocation-free crystals in the film), preventing interdiffusion of substrate chemicals into the film, avoiding excessive shrinkage (and concomitant cracking) during cooling, and of course providing adequate physical strength. All the substrates used so far with HTSC films have limitations in one or more categories.

Improvements to the substrate are as important (and as frequent) as improvements in the film itself. Depending on the application intended, researchers strive to optimize one property or another of the substrate, which obviously involves trade-offs. To illustrate this, consider only a single example:

The Polish Academy of Sciences studied the substrate calcium neodymium aluminate ($CaNdAlO_4$). This substrate has a very low dielectric permittivity ($\varepsilon \cong 20$) and low losses ($\tan \delta \cong 10^{-3}$) in the frequency range of interest for infrared and microwave applications. However, the novel feature is that the material is itself a perovskite, which enhances compatibility with the perovskite HTSCs: the interdiffusion is extremely low. The lattice match is also better than that of most conventional substrates, with discrepancies of only 4.4% with YBCO and 3.4% with BSCCO. There is some degradation of T_c (onset of superconductivity at 87 K, zero resistance at 84 K) but since the operating temperature is 77 K, this is tolerable. The achievement of the Polish Academy was to grow twin-free single crystals of $CaNdAlO_4$, a prerequisite for producing good quality films.

This kind of steady improvement is typical of HTSC research. Prior to the existence of perovskite superconductors, there was no motivation to have perovskite substrates. In the coming years, we expect to see improved materials come along in response to needs created by other new materials.

25.2.3. Commercial Products

The research on HTSC SQUIDs has been carried far enough to earn the term "satisfactory" for temperature and noise. Now the problem is to make them inexpensively. For many applications, the SQUID itself is often not the significant cost factor. Roger Koch, who leads IBM's research team on SQUIDs, has said[7] "If you had free SQUIDs with no noise, you still couldn't improve the market-penetration of SQUIDs, because of many other factors." For example, in biomedical applications (see Chapter 5), where perhaps 200 channels of SQUIDs comprise a scanning system, the cost of a shielded room in which to operate the scanner is about $500,000, far greater than the SQUID system.

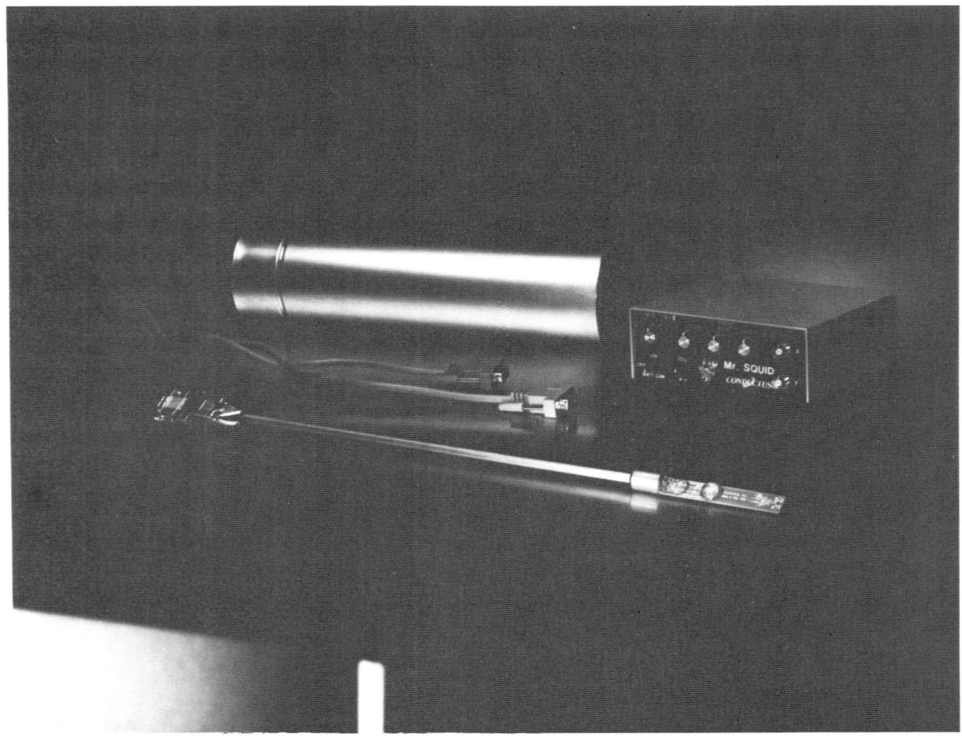

Figure 25.3.   Photograph of the "Mister SQUID" package sold by Conductus, Inc.

Attention is focused on developing a manufacturable SQUID, as contrasted to making a single copy of the best possible SQUID. Typically, a HTSC SQUID costs a few thousand dollars, about the same as a LTSC SQUID. This indicates that attention must turn to solving the engineering problems associated with manufacturing SQUIDs. The experience of IBM is informative: During 1990–92, they learned how to make step edge and SNS junctions by lithography, in analogy with semiconductor manufacturing processes. At IBM, it is no longer important to have a state-of-the-art low-noise SQUID; rather, effort goes into improving reliability. Yield rates have risen, and this is very significant for keeping manufacturing costs down. In 1990, about one in every ten SQUIDs had acceptable characteristics; the rate is much better now.

Several companies are already making practical devices containing HTSC SQUIDs, and we cannot even begin to summarize them. As one example, Conductus, Inc. sells one that is packaged with its electronics, known as the *Mister SQUID*. It operates at 77 K, and is an indicator of better things to come. Figure 25.3 shows the entire product, and Figure 25.4 is a close-up of the SQUID loop in it. The sensitivity is only about $10 \, \mu G$ (10^{-9} T), due to very limited electronics.

At the outset, the *Mister SQUID* is certainly not a money-maker; it sells for $1949 and cannot recover the cost of development this way. But it *is* the beginning of a product line, a foot in the door. It sells mainly to universities who want a classroom demonstration of a

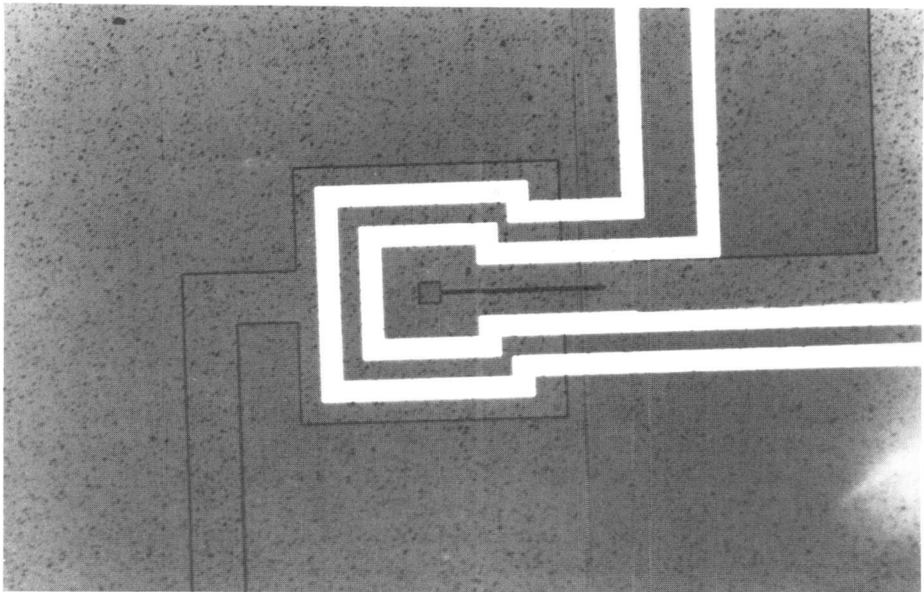

Figure 25.4.   Close-up of the actual SQUID on the probe of Figure 25.3. The bright stripes are gold modulation and coupling coils. The SQUID itself is the dark square washer, 500 μm $\times$ 500 μm, with a hole 30 μm $\times$ 30 μm in the middle. The dark lines that exit the photo (lower left, upper right) are the lead-ins to the SQUID.

SQUID, which, after all, is a macroscopic quantum-mechanical effect. As today's students become familiar with the principles underlying SQUIDs, they will go on to employ SQUIDs routinely in the future.

The *Mister SQUID* product is an excellent example of how instrument makers must walk before they can run. With a sensitivity of 10 μG, the obvious question is "Why bother?" A flux-gate magnetometer does that well, without the requirement of LN$_2$ cooling. But the SQUID principle has the potential to do much better. The poor sensitivity in *Mister SQUID* is because the front-end flux transformer is made of normal metal. To overcome that, Conductus has built a multi-layer device that has a YBCO flux transformer on the same chip with a YBCO SQUID; they report[8] improvement by a factor of 127. It is difficult to deposit so many layers (including insulating layers) on a substrate and still preseve the requisite epitaxial alignment, but the reward in better performance is worth it. Conductus' new product line is called the *iMAG Series*, and went on sale in July 1994. The *iMAG Series* are real instruments, and contain microprocessor-controlled, fiber-optic-linked electronics which can control multiple SQUID channels. Conductus feels that the performance of HTSC SQUIDs is rapidly approaching that of LTSC SQUIDs.

It is impossible to review the full SQUID market, which is changing almost daily. Entire conferences are devoted to the advances in HTSC thin films and related devices, including SQUIDs. It would be presumptuous to assert that we have even hit the highlights with this brief sampling. The intent of this section is merely to show that HTSC SQUIDs are steadily getting better. Therefore, in the foreseeable future, they will reliably deliver high quality measurements of weak magnetic fields. Accordingly, we can now turn our attention to suggesting ways in which these SQUIDs may be used.

25.3. APPLICATIONS OF HTSC SQUIDs

As described in Section 5.2.1, LTSC SQUIDs have already been used in the laboratory to detect the current associated with corrosion.[9] It has been demonstrated that during the corrosion process, the identify of the anode and cathode switch back and forth, (i.e., the current reverses). The magnitude and direction of the current meanders around in a chaotic manner; therefore chemical theorists are now bringing the mathematics of chaos into the task of understanding corrosion processes. This could turn out to be a very fundamental advance in our understanding of an important (and expensive!) fact of industrial life. If so, the extreme sensitivity of SQUIDs will deserve some of the credit. The big question is, can a SQUID sensitive enough to detect corrosion current still function properly in an environment where corrosion is taking place?

Questions of this type are plentiful in factory settings. A second possible application of SQUIDs is in the Non-Destructive Evaluation (NDE) of defects in metals. For example, at a weld junction, the normal grain structure of the metal is disrupted, and there is region of higher dirt concentration on either side of the weld line. (This is where "broken welds" almost always occur.) It is well within the sensitivity reach of SQUIDs to see the variations in magnetic field associated with such collected impurities. Upon inspection, a weld would be rejected if it went beyond a certain limit of magnetic field deviation. The principle is easy enough to grasp, but building a SQUID-based detector that performs reliably in factories where welding takes place is quite another story.

The remainder of this section presents three separate concepts that are unique to superconductors and applicable to control devices. In all cases the economic justification is such that only HTSC SQUIDs could ever be considered; liquid helium refrigeration is far too expensive.

25.3.1. Traffic Control

It is currently the case that some traffic intersections use eddy current detectors as part of a control system to change lights from red to green. A wire loop buried just below the surface of the pavement detects when a car is above it, and advises the traffic-light control system that a car is waiting to enter the intersection. The variable length of many "left-turn arrows" is attributable to such eddy current detection systems. The size of the loop is about the same as the size of the vehicle, and the range of detection is only a few car lengths.

At freeway on-ramps, the short range of eddy current detectors limits their usefulness in controlling entrance of cars into the mainstream. It is desirable to detect approaching cars at much greater range (perhaps 100 meters) in order to properly time the merging of cars at high speed with maximum safety.

SQUIDs offer an alternate way of detecting approaching cars. Imagine a permanent magnet 1 meter across mounted on the front bumper of a car, with a field strength of 0.1 T at the poles. Recalling that dipole fields fall off as $1/r^3$, we can say that 100 meters away the field strength will be approximately 10^{-7} T. Since the earth's field is 400 times bigger, this is not very interesting, except that as the vehicle moves, the field at a fixed observation point will change.

Conventional magnetometers cannot detect such changes, and hence eddy current detectors are the best thing available. What about a SQUID? Consider a SQUID mounted on a traffic light 100 meters from an approaching car. At typical freeway speeds of 25 m/sec, the time rate of change of magnetic field would be $dB/dt = 2.5 \times 10^{-8}$ T/sec. At 50 meters range, $dB/dt = 4 \times 10^{-7}$ T/sec. In the half second after passing the 100 meter point (as a car approaches the intersection), the magnetic field at the intersection increases by 40%. Since LTSC SQUIDs routinely detect 10^{-7} Gauss (= 10^{-11} T), this change constitutes a booming signal.

The implications for traffic control are clear at once. With the ability to detect oncoming traffic 100 meters away, SQUIDs could be used to control merging cars at high speed on-ramps. Traffic lights on urban arteries, now often controlled by timers without any feedback about traffic conditions, could be made much more responsive to the "wavefront" of oncoming traffic. Benefits include safety, fuel savings, time and "hassle" related to stop and go driving. These must be weighed against the cost of operating the SQUID, which is prohibitive for an LTSC device, but may be only the cost of running a cryocooler at 100 K for a TBCCO SQUID.

25.3.2. Positioning

The exceptional precision with which SQUIDs can measure magnetic fields facilitates a variety of controls that have been done (with less accuracy) by other means. The examples of Chapter 5 (corrosion detection, etc.) are candidates for incorporation into control systems. Here is one example drawn from the field of metal working:

Obviously a precise magnetic detector can tell how far away it is from a piece of steel. With this in mind, consider the automobile production step known as *die spotting*. Before pressing a fender, the two dies have to be positioned almost perfectly, or else there will be lumps and creases in the fender. A very exact gap measurement is needed to get it right. If several SQUID detectors feed their data into a computer, the control decisions can be made and executed to properly position the dies.

Will this be cheaper and faster than mechanical or optical techniques for carrying out the same task? It is too early to say. HTSCs raise the possibility of doing it magnetically, but do not guarantee that the economics will make it happen this way.

The catalog of other "thickness" measurements is very long: rolled sheet steel, friction and wear, etc. The measurement of small magnetic fields is a new tool in the hands of process designers.

25.3.3. Inspection

Inspection of metal parts can be improved by precise magnetic measurements. Indeed, a magnetic image of the parts can be formed. To understand this, consider visual inspection of parts: we are focusing optical rays and identifying a pattern in the signals picked up by eye. Magnetic imaging involves many more computational steps, but it *is* possible to sort out data from an array of SQUIDs and reconstruct the source of the magnetic field.

As one example, consider an ideal steel piece passed through a large array of SQUID magnetometers: a certain pattern of signals will be detected as it passes; these signals can be stored in a computer. When subsequent pieces pass, their signal patterns can be compared to the fiducial pattern. Those that deviate beyond a certain amount are rejected. Here we are applying a fairly universal principle of inspection, aided by the remarkable improvement in magnetic detection that SQUIDs confer.

Magnetic inspection devices have large growth potential, because there is a strong economic drive towards 100% inspection of parts. For example, a General Motors plant inspects 100% of steering knuckles, motivated by fear of lawsuits.[10]

Routine maintenance inspection within a plant includes the search for energy wasters, such as pneumatic leaks, stuck steam traps, or other hot spots. *Infrared thermography* is a tool[11] that has proven very helpful for detecting hidden heat leaks. Sensitive magnetic detectors could be used to locate unplanned current leaks. Indeed, the same principles that allow magnetic brain scans (Magneto-EncephaloGraphy) might be used to generate a map of current flow paths in factory equipment.

25.4. MAGNETIC SHIELDING

It is well known that a lot of interesting laboratory techniques do not work in factories. The reason why is obvious when mechanically delicate instruments are compared with the booming, vibrating environment of a forge shop or comparable industrial facility. Less obvious is the impediment posed by ElectroMagnetic Interference (EMI) in a setting where large electric motors or other electrical machinery is present. The fact is that a lot of electronic instruments are rendered useless not by the harshness of the mechanical environment but by the EMI environment. Analog instruments are particularly susceptible to disruption of this sort; consequently most instrument manufacturers strive to digitize their sensor signal at the earliest possible stage. However, any interaction between an industrial process and a sensor creates a signal that is analog at first, and no amount of digital manipulation can reclaim a signal that is lost to noise at its initial stage.

Magnetic shielding is customarily done with *Mu-metal*, an alloy that impedes penetration of a magnetic field to a high degree. A perfect diamagnetic material would provide the best possible shielding, and superconductors act as perfect diamagnets because of the Meissner effect. Liquid helium refrigeration requirements has so far prevented superconductors from being used as practical magnetic shields in this way. For the future, a HTSC cooled by liquid nitrogen might very well prove to be an economically viable magnetic shield. A sheet of material only five microns thick would be very effective. If its transition temperature should be 120 K, it could be cooled to 100 K or 110 K by vapors blown from a nearby 77 K reservoir of LN_2.

One way to characterize magnetic shielding is via the shielding effectiveness of a material. This is defined in analogy with acoustic attenuation through the formula

$$S = -20 \log (H/H_0)$$

where H and H_0 are the measured magnetic fields with and without the material present. An apparatus for determining this parameter of a superconducting material has been built,[12] and it is highly accurate because it suppresses fringing fields, which would otherwise distort the measurement.

The value of magnetic shielding is much greater than first meets the eye. There are certain industrial measurements that are cluttered by stray magnetic fields, and these would be improved, of course. Video recording is a good example from this category. However, there are far more that are never even tried by industry because of the heretofore dreadful magnetic effects. The extremely sensitive measurement capabilities of SQUIDs illustrate this point; they are not used in factories today. However, successful magnetic shielding would open a door to let a large variety of laboratory instruments into the factory: microwave devices, balanced-bridge circuitry, etc.; the list is long. This latter improvement is by far the greater contribution of magnetic shielding via HTSCs—it is a technology that enables other technology to be used.

25.5. DIGITAL CIRCUIT APPLICATIONS

The application of superconductors to computers is the subject of many other studies.[13–14] Their chief advantage in computers is to increase the speed of computations; a second benefit is to reduce the burden of heat removal. An entire book could be written on the topic. Here we only wish to illustrate the general idea that the distinctive properties of HTSCs convey some special advantage. Two entirely different examples of this will suffice to make the point:

25.5.1. Noise-Removing Switch

At Bellcore in 1989, researchers proposed a HTSC switch to remove noise from digital circuits.[15] This is a clever application, because it relies on a nonlinear property of the HTSCs which is not seen in conventional LTSCs.

The underlying principle of this switch is the observation that as current increases, there is at first zero resistance, then (above some I_{c1}) a gradual increase in resistance, and eventually (at some I_{c2}) a sudden rise to full normal resistance. This is in contrast to conventional LTSCs, which show a very sharp change in resistance when the metal goes normal. This switch takes advantage of the region between I_{c1} and I_{c2}, where the resistivity is proportional to I, thus introducing a nonlinear current-voltage relation in place of Ohm's law. This resistive behavior with current is shown in Figure 25.5. This component is placed as a shunt to ground in a test circuit that is deliberately made noisy (see Figure 25.6). The

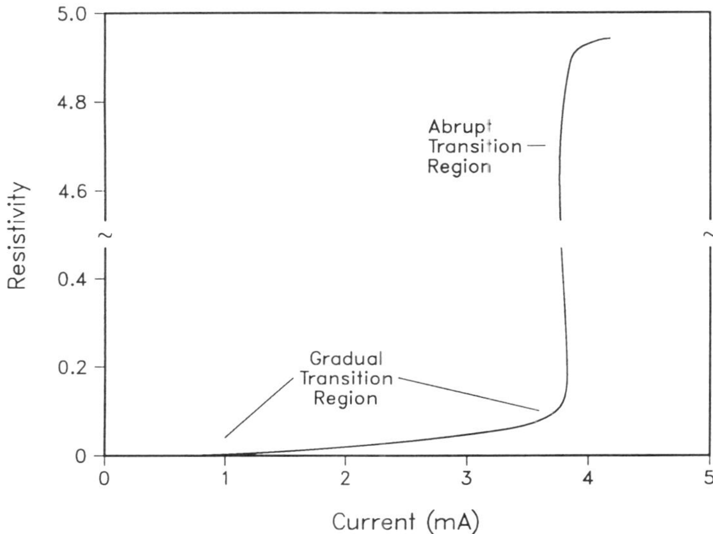

Figure 25.5. Resistivity as a function of current in a YBCO thin film. Resistivity is zero below I_{c1} and rises linearly to I_{c2}.

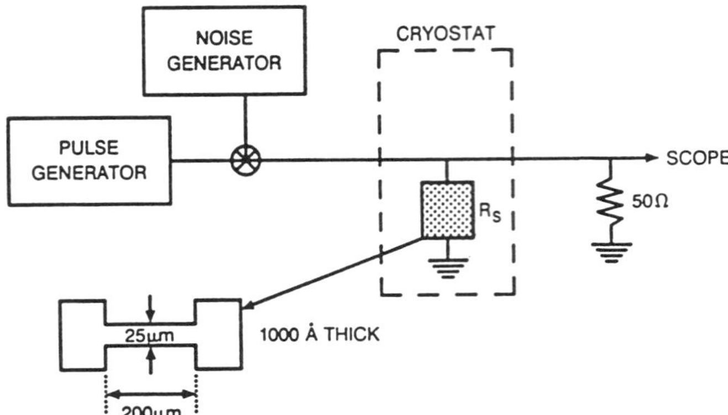

Figure 25.6. Experimental arrangement for the study of fast nonlinear switching of HTSC thin film between the superconducting and dissipating state.

resulting circuit removes small noise (below I_{c1}) but passes major signals with little attenuation. The switching time is below a nanosecond. The device can readily be used to clean up digital signals at a bit rate up to 200 Mb/sec.

This kind of circuit is of great interest for telephone communications. The circuit itself is remarkably simple; what is clever is noticing that because HTSCs have a gradual, nonlinear change in resistivity with increasing current, a device of this type is possible. It is fair to suggest that this particular insight is only one of many applications of the unusual properties of the HTSCs that will gradually be discovered.

25.5.2. Real Time Data Analysis

The field of industrial process control can also take advantage of both the speed and the heat-conduction properties of superconducting electronics. In controlling any process, there are times when simple direct measurements are extremely elusive, and data is contaminated by extraneous factors. ("Drift" of a zero-point is the simplest example of this.) Additional measurements of those factors can be used to correct the primary measurement, thus giving valid information about the process. (For example, in measuring "whiteness" of paper, an optical gauge can be deceived by certain contaminants, but a chemical measurement of titanium dioxide content will clear up the confusion.) Unfortunately, the time scale of such corrections is often slower[16] than the available time for corrective action on the production line. With faster processors, prompt corrections become more plausible.

With processors that dissipate less heat, it will be possible to embed a small dedicated microprocessor into the sensor itself. This will enable sensors to report out calculated quantities; consequently the level of distributed control will be elevated; and the control loops on the process can be tightened. That adds up to more uniform, higher quality product.

Robotics is a particularly good target for industrial applications of superconducting microcomputers. The kind of scanning, pattern recognition and action decisions that go into robot vision would be substantially enhanced by smaller, faster and more compact computers. This is doubly true for mobile robots.

25.6. COMPETING TECHNOLOGY

If superconductors are ever to become "mainstream" electronics, they must achieve cost, performance, reliability, etc., all comparable to silicon technology. Furthermore, "silicon technology" is a moving target. It is important to remember that the user doesn't care about superconductivity; the user merely wants a benign workstation that functions unchecked for five years.

The great advantage of Josephson junction technology is speed; but in silicon, CMOS is still a very strong competitor to superconducting technology. Probably CMOS can run at 1 GHz at 10 K, because the CMOS I–V characteristic sharpens considerably. Further, the cost of CMOS is very competitive: an 8" wafer is processed for somewhere between $400 and $1000, and with low defect density, the yield will be 125 to 200 chips of 1 cm^2 each, costing $24 to $48 each. If the power density limitations will permit, CMOS will contain over 100 million transistors in 1 cm^2. The only real limitation of CMOS is its clock speed, which can't get much above 1 GHz.

The difficulty of market penetration experienced by gallium arsenide indicates what Josephson junction electronics (LTSC or HTSC) can expect. The experience of Tektronix Corporation is representative.[17] At Tektronix, GaAs work began in 1978, and was transferred to a subsidiary in 1984; by 1991 about $60,000,000 was spent. GaAs technology improves steadily, by lowering the defect density and raising the switching speed (up to 70–80 GHz). Still, Tektronix has not yet found the customers, because they don't have enough margin over silicon to get people to absorb the cost of changing.

There are a number of serious obstacles facing HTSCs. Technology transfer from the lab to the manufacturing floor is a long, arduous task. Most corporations are unwilling to place bets on new technologies (like HTSC). Therefore, one has to build things in the research lab and take them all the way to market, just to get the attention of management. Moreover, any deviation from "mainstream" will result in a heavy penalty for the cost of "tools." There is about $ 5 billion worth of software in place now for the basic architecture characteristic of silicon.

The power-dissipation question is significant, too. If 10^8 gates run at 10^9 Hz, the dissipation is 7 kW. However, the CMOS designers would see to it that most of them stay only in standby mode, not all turned on at once. At 77 K, cooling of 6 KW/cm^2 is possible.

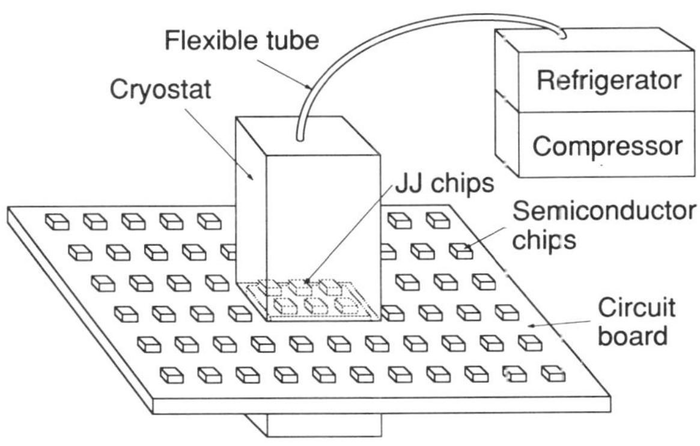

Figure 25.7. Fujitsu Laboratories Josephson–semiconductor hybrid system. (©IEEE.)

Thus CMOS is challenged, but not defeated, by such conditions. It is important to remember that the CMOS folks will do everything they can to hold onto their business.

The idea of hybrid circuits is being explored. Here, superconductors do not replace semiconductors, but cooperate with them. Fujitsu Laboratories in Japan has envisioned a circuit board concept (Figure 25.7) where the superconducting elements would carry out the fastest switching, and conventional CMOS would do the rest. A less ambitious form of hybrid is a Multi-Chip-Module (MCM) from Superconductor Technologies, Inc., in which the active circuit components are CMOS but the interconnecting leads are YBCO.[18] Also, Superconductor Technologies has a very compact cryocooler which uses helium gas to reach 100 K; MCMs that incorporate TBCCO can operate there. Such hybrid circuits may prove to be the optimum marriage between HTSCs and CMOS.

25.7. SUMMARY

This chapter has featured some novel ideas as to how the HTSCs might by used in the future to improve measurement and process control in industry. The basic premise is that entirely new types of very sensitive measurements will become possible in the future, and once certain measurements are possible, control strategies will be constructed around them. The workhorse measurement tool will of course be the SQUID operating at 77 K.

It will be some time before competitive vendors make SQUIDs a practical reality for factories. When that eventually happens, it will promote some exploratory questioning by engineers about what is going on in their production processes, and the *real* advances will emerge from those investigations.

For example, in the dryer section of a paper machine, the sheet builds up a considerable static charge. This charge has always been discarded by draining it to the ground through conducting brushes touching the sheet. However, by asking *why* the sheet charges up, attention is drawn to the current flow during the drying process. Where does that current come from? Can it be measured? What would it tell us? Could we feed that information back to help control the process?

Questions of this type are at the heart of new industrial measurements. In the past, the lack of sensitive magnetic instruments has kept such questions irrelevant, but high-T_c SQUIDs will change the ground rules. Small currents associated with certain physical phenomena can now be observed.

The promise of advances in productivity are real, but the intermediate step of rethinking industrial processes must not be neglected.

REFERENCES

1. R. R. Gatts, R. G. Massey, and J. C. Robertson, *Energy Conservation Program Guide for Industry and Commerce*, National Bureau of Standards Handbook 115 (U.S. Supt. Documents, Washington, DC: 1974).
2. O. Eckstein *et al.*, *The DRI Report on U.S. Manufacturing Industries*, (Data Resources, Inc., Boston: 1984).
3. T. P. Sheahen, Ch. 34 in *Advances in Energy Cost Savings for Industry & Buildings* (Association of Energy Engineers, Atlanta: 1983).
4. See, for comparison, P.R. Gray *et al.*, "Weighted Capacitor Analog/Digital Converting Apparatus and Method", U.S. patent # 4,129,863 (1978).
5. P. Chaudhari, presented at AMSAHTS '90 Conference, Greenbelt, MD (April 2–6, 1990).
6. R. Gross, P. Chaudhari *et al.*, *Appl. Phys. Lett.* **57,** 727 (1990).
7. R. Koch, IBM Corp., private communication.

8. L. P. Lee *et al.*, *Appl. Phys. Lett.* **59**, 3051 (1991).

9. J. G. Bellingham, M. L. A. MacVicar, and M. Nisenoff, IEEE Trans. Magn. **MAG-23** (2), 477 (1987) and J. G. Bellingham *et al.*, *J. Electrochem. Soc.* **133**, 1753 (1986).

10. J. M. Shaw, private communication.

11. T. P. Sheahen, "Using Infrared Thermography for Industrial Energy Conservation", Ch. 6 in *Energy Auditing and Conservation*, Y. Y. Haimes, ed. (Hemisphere Publ. Co.: 1980).

12. Q. Y. Chen *et al.*, *Appl. Phys. Lett.* **57**, 2603 (1990).

13. T. Van Duzer, presented at Federal Conference on Commercial Applications of Superconductivity (Washington, DC, 28 July 1987).

14. A. P. Malozemoff, W. J. Gallagher, and R. E. Schwall, Amer. Chem. Soc. Symposium Series (September 1987).

15. A. Frenkel, T. Venkatesan, *et al.*, *Appl. Phys. Lett.* **53,** 2704 (1988).

16. R. S. Berry *et al.*, *Innovative R&D Opportunities for Energy Efficiency*, report by the Energy Engineering Board, National Research Council (National Academy Press, Washington, DC: 1986).

17. T. Long, Tektronix Corp., presented at Conference on Superconducting Digital Circuits (1991).

18. F. Reynolds, Superconductor Technologies Inc., private communication.

26

High Magnetic Fields

The upper critical magnetic field (H_{c2}) of the new superconductors exceeds 100 T near absolute zero; at 77 K, superconductors with $T_c > 110$ K will sustain 30 T without going normal. However, the highest useful field is the irreversibility field H_{irr}, which is still rather high. Optimistically anticipating future improvements, it is not absurd to imagine coils producing 100 T at 77 K. No magnetic fields this strong have yet been produced, and so it would not be surprising to find exceptional behavior of many materials in such fields. A wide spectrum of applications will open up under these circumstances.

However, there are serious obstacles to generating and using high magnetic fields: First, it is possible now to create fields above 10 T, but that is seldom being done, owing to the general requirement of uniform field.[1] A prominent medical application of (low-temperature) superconducting magnets is magnetic resonance imaging (MRI), described in Chapter 5. As stated there, the magnetic field is required to be extremely uniform, both in space and in time. That requirement will not go away no matter how high T_c or H_0 rises. For MRI applications, a more important design criteria is the mitigation of flux jumping, because it is indispensable for the persistent current to be very stable. Accordingly, we expect to see very conservative magnet designs as HTSCs are introduced. If HTSCs were only to serve the brute force of the industrial high-field market, they would be very limited in their applications.

Second, strong magnetic fields generate forces that distort the coils themselves. (The force acts both radially outward and axially along the coil.) Strains in other superconducting magnets ($H < 10$ T) are a few tenths of a percent; Malozemoff *et al.*[2] showed that in a 20 T solenoid held together by steel, the strains reach 0.6%, which is enormous. Brittle ceramics such as YBCO would break well below this degree of strain. This obstacle may eventually prove to be a show-stopper for ceramic superconducting magnets.

If we assume that research will overcome the obstacles to high magnetic fields, we might then imagine ways to apply such fields. The potential applications are widespread; later in this chapter we give a few imaginative examples to illustrate the diversity of uses. First, however, it is desirable to understand just what we are dealing with.

26.1. ENERGY DENSITY AND MAGNETIC PRESSURE

The energy stored in any magnetic field is the product **B•H**, which in air or vacuum means the energy density is $B^2/2\mu_0$. This implies that as magnetic field strength increases, the stored energy (and hence the stresses) goes up as B^2. Figures 26.1 and 26.2 are plots of

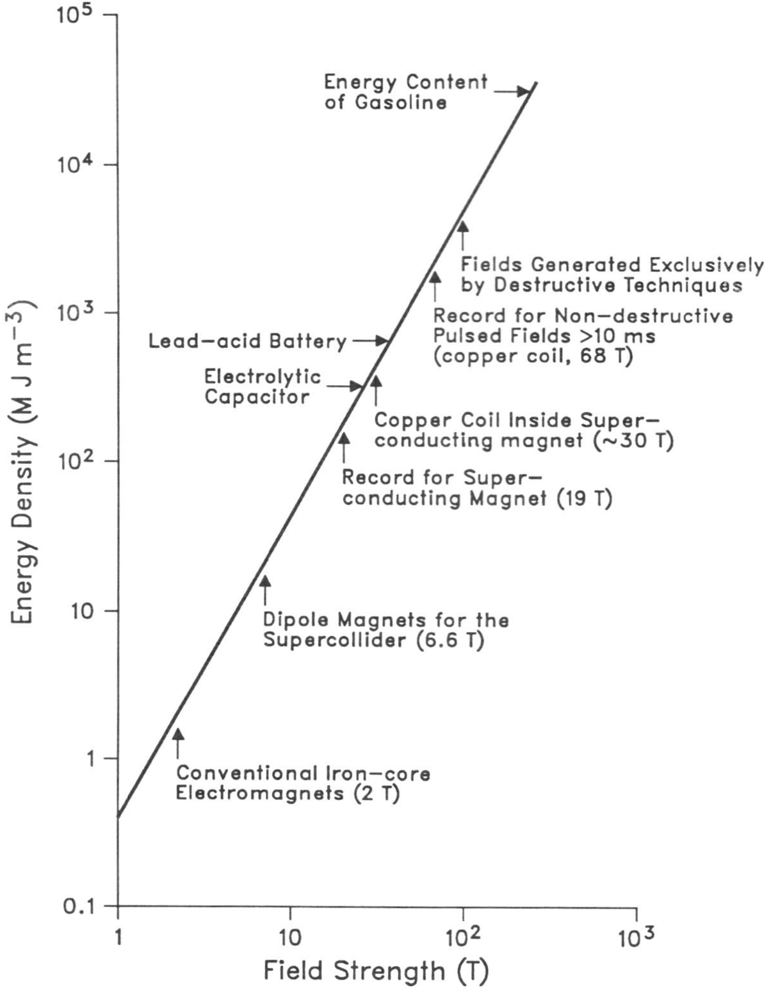

Figure 26.1. Magnetic energy storage per unit volume.

this relationship on log–log paper[3]; they show the energy density and the stresses associated with various field strengths. The energy density is nothing special; we have all handled gasoline before, and lead-acid batteries are unremarkable. However, the magnetic pressure is something else again: enormous pressures are generated by quite modest fields. 1,000 or 10,000 atm can readily be produced by the sorts of fields we are considering here.

The first obvious implication is that a simple design cannot trivially be extended to higher fields. Extra steel is required during construction to counteract the mechanical distortions that tend to occur. Such everyday things as a copper cold finger for cryogenic cooling have to be reconsidered if the field is to exceed 10 T; it could distort badly in the pressure of a 10 T magnet.

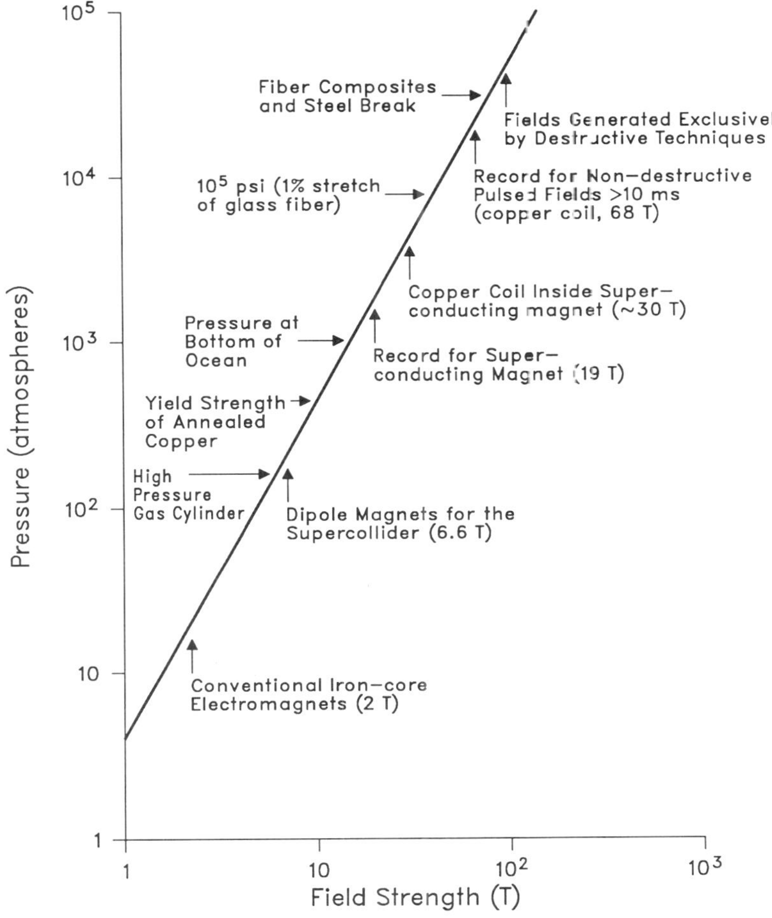

Figure 26.2. Magnetic pressure vs. magnetic field strength.

We saw in Chapter 21 that a substantial fraction of the cost of a SMES is in the structure, and Figure 26.2 is the reason why. When construction costs get high enough, the comparative cost of refrigeration dwindles, and the cost savings associated with operating at 77 K instead of 4 K is of less significance. Thus, although it may at first seem disappointing to have to settle for 20 K operation in order to keep J_c up in a strong magnetic field, the numerical cost is not that great. The fact that magnetic fields above 20 T can be produced at any temperature is what makes the HTSCs attractive.

26.2. HIGH FIELDS USING BSCCO

Shortly after the discovery of HTSCs, the popular press poured extravagant hype on the myriad applications of high magnetic fields. Reality, of course, takes somewhat longer. The best hope for making wire seems to lie with the bismuth family of HTSCs, because its micaceous structure alleviates fabrication difficulties. However, due to flux-pinning limitations, BSCCO is very unlikely to be useful at 77 K in any significant magnetic field.

Nevertheless, BSCCO continues to show promise for use in the low- and intermediate-temperature regimes. For example, Sumitomo collaborated with MIT's National Magnet Lab to measure[4] the J_c values of bismuth compounds in magnetic fields up to 20 T, at low temperatures. Their results (J_c in A/cm^2) are as follows:

Critical Currents (A/cm^2) at Low T and High B

Temperature		0 Tesla	5	10	19.75
4.2 K	$J_c =$	200,000	135,000	110,000	85,600
15		180,000	100,000	80,000	65,000
20		170,000	88,000	72,000	55,000

The persistence of J_c values near 100,000 A/cm^2 in high fields up to 20 K is a very favorable outcome, which supports the hope that copper oxide superconductors will become the preferred means of achieving high magnetic fields. Considering how much energy is dissipated in a conventional nonsuperconducting magnet (the alternative), the refrigeration cost of reaching low or intermediate temperatures may be acceptable. With this possibility in mind, consider some of the applications that might be implemented with very powerful magnets:

26.3. APPLICATIONS TO RESEARCH FACILITIES

To appreciate what is at stake in the world of high magnetic fields, it is appropriate to examine two different kinds of present-day applications that occur in laboratory settings. Both of these use conventional LTSC magnets, that is, niobium-based magnets operating at liquid helium temperatures. The hope for the long term is that someday HTSCs will make these magnet systems simpler and less costly, or perhaps more powerful for the same cost.

26.3.1. Synchrotron Magnets

The field of high-energy physics depends on particle accelerators (colloquially termed atom smashers) as its principal experimental tool. Whatever particle is accelerated (commonly protons or electrons), the goal is to put as much energy as possible into the particles before they strike the target. In the configuration known as a collider, the target is a stream of oppositely charged particles coming head-on at the beam of particles. Most accelerators are circular, with the particles confined in a circular path by bending magnets. On each pass around the circle, they receive a carefully synchronized jolt of energy; hence the name *synchrotron*. Eventually the particles are travelling at very nearly the speed of light, and so their mass increases by relativistic effects.[5]

For a slow-moving object subject to classical mechanics, the energy E is related to the momentum p by $E = p^2/2m$. At relativistic speeds, the relation is revised to

$$E^2 = E_o^2 + (pc)^2$$

where E_o is the rest energy, $m_o c^2$ (c is the speed of light.) At extremely high energies typical of synchrotrons, the rest energy is negligible, and so the energy momentum relation is just $E = pc$. For a particle on a curved trajectory in a magnetic field, the momentum is

$$p = Ber$$

where B is the magnetic field, e is the electronic charge, 1.6×10^{-19} Coulombs, and r is the radius of curvature. Consequently, in a synchrotron there is a close relationship between the strength of the bending magnets and the size of the machine. For a machine of fixed radius, increasing the magnetic field strength linearly increases the energy. In fact, this has been done. Fermilab (Figure 3.19) was originally built in the late 1960s with magnets made of copper wire. During the early 1980s, a second ring was installed in the same tunnel using NbTi magnets. This is pictured in Figure 26.3. This was termed the *energy saver/doubler* because energy was saved by using superconducting magnets, and the final energy of the accelerated protons doubled. Specifically, the initial ring produced 400 GeV protons with B

Figure 26.3. The tunnel of the main accelerator at Fermilab. Visible in the center of the photograph are sections through two rings that run the entire circumference of the tunnel. The upper ring of conventional magnets comprise the fourth stage of particle acceleration. The particle beam is then injected into the Tevatron, which is composed of superconducting magnets (lower ring) designed to accelerate the protons to energies approaching 1 TeV. (Photo courtesy of Fermilab Visual Media Services.)

= 1.8 T. After the new ring was installed the superconducting magnets gave 4.4 T to produce 900 GeV protons. The name *Tevatron* was given to this final configuration.[6]

Many other accelerators around the world today also use superconducting magnets, often cooled to 1.8 K to obtain the maximum possible magnetic field. Niobium technology has been taken as far as it can go. The largest accelerator designed, the SSC (*Superconducting Super Collider*), was to be over 20 miles in diameter, simply because there was no other way to increase the product *Ber.* Obviously, the thought of obtaining 20 or 30 or 100 T from BSCCO has occurred to accelerator designers. However, there are formidable obstacles: beyond the difficulty of making *any* wire from BSCCO, each magnet must be virtually identical to the others, in order to confine and stabilize the beam of charged particles. There is very little tolerance for error in accelerator magnet construction.

No one knows how long the wait will be until uniform and reliable BSCCO magnets appear. For major accelerators already in place (Fermilab, CERN in Geneva, KEK in Japan, etc.) the possibility of someday upgrading their magnets and increasing the energy by a factor of 5 or more makes sense. (Not every accelerator is eligible for such an upgrade; the Advanced Photon Source at Argonne, for example, has no need to increase its magnetic field.) This effectively defines a future niche market: potential vendors of BSCCO wire can calculate today how much they can sell to these high-energy physics laboratories, and anchor their business projections to such forecasts. This can be terribly important in deciding whether to stay the course over a lengthy period of product development punctuated by intermittent setbacks and breakthroughs.

26.3.2. High Magnetic Field Laboratories

Very high magnetic fields are useful in all kinds of materials research. A major national user-facility is being built, partly at Florida State University (FSU) in Tallahassee and partly at Los Alamos National Laboratory in New Mexico. The National Science Foundation is a major sponsor. The intent is to give experimenters a wide choice of magnetic field strengths to work with. In describing this laboratory, we hope to convey some of the complexity associated with producing high fields. The importance of future progress toward HTSC magnets will then be apparant.

The range of scientific disciplines that will make use of the FSU facility is very great—NMR imaging, structural biology, and so on, in the life sciences; polymers and macromolecules, quantum fluids, and highly correlated electron systems in physics; and controlled fusion in engineering.

The first construction project at Florida State was a power supply that outputs up to 40 MW DC. (A 50 MW electrical substation is right next to the facility.) The power supply must be stable to 1 part in 10^4, with ripple-control to 1 in 10^5. In this design, normally the power supply is rated at 400 V, 24 MW; or at 500 V, 32 MW. Happily, early testing of the system showed that it can run continuously at 40 MW—at least for an hour or two. It is conservatively anticipated that power supply losses will be 2 or 3 MW. Furthermore, the cooling system pump losses are about 4 MW, and the cooling system chillers require 6 MW. When operating, the power supply can either drive two 12 MW magnets at once, or one 24 MW magnet, or four magnets drawing 6–8 MW. Four power supplies in parallel cost $1 million each, and related DC distribution-bus and control system costs bring the total to $9.5 million for the entire power supply. That is the heart of the facility, and it drives copper magnets.

The plans for this National High Magnetic Field Laboratory are ambitious, and include five distinct types of magnets to span the full range of interest to researchers:

- Flux compression magnets, producing 100 to 1000 T for 1 to 10 μsec.
- Quasi-continuous magnets that deliver 60 T for 100 msec.
- DC hybrid magnets,[†] giving 45–50 T.
- Bitter and polyhelix resistive magnets yielding 25–35 T DC.
- Superconducting magnets that produce 20–25 T.

A key point here is that if HTSCs someday can be made into large magnets, the range of both the superconducting magnet class and the hybrid magnet class will obviously shift upward. Nb_3Sn, Nb_3Ge, and related compounds cannot exceed 30 T, but BSCCO can. Still, construction of this facility is not waiting upon HTSC developments.

This laboratory is not all at one site. An important collaborator is Los Alamos National Laboratory, who have considerable expertise with very high magnetic fields. (Los Alamos's long experience with controlled fusion enabled them to develop this capability.) The extremely high-power pulsed magnets will be at Los Alamos.

The pulsed-magnet systems are remarkable creations. For example, the 60 T, 100 msec flat-pulse magnet will be driven by a large utility generation system that stores 1.2 GJ. The generator normally runs at 1800 rpm, but its speed drops to 1200 rpm in only 1 sec when the rectifier switches energy into the magnet. The hardware to do all this is truly exceptional. A Florida State newsletter[7] gives this ebullient description:

> Befitting the heaviest object ever moved on New Mexico highways, the stately approach of this behemoth to its final destination was followed daily in the press and marked by heavily reinforced bridges, barking dogs, crying children and stunned motorists. The 1200-ton generator with its 4800-ton inertia block rests on 60 springs to minimize earth tremors when the 212-ton rotor is suddenly decelerated from 1800 rpm to 1260 rpm to produce 600 MJ in one second. . . .

Simultaneously with major construction, smaller-scale tests at Los Alamos have achieved over 60 T for several msec, using a 1 MJ capacitor bank as a pulsed power supply. The time dependence of a magnetic field in a typical test is shown in Figure 26.4. The design objective is to shorten the ramp-up and ramp-down times, and make the flat part as long as possible. Heat generation is the limiting factor.

Already existing is a flux-compression pulsed magnet that relies on an explosion to compress the flux lines. Experiments have already been done up to 200 T, in a pulse that lasts about 5 μsec. The shape of the pulse can be varied by choosing different types of explosives, but it is by no means a flat pulse. The explosion destroys the apparatus and wooden bench, all of which are cheap; within microseconds, the data is gathered, but there is nothing left of the apparatus. This facility is in a New Mexico canyon near Los Alamos, but it is part of the lab. Ultimately, this concept will be used to build a 100–1,000 T flux-compression facility.

Collaborations with other magnetic research centers will be important, too. The MIT National Magnet Lab will continue to operate. They routinely reach 25–30 T, and a hybrid magnet can go to 35 T. The lab in Grenoble, France, has a 20 MW capacity. The Lebedev Institute in Moscow is active, as is Tohoku University in Sendai, Japan. The Japanese

---

[†]A hybrid magnet is composed of a superconducting outer magnet with a copper magnet in its core; the highest fields (at the center) are well above the maximum possible field of the superconductor.

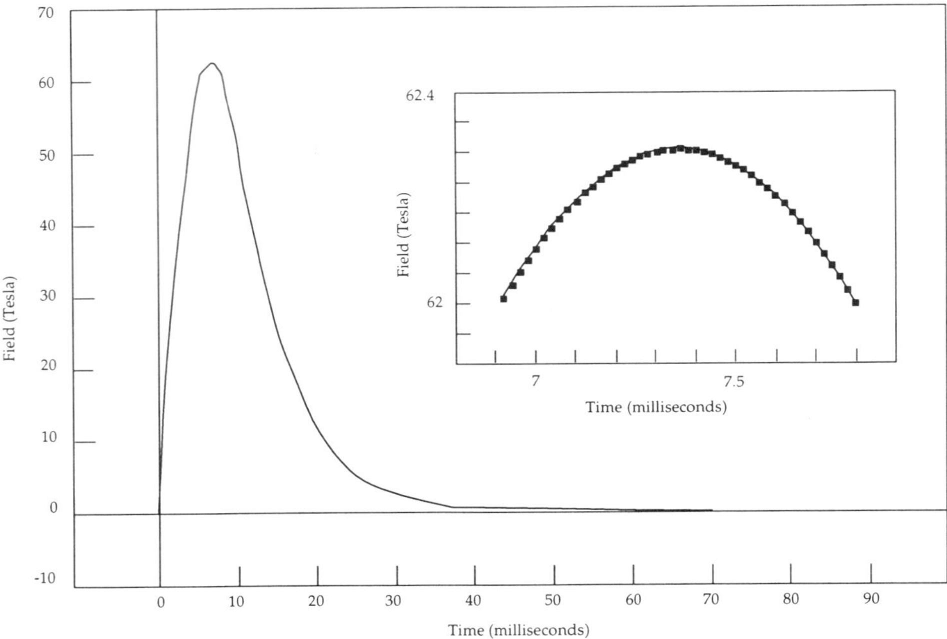

Figure 26.4.   Magnetic field variation with time in the NHMFL pulsed high-field magnet.

National Institute of Iron and Steel is very similar to the Florida lab. Wroclaw, Poland has a 30–50 T capability. Good coordination and scheduling are essential to maximizing the usefulness of this international assemblage of magnet laboratories.

The overall point is that there are many applications of high magnetic fields to research, and this in turn creates a demand for better ways to provide such fields. The prospects for utilizing copper oxide ceramic superconductors, even running at 4 K, are quite optimistic. Like so many other applications of HTSCs, the stumbling block is the difficulty of making wire.

26.4.   MANUFACTURING PROCESSES

The use of high magnetic fields by industry is an easy topic to speculate on, because such fields have never before been routinely available to industry. We can expect factories to introduce any new tool once it is made practical, so high magnetic fields will inevitably be used. Although a great number of possibilities exist, here we cite only a few brief examples, with the intention of encouraging the reader to imagine other examples.

26.4.1.   Forming and Joining Techniques

High magnetic fields have certain established "brute force" uses. When superconducting magnets become commonplace in factories, these applications will become cheaper and more readily accessible. Magnetic separation, discussed in Chapter 4, is a leading example.

One joining technique is the use of the Magnaform process, where a high magnetic field is used to squeeze an aluminum drive shaft tube onto a yoke.[8]

Enough information is available about such processes that economic calculations can be done to determine the relative advantages accruing from HTSCs. Conventional economics will then drive the decision to use or not use these methods. Where superconductors may create surprising new advantages is by enabling a process to be carried out in a different way.

Already plants have fork lift trucks that use the familiar electromagnet to help move metal parts around quickly. Despite being mundane, this is an essential step in most manufacturing processes. When much higher magnetic fields become available, the range of tasks that electromagnets can perform will expand. Armed with this new flexibility of an old tool, manufacturing engineers may create more efficient pathways for materials to flow through a production line. The other side of this coin is the use of magnetic levitation to move objects around the factory, replacing some of the fork lifts.

On the high-tech side, the military is pursuing electromagnetic launchers for kinetic penetrators. Aerospace applications of these are mentioned in Chapter 28. Electromagnetic launchers are essentially nail-drivers of great power. They require magnetic fields of 10–20 T. By using superconductors in these devices, DARPA estimates it can save 50% of the energy and an even bigger fraction of the weight.[9] Although we ordinarily think of a kinetic penetrator as smashing an enemy missile into a million pieces, it is entirely reasonable to imagine it nailing two pieces of metal together. (Rivets serve the same function, but the pieces to be joined must have predrilled holes.) An EM launcher used under precise control may serve very well as a nail gun for some applications. Whether it competes successfully with welding is a downstream question; the point here is that a device developed for military purposes is likely to have commercial applications that alter manufacturing processes.

26.4.2. Metal Casting Applications

The continuous casting of steel is customarily done by pouring molten steel into a containment channel that can withstand high temperatures. However, containment within an intense magnetic field is being developed. (Magnetic "bottles" have been used for 30 years to confine plasmas in fusion power research.) This concept has other applications as well.

In growing single crystals for semiconductor manufacturing, convection within the melt leads to defects in the crystals. By growing crystals in microgravity,[10] where convection is negligible, much better crystals are obtained. (This argues in favor of space commercialization.) But convection can be suppressed on earth by imposing a magnetic field on the growth zone[11]; this "magnetic Czochralski method" produces very high quality crystals. It is fair to ask what effect an even stronger magnetic field might have on solidification processes for steel and other alloys. With the prospect of high magnetic fields in the future, it is timely now to conduct research on this question. A new means of improving and controlling crystal structure may be discovered here.

26.5. MAGNETIC SEPARATION

Magnetic separation is a viable technology: as described in Chapter 4, present-day electromagnets purify kaolin by removing iron oxide. Other comparable uses have been conceptually designed, and Figure 4.8 displays one of these, but economics is often an obstacle. HTSC magnets would improve the economics considerably, and are therefore a goal of many R&D projects, including CRADAs in which National laboratories collaborate with industry.

Because of the severe difficulties encountered in constructing high-field magnets, especially the strain and deformation problems, it may be a wait before high fields become practical. Nevertheless, a creative new idea can be tried out in a factory even today, using low-temperature magnets or hybrid copper/superconductor magnets. If the idea proves feasible, it will add to the market pull driving the development of HTSC magnets. From the point of view of a research agenda, it is entirely reasonable to test new ways to use magnetism, in anticipation of high field magnets being realized eventually.

26.5.1. Lower Ore Grade and Cost of Separation

Magnetic separation provides an example of how superconducting magnets introduce unexpected benefits. In a conventional magnet, the I^2R losses trade off against the ability to separate materials, which increases roughly proportionally to B or I. Beyond a certain point it doesn't pay to separate further. By contrast, a superconducting magnet can run flat out for the same refrigeration cost; therefore, lower grade ores can be treated economically, which in turn extends the raw materials supply at no extra cost.

This point has not been adequately recognized by refiners of ores. Those cost comparisons that have been made simply weigh the cost of refrigeration against the I^2R cost of a copper magnet, assuming the ore grade is held constant. But this is exactly the place where a superconducting magnet makes an unexpected difference. Here we sketch the essential features of this cost analysis. There are a lot of arbitrary constants in the analysis, because different industrial processes contain different parameters, but the general trend is clear: getting a higher B field at no extra cost definitely extends the range of ores that can be tolerated.

The central hypothesis of this model is based on experience with waterborne slurries of ores that are a mixture of magnetic and nonmagnetic materials: the efficiency of separation is proportional to the magnetic field strength. The example of Chapter 4, in which the ore is mostly titanium oxide, but contaminated with small amounts of iron oxide, is representative. The proportionality only holds up to a finite point: depending on details of the geometry, eventually the field reaches a saturation point[†] and so there is no point having a stronger magnet. Meanwhile, it is assumed that ordinary dirt from mining operations (overburden) is removed by other means before the slurry enters the magnetic separator.

How does all this relate to the cost of operations and profitability of the factory? There are several considerations that go into determining how low a grade of ore can be separated cost-effectively. The efficiency of separation, what to do with the waste products, the cost of electricity—all come into play. We can assemble most of the components of the cost equation, as follows: Let the amount of input ore be of weight W and purity grade g ($0 < g < 1$). The revenue from separated final product sold is then $C_0 Wg$, where C_0 is a proportionality factor determined by the marketplace. The cost of operations is the sum of several fixed costs (management, electricity, etc.) plus costs that vary with the ore grade g.

[†]For example, in the Kaolin separator with iron mesh (described in Chapter 4), Eriez discovered that a 5 T magnet performed no better than a 2 T magnet, because once the iron mesh saturated it gained no further ability to attract iron oxide particles.

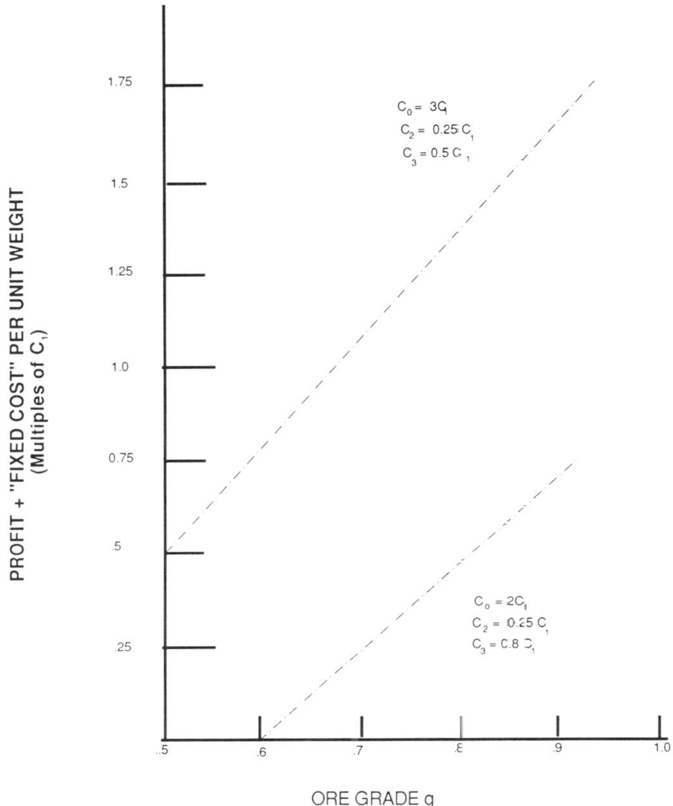

Figure 26.5. Variation of profitability with ore grade, for two different sets of constants.

To analyze this, the first step is to focus on those variable costs. The cost of the original ore depends on the ore grade g; assuming an ideally fair market, we have a linear proportionality $C_1 W g$. There is a cost of pumping the slurry around, $C_2 W$; and a cost of disposing of impurities removed, $C_3 W(1 - g)$. That cost (per ton of input) goes as $(1 - g)$, but the cost per ton of output goes as $(1 - g)/g$. This gets expensive in a hurry as ore quality drops: a 90% ore has 0.11 for this factor, whereas a 50% ore has 1.0—nine times higher. To find the impact of ore grade on profitability, we subtract all the grade-related costs from the revenue stream from sales. The gross profit per unit weight of output is then

$$P / W = g [C_0 - C_1 - C_2/g - C_3 (1 - g)/g] - (\text{fixed costs})/ W \qquad (26.1)$$

Figure 26.5 displays the resulting variation with ore grade when certain numerical choices are made for the constants $\{ C_0 \cdots C_3 \}$. These constants are arbitrarily normalized to C_1 as the unit of money. Two different sets of $\{ C_0 \cdots C_3 \}$ are plotted. Evidently, different choices of these constants would give somewhat different curves, but the general trend is the same in all cases.

The next step is to include the cost of electricity. This is where the choice of magnets makes a big difference. Normally electricity is lumped in with the fixed costs in Equation (26.1). A copper magnet dissipates energy constantly, and its cost goes as I^2R. By contrast, a superconducting magnet loses nothing electrically in its "on" state. Its cost is simply the cost of running the refrigerator, which is determined by the heat leak. In addition, when the magnet is turned on or off, there are eddy current losses during the ramping phase, which boils off coolant. Moreover, the superconducting magnet has a much higher initial cost of installation, which becomes part of the fixed costs of the mill. Clearly, this large baseline cost would drop precipitously if liquid nitrogen could be used.

In the third step, remember that the selling price C_0 is heavily dependent on how well the impurities are cleaned out; a poor job of separation will bring a poor price. Suppose that with current I_1 (and magnetic field B_1), the final purity is only good enough to bring a price of $C_0 = 2C_1$. Then the profit line is the lower (dot-dash) curve on Figure 26.5. But what if running at current I_2 (and magnetic field B_2) would yield purity sufficient to bring a price of $C_0 = 3C_1$? That would shift the profit line to the upper (dashed) curve. In this scenario, the price C_0 depends on the strength of magnetic field and hence on the magnet current, which brings us right back to the operating cost difference between copper and superconducting magnets.

The vertical axis in Figure 26.5 is profit *plus fixed costs*, so it is necessary to select a fixed cost of electricity and subtract it in order to find the profitability of using any particular grade of ore. For a copper wire magnet, let the cost of running at current I_1 be 0.5 C_1, and the cost of running at I_2 be 1.0 C_1. On the other hand, for a superconducting magnet, let the cost of running at either current be 0.8 C_1. By imagining these three horizontal cutoff lines running across Figure 26.5, it becomes clear that:

- At current I_1 in the copper magnet, ore grade must be above 0.86 to yield a profit;
- At current I_2 in the copper magnet, ore grade must be above 0.70;
- In the superconducting magnet (obviously running at I_2), ore grade must exceed 0.62.

If the ore grade is specified ahead of time, the most profitable strategy is to run the superconducting magnet flat out. Of course, this example contains lots of arbitrary constants, but the general message is clear: a superconducting magnet will shift the break-even point downward in ore grade.

The key concept here is that the efficiency of separation increases with B (and hence with I). In all likelihood, other industrial separation opportunities are hiding in this principle. That is, there are probably some magnetic separations that do not now take place because of the high cost of electricity to produce the necessary magnetic field using copper. If a superconducting magnet were available, the economics would change enough to warrant carrying out the separation.

26.6. FUTURE APPLICATIONS

Among the industrial possibilities offered by high magnetic fields, quite a few concepts are waiting in the wings, unable to find application just yet because the field strength required to make them work has not yet been reached. Others technologies are still in the research stage, but if they come true (i.e., device development is eventually successful) they will have their usefulness greatly enhanced by high magnetic fields. To illustrate this point, we present two examples here: one small and one large.

26.6.1. Electromagnetic Acoustic Transducers (EMATs)

One method of measurement that will apply to materials processing is the *electromag-netic acoustic transducer* (EMAT). It will be useful in the steel industry, as well as in many foundry and materials fabrication industries. The EMAT is a remarkably appropriate example of how a laboratory technique, aided by a major innovation in another field (high T_c superconductivity) can be converted to serve industry. We give some detail here, not because EMATs themselves are so special, but to show how the mutual advances in two research areas combine to enable a new technology to be used.

EMATs makes use of the interaction between acoustic waves, magnetic fields, and eddy currents. To understand the principle, first recall that the force on a charge moving in a magnetic field is $\mathbf{F} = q\mathbf{V} \times \mathbf{B}$. Therefore, if a current flows in a circle around a magnetic field line, a force will be created that is perpendicular to both. If the current oscillates with time, so will the force, in which case a small oscillating displacement will propagate through the medium. This is an acoustic wave.

It works the other way, too. If an acoustic wave travels through the medium and encounters a fixed magnetic field, an oscillating circular current (an eddy current) will be set up in the medium. That eddy current can be detected with a nearby induction coil; the eddy current sets up a current in the detection coil. Figure 26.6 is a diagram of this interaction.[12] In this way, the original acoustic wave is converted into a current in an external circuit, then amplified, displayed and analyzed—hence the name, electromagnetic acoustic transducer.

The usefulness of the technique lies in the ability to detect acoustic waves. During the fabrication cycle of various metals,[13] it is helpful to know the grain size, the number of dislocations, and other defects. When an acoustic wave propagates through the metal, Rayleigh scattering will take place at grain boundaries and other defects. Varying the frequency and observing the spectrum of scattered waves tells the grain size and provides information about defects in the crystal lattice.

If the acoustic waves can be detected with sufficient signal/noise ratio, then processing and interpretation of the data can be done by a microprocessor in real time. This means that

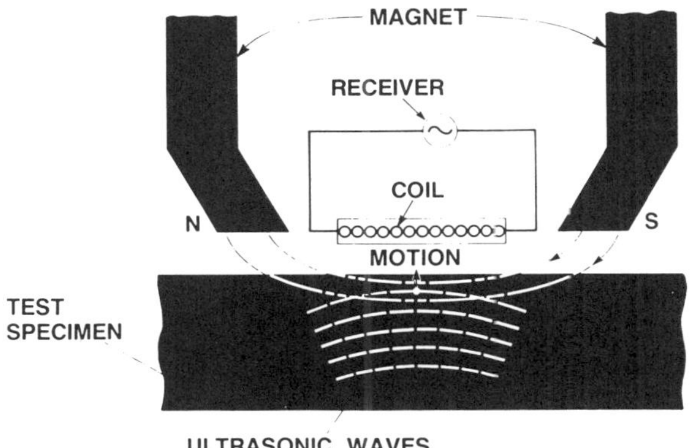

Figure 26.6. Conceptual drawing of the apparatus and interactions in an electromagnetic acoustic transducer (EMAT)

metals can be studied during the heat treating cycle, and corrective action can be taken if the process starts to give undesirable properties.

Until recently, metal alloys were hardened and annealed by two-step processes in which the metal was held at some fixed high temperature for a certain time and then cooled to room temperature over another time period. Choices were rapid quench, air cooling, and so on. With real-time process control, it is easy to follow a complicated time-temperature path. A sensor that gives a look at the grain structure during that path would allow corrections to be made en route. This would in turn reduce waste in metals processing facilities and reduce the likelihood of defective parts ever reaching the customer. All this adds up to higher productivity.

Research laboratories that investigate EMATs and similar devices are developing "enabling technology". Such technology enables an improved process to be realized. The payoff from the research comes in the leverage it gives over an industrial process. When a new sensor is introduced, the widespread benefits are limited only by the imagination of process engineers who find a way to utilize it.

Where does superconductivity come into this? When acoustic waves are scattered, the strength of the detected current is linearly proportional to the strength of the applied magnetic field. Contemporary research experiments test the principle with fields of about 0.5 T; in a factory setting, that signal gets lost in the noise. Having a field of 10–20 T would improve the signal/noise by a factor of 30. Stable and uniform fields of 10 T are reasonable with superconducting magnets.

How soon can we have it? The EMAT technology is making good progress in the laboratory,[14] so the pacing item is the availability of high magnetic fields. In the next decade an EMAT might be successfully tested in a factory using a low-temperature superconducting magnet. That would give some indication of the economic value of the measurement. Assuming the resulting market pull is strong, then after high-temperature superconducting magnets are developed, it would take only about three more years to have a commercial device widely available to factories.

26.6.2. MHD Ship Propulsion

A very large scale example of the use of high magnetic fields is that of magnetohydro-dynamic (MHD) ship propulsion. The starting point here is the same as for the EMAT: the force on a moving charge is $\mathbf{F} = q\mathbf{V} \times \mathbf{B}$. The ions in seawater will experience a force while moving in a magnetic field. To accelerate a ship, the force must be directed along the axis (nominally, the z-axis) of the ship, through a thruster tube, rather like a jet engine. This requires that the flowing current and the magnetic field lie in the cross-section of the ship (the x- and y-, or r- and θ-axes). Obviously, as $\mathbf{B}$ increases, so does the acceleration.

The novel and movie *The Hunt for Red October*[15] popularized the notion of propelling submarines very quickly and silently via this mechanism. Conventional engines and propellers readily give away the position of a submarine, which defeats an essential purpose (stealth) of their existence. Moreover, propeller cavitation limits the speed of conventional ships to 40 mph. A jet-thruster propulsion system is very desirable, and MHD seems to offer that.

This concept has been considered by the U.S. Navy for over 30 years, but it has never been an obvious winner. Design studies have given ambiguous predictions, and the overall efficiency is not calculable with any reliability. The biggest problem is that the level of ionization in seawater is 1 part in 10^{-7} (that's what pH = 7 means), so that only a tiny fraction of the water molecules participate in this force mechanism, and they have to provide all the momentum to move the water relative to the ship. The way around this is to provide a huge magnetic field, which is where superconducting magnets come in. Without superconducting magnets, it would be far too expensive to provide the magnetic field.

The trouble is, huge magnetic fields generate great Lorentz forces, tending to distort the mechanical structure, and therefore massive structural members are required to hold it together. But massive support structures are heavy and impose a severe weight penalty on any ship containing them. One Naval researcher quipped: "The only direction a ship will move under MHD power is straight to the bottom!"[16] To overcome the weight penalty, it is necessary to displace a lot more water, which means the vessel has a greater size and therefore more friction as it moves through the water. That friction in turn penalizes the efficiency of propulsion, and must be taken into account.

Despite all these limitations, researchers agree that experimental tests are warranted. Japan built a surface ship (the *Yamoto 1*) powered by MHD, which moves at 7 mph, thus showing that the principle is valid. At Argonne National Laboratory, a 2 T magnet has been used to accelerate seawater in a pipe that loops through the bore of the magnet[17]; Figure 26.7 shows the apparatus. The tests that were done verified the computational model of the flow and gave credence to numerical predictions that a magnetic field strength between 10 and 20 T would be necessary to power a submarine.[18] Recalling Figure 26.2, it is sobering to note that this magnetic pressure is about equal to the pressure at the bottom of the ocean. Of course, the Argonne laboratory magnet was housed in a building, so weight was no obstacle. The next step is to modify an existing submarine. Any real system for use by the Navy will involve a number of trade-offs.

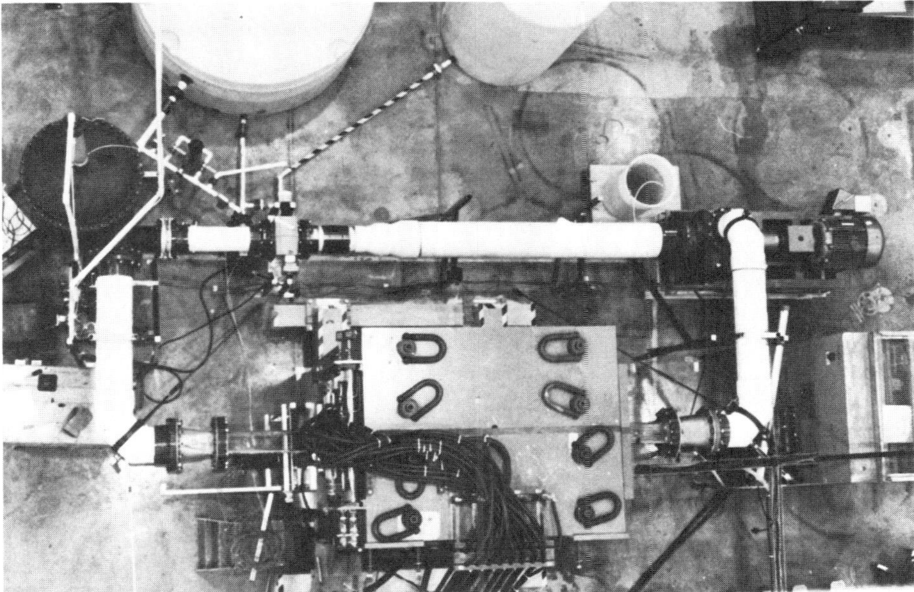

Figure 26.7. Apparatus for MHD test on flowing brine at Argonne National Laboratory. This overhead view shows the rectangular loop in the center, storage tanks at the top, the magnet in the center, with cables leading off to the bottom to the power supply. (Photo courtesy of W. C. Sikes, Newport News Shipbuilding Co.)

Perhaps the biggest worry for the U.S. Navy is the refrigeration system. The vibration and shock environment[19] inherent in battlefield conditions is far more severe than that seen by a commercial fishing vessel. If the refrigerator fails, the magnet shuts down and the ship is dead in the water. Recognizing this weakness, MHD propulsion is not yet considered an option for military systems. However, if HTSCs can run at an intermediate temperature, no longer demanding a helium liquifier on-board ship, then the refrigerator is less of a problem. On the other hand, the brittleness of the HTSCs suggests that vibration and shock would threaten the integrity of the magnet itself.

Considering both the good and the bad features, MHD propulsion is certainly to be considered a legitimate *research* activity. Superconductivity makes it possible to consider this technology, but many other factors will determine whether it is ever practical.

26.7. SUMMARY

The use of HTSCs to achieve high magnetic fields is a major innovation. Whether in laboratory or industrial applications, the prospect of having hitherto unachievable magnetic fields available is one of the most attractive features of this entire field.

Magnets made of BSCCO, running at 4 K or 20 K, are most likely to impact research needs. At present, the upper limit of DC magnetic fields is set by niobium technology, and BSCCO promises to raise that limit considerably—if wire can ever be made and actual magnets fabricated. Major research facilities, such as accelerators and dedicated magnetic-research laboratories, are the likely beneficiaries of such accomplishments.

Progress toward better refrigerators will eventually make intermediate temperatures accessible. As we saw in Chapter 24, a variety of new techniques are under development to attain temperatures between 20 and 50 K. The efficiency (and therefore the economics) is not as good as at 77 K, but if reliability is high, then there will be economically viable applications for such refrigerators.

For industrial applications, the outlook is more speculative because of the requirement to operate at high temperatures. It is easy to cool to 77 K, and a factory manager can plan ahead by keeping a large LN_2 storage tank on the premises, with regularly scheduled refills. To go lower in temperature increases the risk of down-time. Advances in refrigeration may alleviate this risk, but for the present, industrial managers are not willing to bet their production lines on anything that requires cryogenics other than liquid nitrogen.

In thinking about futuristic applications, it is important to retain an engineering perspective whenever a big project is considered. Energy density, magnetic pressure, stress, and strain are all significant factors in trying to build something real out of real materials. The ceramic superconductors are far from ideal materials.

Nevertheless, there is cause for optimism, because the incentive to improve industrial processes is always so great. The cliche "Necessity is the mother of invention" is no less true today, and the combination of advances in sensors, computer controls, and other factors open up many avenues for exploration. In this chapter we have sampled a few specific applications, just to illustrate the more general point that high magnetic fields enable the use of other clever technologies. As shown by the discussion of ore processing, having a magnetic field available where the cost is *not* proportional to the field strength is able to dramatically change the economics of production.

We have made no mention of the secondary effects of having such fields available, and yet this may eventually be very important. Some needs don't even exist until another invention comes along to create them. Recall that in the early days of television, there was no perceived need to watch a show at a time different from its broadcast, because that simply wasn't an option. Yet today many people have bought VCRs for exactly that purpose. The first technology created the need for the second. The applications of superconductivity are likely to follow that path.

REFERENCES

1. E. W. Collings, *Applied Superconductivity, Metallurgy, and Physics of Titanium Alloys* (Plenum Press, New York: 1986).
2. A. P. Malozemoff *et al.*, *Amer. Chem. Soc. Symposium Series* (American Chemical Society, New York: September 1987).
3. P. Berdahl, Lawrence Berkeley Laboratory, private communication.
4. K. Sato *et al.*, *Appl. Phys. Lett.* **57**, 1928 (1990).
5. R. B. Leighton, *Principles of Modern Physics* (McGraw-Hill, New York: 1959).
6. L. M. Lederman, "The Tevatron," *Sci. Am.*, pp. 48–55 (March 1991).
7. *NHMFL Reports*, **1**(2) (National High Magnetic Field Laboratory Office of Educational Media, FSU: 1992).
8. F. Culler, at Federal Conference on Commercial Applications of Superconductivity (28 July 1987).
9. C. Fields, DARPA, testimony before House Science and Technology Committee (7 October 1987).
10. R. J. Naumann and H. W. Herring, *Materials Processing in Space: Early Experiments*, NASA SP-443 (1980).
11. T. P. Sheahen, *Physical Principles of Microgravity Research,* report to NASA by Western Technology, Inc. (1985).
12. H. N. G. Wadley, private communication.
13. G. A. Ayers, in *Proceedings of Conference on Intelligent Processing of Materials and Advanced Sensors*, Orlando, FL, October 1986; edited by H. N. G. Wadley *et al.* (AIME-TMS, Pittsburgh: 1987).
14. H. N. G. Wadley *et al.*, *Phil. Trans. Roy. Soc. London A* **320**, 341 (1986).
15. T. Clancy, *The Hunt for Red October* (Berkeley Press, New York: 1984).
16. M. Superczynski, David Taylor Research Center, private communication.
17. B. F. Picologlou *et al.*, "MHD Seawater Thruster Performance: A Comparison of Predictions with Experimental Results from a 2 T Test Facility," Proc. 11th International Conf. on MHD Electrical Power Generation, Beijing China (October 1992).
18. E. D. Doss and H. K. Geyer, "A Need for Superconducting Magnets for MHD Seawater Propulsion," 25th Intersociety Energy Conversion Engineering Conference, Reno, NV (August 1990).
19. *Federal Research Programs in Superconductivity*, report by FCCSET Committee on Materials, Working Group on Superconductivity (December 1992).

27

Organic Superconductors

The excitement over high-temperature superconductors (HTSCs) of ceramic copper oxides has, to a certain degree, overshadowed a steady path of progress made on organic superconductors. No organic compound has its T_c anywhere near 77 K, but these materials are *layered* superconductors. Therefore, they are in some sense relatives of the HTSCs. This chapter describes the special nature of organic superconductors and compares them with the now-familiar HTSCs. The unusual new compounds based on carbon-60 can also form superconductors.

27.1. HISTORY

Organic superconductors were first proposed by W. A. Little of Stanford University in 1964, who suggested that superconducting behavior could arise from the polarization of metal chains surrounded by organic molecules.[1] Little included some speculations about the possible biological occurrence of such superconductors with T_c at or above room temperature; this motivated a lot of interest in the idea. However, his hypothesis was for one-dimensional superconductivity, and it was later shown that a purely one-dimensional structure is unstable. Little's concept eventually went on the back burner.

About 10 years later, a flurry of activity surrounded an organic molecule known as TTF-TCNQ (tetrathiafulvalene-tetracyanoquinodimethane), which was thought (for a while) to be a superconductor. It turned out to be a conductor to low temperatures, but not a superconductor.[2]

The first true organic superconductor was discovered[3] in 1980: TMTSF (= *tetra-methyltetraselenafulvalene*) forms the compound (TMTSF)₂PF₆ with a superconducting T_c of 0.9 K under a pressure of 12 kbar. Figure 27.1 shows[4] a collection of molecules that form superconductors. In general, they are flat, planar molecules containing sulfur or selenium atoms (among other elements). They form layered compounds with these molecules often standing perpendicular to the layer, like picket-fence boards. At the end of each picket, the junctions between layers of TMTSF molecules are separated by PF_6^- anions. The TMTSF molecules form radical-cations and the PF_6^- are anions, so the two exchange charges much like an ionic crystal. Figure 27.2 is a sketch of the crystal structure of (TMTSF)₂PF₆, viewed down the layers of the organic electron-donor molecule.

After 1980, several more organic superconductors of similar structure were discovered. In all cases, some anion is needed (in the style of Figure 27.2) to effect charge balance in order to obtain metallic properties and (at low temperatures) superconductivity. There are

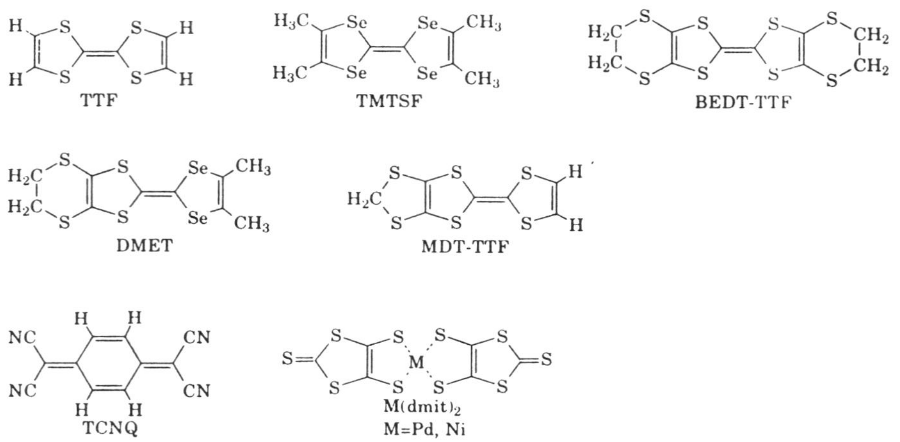

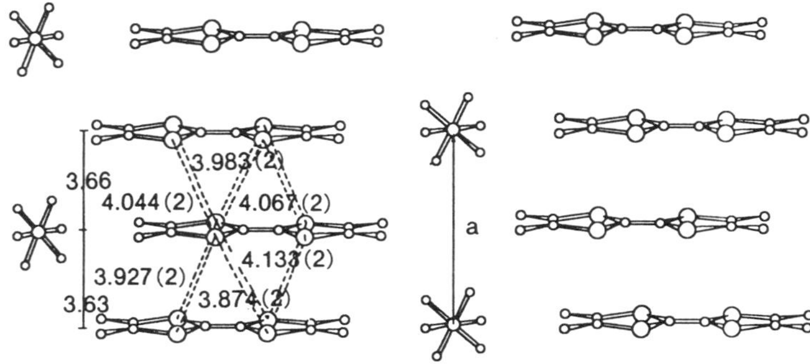

Figure 27.1. Molecular structure of organic compounds that combine to form conductors. (From Ref. 4.)

six different classes of organic superconductors, but most of them yield very low values of T_c. The highest $-T_c$ materials are based on the compound BEDT-TTF [= *bis*(ethylene*di*thio) *tetrathiaful*valene)], shown in the upper right of Figure 27.1. In 1983, the compound (BEDT-TTF)$_2$ ReO$_4$ was found to superconduct at 2.0 K under 4 kbar pressure.[5]

Much attention then turned to this class of compounds. Transition temperatures for BEDT-TTF–based superconductors rose steadily during the 1980s, and many are ambient pressure superconductors. (Usually it is possible to evade the requirement of external pressure by choosing a molecule for the anion that mimics the effect of pressure within the crystal.) By 1988, researchers in Japan[6] reached $T_c = 10.4$ K at ambient pressure, using the Cu(NCS)$_2^-$ anion. This went nearly unnoticed amid the excitement over HTSCs.

27.2. CONTEMPORARY PROGRESS

A useful convention of nomenclature has been adopted: BEDT-TTF is now abbreviated ET. The highest T_c organic superconductors all have the formula (ET)$_2$X.

Figure 27.2. Crystal structure of the organic superconductor (TMTSF)$_2$PF$_6$. This view is of the edge of the TMTSF molecule shown in Figure 27.1. The plane in which superconductivity occurs stands perpendicular to the plane of this drawing, in the y-direction. [From N. Thorup *et al.*, *Acta. Cryst.* **B37**, 1236 (1981).]

In 1989, Japanese researchers slightly changed $(ET)_2Cu(NCS)_2$ by using deuterium in place of some of the hydrogen in ET, and the transition temperature actually rose to 11.0 K—a sort of reverse isotope effect. Then in 1990, Jack Williams and co-workers at Argonne National Laboratory achieved $T_c = 12.8$ K at 0.3 kbar pressure.[7] The new compound is $(ET)_2Cu[N(CN)_2]Cl$. Earlier, the similar compound with Br instead of Cl yielded $T_c = 11.6$ K, at ambient pressure.[8]

What is remarkable in all this is the way T_c rose in the organic superconductors. In only 10 years, T_c increased over a factor of 10. The question immediately springs to mind, Is there another factor of 10 or more possible for these highly complex compounds? No one knows the answer, but experience suggests that it is worth investigating. These organic compounds are layered superconductors, so they are certainly eligible for some of the same surprises that the copper oxides produced.

Whereas the HTSC ceramics are prepared at temperatures near 1000°C, organic superconductors are prepared at ambient temperatures. In fact, if elevated in temperature above 100°C, they decompose, melt or change composition. The *electrocrystallization* synthesis process is generally used to make any organic charge-transfer salt, including the $(ET)_2X$ series. Solutions of the cation and the anion are placed in a container, separated by a porous glass plug (a "frit") that allows ions to pass only when current flows. Applying a small current $(0.1–0.5~\mu A/cm^2)$ causes small crystals of $(ET)_2X$ to form on the anode. Typical crystal masses are 140–280 μg. They are very thin little black crystals about 1 to 2 mm long.

Clearly, at this stage no one regards the organic superconductors as practical materials. However, the properties discussed below make them inherently interesting, and the diversity of possible organic molecules and anion choices remains very large, thus providing great future hopes for these materials.

27.3. ELECTRICAL PROPERTIES

The customary way to detect superconductivity in these crystals is via a magnetic measurement, namely, the shift in resonant frequency of a coil (inside of which the compound is placed) as the temperature decreases below T_c. Figure 27.3 is a typical plot of the superconducting transition.[8,9] Resistivity measurements are made via a four-point method (attaching both current and voltage leads). In the highest T_c sample, $(ET)_2Cu[N(CN)_2]Cl$, resistance measurements performed in the high-pressure apparatus at Sandia gave a midpoint $T_c = 12.8$ K and a width of less than 0.3 K at 0.3 kbar, which is the lowest pressure required to cause any such material to become superconducting.

The upper critical magnetic fields $H_{c2}(T)$ have also been measured[10] for $(ET)_2Cu([N(CN)_2]Br$ and are shown in Figure 27.4. The material is highly anisotropic, so it is no surprise that alignment of H along different crystal axes produces different H vs. T curves. What is surprising is the very steep gradients indicated: -20 T/K for $H\|ac$ and -2.2 T/K for $H\|b$. These numbers are similar to those for HTSCs. Theoretical projection from this slope data gives estimated critical fields at zero temperature: H_{c2} $(T = 0) = 216$ T in the ac-plane and 24 T in the b-direction. But these are no more valid than projections for HTSCs, because the limit where magnetic fields break down the Cooper pairs is only 20 T for organic superconductors.

Estimates of the anisotropy can also be obtained from critical field data, and it is then a short computational step to derive values of the coherence length $\xi = 37$ Å and 4 Å,

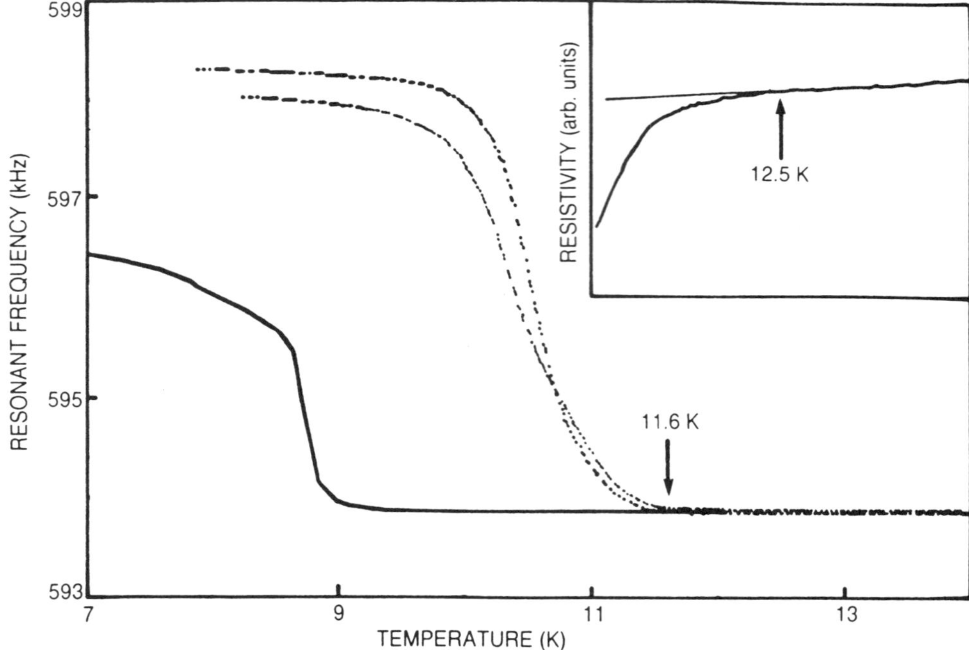

Figure 27.3. Superconducting transition in $(ET)_2Cu[N(CN)_2]Br$ is first seen at 11.6 K by magnetic susceptibility measurements (dotted lines). The vertical axis is the resonant frequency of a coil surrounding the ET salt. The inset at upper right presents the variation of resistivity with temperature for this material. The solid line (at lower left) shows the susceptibility of another member of the ET family, with the anion $Cu(SCN)_2^-$.

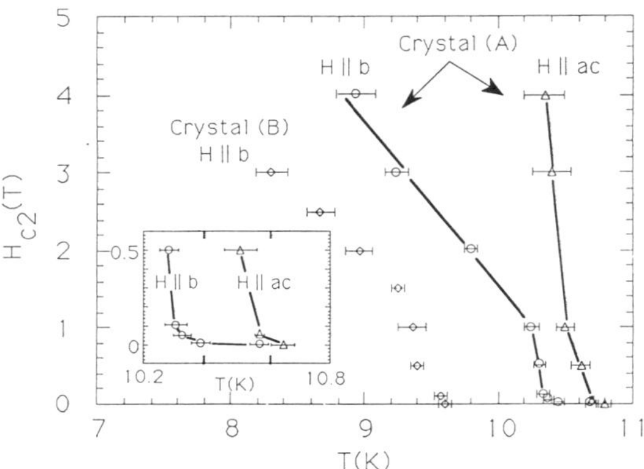

Figure 27.4. Temperature dependence of the magnetically determined upper critical fields of $(ET)_2Cu[N(CN)_2]Br$; for $H\|ac$ and $H\|b$ for an unquenched crystal (A); and $H\|b$ for a quenched crystal (B). Inset: Low-field part of the phase diagram for crystal (A).

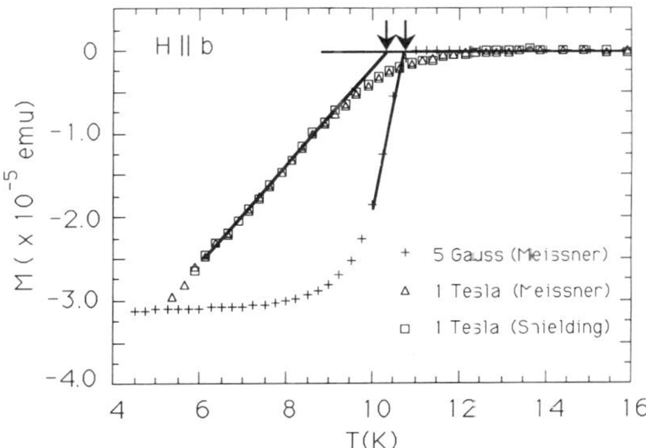

Figure 27.5. Temperature dependence of the magnetization of $(ET)_2Cu[N(CN)_2]Br$ in 1 T field (FC = field cooled; ZFC = zero-field cooled; see Appendix B), for $H\|b$. Also shown (+) are FC measurements in 5 Oe = 0.0005 T. The arrows designate the mean-field T_c for the two magnetic fields.

respectively. These results point to a striking similarity between these organic superconductors and their HTSC counterparts (3–30 Å).

For the organics, the shape of the superconducting transition in an external magnetic field also resembles that of the HTSCs. Figure 27.5 compares the transitions in a field of 5 Oe (= 0.0005 T) and in a field of 1 T. The broadening of the transition seen here is by now a familiar characteristic of YBCO, BSCCO, and so on. Our understanding of this phenomenon in HTSCs attributes the broadening to a dissipative mechanism such as flux flow or Josephson junction weak-link behavior. The similarity in the data warrants a similar speculation for the organic superconductors.

There is one very important difference, however. The exact superposition of triangles and squares in Figure 27.5 indicates that samples cooled in zero field (ZFC) are magnetized the same way as samples cooled with the field already turned on (FC). This implies that the magnetic behavior is reversible, and hence flux pinning must be negligible. That in turn implies that the critical current J_c will be quite small. Although historically J_c values have been estimated from magnetic data using the Bean model, we know from the HTSCs (also highly anisotropic) that this method does not yield reliable true transport values of J_c. It is not uncommon for pure single crystals to have low J_c, but unless some means of flux pinning is eventually discovered for bulk samples of the organics, they will not be useful for current-carrying applications.

Several other measurements have been made on the organic superconductors. For the compound $(ET)_2I_3$, the DC Hall effect has been measured by Murata *et al.*[11] Murata *et al.* carried out[12] similar measurements of the DC Hall effect in $(ET)_2Cu(NCS)_2$. In both cases their data suggests the occurrence of another phase transition around 20 K. Moreover, for the $(ET)_2Cu(NCS)_2$ material, the penetration depth has been deduced from susceptibility measurements by Kanoda *et al.*[13] In that work, the temperature dependence of λ did not follow the simple BCS theory. This outcome suggests that the energy gap is not constant, but varies directionally due to the anisotropy of the superconductor.

It is likely that other electronic properties of the organic superconductors will soon be measured, due to the recent renewal of interest in the field.

27.4. STRUCTURAL PROPERTIES

In a HTSC such as $Y_1Ba_2Cu_3O_7$, the superconductivity takes place in the copper oxide planes (the *ab* crystallographic planes), with relatively little charge motion in the perpendicular *c* direction. The Y atoms provide charge carriers, but otherwise are mere spacers; indeed, Y can be replaced with most of the rare-earth elements without affecting T_c very much. In the organics, the stacking of the ET planes is more complicated than are the CuO_2 planes in HTSCs: In each layer, the ET molecules stick up from the layer like picket-fence boards. Once again the anions are only charge-compensating spacers, and the conductivity is in the organic layer. Superconductivity is likewise greatest in the layer, because of overlapping orbitals associated with sulfur atoms on adjacent ET pickets.

Figure 27.6 is a view looking down on a layer. [Figure 27.6(a) is the β-phase and 27.6(b) is the κ phase, but that distinction is not crucial here.] In either case, the ET molecules are nearly perpendicular to the layer, which is the plane of the figure. The anion layers—$Cu[N(CN)_2]Br^-$, for instance—are not shown because they reside above and below this plane. The sulfur-to-sulfur contact distances are also indicated. A conducting pathway thus zig-zags through the compound, and net current flow can flow anywhere in the layer, which is called the *ac*-plane. The *b*-direction is perpendicular to the organic layer.

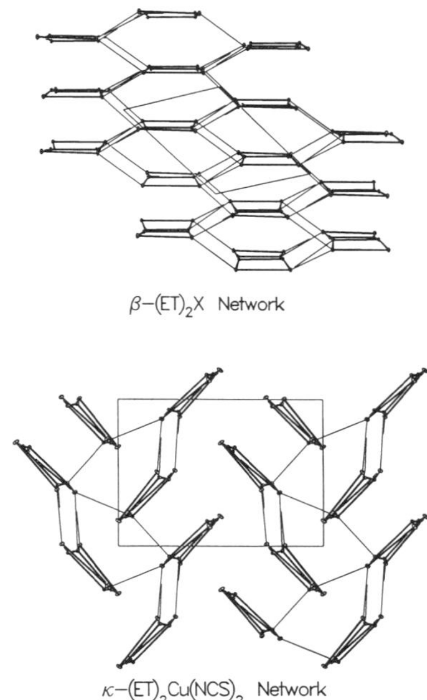

Figure 27.6. Structure of organic superconductors of the form $(ET)_2X$, showing β-phase (top) and κ-phase (bottom). In both cases, the plane of superconductivity is the plane of the page; the ET molecules stick up out of the page like picket fence boards. The unit cell is indicated by the rectangle in the lower drawing. The anions are not shown at all, to reduce clutter. The lines connecting various points on adjacent ET molecules are sulfur-to-sulfur connections, which provide a pathway for superconducting current to zig-zag through the lattice. (Courtesy of J. M. Williams, Argonne National Laboratory.)

Comparing Figures 27.2 and 27.6, the layer of Figure 27.6 slices vertically through Figure 27.2, and conversely. Thus, if Figure 27.6 is a *plan* view, Figure 27.2 is an *elevation*.

The density of these compounds is about 2 g/cm^3. To date, all crystals of the form $(ET)_2X$ are either long thin needles (typically 10–20 mm) or rectangular plates. Only laboratory samples of organic superconductors exist at present. Almost nothing is known of their mechanical strength. There is no effort at this time to fabricate wire, although thin films have been made in Japan. Of course, if the T_c's were higher, measurements of mechanical properties would soon be made.

If a polymer organic superconductor could be made, as contrasted to solid crystals, it might have good mechanical properties. (Conducting polymers like polyacetylene are flexible). However, no organic polymer yet discovered goes superconducting. One polymer superconductor, known as $(SN)_x$ $(T_c = 0.3$ K) contains no carbon and is not usually classified as organic.

27.5. FUTURE EXPECTATIONS

Enough is still unknown about the organic superconductors to keep interest in them high. Much of the research is motivated by a search for higher T_c materials. Because T_c is rising at a rate faster than the HTSCs, there is great expectation for higher T_c's, but practical applications must be deferred until such materials appear.

Meanwhile, since they are two-dimensional, studying the organic superconductors may enhance our knowledge of other layered superconductors.[14] The structural similarity to the HTSCs invites the response "Why not?" to questions about higher T_c organics superconductors. But the extreme anisotropy associated with the unusual structure of $(ET)_2X$ warns us to beware of extrapolating simple theories. Both the oxide superconductors and the organic superconductors are planar, two-dimensional materials. The similarities in certain properties—Fermi level, energy gap, coherence length—support the case for associating the organics with the copper oxides.

The first question asked is always, "Is this a BCS superconductor?" The values of the specific heat jump, the energy gap, and the slope of the critical magnetic field in both organics and copper oxides are too large to fit the BCS weak-coupling model, so strong-coupling superconductivity is a genuine possibility. Organic charge-transfer salts have not been studied extensively in the past,[9] and it is an open question as to what kind of pairing mechanism must be invoked. With progress in the theory of two-dimensional superconductors, the mechanism of electron-pairing in organic compounds will, more than likely, be elucidated.

27.6. CARBON-60 SUPERCONDUCTING COMPOUNDS

Historically, any *allotrope* based on the element carbon has been classed as organic, but a new carbon allotrope stretches that definition. The pure element carbon forms not only graphite and diamond but a soccer-ball shaped molecule containing 60 atoms.[15] Because the structure is a mixture of five-sided and six-sided polygons, reminiscent of the geodesic domes pioneered by architect Buckminster Fuller, the molecule C_{60} has been affectionately named "buckminster-fullerene," or just "buckyballs" for short. There are also higher molecular weight variations such as C_{70}, C_{72}, C_{100}, and so forth, which share many of the same properties. The series of C_{60} compounds with other elements are called *fullerenes*.

Only since 1990 has C_{60} been available[16] to many laboratories, and during this time period a number of different C_{60}-based compounds have been created for the first time. For example, the *fulleranes* are combinations of C_{60} with hydrogen, and it has recently been found that the lowest energy state is achieved with ten hydrogen atoms *inside* the ball of carbon.[17] The possible combinations of this very new form of matter seem almost as unlimited as the field of organic chemistry itself.

Of greatest interest for our purposes is the fact that when C_{60} is combined with alkali elements such as potassium,[18] rubidium,[19] and cesium,[20] superconductivity results. The basic face-centered-cubic structure is one in which eight C_{60} molecules occupy the eight corners of a cubic unit cell, with the alkali atoms centered on the six faces. Sharing with adjacent unit cells contributes $8 \times \frac{1}{8}$ for the C_{60} molecules, and $6 \times \frac{1}{2}$ for the alkali-metal ions. The chemical formula is typically M_3C_{60}, with $M = Rb$, K.

The transition temperatures are in the range 18–33 K for the alkali-doped fullerenes, and the interest in these materials is very high. Ancillary experiments have been conducted to determine properties such as the coherence length (26 Å), penetration depth (4800 Å), critical magnetic field ($H_{c1} = 0.026$ T, $H_{c2} = 78$ T), and so on. These numbers imply type II superconductivity, almost certainly of the BCS mechanism, but with details at least as complex as that of the copper oxide superconductors. Because this is a very fast moving field, most of these results must be termed preliminary. Unfortunately, these materials are extremely unstable in air, burning spontaneously if not kept in an inert atmosphere.

The future of C_{60} superconductivity is a paradigm for the entire field of HTSCs. To the question "How high will T_c go in fullerenes?" Art Ramirez of AT&T Bell Laboratories summarized the outlook[21] as follows:

> I think 33 K is the limit. . . . You can mix potassium and rubidium in any proportion, and T_c scales monotonically with the ratio. If you study these mixtures by x-ray diffraction, you can see that the lattice constant also changes uniformly. T_c is virtually a linear function of lattice constant. That makes sense in terms of BCS theory, because as you increase the lattice constant you also increase the density of states. The highest T_c and longest lattice constant occurs in Cs_2RbC_{60}. Why not Cs_3C_{60}? Because the structure can only be expanded so much, and [then] it goes into another structure. You no longer have face-centered cubic material; you have body-centered cubic or something else. The A_3C_{60} face-centered cubic structure is the only superconducting fullerene we know so far. But if we can find another fullerene and another structure, who knows?

The truly important point here is that carbon buckyballs and the fullerenes are entirely new materials, likely to have a very rich future. The occurrence of superconductivity in such compounds is one more of the many surprises they hold. Future advances in this category of superconductors certainly bear watching.

27.7. SUMMARY

Since the discovery of the copper oxide superconductors with transition temperatures above 77 K, progress on other types of superconductors has gone unnoticed by many observers. Actually, there has been a higher rate of increase in the T_c values of organic superconductors during that same time period. In 1980, the first organic superconductor was discovered, with $T_c = 0.9$ K; today, the highest $T_c = 12.8$ K.

Organic superconductors have a number of similarities to the HTSCs. First of all, the superconductivity takes place in layers in either set of compounds, and the anisotropy is so high as to be effectively two-dimensional superconductivity. Second, the coherence length ξ is extremely small, typically < 10 Å, about the size of a unit cell. (This means that relatively few electrons participate in the cooperative motion required for superconductivity.) Third, the BCS coupling strength suggests that these are strong-coupling superconductors. A number of researchers find in these similarities a motivation to study organic superconductors, in order to improve understanding of the HTSCs.

These compounds form tiny (millimeter-length) single crystals via an electrocrystallization process. They are extremely brittle (much more so than YBCO). The primary goal for organic superconductors at this time is to find compounds with still higher T_c values. When that goal is reached, great thought will be given to their mechanical properties. Practical applications for the organics will await the discovery of higher T_c systems. If an organic superconductor were found with $T_c > 77$ K, applications would be explored immediately.

The molecule C_{60}, known as a buckyball, has been called an entirely new state of matter. Fullerenes (chemicals based on C_{60}) are at best distant relatives of organic molecules, but they too have been shown to form superconductors. The alkali elements K, Rb, and Cs combine with eight buckyballs in a cubic lattice to make compounds such as Rb_2CsC_{60}, with $T_c \approx 30$ K. As research proceeds into the very exceptional properties of various C_{60} compounds, more surprises may be in store.

REFERENCES

1. W. A. Little, *Phys. Rev. A* **134**, 1416 (1964).
2. W. D. Gregory "Superconductivity in Alloys, Compounds, Mixtures, and Organic Materials," *Encyclopedia of Materials Science and Engineering,* Vol. 6, M. Beaver, ed. (Pergamon Press, New York: 1984).
3. D. Jerome *et al.*, *J. Phys. Lett. (Paris)* **41**, L95 (1980).
4. H. Mori, *ISTEC J.* **3**(2), 12 (1990).
5. S. S. P. Parkin *et al.*, *Phys. Rev. Lett.* **50**, 270 (1983).
6. H. Urayama *et al.*, *Chem. Lett.* **55**, 1988 (1988).
7. J. M. Williams *et al.*, *Inorg. Chem.* **29**, 3272 (1990).
8. A. M. Kini *et al.*, *Inorg. Chem.* **29**, 2555 (1990).
9. Physics Today **43**(9), 17–19 (Sept. 1990).
10. W. K. Kwok *et al.*, *Phys. Rev. B* **42**, 8686 (1990).
11. K. Murata *et al.*, *J. Phys. Soc., Japan* **58**, 3469 (1989).
12. K. Murata *et al.*, *Solid State Comm.* **76**, 377 (1990).
13. K. Kanoda *et al.*, *Phys. Rev Lett.* **65**, 1271 (1990).
14. R. L. Greene, "Organic and Oxide Superconductors: A Comparison," in *Organic Superconductivity,* edited by V. Z. Kresin and W. A. Little (Plenum Press, New York: 1990).
15. H. W. Kroto *et al.*, *Nature* **318**, 162 (1985).
16. W. Kratschmer *et al.*, *Nature* **347**, 354 (1990).
17. M. Saunders, *Science* **253**, 330 (1991).
18. A. F. Hebard *et al.*, *Nature* **350**, 600 (1991).
19. M. J. Rosseinsky *et al.*, *Phys. Rev. Lett.* **25**, 2830 (1991).
20. K. Tanigaki *et al.*, *Nature* **352**, 222 (1991).
21. A. Ramirez, interview reported in *Quantum States* (Fall 1992).

28

Aerospace Applications

As we have seen in prior chapters, the capacity of high-temperature superconductors (HTSCs) to sustain great magnetic fields leads to a number of applications where components can be smaller and lighter in weight than in previous embodiments. This provides the motivation to consider employing HTSC-based devices in space vehicles, and so NASA is quite interested in using HTSC. Furthermore, because the ambient temperature in space is usually below 100 K, such concepts can be pursued without including the weight penalty due to refrigerators.

At the conclusion of a book introducing a very new field, perhaps it is permissible to reach outward and upward. Futurists have speculated about space travel for many years, and a key element in that speculation is always wonderful new materials and innovative technology. In that same spirit, this chapter pursues what *might* come true in aerospace technology if HTSC succeeds. After a brief explanation of why NASA's outlook is unique, we describe some of the applications that are within reach of existing technology. After that, some of the more visionary concepts are described—ones that call for magnetic fields far beyond today's technology.

28.1. NASA's PERSPECTIVE

Long-range research in aerospace technology is conducted primarily by NASA, but often in collaboration with National Laboratories or major companies in the aerospace industry. Aerospace technology provides a good example of systems engineering, in which an ensemble of many components must each function correctly for the entire effort to succeed. Because of the long lead-times required between research and final implementation, and because design commitments (once made) are very hard to change downstream, NASA is striving to make sure that technologies being developed now are poised to take advantage of the new superconducting materials.[1]

- *Weight.* There is a very strong economic incentive toward light weight in the aerospace field. The launch costs associated with the space shuttle are so great that it costs above \$10,000 per pound to take raw material into space. This has proved an insurmountable obstacle for all but a very few products (certain medicines are worth \$1 billion per kilogram: if a 1-mg dose will cure you of a fatal disease, you'll gladly pay \$1000 for it). It is a sobering thought to recall the children's story of Rumpelstiltskin, the little gnome who could spin straw into gold. Had Rumpelstiltskin been able to do that on the space shuttle, he would have lost money, because the launch cost of the straw exceeds $700 per ounce.

Clearly, anything that can reduce weight has very great value in the economics of NASA. Although far less critical, for commercial aircraft light weight translates into smaller engine requirements for takeoff and lower fuel consumption in flight.

- *Refrigeration.* NASA seeks improvements in refrigeration systems, which must combine extreme reliability with light weight for optimum performance. They are hoping for major assistance from the HTSCs, because the liquid helium situation is so bleak: it takes 1000 W of refrigeration to cool 1 W of load using LHe; by contrast, it's 140:1 for neon and 30:1 for LN_2. Just by using HTSC lead-in wires to get from 77 K to 4 K, they can save nearly half the heat leak, thus extending the lifetime of their instruments substantially. The Infrared Astronomy Satellite (IRAS) could see out to 200 mm in the infrared; it found a planet around another star. That spacecraft only lasted less than 1 year, but would have gone 2 years with HTSC lead-ins. A more advanced refrigerator is the *Oxford cooler*, which Jet Propulsion Lab has upgraded. NASA is hoping to operate at 65 K, because that is where mercury cadmium telluride detectors work best. HTSC lead-ins would reduce parasitic losses here, too.

28.2. NEAR-TERM APPLICATIONS

NASA has defined four categories of technology into which HTSC applications might fall: power, propulsion, communications/data systems, and sensors. The latter two categories are primarily electronic applications, and researchers are very optimistic about progress in these areas. Following initial research and fabrication of prototypes, it is reasonable to anticipate that commercial development will convert research results into practical devices.

28.2.1. Electronics

Obviously the international race to bring superconducting electronics to practicality will have space applications. Powerful, lightweight on-board computers are very desirable, but there is much more: the use of thin-film HTSC devices for microwave resonators, and so on, is already well into the development stage.[2] Other books cover thin-film electronics in far more detail, so here we briefly mention only a few examples of thin-film sensors that are of direct interest to NASA.

1. NASA hopes to develop a HTSC infrared bolometer to be launched on missions to the outer planets. For example, in 1996, "Cassini" is scheduled to leave for a rendezvous with Saturn in 2002, where it will study the atmosphere of the moon Titan, because it's atmosphere resembles that of the early earth. Assuming an ambient temperature below 92 K, the HTSC bolometer would serve as the detector in a Fourier-transform spectrometer (FTIR), capturing infrared radiation in the range 16 μm–1000 μm. The properties on NASA's wish list for this detector are ambitious: 12-year lifetime in a high vacuum, $D^* > 7 \times 10^9$ in either range $16 < \lambda < 50$ μm or $30 < \lambda < 1000$ μm. The best bet for a "fast" bolometric detector (one with little specific heat) looks like a thin film of YBCO on a diamond substrate, with a buffer layer in between. NASA is hopeful that the more traditional thermopile detector can be replaced with this.

2.   An early flight that includes HTSC hardware is the HTSSE-II experiment of the Naval Research Laboratory.[3] There a low-noise receiver will incorporate several superconducting components, along with cryogenically cooled semiconductor components.[4] Earth-bound uses of such hybrid semiconductor–superconductor technology are so promising that it is only natural to try the same thing in space.

3.   Another electronic application of interest to NASA is an ultrastable space clock in which a HTSC-coated cavity with an extremely high Q may give stability as high as one part in 10^6. Precision timing is terribly important in lengthy space missions.

28.2.2. Levitation

A completely different, but equally possible, application of HTSCs is in the area of levitation, specifically magnetic bearings. The environment of a spacecraft, once deployed, is nearly zero-gravity; on the moon, gravity is $\frac{1}{6}$ that of earth. In either case, the bearings supporting rotating components do not have to sustain large forces. These are perfect opportunities to use HTSC bearing materials.

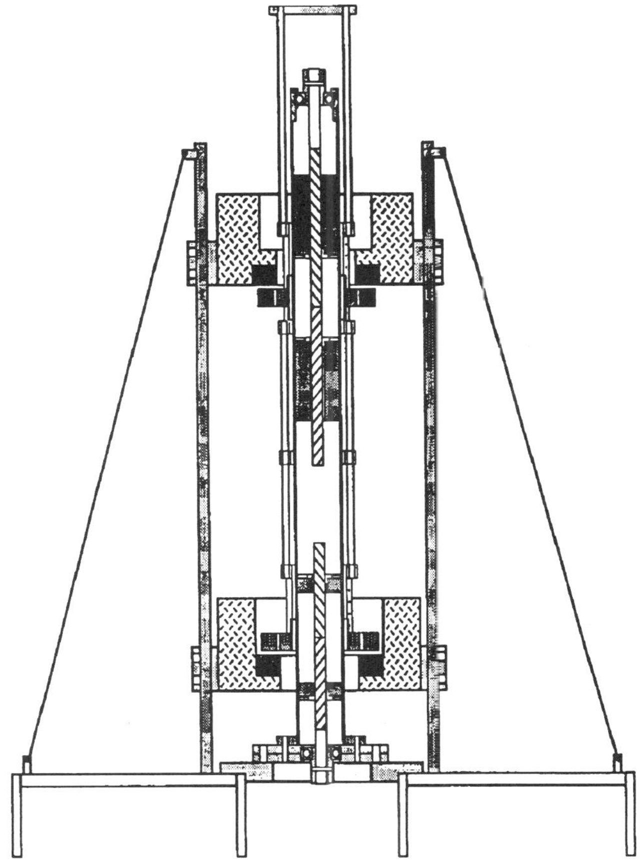

Figure 28.1.   Cross-sectional diagram of TCSUH Lunar telescope mount using superconducting bearings.

As was discussed in Section 20.5, the Meissner effect can be used to advantage in a hybrid combination of a permanent magnet with a superconductor. The Texas Center for Superconductivity at Houston (TCSUH) pioneered this configuration; Figure 20.10 shows the dimpled mountain potential well that creates a stable position. TCSUH also has built some hybrid bearings, and has shown that they are truly frictionless.[5] In a vacuum, they can run forever—as long as the superconductor is kept cold.

The combination of cold and vacuum occurs on the moon, and TCSUH hopes to see these bearings used in an unmanned lunar observatory. Figure 28.1 is a drawing of how a telescope might be supported with such bearings. The conditions of observation are quite different from those on earth: the slower spin rate of the moon implies that the telescope must rotate about 0.1 μm/sec in order to faithfully track stars across the night sky on the moon. On earth, machinery that can control bearing motion with such precision is bulky and massive; it would be very difficult to send it to the moon. This kind of bearing, however, is lightweight to begin with, and the lack of any friction makes position control much easier.

Figure 28.2. Photograph of the mock-up of the Lunar telescope mount.

Just as on any planet, the moon experiences frequent tiny earthquakes (moonquakes?), and so the telescope support apparatus will vibrate. It turns out that YBCO acts as a vibration damper,[6] for the following reason: when a magnetic material moves periodically (vibrates) near YBCO, the hysteresis in the restoring force turns the system into a sink for mechanical vibration energy.[7] This feature promises to make the lunar telescope exceptionally quiet and stable.

This device has reached the mock-up stage, and Figure 28.2 is a photo of it. The entire structure is about 1 m high, with a base ≈ ½ meter. The bearing assembly can support a load of 10 lb, while its own weight is 15 lb. In order to support a central shaft of roughly 7.5 cm diameter, two superconducting bearings must be 15 cm in diameter and 7.5 cm tall. That is a large YBCO sample, especially if melt-processed, but not an impossible goal.

28.3. APPLICATIONS OF HIGH MAGNETIC FIELDS

Sensors and electronics contribute to many industries. However, there are other uses of superconductivity that are unique to the aerospace industry; some can best be termed visionary.[1] Table 28.1 lists several of these.

28.3.1. Electromagnetic Launch

The idea of electromagnetic (EM) launching of objects is much older than high-temperature superconductivity.[8] Aside from rail-guns for kinetic penetrators in weapons research, designs for various EM launchers have been known in the aerospace community for years. Figures 28.3 and 28.4 were drawn long before the discovery of high-temperature superconductivity.

Through earlier studies,[9] EM launch appears to be technically feasible, and (in large-scale applications) economically attractive. Technical barriers are serious but not insurmountable. It is not reasonable to launch people this way, because the initial kick is many times the force of gravity. Therefore, the technique is limited to launching bulk cargo. If there is enough such cargo, the economics are good: 2000 kg per day are enough to break even. What high-temperature superconductivity brings to the party are sharp reductions in refrigeration cost (perhaps 90%), and a definite cut in the cost of energy storage (25%).

The idea of mining valuable minerals from the surface of the moon has been a popular dream so far, but it may be practical someday. If solar cells collect sunlight and a superconducting ring is used to store the energy, a launching device as depicted in Figure 28.2 could be used to lift the valuable materials off the surface and into moon orbit. Recalling the time

Table 28.1. Aerospace Applications of
Superconductivity

Electromagnetic launch
Electromagnetic airport applications
Microwave power transmission
Shielding from high-energy charged particles
Magnetic energy storage for space applications
Re-entry magnetic heat shield
Magnetoplasmadynamic (MPD) propulsion

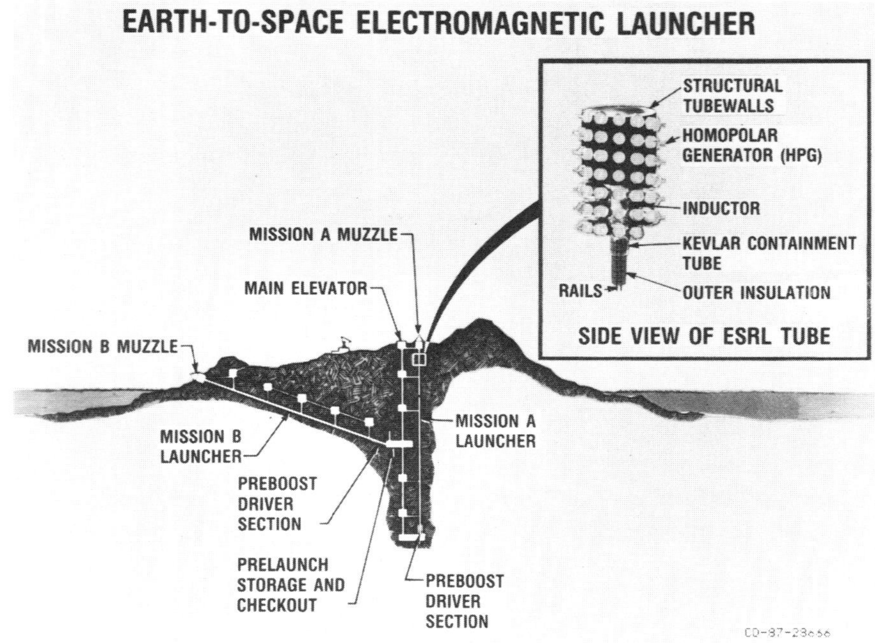

Figure 28.3.　Earth-to-Space ElectroMagnetic Launcher. (Courtesy of NASA Lewis Research Laboratory.)

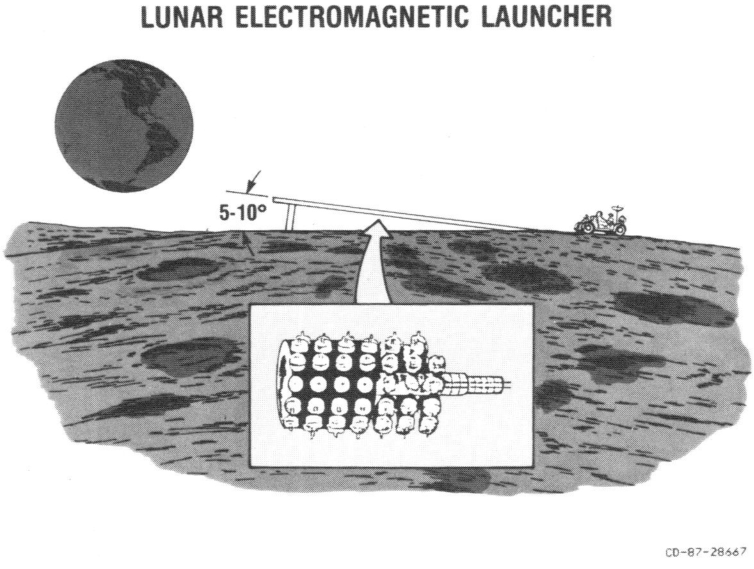

Figure 28.4.　Lunar ElectroMagnetic Launcher. (Courtesy of NASA Lewis Research Laboratory.)

that astronaut Alan Shepard hit a golf ball far along the moon's surface is a good way to remember that it doesn't take as much to overcome the moon's gravity as it does the earth's.

28.3.2. Electromagnetic Airport Applications

Magnetic levitation is not confined strictly to trains; the application to airport runways also may be contemplated. Airplanes could land on much shorter runways if it weren't for the safety requirement of hitting the pavement gently. In a magnetic-levitation runway, the pavement is the very squishy region of strong magnetic fields. Moreover, suddenly switching the polarity of two adjacent magnetic fields provides much better braking than does reversing the thrust of jet engines. The suddenness of stopping is limited by how much deceleration passengers can endure, not by the technology of the stopping mechanism. It is estimated that the amount of airport traffic per unit area can be increased by about a factor of 10 using such techniques, with no decrease in safety.

On the take-off side, an EM launcher gives too hard a kick for people to tolerate, but electromagnetic assist is entirely plausible. Here the engines are running at full throttle as usual, but the take-off acceleration is enhanced by a gentler EM launching device. EM assist reduces the power required from the engines. The benefit here is not to get people moving faster, but to allow large airplanes to have much smaller engines. That downsizing and reduction of weight would save fuel throughout the duration of the flight. An additional benefit would be the reduction of noise and exhaust fumes around the airport.

28.3.3. Microwave Power Transmission

The term *microwave* is used to span a broad range of the electromagnetic spectrum. The lowest frequencies that fit this name are around 1 Ghz, or 10^9 Hz. Somewhat higher frequencies, in the vicinity of 10 Ghz, are widely used in police radar systems. In general, the frequency and wavelength are related via the speed of light: $f\lambda = c = 3 \times 10^8$ m/sec. For example, a frequency of 5 Ghz (5×10^9 cycles/sec) will have a wavelength of 0.06 m = 2.4 in. Frequencies ten times higher are known as millimeter waves; higher still gives submillimeter waves. The far infrared, infrared, and visible regions of the spectrum follow, as frequency increases.

The commonest and most practical microwaves are in the 1–10 Ghz range, having dimensions of inches, typically. Nearly all telephone transmission was carried by microwaves from one mountaintop to the next, until fiber optics were invented. Electronic components and waveguides are not difficult to make in this size. Higher frequencies and shorter wavelengths create more stringent demands on component machining accuracy, and so on. Consequently, for many years there were virtually *no* practical applications of millimeter and submillimeter waves.

Because the size of components scales with the wavelength λ, the weight scales as λ^3. That tells us at once that aerospace applications offer an urgent invitation to push the technology of high-powered, short-wavelength microwaves. It is at this point that superconductivity has something to offer.

The classical means of generating microwaves involves using a high magnetic field to control electron flow in a large vacuum tube. Superconductors can shrink the size and weight of the magnet considerably, and possibly provide extremely large focusing fields with almost arbitrary geometrical arrangement. Theory suggests such focusing fields could enable much shorter wavelength microwave tubes. As long as a power source is available, the shortest possible waves are best. Calculations suggest that the optimum frequency may be 500 GHz or higher, well into the submillimeter range.

The biggest weight savings (and concomitant cost savings) comes in the antenna and rectenna systems. A factor of 100 in λ suggests a factor of a million in weight. In reality, the outcome would be to drive down the weight of the antenna and rectenna to where they are no longer prohibitively heavy components in the overall system design.

28.3.4. Shielding from High-Energy Charged Particles

High-energy charged particles are constantly being given off by the sun; they comprise the solar wind. They have negligible effect on people at the earth's surface, owing to the shielding effect of the atmosphere, but in space astronauts (as well as electronics) must be shielded from them. Typical barrier methods of shielding are heavy and therefore expensive to carry into orbit.

Magnetic shielding is an alternative that may become the method of choice if superconductors can provide high magnetic fields at low cost and weight. Any charged particle entering a magnetic field $\mathbf{B}$ experiences a force perpendicular to its velocity $\mathbf{V}$, $\mathbf{F} = e\mathbf{V} \times \mathbf{B}$, and hence is deflected into a circular path. This is called Larmor precession.[10] The momentum of a particle having energy E and traveling at nearly the speed of light (c) is just E/c. In a magnetic field $\mathbf{B}$, the radius of the trajectory r is determined by requiring this momentum to equal Ber. In the earth's natural magnetic field ($\frac{1}{2}$ G $= 0.5 \times 10^{-4}$ T), the radius of a 1 GeV (1 billion electron-volt) particle is about 100,000 m. Hence, the particles gently spiral in to earth near the north and south magnetic poles, creating the *aurora borealis* via collisions with air molecules as they enter the atmosphere. But in a magnetic field of 10 T (conceivable for a superconducting shielding magnet near a spacecraft), the radius is only 30 cm, or 1 ft. Thus, charged particles could be repelled from the vicinity of a spacecraft rather easily, using superconducting magnets.

28.3.5. Magnetic Energy Storage for Space

Using a superconducting ring to store energy is just as attractive in space as it is on earth. It allows energy to be collected by solar cells during one portion of orbital flight, stored, and spent later. The round-trip efficiency of SMES (90%) compares favorably with other means of storing energy in space.

Of course, earthbound designs for such storage rings achieve optimal performance by going to very large scale, perhaps 1 Km in diameter. To withstand the huge distorting forces, such systems need to be carved out of massive stone. Space applications would be far smaller and lighter. However, the distorting (Lorentz) force would still be present, and without retaining walls the modern ceramic superconductors would break easily. For this reason, application of SMES in space awaits the development of a flexible, stretchable high T_c superconductor.

After the mechanical problems associated with magnetic energy storage are solved, the advantages are significant. Even the most advanced batteries and fuel cells have a round-trip efficiency below 80%, and their mass represents a substantial weight penalty. Higher efficiency of storage and retrieval means that fewer solar cells would be needed for the same power output; therefore, a slight multiplier effect in weight reduction occurs. The ability to charge and discharge rapidly and at a high repetition rate enhances the flexibility of this storage method compared to others.

28.3.6. Re-Entry Magnetic Heat Shield

The same outward pressure that makes it so difficult to hold a magnetic energy storage ring together can be utilized to help decelerate a space shuttle during re-entry. At present, any vehicle re-entering the earth's atmosphere strikes the air molecules hard enough to ionize some of them, and therefore the vehicle is engulfed in an ionized gas (a plasma) during a critical period of re-entry, from about 250,000 ft altitude down to 50,000 ft. The hot plasma can be very damaging to exposed surfaces, and hence the orbiters have a bottom layer of ceramic tiles to dissipate the heat. Lots of tiles have to be replaced after each use. It is desirable to keep the hot plasma away from the surface of the vehicle.

Ions interact with a magnetic field, and therefore if the descending orbiter presents a high enough magnetic field, the ions can be deflected away from the vehicle. Setting up a magnetic shield would alleviate the heat-dissipation load on the vehicle itself. (Some uncharged gas molecules would still get through this shield, and their high speed would heat the vehicle surface; but the overall severity of the heat load would be reduced.)

Figure 28.5 compares the standard form of aerobraking with a comparable magnetic braking configuration. By using compact high-field electromagnets (presumably supercon-ductors), a magnetic field B can be set up in front of the vehicle. The pressure is $P = B^2/2\mu_0$; this equals 4 atm for a 1 T field—enough to drive away a major fraction of the incoming plasma.

If the vehicle is to slow down at all, the energy of motion must be dissipated as heat somewhere. The point of this magnetic braking method is to dissipate the energy in the air, not in the orbiter. By suitably shaping the magnetic field, a shock front can be set up that will cause the ions to recoil throughout a large volume of air. Energy will be transferred to

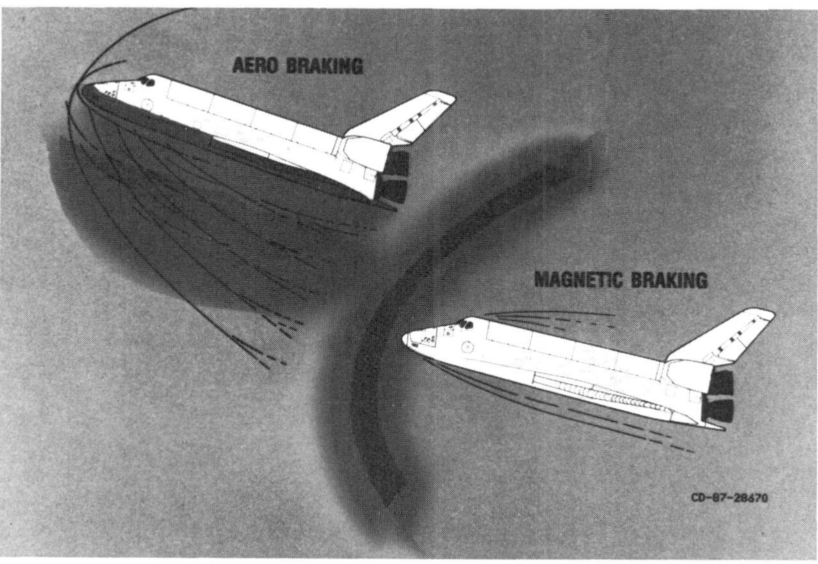

Figure 28.5. Comparison of two types of braking: conventional aerobraking (upper left) vs. magnetic braking (lower right). (Courtesy of NASA Lewis Research Laboratory.)

that air through collisions between ions and neutrals, and much less energy will reach the heat shield of the space shuttle.

The effect of the magnetic shock front is to greatly increase the effective cross-sectional area of the vehicle, so that deceleration can occur rapidly. The early space capsules, with their blunt heat shields, used this principle. Here the blunt surface is made of ionized air molecules, held in place by the magnetic field generated back at the vehicle itself. The heat load is shifted away from the vehicle and into the surrounding air.

The best feature of this entire braking method is the flexibility that accrues to the pilot. The magnetic field strength and shape is under the pilot's control. Just as an automobile driver can depress the brake pedal with different degrees of force, magnetic aerobraking offers a degree of flexibility in braking that has been absent so far in atmospheric re-entry. That in turn means that the final trajectory of re-entry is not absolutely determined at the start of re-entry, above the atmosphere.

There is an application to weaponry here as well. A warhead of a certain shape on a ballistic trajectory normally has an easily calculated re-entry path to a target. A military opponent who spots the warhead on radar can figure rather accurately where it will land some minutes later, and shoot a defensive missile at it. However, if the warhead has on-board a means of producing a strong (and variable) magnetic field during re-entry, its path and final target become harder for the opponent to discern.

Advanced techniques for enabling warheads to maneuver during re-entry have been studied extensively in the past. The day-to-day uncertainties in the atmosphere degrade the reliability of all maneuvering strategies. This method of magnetic shielding is not exempt from that limitation. Still, with on-board sensors feeding back data to control the magnetic field in real time, the method may have promise.

28.3.7. *Magnetoplasmadynamic (MPD) Propulsion*

Despite its elaborate name, the principle underlying MPD propulsion is an everyday occurrence. When an auto-ignition switch is turned to start, current flows through a solenoid, creating a magnetic field, and a conductor is pulled toward the inside of the solenoid (thus closing a switch and causing the starter motor to run). In MPD, ions are drawn toward the interior of a magnetic solenoid, accelerating as they enter the region. Properly positioned anodes and cathodes ensure that the moving ions exit through a nozzle at the back of the rocket. The equal and opposite force of Newton's third law propels the vehicle forward. Figure 28.6(a) shows a typical configuration.

It is noteworthy that Figure 28.6 dates from 1971. The technology to do this is not new. It simply has not yet been a very efficient form of propulsion. Figure 28.6(b) shows why: With magnetic fields below 1 T, the efficiency is below 35%. For low-thrust rockets (verniers, orbit trimming, etc.) other technologies are better.

Assuming successful development of HTSCs, magnetic fields of many tesla may become possible. If the ambient temperature of space makes refrigeration unnecessary, MPD technology may rise to the top of the list. NASA has renewed interest in it,[1] because MPD rockets deliver high specific impulse at low thrust, and operate continuously for a long time.

These characteristics make MPD a candidate technology for long, slow space journeys. For example, a cargo vehicle going from Mars or the moon to earth could use MPD propulsion after an initial lift-off phase. Similarly, the very high fuel efficiency promised by MPD with high magnetic fields recommends this as a means of delivering heavy payloads to other planets.

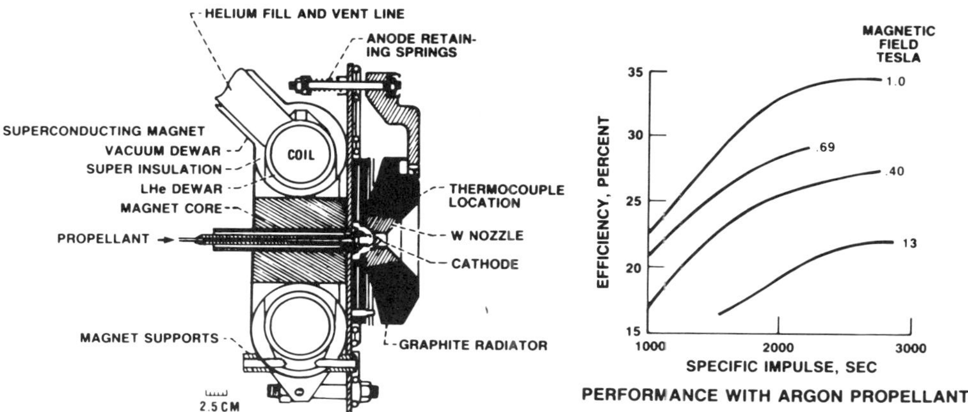

Figure 28.6. Superconducting Magnet MPD thruster, with emphasis on importance of high magnetic fields. (a) cross-sectional diagram of unit; (b) efficiency vs. specific impulse, for various magnetic fields.

28.4. FUTURE EXPECTATIONS

The hope of high magnetic fields makes superconductors attractive to the aerospace industry. There are a lot of good ideas that have sat on the shelf for many years, impeded by the lack of strong magnetic fields. Most categories discussed in this chapter share this limitation. The excessive weight associated with conventional magnets precludes their use in space.

Several of the many applications discussed in Chapters 19–26 may be useful in space. Beyond levitated bearings, another obvious candidate is very small electric motors. Superconducting magnetic energy storage (SMES) is certainly preferable to conventional chemical batteries[11]; indeed, with an array of solar panels on board the spacecraft, SMES might be able to maintain high power to an earth-orbiting satellite during dark periods. HTSC cables, akin to those described in Chapter 19, might be used to carry electric power from a nuclear reactor to the space station in an ambient < 100 K. When radiative cooling alone is adequate, this application is feasible with existing technology.[12]

The requirement of liquid helium refrigeration has kept superconductors from being cost-effective in the past. With ceramic superconductors able to operate at 20–35 K, and new types of refrigerators to reach those temperatures, that obstacle may be overcome. Moreover, whenever superconductors can operate above the ambient temperature in space, engineers will rapidly find other applications at least as ingenious as those described here. Space exploration stands to become a major beneficiary of progress in superconductivity.

REFERENCES

1. D. J. Connolly *et al.*, *Applied Superconductivity* **1**, 1231 (1993).
2. R. F. Leonard *et al.*, *Applied Superconductivity* **1**, 1341 (1993).
3. C. L. Lichtenberg *et al.*, *Applied Superconductivity* **1**, 1313 (1993).
4. W. Chew *et al.*, *Applied Superconductivity* **1**, 1259 (1993).
5. C. K. McMichael *et al.*, *Appl. Phys. Lett.* **60**, 1893 (1992).
6. K. B. Ma *et al.*, *IEEE Trans. Applied Superconductivity* **3**(2), 388 (1993).

7. M. A. Lamb *et al.*, pp. 453–458 in *HTS Materials, Bulk Processing and Bulk Applications*, ed. by C. W. Chu *et al.* (World Scientific, Singapore: 1992) ISB: 981021032-9.

8. F. Chilton, "Mass Driver Theory and History," AIAA Paper 77-533 (May 1977).

9. L. Miller *et al.*, "Preliminary Analysis of Space Mission Applications for ElectroMagnetic Launchers," NASA Report CR-1747-48 (1984).

10. J. A. Stratton, *Electromagnetic Theory*, (McGraw-Hill: 1941).

11. U. M. Eyassa *et al.*, "SMES for Space Applications," European Space Power Conf., Florence, Italy (September 1991).

12. J. R. Hull and I. T. Myers, "HTSCs for Space Transmission Lines," *Proc. ASME Winter Meeting* (December 1989).

Appendix A

Measurement of Critical Current

For practical superconductors, maximizing J_c is the object of a great deal of materials engineering practice. To measure success in this regard, it is necessary to measure the value of J_c in a superconductor.

For low-temperature superconductors (LTSCs), a set of measurement procedures was defined. These are contained in ASTM standard B 714-82. Experience with high-temperature superconductors (HTSCs) has shown that this standard is not applicable because (1) it was intended for longer lengths of wire than are common in HTSC testing, and (2) it is based on the LTSC observation that the current–voltage curve has a very sharp "knee" at the critical current. Figure A.1 illustrates the problem caused by the soft "knee" of the HTSCs. The point at which we call the current *critical* is set by some threshold electric field E_0. A different choice of E_0 results in a different value for J_c. This was never a problem for the LTSCs. Because the properties of the HTSCs differ in this and several other respects from the LTSCs, it is necessary to take a fresh look at this entire measurement task.

A.1. MAGNETIZATION MEASUREMENT OF J_c

For purposes of carrying current, it is the *transport* measurements that matter—how much current actually can move through a wire, grain to grain. However, experimentally it is often more convenient to do magnetic measurements—i.e., the change in magnetic susceptibility—and thus determine J_c as a circulating current within a grain. The customary theoretical connection relating magnetic susceptibility to J_c involves the *Bean critical state model*.[1] In this model, when current flows in a superconductor, it does so first on the surface, with a current density J_c in a shallow layer. To transport more total amps, the shallow layer thickens, still carrying J_c, but over a larger cross-section. As the total current increases, all that happens is that the boundary between J_c and zero moves inward, closer to the center of the superconductor. (When the current decreases, the boundary moves outward.) Figure A.2 displays the profile of the current in a typical slab of superconducting material. By using this model, it is easy to obtain a value of J_c via magnetic measurements.

When an external magnetic field H is applied to a superconductor of characteristic dimension D, that field will set up surface currents having value J_c. The relation $\partial H/\partial x = J_c$ leads at once to the Bean formula $J_c = H/D$. Thus, J_c can be measured magnetically, without having to conduct large quantities of electricity through external leads attached to a sample. This fact led to magnetic measurements becoming the preferred means[2] of determining J_c in NbTi and Nb$_3$Sn. This is known as the magnetization method; it depends for its validity

555

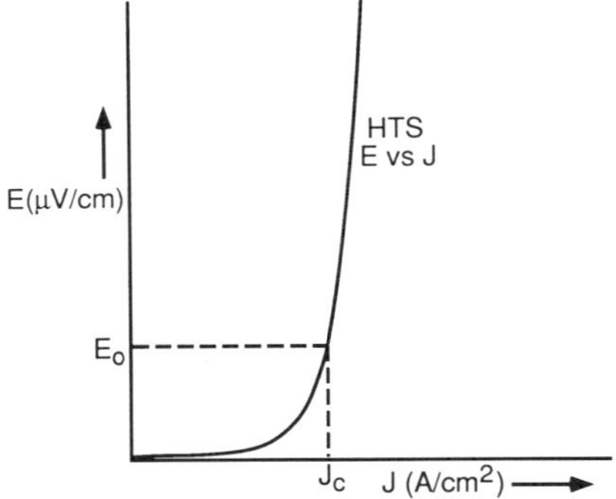

Figure A.1.　Current–voltage relation in a HTSC, showing how the choice of E_0 influences the defined value of J_c.

on both the critical state model (which is well established for conventional superconductors) and on a correct estimate of the diameter of the surface current loops.

Magnetization measurements have the advantage that they eliminate the need for high test currents. Unfortunately, there has been some disagreements between different measurements made on the HTSCs this way. When the superconductor is composed of poorly connected grains, as in YBCO or the other ceramics, the correct diameter is not obvious. Any current confined within one grain sees a very small diameter which may be far smaller

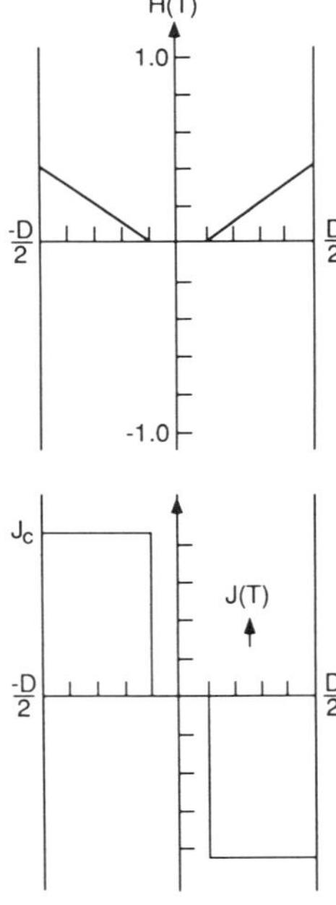

Figure A.2.　Current distribution and magnetic field according to the Bean model.

than the full wire diameter. Then applying the Bean model blindly will give wildly different J_c values in the bulk and in a grain.

A.2. TRANSPORT MEASUREMENT OF J_c

Recognizing that magnetic measurements are inadequate for HTSCs, it becomes necessary to measure the actual current moving down the wire, known as the *transport current*. At first sight, it is conceptually an easy measurement to determine the critical current I_c in a sample, and then just a matter of geometry to divide out the cross-sectional area to get J_c.

Of course, reality is never that simple. Great care must be taken to attach the lead-in wires, to measure the voltage across the sample, to ensure that the current is actually flowing through the presumed cross-section, and to control the ambient magnetic field. The experiment is sketched in Figure A.3; Figure A.4 is a photograph of a typical thin film set-up for J_c measurement. It is important to note the very small scale, typically a few millimeters between electrical contacts.

In the basic transport measurement, current enters one end of the sample and leaves via the other end. At two points a known distance apart, the voltage difference is measured. For small currents, that voltage is zero, but when $J = J_c$ the sample becomes normal and the voltage suddenly rises. Accordingly, increasing J steadily until the voltage jumps is the easy way to determine J_c.

Four types of transport measurements can be distinguished:

- continuous DC ramp in which current is steadily increased while the output of a DC voltmeter is monitored;
- a stepped DC ramp, in which the current pauses at discrete values;
- a DC bias current is applied at some level, while an AC modulation is superimposed (this allows use of lock-in amplifiers with concomitant higher sensitivity);
- pulsed current methods, which prevent overheating but permit transient effects.

Choosing between these competing methods depends on several factors.

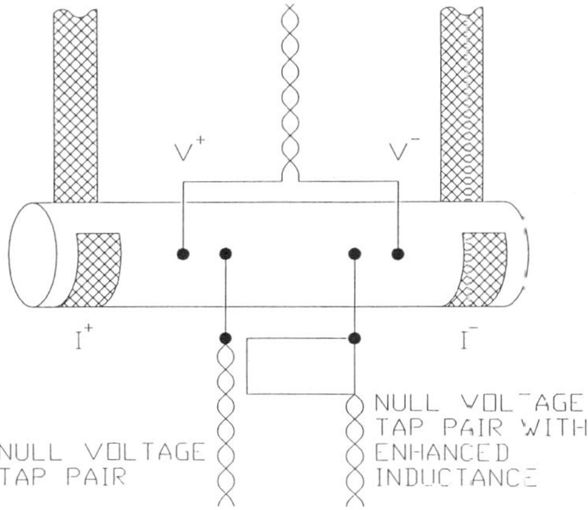

Figure A.3.   Schematic drawing of experimental set up for measuring transport currents. (From Ref. 4.)

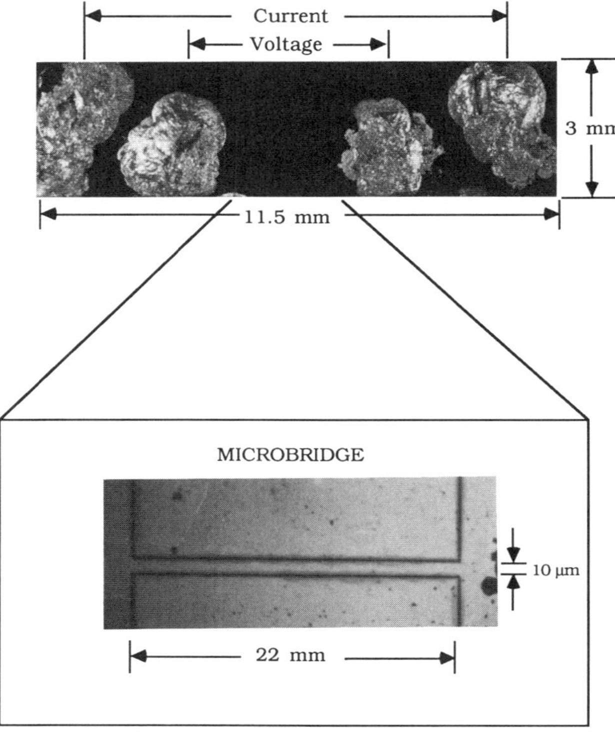

Figure A.4. Photograph of contacts on a small YBCO sample

Of the many complications, one is that when the voltage jumps, tremendous heat dissipation begins and the sample will burn up unless the current is shut off quickly. Also, there is ohmic heating at the contacts where the lead-in wires are soldered to the superconductor. A test circuit that remains on for a long time may yield false measurements of J_c because the sample's temperature may change. For this reason, test durations of 1 sec or less are commonplace; a millisecond test is often used. The trouble with short test durations is that the motion of flux lines takes place on the scale of minutes (flux creep), and critical currents may appear deceptively big because the short test does not allow enough time for flux creep to be noticed. Optimizing the experimental configuration demands compromise between conflicting choices of test duration.

HTSCs are generally so small that the total current flowing in a test is only an amp or two in bulk, and milliamps in a thin film. At the same time, spurious currents are present, caused by thermoelectric power of the lead-in wires running out to room temperature. Meanwhile, voltage probes may be only a millimeter apart, so that one must measure nanovolts accurately to determine the voltage in microvolts per meter. There is much more to a critical current measurement than meets the eye.

For example, in NbTi the National Institute of Standards and Technology lists seven parameters that must be specified with regard to voltage alone: (1) placement of contacts; (2) separation of contacts; (3) noise level; (4) detection criterion (usually 1 μV); (5) inductive voltage; (6) thermoelectric voltage; and (7) meter response. Similarly, the current and magnetic field each have six parameters to be specified, and so on, up to a total of 37 items that can affect a J_c measurement.

A.3. CONTACT HEATING

Accurate transport current measurements in HTSCs has been impeded by the fact that the electrical contacts between the superconductor and the external wires are points of relatively high resistance. In many cases, powerful currents have to be applied for only a fraction of a second in order to prevent overheating and burnup of the contact points. Because J_c declines with increasing temperature, local heating can raise the sample above 77 K, resulting in artificially low J_c.

For copper oxide superconductors, the preferred contact wire is silver, which does not chemically react and steal oxygen from the crystal. AT&T Bell Labs invented a more reliable means of attaching silver lead-in wires to YBCO. Their basic method is to put the silver in *before* sintering the precursor material to make it superconducting. They reported[3] three different methods: (1) silver wires are embedded in the YBCO powder; (2) silver particles totaling 25% by weight are mixed in with the powder, so that the finished product is a granular composite of silver and YBCO; and (3) silver cladding surrounds the precursor powder, which is then rolled, patterned, and etched so that silver stripes remain on the surface—after which the entire unit is sintered.

The biggest near-term consequence of innovations such as this will be that experiments on J_c in HTSCs will not be distorted as much by the extraneous influence of contact heating. Eliminating this nuisance should permit more DC measurements, which are better understood than pulsed measurements.

A.4. PROGRESS TOWARD STANDARDS

Subsequent to the initial confusion associated with determining J_c in the HTSCs, in 1989 DARPA sponsored a meeting of researchers to discuss standard measurement procedures. The consensus appears in Table A.1, which enumerates the 21 most important variables to be reported when doing a HTSC critical current measurement. The message of Figure A.1 is embedded in the measurement guidelines that they set: Keep the voltage taps well separated; use an electric field criterion of 1 μV/cm; do at least one DC transport measurement; use magnetic fields of 0.01, 0.1, and 1 T, and measure J_c at both 77 K and 4 K. The group also issued precautions pertaining to moisture, electrical contacts, solder, mechanical stress, spurious magnetic fields, and the data acquisition system being used.

By 1990, to improve the reliability of measurements coming from various laboratories, the National Institute of Standards and Technology (NIST) began a program to fabricate standard reference materials for HTSCs. NIST hopes to overcome certain nuisance problems (such as contact resistance variability) by building a standard reference device: made of a fiducial compound of YBCO, NIST would attach the contacts, protect it from environmental changes (CO_2, H_2O, etc.), and send it out consecutively to participating laboratories. The intent of this activity is to provide a calibration of the measurement techniques being used in diverse laboratories.

Table 1. DARPA Conference

Parameters for Reporting J_c Results

1. Critical Parameters
   a. Measured I_c
   b. Cross sectional area used for J_c
   c. J_c
   d. Applied magnetic field
   e. Sample temperature
   f. Magnitude of electric field or resistivity criteria

2. Measurement Geometry
   a. Voltage tap separation
   b. Cross sectional dimensions
   c. Overall specimen length
   d. Sample configuration (i.e., straight, coil, hair pin, etc.)
   e. Substrate material
   f. Orientation of magnetic field and specimen current directions with respect to the specimen geometry and with respect to the crystallographic orientation when appropriate
   g. Magnetic field history (magnitude and orientation)
   h. For cladded materials, an estimate of the current that could flow in the cladding (at the criterion used) must be reported

3. Voltage—Current Characteristics
   a. A representative V-I curve
   b. n-value and range over which it was determined
   c. Summary of the V-I data acquisition technique with enough details such that it can be duplicated
   d. Level of interfering voltages

4. Current Contacts
   a. Method of application
   b. Upper limit of contact resistance

5. Specimen Cooling
   a. Type of immersion (liquid or vapor)

Goodrich and Bray of NIST prepared a compendium[4] of all the things that can go wrong. The biggest problem in determining J_c is in choosing the voltage criterion. In conventional LTSCs, the voltage remained essentially zero while current increased until at some point V shot upward, usually proportional to about the 30th power of I. This value of current offered an obvious choice for the designation J_c. The HTSCs show a far softer upward trend in V with increasing I, due to giant flux creep. There is no one particular break point in the I-V curve, and hence little reason to designate one particular value as the critical current. The efforts of NIST hopefully will lead to agreement on standard methods of testing, which will then produce correspondingly more reliable values of J_c.

NIST also built a black box device[5] that simulates the behavior of a HTSC. This permits examination of various techniques for measuring critical current density under controlled conditions. Using this device, NIST established that pulsed currrent measurements, if done correctly, can give almost the same critical current density as DC measurements. However, NIST cautions that good pulsed measurements are difficult to perform correctly.

REFERENCES

1. C. P. Bean, *Rev. Mod. Phys.* **36**, 31 (1964).
2. M. N. Wilson, *Superconducting Magnets* (Oxford University Press: 1983).
3. S. Jin *et al.*, *Appl. Phys. Lett.* **54**, 2605 (1989).
4. L. Goodrich and S. L. Bray, *Cryogenics* **30**, 667 (1990).
5. L. Goodrich *et al.*, *Cryogenics* **33**, 1142 (1993).

Appendix B

Magnetic Measurements Upon Warming or Cooling

Donn Forbes and John R. Clem

The Meissner effect (explained in Chapter 2) is a stronger condition than perfect diamagnetism, because when a superconductor cools through T_c it will expel an existing field. Thus, it is essential to distinguish between data from samples cooled in zero field (ZFC) or cooled in a field already present (FC). But there is a further point: where irreversibility in HTSCs (covered primarily in Chapter 14) is concerned, it makes a difference whether measurements are taken as the sample cools or as the sample warms. This appendix explains why that is so. It is copied with permission from the *Quantum States* newsletter, Fall 1992 edition. It is based on the work of John Clem and Zhidong Hao of the Ames Laboratory at Iowa State University.[1]

When determining the irreversibility line of high-temperature superconductors, the procedure is to make two measurements of magnetization as a function of temperature—first a zero field cooled measurement (ZFC), then a field-cooled measurement. The common assumption is that these two curves diverge at the irreversibility temperature; but this holds *if and only if the field-cooled measurements are taken while the sample is cooling*; i.e., magnetization is measured at successively lower temperatures (field-cooled cooling, or FCC). For convenience, some researchers have made their field-cooled measurements by first cooling the sample back to low temperatures, then taking data while warming (field-cooled warming, or FCW). Unfortunately, experiments conducted in this way have not measured the irreversibility line. This appendix explains what is taking place.

The key point is that the magnetization vs. temperature curve is hysteretic. For field-cooled measurements of magnetization vs. temperature in a constant field, the value of magnetization obtained (at a given temperature) depends on whether that temperature is reached by cooling or by warming. Figure B.1 illustrates the hysteretic nature of the magnetization vs. temperature curve. Figure B.2, data taken on a YBCO thin film by McElfresh and co-workers,[2] confirms the distinct nature of the three curves: ZFC, FCC, and FCW.

The rest of this appendix explains the origin of hysteresis in the magnetization vs. temperature curve with the aid of *sand-hill diagrams* based on the Bean critical state model.[3] Consider a superconducting solid cylinder of radius R (Figure B.3). The sand-hill diagrams (Figures B.4 to B.6) show flux density B inside the cylinder vs. distance from the center r. Each bold line in a sand-hill diagram corresponds to a different temperature. We first discuss these three figures individually, then compare them to show how they confirm the most important features of Figure B.1.

563

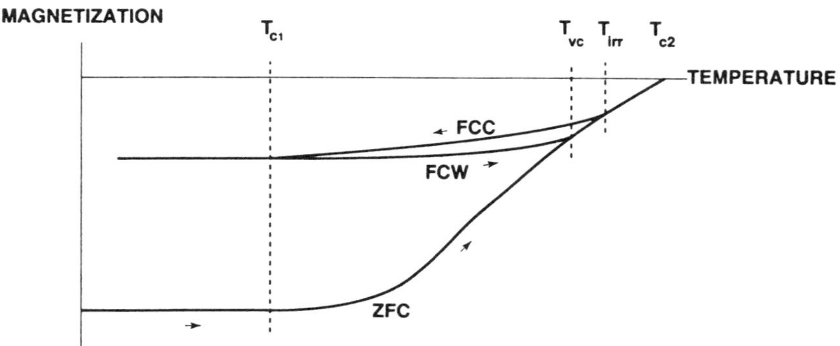

Figure B.1

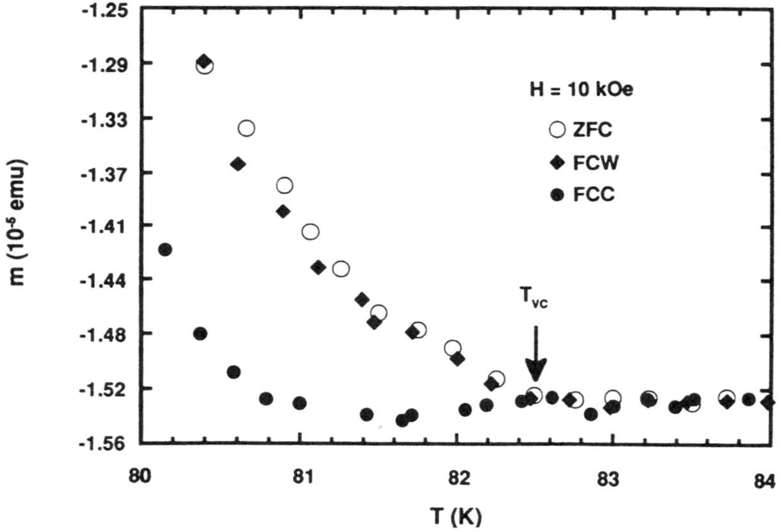

Figure B.2

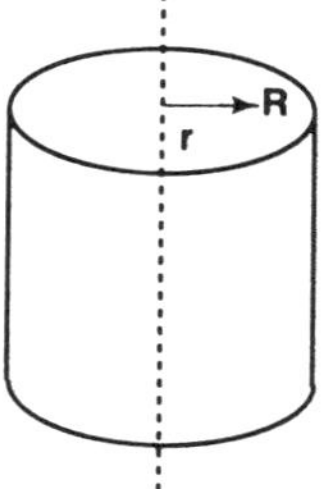

Figure B.3

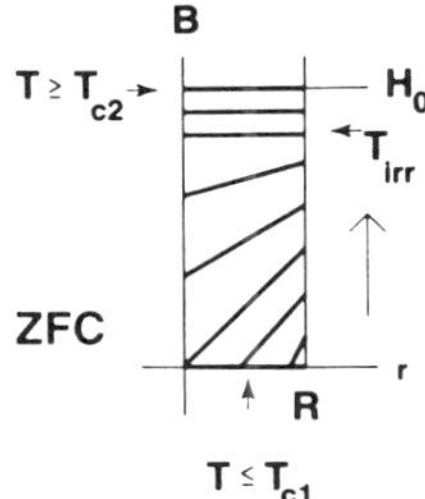

Figure B.4

First, consider the zero field cooled case (Figure B.4): cool the superconductor to some low temperature in zero field, then apply a field H_0 and trace the flux density profile as T increases. (As in Chapter 2, T_{c1} is defined as the temperature at which the cylinder makes the transition from the Meissner state into the mixed state, and T_{c2} as the temperature at which the cylinder makes the transition from the mixed state into the normal state.) When $T \leq T_{c1}$, the cylinder is in the Meissner state, and the flux density is zero everywhere inside the cylinder. At temperatures slightly above T_{c1}, flux begins to penetrate the cylinder—at first, it doesn't reach the center. At all temperatures between T_{c1} and the irreversibility temperature (T_{irr}), there is a gradient of flux density. The slope of this flux density vs. position curve is proportional to J_c at that temperature. Between T_{irr} and T_{c2}, the lines are flat—i.e., the slope is zero, and J_c is effectively zero. At T_{c2} and higher temperatures, the flux density everywhere inside the cylinder is equal to the applied field H_0.

Next consider the field-cooled situation, with data taken while cooling (Figure B.5)— i.e., we turn on a field while the cylinder is nonsuperconducting, then trace the flux density profile as the temperature decreases. For $T > T_{c2}$, the material is in the normal state, and the flux density everywhere inside the sample is equal to the applied field H_0. For $T < T_{c2}$ but $T > T_{irr}$, the distribution of flux density is equal throughout the sample, the slope of the line is zero, and $J_c = 0$. For $T < T_{irr}$, there is a gradient of flux density within the cylinder—higher at the center, lower at the surface. The slope of the line describing flux density vs. position is proportional to $-J_c$. At T_{c1} and lower temperatures, flux remains trapped in the cylinder, and the flux density profile remains constant—zero at the surface of the cylinder, and highest in the center.

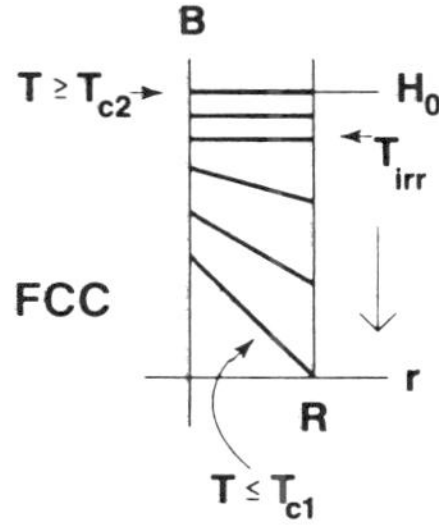

Figure B.5

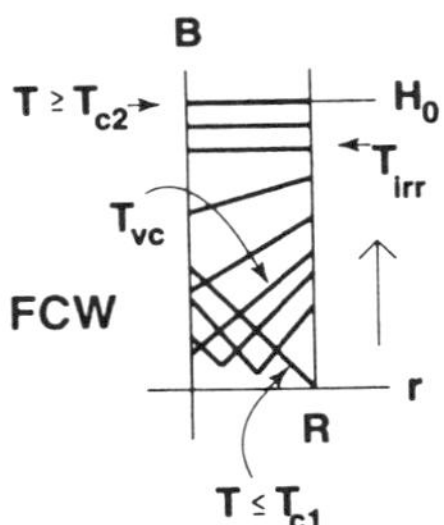

Figure B.6

Finally, we look at the field-cooled situation, with data taken while warming (Figure B.6). In this case, we turn on a field, then cool the cylinder to some $T < T_{c1}$. As we cool the cylinder, the flux density profile changes in exactly the way we traced in the previous paragraph (FCC), but the temperature changes continuously and no measurements are made. For $T < T_{c1}$, we observe the flux density profile with which we ended the FCC discussion: the flux density is zero at the surface, and it increases toward the center. At temperatures slightly greater than T_{c1}, flux begins to enter the sample, and we see a V-shaped minimum in flux density near the surface. As the temperature increases, that V-shaped minimum moves toward the center. At some temperature T_{vc}, the V-shaped minimum progresses all the way to the center of the cylinder. At temperatures between T_{vc} and T_{irr}, the slope of the line is positive and proportional to J_c. At temperatures between T_{irr} and T_{c2}, the flux density is equal everywhere in the sample, the slope is zero, and J_c is effectively zero.

To see why the magnetization vs. T_{irr} curve is hysteretic, compare these three flux density profiles. The key to comparing the profiles is to note that the internal flux density at the surface of the cylinder is not hysteretic. In a constant field, the flux density just inside the surface of the cylinder has only one equilibrium value for each temperature, no matter whether the temperature was ramping upward or downward (assuming there are no surface barriers to flux exit or entry). Therefore, if two sand-hill diagrams show the same value for flux density at the surface of the cylinder, they must correspond to the same temperature, regardless of the mode in which the measurements were taken (ZFC, FCC, or FCW).

By comparing these flux density profiles, we can make four observations that corroborate the validity of the magnetization vs. temperature curves shown in Figure B.1. The measured quantity is magnetization M, which is the difference between the average flux

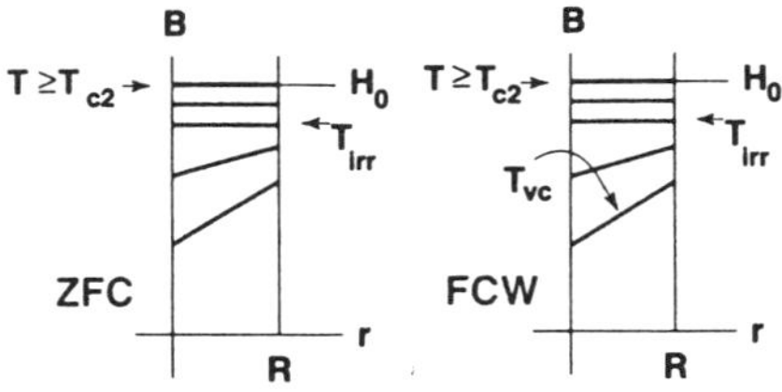

ZFC and FCW, T ≥ T_vc

Figure B.7

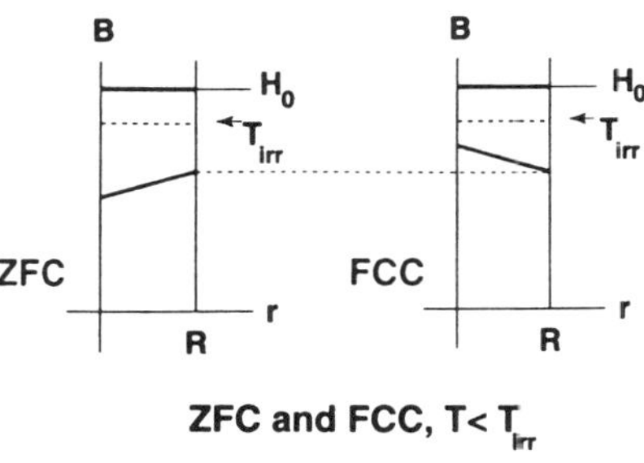

ZFC and FCC, T< T_irr

Figure B.8

density $<B>$ within the sample and the applied field H: $4\pi M = <B> - H$. The magnetization increases as the average flux density within the sample increases.

1. The flux density profile for the ZFC case is exactly the same as the flux density profile for the FCW case at temperatures between T_{vc} and T_{irr} (see Figure B.7). This comparison explains why the ZFC and FCW curves do not diverge at T_{irr}, as some people have assumed—instead, they diverge at T_{vc}, the temperature at which the V–shaped minimum in the FCW flux density profile reaches the center of the cylinder. Researchers who followed the ZFC-FCW experimental protocol actually measured T_{vc}.

2. At temperatures below T_{irr}, the average flux density measured in the ZFC case is less than that measured in the FCC case (Figure B.8). Therefore, the magnetization measured in ZFC mode will be less than the magnetization measured in FCC mode at all $T < T_{irr}$. This explains why the ZFC and FCC magnetization curves diverge at T_{irr}.

The next two observations confirm that for field-cooled measurements, the magnetization vs. temperature curve is hysteretic—i.e., the observed values of magnetization depend on whether data is taken while cooling (FCC) or warming (FCW).

3. Figure B.9 compares the flux density profile in FCC mode with the flux density profile in FCW mode at temperatures between T_{c1} and T_{vc}. Because the average flux density in the cylinder is greater in the FCC mode, higher magnetization will be measured in the FCC mode than in the FCW mode at temperatures between T_{c1} and T_{vc}.

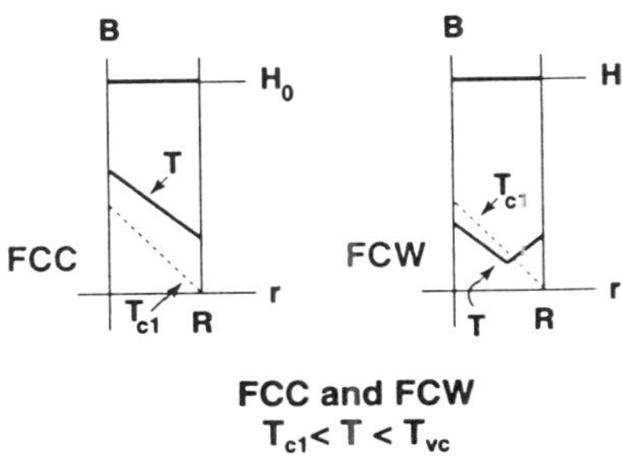

FCC and FCW
$T_{c1} < T < T_{vc}$

Figure B.9

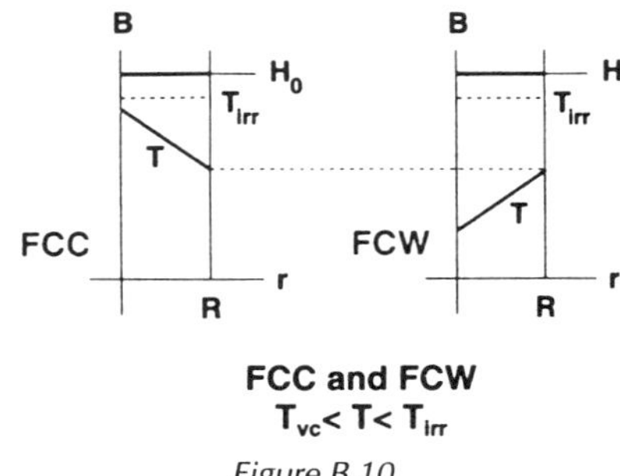

FCC and FCW
$$T_{vc} < T < T_{irr}$$

Figure B.10

4. Figure B.10 compares the flux density profiles in the FCC mode and in the FCW mode in a higher temperature regime, between T_{vc} and T_{irr}. Again, higher magnetization will be found in FCC mode because the average flux density in the cylinder is greater.

REFERENCES

1. J. R. Clem and Z. Hao, *Phys. Rev. B.* **48**, 13774 (1993).
2. J. Deak *et al., Phys. Rev. B* **47**, 8377 (1993).
3. C. P. Bean, *Phys. Rev. Lett.* **8**, 250 (1962).

Glossary

Anisotropy The property of a material by which certain characteristics are different along different directions within the crystal structure. Ceramic superconductors are so highly anisotropic that they are sometimes discussed as "two-dimensional materials."

BCS Theory The explanation of superconductivity in terms of quantum mechanics. It introduced several original concepts, especially that electrons are held together in pairs through an interaction with the lattice vibrations, or phonons. It is very widely accepted today for most superconductors, and is easily the leading candidate to explain high-temperature superconductors.

BSCCO Acronym for bismuth strontium calcium copper oxide, a ceramic superconductor, which can be made into wire.

Brittleness Ceramics tend to be brittle, meaning that they break easily. Unlike copper, which deforms under stress and is readily drawn into wire, a brittle material fractures when stressed beyond a certain point.

Critical Current Density Measured in A/cm^2 and denoted by J_c, the critical current density is the highest amount of electricity that can flow through a superconductor. Any greater current causes superconductivity to vanish, and the material returns to its normal state.

Critical Magnetic Field Measured in tesla and denoted by H_c, the critical magnetic field is the highest value of field (at any given temperature) for which superconductivity remains. Type II superconductors have lower and upper critical fields, H_{c1} and H_{c2}.

Critical Temperature Measured in degrees Kelvin and denoted by T_c, the critical temperature is the highest temperature at which a material remains superconducting.

Energy Gap A key concept in the explanation of superconductivity, by which certain energy levels cannot be occupied; it prevents electrons from exhibiting normal behavior.

Flux Lattice A regular array of magnetic flux lines penetrating a superconductor.

HTSC Acronym for high-temperature superconductor.

LTSC Acronym for low-temperature superconductor.

Magnetic Flux Lines Any magnetic field can be described in terms of its lines of magnetic flux. In many cases of interest, superconducting materials allow magnetic flux lines to penetrate the material, so that superconductivity exists side-by-side with magnetism.

Meissner Effect The property of a superconductor by which *all* magnetic flux lines are forced to stay outside of the superconducting material. This principle is used as the basis for levitation devices.

Normal State   The nonsuperconducting state of a material, in which current flows with electrical resistance, and so on. Materials revert to the normal state either at higher temperatures (above T_c), or in high magnetic fields, or when high current density (above J_c) is passed through them.

Persistent Current Loop   Once started in a loop of superconducting wire, a current flows without running down for indefinitely long times (as long as the material is kept cold).

Type I Superconductor   Most elemental superconductors are type I and have low valves of T_c, H_c, and J_c. Because they carry very little current, these *soft* superconductors are not of interest for practical devices.

Type II Superconductors   Several important alloy superconductors, plus all the high-temperature superconductors, are type II. Here the magnetic field can co-exist with superconductivity, thus allowing high currents to flow. These *hard* superconductors are used to make electric wire and a wide variety of superconducting devices.

TBCCO   Acronym for thallium barium calcium copper oxide, a ceramic superconductor which has $T_c = 125$ K.

YBCO   Acronym for yttrium barium copper oxide, the first substance found to remain superconducting above 77 K, the temperature of liquid nitrogen.

Index

A-15 compounds, 32, 398
Abrikosov lattice, 25, 265
AC losses, 373–393, 404, 413, 425, 441, 456, 472
AC transmission cables, 404
Accelerators, 59–62, 78–79, 518–520
Active Magnetic Regenerator, 492–496
Adiabatic demagnetization, 50–53, 492
Adiabatic limit, 364
Adiabatic stability, 366, 474
Air core motor, 453–463
Air core reactor, 468–469, 473, 479
Air gap magnetic field B_{gmax}, 450–454, 462–463
Allied Signal Corp., 426–427
Alloys, 167–170
Ambient temperature dielectric cables, 407–409, 411
American Superconductor Corp., 8, 130, 218, 321–322, 329, 339–344, 459
Ames Laboratory (Iowa State University), 106, 200, 220, 321, 381, 563
Anisotropy
    of coherence length, 150, 231, 234, 238, 243, 249, 258, 535–537
    of crystal structure, 138, 149, 256–258, 535–539
    of energy gap, 113, 231, 537
    and flux pinning, 279–281
    of HTSCs, 4, 137, 141–144, 231, 239
    of London model, 227, 239
    measurements, 151–152
    of momentum space, 225
    of penetration depth, 150, 537
Annealing, 292–293
Antiferromagnetism, 191–195
Argonne National Laboratory, 10, 127, 144, 151, 200, 214–220, 226, 255, 279, 282, 292, 304, 311, 313, 320, 386, 429, 459, 529, 535
    Advanced Photon Source, 520
    ATLAS, 59–60
    Intense Pulsed Neutron Source (IPNS), 199
Arrhenius plot, 269, 286

Asea Brown Boveri (ABB), 436, 473–474
Astronautics Corporation of America (ACA), 492–496
AT&T Bell Laboratories 117–119, 153, 181, 252, 256, 283–286, 291–316, 319, 322, 540
Atomics International Corp., 422

Babcock-Wilcox Corp., 220, 321
BCS
    energy gap, 25, 97–115, 223–240, 354
    integral equation, 110
    ratio, 111, 237, 240
    and specific heat, 356–357
    theory, 17, 97–115, 223–240
Bean model 15, 28–29, 34, 360, 375, 392, 537, 555–557, 563
Bearings, superconducting magnetic, 11, 425–429, 545–547
BEDT-TTF, 533–534
Bellcore, 509
Beta-gauge, 500
Biomagnetism, 89–94
Biomagnetic Technologies, Inc., 91–95
Bipolar power supply, 76
Bohr magnetron, 52
Bond length, 137
Bond sum rule, 141
Bond valence sum, 189
Bonneville Power Administration, 76–78, 435, 438
Bose glass, 284–288
Bragg condition, 296
Brayton cycle, 45
Brick wall model, 255
Brillouin zone, 99, 104, 236
Brinnell hardness test, 213
Brittleness, 205–210, 214, 329
Broadening of transition, 100–101, 279–280
Brookhaven National Laboratory, 373, 382–392, 404–406
Buckyballs, 539–541
Bulk power, 397–414